RULES FOR FINDING DERIVATIVES

Rule 1. If $f(x) = k$, where k is any constant, $f'(x) = 0$.

Rule 2. If $f(x) = x^n$, where n is any real number and $n \neq 0$, $f'(x) = nx^{n-1}$.

Rule 3. If $f(x) = kg(x)$, then $f'(x) = k \cdot g'(x)$.

Rule 4. If $f(x) = [g(x) \pm h(x)]$, then $f'(x) = [g'(x) \pm h'(x)]$.

Rule 5. If $f(x) = a^{g(x)}$, where $a>0$ and $a \neq 1$, $f'(x) = g'(x)a^{g(x)} \ln a$.

If $f(x) = e^{g(x)}$, then $f'(x) = g'(x)e^{g(x)}$.

Rule 6. If $f(x) = \log_a g(x)$, then $f'(x) = \dfrac{g'(x)}{g(x) \ln a}$.

If $f(x) = \ln g(x)$, then $f'(x) = g'(x)/g(x)$.

Rule 7. If $h(x) = f(x) \cdot g(x)$, then $h'(x) = f(x) \cdot g'(x) + g(x) \cdot f'(x)$.

Rule 8. If $h(x) = \dfrac{f(x)}{g(x)}$, then $h'(x) = \dfrac{g(x) \cdot f'(x) - f(x) \cdot g'(x)}{[g(x)]^2}$

Rule 9. If $y = f(u)$ and $u = g(x)$, then $dy/dx = dy/du \cdot du/dx$.

Rule 10. If $y = [f(x)]^n$, then $y' = n[f(x)]^{n-1} \cdot f'(x)$.

Rule 11. If $f(x) = [g(x)]^{h(x)}$, then $f'(x) = g(x)^{h(x)} \left[h'(x) \cdot \ln g(x) + h(x) \cdot \dfrac{g'(x)}{g(x)}\right]$.

MATH OF FINANCE FORMULAS:

Simple Interest: $I = Prt$

Amount (Future Value) at Simple Interest: $S = P(1+rt)$

Simple Discount: $D = Mdt$

Proceeds on Simple Discount Note: $P = M(1-dt)$

Amount with Compound Interest: $S = P(1+i)^n$

Present Value of Compound Amount: $P = S(1+i)^{-n}$

Effective Interest Rate: $(1+r/m)^m - 1$

Amount with Continuous Compounding: $S = Pe^{rt}$

Effective Rate—with continuous compounding: $e^r - 1$

Amount of an Ordinary Annuity: $S = R\left[\dfrac{(1+i)^n - 1}{i}\right]$

Required Periodic Payment—into Sinking Fund: $R = S\left[\dfrac{i}{(1+i)^n - 1}\right]$

Present Value of an Annuity: $A = R\left[\dfrac{1-(1+i)^{-n}}{i}\right]$

Required Periodic Payment—for Amortizing a Debt: $R = A\left[\dfrac{i}{1-(1+i)^{-n}}\right]$

Applied Mathematics: For Business, Economics, and the Social Sciences

The Irwin Series in Quantitative Analysis for Business
Consulting Editor Robert B. Fetter Yale University

Applied Mathematics: For Business, Economics, and the Social Sciences

Ann J. Hughes
Georgia State University

1983

RICHARD D. IRWIN, INC.
Homewood, Illinois 60430

ISBN 0-256-02829-X

Library of Congress Catalog Card No. 82-83422

Printed in the United States of America

1 2 3 4 5 6 7 8 9 0 K 0 9 8 7 6 5 4 3

Preface

Applied Mathematics: For Business, Economics, and the Social Sciences is designed for a thorough freshman or sophomore level course in mathematics beyond basic algebra, emphasizing topics that are most useful to students in business, economics, and the social sciences. It will fit nicely into either a business curriculum or a liberal arts curriculum. Appropriate topics are covered in sufficient depth for a one-quarter or one-semester course in finite mathematics followed by a one-quarter or one-semester course in calculus; or a three-quarter course can be built around mathematical foundations, model building, linear systems, calculus, and probability. When certain introductory topics are assigned as student self-review, the remaining topics are sufficiently rigorous for a good MBA-level mathematics refresher course. Actually, the choice of topics and the organization of the material, as well as the depth of coverage, make the book readily adaptable to a variety of courses.

The goals of the text are twofold. One goal is to provide a solid foundation in the mathematical tools most useful to students of either a business or a liberal arts curriculum. A collateral goal is to stimulate the students' interest in mathematics by emphasizing the practical usefulness of the techniques being discussed.

The material is presented in such a way as to be easily readable—and, more important, understandable—by the student who is not truly mathematically oriented. While the mathematical content is complete and correct, the basic ideas are developed in a "reasonable," intuitive manner rather than from a formal, highly theoretical point of view. The language—and the symbolism—are kept as simple and as straightforward as possible. Important ideas are explained, illustrated by mathematical examples (which are worked out in detail), shown graphically whenever possible, made meaningful through down-to-earth applications, and then reinforced by extensive exercise sets. Derivations and mathematical proofs are given only where their inclusion adds significant insight into a particular concept. This style of presentation is meant to be inviting and unintimidating; it is not condescending. Our aim is not to sacrifice mathematical rigor—only to make it more palatable.

An abundance and a wide variety of applications that are relevant to the intended audience appear throughout the text, in the examples, and in the exercises. Students will be able continually to see practical uses for the mathematics they are learning. The concepts on which the applications are based are presented in such a way, however, that no previous technical knowledge in these subject areas is necessary. Thus, the book is meaningful to students of varied interests.

Because students tend to learn mathematics more easily when it is administered in small "doses," the textual material is broken down into relatively compact sections. Exercise sets which include not only a variety of applications-type situations but also a plentiful supply of drill-type problems on techniques are located at the end of almost every section. (There are some 2,900 exercises in the text!) These are tightly integrated to that particular section and the concepts it develops. These exercise sets provide an opportunity for the student to pause and practice the mathematical technique to which he has just been introduced until he feels comfortable with it, or to consider the wide variety of real-world applications for that mathematical technique. Then he is ready to proceed to another concept. It is this logical, and methodical, transition from the mastery of one concept to the next that enables students to grasp "the big *mathematics* picture."

Surely there are many equally valid ways of ordering the topics included in courses for which this text is intended. With this in mind, we have attempted to reserve for the instructor maximum flexibility in designing a course to meet the needs of his particular students. The organization of topics, their manner and depth of coverage—within the book as a whole and within the individual chapters—makes the text readily adaptable to a variety of courses.

The only prerequisite assumed is basic algebra. However, because much of this material is forgotten through non-use, the fundamentals of *algebra* are discussed in Chapter 1. This material can be treated in a systematic way in class or assigned as individual student reference, as needed. Chapter 2 presents the basic concepts of *set theory* and establishes the foundational structure for a study of *probability* (Chapters 16 and 17). *Matrix algebra* (in Chapter 3) merits attention not only because of the existence of many important matrix models but also because it is an important tool in the manipulation of *linear systems* (Chapter 6).

Mathematical *model building* is introduced early (in Chapter 4) and is emphasized throughout the text. A discussion of *linear functions* (Chapter 5) and *systems of linear functions* (Chapter 6) is followed by two chapters on *linear programming.* Chapter 7 introduces the linear programming model and develops a graphical solution procedure; Chapter 8 develops the Simplex algorithm. *Nonlinear functions,* including exponential and logarithmic functions, are covered in Chapter 9. Chapters 10 through 15 provide a comprehensive treatment of the *calculus.*

Beginning with *limits and continuity,* the discussion moves from *derivative rules* and *applications* to *integral rules* and *applications* and concludes with a look at the calculus of *multivariate functions.* The important topic of *mathematics of finance* is included in Chapter 18.

Basic chapters of the text are sufficiently elementary so as to be within the scope of students with somewhat limited mathematical backgrounds. The maturity attained in these chapters enables the student to proceed, without undue difficulty, to the more advanced topics developed in the chapters which follow.

Acknowledgments

I am deeply indebted to the many people who have encouraged and assisted me in producing this book. My sincere thanks go to Gerald E. Rubin, Marshall University, and Arthur B. Kahn, University of Baltimore, who generously prepared detailed reviews of the manuscript; to William D. Blair, Northern Illinois University; Leonard Fountain, San Diego State University; Terry D. Lenker, Central Michigan University; Peter Rice, University of Georgia; John C. Shannon, Suffolk University; and Dalton Tarwater, Texas Tech University, for their perceptive criticisms and suggestions; and to my students for their good-natured and ever-so-helpful comments and complaints as we struggled through rough drafts of the text. And my very special thanks to my daughter Robyn for her continued patience, understanding, and faith.

Ann J. Hughes

Contents

C·H·A·P·T·E·R 1

A Review of the Basic Concepts of Mathematics

We, who are known to our creditors, and to the tax collector, by our social security number rather than by our name, who negotiate wage contracts based on a "cost-of-living" index, who carry a hand-held calculator in our briefcase, take for granted the existence and use of numbers in our lives. Consciously or subconsciously, we use and associate with numbers in almost everything we do, whether paying our income taxes, dialing the phone, or adjusting the speed of our automobile to the legal limit. Numbers, and mathematics, the body of formalized rules for manipulating numbers, have become an essential part of our everyday thought and language.

Regrettably, people tend to take one of two extreme attitudes towards numbers. They either view numbers and mathematical expression as being a dreadfully complex language, one to be avoided at all costs; or they perceive mathematical expression as the ultimate in beauty and goodness, the essence of all truth. Both extreme attitudes are devoid of real understanding. Useful as they are—and they *are* indispensable to modern man—there is nothing magical, mystical, demonic, or sacrosanct about either numbers or mathematics. Neither are reality; both are artificial constructs of humanity, developed by persons to aid them in understanding a portion of reality.

Although the origin of numbers is lost in antiquity, we do know that their existence predates the time of the earliest written records. And, while one can only speculate about the first use of numbers, it is not difficult to picture how this might have developed. Prehistoric men and women perhaps had their first very limited sense of number or quantity when collecting food or fuel, and this was probably only comparable to our "Is this enough?" or "Do we need more than this?" rather than a more specific "How much?" As civilization developed, ways of using numbers became more astute.

Tallying was very likely one of the devices people used when they began to take a more extensive interest in the size of their collections of animals or other valuable objects. In tallying, instead of trying to assess a collection as a whole, people took the objects forming the collection one at a time and placed a tally to correspond to each item in the collection. It seems reasonable to hypothesize that shepherds, for example, may have developed the concept of tallying in order to verify that the same number of sheep returned to the fold at night as went to the fields in the morning. This may have been accomplished by placing a pebble in a pouch in the morning for each sheep that went out to the pasture and removing a pebble for each sheep that returned in the evening.

It must soon have become apparent that three pebbles could represent one's three children, or three fish, or three of the neighbors' sheep as well as one's own three sheep. The result would have been a mental symbol for three which, in fact, turned out to be a splendid way of remembering a certain quantity of any object. In time, people must have realized that this mental symbol for three could be considered as an entity unto itself rather than just a property of some collection of objects. This evolution, or one similar, led to the concept of a numeral.

Next, people must have started using words to represent quantity. For example, relics indicate that one ancient culture used the word that meant "wing" to represent two objects and the word that meant "hand" for five objects. The appearance of such number-names in various primitive languages shows that people very early began to organize their ideas about numbers and quantity. Eventually, the number-names were arranged in a certain sequence, making it possible to count, and various peoples in many parts of the world did in fact develop formal counting systems.

Even before 3000 B.C. the peoples of ancient Babylonia, China, and Egypt had developed practical systems of mathematics. They were using written symbols to represent numbers, and they performed simple arithmetic operations. This knowledge was useful to them in mercantile transactions and in government. They also developed a practical geometry helpful in agriculture and engineering. The ancient Egyptians knew how to survey their fields and to make the intricate measurements necessary in the construction of huge pyramids. This earliest body of mathematical procedures was developed as it was needed and was applied only to practical problems; it was *applied,* rather than *pure, mathematics.*

The Greeks, between 600 and 300 B.C., became the first people to separate mathematics from practical problems by making it into an abstract exploration of space. Their study was based on logical reasoning, on the use of points, lines, circles, or other simple figures, rather than upon phenomena observed in nature. The famous Pythagorean Theorem was developed; and Euclid, with his book *The Elements,* organized geometry as a single system of mathematics.

By 450 B.C. Greek mathematicians recognized irrational numbers. In 225 B.C. Archimedes invented processes that foreshadowed those used in integral calculus. Ptolemy (A.D. 150) developed trigonometry, and Diophantus (A.D. 275) with his works on numbers in equations earned for himself the title "Father of Algebra."

Although the Middle Ages were characterized by little progress in the development of mathematical concepts, with the Renaissance came exciting advances in this area. Renewed interest in mathematics stemmed in part from the fact that the exploration of new lands and continents required better mathematics of navigation. And, in a similar way, the growth of business and commerce demanded better mathematics of banking and finance. But interest also grew in pure mathematics. Copernicus contributed significantly to the theory of mathematics through his work in astronomy. Napier invented logarithms. Galileo expanded mathematical knowledge through his studies of the stars and planets. Descartes invented analytic geometry; Pascal and Fermat invented mathematical theory of probability; Newton and Leibniz originated calculus.

Over the years, many great mathematical minds—such as Abraham de Moivre, Leonhard Euler, Marquis de Laplace, Karl Friedrich Gauss, and Albert Einstein to name only a few—have developed and added to our present consciousness of numbers. And important advances continue to be made in the field of mathematics. Certainly the advent of the electronic computer has increased immeasurably our ability to manipulate vast quantities of numerical data. While mathematical methods have long been an integral part of the study of engineering and other physical sciences, recognition of the usefulness of these methods in analyzing relationships existent in a business, or economic, environment, and in the social sciences, is becoming more and more widespread.

This chapter is designed to review some of the basic properties of numbers and mathematics. The concepts presented here will not be new to you; you will have encountered each of them in some previous math course. However, your recollection of the concept, or your finesse in using the technique, may have dimmed from nonuse.

1.1 TODAY'S SYSTEM OF COMPLEX NUMBERS

A fundamental concept of mathematics is that of a SET, where a set is a WELL-DEFINED COLLECTION OF DISTINCT OBJECTS. The numbers that we work with are often considered in groups, or sets.

The basic set of numbers, known as the NATURAL or COUNTING NUMBERS, comprises the POSITIVE INTEGERS, 1, 2, 3, ... These positive integers meet the need for enumerating objects or possessions.

Mathematically, this set is *closed* under the operation of addition; that is, if any two positive integers are added together, the result will be another positive integer. Thus, the set of numbers which comprises the NEGATIVES OF THE NATURAL NUMBERS was developed only as people recognized the need for an operation of negative addition, or "subtraction." These two sets—the natural numbers and the negatives of the natural numbers—along with that all-important discovery, the number ZERO, form the set of INTEGERS: . . . , −3, −2, −1, 0, 1, 2, 3, ...

The set of integers is closed under both addition and subtraction, and under multiplication as well. Any integer added to an integer, subtracted from an integer, or multiplied times an integer, will yield a result which is itself an integer. But the set of integers is not closed under the mathematical operation of division. Hence, the need arose to further extend the system of numbers, and the set of NONINTEGER NUMBERS emerged. The integers and the nonintegers taken together became labeled RATIONAL NUMBERS.

A RATIONAL NUMBER is defined as THE RATIO OF TWO INTEGERS, as p/q, with q being nonzero. Thus, 3/2, −4/7, 13/31 are all rational numbers. If $q = 1$, the result is an integer. Otherwise, the result may be either a TERMINATING DECIMAL, as $3/2 = 1.5$, or a NONTERMINATING, REPEATING DECIMAL, as $4/3 = 1.33\overline{3}$.

It would seem that the matter might end here, but not so. Pythagoras discovered that $\sqrt{2}$ is not a rational number, since no two integers p and q can be found such that $p/q = \sqrt{2}$. Because it was not rational

Figure 1.1

Today's System of Complex Numbers

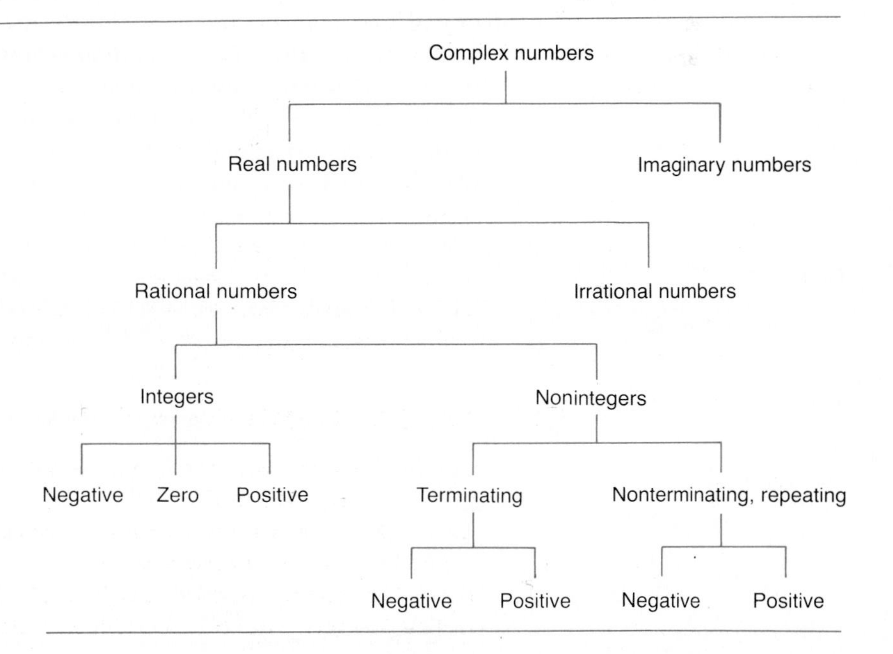

(*not* the ratio of two integers), it was labeled an IRRATIONAL NUMBER. Eventually, other irrational numbers were discovered; among these are $\sqrt{3}$, $\sqrt{5}$, and $\pi = 3.14159\ldots$ of πr^2 fame. These all result in NONTERMINATING, NONREPEATING, DECIMALS. The set of rational numbers along with the set of irrational numbers forms the set of REAL NUMBERS.

Last, there is the set of IMAGINARY NUMBERS. Consider $\sqrt{-1}$. No real number exists which is the square root of a negative number. These kinds of numbers are called "imaginary" numbers, but the descriptive term does not here have the usual colloquial meaning. These numbers play a fundamental role in science and engineering. The real and the imaginary numbers taken together form the set of COMPLEX NUMBERS.

1.2 THE REAL NUMBER LINE

Real numbers can be represented geometrically as points on an infinitely long straight line called a REAL NUMBER LINE, or simply NUMBER LINE. For each and every real number there exists a corresponding, and unique, point on the number line. Conversely, each point on the line represents a unique real number.

A point on the line is chosen as a point of reference and represents the number zero; this point is termed the ORIGIN. Then a standard measure of distance, or UNIT DISTANCE, is selected and successively marked off both to the right and to the left of the origin. The ALGEBRAIC VALUE of a number is its DIRECTED DISTANCE, or SIGNED DISTANCE, from the origin. That is, algebraic values are either positive or negative. Points to the right of the origin represent positive numbers, while points to the left of the origin represent negative numbers. Thus, the point that is one unit distance to the right of the origin corresponds to the number +1; the point that is one half unit distance to the left of the origin corresponds to the number $-1/2$; and so on. The ABSOLUTE VALUE of a number is its NONDIRECTED DISTANCE from

Figure 1.2

The Real Number Line

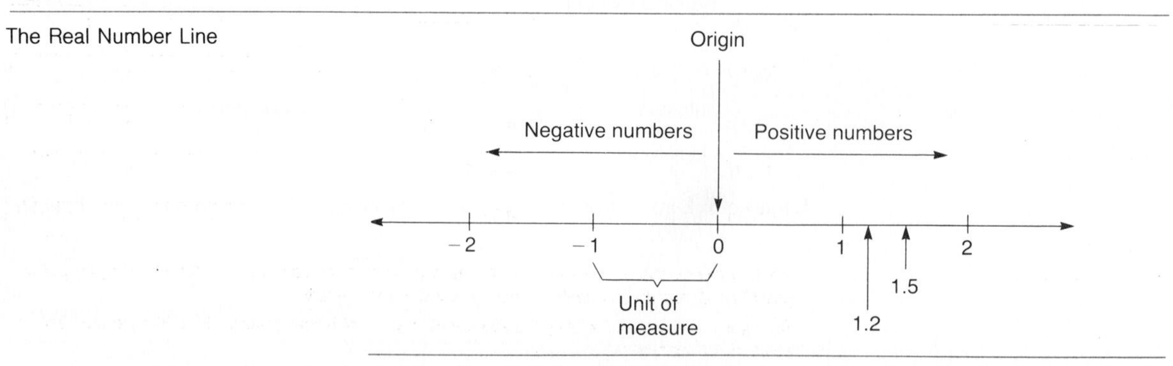

the origin. That is, absolute values are neither positive nor negative. The symbol $|a|$ is used to represent the absolute value of a.

Identity and Rank Order Properties of Numbers

For any real numbers, a, b, and c,

1. If $a = b$ then $b = a$.
2. If $a = b$ and $b = c$, then $a = c$.

If a is not equal to b, symbolized $a \neq b$, we write $a < b$ to signify that a is *less than* b (meaning that a is to the left of b on the number line). Or we write $a > b$ to signify that a is *greater than* b (meaning that a is to the right of b on the number line).

3. Precisely one of the properties $a < b$, $a = b$, or $a > b$ holds.
4. If $a < b$ and $b < c$, then $a < c$.

Intervals on the Real Number Line

Given two real numbers, a and b, with $a < b$, we define the following types of intervals on the real number line.

> **A closed interval,** denoted $[a,b]$, includes all real numbers x from a to b, *inclusive,* on the number line. An OPEN INTERVAL, denoted (a,b), consists of all real numbers x between a and b, *exclusive* of both a and b.

In addition, many HALF-OPEN, or HALF-CLOSED, INTERVALS are defined. These are summarized in Table 1.1.

Table 1.1

Intervals on the Real Number Line (for any two real numbers, a and b, with $a < b$)

Interval notation	Inequality notation	Representation on the real number line
$[a,b]$	$a \leq x \leq b$	
(a,b)	$a < x < b$	
$[a,b)$	$a \leq x < b$	
$(a,b]$	$a < x \leq b$	
$[a,\infty)$	$a \leq x$	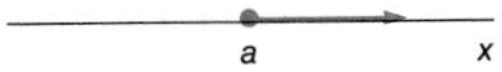
(a,∞)	$a < x$	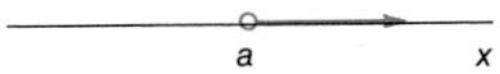
$(-\infty,a]$	$a \geq x$	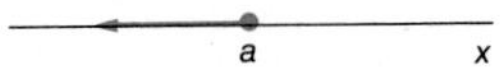
$(-\infty,a)$	$a > x$	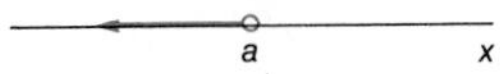

Note: A closed dot is used to indicate that the point is included in the interval; an open dot is used to indicate that the point is not included in the interval.

The symbol ∞ is read "infinity." Because infinity is not a real number, it cannot be included in a set of real numbers.

1.3 ALGEBRAIC OPERATIONS ON REAL NUMBERS

The basic algebraic operation performed on real numbers is ADDITION. The concept of addition is extended to include the operations of SUBTRACTION, MULTIPLICATION, and DIVISION.

Addition and Multiplication on Real Numbers

The following properties of real numbers with respect to addition and multiplication ensure that these operations yield results which are internally consistent.

The closure properties If a and b are real numbers, then their sum, $a + b$, or their product, ab, is also a real number.

The commutative properties When adding or multiplying two real numbers, a and b, the order in which they are added, or multiplied, is immaterial; that is,

$$a + b = b + a \qquad \text{and} \qquad ab = ba$$

The associative properties In addition or multiplication, numbers may be grouped in any order; that is,

$$a + (b + c) = (a + b) + c \qquad \text{and} \qquad a(bc) = (ab)c$$

The identity properties The quantity *zero* is called the ADDITIVE IDENTITY. Added to any other real number, zero does not change the number; that is

$$a + 0 = 0 + a = a$$

The quantity *one* is called the MULTIPLICATIVE IDENTITY. One multiplied times any other real number does not change the other number; that is,

$$(a) \cdot (1) = (1) \cdot (a) = a$$

The inverse properties For each real number a there is a unique real number, denoted $-a$ and called the ADDITIVE INVERSE of a, or the NEGATIVE of a, with the property that

$$a + (-a) = (-a) + a = 0$$

If a is any *nonzero* real number, there is a unique real number, denoted $1/a$ and called the MULTIPLICATIVE INVERSE of a, or the RECIPROCAL of a, with the property that

$$a \cdot (1/a) = (1/a) \cdot a = 1$$

The multiplicative inverse of a is also often represented by a^{-1}.

Zero does not have a multiplicative inverse.

The distributive property (of multiplication over addition) If a, b, and c are real numbers, then

$$a(b + c) = ab + ac$$

Explain

Example 1

$$(8)(2+4) = (8)(2) + (8)(4)$$
$$(8)(6) = 16 + 32$$
$$48 = 48$$

[]

The distributive property holds for any number of terms.

Rules of algebraic signs in addition If two numbers having the same algebraic sign are to be added, add their absolute values and use their common sign as the sign of the sum. If two numbers having opposite algebraic signs are to be added, find the difference in their absolute values and use the sign of the larger absolute number as the sign of the sum.

Rules of algebraic signs in multiplication If the two real numbers being multiplied have the same algebraic sign, their product is positive. If the two numbers being multiplied have opposite algebraic signs, their product is negative.

A zero product The product of any two nonzero real numbers is also nonzero; that is, if $a \neq 0$ and $b \neq 0$, then $ab \neq 0$. The only way a product of zero can be obtained is for at least one of the factors of the product to be zero; that is, if $ab = 0$, then either $a = 0$ or $b = 0$ or both $a = 0$ and $b = 0$.

Division on Real Numbers

We use the terminology

$$\text{DIVIDEND} \div \text{DIVISOR} = \text{QUOTIENT}$$

and define division in terms of multiplication by the multicative inverse, or reciprocal, of the divisor. If a and b are any real numbers, where $b \neq 0$, the quotient $a \div b$ is given by

$$a \div b = a(1/b) = a/b$$

Thus, the basic properties of multiplication within the real number system extend to the operation of division.

The role of zero in division Division into zero is permitted, it yields a quotient of zero. DIVISION BY ZERO, however, IS NOT DEFINED. That is, if a is any nonzero real number,

$$0/a = 0 \qquad \text{but} \qquad a/0 \text{ is not defined}$$

(Recall that zero does not have a multiplicative inverse; hence, multiplication by the multiplicative inverse of zero is not possible.)

Rules of algebraic signs in division If the dividend and divisor have the same algebraic sign, the quotient is positive. If the dividend and the divisor have opposite signs, the quotient is negative.

Sequences of Operations

Algebraic operations are commonly used in a sequence, as

$$5 + (3 \times 4) - (6 \div 2)$$

When the operations indicated in any such sequence are carried out, unless otherwise specified, ALL MULTIPLICATIONS AND DIVISIONS ARE PERFORMED FIRST, FOLLOWED BY ADDITIONS AND SUBTRACTIONS, generally working from left to right.

The sequence in which the operations are performed can be altered through the use of parentheses ( ), brackets [ ], and braces { }. Operations within these SYMBOLS OF GROUPING are carried out before operations outside the symbols are performed. If several sets of symbols of grouping are needed, we shall usually enclose parentheses within brackets, and brackets within braces. However, it is becoming more and more conventional to use only parentheses as symbols of grouping. Thus, we may have parentheses within parentheses within parentheses. At any rate, operations within the *innermost* set of grouping symbols are carried out first. When these symbols are removed, the operations within the remaining innermost set of such symbols are carried out. This continues until all such symbols are removed.

Example 2

$$(3 \times 5) + (3 \times 4) - (6 \div 2)$$

is solved by

$$\begin{aligned}&(3 \times 5) + (3 \times 4) - (6 \div 2)\\&= 15 + 12 - 3\\&= 24\end{aligned}$$

$$3\{[(5 + 3) \times 4 - 6] \div 2\}$$

is solved by

$$\begin{aligned}&3\{[8 \times 4 - 6] \div 2\}\\&= 3\{26 \div 2\}\\&= 3\{13\}\\&= 39\end{aligned}$$

[ ]

Clearly, symbols of grouping are a very important part of any mathematical statement.

EXERCISES

Reduce each of the following to a single number.

1. $-5 + 3$
2. $5 + (-3)$
3. $(-5) \times 3$
4. $4 \times (-2) \times (-5)$
5. $(-10) \div (-5)$
6. $7 - 3 \times 4 - 1$
7. $7 - 3 \times (4 - 1)$
8. $2(5 - 1) + 3(6 + 3)$
9. $3[1 - 2(5 - 3) + 8 \div (4 - 2) - (-2 + 3) \times 4]$
10. $\{3 - [2 \times (4 + 1) \div 5]\} \times \{6 \times (4 - 2) + 3 + [(3 \times 5) \div (4 + 1)]\}$

1.4 WORKING WITH FRACTIONS

If a and b are integers, with $b \neq 0$, then a/b is called a FRACTION (or RATIONAL NUMBER). We use the terminology

$$\frac{\text{NUMERATOR}}{\text{DENOMINATOR}}$$

to refer to the parts of the fraction.

Addition and Subtraction of Fractions

To add (or subtract) two fractions which have the same denominator, add (or subtract) the numerators and retain the denominator that is common to both; that is,

$$\frac{a}{b} + \frac{c}{b} = \frac{a+c}{b}$$

To add (or subtract) two fractions which have different denominators, multiply both the numerator and the denominator of each fraction by an appropriate number to make the denominators alike, and then apply the rule stated above; that is,

$$\frac{a}{b} + \frac{c}{d} = \frac{ad}{bd} + \frac{bc}{bd} = \frac{ad+bc}{bd}$$

Example 3

a. $\frac{2}{5} + \frac{1}{5} = \frac{2+1}{5} = \frac{3}{5}$ *b.* $\frac{2}{5} + \frac{3}{7} = \frac{7 \times 2}{7 \times 5} + \frac{5 \times 3}{5 \times 7} = \frac{14+15}{35} = \frac{29}{35}$ []

Multiplication and Division of Fractions

The product of two fractions is found by dividing the product of their numerators by the product of their denominators; that is,

$$\frac{a}{b} \times \frac{c}{d} = \frac{ac}{bd}$$

The quotient of two fractions is found by inverting the fraction in the denominator and proceeding as with multiplication; that is,

$$\frac{a}{b} \div \frac{c}{d} = \frac{a}{b} \times \frac{d}{c} = \frac{ad}{bc}$$

Example 4

a. $\frac{3}{8} \times \frac{7}{5} = \frac{3 \times 7}{8 \times 5} = \frac{21}{40}$ *b.* $\frac{3}{8} \div \frac{2}{5} = \frac{3}{8} \times \frac{5}{2} = \frac{15}{16}$ []

Equal, or Equivalent, Fractions

If a, b, c, and d are integers with $b \neq 0$ and $d \neq 0$, two fractions a/b and c/d are EQUAL, or EQUIVALENT, if and only if $ad = bc$.

The value of a fraction does not change when *both* its numerator and denominator are multiplied, or divided, by the same nonzero factor.

If a, b, and c are integers and $c \neq 0$, then

$$\frac{a}{b} = \frac{a}{b} \times \frac{c}{c} = \frac{ac}{bc} \qquad \frac{ac}{bc} = \frac{ac}{bc} \div \frac{c}{c} = \frac{a}{b}$$

Example 5 *a.* $\frac{3}{5}=\frac{3\times7}{5\times7}=\frac{21}{35}$ *b.* $\frac{15}{25}=\frac{15\div5}{25\div5}=\frac{3}{5}$ []

A fraction is said to be in REDUCED FORM (or in LOWEST TERMS) when the numerator and the denominator have no common factor. To change a fraction to an equivalent fraction in reduced form, divide both the numerator and the denominator by all of their common factors. The fraction 15/25 is not in reduced form; a fraction equivalent to 15/25, but in reduced form, is 3/5.

EXERCISES

State as a single fraction in reduced form.

11. $\frac{4}{6}+\frac{1}{6}$
12. $\frac{5}{7}\div\frac{2}{7}$
13. $\frac{3}{7}-\frac{2}{9}$
14. $\frac{-4}{3}\times\frac{-1}{6}$
15. $\frac{3}{5}+\frac{2}{3}+\frac{1}{2}$
16. $\frac{7/11}{-4/7}$
17. $\frac{5}{8}-\frac{4}{5}\div\frac{2}{3}$
18. $\frac{2}{5}\left(\frac{7}{12}-\frac{5}{6}\right)$
19. $\frac{3}{4}\times\frac{2}{5}\times\frac{1}{6}$
20. $\frac{1}{2}+\frac{2}{3}\times\frac{3}{4}-\frac{4}{5}$

1.5 POWERS AND ROOTS OF REAL NUMBERS

If n is a positive integer, then a^n represents the product

$$\underbrace{a\cdot a\cdot a\cdot a\cdot\ldots\cdot a}_{n\text{ times}}=a^n$$

n is called the EXPONENT of a, and a^n is itself called the nTH POWER OF a.

Example 6

a. $4^2=(4)\cdot(4)=16$

b. $(1/2)^3=(1/2)\cdot(1/2)\cdot(1/2)=1/8$

c. $2^5=(2)\cdot(2)\cdot(2)\cdot(2)\cdot(2)=32$

d. $(-3)^3=(-3)\cdot(-3)\cdot(-3)=-27$ []

If n is a positive integer greater than 1, and if $a^n = b$, then a is called the nTH ROOT of b. In particular, if $a^2 = b$, than a is called a SQUARE ROOT of b; if $a^3 = b$, then a is called a CUBE ROOT of b. An nth root of b is symbolized.

$$\sqrt[n]{b}=a$$

The symbol $\sqrt{\ }$ is called the RADICAL, b the RADICAND, and n is the INDEX of the radical. If $n = 2$, it is customary to omit the index and simply write $\sqrt{b}$. If n is any value other than 2, the index should be explicitly stated.

Both even and odd roots are defined for positive numbers. The odd root of a negative number is also defined. However, the even root of a negative number is not defined in the real number system.

Example 7

a. $\sqrt[3]{-8} = -2$ since $(-2)^3 = (-2)\cdot(-2)\cdot(-2) = -8$.

b. $\sqrt{-8}$ is not defined since there exists no real number that when multiplied times itself yields the product -8. []

Every positive number b has two square roots, one root being the negative of the other. For example, both $+4$ and -4 are square roots of 16 since both $(+4)^2 = 16$ and $(-4)^2 = 16$. Every real number has exactly one real cube root.

Laws of Exponents

The following are the basic definitions and properties of exponents.

Law	*Example*
1. $0^n = 0$ if $n > 0$	$0^2 = 0$; $0^3 = 0$; $0^8 = 0$
2. $a^0 = 1$ if $a \neq 0$	$5^0 = 1$; $(1/3)^0 = 1$; $(-2)^0 = 1$
3. $a^{-n} = 1/a^n$ if $a \neq 0$	$3^{-2} = 1/3^2 = 1/9$; $(-4)^{-3} = 1/(-4)^3 = 1/-64 = -1/64$
4. $a^{m/n} = (\sqrt[n]{a})^m = \sqrt[n]{a^m}$ if the root is defined	$4^{(3/2)} = (\sqrt{4})^3 = 2^3 = 8$ $8^{(2/3)} = \sqrt[3]{8^2} = \sqrt[3]{64} = 4$ $(-4)^{(3/2)} = \sqrt{(-4)^3} = \sqrt{-64} =$ undefined
5. $a^m \cdot a^n = a^{m+n}$	$(4^3)\cdot(4^{-2}) = 4^{3-2} = 4^1 = 4$ $(-6)^{(1/2)}(-6)^{(3/2)} = (-6)^{(1/2)+(3/2)} = (-6)^2 = 36$
6. $(a^m)^n = a^{mn}$	$(3^3)^2 = 3^6 = 729$ $(4^{1/5})^{-10} = 4^{-2} = 1/4^2 = 1/16$
7. $(a^m)\cdot(b^m) = (ab)^m$	$(5^2)\cdot(6^2) = (5 \times 6)^2 = 30^2 = 900$ $(2^{1/2})(8^{1/2}) = (2 \times 8)^{1/2} = 16^{1/2} = \sqrt{16} = 4$
8. $\dfrac{a^m}{a^n} = a^{m-n}$ if $a \neq 0$	$\dfrac{(-7)^5}{(-7)^2} = (-7)^{5-2} = (-7)^3 = -343$ $\dfrac{81^{1/8}}{81^{3/8}} = (81)^{(-1/4)} = \dfrac{1}{81^{1/4}} = \dfrac{1}{\sqrt[4]{81}} = \dfrac{1}{3}$
9. $\dfrac{a^m}{b^m} = (a/b)^m$ if $b \neq 0$	$6^2/2^2 = (6/2)^2 = 3^2 = 9$ $4^7/8^7 = (4/8)^7 = (1/2)^7 = 1/128$

Computations involving roots are greatly facilitated by the following properties of radicals.

Product rule for radicals If a and b are positive real numbers, then

$$\sqrt[n]{ab} = (\sqrt[n]{a})(\sqrt[n]{b})$$

Quotient rule for radicals If a and b are positive real numbers, then

$$\sqrt[n]{a/b} = \frac{\sqrt[n]{a}}{\sqrt[n]{b}}$$

CAUTION!

$$-x^2 \neq (-x)^2$$
$$4x^n \neq (4x)^n$$
$$(x+y)^n \neq x^n + y^n$$
$$\sqrt[n]{x+y} \neq \sqrt[n]{x} + \sqrt[n]{y}$$

EXERCISES

Evaluate each of the following.

21. 0^4
22. 4^0
23. 4^{-2}
24. $4^{1/2}$
25. $4 \cdot 2^4$
26. $(4 \times 2)^4$
27. $(5+3)^2$
28. $2^2 \times 2^3$
29. $2^2/2^3$
30. $2^2 + 2^3$
31. $5^{1/2} \times 5^{3/2}$
32. $5^0 + 5^1$
33. 8^{-3}
34. $8^{1/3}$
35. $(-8)^{-1/3}$
36. $8^{2/3}$
37. $81^{1/2}$
38. $2^{-3} \times 2^4$
39. $9^{-3/2} \times 9^{5/2} \div 9^{1/2}$
40. $(-3)^{-3} \div (-5)^{-2}$
41. $[(-1)^3]^3$
42. $[(-8)^{2/3}]^{1/2}$
43. $5^3/25^3$
44. $\sqrt[3]{27/125}$
45. $\sqrt[3]{375}$
46. $(1000/27)^{1/3}$
47. $\dfrac{25^{3/4}}{25^{1/4}}$
48. $\dfrac{2^{3/4}}{2^{1/2}}$
49. $2^{1/2} \times 2^{1/2}$
50. $(2^{1/2})^3$

1.6 ALGEBRAIC EXPRESSIONS

An ALGEBRAIC EXPRESSION is a mathematical statement indicating that numerical quantities are to be combined by the operations of addition, subtraction, multiplication, division, or extraction of roots. The quantities may be either CONSTANTS or VARIABLES. A constant is a quantity that remains unchanged (holds *constant*) in a given problem; a variable is a quantity that can assume different values in a given problem. Variables are generally represented in an algebraic expression by letters such as x, and y, or z. Constants are generally written as a numeral, but they, too, may sometimes be represented by an alphabetical character.

Each of the following is an algebraic expression.

$4x^3$ $\quad 3x + 2y$ $\quad 9 - x + y - z$ $\quad ax^2 + bx + c$ $\quad 5x^2y - 3xy^2$

Algebraic expressions are composed of MONOMIALS, or TERMS, where a monomial is the product of a nonzero constant and nonnegative integer powers of the variables, as $3x^2$, $5xy$, or $-2x^2yz$. Thus, the terms may be composed of two or more FACTORS. The constant factor, such as 6 in the term $6x$, is referred to as the COEFFICIENT of the variable; but a constant term standing alone, as the 9 in $9 - x + y - z$, is referred to as a CONSTANT TERM. The VARIABLE PART of a term may consist of one variable or the product of two or more variables, as x^2, xy, or x^2yz. Algebraic expressions composed of more than one term are often called MULTINOMIALS.

Addition and Subtraction of Algebraic Expressions

We add or subtract multinomials by combining LIKE TERMS, where like terms are terms which have exactly the same variable part; they differ only in their numerical coefficients. The terms $4x$ and $-5x$ are like terms since both have the same variable part, namely x; they differ only in the value of the coefficients, 4 and -5. The terms $3x^2$ and $5x$ are not like terms since one variable part is x and the other variable part is x^2.

Like terms are combined by summing their coefficients (using the rules for summing real numbers) and bringing forward the variable part of the term. It is the distributive law that enables us to combine terms in this manner.

Example 8 *a.* $2x + 3x = (2 + 3)x = 5x$ *b.* $7x^2y + 8x^2y = (7 + 8)x^2y = 15x^2y$ []

When algebraic expressions are added or subtracted, *only like terms may be combined.* Terms which are not like terms cannot be combined into a simpler form.

Removing parentheses If the various expressions to be summed are enclosed in parentheses, and each set of parentheses is preceded by a positive (plus) sign, the parentheses may simply be removed without affecting the value of any of the terms.

One algebraic expression which is to be subtracted from another algebraic expression is generally contained within a set of parentheses with a negative (minus) sign preceding it. The negative sign before the parentheses affects the entire expression. Thus, the parentheses can be removed only by changing the algebraic sign of each term within the parentheses. (We are actually multiplying each by -1.) Then we proceed as with addition.

Example 9

$$\begin{aligned}(4x^2y - 2xy^2 + 3x - y) - (x^2y - 4xy^2 + x - 2y) \\ &= 4x^2y - 2xy^2 + 3x - y - x^2y + 4xy^2 - x + 2y \\ &= (4x^2y - x^2y) + (-2xy^2 + 4xy^2) + (3x - x) + (-y + 2y) \\ &= 3x^2y + 2xy^2 + 2x + y\end{aligned}$$

[]

Multiplication of Algebraic Expressions

Multiplication of algebraic expressions is accomplished by multiplying the terms. Then the product should be simplified as much as possible by combining like terms.

Multiplying two terms To multiply two terms, multiply their coefficients using the laws of real numbers. Then multiply their variable parts using the rules of exponents.

Example 10

a. $(3x)\cdot(4x) = (3 \times 4)\cdot(x \times x) = 12x^2$

b. $(3x^2y)\cdot(6xy) = (3 \times 6)\cdot(x^2 \times x)\cdot(y \times y) = 18x^3y^2$ []

Multiplying two expressions To multiply two expressions, multiply each term of one expression by every term in the other expression.

Example 11

a.
$$\begin{aligned}(x+3)(x^2+4) &= (x)(x^2+4) + (3)(x^2+4) \\ &= (x)(x^2) + (x)(4) + (3)(x^2) + (3)(4) \\ &= x^3 + 3x^2 + 4x + 12\end{aligned}$$

b.
$$\begin{aligned}(6x^2 - 2x)(x^3 + 2x^2 - 4x) &= 6x^2(x^3 + 2x^2 - 4x) - 2x(x^3 + 2x^2 - 4x) \\ &= (6x^2)(x^3) + (6x^2)(2x^2) + (6x^2)(-4x) + (-2x)(x^3) \\ &\quad + (-2x)(2x^2) + (-2x)(-4x) \\ &= 6x^5 + 12x^4 - 24x^3 - 2x^4 - 4x^3 + 8x^2\end{aligned}$$

Then, combining like terms, the product is simplified to equal

$$6x^5 + 10x^4 - 28x^3 + 8x^2$$

[]

EXERCISES

Perform the indicated operations and simplify.

51. $(8x^2 + 6y^2) - (4x^2 - 3y^2)$
52. $(4x^2 + 5x + 6) - (x^2 - 2x + 4)$
53. $(3x + 5y) - (2x + 4y) + (x - y)$
54. $(x^3 + 7x^2 + 5x - 8) + (x^2 + 8)$
55. $[(1/2)x^2 + (1/4)x + (1/8)] - [(3/2)x^2 + (1/4)]$
56. $(x^2 + 1) - (x^2 + 3) + (2x^2 + 7) - (3x^2 - 1)$
57. $(x^{10} + x^6 + 3x^2) + (5x^{10} - 3x^6 + 2x^2)$
58. $[(1/x) + (1/y)] + [(2/x) + (5/y)]$

Perform the indicated operations and simplify the result.

59. $3(x-5)$
60. $4(x+1/4)$
61. $x^2(x+4)$
62. $(x-1)(x+2)$
63. $(x+3)(x+4)$
64. $(x-2)(x^2+2x+3)$
65. $(x+1)(x^2-x+5)$
66. $(x-1)(x^4+x^2+8)$
67. $(x^2-1)(2x^2+3x+4)$
68. $(x^2+x+1)^2$
69. $(x-1)^3$
70. $(x+y)(2x+3y)(x-5)$

Perform the indicated operations and simplify the result.

71. $x\{3(x+1)(x-3)+2[x+4(x-7)]\}$
72. $x(2x+y)-3(x-4y)^2$
73. $x-\{5+3[(x+1)(x-1)+6x]\}$
74. $4\{2(x+3)-x[5-(x+4)]\}$
75. $2\{x+5[(3+x)(1-x)-4]-(2x+1)(x+4)\}$

Factoring Algebraic Expressions

When two or more expressions are multiplied together, the expressions are termed FACTORS of the product. The name FACTORING is given to THE PROCEDURE OF IDENTIFYING THE FACTORS OF AN EXPRESSION AND RESTATING THE EXPRESSION AS THE PRODUCT OF ITS FACTORS. When we write $x^2+2x-3=(x+3)(x-1)$, we have factored the original expression. The procedure is based on use of the distributive law in reverse.

The following rules aid in the factoring process.

Rule 1	$ax+ay+az=a(x+y+z)$	(common factor)
Rule 2	$ax+by+bx+ay=(a+b)(x+y)$	(grouping of terms)
Rule 3	$x^2+(a+b)x+ab=(x+a)(x+b)$	
Rule 4	$acx^2+(ad+bc)x+bd=(ax+b)(cx+d)$	
Rule 5	$x^2+2ax+a^2=(x+a)^2$	(perfect square)
Rule 6	$x^2-2ax+a^2=(x-a)^2$	(perfect square)
Rule 7	$x^2-a^2=(x+a)(x-a)$	(difference of two squares)
Rule 8	$x^3+a^3=(x+a)(x^2-ax+a^2)$	(sum of two cubes)
Rule 9	$x^3-a^3=(x-a)(x^2+ax+a^2)$	(difference of two cubes)

When factoring, always factor as completely as possible.

Example 12

a. $6x+4y+2z=2(3x+2y+z)$ (by Rule 1)

b.
$$\begin{aligned} x^3+3x-4x^2-12 &= (x^3+3x)+(-4x^2-12)\\ &= x(x^2+3)-4(x^2+3)\\ &= (x^2+3)(x-4) \end{aligned}$$
(by Rule 2)

c. $x^2 + x - 6 = x^2 + (3-2)x + (3)(-2)$ — how?
$= (x+3)(x-2)$ (by Rule 3)

d. $8x^2 + 14x + 3 = (2)(4)x^2 + [(2)(1) + (3)(4)]x + (1)(3)$
$= (2x+3)(4x+1)$ (by Rule 4)

e. $x^2 + 10x + 25 = (x+5)^2$ (by Rule 5)

f. $x^2 - 10x + 25 = (x-5)^2$ (by Rule 6)

g. $x^2 - 16 = (x+4)(x-4)$ (by Rule 7)

h. $x^3 + 8 = (x+2)(x^2 - 2x + 4)$ (by Rule 8)

i. $x^3 - 1 = (x-1)(x^2 + x + 1)$ (by Rule 9) [ ]

EXERCISES

Factor each expression as completely as possible.

76. $8x + 6$
77. $16x^3y^2z + 8x^2yz - 4xyz^2$
78. $x^3 + 2x^2 + 4x + 8$
79. $x^2 + 7x + 12$
80. $x^2 + x - 12$
81. $6 - x - x^2$
82. $6x^2 + 7x - 20$
83. $30x^2 - 7x - 2$
84. $x^2 + 6x + 9$
85. $x^2 + 8x + 16$
86. $x^2 - 4x + 4$
87. $x^3 - 6x^2 + 9x$
88. $x^2 - (4/3)x + (4/9)$
89. $x^2 - 4$
90. $9x^2 - (1/9)$
91. $x^2 - 25$
92. $x^4 - 1$
93. $8x^3 - 1$
94. $27 + 8x^3$
95. $x^3 + 64$
96. $x^{2/3}y - 9x^{8/3}y^3$
97. $x^3 - 4x + 8 - 2x^2$
98. $x^{1/4} - x^{5/4}$
99. $x^8 - 1$

1.7 FURTHER WORK WITH EXPRESSIONS IN FRACTIONAL FORM

A RATIONAL EXPRESSION is the ratio of two algebraic expressions. Examples of such expressions are

$$\frac{5}{x+1}, \quad \frac{x+2}{x-2}, \quad \frac{1+x}{6xyz}, \quad \text{and} \quad \frac{x^2+2x+3}{x-4}$$

The rules governing mathematical operations on real numbers written as fractions extend to these same operations performed on algebraic expressions which are in fractional form.

Addition and Subtraction of Rational Expressions

To add (or subtract) two expressions each of which is in fractional form and which have the same denominator, add (or subtract) the terms in the numerators and bring forward the common denominator. If the two fractions have different denominators, use the fundamental principles of fractions to obtain the same denominator for each expression. Then add (or subtract) the terms in the numerators and place over the common denominator.

Example 13

a. $$\frac{x}{x+1} - \frac{5}{x+1} = \frac{x-5}{x+1}$$

b. $$\frac{2x}{x(x-1)} + \frac{4x+1}{x^2} = \frac{x}{x} \cdot \frac{2x}{x(x-1)} + \frac{x-1}{x-1} \cdot \frac{4x+1}{x^2}$$
$$= \frac{2x^2}{x^2(x-1)} + \frac{4x^2-3x-1}{x^2(x-1)}$$
$$= \frac{6x^2-3x-1}{x^2(x-1)}$$ []

Multiplying Rational Expressions

To multiply two expressions which are in fractional form, multiply the numerators and then multiply the denominators.

Example 14

$$\frac{x}{x+2} \cdot \frac{x^2+1}{x+3} = \frac{x(x^2+1)}{(x+2)(x+3)} = \frac{x^3+x}{x^2+5x+6}$$

(Actually, it may be desirable to leave both the numerator and the denominator in factored form.) []

Dividing Rational Expressions

To divide two expressions which are written as fractions, invert the denominator fraction and proceed as with multiplication.

Example 15

$$\frac{(x+4)/x}{(x-5)/3} = \frac{x+4}{x} \cdot \frac{3}{x-5} = \frac{3(x+4)}{x(x-5)}$$ []

Dividing one term into another To divide one term by another term, divide the coefficients using the laws of real numbers. Then divide the variables using the rules of exponents.

Example 16

a. $$\frac{18x^6}{6x^2} = \left(\frac{18}{6}\right)\left(\frac{x^6}{x^2}\right) = 3x^4$$

b. $$\frac{8x^3y^4z}{4xy^2z} = \left(\frac{8}{4}\right)\left(\frac{x^3}{x}\right)\left(\frac{y^4}{y^2}\right)\left(\frac{z}{z}\right) = 2x^2y^2$$ []

Dividing an expression by a single term To divide an algebraic expression by a single term, divide each term of the expression by the term and algebraically sum the individual quotients.

Example 17

$$\frac{6x^2y - 15xy + 5xy^2}{xy} = \frac{6x^2y}{xy} - \frac{15xy}{xy} + \frac{5xy^2}{xy} = 6x - 15 + 5y$$ []

When the divisor is a factor of the dividend If the dividend can be factored in such a way that the divisor is one of the factors, then the quotient can be found by "cancelling" the common factors found in both dividend and divisor.

Example 18 Divide $x^2 + 7x + 12$ by $x + 4$.
The dividend can be factored, as

$$x^2 + 7x + 12 = (x + 4)(x + 3)$$

Then

$$\frac{x^2 + 7x + 12}{x + 4} = \frac{\cancel{(x + 4)}(x + 3)}{\cancel{(x + 4)}} = (x + 3)$$

[]

CAUTION! Cancellation is permissible *only when the factors to be cancelled are factors of the entire numerator and the entire denominator.*

$$\frac{a + b}{c + b} \neq \frac{a}{c}$$

The *bs* cannot be cancelled because they are not factors of the numerator and the denominator.

Dividing by expressions which contain fractions Division of fractions requires that both the numerator and the denominator be expressed as simple fractions before division takes place.

Example 19

$$\frac{2 - (1/x)}{3/x^2} = \frac{\frac{2x}{x} - \frac{1}{x}}{\frac{3}{x^2}} = \frac{\frac{2x - 1}{x}}{\frac{3}{x^2}} = \frac{2x - 1}{x} \cdot \frac{x^2}{3} = \frac{x(2x - 1)}{3}$$

[]

EXERCISES

Perform the indicated opertions and simplify the result.

100. $\frac{2}{x} + \frac{3}{2x}$

101. $\frac{2}{x} + \frac{1}{x + 1}$

102. $\frac{4}{x - 1} - \frac{3x}{x - 1}$

103. $\frac{2}{x - 2} + \frac{5}{x + 1}$

104. $\frac{2a}{a - b} - \frac{3b}{a + b}$

105. $\frac{4}{3x} + \frac{4x - 1}{2}$

106. $\frac{5}{x + 1} + 3$

107. $\frac{5 + x}{x^2} + \frac{x^2}{5 + x}$

108. $\frac{5 + x}{x^2} + \frac{3 + x}{x + 1}$

109. $\frac{x + 2}{3x} - \frac{5}{x} + 1$

110. $\frac{3}{x} \times \frac{5}{x}$

111. $\frac{x}{y} \times \frac{y - 1}{x - 1}$

112. $\frac{4x}{y} \times \frac{5y}{xz}$

113. $\frac{3y}{x + 1} \times \frac{2x}{4y - 1}$

114. $\dfrac{x+1}{3x} \times \dfrac{x-1}{5}$

115. $\dfrac{5}{3-x} \times \dfrac{x-3}{7}$

116. $\dfrac{x-2}{3} \div \dfrac{4}{x-2}$

117. $\dfrac{x}{x+1} \div \dfrac{y}{xy+y^2}$

118. $\dfrac{2x+8}{3} \div \dfrac{x+4}{9x-6}$

119. $\dfrac{3+(2/x)}{x-(3/x^2)}$

120. $\dfrac{x}{y} \div \dfrac{2x^2-x}{3y} + 4x$

121. $\dfrac{x^2-x-2}{2} \div \dfrac{x-2}{4x^2-6x}$

122. $\dfrac{1}{2x-y} \div \dfrac{y-2x}{2} \times \dfrac{3}{x+1}$

123. $\dfrac{x+1}{x-1} \times \dfrac{x-1}{3x+6} \div \dfrac{x+2}{3}$

Find the indicated quotient.

124. $\dfrac{10x^2}{5x}$

125. $\dfrac{12x^2y^3}{4x^3y^2}$

126. $\dfrac{8x^3-4x^2}{2x^2}$

127. $\dfrac{6x^6+15x^4-3x^2}{3x}$

128. $\dfrac{x^2-6x+8}{x-2}$

129. $\dfrac{x^4-x^2-12}{x+2}$

1.8 EQUATIONS

An EQUATION is a mathematical statement of equality between two expressions. Equations may involve one or more variables. Examples of equations in one variable are $2x - 1 = 0$ and $x^2 = 4$, while $x + y = 6$ and $x + 1 = y + 2$ are equations in two variables. An equation in three variables is $x + y = 10 - 2z$.

Two kinds of equations, IDENTITIES and CONDITIONAL EQUATIONS, should be distinguished. An IDENTITY is an equation which is true for all allowable values of the variables involved, as

$$3x + 1 = \frac{6x+2}{2} \qquad \text{and} \qquad 3(x+y) = 3x + 3y$$

Any values assigned to the variables will make the two sides of the equations equal.

A CONDITIONAL EQUATION is true for only a limited number of values of the variables. The equation

$$x + 2 = 5$$

becomes a true statement for only one value of x—that is, $x = 3$.

If an equation contains only one variable, any value of that variable which makes the equation true is called a SOLUTION, or ROOT, of the equation. The solution of the equation $x + 2 = 5$ is $x = 3$. The solution of the equation $2x = 8$ is $x = 4$.

A solution for an equation in two variables, such as x and y, is any ORDERED PAIR of values (x,y) that yields a correct statement when substituted for x and y, respectively, in the equation. For example ($x = 2$, $y = 5$), or, more compactly, (2,5), is a solution to the equation $3x + y = 11$ since substitution of 2 for x and 5 for y yields $3(2) + 5 = 11$, a correct statement. Note that (1,8) also represents a solution to the same equation since $3(1) + 8 = 11$. There are, in fact, many ordered pairs of values for the variables that represent solutions. All such ordered pairs are said to be members of the SOLUTION SET for the equation. Among these are (0,11), (1/3, 10), (−1,14), and (10,−19). Needless to say, there are many, many ordered pairs that do not represent solutions to this equation.

The solution to an equation in three variables—say x, y, and z—would be any ORDERED TRIPLE (x,y,z) of values which would, when substituted into the equation, yield a correct statement.

Finding the Solution of an Equation

The process of finding the solution, or solutions, of an equation, if any exist, is referred to as SOLVING THE EQUATION. The exact procedure followed in solving an equation depends upon the nature of the equation. Of fundamental importance to the procedure of solving any equation, however, are mathematical operations which allow us to transform an equation into an equivalent equation.

Two equations are said to be EQUIVALENT if and only if they have the same solution sets.

$$3x + 1 = 10 \qquad \text{and} \qquad 3x = 9$$

are equivalent equations since they both have the same solution, namely $x = 3$. In much the same manner,

$$2x + y = 5 \qquad \text{and} \qquad 4x + 2y = 10$$

are equivalent equations since each ordered pair (x,y) that satisfies one also satisfies the other.

The following OPERATIONS OF EQUIVALENCE can be applied to one equation to obtain an equivalent equation.

1. The same constant may be added to, or subtracted from, both sides of an equation.
2. Both sides of an equation may be multiplied, or divided, by the same nonzero constant.
3. A term which appears on either side of the equation may be added to, or subtracted from, both sides of the equation. (We speak of transposing the term.)

Let us illustrate these operations. First, by subtracting the constant 1 from both sides of $3x + 1 = 10$, we obtain an equivalent equation $3x = 9$. Or, by multiplying each side of the equation $2x + y = 5$ by the constant 2, we obtain an equivalent equation $4x + 2y = 10$. Lastly, by subtracting x from both sides of the equation $x + y = 15$, we obtain $y = 15 - x$, which is an equivalent equation.

We attempt to use the equivalency operations in such a sequence that we isolate the variable on the left side of the equation and its solution value on the right side of the equation.

Example 20 Find the real number x for which $3x - 2 = 13$ is a true statement.

Using the operations listed above, we proceed as follows:

$3x - 2 = 13$

$3x - 2 + 2 = 13 + 2$ (by adding the constant 2 to each side)

$3x = 15$ (by collecting terms)

$(3x)(1/3) = (15)(1/3)$ (by multiplying each side by the constant 1/3)

$x = 5$ (by collecting terms)

Hence, the value of x which makes the equation a true statement is $x = 5$. [ ]

Example 21 Find two ordered pairs of values (x,y) which are members of the solution set of the equation $2x + y = 12$.

There will generally be many solutions to an equation in two variables. To determine these solutions we may select an allowable value of one variable and substitute this in the equation to obtain the corresponding value of the other variable. The process may be facilitated by isolating one variable on the left side of the equation. This can be accomplished by transposing terms containing the other variable to the right side of the equation.

Thus, by adding $-2x$ to each side of the equation $2x + y = 12$, we obtain the equivalent equation

$$y = 12 - 2x$$

Now if we arbitrarily select $x = 0$, we find that

$$y = 12 - 2(0) = 12$$

Thus, (0,12) is an ordered pair that represents a solution to the equation. Or, if we arbitrarily select $x = 1$, we find that

$$y = 12 - 2(1) = 10$$

so that (1,10) is another ordered pair that represents a solution to the equation. [ ]

Additional Solution Procedures

There are operations in addition to the three stated above that we sometimes use to solve equations. Let us add these to the list.

4. Both sides of an equation may be multiplied, or divided, by a nonzero expression involving the variable.
5. Both sides of an equation may be raised to the same power.

While these last two operations generally provide satisfactory results, we should be cautious about using them since *they do not necessarily result in equivalent equations.* Consider the equation $x = 1$, whose only solution is $x = 1$. If both sides of this equation are multiplied by the variable x (Operation 4), the result is $x^2 = x$, which has two solutions, $x = 1$ and $x = 0$. Because they do not have the same solution sets, the equations $x = 1$ and $x^2 = x$ are not equivalent equations.

Or, consider the equation $x = 3$. If both sides of this equation are squared (Operation 5), the result is $x^2 = 9$, which is a true statement for $x = +3$ and for $x = -3$ as well. Thus, $x = 3$ and $x^2 = 9$ are not equivalent equations.

So, whereas Operations 1 through 3 guarantee that the equations are equivalent, Operations 4 and 5 carry no such guarantee. We should always verify that any "solution" produced by using one of these latter operations actually satisfies the *original* equation by substituting the values back into that equation.

Example 22 Solve the equation

$$\frac{4}{x-3} = \frac{3}{x+2}$$

To solve a fractional equation, we first write in a form free of fractions. Multiplying both sides by $(x - 3)(x + 2)$, we obtain

$$(x-3)(x+2) \cdot \frac{4}{(x-3)} = (x-3)(x+2) \cdot \frac{3}{(x+2)}$$

$$(x+2)4 = (x-3)3$$

$$4x + 8 = 3x - 9$$

$$4x - 3x = -9 - 8$$

$$x = -17$$

Because we multiplied each side of the equation by an expression involving the variable, we must verify that the solution obtained actually satisfies the original equation. We thus substitute

$$\frac{4}{-17-3} = \frac{3}{-17+2}$$

$$\frac{4}{-20} = \frac{3}{-15}$$

$$\frac{-1}{5} = \frac{-1}{5}$$

Both sides of the equation are equal, indicating that $x = -17$ is, indeed, a solution of the original equation. [ ]

Finding the Solution by Factoring

Whenever the product of two or more quantities is zero, at least one of the quantities must be zero. Because of this principle, factoring can be very effectively used to find the solutions to many equations.

Example 23 Find the solutions to the equation $x^2 + 3x + 2 = 0$.

The left side of the equation factors as $x^2 + 3x + 2 = (x + 1)(x + 2)$ Thus,

$$x^2 + 3x + 2 = (x + 1)(x + 2) = 0$$

Now, if their product equals zero, either $(x + 1) = 0$ *or* $(x + 2) = 0$. Solving these equations

$$x + 1 = 0 \qquad \text{and} \qquad x + 2 = 0$$
$$x = -1 \qquad\qquad\qquad x = -2$$

we obtain the solutions, $x = -1$ and $x = -2$, for the original equation. []

EXERCISES

Find two members of the solution set for each of the following equations.

130. $x + y = 25$
131. $2x + 3y = 13$
132. $3y = 6x + 12$

Solve the following equations in one variable.

133. $x - 3 = 7$
134. $2x + 3 = 15$
135. $2(x + 1) = 12$
136. $4 - x = 2x + 1$
137. $\dfrac{4}{x} = 16$
138. $\dfrac{2x - 5}{4} = \dfrac{1}{2}$
139. $\dfrac{1}{x} + \dfrac{2}{7} = \dfrac{5}{7}$
140. $\dfrac{2x - 5}{4x + 1} = 6$
141. $\sqrt{x + 11} = 4$
142. $x + 2 = 2\sqrt{4x - 7}$
143. $\dfrac{9}{x - 4} = \dfrac{3x}{x - 4}$
144. $\dfrac{1}{x^2 - 9} = \dfrac{3x}{x + 3} + \dfrac{1}{x - 3}$
145. $(x - 3)^{3/2} = 8$
146. $4x^2(x - 3) - 9(x - 3) = 0$
147. $x^2 - x = 20$
148. $\dfrac{3x + 2}{x - 3} - \dfrac{3x - 1}{x + 2} = \dfrac{5}{x^2 - x - 6}$
149. $x^2 - 8x = 0$
150. $x^2 - 2x - 3 = 0$

1.9 INEQUALITIES AND THEIR SOLUTION SETS

An INEQUALITY is an expression relating the condition that either two quantities are not equal or that they may or may not be equal. The first is a STRICT INEQUALITY and involves the conditions "greater than" or "less than"; the latter involves the conditions "equal to or greater than" or "equal to or less than." Examples of inequalities are

$$3x + y \geq 15 \qquad x \leq y + 5 \qquad x^2 + 3 > 12 \qquad x + y < x - 5$$

Properties of Inequalities

For any real numbers a, b, and c, the following properties of inequalities hold.

1. If $a > b$, then $a + c > b + c$. (*The direction of the inequality remains unchanged if the same constant is added to both sides of the inequality.*)
2. If $a > b$ and c is positive, then $ac > bc$. (*The direction of the inequality remains unchanged if both sides of the inequality are multiplied by the same positive constant.*)
3. If $a > b$ and c is negative, then $ac < bc$. (*The direction of the inequality is reversed if both sides are multiplied by the same negative constant.*)

Solving Inequalities

To "solve" an inequality means to find all values for the variables that will make the statement true. These values constitute the SOLUTION SET for the inequality. Inequalities, like equations, are solved by obtaining a series of equivalent inequalities until we obtain an inequality with an obvious solution set.

Example 24 Find the solution set for the inequality $4 + 3x \leq 6 + x$.

We proceed as follows:

$4 + 3x \leq 6 + x$	
$4 + 3x - x \leq 6 + x - x$	(by adding $-x$ to each side)
$4 + 2x \leq 6$	(by collecting terms)
$4 + 2x - 4 \leq 6 - 4$	(by adding -4 to each side)
$2x \leq 2$	(by collecting terms)
$(2x)(1/2) \leq (2)(1/2)$	(by multiplying each side by $1/2$)
$x \leq 1$	(by collecting terms)

Figure 1.3

Solution Set for the Inequality $4 + 3x \leq 6 + x$

The solution set for the inequality consists of all real x such that $x \leq 1$ (see Figure 1.3). []

Example 25 Find the solution set for the inequality $2x + 8 < -4 + 5x$.

We proceed as follows:

$2x + 8 < -4 + 5x$	
$2x + 8 - 5x < -4 + 5x - 5x$	(by adding $-5x$ to each side)
$-3x + 8 < -4$	(by collecting terms)
$-3x + 8 - 8 < -4 - 8$	(by adding -8 to each side)
$-3x < -12$	(by collecting terms)
$(-3x)(-1/3) > (-12)(-1/3)$	(by multiplying both sides by $-1/3$, thereby reversing the direction of the inequality)
$x > 4$	(by collecting terms)

Figure 1.4

Solution Set for the Inequality $2x + 8 < -4 + 5x$

The inequality $2x + 8 < -4 + 5x$ is true for all $x > 4$ (see Figure 1.4). []

Example 26 Find the solution set for

$$\frac{3x}{x+4} > 2$$

To solve the inequality, we will multiply both sides by $x + 4$. Now, the quantity $(x + 4)$ may be positive, or it may be negative. And the two cases must be considered separately.

If $(x+4) > 0$ (or $x > -4$)

$$\frac{3x}{x+4}(x+4) > (2)(x+4)$$

$$3x > 2x + 8$$

$$x > 8$$

The result, $x > 8$, is compatible with the condition $x > -4$. This solution is, thus, $x > 8$.

If $(x+4) < 0$ (or $x < -4$)

$$\frac{3x}{x+4}(x+4) < (2)(x+4)$$

$$3x < 2x + 8$$

$$x < 8$$

The result, $x < 8$, is compatible with the condition $x < -4$ *only if the interval* $-4 \le x < 8$ *is excluded.* This solution, therefore, is $x < -4$.

Figure 1.5

Solution Set for the Inequality $\frac{3x}{x+4} > 2$

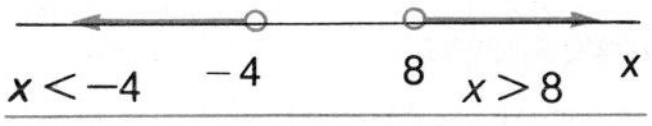

$x<-4$ -4 8 $x>8$ x

The graph of the solution set is shown in Figure 1.5. It consists of two nonoverlapping OPEN RAYS, consisting of all points on the real number line exterior to the closed interval $-4 \le x \le 8$. []

EXERCISES

Find the solution set for each of the following inequalities. Picture the solution set on the real number line.

151. $3x + 5 \le 4$
152. $4x + 3 > 2x - 1$
153. $5 - 3x \ge 6$
154. $-2x + 7 \le 5$
155. $6 - 3x \le 2x - 4$
156. $5x + 4 \ge 3x - 6$
157. $(3x - 2)^2 < 9x^2 + 7$
158. $7x - 2 < \frac{4x-2}{3}$
159. $\frac{4x}{x-4} > 3$
160. $\frac{3}{1-2x} < 1$

1.10 SCIENTIFIC NOTATION

It is generally quite cumbersome to write very large or very small numbers in standard decimal notation or to make computations using such numbers. We may find it more convenient to represent such numbers in SCIENTIFIC NOTATION, that is, as the product of a number between 1 and 10 and a power of 10.

Example 27

$4 = 4 \times 10^0$	$0.4 = 4 \times 10^{-1}$
$45 = 4.5 \times 10^1$	$0.045 = 4.5 \times 10^{-2}$
$456 = 4.56 \times 10^2$	$0.0045 = 4.5 \times 10^{-3}$
$45600 = 4.56 \times 10^4$	$0.000045 = 4.5 \times 10^{-5}$
$456{,}000{,}000 = 4.56 \times 10^8$	$0.0000000045 = 4.5 \times 10^{-9}$ []

Note that the power of 10 corresponds to the number of places the decimal point is moved in order to form a number between 1 and 10. This power is positive if the decimal point was moved to the left and negative if it was moved to the right.

EXERCISES

Write each number in scientific notation.

161. 351,000,000
162. −0.000000793
163. 53,400,000
164. 74,000
165. 0.0000821
166. −371,000
167. 1500
168. −0.0033
169. 0.014

1.11 LOGARITHMS

Essentially, A LOGARITHM IS AN EXPONENT.

> *Given some positive constant* a, *where* a ≠ 1, *the unique value of* x *associated with each positive value of* y *such that* $y = a^x$ *is called the LOGARITHM OF* y *TO THE BASE* a. *This statement is symbolized by*
>
> $$\log_a y = x$$

If x is the logarithm of y to the base a, then x is the power to which a must be raised to obtain y. The logarithm is defined only for base $a > 0$, and $a \neq 1$.

Example 28

a. Since $3^2 = 9$, $\log_3 9 = 2$

b. Since $5^0 = 1$, $\log_5 1 = 0$

c. Since $4^1 = 4$, $\log_4 4 = 1$

d. Since $(25)^{1/2} = 5$, $\log_{25} 5 = 1/2$

e. Since $(1/2)^{-1} = 2$, $\log_{1/2} 2 = -1$

f. Since $(2)^{-3} = 1/8$, $\log_2 (1/8) = -3$ []

Any positive number has a unique logarithm to a definite base other than one. Because the base a is positive, $y = a^x$ is always positive. Consequently, $\log_a y$ makes sense only if y is positive. THE LOGARITHM OF ZERO OR A NEGATIVE NUMBER IS NOT DEFINED.

Common Logarithms and Natural Logarithms

Although the base of the log system may be any positive number not equal to one, two bases are more frequently used than any others. Logarithms using the base 10 are known as COMMON LOGARITHMS and are often denoted simply as "log" x without any explicit mention of the base. For example, $log\ 5.64$ is automatically interpreted as $log_{10} 5.64$.

NATURAL LOGARITHMS use the familiar base $e = 2.71828\ldots$ and are often symbolized $ln\ x$. Thus, $ln\ 27$ is automatically interpreted as $log_e 27$.

If the base of the system of logarithms is anything other than 10 or e, the specific value must be denoted, as $log_3 x$, or $log_5 7$.

Even the inexpensive hand-held calculators frequently have "$log\ x$" and/or "$ln\ x$" keys, which greatly simplify the process of determining the logarithm of a number. However, in case such a calculator is not available for use, Appendix Table I is a table of natural logarithms.

Change of base rule The following rule allows us to convert from a system of logarithms using one base to an equivalent system using another base.

Change of base rule

$$log_a y = \frac{log_b y}{log_b a} = \frac{1}{log_y a} \quad (1.1)$$

Example 29

a. $$log_3 15 = \frac{ln\ 15}{ln\ 3} = \frac{2.708050}{1.098612} = 2.464974$$

b. $$log_3 e = \frac{ln\ e}{ln\ 3} = \frac{1}{1.098612} = 0.910239$$

c. $$log\ e = \frac{ln\ e}{ln\ 10} = \frac{1}{2.302585} = 0.434294$$

[]

Laws Governing Logarithms

The laws governing exponents may be translated into corresponding laws governing logarithms.

Rules governing logarithms *For any base* $a > 0$, $a \neq 1$, *and positive real numbers* m *and* n.

Rule 1 $log_a 1 = 0$
Rule 2 $log_a a = 1$
Rule 3 $log_a a^x = x$
Rule 4 $log_a(m \cdot n) = log_a m + log_a n$
Rule 5 $log_a(m/n) = log_a m - log_a n$
Rule 6 $log_a(m^n) = n \cdot log_a m$
Rule 7 $log_a \sqrt[n]{m} = log_a(m^{1/n}) = (log_a m)/n$ (*for* n *a positive integer*)

Logarithmic and Exponential Transformations

A LOGARITHMIC TRANSFORMATION involves transforming both sides of an equation to logarithms. An EXPONENTIAL TRANSFORMATION involves expressing both sides of an equation in exponential form. Because the same operation is performed on both sides of the equation, the integrity of the equation is not violated.

Example 30

Find x given $5^{2x} = 259$.

We take the natural logarithm of each side of the equation, as

$ln\ 5^{2x} = ln\ 259$	(by taking the natural log of each side)
$(2x)(ln\ 5) = ln\ 259$	(by Rule 6)
$2x = \dfrac{ln\ 259}{ln\ 5}$	(by dividing each side by ln 5)
$2x = \dfrac{5.556828}{1.609438}$	(by using hand-held calculator or Appendix Table I)
$2x = 3.452651$	
$x = 1.726326$	(by dividing each side by 2) []

Example 31

Find x given $ln(2x + 3) = 8$.

We begin by expressing each side of the equation as a power of the base e, as

$$e^{ln(2x+3)} = e^8$$

Now, because $ln(2x + 3)$ is the power to which we would have to raise e in order to get $2x + 3$, $e^{ln(2x+3)} = 2x + 3$. Hence, we continue by writing

$2x + 3 = e^8$	
$2x + 3 = 2980.957987$	(by using hand-held calculator or Appendix Table II)
$2x = 2977.957987$	(by adding -3 to each side)
$x = 1488.978994$	(by dividing each side by 2) []

EXERCISES

Rewrite each of the following in logarithmic notation; that is, in the format $\log_a y = x$.

170. $(16)^{1/2} = 4$
171. $4^3 = 64$
172. $(243)^{0.2} = 3$
173. $5^{-2} = 1/25$
174. $e^2 = 7.389056$
175. $(1/4)^{-1} = 4$
176. $e^0 = 1$
177. $e^{2.302585} = 10$
178. $(10)^3 = 1,000$

Rewrite each of the following in exponential notation; that is, in the format $y = a^x$. *Then solve for* x.

179. $\log_5 x = -2$
180. $\log_{1/2} 1/8 = x$
181. $\log_x 9 = 2$
182. $\log_{32} 2 = x$
183. $\ln x = 3$
184. $\log_{25} 1 = x$
185. $\log_3 x = -1$
186. $\log_x 2 = 1/2$
187. $\log_4 64 = x$

Solve each of the following expressions for x.

188. $10^x = 3$
189. $2 = (1 + x)^5$
190. $e^x = 14$
191. $e^{2x} = 33$
192. $25.7 = e^{0.4x}$
193. $8.98 = e^{-0.1x}$
194. $2^{-3x} = 1/64$
195. $5 = 1.14^x$
196. $(3^{x+1})^2 = 3^{12}$
197. $4^{-x} = 0.0156$
198. $3^{4-2x} = 27^{2x}$
199. $e^{x^2} = 6$
200. $10^{x^2+1} = 1{,}806$
201. $\ln 5 + \ln x = 2.7081$
202. $x = \ln e^3$
203. $x = \log_5 5^3$
204. $\log 4x - \log(x + 3) = \log 1$
205. $\ln e^{3+x} = 8$
206. $\log_2 16 = x + 3$
207. $\log(x + 1) - \log(2x) = \log 3$
208. $\ln(3x + 5) = 11$
209. $\ln x^2 = 15$
210. $\log(x + 4) = 1.5$
211. $\log(x^2 + 1) = 5$
212. $\log x = (2/3)\log 8 + (1.2)\log 9 - \log 6$

CHAPTER 2 A Review of Set Theory and Counting

A PREVIEW of questions you will be able to answer after studying this chapter:

Records indicate that of the 300 persons attending an office-automation conference, 120 visited Exhibit A, 115 visited Exhibit B, 87 visited Exhibit C, 30 visited Exhibits A and B, 27 visited Exhibits B and C, 22 visited Exhibits A and C, and 12 visited Exhibits A, B, and C. How many persons attending the conference visited none of the exhibits? How many persons attending the conference visited only one of the exhibits? How many persons attending the conference visited at least one of the exhibits? How many persons attending the conference visited exactly two of the exhibits? How many persons attending the conference visited at least two of the exhibits?

A customer considering the purchase of a new automobile is told that the standard model is available in six colors, two types of transmissions, and five different accessory packages, while the sports model is available in seven colors, three types of transmissions, and four different accessory packages. How many choices of different automobiles does the customer have?

A salesperson is needed in each of five different sales territories. If 11 equally qualified persons apply for the jobs, in how many ways can the jobs be filled?

In how many ways can four identical motorcycles and three identical wheelchairs be displayed in a row in a display window?

Karen is taking a 15-question essay examination. If she must answer only 12 of the 15 questions, in how many different ways can she take the examination?

Pedro's Pizza Parlor offers 11-inch, 13-inch, and 17-inch pizzas with choices of cheese, sausage, pepperoni, mushrooms, onions, and peppers or any combination thereof, and thick or crispy crust. How many different pizzas are available?

The 24 students of an operations management class are assigned to one of four teams, each team consisting of six members. How many ways can the teams be set up? How many distinct teams are possible if, after the groups of six are designated, a president, vice-president, treasurer, and production manager are elected for each team?

Set theory serves as the foundation for modern mathematics. Not only does it provide the vehicle for developing precise definitions for the important concepts of relations and functions, it also serves as a thoroughly developed arithmetic for manipulating groups of objects. Thus, set theory aids in the analysis of a significant number of problem areas in the business environment which are not adaptable to conventional algebraic techniques. A knowledge of the fundamental concepts of set theory can pave the way to a clearer understanding of probability and the methods of statistical inference.

2.1 THE CONCEPT OF A SET AND ITS ELEMENTS

A SET is a WELL-DEFINED COLLECTION OF DISTINCT OBJECTS. We are all familiar with such notions as a "set" of dishes, a "set" of golf clubs, or a "set" of encyclopedia volumes. All books published in 1980, the members of the graduating class of 1931 of Holyoke College, the oceans of the world, the set of integers between 1 and 1,000, the automobiles manufactured in Detroit in 1982, the cards in an ordinary bridge deck—all these also constitute sets. And the objects contained in a set need not be as concrete as the examples mentioned. Abstract concepts as well—such as all positive integers, all points in the interval $[a,b]$ on a straight line, and all nonnegative rational numbers—may be the objects found in a set. The items comprising a set, then, may be of any sort: people, things, geographical locations, geometrical figures, outcomes of experiments. Each object in a set is called an ELEMENT or MEMBER of the set.

In order to form a set, the collection of objects must meet these two requirements. First, *the aggregate must be well defined.* The individual items must have a common characteristic or characteristics that cause them to belong to a particular set. A rule or method must exist by which it is possible to determine for any object whatsoever whether or not that object is a member of the set in question. The set "all subscribers to *Atlanta Magazine* as of January 1, 1983" is a well-defined set; the set of "worthwhile" books published by Tripleday Publishing Company is not well defined.

Second, *the elements of a set are distinct; no object in a set is counted twice.* When one is listing the elements of a set, once an object is recorded, it is not repeated. The set of letters in the word *Miami,* then, is not a set containing five letters but is, instead, a set of three distinct letters: *m, i,* and *a.* The sequence in which the elements are listed when they are enumerated is immaterial.

Set Notation

Usually capital letters such as A, B, X, and Y are used to denote sets, while lowercase letters such as a, b, x, and y are used to represent the individual elements of a set.

Sets may be described in either one of two ways:

1. *The roster, or listing, method.* All elements of the set are listed, separated by commas, and enclosed in braces { }.
2. *The rule, or defining-property, method.* The rule which can be used to determine whether or not an object belongs to the set is stated and enclosed in braces.

Thus, the set A, consisting of the integers between 5 and 10, may be written

$$A = \{6, 7, 8, 9\}$$

This notation is read, "The set A whose elements are 6, 7, 8, and 9." Or, the same set may be denoted by the defining-property method as

$$A = \{a \mid a \text{ is an integer and is between 5 and 10}\}$$

This notation would be read, "A is the set of all a's such that a is an integer between 5 and 10."

Elements of a Set

In set notation, the symbol $\in$ means "is an element of," or "belongs to," or "is a member of" a set. Contrarily, the symbol $\notin$ means "is not an element of" or "does not belong to" a set.

Example 1

Let the set $X = \{x | x$ is a positive integer less than 10 and x is exactly divisible by 4$\}$. Then $8 \in X$ but $7 \notin X$. []

Example 2

Let the letter a represent Mr. Anderson and the letter B represent the set of directors on the board of First City National Bank. Then $a \in B$ indicates that Mr. Anderson is a member of the Board of Directors of First City National Bank; $a \notin B$ indicates that Mr. Anderson is not a member of the Board of Directors of First City National Bank. []

Finite and Infinite Sets

If a set has a definable number of elements, it is called a FINITE SET. It is quite possible for the set to contain a fantastically large number of elements and still be a finite set. Whenever it is theoretically possible for the elements to be enumerated in some sequence and then counted one by one until the final element is reached, the set is finite.

If the number of elements in a set is without limit, the set is said to be an INFINITE SET. A simple example of an infinite set is the set of positive integers 1, 2, 3, . . . , the counting, or natural, numbers.

An infinite set is said to be DENUMERABLY or COUNTABLY INFINITE if its members, although unlimited in number, can be arranged in a sequence so that every member has a definite place in the sequence and any two elements are separated from each other by some nonzero distance.

Other infinite sets cannot be put in such a sequence—for example, the set of irrational numbers in the interval [0,1]. Such sets are known as NONDENUMERABLY, or UNCOUNTABLY, INFINITE SETS.

Finite and countably infinite sets are often referred to as DISCRETE SETS. A CONTINUOUS SET is an uncountably infinite set.

Equal Sets

Two sets, A and B, are said to be EQUAL if and only if they each contain exactly the same elements. Equality is symbolized by writing $A = B$ or $B = A$. If either of the sets has at least one element that does not belong to the other set, the two sets are not equal. This inequality is symbolized by writing $A \neq B$ or $B \neq A$.

Universal Sets

In any analysis where set theory is employed, a basic set which contains all elements to be considered in that investigation is tacitly assumed to exist. This all-encompassing master set is called the UNIVERSAL SET or the UNIVERSE and is denoted by the symbol $\mathcal{U}$. All other sets considered in the investigation are assumed to be defined on this basic set.

Note that a different universal set is defined for each different problem or investigation.

The Null Set

The set that contains no elements whatever is called the NULL or EMPTY SET and is denoted by the symbol ϕ, or by $\{\}$. The set of all joggers who regularly traverse 10 miles in one minute, the set of all football players weighing more than 2,000 pounds, the set of all employees who have worked for the city government for more than 150 years are examples of empty sets.

Because any two null sets are equal, it is logical to refer to *the* null set.

2.2 SUBSETS

If all the elements of set A are also elements of set B, A is called a SUBSET of B. The relationship is symbolized $A \subset B$, read "A is a subset of B." Although $A \subset B$ indicates that every element belonging to set A is also an element of B, all elements in B may or may not be included in A for the statement "A is a subset of B" to be true.

The symbol $\not\subset$ is used to indicate that one set is not a subset of another set.

Example 3

Given $A = \{1, 2\}$, $B = \{1, 2, 3\}$ and $C = \{2, 3, 4\}$. The set A is a subset of the set B, but A is not a subset of C. That is, $A \subset B$ but $A \not\subset C$. []

Every set is a subset of itself; the null set is a subset of every set.

Venn Diagrams

When considering a universal set and its subsets, it is often helpful to make a geometrical representation of these sets and the relationships between them. VENN DIAGRAMS are used to illustrate sets pictorially. A large rectangle is commonly employed to symbolize the universal set $\mathcal{U}$, while circles or ovals or other simple shapes are drawn inside the rectangle to depict subsets of $\mathcal{U}$. The only stipulation is that the symbols used to represent the subsets must lie entirely within the box used to represent the universal set. The size or shape of the configurations has no direct bearing on the number of elements of the set and subsets.

Figure 2.1 shows the subsets A, B, and C defined on a universal set $\mathcal{U}$ and illustrates that $A \subset B$, $A \not\subset C$, and $B \not\subset C$.

Number of Subsets of a Set

Once the universal set $\mathcal{U}$ has been defined in a particular analysis, all sets which can be formed from the elements of $\mathcal{U}$ are known as subsets of $\mathcal{U}$. The total number of possible subsets depends on the number of elements of $\mathcal{U}$.

> A set with n elements has 2^n possible subsets.

Figure 2.1

$A \subset B$, $A \not\subset C$, and $B \not\subset C$

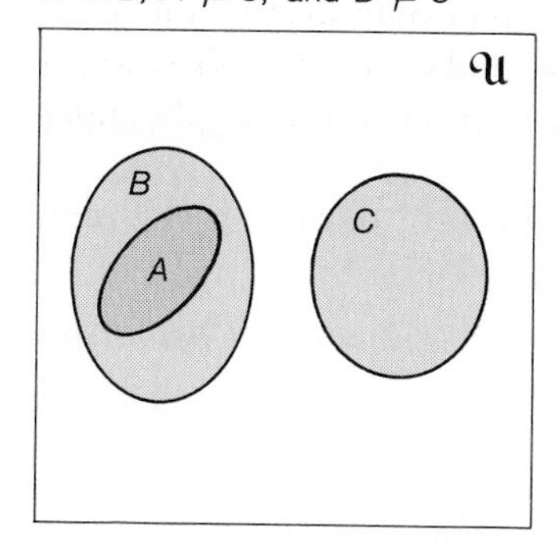

Thus, a set with 3 elements has $2^3 = 8$ possible subsets; a set with 10 elements has $2^{10} = 1{,}024$ subsets.

The set of all possible subsets of $\mathscr{U}$ is called the POWER SET OF $\mathscr{U}$. The power set always consists of these subsets:

Number of subsets	*Number of elements in subsets*
1	Null set
$\binom{n}{1}$	One-element subsets
$\binom{n}{2}$	Two-element subsets
$\vdots$	
$\binom{n}{n-1}$	$(n-1)$-element subsets
1	The set itself

where the notation $\binom{n}{x} = \frac{n!}{x!(n-x)!}$. The symbol $n!$ (read "n factorial") is defined by $n! = n \cdot (n-1) \cdot (n-2) \cdot \ldots \cdot (3) \cdot (2) \cdot (1)$.

Table 2.1 gives an example of subsets formed from a universal set.

Table 2.1

The Subsets of a Universal Set

Universal set	Number of subsets	Subsets
$\mathscr{U} = \{\ \}$	$2^0 = 1$	$\{\ \}$
$\mathscr{U} = \{a\}$	$2^1 = 2$	$\{\ \}\ \{a\}$
$\mathscr{U} = \{a,b\}$	$2^2 = 4$	$\{\ \}, \{a\}, \{b\}, \{a,b\}$
$\mathscr{U} = \{a,b,c\}$	$2^3 = 8$	$\{\ \}, \{a\}, \{b\}, \{c\}, \{a,b\}, \{a,c\}, \{b,c\}, \{a,b,c\}$
$\mathscr{U} = \{a,b,c,d\}$	$2^4 = 16$	$\{\ \}, \{a\}, \{b\}, \{c\}, \{d\}, \{a,b\}, \{a,c\}, \{a,d\}, \{b,c\}, \{b,d\}, \{c,d\}, \{a,b,c\}, \{a,b,d\}, \{a,c,d\}, \{b,c,d\}, \{a,b,c,d\}$

EXERCISES

For each of the following groups of sets, determine which are equal and which are not equal.

1. $A = \{11, 13, 15\}$; $B = \{11, 12, 13, 14, 15\}$; $C = \{15, 11, 13\}$; $D = \{15, 14, 13, 12, 11\}$
2. $R = \{r \mid r$ is a member of the Board of Directors of X Company$\}$
   $S = \{s \mid s$ is an employee of X Company$\}$
   $T = \{t \mid t$ is a customer of X Company$\}$

3. $F = \{x|x^2 - 2x = 0\}$
   $G = \{x|\ 3 - x = 1\}$
   $H = \{4\}$
4. $A = \{x|2x^2 + 5x + 2 = 0\}$; $B = \{x|2x^3 + 5x^2 + 2x = 0\}$
5. $K = \{a,e,i,o,u\}$; $L = \{e,i,j,a,u\}$; $M = \{v|v$ is a vowel in the English alphabet$\}$
6. $A = \{\ \}$; $B = \{0\}$; $C = \phi$

Given the universal set $\mathscr{U} = \{\text{a,b,c}\}$, *which of the sets listed below are subsets of* $\mathscr{U}$?

7. $\{c,b,a\}$
8. $\{b\}$
9. $\{\ \}$
10. $\{a,b\}$
11. $\{a,b,c,d\}$
12. $\{b,d\}$

Given A = $\{1,2,3,4,5,6,7\}$, *which of the following statements are true and which are false?*

13. $4 \in A$
14. $-10 \in A$
15. $\{6\} \subset A$
16. $\{6\} \not\subset A$
17. $\{1,2,3\} \in A$
18. $\{1,2,3\} \subset A$
19. $\{1,2,3\} \not\subset A$
20. $\phi \in A$
21. $\phi \subset A$
22. $\phi \not\subset A$

Given $\mathscr{U} = \{2,4,6,8\}$ *and* A = $\{2,4\}$, *which of the following statements are true? Explain. Rewrite using correct notation where needed.*

23. $A \subset A$
24. $A \subset \mathscr{U}$
25. $\mathscr{U} \subset A$
26. $\phi \subset \mathscr{U}$
27. $\phi \subset \phi$
28. Given the universal set $\mathscr{U} = \{3, 5, 7, 9\}$, list the power set of $\mathscr{U}$.

Given $\mathscr{U} = \{\text{a,b,c}\}$ *and using the notation* 2^u *to represent the power set of* $\mathscr{U}$, *explain the following statements.*

29. $\phi \notin \mathscr{U}$ but $\phi \in 2^u$
30. $a \in \mathscr{U}$ but $a \notin 2^u$
31. $\{a,b\} \notin \mathscr{U}$ but $\{a,b\} \in 2^u$

Given A, B, *and* C, *defined on a universal set* $\mathscr{U}$, *use Venn diagrams to picture each of the following.*

32. $A \subset B,\ A \not\subset C,\ B \not\subset C$
33. $A \subset B,\ A \subset C,\ B \subset C$
34. $C \subset A,\ A \subset B,\ B \not\subset C$
35. $C \not\subset A,\ C \not\subset B,\ A \not\subset B$

2.3 CARTESIAN PRODUCT OF SETS

An ORDERED PAIR is a pair of objects for which we distinguish the sequence in which the objects are considered. The notation (a,b) is used to represent an ordered pair in which a is the first component and b is the second component. The ordered pair (a,b) is quite different from the set $\{a,b\}$ containing the two elements a and b. In the set $\{a,b\}$ there is no "first component" because the order in which the elements of a set are listed is immaterial. Therefore, although the set $\{a,b\}$ is equivalent to the set $\{b,a\}$, the ordered pair (a,b) is not equal to the ordered pair (b,a). Two ordered pairs are equal if and only if their first components are the same and their second components are the same.

Whenever we have two sets we can form ordered pairs by taking the first component from the elements of one set and the second component from the elements of the second set. If A and B are two sets, the set of all ordered pairs in which the first component is taken from A and the second component from B is called the CARTESIAN PRODUCT of A cross B (named for the mathematician René Decartes) and is denoted $A \times B$, usually read "A cross B." In symbolic notation,

$$A \times B = \{(a,b) | a \in A \text{ and } b \in B\}$$

If A and B are each finite sets, A consisting of m elements $a_1, a_2, \ldots, a_m$, and B consisting of n elements $b_1, b_2, \ldots, b_n$, $A \times B$ is a set consisting of the following mn elements:

$$\begin{matrix} (a_1,b_1) & (a_1,b_2) & \cdots & (a_1,b_n) \\ (a_2,b_1) & (a_2,b_2) & \cdots & (a_2,b_n) \\ \vdots & \vdots & & \vdots \\ (a_m,b_1) & (a_m,b_2) & \cdots & (a_m,b_n) \end{matrix}$$

If the first element of the ordered pairs is taken from set B and the second element from set A, the Cartesian product set is of B cross A, denoted $B \times A$.

Example 4 If set A represents the outcomes on the toss of a coin, $A = \{H,T\}$ where H denotes heads and T denotes tails. Suppose that set $B = \{1,2,3,4,5,6\}$, the possible outcomes of the toss of a die. The following sets are a few of the Cartesian product sets which might be formed.

$$A \times B = \{(H,1),(H,2),(H,3),(H,4),(H,5),(H,6),(T,1),(T,2),(T,3),(T,4),(T,5),(T,6)\}$$

$$B \times A = \{(1,H),(1,T),(2,H),(2,T),(3,H),(3,T),(4,H),(4,T),(5,H),(5,T),(6,H),(6,T)\}$$

$$A \times A = \{(H,H),(H,T),(T,H),(T,T)\}$$

[]

This concept can be extended to the Cartesian product of n sets; that is, a set of n ordered elements may be formed. If A and B and C are sets, several such constructions may be made. The Cartesian product $A \times B$ may be taken to form a new set, which may then be combined with C to form $(A \times B) \times C$. Or B and C could be combined to form $B \times C$, and then a second combination $(B \times C) \times A$ could be made, and so on.

Number of Elements in the Cartesian Product Set

Let the number of elements in set A be $n(A) = m$ and the number of elements in set B be $n(B) = n$. In the Cartesian product set $A \times B$, the first element from A will be paired with n elements from B; then the second element from A will be paired with n elements from B; until m elements from A have been paired with n elements from B. The number of elements in the Cartesian product set $A \times B$ will be

$$n(A \times B) = n(A) \cdot n(B)$$

This concept can be extended to the Cartesian product of any number of finite sets.

Relations

A RELATION between set A and set B, denoted R, is any subset of the Cartesian product set $A \times B$. The number of relations in any Cartesian product set depends upon the number of ordered pairs in that particular product set. If the number of ordered pairs is p, the number of relations is 2^p.

Example 5 If $A = \{a_1,a_2\}$ and $B = \{b_1,b_2\}$, the Cartesian product set $A \times B = Y$ consists of $(2) \cdot (2) = 4$ ordered pairs, as follows:

$$Y = A \times B = \{(a_1,b_1),(a_1,b_2),(a_2,b_1),(a_2,b_2)\}$$

Any subset of ordered pairs from this Cartesian product set is a relation. Here there are $2^4 = 16$ relations, as follows:

$R_1 = \{(a_1,b_1),(a_1,b_2),(a_2,b_1),(a_2,b_2)\}$ $R_2 = \{(a_1,b_1),(a_1,b_2),(a_2,b_1)\}$
$R_3 = \{(a_1,b_1),(a_1,b_2),(a_2,b_2)\}$ $R_4 = \{(a_1,b_1),(a_2,b_1),(a_2,b_2)\}$
$R_5 = \{(a_1,b_2),(a_2,b_1),(a_2,b_2)\}$ $R_6 = \{(a_1,b_1),(a_1,b_2)\}$
$R_7 = \{(a_1,b_1),(a_2,b_1)\}$ $R_8 = \{(a_1,b_1),(a_2,b_2)\}$
$R_9 = \{(a_1,b_2),(a_2,b_1)\}$ $R_{10} = \{(a_1,b_2),(a_2,b_2)\}$
$R_{11} = \{(a_2,b_1),(a_2,b_2)\}$ $R_{12} = \{(a_1,b_1)\}$
$R_{13} = \{(a_1,b_2)\}$ $R_{14} = \{(a_2,b_1)\}$
$R_{15} = \{(a_2,b_2)\}$ $R_{16} = \phi$ []

Interest is usually focused on only a relatively few of the many possible relations.

Example 6 A white die and a black die are tossed. Let B represent the possible outcomes for the black die and W the possible outcomes for the white die. Then $B = W = \{1,2,3,4,5,6\}$. In the Cartesian product set $B \times W$ there are $(6) \cdot (6) = 36$ elements which are ordered pairs. The product set can be noted symbolically as

$$X = B \times W = \{(b,w) | b \in B \text{ and } w \in W\}$$

There are 2^{36} possible relations. Specific examples of these relations which might be of special interest are:

$$R_1 = \{(b,w) | b = w \text{ and } (b,w) \in B \times W\}$$

The ordered pairs of this relation are:

$$R_1 = \{(1,1),(2,2),(3,3),(4,4),(5,5),(6,6)\}$$

Or, we might be especially concerned with the relation

$$R_2 = \{(b,w) | b > w \text{ and } (b,w) \in B \times W\}$$

whose elements are the ordered pairs

$$R_2 = \{(2,1),(3,1),(3,2),(4,1),(4,2),(4,3),(5,1),(5,2),(5,3),(5,4),(6,1),(6,2),(6,3),(6,4),(6,5)\}$$

[]

The DOMAIN of the relation R is the set of all first elements of the ordered pairs in R. The RANGE of a relation R is the set of all second elements of the ordered pairs in R.

Functions

A FUNCTION is a special case of a relation. *Any* subset of $A \times B$ is a relation. The relation is a FUNCTION FROM A TO B if TO EVERY ELEMENT OF SET A THERE IS ASSIGNED EXACTLY ONE ELEMENT FROM SET B. In other words, if each element of the domain is associated with one and only one element from the range, the association is said to be a function. Note, then, that the number of ordered pairs in a function is equal to the number of elements in set A, the set that furnishes the first component of the ordered pairs.

Example 7

We have seen in Example 5 that where $A = \{a_1,a_2\}$ and $B = \{b_1,b_2\}$ there are 16 relations (or subsets) possible on the Cartesian product set $A \times B$. Of these relations, only four conform to the definition of a function. These four functions are

$$R_7 = \{(a_1,b_1),(a_2,b_1)\} \qquad R_8 = \{(a_1,b_1),(a_2,b_2)\}$$
$$R_9 = \{(a_1,b_2),(a_2,b_1)\} \qquad R_{10} = \{(a_1,b_2),(a_2,b_2)\}$$

Each of these functions consists of two ordered pairs, or $n(A)$ ordered pairs. a_1 appears as the first element once and a_2 appears as the first element once in each function. No two distinct ordered pairs of a function have the same first coordinate. []

EXERCISES

Given A = {4,5}, B = {3,6,9}, *and* C = {7,8}, *construct the following cross-product sets:*

36. $A \times B$
37. $A \times C$
38. $C \times B$
39. $C \times B \times A$
40. $C \times C \times C$
41. $B \times B$
42. If the Cartesian product $A \times B = \{(1,2),(1,3),(1,4),(3,2),(3,3),(3,4),(5,2),(5,3),(5,4)\}$, what are the elements of set A? Of set B?

If A *has three elements,* B *has four elements, and* C *has five elements, how many elements will be found in each of the following product sets?*

43. $A \times B$
44. $B \times C$
45. $C \times A$
46. $A \times B \times C$
47. $A \times A$
48. $C \times C \times C$

Let A = {2,4}, B = {3,4}, *and* C {1,2,3} *where the universal set* $\mathcal{U}$ = {1,2,3,4}. *List the elements in each of the following sets:*

49. $A \times C$
50. $A \times B$
51. $C \times A$
52. $A \times B \times C$

Let A = {1,2}, B = {4,6}, *and C* = {5}. *List the elements in each of the following sets:*

53. The power set of *A*.
54. The power set of *B*.
55. $A \times B$
56. $B \times C$
57. $(A \times B) \times (B \times C)$
58. $(A \times C) \times (B \times C)$

For the set A = {H,T}, *find a subset of* A × A *for which*

59. No *H* occurs.
60. One *H* occurs.
61. Two *H*s occur.

For the set A = {0,1,2,3}, *and* B = {2,4,6}, *find a subset of* A × B *for which*

62. The sum of the elements in each ordered pair is 6.
63. The two items of the ordered pair are the same.
64. The sum of the two elements of the ordered pair is an odd number.
65. The sum of the elements of the ordered pair is greater than 4.
66. The sum of the elements of the ordered pair is less than or equal to 5.

Let X = {1,2,3,4} *and* Y = {2,3,4,5}. *List the ordered pairs in each of the following relations. Are any of these relations also functions?*

67. $R_1 = \{(x,y) | y \leq x + 1\}$
68. $R_2 = \{(x,y) | y \geq x + 1\}$
69. $R_3 = \{(x,y) | y = x + 1\}$

Given A = {2, 4, 6} *and* B = {3,5}, *list the ordered pairs in each function defined on*

70. $A \times B$
71. $A \times A$
72. $B \times A$

2.4 OPERATIONS ON SETS

Just as numbers can be combined by the basic mathematical operations of addition, subtraction, multiplication, and division to form a new number, so, too, can sets be combined to form a new set. All of the sets involved in the combination are subsets of the same universal set; the newly formed set will also be a subset of the same universal set.

The basic operations on sets are (1) COMPLEMENT OF SETS, (2) SET INTERSECTION, and (3) SET UNION.

Complement of Sets

> The COMPLEMENT of set A with respect to the universal set $\mathcal{U}$ is the set containing all elements of $\mathcal{U}$ that are not in A.

Figure 2.2

A^c, the Complement of A

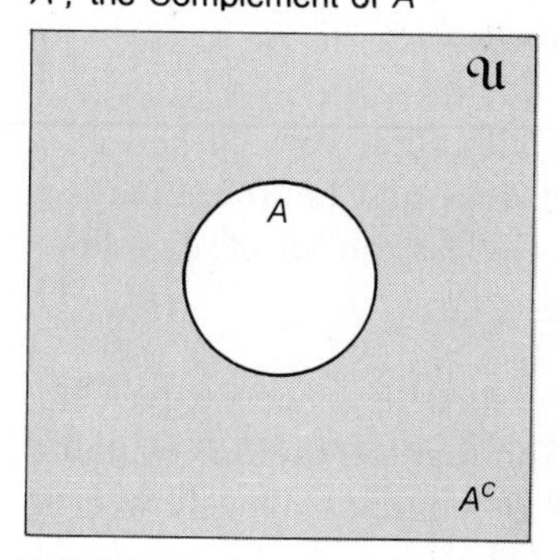

Suppose that the universal set $\mathcal{U}$ has as its elements all 26 letters of the English alphabet. If A is a subset of $\mathcal{U}$ which contains all the vowels in the English alphabet, then all the consonants form another subset, also a subset of $\mathcal{U}$, which is known as the COMPLEMENT OF A WITH RESPECT TO $\mathcal{U}$. The symbol A^c, read "not A," or "the complement of A," is used to represent the complement of A. (See Figure 2.2). The relationship may be symbolized

$$A^c = \{x \mid x \in \mathcal{U} \text{ and } x \notin A\}$$

Example 8 The complement of the set of all even integers with respect to the universal set of all integers is the set of all odd integers. []

Example 9 The complement of the set of employees of the XYZ Company who are 45 years old or older with respect to the universal set of all employees of XYZ Company is the set whose elements are those employees of XYZ Company who are less than 45 years of age. []

The complement of the universal set $\mathcal{U}$ with respect to itself is the empty set ϕ; the complement of the empty set ϕ is the universal set $\mathcal{U}$.

Intersection

> The INTERSECTION of two sets A and B, denoted $A \cap B$, is the set of elements that belong to both A and B. Symbolically,
>
> $$A \cap B = \{x \mid x \in A \text{ and } x \in B\}$$

The intersection of two sets is shown in Figure 2.3.

Figure 2.3

$A \cap B$ (the intersection of A and B is shown as the shaded area)

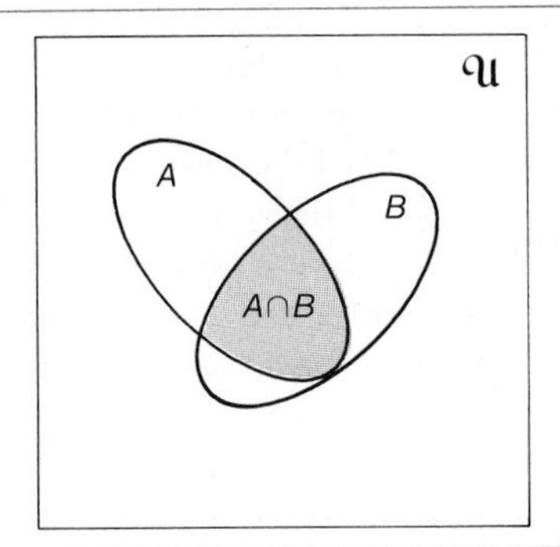

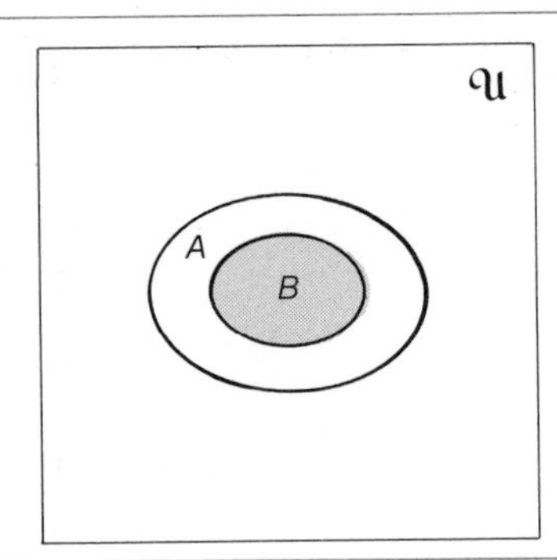

Example 10 If $A = \{2,4,6,8\}$ and $B = \{3,4,7\}$, then $A \cap B = \{4\}$. [ ]

Example 11 If the set A is a set whose elements are all yellow cars parked in a particular parking lot and the elements of set B are all Chevrolets parked on that lot, the intersection of A and B, $A \cap B$, is the set of all yellow Chevrolets on the parking lot. [ ]

Example 12 If the set A consists of all yellow cars parked in a certain parking lot and set B contains all Chevrolets parked on that lot, then B^c contains all cars that are not Chevrolets and $A \cap B^c$ contains all yellow cars except the yellow Chevrolets. (See Figure 2.4) [ ]

The notion of intersection can easily be generalized to situations involving more than two sets. Thus, the intersection of the sets A_1, A_2, $\ldots$, A_n, written $A_1 \cap A_1 \cap \ldots \cap A_n$, is the set of elements common to all the sets $A_1, A_2, \ldots, A_n$.

Example 13 Defined on the universal set $\mathscr{U}$ whose elements are all members of the labor force are set A, whose elements are the employees of XYZ Company; set B, whose elements are all female members of the labor force; and set C, whose elements are all members of the labor force who are under 25 years of age. The intersection of these sets, $A \cap B \cap C$, would be the set of female employees of XYZ Company who are less than 25 years old. [ ]

Figure 2.4

$A \cap B^c$

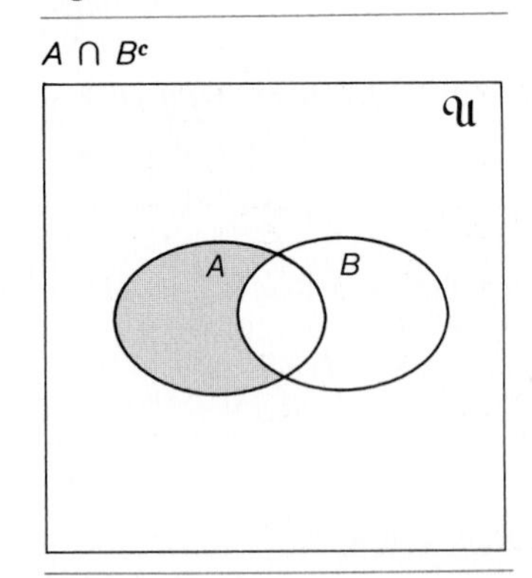

Disjoint Sets If two sets A and B have no elements in common, $A \cap B = \phi$, and the sets are said to be DISJOINT or MUTUALLY EXCLUSIVE. In a Venn diagram, such as the one shown in Figure 2.5, disjoint sets are shown as having no overlapping area. (The term CONJOINT is used to describe two sets that are not disjoint.)

Figure 2.5

$A \cap B = \phi$ (mutually exclusive or disjoint sets)

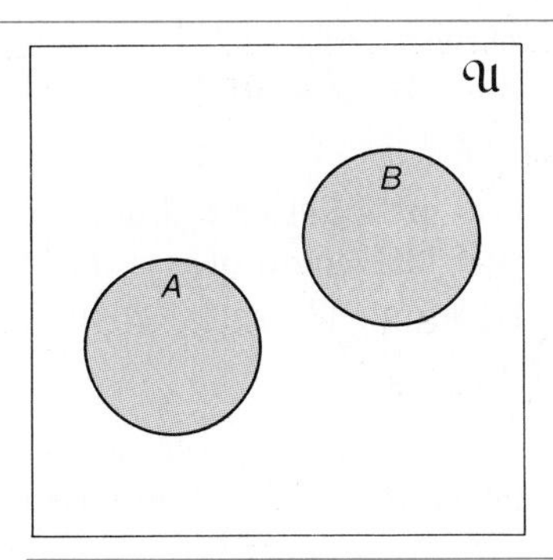

Example 14 If the universal set $\mathcal{U}$ consists of the 52 cards in a deck of ordinary bridge cards and if these subsets have been defined

$$R = \{\text{red cards}\} \qquad \text{and} \qquad B = \{\text{black cards}\}$$

then $R \cap B = \phi$. Sets R and B are mutually exclusive, or disjoint. [ ]

Union Figure 2.6 illustrates the union of A and B which is defined as follows.

> **The UNION of A and B** (denoted $A \cup B$) when A and B are two sets defined on a universal set $\mathcal{U}$, contains only those elements that belong either to set A or to set B or to both.
>
> $$A \cup B = \{x | x \in A \text{ or } x \in B\}$$

Example 15 If $A = \{1,2,3,4\}$ and $B = \{2,4,6,8\}$, then $A \cup B = \}1,2,3,4,6,8\}$. [ ]

Figure 2.6

$A \cup B$ (the union of A and B is shown as the shaded area)

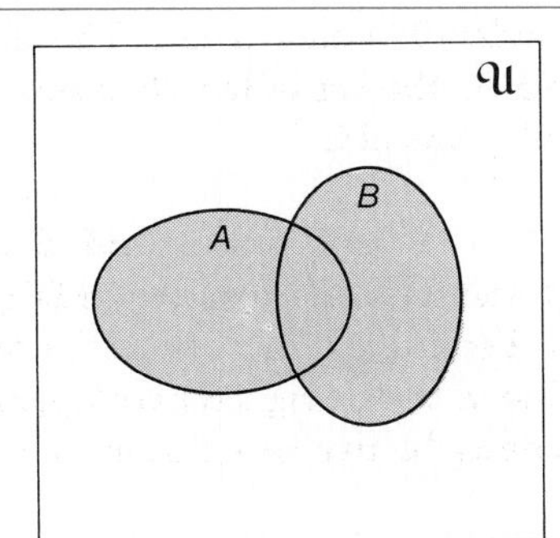

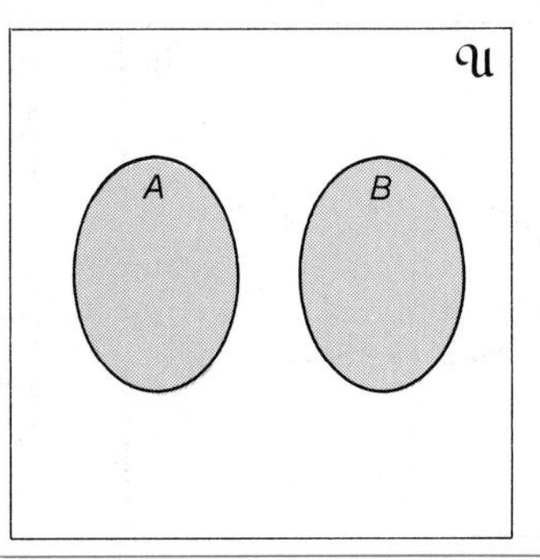

Example 16 If set A contains all the yellow cars on a parking lot and set B contains all the Chevrolets on that lot, the union of A and B, $A \cup B$, contains all the yellow cars plus the Chevrolets of other colors on the parking lot. []

The notion of union can also be extended to cases involving more than two sets. The union of sets $A_1, A_2, \ldots, A_n$, noted $A_1 \cup A_2 \cup \ldots \cup A_n$, is the set of elements which are in at least one of the sets $A_1, A_2, \ldots, A_n$.

Example 17 Consider the following sets defined on the universal set $\mathcal{U}$ whose elements are all the residents of Gritty City, Pennsylvania,

$$A = \{x | x \text{ is a college professor}\}$$
$$B = \{x | x \text{ is a married person}\}$$
$$C = \{x | x \text{ is under 35 years of age}\}$$

Then the set $A \cup B \cup C$ represents all residents of Gritty City who are either college professors or married or under 35 years of age. (See Figure 2.7A.)

The set of residents who are married college professors and are under 35 years of age is denoted $A \cap B \cap C$. (See Figure 2.7B.)

Figure 2.7

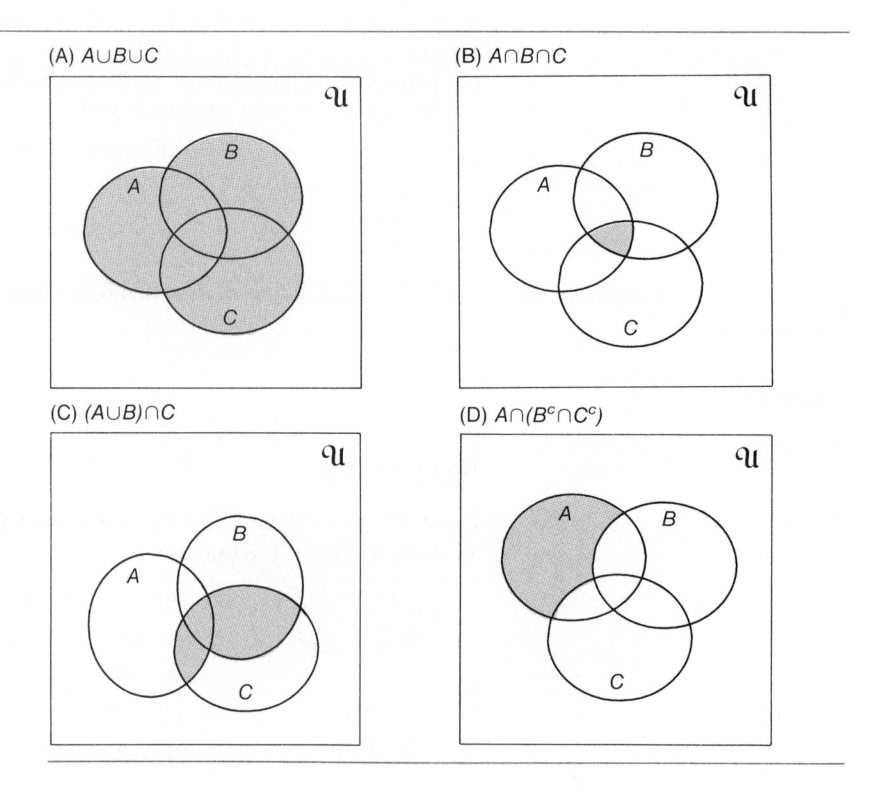

Figure 2.8

A Group of Mutually Exclusive and Exhaustive Sets Form a PARTITION of the Universal Set

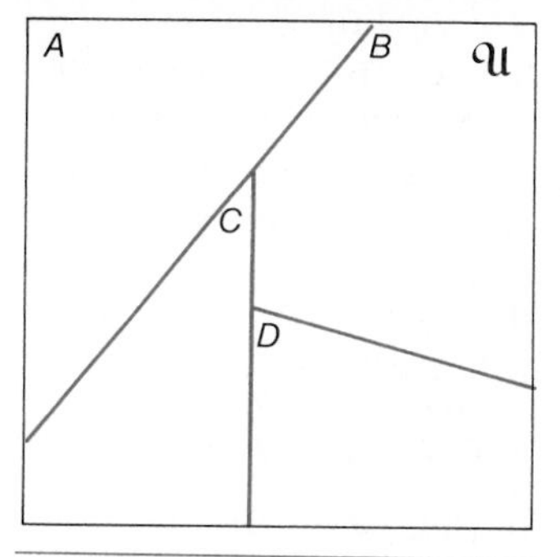

The set of residents who are either college professors or are married and are also under 35 years of age can be depicted by the notation $(A \cup B) \cap C$. (See Figure 2.7C.)

Or the set of residents who are college professors who are not married but are over 35 years of age can be symbolized $A \cap (B^c \cap C^c)$. (See Figure 2.7D.)

[]

A collection of subsets is said to be EXHAUSTIVE if its union contains each and every element in the universal set on which they are defined. A group of sets which are mutually exclusive and exhaustive is called a PARTITION. Such a partition is shown in Figure 2.8.

Example 18

Given the universal set $\mathcal{U} = [1,2,3,4,5]$, the collection of subsets $A = \{1,2\}$, $B = \{3\}$, and $C = \{4,5\}$ forms a partition of $\mathcal{U}$.

A universal set may be partitioned in different ways, of course. For instance, the collection of subsets $D = \{1\}$, $E = \{3,5\}$ and $F = \{2,4\}$ also forms a partition of the universal set given above.

[]

Figure 2.9

$A = (A \cap B) \cup (A \cap B^c)$ and $(A \cup B) = (A \cap B^c) \cup (A \cap B) \cup (A^c \cap B)$

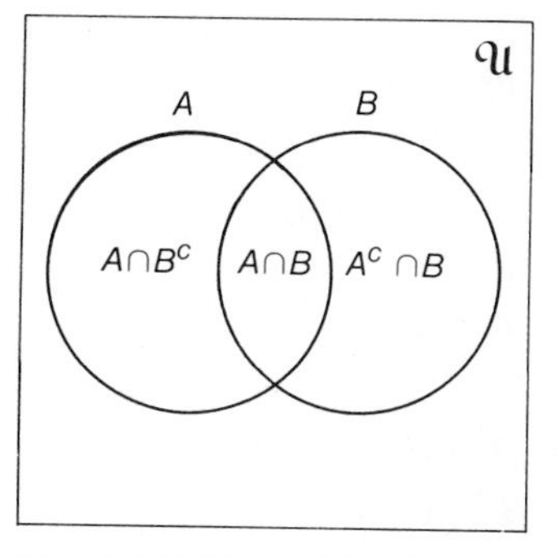

Given two sets, A and B, which are not disjoint, the set A can be partitioned into two disjoint subsets $(A \cap B)$ and $(A \cap B^c)$. Furthermore, the union of the two sets, $A \cup B$, may be partitioned into three disjoint subsets $(A \cap B^c)$, $(A \cap B)$, and $(A^c \cap B)$, as illustrated in Figure 2.9.

EXERCISES

Given $\mathcal{U} = \{1,2,3,4,5,6\}$, $A = \{1,2,3,4\}$, and $B = \{3,4,5,6\}$, list the elements in each of the following sets.

73. A^c
74. B^c
75. $A \cap B$
76. $A \cup B$
77. $A \cup \phi$
78. $\phi \cap B$
79. $\mathcal{U} \cup A$
80. $B \cap B^c$
81. $A^c \cap B$
82. $A^c \cup B$
83. $A^c \cap B^c$
84. $(A \cap B)^c$
85. $(A \cup B)^c$
86. $A \cup (A^c \cap B)$
87. $(A \cap B^c) \cup (A^c \cap B)$

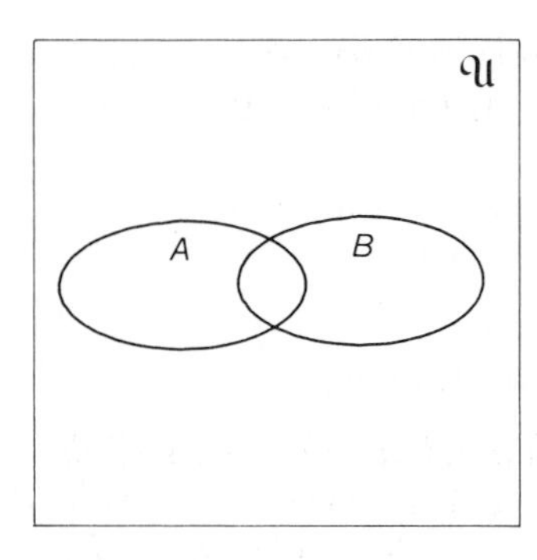

Use Venn diagrams similar to the one given at right to illustrate each of the following sets.

88. A^c
89. B^c
90. $A \cap B$
91. $A \cup B$
92. $(A \cap B)^c$
93. $(A \cup B)^c$
94. $A^c \cup B$
95. $A^c \cap B$
96. $A \cup B^c$
97. $A \cap B^c$
98. $A^c \cup B^c$
99. $A^c \cap B^c$

Let $\mathscr{U}$ be the set of all United States senators, E the set of senators whose states are east of the Mississippi River, and D the set of senators who are Democrats. Explain each of the following.

100. E^c
101. D^c
102. $E \cup D$
103. $E \cap D$
104. $E^c \cup D$
105. $E^c \cap D$
106. $E \cup D^c$
107. $E \cap D^c$
108. $E^c \cap D^c$
109. $E^c \cup D^c$
110. $(E \cap D)^c$
111. $(E \cup D)^c$

Given $\mathscr{U} = \{1,2,3,4,5,6,7,8\}$, $A = \{1,2,3,4\}$, $B = \{1,2,5,6\}$, and $C = \{1,3,5,7\}$, list the elements in each of the following sets.

112. A^c
113. B^c
114. C^c
115. $A \cup B$
116. $A \cup C$
117. $B \cup C$
118. $A \cap B$
119. $A \cap C$
120. $B \cap C$
121. $A^c \cup B^c$
122. $B^c \cup B$
123. $C \cup B^c$
124. $A \cup B \cup C$
125. $A \cap B \cap C$
126. $A \cup (B \cap C)$
127. $(A \cap B) \cup C$
128. $A \cap (B \cup C)$
129. $(A \cup B) \cap C$
130. $A^c \cup B \cup C$
131. $A \cup B^c \cup C$
132. $A \cup B \cup C^c$
133. $A^c \cap B \cap C$
134. $A \cap B^c \cap C$
135. $A \cap B^c \cap C^c$
136. $(A \cup B \cup C)^c$
137. $(A \cap B \cap C)^c$
138. $A^c \cup B^c \cup C^c$
139. $(A \cup B) \cap (A \cup C)$
140. $(A \cap B) \cup (A \cap C)$

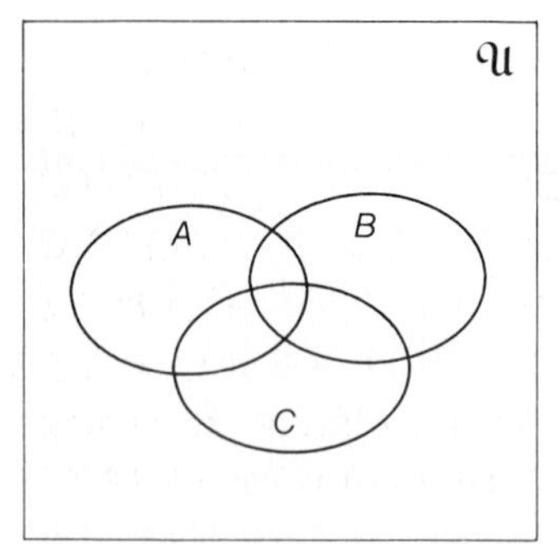

Using Venn diagrams similar to the one shown at right, illustrate each of the following sets.

141. $A \cap B \cap C$
142. $A \cup B \cup C$
143. $A \cap B \cap C^c$
144. $A \cap B^c \cap C$
145. $A^c \cap B \cap C$
146. $A \cap B^c \cap C^c$
147. $A^c \cap B^c \cap C^c$
148. $(A \cap B \cap C)^c$
149. $A \cup B \cup C^c$
150. $A \cup B^c \cup C$

Using Venn diagrams, show that

151. $A \cap (A \cup B) = A$

152. $A \cup (A \cap B) = A$

153. $A \cap (A \cap B) = A \cap B$

154. $(A \cup B)^c = A^c \cap B^c$

155. $A \cup (B \cup C) = (A \cup B) \cup C$

156. $A \cap (B \cap C) = (A \cap B) \cap C$

157. $A \cup (B \cap C) = (A \cup B) \cap (A \cup C)$

158. $(A \cap B) \cup (A \cap B^c) = A$

Draw a series of Venn diagrams to show the following subsets of the universal set of all employees of a firm. (Use S *to represent the set of all salespersons,* M *to represent the set of all male employees, and* A *to represent the set of all employees who are over 45 years of age.)*

159. The set of all salespersons.
160. The set of all employees over 45 years of age.
161. The set of all male employees.
162. The set of all salespersons who are over 45 years of age.
163. The set of all female employees.
164. The set of all male employees who are not salespersons.
165. The set of all male salespersons under 45 years of age.
166. The set of all female salespersons.
167. The set of all female employees, under 45 years of age, who are not salespersons.
168. The set of all female salespersons who are under 45 years of age.

Let $\mathcal{U}$ = {s|s *is a student at Tree-of-Knowledge College*} *with the following subsets defined on* $\mathcal{U}$:

$A = \{a \mid a$ is a student taking an accounting course$\}$
$B = \{b \mid b$ is a student taking a biology course$\}$
$C = \{c \mid c$ is a student taking a chemistry course$\}$

Describe each of the following sets in words.

169. A^c

170. B^c

171. C^c

172. $A \cap B$

173. $B \cap C$

174. $A \cap C$

175. $A \cup B$

176. $A \cup C$

177. $A \cap B \cap C$

178. $A \cup B \cup C$

179. $A \cup B \cup C^c$

180. $A \cup B^c \cup C$

181. $A^c \cup B \cup C$

182. $A \cap B \cap C^c$

183. $A \cap B^c \cap C$

184. $A^c \cap B \cap C$

185. $A \cap B^c \cap C^c$

186. $A^c \cap B^c \cap C^c$

187. $A \cap (B \cup C)$

188. $A \cup (B \cap C)$

189. $C \cup (A \cap B)$02

190. Given that A, B, and C are subsets of a universal set $\mathcal{U}$, arrange the sets listed below in such an order that each set is a subset of the sets that follow.

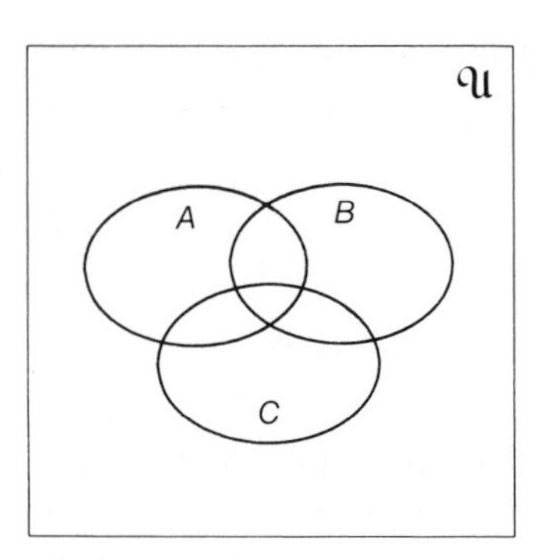

$\mathscr{U}$	A	ϕ	$A \cap B$
$A \cup B$	$A \cup (B \cap C)$	$A \cap (B \cup C)$	$A \cup B \cup C$
$A \cap B \cap C$			

A, B, and C are subsets of the universal set $\mathscr{U}$ as shown in the Venn diagram at right.
Partition each of the following:

191. A	192. $\mathscr{U}$	193. $A \cup C$
194. $A \cap B$	195. $B \cup C$	196. $A \cup B \cup C$

2.5 NUMBER OF ELEMENTS IN GROUPS OF FINITE SETS

The number of elements in any finite set A may be denoted $n(A)$. For instance, if $A = \{1,2,3,4\}$, then $n(A) = 4$.

We are frequently concerned with the number of elements in various combinations of finite sets. The following observations will be helpful in such situations.

1. The null set contains no elements; that is,

$$n(\phi) = 0$$

2. A nonempty set cannot have a negative number of elements; that is,

$$n(A) > 0 \qquad \text{if } A \text{ is not empty.}$$

3. If A and B are two disjoint sets, they have no elements in common, and the set $A \cap B$ is the empty set; that is,

$$n(A \cap B) = 0 \qquad \text{if } A \text{ and } B \text{ are disjoint sets.}$$

4. If A and B are two disjoint sets, the number of elements in $A \cup B$ is equal to the number of elements in A plus the number of elements in B; that is,

$$n(A \cup B) = n(A) + n(B) \qquad \text{if } (A \cap B) = \phi. \tag{2.1}$$

5. For any two sets A and B, the number of elements in $A \cup B$ is equal to the number of elements in A plus the number of elements in B minus the number of elements that are common to both sets; that is,

$$n(A \cup B) = n(A) + n(B) - n(A \cap B) \tag{2.2}$$

6. For any two sets A and B that are not disjoint, the set A can be partitioned into two disjoint sets $A \cap B$ and $A \cap B^c$. (See Figure 2.10.) Thus,

$$n(A) = n(A \cap B) + n(A \cap B^c) \tag{2.3}$$

Figure 2.10

$A = (A \cap B) \cup (A \cap B^c)$; therefore, $n(A) = n(A \cap B) + n(A \cap B^c)$

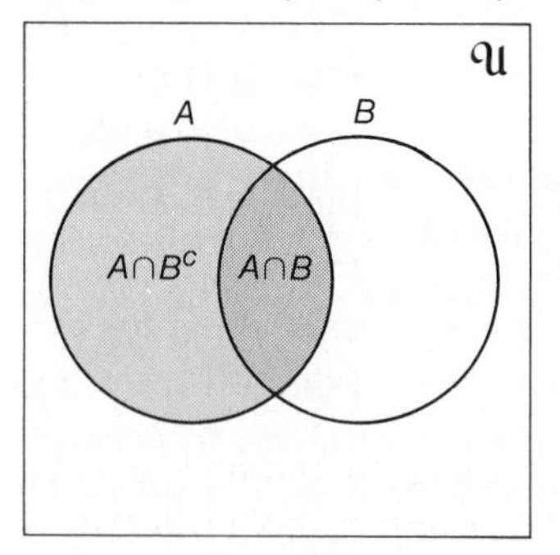

Figure 2.11

$\mathscr{U} = (A \cup B) \cup (A \cup B)^c$; therefore, $n(\mathscr{U}) = n(A \cup B) + n(A \cup B)^c$

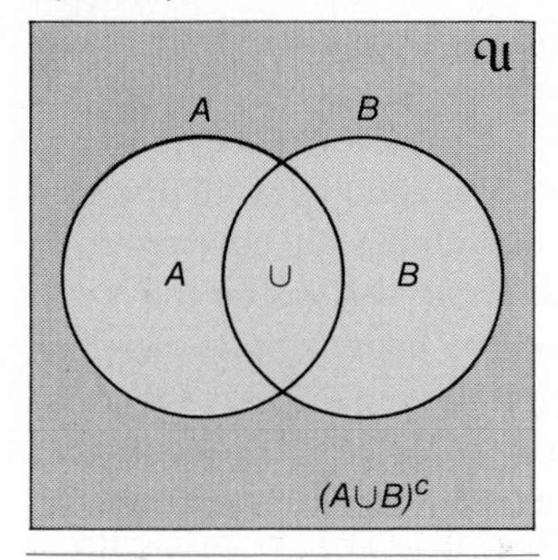

7. For any universal set $\mathscr{U}$ on which the sets A and B are defined, the universal set can be partitioned into two disjoint sets $A \cup B$ and $(A \cup B)^c$. (See Figure 2.11.) Thus,

$$n(\mathscr{U}) = n(A \cup B) + n(A \cup B)^c \tag{2.4}$$

8. The set $(A \cup B)^c$ and the set $(A^c \cap B^c)$ are equal because they consist of precisely the same elements. (See Figure 2.12.) Thus

$$n(A^c \cap B^c) = n(A \cup B)^c$$

And, because $n(\mathscr{U}) = n(A \cup B) + n(A \cup B)^c$,

$$n(A^c \cap B^c) = n(\mathscr{U}) - n(A \cup B) \tag{2.5}$$

Figure 2.12

A^C

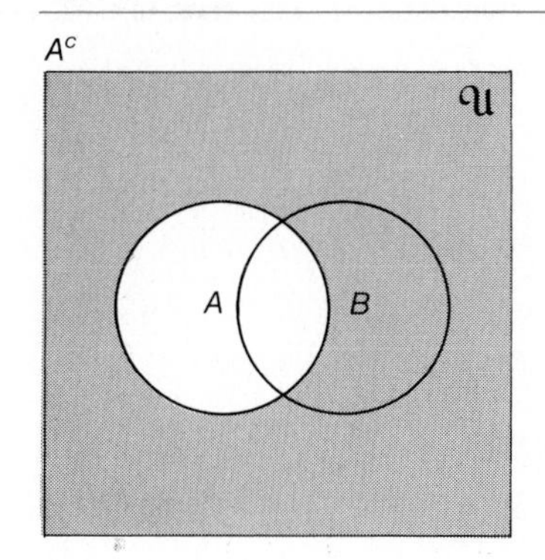

B^C

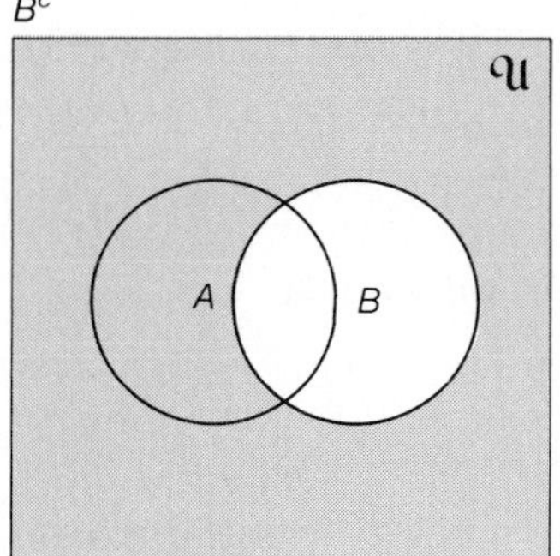

$(A^C \cap B^C) = (A \cup B)^C$

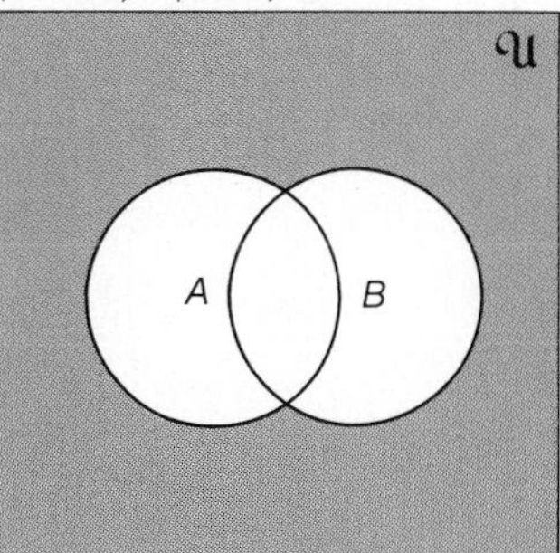

Example 19 Given $A = \{1,2,3,4,5\}$ and $B = \{6,7,8\}$, then $n(A) = 5$ and $n(B) = 3$. We see that A and B have no elements in common; thus, $A \cap B = \phi$ and $n(A \cap B) = 0$. Also, $A \cup B = \{1,2,3,4,5,6,7,8\}$ and $n(A \cup B) = n(A) + n(B) = 5 + 3 = 8$. []

Example 20 Given $A = \{2,3,4,5\}$ and $B = \{2,4,6\}$, then $n(A) = 4$ and $n(B) = 3$. In this case, $A \cap B = \{2,4\}$ and $n(A \cap B) = 2$.

The set $A \cup B = \{2,3,4,5,6\}$ and $n(A \cup B) = 5$. When A and B are not disjoint, the number of elements in $A \cup B$ is *not* the sum of $n(A)$ and $n(B)$ but is instead, $n(A \cup B) = n(A) + n(B) - n(A \cap B)$. Here, $5 = 4 + 3 - 2$. []

Venn diagrams are most useful in determining the number of elements in combinations of finite sets, as the following example will show.

Example 21 On a recent Sunday morning 325 persons stopped by a newsstand. Of these, 185 bought the *New York Times,* 150 bought the *Washington Post,* and 95 bought both papers. How many persons did not purchase either newspaper? How many persons purchased the *Times* but not the *Post*? How many persons purchased the *Post* but not the *Times*? How many persons purchased at least one of the newspapers?

Figure 2.13

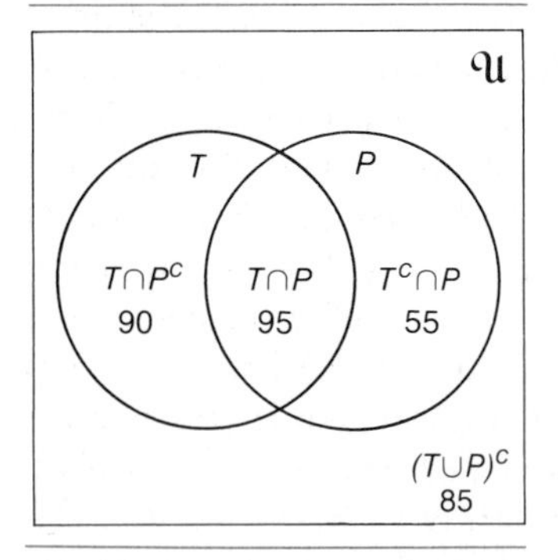

The Venn diagram in Figure 2.13 will help us to answer the questions which have been posed. First, let us define set T as containing all purchasers of the *New York Times* and set P as containing all purchasers of the *Washington Post.* Then we carefully label the regions on the Venn diagram. Using the information that 95 persons bought both papers—that is, $n(T \cap P) = 95$—we place this number in the region corresponding to $T \cap P$. Next, because $n(T) = n(T \cap P) + n(T \cap P^c)$, it follows that $n(T \cap P^c) = 185 - 95 = 90$; and we enter this number in the appropriate region. Also, $n(P) = n(T \cap P) + n(T^c \cap P)$ so that $n(T^c \cap P) = 150 - 95 = 55$. Again we enter this number in the appropriate region.

To determine the number of persons who do not buy either newspaper, we partition the universal set $\mathscr{U}$ into two disjoint sets; that is,

$$\mathscr{U} = (T \cup P) \cup (T \cup P)^c$$

Thus, we have

$$n(\mathscr{U}) = n(T \cup P) + n(T \cup P)^c$$

Further, we see that

$$(T \cup P) = (T \cap P^c) \cup (T \cap P) \cup (T^c \cap P)$$

so that, in this case,

$$n(T \cup P) = 90 + 95 + 55 = 240$$

From this we are able to obtain

$$n(T \cup P)^c = 325 - 240 = 85$$

In summary, we have determined that 85 people did not purchase either newspaper, 90 purchased the *Times* only, 55 purchased the *Post* only. In addition, $90 + 55 = 145$ purchased exactly one of these papers; $325 - 85 = 240$ purchased at least one of these papers. [ ]

Example 22 In a survey of 52 students, the following information was obtained:

25 were enrolled in an economics course.
22 were enrolled in a marketing course.
20 were enrolled in a history course.
13 were enrolled in an economics course and a marketing course.
10 were enrolled in a marketing course and a history course.
8 were enrolled in an economics course and a history course.

5 were enrolled in an economics course, a marketing course, and a history course.

How many students were enrolled in exactly one of the courses? in exactly two of the courses? in at least one of the courses? in none of the courses? How many students were enrolled in an economics course only? in a marketing course only? in a history course only?

Letting E represent the set of students enrolled in an economics course, M the set of students enrolled in a marketing course, and H the set of students enrolled in a history course, we construct a Venn diagram like the one shown in Figure 2.14. We begin with the information that five students are enrolled in all three courses—that is $n(E \cap M \cap H) = 5$—and enter this number in the appropriate region of the diagram. Because a total of eight students are enrolled in economics and history courses, this leaves $8 - 5 = 3$ who are enrolled in economics and history but not marketing. Similarly, because a total of 10 students are enrolled in marketing and history courses, $10 - 5 = 5$ are enrolled in marketing and history but not economics. And $13 - 5 = 8$ is the number of students enrolled in economics and marketing but not history. These numbers are placed in the appropriate regions of the Venn diagram.

Because the total number of students enrolled in a history course is 20, the number of students enrolled in history only is $20 - 5 - 5 - 3 = 7$. Similarly, because the total number of students enrolled in a marketing course is 22, then the number taking marketing only is $22 - 8 - 5 - 5 = 4$. Thus, the number of students taking economics only is 25 −

Figure 2.14

I. $E \cap M \cap H$
II. $E \cap H \cap M^c$
III. $M \cap H \cap E^c$
IV. $E \cap M \cap H^c$
V. $H \cap E^c \cap M^c$
VI. $M \cap E^c \cap H^c$
VII. $E \cap M^c \cap H^c$
VIII. $(E \cup M \cup H)^c$

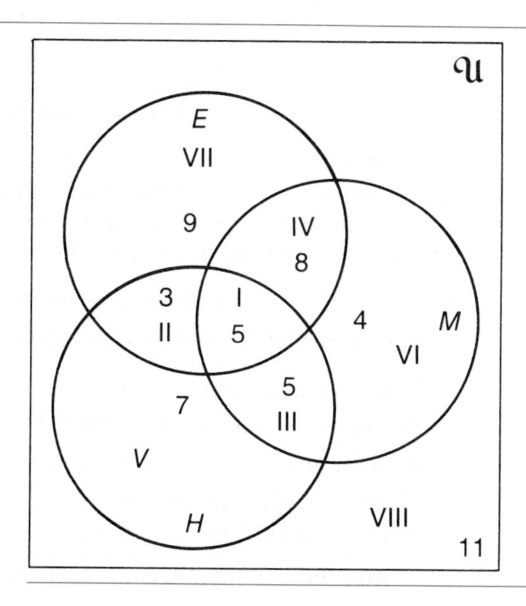

$8 - 5 - 3 = 9$. These numbers are placed in the appropriate regions of the Venn diagram.

Finally, the number of students not taking any of the courses is given by $52 - 9 - 8 - 5 - 3 - 5 - 4 - 7 = 11$.

Now we are able to answer the questions raised. The number of students enrolled in exactly one of the courses is given by $9 + 7 + 4 = 20$. The number of students enrolled in exactly two of the courses is $8 + 3 + 5 = 16$. The number of students enrolled in at least one of the courses is $9 + 8 + 4 + 3 + 5 + 5 + 7 = 41$. The number of students enrolled in none of the courses is 11. []

EXERCISES

Let $A = \{1,2,3,4\}$, $B = \{4,5,6\}$, *and* $C = \{3,5,7,9,11\}$. *How many elements are in each of the following (where* $\mathcal{U} = \{1,2,3,4,5,6,7,8,9,10,11,12\}$*)?*

197. $A \cup B$
198. $A \cap B$
199. $A \cup C$
200. $A \cap C$
201. $B \cup C$
202. $B \cap C$
203. $A \cup B^c$
204. $A \cup C^c$
205. $B \cup C^c$
206. $A \cap B^c$
207. $A \cap B \cap C$
208. $A \cup B \cup C$
209. $(A \cup B)^c$
210. $A^c \cap B^c$
211. $A^c \cup B^c \cup C^c$

Given A *and* B *are subsets of* $\mathcal{U}$, *where* $n(\mathcal{U}) = 250$, $n(A) = 120$, $n(B) = 90$, *and* $n(A \cap B) = 40$, *determine the following:*

212. $n(A^c)$
213. $n(B^c)$
214. $n(A^c \cap B)$
215. $n(A \cap B^c)$
216. $n(A \cup B)$
217. $n(A^c \cap B^c)$

Given A *and* B *are subsets of* $\mathcal{U}$, *where* $n(\mathcal{U}) = 105$, $n(A) = 55$, $n(B) = 50$, *and* $n(A \cap B) = 15$, *determine the following:*

218. $n(A^c)$
219. $n(B^c)$
220. $n(A^c \cap B)$
221. $n(A \cap B^c)$
222. $n(A \cup B)$
223. $n(A^c \cup B^c)$
224. $n(A^c \cup B)$
225. $n(A \cup B^c)$
226. $n(A^c \cap B^c)$

A survey of 110 automobiles inspected at an automobile safety inspection center revealed that 42 had defective brakes while 53 had defective tires and 15 had both defective brakes and defective tires.

227. How many of the automobiles had defective tires but not defective brakes?
228. How many of the automobiles had defective brakes but not defective tires?
229. How many of the automobiles had at least one or the other, defective tires or defective brakes?

230. How many of the automobiles had neither defective tires nor defective brakes?

A survey of 300 white-collar workers showed that 155 regularly watched the six o'clock news on television and that 185 regularly watched the late-night news on television, while 60 regularly watched both newscasts.

231. How many of the workers watched the six o'clock news but not the late-night news?
232. How many of the workers watched the late-night news but not the six o'clock news?
233. How many of the workers watched exactly one of the newscasts?
234. How many of the workers watched at least one of the newscasts?
235. How many of the workers did not watch either of the newscasts?

In a group of exactly 1,000 customers entering a store during the day, it has been reported that there were exactly

420 customers who were women.
525 customers who had charge accounts at the store.
325 customers who made a purchase in the store.
40 women customers who had charge accounts but did not make a purchase.
150 customers who made a purchase and had charge accounts.
30 women who made a purchase but did not have a charge account.
50 women who had a charge account and made a purchase.

236. Use a Venn diagram to determine whether or not this report is consistent.

In a survey of 100 local union members, the number of members who had attended the last three monthly union meetings was found to be: January only, 10; January but not February, 25; January and March, 20; January, 40; March, 45; March and February, 10; none of the three meetings, 25; all three meetings, 5.

237. How many members attended the February meeting?
238. How many members attended two consecutive meetings?
239. How many members attended the March meeting but did not attend the February meeting?
240. How many members attended the January and February meetings but did not attend the March meeting?
241. How many members attended the February meeting only?

Records indicate that of the 300 persons attending an office automation conference,

120 visited Exhibit A.
115 visited Exhibit B.
87 visited Exhibit C.
30 visited Exhibits A and B.
27 visited Exhibits B and C.
22 visited Exhibits A and C.
12 visited Exhibits A, B, and C.

242. How many persons attending the conference visited none of the exhibits?
243. How many persons attending the conference visited only one of the exhibits?
244. How many persons attending the conference visited at least one of the exhibits?
245. How many persons attending the conference visited exactly two of the exhibits?
246. How many persons attending the conference visited at least two of the exhibits?

A consumer research study of 500 heads of household in a certain market area indicates the following about the purchasing plans of the household for the next 12 months:

175 plan to purchase a new stove.
190 plan to purchase a new refrigerator.
170 plan to purchase a new washing machine.
65 plan to purchase both a stove and a refrigerator.
55 plan to purchase both a stove and a washing machine.
80 plan to purchase both a refrigerator and a washing machine.
25 plan to purchase a stove, a refrigerator, and a washing machine.

247. How many households plan to purchase exactly one of the appliances?
248. How many households plan to purchase exactly two of the appliances?
249. How many households plan to purchase none of the appliances?

The reactions of a number of customers to a new product are tabulated at right.

	Liked	Disliked
Single:		
Men	16	15
Women	22	4
Married:		
Men	19	12
Women	24	5

Letting L *denote the set of all customers who liked the product,* S *the set of single customers, and* M *the set of men, define in words the following sets. How many elements are there in each of the sets?*

250. M^c
251. L
252. $M \cap L \cap S$
253. $(M \cup S)^c$
254. $M \cap L^c \cap S$
255. $M^c \cap S^c \cap L$

2.6 A FUNDAMENTAL PRINCIPLE OF COUNTING (THE MULTIPLICATION RULE)

We are not always able to determine how many elements are in a universal set or one of its subsets through such simple and straightforward counting procedures as tabulating the number of students enrolled in a class, the number of responses to a questionnaire, or the number of automobiles passing a certain tollgate over a holiday weekend. In many instances the universal set and its subsets, or a sample space and the events defined on it, are of such a nature that special rules for counting are needed in determining the number of elements involved. Consider the universal set consisting of all sums of money possible from a penny, a nickel, a dime, a quarter, a half-dollar, and a silver dollar; or all possible lineups when the six newly elected city councilmen are asked to stand together in a row for a picture-taking session; or the number of 5-card hands possible from a standard deck of 52 playing cards; or the number of different routes a hitchhiker may take traveling from Haleyville to Tipton with intermediate stops first at Jonesboro and then at Crawford, given that there are four routes from Haleyville to Jonesboro, five routes from Jonesboro to Crawford, and three routes from Crawford to Tipton.

So, you see, it is important that we become familiar with some techniques which can be used for counting various collections and arrangements of objects. While a great variety of counting procedures is available, only those most frequently used will be discussed here.

The following procedure is a basic strategy for counting: Divide an entire selection process S into a finite sequence of selections $s_1, s_2, \ldots, s_k$ such that the ways of making any selection s_i do not depend upon the results of the previous selections in the sequence. The number of ways of realizing S will be the product of the number of ways of realizing the individual selections in the sequence.

> If a selection operation can be performed in n_1 ways, and if for each of these a second operation can be performed in n_2 ways, and for each of these two a third operation can be performed in n_3 ways, and so on, the entire sequence of k operations can be performed in this number of ways:
>
> $$n(S) = (n_1) \cdot (n_2) \cdot \ldots \cdot (n_k) \qquad \textbf{(2.6)}$$

Example 23 A construction company builds two styles of homes—colonial and contemporary—and uses three colors of paint—gray, green, and white. How many different-looking houses can be built?

In selecting a style of house, there are *two* possibilities and, for each of these, there are *three* possible color choices; hence, there are in all

$$n(S) = (2) \cdot (3) = 6$$

possible different-looking houses.

A "tree diagram" may be useful in providing a list of these possibilities, as shown in Figure 2.15. []

Figure 2.15

Tree Diagram for Process of Selecting Style and Color of House

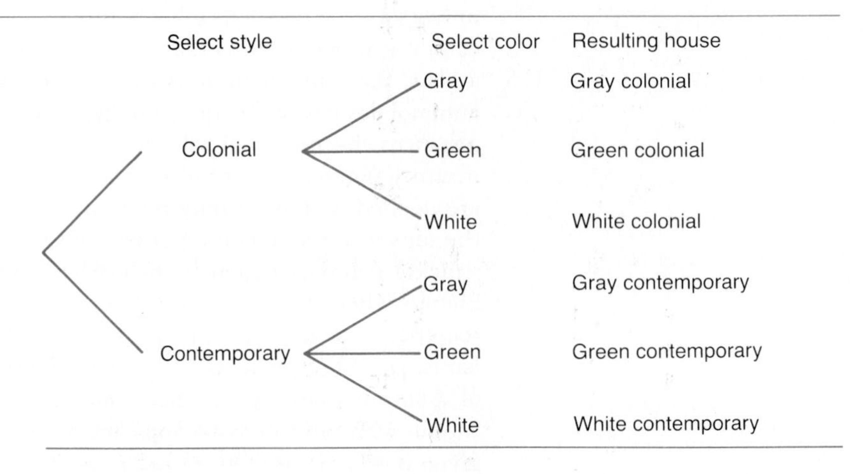

Example 24 The letters A, B, and C are used as prefixes for inventory numbers.

(*a*) How many different one-letter prefixes can be made?

There are only three choices:

$$A \qquad B \qquad C$$

(*b*) How many different two-letter prefixes are possible, if there are no restrictions on the use of the letters?

We must fill in two positions in the prefix; there are three ways of filling in the first position and then three ways of filling in the second position, as

$$\frac{(3 \text{ ways})}{1st\ letter} \times \frac{(3 \text{ ways})}{2nd\ letter} = 9 \text{ possible two-letter prefixes}$$

Using a tree diagram, as shown in Figure 2.16, we are able to compile a list of these possible two-letter prefixes.

(*c*) How many different three-letter prefixes are possible if there are no restrictions placed on the use of the letters?

Now we must fill in three positions in the prefix and have three possi-

Figure 2.16

Tree Diagram for Process of Selecting Letters *A*, *B*, *C* for a Two-Letter Prefix

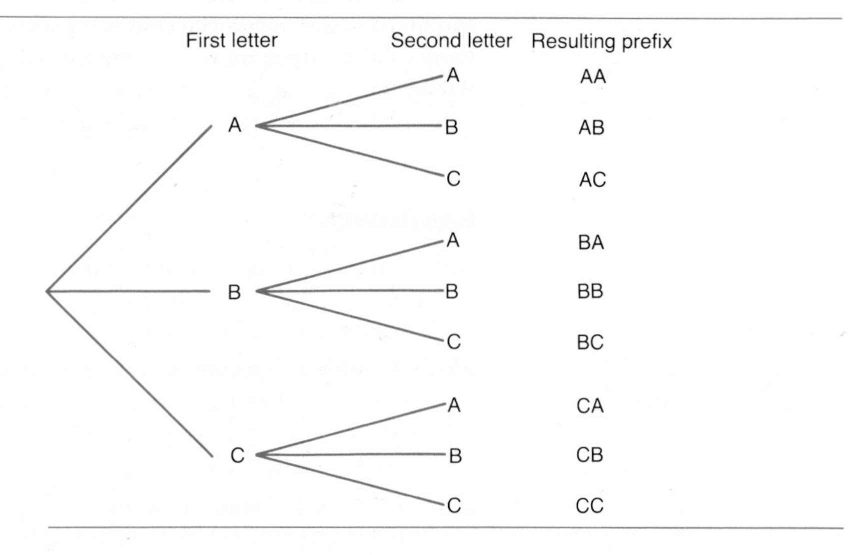

bilities for each position. Hence, the number of distinct three-letter prefixes possible is

$$\frac{(3 \text{ ways})}{1st\ letter} \times \frac{(3 \text{ ways})}{2nd\ letter} \times \frac{(3 \text{ ways})}{3rd\ letter} = 27 \text{ possible three-letter prefixes}$$

These three-letter prefixes can be enumerated by visually extending the tree diagram shown in Figure 2.16. They are

AAA	*ABA*	*ACA*	*BAA*	*BBA*	*BCA*	*CAA*	*CBA*	*CCA*
AAB	*ABB*	*ACB*	*BAB*	*BBB*	*BCB*	*CAB*	*CBB*	*CCB*
AAC	*ABC*	*ACC*	*BAC*	*BBC*	*BCC*	*CAC*	*CBC*	*CCC*

(*d*) How many prefixes are possible if either one-letter, two-letter, or three-letter prefixes are allowed?

The total number of prefixes of either one-, two-, or three- letters is the sum of the number of one-letter and the number of two-letter and the number of three-letter possibilities, as

$$\frac{3}{\substack{number\ of\\ one\text{-}letter\\ prefixes}} + \frac{9}{\substack{number\ of\\ two\text{-}letter\\ prefixes}} + \frac{27}{\substack{number\ of\\ three\text{-}letter\\ prefixes}} = \frac{39}{\substack{total\ number\ of\\ distinct\\ prefixes}}$$

[]

Example 25 A plant cafeteria offers a choice of two soups, five sandwiches, three desserts, and three drinks. How many different lunches—each consisting of a soup, a sandwich, a dessert, and a drink—are possible?

A soup can be chosen in two different ways; for each of these, a sandwich can be chosen in five different ways; for each soup-sandwich combination, a dessert can be selected in three ways; and, finally, a drink can be picked in three different ways. Thus, the total number of different lunches of a soup, a sandwich, dessert, and drink possible at the cafeteria would be

$$n(S) = (2) \cdot (5) \cdot (3) \cdot (3) = 90$$ []

EXERCISES

256. How many distinct three-digit numbers can be formed from the digits 1, 3, 5, 7, 9 if there is no restriction on the number of times each digit can be used?
257. A fashion designer works with six dress patterns and five types of fabric. How many different-looking dresses can he produce?
258. Three dice—one red, one blue, and one green—are tossed. In how many different ways may they fall?
259. How many different license plates can be made where the first two positions are letters (selected from the 26 letters of the English alphabet) and the third, fourth, and fifth positions are numbers (selected from the 10 digits 0 through 9)? No restrictions are placed on the use of the letters or the numbers.
260. A cafeteria offers seven different salads, six different main courses, 12 different desserts, and three different drinks. How many different lunches—each consisting of a salad, a main course, a dessert, and a drink—are possible?
261. A municipal planning committee is to be set up, with one member from the chamber of commerce, one member from the city council, and one member from the League of Women Voters. If there are 13 candidates from the chamber of commerce, eight candidates from the city council, and five candidates from the League of Women Voters, how many different planning committees are possible?
262. How many different one-, two-, or three-letter Greek fraternity names are possible if there are 26 letters in the Greek alphabet and there are no restrictions on the use of the letters?
263. A hitchhiker plans to travel from Haleyville to Tipton with intermediate stops first at Jonesboro and then at Crawford. There are four routes from Haleyville to Jonesboro, five routes from Jonesboro to Crawford, and three routes from Crawford to Tipton. How many different routes could the hitchhiker take from Haleyville to Tipton?
264. A customer considering the purchase of a new automobile is told that the standard model is available in six colors, two types of transmissions, and five different accessory packages, while the sports model is available in seven colors, three types of transmis-

sions, and four different accessory packages. How many choices of different automobiles does the customer have?

265. A system has eight switches, each of which may be either open or closed. The state of the system is described by indicating for each switch whether it is open or closed. How many different states of the system are there?

A student owns five pair of slacks, six sports coats, eight shirts, and four sweaters.

266. How many different outfits can be put together if an outfit consists of a pair of slacks, a sports coat, and a shirt?
267. How many different outfits can be put together if an outfit consists of a pair of slacks, a sports coat, and a sweater?
268. How many different outfits can be put together if an outfit consists of a pair of slacks, a sports coat, a shirt, and a sweater?
269. How many different ways can a student answer a quiz with 10 multiple-choice questions if each question has three possibilities?

How many four-digit numbers may be formed from the digits 1, 2, 3, 4, 5, 6, 7

270. If there are no restrictions placed on the use of the digits?
271. If each digit can be used only once?
272. If the number formed must be less than 5,000 but there are no other restrictions?
273. If the number formed must be exactly divisible by 5 but no other restrictions are placed on the use of the digits?

2.7 PERMUTATIONS

Frequently we are concerned with a universal set that has as its elements all possible ORDERS or ARRANGEMENTS of a group of objects. We are interested not only in the composition of the group of objects but also in the sequence in which the objects appear. Such an ORDERED ARRANGEMENT of items, selected without replacement from one finite collection of items, is called a PERMUTATION.

Permutations: n Objects Taken n at a Time

An ordered arrangement of all n distinct elements of a set is called a *permutation of* n *distinguishable objects taken* n *at a time.*

Any one of the n elements in the basic group could be chosen to fill the first position in the arrangement. For the second position, there are only $(n - 1)$ items to select from. Hence, the first two positions can be filled in $(n) \cdot (n - 1)$ ways. Similarly, the third position can be filled in $(n - 2)$ ways, and the fourth in $(n - 3)$ ways, and so on until the nth position can be filled in only one way. Therefore, the total number of permutations of n distinct objects taken n at a time is equal to $(n) \cdot (n-1) \cdot (n-2) \cdot (n-3) \cdot \ldots \cdot (1) = n!$

> The number of permutations of n distinct objects taken n at a time, commonly denoted $P(n,n)$, is
>
> $$P(n,n) = (n) \cdot (n-1) \cdot (n-2) \cdot \ldots \cdot (1) = n! \qquad (2.7)$$

Example 26 Consider the three letters X, Y, and Z. The possible permutations of all three letters are:

XYZ	YZX	ZXY
XZY	YXZ	ZYX

That there are six distinct sequences of these letters could have been determined by

$$P(3,3) = 3! = (3) \cdot (2) \cdot (1) = 6.$$ []

Note that in a permutation all selections are made from the same basic set and that no element from this basic set may be used more than once!

Example 27 A bookstore is promoting a 12-volume set of mystery novels. In how many different ways can these books be arranged on a shelf?

The 12 volumes may be arranged in

$$P(12,12) = 12! = 479{,}001{,}600$$

different sequences on a display shelf. []

Example 28 How many different batting orders are possible for a Little League baseball team consisting of nine players?

The number of permutations of nine distinct objects taken nine at a time is

$$P(9,9) = 9! = 362{,}880$$

How many different batting orders are possible if the pitcher must bat last?

Now there remain only eight batting positions to fill and these may be filled in

$$P(8,8) = 8! = 40{,}320$$

ways. In each of these batting sequences, the pitcher bats last. []

Example 29 How many different ways can six workers be assigned to six work desks which are arranged in a row in the work center?

There are

$$P(6,6) = 6! = 720$$

different ordered arrangements of the six workers.

How many ways are there to assign the workers to the desks if Eric and Fred refuse to work at adjacent desks?

There are five locations within the row of six positions where the two workers might be stationed adjacent to one another. Then within these two work stations, the two workers could be arranged in $2! = 2$ ways, *EF* or *FE*. Thus, there are $5 \times 2! = 10$ arrangements involving Eric and Fred which would have to be eliminated from consideration. These are as follows:

E	*F*	?	?	?	?		*F*	*E*	?	?	?	?
?	*E*	*F*	?	?	?		?	*F*	*E*	?	?	?
?	?	*E*	*F*	?	?		?	?	*F*	*E*	?	?
?	?	?	*E*	*F*	?		?	?	?	*F*	*E*	?
?	?	?	?	*E*	*F*		?	?	?	?	*F*	*E*

Note that many more than 10 arrangements are involved, however, since for each arrangement that is listed above, the blank spaces may be filled in $4! = 24$ different ways. Hence, $24 \times 10 = 240$ of the total number of possible ways of arranging the six workers would not be acceptable. Then the number of acceptable arrangements becomes

$$6! - [5 \times 2! \times 4!] = 480$$

[]

Permutations: *n* Objects Taken *r* at a Time

An arrangement of r objects selected from a group consisting of n distinct objects, given that $r < n$, where the sequence of the objects is of consequence, is called a *permutation of* n *objects taken* r *at a time.*

Again, any one of the n elements in the basic group could be chosen to fill the first position in the arrangement. For the second position, there are $(n - 1)$ elements to select from. The third position can be filled in $(n - 2)$ ways, and so on, until the rth position can be filled in $(n - r + 1)$ ways. Therefore, the total number of permutations of n distinct objects taken r at a time is given by

$$(n) \cdot (n-1) \cdot (n-2) \cdot \ldots \cdot (n-r+1)$$

If this expression is multiplied by $(n - r)!/(n - r)!$, the result is

$$\frac{(n) \cdot (n-1) \cdot (n-2) \cdot \ldots \cdot (n-r+1)(n-r)!}{(n-r)!} = \frac{n!}{(n-r)!}$$

The number of permutations of n distinct objects taken r at a time, commonly denoted $P(n,r)$, is given by

$$P(n,r) = \frac{n!}{(n-r)!} = (n) \cdot (n-1) \cdot (n-2) \cdot \ldots \cdot (n-r+1) \qquad (2.8)$$

Example 30

How many three-digit numbers in which no digit is repeated can be made with the nine digits 1 through 9?

We note that the three positions in the number can be filled as follows

$$\frac{(9 \text{ ways})}{1st\ position} \times \frac{(8 \text{ ways})}{2nd\ position} \times \frac{(7 \text{ ways})}{3rd\ position} = 504 \text{ ways}$$

Or, using Rule 2.8 above, we determine that

$$P(9,3) = \frac{9!}{6!} = \frac{(9) \cdot (8) \cdot (7) \cdot \cancel{(6!)}}{\cancel{(6!)}} = (9) \cdot (8) \cdot (7) = 504$$

different three-digit numbers could be formed from the nine digits. [ ]

Example 31

How many ways could a president, vice president, secretary, and treasurer of an executive committee be selected from the 11 members of the committee?

The number of permutations of 11 objects taken 4 at a time is given by

$$P(11,4) = 11!/(11-4)! = 11!/7! = (11) \cdot (10) \cdot (9) \cdot (8) = 7{,}920$$

Thus, there are 7,920 different ways of selecting the four officers from the committee of 11 people. [ ]

Permutations: Like Objects

The formulas for permutations given previously are valid only when all of the objects in the basic group from which the selections are made are completely distinguishable. In those situations where some of the objects are alike, another method of finding the number of permutations is required.

Let us assume that a set of n objects consists of only two different types of items. There are n_1 indistinguishable objects of one type and n_2 indistinguishable objects of a second type where $n_1 + n_2 = n$. If all n objects were different, there would be $n!$ permutations. Now the n_1 objects can be arranged in $n_1!$ different ways, while the n_2 objects can be arranged in $n_2!$ different ways. Thus, the number of permutations of all n objects taken together is reduced from $n!$ to $n!/(n_1!n_2!)$.

This concept of permutations of like objects may be extended to any number of groups of like objects.

The number of distinct permutations of n things when n_1 are indistinguishable and of one type, n_2 are indistinguishable and of a second type, . . . , and n_k are indistinguishable and of a kth type, is given by

$$\frac{n!}{n_1!n_2! \ldots n_k!} \tag{2.9}$$

where $\sum_{n=1}^{i} n_i = n$

Example 32 In how many different ways can two identical white refrigerators and two identical yellow refrigerators be arranged in a row across a showroom floor?

If we used red paint to mark the numbers 1 and 2 on the front of the two white refrigerators, we would be able to distinguish between the arrangement W_1W_2YY and W_2W_1YY. However, without the red numbers, the arrangements $WWYY$ and $WWYY$ are indistinguishable. Thus, the total number of permutations of distinct objects must be reduced accordingly. Indeed, the total number of permutations must be reduced by the number of permutations possible using the two white refrigerators; that is, 2! And a similar adjustment must be made because the two yellow refrigerators are indistinguishable.

In the final analysis, the total number of permutations of these four objects is

$$4!/(2!2!) = 6$$

These arrangements are

$WWYY$ $WYWY$ $WYYW$ $YYWW$ $YWYW$ $YWWY$ [ ]

Example 33 The word *statistics* contains 10 letters. However, the three *s*'s are alike, the three *t*'s are alike, and the two *i*'s are alike. The number of permutations of the letters in the word is, therefore, not 10! but rather,

$$\frac{10!}{3!3!2!1!1!} = 50{,}400$$ [ ]

EXERCISES

274. Four different soft drinks—*A*, *B*, *C*, and *D*—are ranked in order of preference by independent taste-testers. How many distinct rankings (ruling out ties) are possible?

275. How many five-digit numbers can be formed from the digits 1, 2, 3, 4, and 5, if each digit can be used only once in a number?

276. In how many ways that are incorrect can a 22-volume encyclopedia set be arranged on a bookshelf?

277. From the 2,500 lottery tickets which are sold, 3 tickets are to be selected for first, second, and third prizes. How many possible outcomes are there?

278. How many three-letter codes can be formed from the letters *G*, *K*, *Q*, *R*, and *Z*, using each letter only once?

279. In how many different ways can six students line up at a bar?

280. In how many different ways can five people be seated in a room containing seven chairs?

281. In how many ways can seven guest speakers be seated in a row on a lecture platform?

282. In how many ways can seven speakers be seated in a row on a lecture platform if the principal speaker must be seated in the center?

283. A salesperson is needed in each of five different sales territories. If 11 equally qualified persons apply for the jobs, in how many ways can the jobs be filled?

The letters W, X, Y, *and* Z *are used as prefixes for inventory numbers.*

284. How many different four-letter prefixes can be formed if there are no restrictions on the use of the letters?

285. How many different four-letter prefixes can be formed if each letter can be used only once in each prefix?

286. How many different prefixes are possible if the prefix is to consist of only three letters and no letter can be used more than once?

287. How many prefixes are possible if either one-, two-, three-, or four-letter prefixes are allowed but a letter cannot be repeated within a prefix?

Five married couples have 10 seats in a row for a certain opera performance.

288. In how many different ways can they be seated?

289. In how many ways can they be seated if man and wife sit in adjacent seats?

290. In how many ways can they be seated if all the men sit together and all the women sit together?

291. How many ways may the group photograph of the nine officers of the Grant Company be taken if the officers are to stand in a line?

292. How many ways may the photograph of the nine officers of the Grant Company be taken if the president must stand in the center of the line, with the two vice presidents on either side of the president?

293. There are five different computer-programming projects to be assigned to 10 different programmers, with 1 programmer per project. How many different assignments are possible?

During a sale, a clerk in a bookshop wants to arrange in a display area two different travel books, three different autobiographies, four different cookbooks, and five different novels. How many different arrangements (in a row) are possible if

294. The books may be arranged in any order?
295. The two travel books must be placed one at each end of the row?
296. The books of each type must all stand together, novels on the right, cookbooks next, followed by autobiographies, and then the travel books?
297. The books of each type must all stand together but the types can be placed in any sequence?
298. In how many different ways may six color-coded wires be connected to six color-coded terminals?
299. In how many ways can four identical motorcycles and three identical wheelchairs be displayed in a row in a display window?
300. How many distinguishable arrangements of the letters in the word *Mississippi* are possible?
301. How many different signals can be made by arranging four red flags, five blue flags, and three yellow flags in a row?

2.8 COMBINATIONS

If, when we select a finite number of distinct objects from a given set, we are interested in the membership of the group selected with *no regard to the order in which the objects are selected or arranged,* the group of objects is known as a COMBINATION.

The formula for permutations will yield the number of distinct arrangements of r things that can occur when there are n different objects to select from. With combinations, however, interest is not in the number of possible sequences but only in the number of possible groupings. In forming the permutations, we may think first of a group of r objects selected from a set of n objects—label this group $C(n,r)$—and then think of the number of different arrangements that can be made with the objects selected. There are $r!$ such distinct arrangements of the r objects. Thus, the number of permutations is

$$P(n,r) = C(n,r) \times r!$$

It follows that the number of ways the group could have been selected without regard to sequence—which we have labeled $C(n,r)$—equals

$$C(n,r) = \frac{P(n,r)}{r!} = \frac{n!}{(n-r)!} \div r! = \frac{n!}{r!(n-r)!}$$

The number of COMBINATIONS of n distinct objects taken r at a time, denoted $C(n,r)$ or, perhaps more commonly by $\binom{n}{r}$, is given by

$$C(n,r) = \binom{n}{r} = \frac{n!}{r!(n-r)!} \qquad \textbf{(2.10)}$$

The notation $\binom{n}{r}$ is read "n above r."

Example 34 How many permutations and how many combinations can be formed from the three letters A, B, and C if all three letters are used?

The possible permutations of this set of three letters are the following six arrangements:

ABC *ACB* *BAC* *BCA* *CAB* *CBA*

If sequence or arrangement is disregarded, there is but one combination of the set of three letters—the combination *ABC* (which can, of course, be written in any one of the six ways listed). []

Example 35 The number of ways a committee of 4 persons can be selected from a group of 11 persons is equal to

$$C(11,4) = \binom{11}{4} = \frac{11!}{4!(11-4)!} = 330$$

Compare this result with that of a previous example where specific officers—a president, a vice president, secretary, and a treasurer—were selected from a group of 11 persons. []

Example 36 How many distinct poker hands consisting of 5 cards are possible from a standard deck of 52 playing cards?

The number of possible distinct 5-card hands that can be dealt from the deck of 52 cards is

$$C(52,5) = \binom{52}{5} = \frac{52!}{5!(52-5)!} = 2{,}598{,}960$$ []

Example 37 Angelo's Pizza Parlor offers a plain cheese pizza to which can be added any number of five different toppings: sausage, pepperoni, mushrooms, onions, and green peppers. How many different pizzas can be ordered?

Because the set of five toppings has $2^5 = 32$ different subsets, there are 32 different pizzas that can be ordered, as follows

With no toppings (plain cheese)	$\frac{5!}{0!5!} = 1$
With one topping	$\frac{5!}{1!4!} = 5$
With two toppings	$\frac{5!}{2!3!} = 10$
With three toppings	$\frac{5!}{3!2!} = 10$
With four toppings	$\frac{5!}{4!1!} = 5$
With five toppings	$\frac{5!}{5!0!} = 1$
	32

[]

Sometimes we have difficulty in deciding which of the two quantities—permutations or combinations—is required for a particular analysis. The key lies in whether or not the sequence of occurrence, or the arrangement, of the items is important. If we need to take into account the sequence, we want permutations. If sequence is immaterial and we are interested only in the composition of the group, we want combinations.

Both the permutation and combinations formulas presume selection without replacement from one basic group of objects. If selection is made with replacement, or if selections are made from several different sets of objects, the fundamental multiplicative rule for counting will have to be used to determine the total number of possible outcomes.

Joint Selections

We are often concerned with the number of possible combinations when items are selected from more than one basic group.

> The number of possible ways to simultaneously select r_1 elements from a set containing n_1 different elements, r_2 elements from a set containing n_2 different elements, and so on until r_k elements are selected from a set of n_k different elements, is given by
>
> $$\binom{n_1}{r_1} \cdot \binom{n_2}{r_2} \cdot \cdots \cdot \binom{n_k}{r_k} \qquad \textbf{(2.11)}$$

Example 38

A ballet troupe has as its repertoire four variations from *Swan Lake,* five variations from *The Nutcracker,* and three variations from *Coppelia.* How many different programs, each consisting of two variations from each ballet, can they offer?

The number of different programs possible is

$$\binom{4}{2} \cdot \binom{5}{2} \cdot \binom{3}{2} = \left(\frac{4!}{2!2!}\right) \cdot \left(\frac{5!}{2!3!}\right) \cdot \left(\frac{3!}{2!1!}\right) = (6) \cdot (10) \cdot (3) = 180$$

[]

Example 39 In a certain firm a planning committee composed of three members from the sales department, three members from the production department, and two members from the controller's office is to be appointed. How many different committees are possible given that six members of the sales department, seven members of the production department and four members of the controller's office are qualified to serve on the planning committee?

The number of possible planning committees is given by

$$\binom{6}{3}\binom{7}{3}\binom{4}{2} = \left(\frac{6!}{3!3!}\right)\left(\frac{7!}{3!4!}\right)\left(\frac{4!}{2!2!}\right) = (20)\cdot(35)\cdot(6) = 4200$$ []

Partitions At times, interest is centered upon the number of ways of placing each member of a set of n objects into one and only one of k subsets or cells. A PARTITION is achieved if the intersection of every possible pair of the k subsets is the empty set (that is, there is no overlapping of subsets) and if the union of all the k subsets gives the original set. The order or arrangement of the items within each subset is immaterial.

> The number of ways of PARTITIONING a set of n objects into k cells with n_1 objects in the first cell, n_2 objects in the second cell, and so on, is given by
>
> $$\frac{n!}{n_1!n_2!\ldots n_k!} \qquad (2.11)$$
>
> where $\sum_{i=1}^{k} n_i = n$

Example 40 What is the possible number of partitions of the set $S = \{1,2,3,4,5\}$ into two cells with the first cell containing three elements and the second cell containing two elements?

The number of such partitions is given by

$$5!/3!2! = 10$$

These 10 partitions are:

{(1,2,3),(4,5)}	{(1,3,4),(2,5)}	{(2,3,4),(1,5)}
{(1,2,4),(3,5)}	{(1,3,5),(2,4)}	{(2,4,5),(1,3)}
{(1,2,5),(3,4)}	{(1,4,5),(2,3)}	{(2,3,5),(1,4)}
		{(3,4,5),(1,2)}

[]

Example 41 In how many different ways can 12 sales trainees be assigned to three sales territories if 5 are to be assigned to Territory E, 4 to Territory N, and 3 to Territory W?

The number of such partitions is

$$12!/(5!4!3!) = 27{,}720$$ []

Example 42 In how many different ways can seven engineers be assigned to two projects if four engineers are needed for Project A and three engineers are needed for Project B?

The number of such partitions is

$$7!/(4!3!) = 35$$ []

EXERCISES

302. A display manager wishes to place five different colors of carpet in a display window. If there are nine colors from which to choose, how many combinations of five colors are possible?
303. An engineer has four problems to be solved on an electronic computer. Each problem will require approximately the same amount of time, but there is time available only to solve three problems. What choices does the engineer have?

A three-person committee is to be formed from a list of seven persons.

304. How many possible committees can be formed?
305. How many different committees are possible if, after the group of three is selected, one member is chosen to be the spokesperson?

306. A team of two salespersons is to be selected from the six-person sales force. How many different teams of two are possible?
307. A contractor needs to hire six additional painters. If 14 equally qualified painters applied for the jobs, how many different ways can the 6 be selected?
308. Kevin is taking a 15-question essay examination. If he must answer only 12 of the 15 questions, in how many different ways can he take the examination?
309. A machine shop has five work orders that require lathe work and three work orders that require milling work. If the shop has two lathes and three milling machines, in how many different ways can the work orders be assigned to the machines?

A firm has an inventory of six identical generators in stock, two of which are known to be defective.

310. How many different groups of three generators can be selected from the stock?
311. How many different groups of three generators can be selected that do not include the defective generators?
312. How many different groups of three generators can be selected that contain exactly one defective item?
313. How many different groups of three generators can be selected that contain at least one defective item?

Ten employees from the western regional office and 12 employees from the eastern regional office have applied for four vacancies in the home office. All applicants are equally qualified. In how many ways can the four vacancies in the home office be filled if

314. All four are chosen from the western regional office?
315. All four are chosen from the eastern regional office?
316. Two are chosen from the eastern regional office and two from the western regional office?
317. No restrictions are placed on the selection process?

A special committee of 4 is to be chosen from the 18-member governing committee of the Benevolent Brotherhood, local chapter. The special committee is to attend the national conference in Philadelphia.

318. How many four-person committees are possible?
319. How many 4-person committees are possible if 2 of the 18 abhor each other and refuse to go to the meeting together?
320. How many different sums of money are possible from a penny, a nickel, a dime, a quarter, a half-dollar, and a silver dollar, assuming that at least one coin must be used?

Janet wishes to spend her spring holidays reading. There are six classics she would like to read, but time may prevent her reading all six.

321. How many ways can she select two books from the six?
322. How many ways can she select three books from the six?
323. How many ways can she select at least two books from the six?
324. From three red, four purple, and six pink costumes, how many selections of six costumes are possible if each color must be represented twice?
325. Twelve boys and nine girls have signed up to play in the mixed-doubles matches at the tennis tournament. How many different matches are possible where two teams, each consisting of one boy and one girl, are required for each match?

Hogan-Diaz Ice Cream Parlor offers 13 flavors of ice cream and a choice of five toppings.

326. How many different serving choices are possible if a serving consists of one scoop of ice cream and as many toppings as desired?
327. How many different serving choices are possible if a serving consists of two scoops of ice cream and as many toppings as desired?
328. Pedro's Pizza Parlor offers 11-inch, 13-inch, and 17-inch pizzas with choice of cheese, sausage, pepperoni, mushrooms, onions,

and peppers or any combination thereof, and thick or crispy crust. How many different pizzas are available?

329. In how many ways can a group of 10 salespersons and four typists be divided into two groups consisting of 5 salespersons and two typists each?

330. A firm has 10 salespersons. In how many ways can they be assigned to two territories with five in each territory? with seven in one territory and three in the other?

331. A firm has 10 salespersons. In how many ways can the salesmen be assigned to three territories with three men in one territory, three men in a second territory, and four men in the third territory?

Ten programmers from the computer department of a firm are taking a two-week refresher course at State University and will be housed in three rooms in the school dormitory.

332. In how many ways can the 10 be assigned to the three dormitory rooms where three will stay in each of two rooms and four will stay in the third room?

333. In how many ways can the 10 be assigned to the three dormitory rooms if one pair of programmers demand to room together, three will stay in each of two rooms, and four will stay in the third room?

The 24 students of an operations management class are assigned to one of four teams, each team consisting of six members.

334. How many ways can the teams be set up?

335. How many distinct teams are possible if, after the groups of six are designated, a president, vice president, treasurer, and production manager are elected for each team?

336. In how many ways can 12 books be evenly distributed among four persons?

C·H·A·P·T·E·R 3 Matrix Algebra

A PREVIEW of problems you will be able to work after studying this chapter:

Modular Office Systems offers for sale three models of complete office filing systems. These models consist of various combinations of filing subassemblies A, B, and C. Model I requires three subassemblies A and two subassemblies B. Model II requires one each of subassemblies A and B and four of subassembly C. Model III requires two each of subassemblies B and C.

The subassemblies are, in turn, composed of various parts, 1, 2, 3, and 4. Subassembly A requires two units each of parts 1 and 3 and three units of part 2. Subassembly B requires one each of parts 1 and 2 and five units of part 4. Subassembly C requires six units of part 1, three units each of 2 and 3, and one unit of part 4.

The firm has just received an order for three units of model I, five units of model II, and two units of model III.

Determine the number of units of each type of part required to fill the order.

Five different brands of coffee are being compared. The taster indicates these preferences: Brand A preferred over B and D; Brand B preferred over C and D; Brand C preferred over A; Brand D preferred over E; Brand E preferred over A, B, and C. Is there a preferred brand based on one-stage dominances? Is there a preferred brand based on one-stage and two-stage dominances?

Use the following array of numbers

$$\begin{bmatrix} 0.1 & 0.1 & 0 \\ 0 & 0.05 & 0.2 \\ 0.05 & -0.05 & 0.1 \end{bmatrix}$$

to decode the message: 2.9, 6.15, 2.65, 3.5, 6.15, 5.65, 2.5, 2.35, 0.55, 4.1, 2.15, −0.25, 2, 1.75, 3.2.

Long recognized for their usefulness in almost every branch of science and engineering, matrices and the algebra of matrices are assuming an increasingly important role among the modern techniques used in the mathematical analysis of social and economic problems. A matrix is, in essence, a tool for organizing vast quantities of data. Then, by the theories of matrix algebra, these arrays of data, or matrices, may be treated as single entities and subjected to various types of mathematical operations. Actually the idea of a matrix is not new. The English mathematician Arthur Cayley (1812–95) is credited with first introducing the concept of matrices in 1858. The fairly recent, but very strong, resurgence of interest in this technique undoubtedly stems from the advent of high-speed computers and their amazing capabilities for performing routine, but laborious, mathematical calculations on large volumes of numbers.

The student of matrix algebra will be examining one of the most useful and important, and also one of the most interesting, branches of mathematics.

3.1 WHAT IS A MATRIX?

A matrix may well be as much an innovation in language as an innovation in mathematics. The underlying idea is to be able to consider an entire array of numbers as a single unit, denoted by one name, or symbol. When this is done, the statement of complex mathematical relationships is greatly simplified.

A MATRIX is A RECTANGULAR ARRAY OF REAL NUMBERS ARRANGED IN m ROWS AND n COLUMNS. A boldface capital letter is used to symbolize the entire matrix, as

$$\mathbf{A} = \begin{bmatrix} a_{11} & a_{12} & \cdots & a_{1n} \\ a_{21} & a_{22} & \cdots & a_{2n} \\ \vdots & & \vdots & \vdots \\ a_{m1} & a_{m2} & & a_{mn} \end{bmatrix}$$

Each number appearing in the array is said to be an ELEMENT, or COMPONENT, of the matrix. Elements of a matrix are designated using a lowercase form of the same letter used to symbolize the matrix itself. These letters are subscripted, as a_{ij}, to give the row and column location of the element within the array. The first subscript always refers to the row location of the element; the second subscript always refers to its column location. Thus, component a_{ij} is the component located at the intersection of the ith row and the jth column.

The number of rows, m, and the number of columns, n, of the array give its ORDER, or its DIMENSIONS, $m \times n$ (read "m by n"). The abbreviated notation

$$\mathbf{A}_{m \times n} \qquad \text{or} \qquad [a_{ij}]_{(m \times n)}$$

is often used to represent the matrix and specify its order.

Example 1 The following are examples of matrices.

$$\mathbf{A} = \begin{bmatrix} 1 & 7 \\ 5 & 3 \\ 4 & -2 \end{bmatrix}$$

This is a 3×2 matrix.
Element $a_{12} = 7$
Element $a_{21} = 5$

$$\mathbf{B} = \begin{bmatrix} 1 & 2 & -1 \\ 3 & 5 & 6 \end{bmatrix}$$

This is a 2×3 matrix.
Element $b_{13} = -1$
Element $b_{23} = 6$

$$\mathbf{C} = \begin{bmatrix} 1 & 9 & 8 & 5 \\ -2 & 3 & -7 & 6 \\ -8 & -5 & 4 & 0 \\ 7 & -6 & 2 & -1 \end{bmatrix}$$

This is a 4×4 matrix.
Element $c_{33} = 4$
Element $c_{44} = -1$
Element $c_{31} = -8$ []

Matrices provide a most convenient vehicle for organizing and storing large quantities of data. Because the basic idea is to *organize* the data, we cannot overemphasize the importance of the location of each number within the matrix. It is not simply a matter of putting numbers into rows and columns; each row-column location within each matrix carries with it a special interpretation.

Example 2 **A production matrix** The Fantastic Furniture Company manufactures three styles of sofas and chairs: early American, traditional, and country French. During the past month, the company produced 10 early American sofas, 15 traditional sofas, and 12 country French sofas, along with 22 early American chairs, 28 traditional chairs, and 18 country French chairs.

To organize the data, we might first tabulate the numbers as follows:

	Units Produced in January		
	Early American	Traditional	Country French
Sofas	10	15	12
Chairs	22	28	18

When the pattern in which the information is to be recorded is clearly defined in advance, the numbers may be presented in a simple rectangular array, as

$$\mathbf{Q} = \begin{bmatrix} 10 & 15 & 12 \\ 22 & 28 & 18 \end{bmatrix}$$

The numbers on the first row are understood to represent production of sofas, while those on the second row represent chairs. The numbers in each column refer to a particular style of furniture: early American in the first column, traditional in the second column, and country French in the third column. The individual elements in $\mathbf{Q}$ are, thus, defined q_{ij} = number of units produced, in January, of product type i in style j. []

Example 3 **A transportation cost matrix** A manufacturer of automobile tires has manufacturing facilities in Akron, Canton, and Dover. Tires produced in each of these plants are shipped to four different warehouses, which are located in Stowe, Troy, Utica, and Wayne. The tires are distributed from the warehouses to wholesale and retail outlets. The available shipping routes from plants to warehouses are as follows:

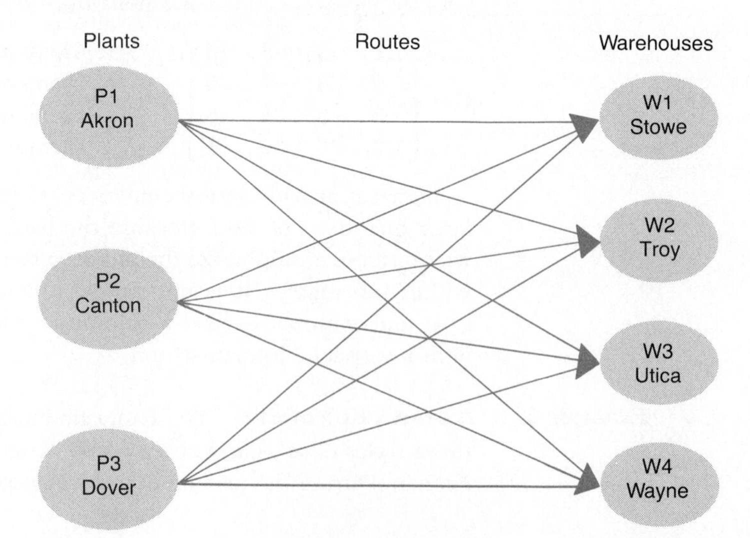

The shipping costs for the various routes are:

From	To	Cost per item
P1—Akron	W1—Stowe	$2.40
P1—Akron	W2—Troy	3.60
P1—Akron	W3—Utica	4.00
P1—Akron	W4—Wayne	3.80
P2—Canton	W1—Stowe	1.90
P2—Canton	W2—Troy	2.10
P2—Canton	W3—Utica	1.80
P2—Canton	W4—Wayne	3.20
P3—Dover	W1—Stowe	2.90
P3—Dover	W2—Troy	3.10
P3—Dover	W3—Utica	1.40
P3—Dover	W4—Wayne	2.30

This cost schedule may be meaningfully displayed in a 3 × 4 matrix, using manufacturing plants as the row labels and warehouses as the column headings, as

$$\begin{array}{c} \\ \mathbf{C} = \end{array} \begin{array}{c} \begin{matrix} \text{W1} & \text{W2} & \text{W3} & \text{W4} \end{matrix} \\ \begin{bmatrix} 2.40 & 3.60 & 4.00 & 3.80 \\ 1.90 & 2.10 & 1.80 & 3.20 \\ 2.90 & 3.10 & 1.40 & 2.30 \end{bmatrix} \end{array} \begin{array}{c} \\ \text{P1} \\ \text{P2} \\ \text{P3} \end{array}$$

where c_{ij} = cost, in dollars, per item shipped from plant i to warehouse j.

Note the importance of the positions within the matrix. All entries in row 1 are costs from the Akron plant; all entries in column 1 are costs to the warehouse at Stowe. The location of each number in the matrix tells us what that number represents.

What route is associated with c_{12}? with c_{23}? with c_{31}? []

EXERCISES

State the dimensions of the following matrices.

1. $\begin{bmatrix} 8 & 7 & 6 \\ 5 & 1/2 & -3 \end{bmatrix}$

2. $\begin{bmatrix} 2 & 7 \\ 0 & 4 \\ 3 & 5 \end{bmatrix}$

3. $\begin{bmatrix} 1/3 & 1/4 \\ 1/2 & 1/8 \end{bmatrix}$

4. $\begin{bmatrix} 4 & 4 & -1 \\ 3 & 0 & 9 \\ 2 & 8 & 10 \end{bmatrix}$

5. $\begin{bmatrix} 7 & 8 & 4 & 2 \\ 4 & 6 & 1 & 3 \end{bmatrix}$

6. $\begin{bmatrix} 9 & 0 \\ 3 & 7 \\ 5 & -1 \\ 6 & -2 \end{bmatrix}$

If

$$\mathbf{A} = \begin{bmatrix} 13 & 2 & -4 \\ -2 & 9 & 10 \\ 1/2 & 7 & 5 \\ 8 & 0 & 6 \end{bmatrix}$$

Give the following elements:

7. a_{12} 8. a_{21} 9. a_{23} 10. a_{32} 11. a_{41}

What is the row-column location in **A** *of each of the following numbers?*

12. −4 13. 9 14. 5 15. 6 16. 1/2

Write the 5 × 5 matrix in which

17. $$a_{ij} = \begin{cases} 0 & \text{if } i = j \\ 1 & \text{if } i < j \\ 2 & \text{if } i > j \end{cases}$$

18. $$a_{ij} = \begin{cases} i & \text{if } i > j \\ j & \text{if } j > i \\ 0 & \text{if } i = j \end{cases}$$

19. $a_{ij} = i + j$ 20. $a_{ij} = i/j$

21. A survey was conducted to accumulate information about the educational level and the political affiliation of residents in a certain geographical area. The following information was obtained:

    225 respondents were Republicans with a college degree.
    345 respondents were Democrats with a college degree.
    110 respondents were independents with a college degree.
    130 respondents were Republicans without a college degree.
    212 respondents were Democrats without a college degree.
    48 respondents were independents without a college degree.

    Display this information in matrix format using columns for educational level and rows for political affiliation. Label the rows and columns. Using **R** to symbolize the matrix, define the meaning of location r_{ij}.

22. Pliable Plastics, Inc., produces laundry baskets, flowerpots, and food-storage containers, among other things. The company operates four factories, each with different production capabilities. Factory A produces 2,000 laundry baskets, 2,500 flowerpots, and 800 food-storage containers each day. At Factory B, 1,000 laundry baskets, 1,250 flowerpots, and 3,000 food-storage containers are produced daily. Factory C has a daily output of 500 laundry baskets, 800 flowerpots, and 1,200 food-storage containers. Factory D produces no laundry baskets but has a daily production of 1,800 flowerpots and 600 food-storage containers.

    Display this information in matrix format, associating with each row a factory and with each column a product. Label rows and columns of the matrix. Using **Q** to represent the matrix, define the meaning of q_{ij}.

23. A manufacturing firm sells items it produces to both wholesalers and retailers, but uses a different discount policy for the two types of accounts. Wholesalers receive a 25 percent discount on both

Product A and Product B but receive a 40 percent discount on Product C. Retailers receive a 15 percent discount on Product A, 10 percent on Product B, and 5 percent on Product C.

Organize these data in a 3 × 2 matrix. Label the rows and columns. Using **D** to symbolize the matrix, define the meaning of element d_{ij}.

24. A firm needs additional machinery for use in the production of a new product it plans to market. The machinery could either be purchased outright or leased. The desirability of leasing versus purchasing depends upon the level of demand for the new product during the coming season. If the machinery is purchased, and demand turns out to be low, the firm will realize a net profit of only $20,000; if demand turns out to be at a medium level, a net profit of $30,000 will be realized should the firm purchase the machinery; or if the machinery is purchased and demand turns out to be high, a net profit of $45,000 will result. On the other hand, if the machinery is leased and demand is low, a net profit of $25,000 will be realized. If the machinery is leased, profits of $27,000 and $30,000 will result from medium and high levels of demand, respectively.

    Express these payoffs in matrix format using the rows to represent possible COURSES OF ACTION (that is, [1] purchase machinery or [2] lease machinery) and the columns to represent MARKET CONDITIONS (that is, [1] low demand, [2] medium demand, and [3] high demand). Using **P** to symbolize the matrix, define the meaning of element p_{ij}.

3.2 SPECIAL TYPES OF MATRICES

Matrices that consist of just one row or just one column may be referred to as VECTORS. More specifically, any $m \times 1$ matrix may be called a COLUMN VECTOR and any $1 \times n$ matrix may be called a ROW VECTOR.

Example 4

$$\mathbf{V} = [-1 \quad -6 \quad 2]$$

is a 1 × 3 matrix or a three-component ROW VECTOR.

$$\mathbf{W} = \begin{bmatrix} 1 \\ 0 \\ 3 \\ 2 \end{bmatrix}$$

is a 4 × 1 matrix or a four-component COLUMN VECTOR. [ ]

A matrix that has the same number of rows as it has columns is called a SQUARE MATRIX or an nth-ORDER MATRIX.

Example 5

$$\mathbf{A} = \begin{bmatrix} 1 & 6 \\ 9 & 12 \end{bmatrix}$$

is a 2×2, or a 2nd-order, matrix.

$$\mathbf{B} = \begin{bmatrix} 0 & -1 & 4 \\ 3 & 5 & 2 \\ 6 & 1 & 7 \end{bmatrix}$$

is a 3×3, or 3rd-order, matrix. []

The elements $a_{11}, a_{22}, a_{33}, \ldots, a_{nn}$ of a square matrix are said to form the MAIN, or PRIMARY, DIAGONAL of the matrix. A square matrix in which all of the primary diagonal entries are ones and all of the off-diagonal entries are zeros is called an IDENTITY MATRIX. The identity matrix, symbolized $\mathbf{I}_n$, is used in matrix algebra in much the same way as the number one is used in regular algebra.

Example 6 Each of the following is an IDENTITY MATRIX.

$$\mathbf{I}_2 = \begin{bmatrix} 1 & 0 \\ 0 & 1 \end{bmatrix} \qquad \mathbf{I}_3 = \begin{bmatrix} 1 & 0 & 0 \\ 0 & 1 & 0 \\ 0 & 0 & 1 \end{bmatrix} \qquad \mathbf{I}_4 = \begin{bmatrix} 1 & 0 & 0 & 0 \\ 0 & 1 & 0 & 0 \\ 0 & 0 & 1 & 0 \\ 0 & 0 & 0 & 1 \end{bmatrix}$$

Note that *each identity matrix is a square matrix.* []

A ZERO MATRIX is a matrix that has zero for every entry. The zero matrix is generally denoted $\mathbf{0}_{m \times n}$ and is used in matrix algebra in much the same way that the number zero is used in regular algebra.

The TRANSPOSE of an $m \times n$ matrix $\mathbf{A}$, denoted $\mathbf{A}^t$, is the $n \times m$ matrix whose rows are the columns in $\mathbf{A}$ (in the same order) and whose columns are the rows in $\mathbf{A}$ (in the same order).

Example 7

$$\text{If } \mathbf{A} = \begin{bmatrix} 1 & 2 & 3 \\ 4 & 5 & 6 \end{bmatrix} \qquad \text{then } \mathbf{A}^t = \begin{bmatrix} 1 & 4 \\ 2 & 5 \\ 3 & 6 \end{bmatrix}$$

Note that $a_{ij}{}^t = a_{ji}$. []

The transpose of a row vector is a column vector and the transpose of a column vector is a row vector.

Two matrices, $\mathbf{A}$ and $\mathbf{B}$, are said to be EQUAL only if they are of the same dimensions and if each element in $\mathbf{A}$ is identical to its *corresponding element* in $\mathbf{B}$; that is, if and only if $a_{ij} = b_{ij}$ for every pair of subscripts i and j. If $\mathbf{A} = \mathbf{B}$, then $\mathbf{B} = \mathbf{A}$; or if $\mathbf{A} \neq \mathbf{B}$, then $\mathbf{B} \neq \mathbf{A}$.

Example 8

$$\mathbf{A} = \begin{bmatrix} 1 & 2 \\ 3 & 4 \end{bmatrix} \qquad \text{is equal to } \mathbf{B} = \begin{bmatrix} 1 & 2 \\ 3 & 4 \end{bmatrix}$$

However,

$$\mathbf{A} = \begin{bmatrix} 1 & 2 \\ 3 & 4 \end{bmatrix} \qquad \text{is not equal to } \mathbf{C} = \begin{bmatrix} 4 & 3 \\ 2 & 1 \end{bmatrix}$$

Even though they contain the same set of numerical values, **A** and **C** are not equal because their corresponding elements are not equal; that is, $a_{11} \neq c_{11}$, and so on.

It is also the case that

$$\mathbf{A} = \begin{bmatrix} 1 & 2 \\ 3 & 4 \end{bmatrix} \qquad \text{and } \mathbf{D} = \begin{bmatrix} 1 & 2 \\ 3 & 4 \\ 5 & 6 \end{bmatrix}$$

are not equal since they are not of the same dimension. []

EXERCISES

25. ABCD Sales, Inc., has received an order for 600 units of Product A, 450 units of Product B, 100 units of Product C, and 40 units of Product D. Express this information in row vector format. Using **Q** to symbolize the vector, define its elements as "$q_j = \ldots$".
26. The per-unit selling prices of the products sold by ABCD Sales, Inc., are six, eight, four, and five dollars for Products *A, B, C,* and *D,* respectively. Express this information in column vector format. Using **P** to symbolize the vector, define its elements as "$p_i = \ldots$".

Given

$$\mathbf{A} = \begin{bmatrix} 1 & 0 \\ 0 & 1 \end{bmatrix} \qquad \mathbf{B} = \begin{bmatrix} 0 & 2 & 1 \\ 7 & 0 & 5 \\ 9 & 8 & 0 \end{bmatrix} \qquad \mathbf{C} = \begin{bmatrix} -1 & 0 & 0 & 0 \\ 0 & -1 & 0 & 0 \\ 0 & 0 & -1 & 0 \\ 0 & 0 & 0 & -1 \end{bmatrix}$$

$$\mathbf{D} = [8 \quad -9 \quad 7] \qquad \mathbf{E} = \begin{bmatrix} 5 \\ 1/5 \\ 25 \end{bmatrix} \qquad \mathbf{F} = [0 \quad 0 \quad 0 \quad 0]$$

$$\mathbf{G} = \begin{bmatrix} 1 & 0 & 1 \\ 0 & 1 & 0 \end{bmatrix} \qquad \mathbf{H} = \begin{bmatrix} 7 & 0 \\ 0 & 7 \end{bmatrix} \qquad \mathbf{K} = \begin{bmatrix} 0 & 0 & 0 \\ 0 & 0 & 0 \end{bmatrix}$$

27. Which matrices, if any, are square matrices?
28. Which matrices, if any, are row vectors?
29. Which matrices, if any, are column vectors?
30. Which matrices, if any, are zero matrices?
31. Which matrices, if any, are identity matrices?
32. What are the elements on the primary diagonal of matrix **B**? of matrix **H**?
33. Find $\mathbf{A}^t$, $\mathbf{G}^t$, $\mathbf{E}^t$, and $\mathbf{D}^t$.
34. Show that $(\mathbf{B}^t)^t = \mathbf{B}$.

Test the following pairs of matrices for equality. If they are not equal, give a reason.

35. $\mathbf{A} = \begin{bmatrix} 2 & 3 & 4 \\ 5 & 6 & 7 \end{bmatrix}$ and $\mathbf{B} = \begin{bmatrix} 5 & 6 & 7 \\ 2 & 3 & 4 \end{bmatrix}$

36. $\mathbf{A} = \begin{bmatrix} 2 & 3 & 4 \\ 5 & 6 & 7 \end{bmatrix}$ and $\mathbf{B} = \begin{bmatrix} 2 & 5 \\ 3 & 6 \\ 4 & 7 \end{bmatrix}$

37. $\mathbf{A} = \begin{bmatrix} 2 & 3 \\ 4 & 5 \end{bmatrix}$ and $\mathbf{B} = \begin{bmatrix} -2 & -3 \\ -4 & -5 \end{bmatrix}$

38. $\mathbf{A} = [4 \quad 5]$ and $\mathbf{B} = \begin{bmatrix} 4 \\ 5 \end{bmatrix}$

3.3 MATRIX ADDITION

Two matrices of the same dimensions are said to be CONFORMABLE FOR ADDITION. The addition is performed by adding corresponding elements from the two matrices and entering the result in the same row-column position of a new matrix.

> If **A** and **B** are two matrices, each of size $m \times n$, then the SUM of **A** and **B** is the $m \times n$ matrix **C** whose elements are
>
> $$c_{ij} = a_{ij} + b_{ij} \qquad \text{for } i = 1, 2, \ldots, m \quad j = 1, 2, \ldots, n \tag{3.1}$$

Example 9 (*a*) $\begin{bmatrix} 1 & 2 \\ -1 & 0 \end{bmatrix} + \begin{bmatrix} 7 & 14 \\ 0 & 3 \end{bmatrix} = \begin{bmatrix} (1+7) & (2+14) \\ (-1+0) & (0+3) \end{bmatrix} = \begin{bmatrix} 8 & 16 \\ -1 & 3 \end{bmatrix}$

(*b*) $[5 \quad 8 \quad 2] + [3 \quad 0 \quad 1] = [(5+3) \quad (8+0) \quad (2+1)] = [8 \quad 8 \quad 3]$

(*c*) $\begin{bmatrix} 6 & 8 \\ 3 & 4 \\ -9 & 1 \end{bmatrix} + \begin{bmatrix} 7 & 11 & 12 \\ 0 & 3 & 7 \end{bmatrix}$ *is not defined* because the two matrices are not of the same dimension []

Example 10 Theater-on-the-Square is presenting a special production of *Faust*. Ticket sales for the afternoon matinee were as follows: 331 orchestra seats, 427 loge seats, and 642 balcony seats. For the night performance, 282 orchestra seat tickets, 438 loge seat tickets, and 834 balcony seat tickets were sold. Use matrix methods to determine total ticket sales, by type, for the two performances.

We set up two vectors as follows:

$$\mathbf{M} = [331 \quad 427 \quad 642] \qquad \text{and} \qquad \mathbf{N} = [282 \quad 438 \quad 834]$$

Then, total ticket sales, by type, is given by

$$\mathbf{S} = \mathbf{M} + \mathbf{N} = [613 \quad 865 \quad 1{,}476] \qquad \square$$

Laws of Matrix Addition

The operation of adding two matrices that are conformable for addition has these basic properties.

1. $\mathbf{A} + \mathbf{B} = \mathbf{B} + \mathbf{A}$ (The commutative law of matrix addition)
2. $(\mathbf{A} + \mathbf{B}) + \mathbf{C} = \mathbf{A} + (\mathbf{B} + \mathbf{C})$ (The associative law of matrix addition)

EXERCISES

Perform the indicated matrix operations. If the operation is not defined, give the reason.

39. $\begin{bmatrix} 1 & 2 \\ 3 & 4 \end{bmatrix} + \begin{bmatrix} 5 & 6 \\ 7 & 8 \end{bmatrix}$

40. $\begin{bmatrix} 4 & 3 & 6 \\ 5 & 0 & -1 \end{bmatrix} + \begin{bmatrix} 2 & -1 & -3 \\ 0 & 6 & 5 \end{bmatrix}$

41. $\begin{bmatrix} 8 & 9 \\ 3 & 4 \\ 5 & -2 \end{bmatrix} + \begin{bmatrix} 3 & 0 \\ 1 & -5 \\ 1 & 1 \end{bmatrix}$

42. $\begin{bmatrix} 2 & 3 & 4 \\ 1 & 0 & 6 \end{bmatrix} + \begin{bmatrix} 8 & 7 \\ 9 & -3 \\ 0 & 1 \end{bmatrix}$ not defined

43. $\begin{bmatrix} 4 & 3 \\ 5 & 0 \end{bmatrix} + \begin{bmatrix} 8 & 1 & 7 \\ 3 & 2 & 5 \\ -1 & 4 & 0 \end{bmatrix}$

44. $[1 \quad 2 \quad 3] + [4 \quad 6 \quad 8]$

45. $\begin{bmatrix} 5 \\ 3 \end{bmatrix} + \begin{bmatrix} 2 \\ 1 \end{bmatrix}$

46. $\begin{bmatrix} 2 & 3 \\ 4 & 3 \end{bmatrix} + \begin{bmatrix} 5 & 1 \\ 0 & 5 \end{bmatrix} + \begin{bmatrix} 7 & 8 \\ 9 & -3 \end{bmatrix}$

47. $\begin{bmatrix} 1/2 & 1/4 \\ 1/3 & 1/5 \end{bmatrix} + \begin{bmatrix} 1/6 & 1/8 \\ 1/7 & 1/9 \end{bmatrix}$

48. July production at Fantastic Furniture Company's plant is summarized in matrix **J** as

$$\mathbf{J} = \begin{array}{c} \text{Early American} \quad \text{Traditional} \quad \text{Country French} \\ \begin{bmatrix} 10 & 15 & 12 \\ 22 & 28 & 18 \end{bmatrix} \end{array} \begin{array}{l} \text{Sofas} \\ \text{Chairs} \end{array}$$

Matrix **A** gives August production as

$$\mathbf{A} = \begin{array}{c} \text{Early American} \quad \text{Traditional} \quad \text{Country French} \\ \begin{bmatrix} 12 & 17 & 8 \\ 20 & 30 & 12 \end{bmatrix} \end{array} \begin{array}{l} \text{Sofas} \\ \text{Chairs} \end{array}$$

What was the total production, by product and style, for the two months?

49. During the fall semester, Sylvester Student received one A, two Bs, one C, and one D. During the spring semester his grades were one A, one B, two Cs, and one F. But during the summer session

Sylvester's grades were two As and 1 B. Use row vectors to show Sylvester's grades by session and matrix addition to summarize his grades for the year.

50. Honest John operates two used-car lots. End-of-the-year inventory shows that there are 32 sedans on the Northside lot: 15 of these are Chryslers, 9 are Fords, and 8 are Chevrolets. There are also 11 Chrysler compacts and 5 Chrysler wagons on the lot, along with 6 Ford compacts, 2 Ford wagons and 4 Chevrolet compacts. Organize the Northside lot inventory information in matrix format, categorizing the vehicles by make—Chrysler, Ford, or Chevrolet—and style—compact, sedan, or station wagon.

    Inventory records indicate that there are on the West End lot 21 Chrysler sedans, 12 Ford sedans, and 7 Chevrolet sedans, as well as 15 Chrysler, 10 Ford and 8 Chevrolet compacts. There are also 4 Chrysler wagons and 6 Chevrolet wagons on the West End lot. Organize the inventory for this lot in matrix format, categorizing the vehicles by make and style as before.

    Use matrix addition to determine the total number of vehicles of each type on hand at the end of the year.

Given,

$$\mathbf{A}=\begin{bmatrix}2 & 1\\4 & 1\end{bmatrix} \qquad \mathbf{B}=\begin{bmatrix}1 & 3\\1 & 2\end{bmatrix} \qquad \text{and} \qquad \mathbf{C}=\begin{bmatrix}4 & 1\\1 & 8\end{bmatrix}$$

51. Using **A** and **B**, verify the commutative law of matrix addition.
52. Using **A**, **B**, and **C**, verify the associative law of matrix addition.
53. Show that the transpose of two matrices is the same as the sum of the two transposed matrices; that is, show that $(\mathbf{A}+\mathbf{B})^t = \mathbf{A}^t + \mathbf{B}^t$.

3.4 MULTIPLICATION BY A CONSTANT

A matrix can be multiplied by a constant by multiplying each component in the matrix by the constant. The result is a new matrix of the same dimensions as the original matrix.

> If k is any real number and **A** is an $m \times n$ matrix, then the product $k\mathbf{A}$ is defined to be the matrix whose components are given by k times the corresponding component of **A**; that is,
>
> $$k\mathbf{A} = [ka_{ij}]_{(m\times n)} \tag{3.2}$$

Example 11 (*a*) If $\mathbf{V} = [1 \quad 2 \quad 9]$, then $5\mathbf{V} = [(5\times 1) \quad (5\times 2) \quad (5\times 9)]$
$= [5 \quad 10 \quad 45]$

(*b*) If

$$\mathbf{B} = \begin{bmatrix} 9 & -3 \\ 6 & 1 \\ 12 & 4 \end{bmatrix} \quad \text{then}$$

$$(1/3)\mathbf{B} = \begin{bmatrix} (1/3)\times 9 & (1/3)\times -3 \\ (1/3)\times 6 & (1/3)\times 1 \\ (1/3)\times 12 & (1/3)\times 4 \end{bmatrix} = \begin{bmatrix} 3 & -1 \\ 2 & 1/3 \\ 4 & 4/3 \end{bmatrix}$$ []

Example 12 The material needed to construct one deluxe doghouse is 3 pounds of nails, 12 two-by-fours, 4 two-by-sixes, 3 sheets of plywood, 2 cartons of roofing, and 1 square yard of carpeting. This information can be expressed as the row vector

$$\mathbf{M} = [3 \quad 12 \quad 4 \quad 3 \quad 2 \quad 1]$$

Then the materials required in the construction of six such deluxe doghouses is obtained by

$$6\mathbf{M} = 6[3 \quad 12 \quad 4 \quad 3 \quad 2 \quad 1] = [18 \quad 72 \quad 24 \quad 18 \quad 12 \quad 6]$$ []

Example 13 Reverberating Sound Company has used, during the last year, the following price schedule on three models of stereo systems to wholesalers and retailers.

	Wholesalers	Retailers	
P =	$348	$402	System I
	$460	$500	System II
	$490	$550	System III

Now, because of increased production costs, prices on all models, to both wholesalers and retailers, are being increased 15 percent. Determine the new price schedule.

We note that $\mathbf{N} = \mathbf{P} + 0.15\mathbf{P} = 1.15\mathbf{P}$ and compute

$$\mathbf{N} = 1.15\begin{bmatrix} 348 & 402 \\ 460 & 500 \\ 490 & 550 \end{bmatrix} = \begin{bmatrix} \$400.20 & \$462.30 \\ \$529.00 & \$575.00 \\ \$563.50 & \$632.50 \end{bmatrix}$$

Wholesalers	Retailers	
$400.20	$462.30	System I
$529.00	$575.00	System II
$563.50	$632.50	System III

[]

Factoring a Constant out of a Matrix It is often convenient to factor a number out of a matrix. We can approach this problem as if it were the reverse operation of multiplying by a constant.

Example 14 Consider the matrix

$$\begin{bmatrix} 2/3 & -3/4 & 1/3 \\ 0 & 5/12 & 7/12 \\ -2/3 & 1 & 1/6 \end{bmatrix}$$

which we would like to rewrite as a constant (which may be a fraction) times a matrix of whole numbers.

It is usually best to find a common denominator for the entries in the matrix and write all the entries as fractions using this common denominator. Here we use 12. Then we can factor out 1/12, as follows:

$$\begin{bmatrix} 2/3 & -3/4 & 1/3 \\ 0 & 5/12 & 7/12 \\ -2/3 & 1 & 1/6 \end{bmatrix} = \begin{bmatrix} 8/12 & -9/12 & 4/12 \\ 0 & 5/12 & 7/12 \\ -8/12 & 12/12 & 2/12 \end{bmatrix} = 1/12 \begin{bmatrix} 8 & -9 & 4 \\ 0 & 5 & 7 \\ -8 & 12 & 2 \end{bmatrix} \quad []$$

Laws of Scalar Multiplication

The operation of multiplying a matrix by a constant (a SCALAR) has these basic properties. If x and y are real numbers and **A** and **B** are $m \times n$ matrices, conformable for addition, then

1. $x\mathbf{A} = \mathbf{A}x$
2. $(x + y)\mathbf{A} = x\mathbf{A} + y\mathbf{A}$
3. $x(\mathbf{A} + \mathbf{B}) = x\mathbf{A} + x\mathbf{B}$
4. $x(y\mathbf{A}) = (xy)\mathbf{A}$

EXERCISES

Calculate the following products:

54. $3\begin{bmatrix} -1 & 2 & 5 \\ 0 & 4 & 8 \\ 3 & 7 & 3 \end{bmatrix}$

55. $-4\begin{bmatrix} 9 & 4 \\ 2 & 3 \\ 0 & 6 \end{bmatrix}$

56. $1/2\begin{bmatrix} 1 & 2 \\ 5 & -6 \\ 0 & 3 \end{bmatrix}$

57. $0.7\begin{bmatrix} 9 & 6 \\ 1 & 3 \end{bmatrix}$

58. $a\begin{bmatrix} 8 \\ 3 \end{bmatrix}$

59. $-2/3[3 \quad -6 \quad 3/2]$

What number must the first matrix be multiplied by in order to obtain the second matrix?

60. $\begin{bmatrix} 1 & -2 \\ 3 & 0 \end{bmatrix}, \begin{bmatrix} 5 & -10 \\ 15 & 0 \end{bmatrix}$

61. $\begin{bmatrix} 4 & -2 \\ 6 & 3 \\ 8 & 5 \end{bmatrix}, \begin{bmatrix} 2 & -1 \\ 3 & 3/2 \\ 4 & 5/2 \end{bmatrix}$

62. $[8 \quad 3 \quad 12], [16/3 \quad 2 \quad 8]$

63. $\begin{bmatrix} 9 \\ -3 \\ 7 \end{bmatrix}, \begin{bmatrix} -6.3 \\ 2.1 \\ -4.9 \end{bmatrix}$

Write each of the following matrices as the product of a fraction and a matrix of integers.

64. $\begin{bmatrix} 1/3 & 5/18 \\ 2/9 & -2/3 \\ -5/6 & 1/9 \end{bmatrix}$

65. $\begin{bmatrix} 1/4 & 5/12 \\ 2/3 & 5/6 \end{bmatrix}$

66. $\begin{bmatrix} 0 & 2/3 & 4/5 \\ 1/3 & 2 & -2/15 \end{bmatrix}$

67. $[9/2 \quad 7/4 \quad 5/8]$

Given

$$\mathbf{A} = \begin{bmatrix} 0 & 4 \\ 1 & -2 \end{bmatrix} \qquad \mathbf{B} = \begin{bmatrix} 5 & 8 \\ 3 & 1 \end{bmatrix} \qquad x = 4, \text{ and } y = 1/3$$

68. Verify that $x\mathbf{A} = \mathbf{A}x$.
69. Verify that $(x + y)\mathbf{A} = x\mathbf{A} + y\mathbf{A}$.
70. Verify that $x(\mathbf{A} + \mathbf{B}) = x\mathbf{A} + x\mathbf{B}$.
71. Verify that $x(y\mathbf{A}) = (xy)\mathbf{A}$.
72. Home Improvement Stores, Inc., sells guttering, installed, at the following prices (in dollars per linear foot):

	5″	6″	
P =	1.59	1.89	Galvanized steel
	1.99	2.19	Aluminum

During August all prices are reduced by 10 percent. Develop the new price schedule by matrix operations.

Daily output at Precision Manufacturing Company is as follows:

	Economy model	Standard model	Deluxe model	
Q =	36	42	15	Product A
	42	58	12	Product B
	15	24	8	Product C

73. If the plant operates five days a week, what is the weekly output?
74. New production-line machinery is available that would allow output to be increased by 20 percent, across the board. What, then, would be the weekly output?
75. Schwartz Construction Company builds small-scale, barnlike storage sheds. The 9-by-12-foot model requires 8 units of lumber, 5 units of roofing, and 3 units of paint. The 7-by-10-foot model requires 7 units of lumber, 4 units of roofing, and 2.5 units of paint. What materials will be required for the construction of three 9-by-12 models and five 7-by-10 models?

3.5 THE NEGATIVE OF A MATRIX AND SUBTRACTION

If **A** is any matrix, then the NEGATIVE OF **A**, denoted $-\mathbf{A}$, is the matrix whose elements are the negatives of the corresponding entries in **A**. The negative of the matrix

$$\mathbf{A} = \begin{bmatrix} 5 & -3 & 12 \\ -1 & 7 & 4 \end{bmatrix}$$

is obtained by scalar multiplication, as

$$-\mathbf{A} = (-1)\mathbf{A} = \begin{bmatrix} -5 & 3 & -12 \\ 1 & -7 & -4 \end{bmatrix}$$

If **A** is any matrix and **0** is the zero matrix with the same dimensions as **A**, then it is evident that

$$\mathbf{A} + (-\mathbf{A}) = \mathbf{0}$$

We can now define SUBTRACTION OF MATRICES in a manner similar to subtraction of real numbers; that is, for matrices **A** and **B**, both of the same dimensions, we define

$$\mathbf{A} - \mathbf{B} = \mathbf{A} + (-\mathbf{B})$$

Example 15 Given

$$\mathbf{A} = \begin{bmatrix} 1 & 2 \\ 0 & -1 \end{bmatrix} \quad \text{and} \quad \mathbf{B} = \begin{bmatrix} 2 & -3 \\ -1 & 7 \end{bmatrix}$$

then

$$-\mathbf{B} = (-1)\mathbf{B} = \begin{bmatrix} -2 & 3 \\ 1 & -7 \end{bmatrix}$$

and

$$\mathbf{A} - \mathbf{B} = \mathbf{A} + (-\mathbf{B}) = \begin{bmatrix} 1 & 2 \\ 0 & -1 \end{bmatrix} + \begin{bmatrix} -2 & 3 \\ 1 & -7 \end{bmatrix} = \begin{bmatrix} -1 & 5 \\ 1 & -8 \end{bmatrix}$$ []

Actually, we do not ordinarily think of subtraction as the addition of the negative matrix; we generally think in the more direct terms of simply "subtracting" corresponding elements.

Example 16 On September 1, the inventory at the Fantastic Furniture Company's warehouse was as follows:

	Early American	Traditional	Country French	
B =	116	125	80	Sofas
	190	110	130	Chairs

Shipments were made from the warehouse to retail outlets during the month of September as follows:

	Early American	Traditional	Country French	
S =	102	95	48	Sofas
	135	75	89	Chairs

No new shipments of mechandise were received at the warehouse during the month. What was the ending inventory?

The ending inventory can be determined by subtracting from the beginning inventory the shipments made from the warehouse during the month, as

$$\mathbf{E}=\mathbf{B}-\mathbf{S}=\begin{bmatrix}116 & 125 & 80\\ 190 & 110 & 130\end{bmatrix}-\begin{bmatrix}102 & 95 & 48\\ 135 & 75 & 89\end{bmatrix}=\begin{bmatrix}14 & 30 & 32\\ 55 & 35 & 41\end{bmatrix} \quad []$$

Solving a Matrix Equation Suppose that **A** and **B** are known matrices. It is also known that the matrix **X** is related to **A** and **B** as follows:

$$\mathbf{X}+\mathbf{A}=\mathbf{B}$$

How can we solve the matrix equation for the unknown matrix **X**? We add the matrix $-\mathbf{A}$ to each side of the matrix equation to obtain

$$\mathbf{X}+\mathbf{A}+(-\mathbf{A})=\mathbf{B}+(-\mathbf{A})$$
$$\mathbf{X}=\mathbf{B}-\mathbf{A}$$

Example 17 Given

$$\mathbf{A}=\begin{bmatrix}1 & 0\\ 1 & 2\end{bmatrix} \quad \text{and} \quad \mathbf{B}=\begin{bmatrix}3 & 4\\ 6 & 9\end{bmatrix}$$

and the matrix equation

$$\mathbf{X}+\mathbf{A}=\mathbf{B}$$

solve for the matrix **X**.

We find that

$$\mathbf{X}=\mathbf{B}-\mathbf{A}$$
$$\mathbf{X}=\begin{bmatrix}3 & 4\\ 6 & 9\end{bmatrix}-\begin{bmatrix}1 & 0\\ 1 & 2\end{bmatrix}=\begin{bmatrix}2 & 4\\ 5 & 7\end{bmatrix} \quad []$$

Example 18 Given

$$\mathbf{A}=\begin{bmatrix}1 & 2 & 3\\ 0 & 1 & 2\\ 0 & 0 & 1\end{bmatrix} \quad \mathbf{B}=\begin{bmatrix}4 & 0 & 4\\ 0 & 4 & 0\\ 4 & 0 & 4\end{bmatrix}$$

and the matrix equation

$$3\mathbf{X}+\mathbf{B}=2(\mathbf{A}+\mathbf{X})$$

we solve for the unknown matrix **X** as follows:

$$3\mathbf{X}+\mathbf{B}=2\mathbf{A}+2\mathbf{X}$$
$$3\mathbf{X}-2\mathbf{X}=2\mathbf{A}-\mathbf{B}$$
$$\mathbf{X}=2\mathbf{A}-\mathbf{B}$$

That is,

$$\mathbf{X}=2\begin{bmatrix}1 & 2 & 3\\ 0 & 1 & 2\\ 0 & 0 & 1\end{bmatrix}-\begin{bmatrix}4 & 0 & 4\\ 0 & 4 & 0\\ 4 & 0 & 4\end{bmatrix}=\begin{bmatrix}-2 & 4 & 2\\ 0 & -2 & 4\\ -4 & 0 & -2\end{bmatrix} \quad []$$

EXERCISES

Given

$$\mathbf{A}=\begin{bmatrix}4 & 1 & 6\\ 5 & -1 & 0\\ 2 & 9 & 0\end{bmatrix} \qquad \mathbf{B}=\begin{bmatrix}3 & 5 & 3\\ 4 & 0 & 1\\ 0 & 0 & -2\end{bmatrix} \qquad \mathbf{C}=\begin{bmatrix}2 & 0 & 2\\ 0 & 1 & -3\\ 1 & 0 & 4\end{bmatrix}$$

determine the result of the following operations:

76. $\mathbf{A}-\mathbf{B}$
77. $\mathbf{B}-\mathbf{A}$
78. $\mathbf{A}-(\mathbf{B}+\mathbf{C})$
79. $(\mathbf{A}-\mathbf{B})-\mathbf{C}$
80. $3\mathbf{A}-6(\mathbf{B}-\mathbf{C})$
81. $3(\mathbf{A}-2\mathbf{B}+2\mathbf{C})$

Given

$$\mathbf{A}=\begin{bmatrix}4 & 0\\ 2 & -3\end{bmatrix} \qquad \mathbf{B}=\begin{bmatrix}-1 & 1/2\\ 5 & 0\end{bmatrix} \qquad \mathbf{C}=\begin{bmatrix}1 & 1\\ 0 & -2\end{bmatrix}$$

solve each of the following matrix equations for the matrix $\mathbf{X}$:

82. $2\mathbf{X}=\mathbf{A}$
83. $\mathbf{X}+\mathbf{A}=\mathbf{B}$
84. $\mathbf{X}-\mathbf{A}=0$
85. $\mathbf{X}+(1/2)\mathbf{B}=\mathbf{A}$
86. $\mathbf{X}+\mathbf{A}=\mathbf{B}+\mathbf{C}$
87. $(0.5)(\mathbf{X}+\mathbf{B})=3(\mathbf{X}+\mathbf{A})+\mathbf{C}$
88. $3(\mathbf{X}+\mathbf{A})=2(\mathbf{X}-\mathbf{B})$
89. $0.25(\mathbf{X}-\mathbf{C})=2\mathbf{A}+4(\mathbf{B}-\mathbf{C})$
90. In 1982 the full-time enrollment at City College was as follows:

    426 undergraduates and 186 graduate students in the school of business, 581 undergraduates and 306 graduate students in the school of engineering, and 325 undergraduates and 142 graduate students in the school of arts. In 1983 there were 491 undergraduates in the school of business, 535 in the school of engineering, and 323 in the school of arts, along with 203 graduate students in the school of business, 298 in the school of engineering, and 149 in the school of arts.

    Use matrix methods to determine the change in enrollment from 1982 to 1983 by school and level (undergraduate or graduate student).

3.6 MULTIPLICATION OF MATRICES

Let us begin by considering the multiplication of two vectors, one a row vector and the other a column vector, but each with the same number of components.

> The product of an n-component row vector $\mathbf{A}$ and an n-component column vector $\mathbf{B}$ is the number $\mathbf{C}$, defined as
>
> $$\mathbf{C}=\mathbf{AB}=[a_1b_1+a_2b_2+\cdots+a_nb_n] \qquad (3.3)$$

Example 19 Given

$$\mathbf{A} = [1 \quad 2 \quad 3] \qquad \text{and} \qquad \mathbf{B} = \begin{bmatrix} 4 \\ 5 \\ 6 \end{bmatrix}$$

the product $\mathbf{AB} = \mathbf{C}$ is

$$\mathbf{C} = \mathbf{AB} = [1 \quad 2 \quad 3] \cdot \begin{bmatrix} 4 \\ 5 \\ 6 \end{bmatrix} = [(1 \times 4) + (2 \times 5) + (3 \times 6)]$$

$$= [4 + 10 + 18] = [32] \quad \square$$

Note these important things about the procedure. A ROW IS MULTIPLIED TIMES A COLUMN. In an orderly progression, the first element in the row is multiplied times the first element in the column, the second element in the row times the second element in the column, and so on until the nth row element is multiplied times the nth column element. These products are then summed to obtain the single number that is the product of the two vectors. Clearly, the procedure can be executed only if the row vector contains the same number of components as the column vector.

Example 20 Vanassa purchased 6 cases of soft drinks, 18 pounds of hot dogs, 12 packages of hot-dog buns, 3 jars of mustard, and 4 jars of relish. Soft drinks cost \$5 a case, hot dogs cost \$2.25 a pound, buns cost \$0.90 a package, mustard \$1.90 a jar, and relish \$1.75 a jar. Use vector multiplication to determine the total amount Vanassa spent.

We set up a price vector and a quantity vector (one must be a row vector and the other a column vector), as

$$\mathbf{C} = \mathbf{PQ} = [5 \quad 2.25 \quad 0.90 \quad 1.90 \quad 1.75] \begin{bmatrix} 6 \\ 18 \\ 12 \\ 3 \\ 4 \end{bmatrix} = \$94.00 \quad \square$$

Let us continue by multiplying an m-component row vector $\mathbf{A}$ times a $m \times n$ matrix $\mathbf{B}$. The product will be an n-component row vector, the first component being the product of row vector $\mathbf{A}$ and the first column in $\mathbf{B}$, the second component being the product of the row vector $\mathbf{A}$ and the second column in $\mathbf{B}$, and so on.

$$\mathbf{A} = [a_1 \quad a_2 \ldots a_m] \qquad \text{and} \qquad \mathbf{B} = \begin{bmatrix} b_{11} & b_{12} & \cdots & b_{1n} \\ b_{21} & b_{22} & \cdots & b_{2n} \\ \vdots & \vdots & \vdots & \vdots \\ b_{m1} & b_{m2} & \cdots & b_{mn} \end{bmatrix}$$

$$\mathbf{C} = \mathbf{AB} = [(a_1b_{11} + a_2b_{21} \cdots + a_mb_{m1}), (a_1b_{12} + a_2b_{22} + \cdots + a_mb_{m2}), \ldots, (a_1b_{1n} + a_2b_{2n} + \cdots + a_mb_{mn})]$$

Example 21 Given

$$\mathbf{A} = [1 \quad 2 \quad 3] \qquad \text{and} \qquad \mathbf{B} = \begin{bmatrix} 4 & 5 \\ 6 & 7 \\ 8 & 9 \end{bmatrix}$$

the product $\mathbf{C} = \mathbf{AB}$ is computed

$$\mathbf{C} = \mathbf{AB} = \left[[1 \quad 2 \quad 3] \cdot \begin{bmatrix} 4 \\ 6 \\ 8 \end{bmatrix} \qquad [1 \quad 2 \quad 3] \cdot \begin{bmatrix} 5 \\ 7 \\ 9 \end{bmatrix} \right]$$
$$= [(1 \times 4 + 2 \times 6 + 3 \times 8) \qquad (1 \times 5 + 2 \times 7 + 3 \times 9)]$$
$$= [40 \quad 46]$$

Now let us multiply two $m \times n$ matrices. If **A** and **B** are two matrices, *the product* **AB** *is defined if and only if the number of columns in* **A** *is equal to the number of rows in* **B**. If this requirement is met, **A** is said to be CONFORMABLE to B FOR MULTIPLICATION. The matrix resulting from the multiplication has dimensions equivalent to the number of rows in **A** and the number of columns in **B**. Thus, for example, a 2×3 matrix can be multiplied by a 3×4 matrix and the product matrix will be of dimensions 2×4. However, multiplication of a 3×4 matrix by a 2×2 matrix is not defined.

The following schematic illustrates the dimensional requirements for the multiplication $\mathbf{C} = \mathbf{AB}$.

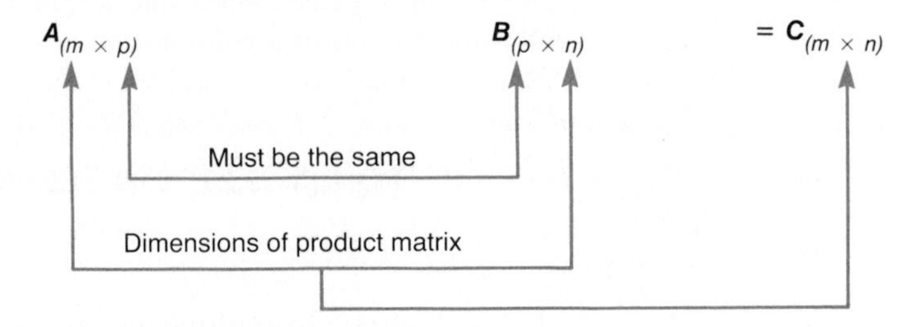

Let us explain how the multiplication is executed; and as we do, the reasons for these restrictions will become apparent. Given **A** is an $m \times p$ matrix and **B** a $p \times n$ matrix, then the product $\mathbf{AB} = \mathbf{C}$ is the $m \times n$ matrix defined as

$$\mathbf{AB} = \mathbf{C}$$

$$\begin{bmatrix} a_{11} & a_{12} & \cdots & a_{1p} \\ a_{21} & a_{22} & \cdots & a_{2p} \\ & & \vdots & \\ a_{m1} & a_{m2} & \cdots & a_{mp} \end{bmatrix} \cdot \begin{bmatrix} b_{11} & b_{12} & \cdots & b_{1n} \\ b_{21} & b_{22} & \cdots & b_{2n} \\ & & \vdots & \\ b_{p1} & b_{p2} & \cdots & b_{pn} \end{bmatrix} = \begin{bmatrix} c_{11} & c_{12} & \cdots & c_{1n} \\ c_{21} & c_{22} & \cdots & c_{2n} \\ & & \vdots & \\ c_{m1} & c_{m2} & \cdots & c_{mn} \end{bmatrix}$$

where

$$c_{ij} = (i\text{th row from } \mathbf{A}) \cdot (j\text{th column from } \mathbf{B})$$
$$= a_{i1}b_{1j} + a_{i2}b_{2j} + \cdots + a_{ip}b_{pj} \quad (3.4)$$

The components of the product matrix **C** = **AB** are formed by the sum of the products of the rows of the first matrix, **A**, and the columns of the second matrix, **B**. The sum of the products of the first row in **A** and the first column in **B** gives the row one-column one component of **C**. Similarly, the sum of the products of the first row in **A** and the second column in **B** gives the row one-column two component of **C**. This procedure is continued until all rows of **A** have been multiplied times each column of **B**.

Example 22 Given

$$\mathbf{A} = \begin{bmatrix} 2 & 4 \\ 3 & 1 \end{bmatrix} \qquad \mathbf{B} = \begin{bmatrix} 5 & 6 & 1 \\ 7 & 8 & 3 \end{bmatrix}$$

then

$$\mathbf{C} = \mathbf{AB} = \begin{bmatrix} [2 \quad 4] \cdot \begin{bmatrix} 5 \\ 7 \end{bmatrix} & [2 \quad 4] \cdot \begin{bmatrix} 6 \\ 8 \end{bmatrix} & [2 \quad 4] \cdot \begin{bmatrix} 1 \\ 3 \end{bmatrix} \\ [3 \quad 1] \cdot \begin{bmatrix} 5 \\ 7 \end{bmatrix} & [3 \quad 1] \cdot \begin{bmatrix} 6 \\ 8 \end{bmatrix} & [3 \quad 1] \cdot \begin{bmatrix} 1 \\ 3 \end{bmatrix} \end{bmatrix}$$
$$= \begin{bmatrix} (2 \times 5 + 4 \times 7) & (2 \times 6 + 4 \times 8) & (2 \times 1 + 4 \times 3) \\ (3 \times 5 + 1 \times 7) & (3 \times 6 + 1 \times 8) & (3 \times 1 + 1 \times 3) \end{bmatrix}$$
$$= \begin{bmatrix} 38 & 44 & 14 \\ 22 & 26 & 6 \end{bmatrix}$$

[]

The fact that a row is multiplied times a column does not preclude the multiplication of a column vector times a row vector, provided they have the same number of components.

Example 23 Given

$$\mathbf{A} = \begin{bmatrix} 5 \\ 6 \end{bmatrix} \qquad \text{and} \qquad \mathbf{B} = [2 \quad 4]$$

the product **C** = **AB** is

$$\mathbf{C} = \mathbf{AB} = \begin{bmatrix} (5 \times 2) & (5 \times 4) \\ (6 \times 2) & (6 \times 4) \end{bmatrix} = \begin{bmatrix} 10 & 20 \\ 12 & 24 \end{bmatrix}$$

[]

Example 24 A firm manufacturing office furniture finds that it has the following variable costs, in dollars:

	Desk	Chair	Table	Cabinet	
V =	50	20	15	25	Material
	30	15	12	15	Labor
	30	15	8	20	Overhead

That is, the manufacture of each desk requires materials costing \$50, labor costing \$30, and overhead of \$30. Each chair requires material costing \$20, and so on.

An order for 5 desks, 6 chairs, 4 tables, and 12 cabinets has just been received. This information is organized in a column vector, as

$$\mathbf{Q} = \begin{bmatrix} 5 \\ 6 \\ 4 \\ 12 \end{bmatrix} \begin{matrix} \text{Desks} \\ \text{Chairs} \\ \text{Tables} \\ \text{Cabinets} \end{matrix}$$

The total material, total labor, and total overhead cost associated with filling this order can be computed as follows:

$$\mathbf{C} = \mathbf{VQ} = \begin{bmatrix} 50 & 20 & 15 & 25 \\ 30 & 15 & 12 & 15 \\ 30 & 15 & 8 & 20 \end{bmatrix} \begin{bmatrix} 5 \\ 6 \\ 4 \\ 12 \end{bmatrix} = \begin{bmatrix} 730 \\ 468 \\ 512 \end{bmatrix} \begin{matrix} \text{Material} \\ \text{Labor} \\ \text{Overhead} \end{matrix}$$

Note that the computations made to determine that total material cost is \$730 were as follows:

\$50 material per desk	×	5 desks	=	\$250
\$20 material per chair	×	6 chairs	=	120
\$15 material per table	×	4 tables	=	60
\$25 material per cabinet	×	12 cabinets	=	300
				\$730

Similar computations yielded the total labor cost of \$468 and the total overhead cost of \$512. []

Note how important it is that the information be arranged in the rows and columns of the matrices so that multiplication is not only possible from a mathematical standpoint but also logical in the context of the problem situation. When considering a specific matrix multiplication problem, we can set up an "information-content" schematic, in much the same way that we set up a schematic for dimensions, and use this to determine if a given order of multiplication will be meaningful. That is,

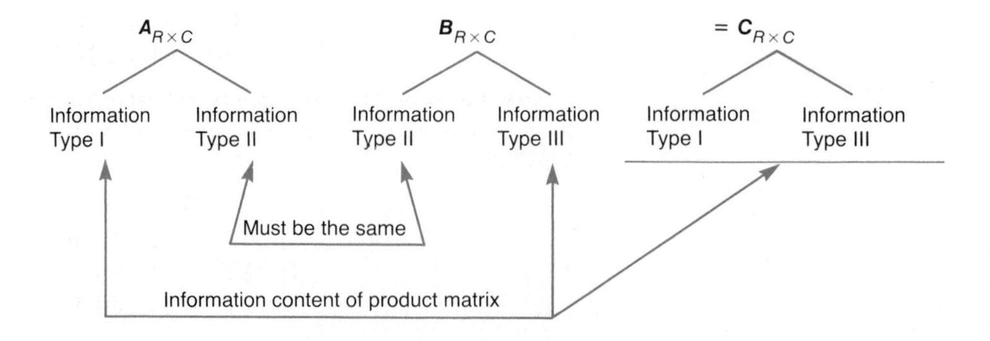

Example 25 The Callavette Company operates plants in Boise and Cairo where two types of hand-held calculators are manufactured. The manufacturing time required, in hours, for each type of calculator and the hourly manufacturing cost, in dollars, in each plant are given below.

$$\mathbf{A} = \begin{array}{c} \text{Assembly} \quad \text{Packaging} \\ \begin{bmatrix} 1.0 & 0.6 \\ 1.2 & 0.8 \end{bmatrix} \end{array} \begin{array}{l} \text{Financial} \\ \text{Scientific} \end{array} \qquad \mathbf{B} = \begin{array}{c} \text{Boise} \quad \text{Cairo} \\ \begin{bmatrix} 9 & 7 \\ 6 & 4 \end{bmatrix} \end{array} \begin{array}{l} \text{Assembly} \\ \text{Packaging} \end{array}$$

The total cost of manufacturing each type of calculator in each plant is given by $\mathbf{C} = \mathbf{AB}$. Note that in this multiplication the information content of the matrices is

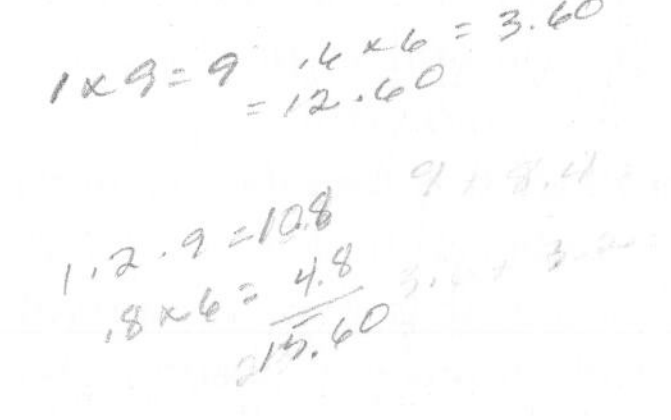

A	**B**	=	**C**
$R \times C$	$R \times C$		$R \times C$
Type calculator / Department	Department / Plant		Type calculator / Plant

We obtain

$$\mathbf{C} = \mathbf{AB} = \begin{array}{c} \text{Boise} \quad \text{Cairo} \\ \begin{bmatrix} 12.60 & 9.40 \\ 15.60 & 11.60 \end{bmatrix} \end{array} \begin{array}{l} \text{Financial} \\ \text{Scientific} \end{array}$$

[]

Example 26 **Parts explosion problem** A manufacturer of microcomputers produces three final models, I, II, and III, which are made from three subassemblies, A, B, and C. Furthermore, each subassembly is made from parts 1, 2, 3, and 4.

PARTS (1, 2, 3, 4) → SUBASSEMBLIES (A, B, C) → MODEL (I, II, III)

Model I requires two subassemblies A and two subassemblies C. Model II requires one each of subassemblies A and C and two subassemblies B. Model III requires four subassemblies B and one subassembly C. This information can be organized in matrix format, as

$$\mathbf{S} = \begin{array}{c} \text{A} \quad \text{B} \quad \text{C} \\ \begin{bmatrix} 2 & 0 & 3 \\ 1 & 2 & 1 \\ 0 & 4 & 1 \end{bmatrix} \end{array} \begin{array}{l} \text{I} \\ \text{II} \\ \text{III} \end{array}$$

In a similar manner, the number of each different type of part required for each subassembly can be organized in matrix format. This information is as follows:

$$\mathbf{P} = \begin{array}{c} 1 \quad 2 \quad 3 \quad 4 \\ \begin{bmatrix} 1 & 2 & 0 & 1 \\ 3 & 1 & 4 & 0 \\ 0 & 3 & 1 & 2 \end{bmatrix} \end{array} \begin{array}{l} \text{A} \\ \text{B} \\ \text{C} \end{array}$$

The matrix **SP**, obtained by multiplying the subassembly matrix **S** times the parts matrix **P**, gives the number of units of each part required in the production of one unit of each of the final products.

$$\mathbf{SP} = \begin{bmatrix} 2 & 0 & 3 \\ 1 & 2 & 1 \\ 0 & 4 & 1 \end{bmatrix} \times \begin{bmatrix} 1 & 2 & 0 & 1 \\ 3 & 1 & 4 & 0 \\ 0 & 3 & 1 & 2 \end{bmatrix} = \begin{array}{c} \begin{matrix} 1 & 2 & 3 & 4 \end{matrix} \\ \begin{bmatrix} 2 & 13 & 3 & 8 \\ 7 & 7 & 9 & 3 \\ 12 & 7 & 17 & 2 \end{bmatrix} \end{array} \begin{matrix} \text{I} \\ \text{II} \\ \text{III} \end{matrix}$$

Notice that in multiplying row 1 of **S** times column 1 of **P**, these computations are made: [(number of units of subassembly A in one unit of product I) × (number of units of part 1 in one unit of subassembly A)] + [(number of units of subassembly B in one unit of product I) × (number of units of part 1 in one unit of subassembly B)] + [(number of units of subassembly C in one unit of product I) × (number of units of part 1 in one unit of subassembly C)] = number of units of part 1 in each unit of product 1. In matrix multiplication, it is most important that we always verify that row-column multiplications give *meaningful results* in the context of the problem situation.

Other numbers in the **SP** matrix have a similar meaning. Thus, from the first row of this matrix we read that each unit of Model I requires 2 units of part 1, 13 units of part 2, 3 units of part 3, and 8 units of part 4. The entries on the second row of the matrix give the number of units of each type of part required in the manufacture of one unit of Model II. The third row gives the number of units of each part required by one unit of Model III.

Now, the firm has just received an order for six units of Model I, three units of Model II, and four units of Model III and needs to know the number of units of each type of part required to fill the order. An order quantity vector can be established as

$$\begin{matrix} & \text{I} & \text{II} & \text{III} \\ \mathbf{Q} = & [6 & 3 & 4] \end{matrix}$$

By multiplying **Q** times **SP**, we obtain the desired information, as follows:

$$\mathbf{Q(SP)} = \begin{array}{c} \begin{matrix} \text{I} & \text{II} & \text{III} \end{matrix} \\ [6 \quad 3 \quad 4] \end{array} \begin{array}{c} \begin{matrix} 1 & 2 & 3 & 4 \end{matrix} \\ \begin{bmatrix} 2 & 13 & 3 & 8 \\ 7 & 7 & 9 & 3 \\ 12 & 7 & 17 & 2 \end{bmatrix} \end{array} \begin{matrix} \text{I} \\ \text{II} \\ \text{III} \end{matrix} = \begin{array}{c} \begin{matrix} 1 & 2 & 3 & 4 \end{matrix} \\ [81 \quad 127 \quad 113 \quad 65] \end{array}$$

We see, thus, that 81 units of part 1, 127 units of part 2, 113 units of part 3, and 65 units of part 4 will be required to manufacture the items included in the order.

The latest price catalog shows that each unit of part 1 costs $4.50, each unit of part 2 costs $9.00, each unit of part 3 costs $7.25, and each unit of part 4 costs $12.00. With this information we can compute the total parts cost for the order. Costs per unit are set up in a vector, as

$$\mathbf{C} = \begin{bmatrix} 4.50 \\ 9.00 \\ 7.25 \\ 12.00 \end{bmatrix}$$

Then, total cost is determined by

$$\mathbf{T} = (\mathbf{QSP})\mathbf{C} = [81 \quad 127 \quad 113 \quad 65] \cdot \begin{bmatrix} 4.50 \\ 9.00 \\ 7.25 \\ 12.00 \end{bmatrix} = [3106.75]$$

Parts needed to fill the order will have a total cost of $3,106.75. []

Example 27 **Depreciation of capital assets** During the year, Little Brothers, Inc., purchased three pieces of capital equipment costing $45,000, $80,000, and $50,000, respectively. Accountants for the firm have chosen to use a procedure (known as the SUM-OF-THE-YEAR'S-DIGITS METHOD) for depreciating these assets which is often used for those capital items that lose more value annually at the beginning of their useful life than at the end (as do automobiles, for instance). By this procedure, each year's depreciation charge for an asset will be a ratio of the cost of the asset minus its salvage value. The ratios for each year will be

$$\frac{n - j + 1}{(n)(n + 1)/2}$$

where n is the useful life of the asset, in years, j is the current year, and $1 \leq j \leq n$.

Each of the items of equipment purchased by Little Brothers has an estimated useful life of four years and estimated salvage value at the end of its useful life of 20 percent of its original cost. Thus, depreciation will be computed on 80 percent of the original cost using annual rates of 4/10, 3/10, 2/10, and 1/10 of this amount. Note that these ratios were computed as follows: For the first year, given $n = 4$ and $j = 1$

$$\frac{4 - 1 + 1}{(4)(4 + 1)/2} = 4/10$$

The denominator of the ratio remains 10 throughout the other computations; the numerator changes as j changes from 1 to 2 to 3 to 4.

A depreciation schedule can be prepared using matrix multiplication. First, we define a row vector $\mathbf{C}$ where c_j = original cost, in dollars, of asset j so that

$$\mathbf{C} = [45{,}000 \quad 80{,}000 \quad 50{,}000]$$

Next we define a scalar s which is the estimated salvage value of the assets at the end of their useful life, as a percentage of their cost; thus, $s = 0.2$. Finally, we define a column vector **R** where r_i = the annual depreciation ratio in year i.

Then the model

$$\mathbf{D} = \mathbf{R}[(1 - s)\mathbf{C}]$$

yields a matrix **D** whose entries d_{ij} give the depreciation in year i on asset j.

The Little Brothers depreciation schedule for assets A, B, and C, will be computed

$$\mathbf{D} = \begin{bmatrix} 4/10 \\ 3/10 \\ 2/10 \\ 1/10 \end{bmatrix} \cdot \{(1 - 0.2)[45{,}000 \quad 80{,}000 \quad 50{,}000] = \begin{array}{c} \begin{array}{ccc} \text{Asset A} & \text{Asset B} & \text{Asset C} \end{array} \\ \begin{bmatrix} 14{,}400 & 25{,}600 & 16{,}000 \\ 10{,}800 & 19{,}200 & 12{,}000 \\ 7{,}200 & 12{,}800 & 8{,}000 \\ 3{,}600 & 6{,}400 & 4{,}000 \end{bmatrix} \end{array} \begin{array}{l} \text{Year 1} \\ \text{Year 2} \\ \text{Year 3} \\ \text{Year 4} \end{array}$$

The depreciation amounts for assets A, B, and C for year 1 are given on the first row of the matrix; that is, in the first year, the depreciation charge on asset A will be $14,400, on asset B will be $25,600, and on asset C will be $16,000. Depreciation amounts for year 2 are given on row 2 and so on.

Over its useful life, depreciation charges on asset A will be $14,000 in the first year, $10,800 in the second year, $7,200 in the third year, and $3,600 in the last year, making a total of $36,000 over the four-year period, which is 80 percent of the cost of the item. Other columns in the matrix give similar information about assets B and C. []

EXERCISES

Perform the following matrix multiplications, if possible. If the indicated multiplication is not defined, give the reason.

91. $\begin{bmatrix} 1 & 2 \\ 3 & 3 \end{bmatrix} \cdot \begin{bmatrix} 4 & 2 \\ 3 & 1 \end{bmatrix}$

92. $[1 \quad 2 \quad 3] \cdot \begin{bmatrix} 4 \\ 2 \\ 5 \end{bmatrix}$

93. $\begin{bmatrix} 1 & 2 \\ 6 & 3 \\ 5 & 0 \end{bmatrix} \cdot \begin{bmatrix} 1 & 6 & 8 \\ 3 & 9 & 4 \end{bmatrix}$

94. $\begin{bmatrix} 2 & 3 & 4 \\ 5 & 1 & 0 \end{bmatrix} \cdot \begin{bmatrix} 6 & 8 & 9 \\ 0 & -1 & 4 \end{bmatrix}$

95. $\begin{bmatrix} 2 \\ 3 \end{bmatrix} \cdot [5 \quad 6]$

96. $\begin{bmatrix} 4 & 6 & 8 \\ 1 & 3 & 2 \end{bmatrix} \cdot \begin{bmatrix} 1 & 4 \\ 3 & 1 \\ 2 & 5 \end{bmatrix}$

97. $\begin{bmatrix} 1 & -2 \\ 4 & -5 \end{bmatrix} \cdot \begin{bmatrix} 8 & -6 \\ 5 & -3 \end{bmatrix}$

98. $\begin{bmatrix} 1/2 & 1/3 \\ 1/6 & 2/5 \end{bmatrix} \cdot \begin{bmatrix} 3/5 & 2/7 \\ 1/7 & 2/3 \end{bmatrix}$

99. $\begin{bmatrix} 1 & 2 & 3 \\ 4 & 5 & 6 \\ 7 & 8 & 9 \end{bmatrix} \cdot \begin{bmatrix} 0 & 1/2 & 1/2 \\ 1 & 0 & 1 \\ 3 & -2 & 5 \end{bmatrix}$

100. $\begin{bmatrix} 0.4 & 0.8 \\ 0.3 & 0.5 \end{bmatrix} \cdot \begin{bmatrix} 0.1 & 0.4 & -0.9 \\ 0.8 & 0.3 & 0.6 \end{bmatrix}$

101. Fantastic Furniture Company has an inventory of 14 early American sofas, 30 traditional sofas, and 42 country French sofas, along with 55 early American, 35 traditional, and 41 country French chairs. All sofas are valued at $500 each, and all chairs are valued at $200 each. Determine the dollar value of the inventory, by style of furniture.

102. A manufacturer produces two products, product 1 and product 2. One unit of product 1 requires three parts of type A and two parts of type B. One unit of product 2 requires one type A part, four type B parts, and two type C parts. Display this information in a 2 × 3 parts matrix, labeled **P**.

The company has received an order for 250 units of product 1 and 325 units of product 2. Set up a row vector displaying this information and label the vector **Q**.

Find, by matrix multiplication, the number of parts of each type required to fill the order.

103. An investor holds 100 shares of stock A, 50 shares of stock B, 500 shares of stock C, and 80 shares of stock D. The dividend rate, in dollars, on each share of these stocks is as follows: $35 on stock A, $8.50 on stock B, $3.25 on stock C, and $250 on stock D.

Find the total dividend income for this investor.

104. An investor owns 10 shares of preferred stock and 100 shares of common stock in Amalgamated, Inc., along with 15 shares of preferred and 50 shares of common in Consolidated Corporation, and 5 shares of preferred and 200 shares of common in Integrated Products. Amalgamated has declared an annual dividend of $50 per share of preferred and $10 per share of common. Consolidated has declared dividends of $100 and $5 on preferred and common stock, respectively, and Integrated has declared dividends of $60 per share of preferred and $12 per share of common.

Determine the total dividend income realized from the stocks held by this investor.

Home Builders, Inc., is planning a new 120-home subdivision. They will construct three kinds of houses—single-story, tri-level, and two-story—with a choice of two exteriors—colonial and contemporary. Plans call for 10 single-story colonials, 10 tri-level colonials, and 20 two-story colonials, along with 30 single-story, 30 tri-level, and 20 two-story contemporaries.

The quantities of each of the primary building materials used depends largely on the exterior of the house. These quantities are

4/x 2

	Colonial	Contemporary
Concrete (cu. yds)	20	40
Lumber (1,000 bd. ft.)	4	6
Brick (in 1,000s)	15	5
Roofing (100 sq. ft.)	3	4

Concrete costs $35 per cubic yard, lumber is $280 per 1,000 board feet, brick is $140 per 1,000, and roofing is $75 per 100 square feet.

105. What quantities of each of the four primary types of building material must be ordered to complete the subdivision?
106. What is the total cost for the building materials for the subdivision?
107. What is the total material cost for each type of house?

Modular Office Systems offers for sale three models of complete office filing systems. These models consist of various combinations of filing subassemblies, A, B, and C. Model I requires three subassemblies A and two subassemblies B. Model II requires one each of subassemblies A and B and four of subassembly C. Model III requires two each of subassemblies B and C.

The subassemblies are, in turn, composed of various parts, 1, 2, 3, and 4. Subassembly A requires two units each of parts 1 and 3 and three units of part 2. Subassembly B requires one each of parts 1 and 2, and five units of part 4. Subassembly C requires six units of part 1, three units each of 2 and 3, and one unit of part 4.

The firm has just received an order for three units of model I, five units of model II, and two units of model III.

108. Determine the number of units of each type of part required to fill the order.
109. Each part 1 costs $9, each part 2 costs $4, each part 3, $5, and each part 4, $10. Determine the total parts cost for the order.
110. Each model I sells for $500, each model II sells for $650, and each model III sells for $750. Determine the profit per model.
111. Determine the total selling price for the order.

During the past tax year, Summit, Inc., purchased four units of capital equipment costing $60,000, $90,000, $115,000, and $35,000, respectively. Each item has an estimated useful life of five years and an estimated salvage value at the end of its useful life of 25 percent of the original cost.

112. Use matrix methods and set up a depreciation schedule for each unit of equipment for each of the five years, assuming that the sum-of-the-year's-digits method of depreciation is to be used.

3.7 SPECIAL PROPERTIES OF MATRIX MULTIPLICATION

While matrix multiplication is like regular multiplication in many respects, there are some critically important differences between the laws governing the multiplication of real numbers and the laws governing the multiplication of matrices. First, let us note two similarities.

Associative and Distributive Laws Apply

Both the associative and the distributive laws of ordinary algebra apply to matrix multiplication. Given three matrices, **A**, **B**, and **C**, which are conformable for multiplication,

1. $\mathbf{A}(\mathbf{BC}) = (\mathbf{AB})\mathbf{C}$ (The associative law)
2. $\mathbf{A}(\mathbf{B} + \mathbf{C}) = \mathbf{AB} + \mathbf{AC}$ (The distributive law)

Commutative Law Breaks Down

On the other hand, the COMMUTATIVE LAW OF MULTIPLICATION does not apply to matrix multiplication. For any two real numbers x and y, the product xy is always identical to the product yx. But for two matrices **A** and **B**, it is not generally true that **AB** equals **BA**. (In the product **AB**, we say that **B** is PREMULTIPLIED by **A** and that **A** is POSTMULTIPLIED by **B**.) Indeed, in many instances for two matrices **A** and **B**, the product **AB** may be defined while the product **BA** is not defined, or vice versa.

Example 28

Given

$$\mathbf{A} = \begin{bmatrix} 2 & 7 \\ 3 & 0 \end{bmatrix} \qquad \mathbf{B} = \begin{bmatrix} 5 & 1 \\ 2 & 4 \end{bmatrix}$$

$$\mathbf{AB} = \begin{bmatrix} 2 & 7 \\ 3 & 0 \end{bmatrix} \cdot \begin{bmatrix} 5 & 1 \\ 2 & 4 \end{bmatrix} = \begin{bmatrix} (2 \times 5 + 7 \times 2) & (2 \times 1 + 7 \times 4) \\ (3 \times 5 + 0 \times 2) & (3 \times 1 + 0 \times 4) \end{bmatrix} = \begin{bmatrix} 24 & 30 \\ 15 & 3 \end{bmatrix}$$

$$\mathbf{BA} = \begin{bmatrix} 5 & 1 \\ 2 & 4 \end{bmatrix} \cdot \begin{bmatrix} 2 & 7 \\ 3 & 0 \end{bmatrix} = \begin{bmatrix} (5 \times 2 + 1 \times 3) & (5 \times 7 + 1 \times 0) \\ (2 \times 2 + 4 \times 3) & (2 \times 7 + 4 \times 0) \end{bmatrix} = \begin{bmatrix} 13 & 35 \\ 16 & 14 \end{bmatrix}$$

$$\mathbf{AB} \neq \mathbf{BA}$$

[ ]

Example 29

Given a 2 × 2 matrix **A**, and a 2 × 3 matrix **B**, **AB** is defined and will be a 2 × 3 matrix; however, **BA** is not defined since the number of columns in **B** is not equal to the number of rows in **A**. [ ]

We must note that in some special cases, **AB** does equal **BA**. In such special cases, **A** and **B** are said to COMMUTE. However, these are the exceptions, not the rule. Most often, the commutative law fails for matrices, and right-hand multiplication and left-hand multiplication must be carefully distinguished. Products cannot be arranged in arbitrary order.

Other "Breakdowns"

Another unusual property of matrix multiplication is that the product of two matrices can be the zero matrix even though neither of the two matrices themselves is zero! We cannot conclude from the result $\mathbf{AB} = \mathbf{0}$ that at least one of the matrices **A** or **B** is a zero matrix.

Also, we cannot, in matrix algebra, necessarily conclude from the result $\mathbf{AB} = \mathbf{AC}$ that $\mathbf{B} = \mathbf{C}$, even if $\mathbf{A} \neq \mathbf{0}$. Thus, the CANCELLATION LAW does not hold, in general, in matrix multiplication.

These anomalies are illustrated in the next set of exercises.

Powers of Matrices

Many mathematical models involve positive integral powers of square matrices. If **A** is a square matrix, then

$$\begin{aligned} \mathbf{A}^1 &= \mathbf{A} \\ \mathbf{A}^2 &= \mathbf{A}\cdot\mathbf{A} \\ \mathbf{A}^3 &= \mathbf{A}\cdot\mathbf{A}\cdot\mathbf{A} = \mathbf{A}\cdot\mathbf{A}^2 \\ &\vdots \\ \mathbf{A}^{n+m} &= \mathbf{A}^n\cdot\mathbf{A}^m \end{aligned}$$

Example 30 Given

$$\mathbf{A} = \begin{bmatrix} 4 & 5 \\ -3 & 2 \end{bmatrix}$$

$$\mathbf{A}^2 = \mathbf{A}\cdot\mathbf{A} = \begin{bmatrix} 4 & 5 \\ -3 & 2 \end{bmatrix} \cdot \begin{bmatrix} 4 & 5 \\ -3 & 2 \end{bmatrix} = \begin{bmatrix} 1 & 30 \\ -18 & -11 \end{bmatrix}$$ [ ]

Multiplication by I and O

The product of any matrix **A** and a conformable identity matrix **I** will be equal to the matrix **A**; that is

$$\mathbf{AI} = \mathbf{IA} = \mathbf{A}$$

Because of this property, the identity matrix is called the IDENTITY ELEMENT FOR MATRIX MULTIPLICATION.

The product of any matrix **A** with a conformable zero matrix, **O**, is a zero matrix; that is

$$\mathbf{A}_{m\times n}\cdot\mathbf{O}_{n\times p} = \mathbf{O}_{m\times p} \qquad \text{and} \qquad \mathbf{O}_{r\times k}\cdot\mathbf{A}_{k\times s} = \mathbf{O}_{r\times s}$$

Using a Summing Vector

A SUMMING VECTOR (which is also known as a UNIT VECTOR) is a vector with a "+1" for each of its elements, as

$$[1 \;\; 1 \;\; \ldots . \;\; 1] \qquad \text{or} \qquad \begin{bmatrix} 1 \\ 1 \\ \vdots \\ 1 \end{bmatrix}$$

Note what happens when a matrix **A** is *premultiplied* by a conformable summing row vector. Given

$$\mathbf{A} = \begin{bmatrix} 2 & 5 \\ 3 & 6 \\ 4 & 7 \end{bmatrix}$$

then

$$[1 \quad 1 \quad 1] \cdot \begin{bmatrix} 2 & 5 \\ 3 & 6 \\ 4 & 7 \end{bmatrix} = [(1 \times 2 + 1 \times 3 + 1 \times 4) \quad (1 \times 5 + 1 \times 6 + 1 \times 7)] = [9 \quad 18]$$

The components in the vector resulting from the multiplication are the sums of the column elements in **A**!

Now, note what happens when a matrix **A** is *postmultiplied* by a conformable summing column vector.

$$\begin{bmatrix} 2 & 5 \\ 3 & 6 \\ 4 & 7 \end{bmatrix} \cdot \begin{bmatrix} 1 \\ 1 \end{bmatrix} = \begin{bmatrix} (2 \times 1 + 5 \times 1) \\ (3 \times 1 + 6 \times 1) \\ (4 \times 1 + 7 \times 1) \end{bmatrix} = \begin{bmatrix} 7 \\ 9 \\ 11 \end{bmatrix}$$

The components of the vector resulting from this multiplication are the sums of the row elements in **A**!

EXERCISES

Given

$$\mathbf{A} = \begin{bmatrix} 0 & 4 \\ 4 & 0 \end{bmatrix}, \mathbf{B} = \begin{bmatrix} 2 & 1 \\ 1 & -2 \end{bmatrix}, \mathbf{C} = \begin{bmatrix} 5 & 3 \\ 1 & 6 \end{bmatrix}$$

Use matrix computations to determine which of the following statements are correct and which are incorrect.

113. $\mathbf{A(B+C)=AB+AC}$
114. $\mathbf{(B+C)A=BA+CA}$
115. $\mathbf{(A+B)C=AC+BC}$

Given

$$\mathbf{A} = \begin{bmatrix} 1 & 2 \\ 3 & 4 \end{bmatrix} \qquad \mathbf{B} = \begin{bmatrix} 1 & -1 \\ 0 & 1 \end{bmatrix} \qquad \text{and} \qquad k = 5$$

calculate

116. **AB**
117. **BA**
118. **(AB)A**
119. **A(BA)**
120. **B(AA)**
121. **A(AB)**
122. **(BA)B**
123. **B(AB)**
124. **B(BA)**
125. k**(AB)**
126. (k**A)B**
127. **A(**k**B)**

For each set of matrices listed below, show that it is not necessarily the case that **AB** *equals* **BA** *in matrix multiplication.*

128. $\mathbf{A} = \begin{bmatrix} 1 & 0 \\ 2 & 1 \end{bmatrix} \qquad \mathbf{B} = \begin{bmatrix} 2 & 4 \\ 2 & -5 \end{bmatrix}$

129. $\mathbf{A} = [1 \quad 2 \quad 4] \qquad \mathbf{B} = \begin{bmatrix} 3 \\ 6 \\ 9 \end{bmatrix}$

130. $\mathbf{A} = \begin{bmatrix} 2 & 3 \\ 5 & 1 \end{bmatrix}$ $\quad$ $\mathbf{B} = \begin{bmatrix} 4 & 3 & -2 \\ 7 & 0 & 1 \end{bmatrix}$

131. Given

$$\mathbf{A} = \begin{bmatrix} 1 & 3 \\ -2 & -6 \end{bmatrix}, \qquad \mathbf{B} = \begin{bmatrix} 4 & 5 \\ 2 & 3 \end{bmatrix} \qquad \mathbf{C} = \begin{bmatrix} 1 & 2 \\ 3 & 4 \end{bmatrix}$$

compute **AB** and **AC**. Is there anything unusual about these results? Can you explain?

132. Compute **AB**, given

$$\mathbf{A} = \begin{bmatrix} 5 & 0 & 0 \\ 4 & 0 & 0 \\ 3 & 0 & 0 \end{bmatrix} \qquad \mathbf{B} = \begin{bmatrix} 0 & 0 & 0 \\ 3 & 8 & 3 \\ 6 & -1 & 4 \end{bmatrix}$$

Explain the anomaly. Would you expect this same type of result from multiplication in regular algebra?

Given

$$\mathbf{A} = \begin{bmatrix} 1 & 3 \\ -2 & 6 \end{bmatrix} \qquad \text{and} \qquad \mathbf{B} = \begin{bmatrix} 0 & 5 \\ 3 & 1 \end{bmatrix}$$

determine which of the following statements (if any) are true.

133. $(\mathbf{AB})^t = \mathbf{A}^t\mathbf{B}^t$ $\quad$ 134. $(\mathbf{AB})^t = \mathbf{B}^t\mathbf{A}^t$ $\quad$ 135. $(\mathbf{AB})^t = (\mathbf{BA})^t$

136. Given an n-component row vector **X**, show that

$$\mathbf{XX}^t = \sum_{i=1}^{n} x_i^2$$

Given

$$\mathbf{A} = \begin{bmatrix} 1 & -1 \\ 2 & 1 \end{bmatrix} \qquad \text{and} \qquad \mathbf{B} = \begin{bmatrix} 2 & -1 \\ 0 & 3 \end{bmatrix}$$

137. Find $\mathbf{A}^2$, $\mathbf{A}^4$, $\mathbf{A}^6$, $\mathbf{A}^8$, $\mathbf{A}^{16}$

138. Show by computation that $(\mathbf{A} + \mathbf{B})(\mathbf{A} + \mathbf{B}) \neq \mathbf{A}^2 + 2\mathbf{AB} + \mathbf{B}^2$. Can you explain why this familiar identity of scalar algebra does not apply in matrix algebra?

139. Determine, by computation, whether $(\mathbf{A} + \mathbf{B})(\mathbf{A} - \mathbf{B})$ equals $\mathbf{A}^2 - \mathbf{B}^2$.

140. Determine, by computation, which of the following are equal:

$\mathbf{ABAB}$ and $\mathbf{A}^2\mathbf{B}^2$ $\quad$ or $\quad$ $\mathbf{AABB}$ and $\mathbf{A}^2\mathbf{B}^2$

141. What is unusual about $\mathbf{A}^2$ when

$$\mathbf{A} = \begin{bmatrix} ab & b^2 \\ -a^2 & -ab \end{bmatrix}$$

Given

$$\mathbf{A} = \begin{bmatrix} 1 & 0 & 3 \\ 0 & 5 & 6 \\ 7 & 1 & 0 \end{bmatrix}$$

compute (*where* **I** *is the* 3×3 *identity matrix*)

142. **AI** 143. **IA** 144. **(AI)I**

What do these results make you suspect about the effect of multiplying by **I**?

Compute (*where* **O** *is the* 3×3 *zero matrix and* **I** *is the* 3×3 *identity matrix*)

145. **AO** 146. **OA** 147. **O(AI)**

What do these results make you suspect about the effect of multiplying by **O**?

Given

$$\mathbf{A} = \begin{bmatrix} 10 & 5 & 2 & 6 \\ 9 & -4 & 3 & 1 \\ 5 & 0 & 7 & -1 \\ 7 & -9 & 0 & 8 \end{bmatrix}$$

148. Show how a summing vector can be used to determine the ROW TOTALS in **A**.
149. Show how a summing vector can be used to determine the COLUMN TOTALS in **A**.

Global Insurance Company has four salespeople working in Hilltown. The number of policies sold during the last month is given in matrix **A**, *as*

	O'Malley	Caponi	Steinberg	Deutsch	
A =	8	7	6	8	Automobile
	6	9	11	5	Life
	4	3	2	0	Health
	0	2	1	3	Homeowners

150. Let $\mathbf{S} = [1 \quad 1 \quad 1 \quad 1]$. Find **SA** and interpret its elements.
151. Find $\mathbf{AS}^t$ and interpret its elements.

3.8 MATRICES AND DIRECTED GRAPHS

Sociologists and psychologists have investigated extensively many types of binary relationships which exist in the various kinds of human communications networks. The set of communications that exists may be exclusively one-way (a DOMINANCE RELATION), exclusively two-way (a PERFECT COMMUNICATIONS RELATION), or a mixture of both. Examples of the dominance relation are:

Person A dominates person B.

Person A likes person B.

Radio A has a one-way communications link with radio B.

Team A defeated team B in soccer.

Examples of the perfect communications relation are:

Person A and person B are mutual friends.

City A and city B are connected by highway.

Nation A and nation B maintain diplomatic relations.

Many of the models which have been developed make extensive use of matrices for organizing data and facilitating computations.

A Matrix Model for Leadership Identification

A DOMINANCE RELATION is a relation such that in every pair of people, one either dominates (or has influence over) the other or is dominated by (is influenced by) the other. Within a group of such people, it is often desirable to determine who is "leader" in the sense that he or she has the strongest influences over the others.

Five city planners who are working together on a special project are being studied to determine the influence exerted by each of them over the other members of the group. The members of the group are paired off so that it can be determined for each possible pair which one has direct influence over the other; it is found that

A has direct influence over D but is directly influenced by B, C, and E.

B has direct influence over A, C, and D, but is directly influenced by E.

C has direct influence over A but is directly influenced by B, D, and E.

D has direct influence over C and E but is directly influenced by A and B.

E has direct influence over A, B, and C but is directly influenced by D.

These results can be pictured using an analytical tool called a DIRECTED GRAPH or DIGRAPH. When a digraph is used to represent the dominance relation, it is also called a DOMINANCE DIGRAPH. The persons (or other entities) are represented by points called VERTICES. These are connected by *directed lines,* or arrows, with the point

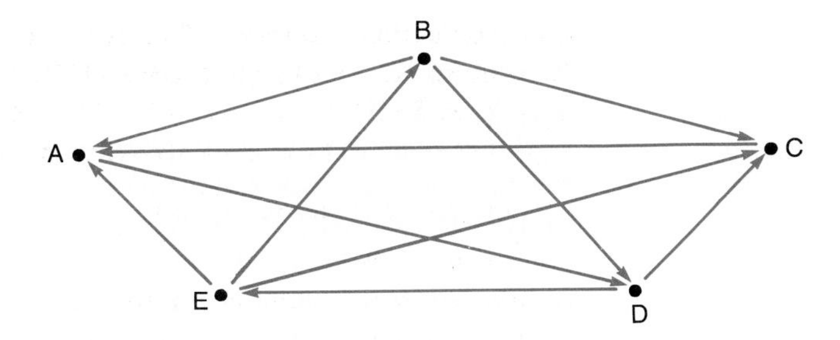

of the arrow toward the member of the pair who is influenced by the other. Thus, A → B indicates that B is influenced by A, and B → A indicates that B has influence over A. In a dominance digraph, for each pair of points, either $P_i \to P_j$ or $P_j \to P_i$, but not both.

The matrix representation of a dominance digraph consists only of 1s and 0s where $m_{ij} = 1$ if i has direct influence over j and $m_{ji} = 0$ if i does not have direct influence over j. Note that $m_{ij} = 1$ implies $m_{ji} = 0$ and $m_{ij} = 0$ implies $m_{ji} = 1$. Such a matrix—one that contains only 1s and 0s—is called an INCIDENCE MATRIX.

For the team of city planners, the incidence matrix is as follows:

$$\begin{array}{c} \\ A \\ B \\ C \\ D \\ E \end{array} \begin{array}{c} \begin{array}{ccccc} A & B & C & D & E \end{array} \\ \begin{bmatrix} 0 & 0 & 0 & 1 & 0 \\ 1 & 0 & 1 & 1 & 0 \\ 1 & 0 & 0 & 0 & 0 \\ 0 & 0 & 1 & 0 & 1 \\ 1 & 1 & 1 & 0 & 0 \end{bmatrix} \end{array} = \mathbf{M}$$

By multiplying the incidence matrix times a row-summing vector, we obtain the number of ONE-STAGE DOMINANCES of each person. For example

$$\begin{bmatrix} 0 & 0 & 0 & 1 & 0 \\ 1 & 0 & 1 & 1 & 0 \\ 1 & 0 & 0 & 0 & 0 \\ 0 & 0 & 1 & 0 & 1 \\ 1 & 1 & 1 & 0 & 0 \end{bmatrix} \cdot \begin{bmatrix} 1 \\ 1 \\ 1 \\ 1 \\ 1 \end{bmatrix} = \begin{bmatrix} 1 \\ 3 \\ 1 \\ 2 \\ 3 \end{bmatrix}$$

shows that A dominates one person (D) in one stage; C also dominates one person (A) in one stage; D dominates two persons (C and E); while B and E each have one-stage dominance over three persons (B over A, C, and D, and E over A, B, and C).

If any one person dominates everyone else in the group in one stage, sociologists term that person a CONSENSUS LEADER of the group. If there is no group member who dominates everyone else in one stage, then the person who dominates the most people in one stage is called the LEADER. Notice that among the five city planners, there are two persons, B and E, who dominate the most people in one stage. It is not immediately clear which one of these two has the most influence over the others.

When there is a tie for the largest number of one-stage dominances, the idea of two-stage dominances is introduced. The notion of TWO-STAGE DOMINANCE involves the idea that if A dominates B and B dominates C (each in one stage), then A dominates C in two stages; that is, using the notation of the dominance digraph, if in one stage A → B and B → C, then in two stages A → B → C. The square $\mathbf{M}^2$ of the incidence matrix **M** is used to determine two-stage dominances. Here

$$\mathbf{M}^2 = \mathbf{M} \cdot \mathbf{M} = \begin{bmatrix} 0 & 0 & 0 & 1 & 0 \\ 1 & 0 & 1 & 1 & 0 \\ 1 & 0 & 0 & 0 & 0 \\ 0 & 0 & 1 & 0 & 1 \\ 1 & 1 & 1 & 0 & 0 \end{bmatrix} \cdot \begin{bmatrix} 0 & 0 & 0 & 1 & 0 \\ 1 & 0 & 1 & 1 & 0 \\ 1 & 0 & 0 & 0 & 0 \\ 0 & 0 & 1 & 0 & 1 \\ 1 & 1 & 1 & 0 & 0 \end{bmatrix} = \begin{bmatrix} 0 & 0 & 1 & 0 & 1 \\ 1 & 0 & 1 & 1 & 1 \\ 0 & 0 & 0 & 1 & 0 \\ 2 & 1 & 1 & 0 & 0 \\ 2 & 0 & 1 & 2 & 0 \end{bmatrix}$$

From $\mathbf{M}^2$ we see that A has two-stage dominance over C and E. We can trace in matrix **M** that A has one-stage dominance over D and D has one-stage dominance over C; hence, in two stages, A → D → C. Also, D has one-stage dominance over E, so that A → D → E. The entry "2" in the row D column A cell of $\mathbf{M}^2$ indicates that D has two-stage dominance over A in two different ways; that is, D → C → A, and D → E → A. Thus, the entries m_{ij} in $\mathbf{M}^2$ give the number of two-stage dominances of person i over person j. If need be, $\mathbf{M}^3$ could be used to determine the number of three-stage dominances and $\mathbf{M}^n$ the number of n-stage dominances.

The sum $\mathbf{M} + \mathbf{M}^2$ gives the number of one- or two-stage dominances. For instance

$$\mathbf{M} + \mathbf{M}^2 = \begin{bmatrix} 0 & 0 & 0 & 1 & 0 \\ 1 & 0 & 1 & 1 & 0 \\ 1 & 0 & 0 & 0 & 0 \\ 0 & 0 & 1 & 0 & 1 \\ 1 & 1 & 1 & 0 & 0 \end{bmatrix} + \begin{bmatrix} 0 & 0 & 1 & 0 & 1 \\ 1 & 0 & 1 & 1 & 1 \\ 0 & 0 & 0 & 1 & 0 \\ 2 & 1 & 1 & 0 & 0 \\ 2 & 0 & 1 & 2 & 0 \end{bmatrix} = \begin{bmatrix} 0 & 0 & 1 & 1 & 1 \\ 2 & 0 & 2 & 2 & 1 \\ 1 & 0 & 0 & 1 & 0 \\ 2 & 1 & 2 & 0 & 1 \\ 3 & 1 & 2 & 2 & 0 \end{bmatrix}$$

The sum of row i in $\mathbf{M} + \mathbf{M}^2$ gives the total number of ways person i can dominate other members in the group in either one or two stages. We can easily compute, then, that A has three such one- or two-stage dominances over other members of the group, B has seven such dominances, C two, D six, and E eight.

Since no one member had more one-stage dominances, that member with the largest number of either one- or two-stage dominances might be considered to be the leader of the group. Here that would be person E, who might be expected to exert more influence over the group than any other one member.

A Perfect Communications Model

The COMMUNICATIONS MODELS assume a collection of elements—people, houses, or cities, and so on—and some mode of communication—telephone lines, highways, or the like—between them. In the PERFECT COMMUNICATIONS models, communication is two-way exclusively. Thus, if A communicates with B, then B, of necessity, communicates with A. Moreover, two elements may or may not communicate with each other. If they do communicate, there may be more than one communications line connecting them.

Assume four cities, A, B, C, and D, with connecting highways as shown below. In the digraph a double arrow is used to denote communication between two points, as A ↔ B. Multiple double arrows between two points indicate multiple two-way communications links between those points. Thus, the digraph shows that, for instance, city A and city B are connected by one highway; city B and city C have two connecting highways; city A and city C have no highway connecting them directly.

The matrix representation of a perfect communications digraph will have as its elements m_{ij}, the number of direct communications links between entity i and entity j. It will be a SYMMETRIC MATRIX in that $m_{ij} = m_{ji}$ for all i and j. Entities are not assumed to have a direct communications link with themselves, so that the entries on the primary diagonal are all zeros. This matrix is not necessarily an incidence matrix because it may well have entries which are not 1s or 0s.

For the four cities, the matrix is as follows:

$$\begin{array}{c} \\ A \\ B \\ C \\ D \end{array} \begin{array}{c} \begin{array}{cccc} A & B & C & D \end{array} \\ \begin{bmatrix} 0 & 1 & 0 & 1 \\ 1 & 0 & 2 & 1 \\ 0 & 2 & 0 & 0 \\ 1 & 1 & 0 & 0 \end{bmatrix} \end{array} = \mathbf{M}$$

The matrix $\mathbf{M}^2$ gives the number of lines of communication between two entities, passing through exactly one other entity. Thus,

$$\mathbf{M}^2 = \begin{bmatrix} 2 & 1 & 2 & 1 \\ 1 & 6 & 0 & 1 \\ 2 & 0 & 4 & 2 \\ 1 & 1 & 2 & 2 \end{bmatrix}$$

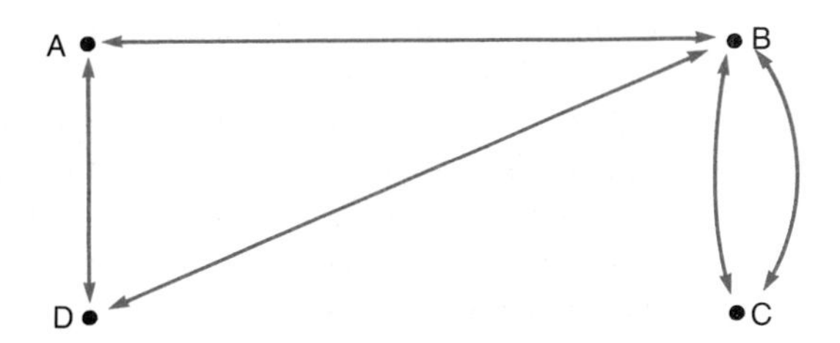

tells us that, even though there is no direct communications link between A and C, there are two ways A can communicate with C in two stages. Note, too, that any city can communicate with itself in two stages and often in several different ways. These are called "feedback paths." The number of feedback paths for each element is given on the primary diagonal. The 6 on the primary diagonal in the second row and second column indicates, for instance, that there are six two-stage feedback paths for B.

The matrix $\mathbf{M}^3$ will give the number of three-stage communications links between pairs of elements; the matrix $\mathbf{M}^k$ gives the number of k-stage communications links. The sum $\mathbf{M} + \mathbf{M}^2 + \ldots + \mathbf{M}^k$ gives the number of one-, two-, . . . , or k-stage communications links between each pair of elements.

EXERCISES

Albert, Barney, Clarence, and Dave are members of a street gang. Albert dominates Clarence, Barney dominates Albert and Dave, Clarence dominates Barney, and Dave dominates Albert and Clarence.

152. Sketch a digraph of the dominance relationships existing among the members of the gang.
153. Set up the incidence matrix. Use a summing vector to determine the total number of dominances by each member. Is there a consensus leader? Is there a leader?
154. Use $\mathbf{M}^2$ to determine the number of two-stage dominances.
155. Is there a leader in the sense that one person has the largest number of one- or two-stage dominances?

Six debate teams attending a conference are involved in a round-robin tournament in which each team must debate every other team once. In each match, one team wins and the other loses; there are no ties. The results are as follows:

Team A won over B, C, D, and E but lost to F.

Team B won over C, D, and F but lost to A and E.

Team C won over F but lost to A, B, D, and E.

Team D won over C and F but lost to A, B, and E.

Team E won over B, C, D, and F but lost to A.

Team F won over A but lost to B, C, D, and E.

156. Construct a digraph to represent the results of the tournament.
157. Set up the incidence matrix. Determine the total number of wins for each team.

158. Is there a winner in the sense that one team won more matches than did any other one team?

159. Use the concept of two-stage dominances (and three-stage dominances, if necessary) to determine who should be declared winner.

Another model that utilizes a dominance digraph is the "paired comparisons" model. Here a person is to select his or her favorite, or preferred, item from a group of items. The procedure is to work with each possible pair of items from the group, selecting the preferred item of the two. The results of such paired comparisons are used to determine the overall favorite.

Five different brands of coffee are being compared. The taster indicates these preferences:

Brand A preferred over B and D.
Brand B preferred over C and D.
Brand C preferred over A.
Brand D preferred over E.
Brand E preferred over A, B, and C.

160. Set up the appropriate incidence matrix.

161. Is there a preferred brand based on one-stage dominances?

162. Use one- and two-stage dominances to determine whether there is a preferred brand.

The following digraph represents the telephone lines connecting five mining outposts.

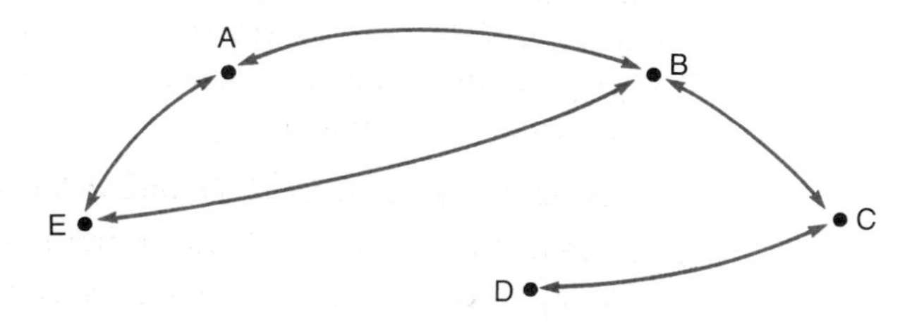

163. Write the matrix **M** representing the digraph.

164. How many one- or two-stage communications links are there between each pair of outposts?

165. In how many ways can each outpost get information back to itself through exactly one other outpost?

3.9 THE MULTIPLICATIVE INVERSE OF A MATRIX

Given a square matrix **A**, the MULTIPLICATIVE INVERSE (or, simply, INVERSE) of **A** is that square matrix, usually denoted $\mathbf{A}^{-1}$, such that

$$\mathbf{A}\mathbf{A}^{-1} = \mathbf{A}^{-1}\mathbf{A} = \mathbf{I}$$

Not all matrices have an inverse. In order for a matrix to have an inverse, the matrix must, first of all, be a square matrix. Still, not all square matrices have inverses. If a matrix has an inverse, it is said to be INVERTIBLE or NONSINGULAR. A matrix that does not have an inverse is said to be SINGULAR. An invertible matrix will have only one inverse; that is, if a matrix does have an inverse, that inverse will be unique. Hence, we speak of *the* inverse of matrix **A**.

Finding the Inverse of a Matrix

Let us begin by considering a tabular format where the square matrix **A** is AUGMENTED with an identity matrix of the same order, as

$$[\mathbf{A} \mid \mathbf{I}]$$

Now, if the inverse matrix $\mathbf{A}^{-1}$ were known, we could multiply the matrices on each side of the vertical line by $\mathbf{A}^{-1}$, as

$$[\mathbf{A}^{-1}\mathbf{A} \mid \mathbf{A}^{-1}\mathbf{I}]$$

Then, because $\mathbf{A}^{-1}\mathbf{A} = \mathbf{I}$ and $\mathbf{A}^{-1}\mathbf{I} = \mathbf{A}^{-1}$, we would have

$$[\mathbf{I} \mid \mathbf{A}^{-1}]$$

We do not follow this procedure, because the inverse is not known at this juncture; we are trying to determine the inverse. We instead employ a set of permissible row operations on the augmented matrix $[\mathbf{A} \mid \mathbf{I}]$ to transform **A** on the left side of the vertical line into an identity matrix. As the identity matrix is formed on the left of the vertical line, the inverse of **A** is formed on the right side. The allowable manipulations are called ELEMENTARY ROW OPERATIONS.

Operations permitted on the rows of a matrix are called ELEMENTARY ROW OPERATIONS. They are as follows:

Type I. Any pair of rows in a matrix may be interchanged. (*EXCHANGE operation*)

Type II. A row can be multiplied by any nonzero real number. (*MULTIPLY operation*)

Type III. A multiple of any row can be added to any other row. (*ADD-A-MULTIPLE operation*)

While the elementary row operations may be applied in a variety of sequences, we recommend the following procedure.

Procedure recommended for transforming [A|I] into [I|A⁻¹]
Working one column at a time and using the appropriate row operations, first get the 1 into the correct location (with respect to the identity matrix) within that column. Then use this 1 to obtain zeros in all other locations within that column. Only after one column is in the correct format should you move to the next column.

The 1 can be obtained in a cell of the matrix by multiplying by the reciprocal of the number presently in that cell (row operation type II). The zero can be obtained in a cell by using an appropriate multiple of the row with the 1 in this column (row operation type III).

Example 31 Find the inverse, if it exists, of the matrix

$$\mathbf{A} = \begin{bmatrix} 4 & 5 \\ -2 & 6 \end{bmatrix}$$

We begin by augmenting the matrix **A** with the second-order identity matrix, as

$$\left[\begin{array}{cc|cc} 4 & 5 & 1 & 0 \\ -2 & 6 & 0 & 1 \end{array}\right]$$

Now we will, working one cell at a time, apply the elementary row operations to this augmented matrix in such a way as to obtain an identity matrix on the left side of the vertical line.

The element in cell a_{11} must be converted to a 1. Row operation type II can be employed to accomplish this. We multiply the first row (R_1)—the *entire* row—by the reciprocal of the number presently in this cell; that is, by 1/4.

$$(1/4)[4 \quad 5 \quad 1 \quad 0] = [1 \quad 5/4 \quad 1/4 \quad 0]$$

The result is the "new" row 1; it replaces the "old" row 1. We write the adjusted tableau as

$$\left[\begin{array}{cc|cc} 1 & 5/4 & 1/4 & 0 \\ -2 & 6 & 0 & 1 \end{array}\right] \quad R_1 \leftarrow (1/4)R_1$$

This notation is read *Row 1 is replaced by 1/4 times Row 1.*

Next, we must have a zero in the a_{21} cell of the tableau. We can obtain a zero in this position through the use of row operation type III; that is, by adding 2 times row 1 to row 2, as

$$\begin{aligned}[-2 \quad 6 \quad 0 \quad 1] + 2[1 \quad 5/4 \quad 1/4 \quad 0] &= [-2 \quad 6 \quad 0 \quad 1] \\ &\quad + [2 \quad 5/2 \quad 1/2 \quad 0] \\ &= [0 \quad 17/2 \quad 1/2 \quad 1]\end{aligned}$$

The adjusted tableau now reads

$$\left[\begin{array}{cc|cc} 1 & 5/4 & 1/4 & 0 \\ 0 & 17/2 & 1/2 & 1 \end{array}\right] \quad R_2 \leftarrow R_2 + 2R_1$$

The first column has been transformed into the identity-matrix format. So, we move to the second column. A 1 is required in the a_{22} cell. We will again accomplish this through row operation type II, by multiplying the second row by the reciprocal of the number presently in this cell. That is, we multiply the second row by 2/17, as

$$(2/17)[0 \quad 17/2 \quad 1/2 \quad 1] = [0 \quad 1 \quad 1/17 \quad 2/17]$$

and enter this on the second row of the augmented tableau.

The tableau now has these entries:

$$\left[\begin{array}{cc|cc} 1 & 5/4 & 1/4 & 0 \\ 0 & 1 & 1/17 & 2/17 \end{array}\right] \quad R_2 \leftarrow (2/17)R_2$$

Finally, a zero must appear in the a_{12} cell. We obtain the zero by adding −5/4 times the second row to the existing first row, as

$$\begin{aligned} [1 \quad 5/4 \quad 1/4 \quad 0] + (-5/4)[0 \quad 1 \quad 1/17 \quad 2/17] &= [1 \quad 5/4 \quad 1/4 \quad 0] \\ &\quad + [0 \quad -5/4 \quad -5/68 \quad -10/68] \\ &= [1 \quad 0 \quad 3/17 \quad -5/34] \end{aligned}$$

When this result is entered in the tableau, it reads

$$\left[\begin{array}{cc|cc} 1 & 0 & 3/17 & -5/34 \\ 0 & 1 & 1/17 & 2/17 \end{array}\right] \quad R_1 \leftarrow R_1 + (-5/4)R_2$$

We have now achieved an identity matrix on the left side of the vertical line in the augmented matrix. The inverse, $\mathbf{A}^{-1}$, appears on the right side of the line. That is,

$$\mathbf{A}^{-1} = \begin{bmatrix} 3/17 & -5/34 \\ 1/17 & 2/17 \end{bmatrix}$$

We can verify that this is, indeed, the inverse of $\mathbf{A}$ through matrix multiplication for $\mathbf{AA}^{-1} = \mathbf{I}$, and

$$\begin{bmatrix} 4 & 5 \\ -2 & 6 \end{bmatrix} \cdot \begin{bmatrix} 3/17 & -5/34 \\ 1/17 & 2/17 \end{bmatrix} = \begin{bmatrix} 1 & 0 \\ 0 & 1 \end{bmatrix}$$

Observe also that $\mathbf{A}^{-1}\mathbf{A} = \mathbf{I}$. []

Example 32 Find the inverse, if it exists, of the matrix

$$\mathbf{A} = \begin{bmatrix} 2 & 0 & 1 \\ 3 & 5 & 2 \\ 4 & 1 & 3 \end{bmatrix}$$

We begin by setting up the augmented matrix $[\mathbf{A} | \mathbf{I}]$, as

$$\left[\begin{array}{ccc|ccc} 2 & 0 & 1 & 1 & 0 & 0 \\ 3 & 5 & 2 & 0 & 1 & 0 \\ 4 & 1 & 3 & 0 & 0 & 1 \end{array}\right]$$

Starting in column 1, we use elementary row operation type II to obtain a 1 in cell a_{11}, as

$$(1/2)\cdot[2\quad 0\quad 1\quad 1\quad 0\quad 0]=[1\quad 0\quad 1/2\quad 1/2\quad 0\quad 0]$$

With this new row 1, the augmented matrix appears as

$$\left[\begin{array}{ccc|ccc} 1 & 0 & 1/2 & 1/2 & 0 & 0 \\ 3 & 5 & 2 & 0 & 1 & 0 \\ 4 & 1 & 3 & 0 & 0 & 1 \end{array}\right] \quad R_1 \leftarrow (1/2)R_1$$

Continuing with column 1, we must obtain zeros in cells a_{21} and a_{31}. This we can do by using elementary row operation type III and multiples of the row which has a 1 in this column. For instance,

$$\begin{array}{ccccc} \text{(old row 2)} & - & \text{3(row 1)} & = & \text{(new row 2)} \\ [3\ 5\ 2\ 0\ 1\ 0] & - & 3[1\ 0\ 1/2\ 1/2\ 0\ 0] & = & [0\ 5\ 1/2\ -3/2\ 1\ 0] \end{array}$$

Then

$$\begin{array}{ccccc} \text{(old row 3)} & - & \text{4(row 1)} & = & \text{(new row 3)} \\ [4\ 1\ 3\ 0\ 0\ 1] & - & 4[1\ 0\ 1/2\ 1/2\ 0\ 0] & = & [0\ 1\ 1\ -2\ 0\ 1] \end{array}$$

These operations have transformed column 1 into the identity-matrix format; the augmented matrix now appears as

$$\left[\begin{array}{ccc|ccc} 1 & 0 & 1/2 & 1/2 & 0 & 0 \\ 0 & 5 & 1/2 & -3/2 & 1 & 0 \\ 0 & 1 & 1 & -2 & 0 & 1 \end{array}\right] \quad \begin{array}{l} R_2 \leftarrow R_2 - 3R_1 \\ R_3 \leftarrow R_3 - 4R_1 \end{array}$$

We move over to column 2. If we interchange rows 2 and 3 (and by elementary row operation type I we are allowed to do this), we obtain the required +1 on the primary diagonal position of column 2 without disturbing the required format of column 1. We have

$$\left[\begin{array}{ccc|ccc} 1 & 0 & 1/2 & 1/2 & 0 & 0 \\ 0 & 1 & 1 & -2 & 0 & 1 \\ 0 & 5 & 1/2 & -3/2 & 1 & 0 \end{array}\right]$$

Then

$$\begin{array}{ccccc} \text{(old row 3)} & - & \text{5(row 2)} & = & \text{(new row 3)} \\ [0\ 5\ 1/2\ -3/2\ 1\ 0] & - & 5[0\ 1\ 1\ -2\ 0\ 1] & = & [0\ 0\ -9/2\ 17/2\ 1\ -5] \end{array}$$

and both column 1 and column 2 are now in the required format, as

$$\left[\begin{array}{ccc|ccc} 1 & 0 & 1/2 & 1/2 & 0 & 0 \\ 0 & 1 & 1 & -2 & 0 & 1 \\ 0 & 0 & -9/2 & 17/2 & 1 & -5 \end{array}\right] \quad R_3 \leftarrow R_3 - 5R_2$$

And we turn our attention to column 3. Notice that we cannot now interchange rows 2 and 3 as a means of obtaining a +1 in cell a_{33} since to do so would destroy the identity-matrix format of column 2. Instead, we obtain the required +1 in this location by

$$-(2/9)[0\quad 0\quad -9/2\quad 17/2\quad 1\quad -5]=[0\quad 0\quad 1\quad -17/9\quad -2/9\quad 10/9]$$

and we write

$$\left[\begin{array}{ccc|ccc} 1 & 0 & 1/2 & 1/2 & 0 & 0 \\ 0 & 1 & 1 & -2 & 0 & 1 \\ 0 & 0 & 1 & -17/9 & -2/9 & 10/9 \end{array}\right] \quad R_3 \leftarrow (-2/9)R_3$$

The zero elements are obtained in cells a_{13} and a_{23} as follows:

$$\begin{array}{c} \text{(old row 1)} \quad - \quad (1/2)\text{(row 3)} \\ [1 \;\; 0 \;\; 1/2 \;\; 1/2 \;\; 0 \;\; 0] - (1/2)[0 \;\; 0 \;\; 1 \;\; -17/9 \;\; -2/9 \;\; 10/9] \\ = \quad \text{(new row 1)} \\ = [1 \;\; 0 \;\; 0 \;\; 13/9 \;\; 1/9 \;\; -5/9] \end{array}$$

and

$$\begin{array}{c} \text{(old row 2)} \quad - \quad \text{(row 3)} \\ [0 \;\; 1 \;\; 1 \;\; -2 \;\; 0 \;\; 1] - [0 \;\; 0 \;\; 1 \;\; -17/9 \;\; -2/9 \;\; 10/9] \\ = \quad \text{(new row 2)} \\ = [0 \;\; 1 \;\; 0 \;\; -1/9 \;\; 2/9 \;\; -1/9] \end{array}$$

The completed augmented matrix is

$$\left[\begin{array}{ccc|ccc} 1 & 0 & 0 & 13/9 & 1/9 & -5/9 \\ 0 & 1 & 0 & -1/9 & 2/9 & -1/9 \\ 0 & 0 & 1 & -17/9 & -2/9 & 10/9 \end{array}\right] \quad \begin{array}{l} R_1 \leftarrow R_2 - (1/2)R_3 \\ R_2 \leftarrow R_2 - R_3 \end{array}$$

The inverse of **A** is

$$\mathbf{A}^{-1} = \begin{bmatrix} 13/9 & 1/9 & -5/9 \\ -1/9 & 2/9 & -1/9 \\ -17/9 & -2/9 & 10/9 \end{bmatrix}$$

Again, as verification of our math computations, we could compute $\mathbf{AA}^{-1} = \mathbf{I}$. []

If, at any stage in the process just outlined, we find it to be impossible to transform **A** into **I** through a series of permissible row operations, then **A** does not have an inverse.

Example 33 Find the inverse, if it exists, of the matrix

$$\mathbf{A} = \begin{bmatrix} 1 & 2 \\ 2 & 4 \end{bmatrix}$$

We begin with the augmented matrix

$$\left[\begin{array}{cc|cc} 1 & 2 & 1 & 0 \\ 2 & 4 & 0 & 1 \end{array}\right]$$

We already have $a_{11} = 1$, so we obtain a zero in location a_{21} through row operation type III, as

$$[2 \;\; 4 \;\; 0 \;\; 1] - 2[1 \;\; 2 \;\; 1 \;\; 0] = [0 \;\; 0 \;\; -2 \;\; 1]$$

The adjusted matrix now reads

$$\left[\begin{array}{cc|cc} 1 & 2 & 1 & 0 \\ 0 & 0 & -2 & 1 \end{array}\right]$$

Row 2 of this matrix has only zero entries to the left of the vertical line. There are no permissible row operations that will allow us to obtain a +1 in the a_{22} position and maintain the zero in the a_{21} position! We cannot transform the **A** portion of the augmented matrix into an identity; hence, **A** does not have an inverse. [ ]

EXERCISES

Find the inverse of each of the following matrices, is possible. Check your answers using matrix multiplication.

166. $\begin{bmatrix} 1 & 2 \\ 3 & 7 \end{bmatrix}$

167. $\begin{bmatrix} 1/2 & 1/2 \\ 1 & 0 \end{bmatrix}$

168. $\begin{bmatrix} 2 & -3 \\ 4 & 5 \end{bmatrix}$

169. $\begin{bmatrix} 6 & 7 \\ 1 & 3 \end{bmatrix}$

170. $\begin{bmatrix} 8 & 5 \\ 4 & -1 \end{bmatrix}$

171. $\begin{bmatrix} 2 & 3 \\ 1 & 2 \end{bmatrix}$

172. $\begin{bmatrix} 0.2 & 0.3 \\ 0.5 & 1.4 \end{bmatrix}$

173. $\begin{bmatrix} 6 & 12 \\ 3 & -4 \end{bmatrix}$

174. $\begin{bmatrix} 4 & -3 \\ 8 & -6 \end{bmatrix}$

175. $\begin{bmatrix} 2 & 1 & 0 \\ 1 & 0 & 2 \\ 3 & 1 & 1 \end{bmatrix}$

176. $\begin{bmatrix} 1 & 2 & 4 \\ 9 & 6 & 1 \\ 8 & 0 & 9 \end{bmatrix}$

177. $\begin{bmatrix} 6 & 5 & 4 \\ 3 & 2 & -1 \\ 1 & 3 & 2 \end{bmatrix}$

178. $\begin{bmatrix} 2 & 0 & 1 \\ 2 & 1 & 2 \\ 4 & 1 & 3 \end{bmatrix}$

179. $\begin{bmatrix} 2 & 6 & 0 \\ 1 & 2 & 0 \\ 3 & 8 & 1 \end{bmatrix}$

180. $\begin{bmatrix} 2 & 0 & 1 \\ 0 & -1 & 3 \\ 1 & 5 & -6 \end{bmatrix}$

181. $\begin{bmatrix} 4 & 3 & 1 \\ 8 & 6 & 2 \\ 1 & 0 & 5 \end{bmatrix}$

182. $\begin{bmatrix} 3 & 2 & 0 & 1 \\ 0 & 1 & 4 & 5 \\ 1 & -1 & 3 & 0 \\ 2 & 0 & 1 & 4 \end{bmatrix}$

183. $\begin{bmatrix} 1 & 0 & 3 & 0 \\ 0 & 2 & 4 & 1 \\ 1 & -1 & 0 & 6 \\ 0 & 1 & 1 & 9 \end{bmatrix}$

184. In the special case that **A** has an inverse, we define for positive integers k

$$\mathbf{A}^{-k} = (\mathbf{A}^{-1})^k$$

Given this definition and the property that $\mathbf{A}^{m+n} = \mathbf{A}^m \cdot \mathbf{A}^n$, show that it is logical to define

$$\mathbf{A}^0 = \mathbf{I}$$

Given

$$\mathbf{A} = \begin{bmatrix} 3 & 7 \\ 2 & 5 \end{bmatrix} \qquad \mathbf{B} = \begin{bmatrix} 2 & 3 \\ 1 & 2 \end{bmatrix} \qquad \text{and} \qquad k = 4$$

185. Show that $(\mathbf{A}^{-1})^{-1} = \mathbf{A}$
186. Show that $(k\mathbf{A})^{-1} = (1/k)\mathbf{A}^{-1}$ if $k \neq 0$
187. Show that $(\mathbf{A}^t)^{-1} = (\mathbf{A}^{-1})^t$
188. Show that $(\mathbf{AB})^{-1} = \mathbf{B}^{-1}\mathbf{A}^{-1}$

3.10 MATRICES AND CRYPTOGRAPHY

CRYPTOGRAPHY, the putting of messages into code or cipher, dates back to the time of the ancient Greeks. The Spartans are said to have wound a belt in a spiral around a cane, written the message along the length of the cane, and unwound the belt. No one could read the message unless he had a cane of the exact size. Cardinal Richelieu, in the 1600s, would place a card with holes prepunched in it (a grille) over a sheet of paper, write his secret message in the holes, and then fill in the rest of the paper to look like an innocent letter. Only a person with an identical grille could read the message. Today matrices and their inverses are being used very successfully to encode and decode secret messages.

Let us see how this might work using the important message "NOON TUESDAY." We shall simply number the letters of the English alphabet as, a = 1, b = 2, . . . , z = 26. Punctuation is ignored but we shall let the number 27 correspond to a space between words. (If numbers were to be encoded, we could continue with zero = 28, 1 = 29, and so on.) Thus, our message is coded

N	O	O	N	—	T	U	E	S	D	A	Y
14	15	15	14	27	20	21	5	18	4	1	25

Next, we separate the numbers into groups of three and write these as column vectors, as

$$\mathbf{V}_1 = \begin{bmatrix} 14 \\ 15 \\ 15 \end{bmatrix}, \mathbf{V}_2 = \begin{bmatrix} 14 \\ 27 \\ 20 \end{bmatrix}, \mathbf{V}_3 = \begin{bmatrix} 21 \\ 5 \\ 18 \end{bmatrix}, \mathbf{V}_4 = \begin{bmatrix} 4 \\ 1 \\ 25 \end{bmatrix}$$

Then we choose any 3×3 matrix that has an inverse. We can use

$$\mathbf{A} = \begin{bmatrix} 3 & 3 & 6 \\ 0 & 1 & 2 \\ 5 & 3 & 2 \end{bmatrix} \qquad \text{whose inverse is } \mathbf{A}^{-1} = \begin{bmatrix} 1/3 & -1 & 0 \\ -5/6 & 2 & 1/2 \\ 5/12 & -1/2 & -1/4 \end{bmatrix}$$

Each of the column vectors is premultiplied by the matrix **A**, as

$$\mathbf{AV}_1 = \begin{bmatrix} 3 & 3 & 6 \\ 0 & 1 & 2 \\ 5 & 3 & 2 \end{bmatrix} \cdot \begin{bmatrix} 14 \\ 15 \\ 15 \end{bmatrix} = \begin{bmatrix} 177 \\ 45 \\ 145 \end{bmatrix}$$

and

$$\mathbf{AV}_2 = \begin{bmatrix} 243 \\ 67 \\ 191 \end{bmatrix}, \mathbf{AV}_3 = \begin{bmatrix} 192 \\ 43 \\ 158 \end{bmatrix}, \mathbf{AV}_4 = \begin{bmatrix} 165 \\ 51 \\ 73 \end{bmatrix}$$

The message is transmitted simply as 177, 45, 145, 243, 67, . . . , 73. Our compatriot, in possession of the inverse, $\mathbf{A}^{-1}$, decodes by separating the numbers into three-component column vectors and premultiplying each by $\mathbf{A}^{-1}$, as

$$\mathbf{A}^{-1}\mathbf{V}_1 = \begin{bmatrix} 1/3 & -1 & 0 \\ -5/6 & 2 & 1/2 \\ 5/12 & -1/2 & -1/4 \end{bmatrix} \cdot \begin{bmatrix} 177 \\ 45 \\ 145 \end{bmatrix} = \begin{bmatrix} 14 \\ 15 \\ 15 \end{bmatrix} = \begin{bmatrix} \text{N} \\ \text{O} \\ \text{O} \end{bmatrix}$$

and

$$\mathbf{A}^{-1}\mathbf{V}_2 = \begin{bmatrix} 1/3 & -1 & 0 \\ -5/6 & 2 & 1/2 \\ 5/12 & -1/2 & -1/4 \end{bmatrix} \cdot \begin{bmatrix} 243 \\ 67 \\ 191 \end{bmatrix} = \begin{bmatrix} 14 \\ 27 \\ 20 \end{bmatrix} = \begin{bmatrix} \text{N} \\ - \\ \text{T} \end{bmatrix}$$

and so on.

The coding procedure, although a relatively simple one, is actually very difficult to "break," especially when large matrices are used.

EXERCISES

189. Given the code matrix

$$\mathbf{A} = \begin{bmatrix} 1 & 0 & 3 & 0 \\ 2 & -1 & 0 & 6 \\ 3 & 0 & 2 & 0 \\ 0 & 1 & 0 & 1 \end{bmatrix}$$

use its inverse to decode the following message:

82, 42, 99, 24, 40, 29, 85, 28, 84, 22, 63, 12,
82, 97, 57, 45, 48, 33, 67, 32, 86, 114, 69, 36,
89, 167, 78, 31, 66, 160, 65, 47, 4, 129, 5, 48

Using the matrix

$$\mathbf{A} = \begin{bmatrix} 0.1 & 0.1 & 0 \\ 0 & 0.05 & 0.2 \\ 0.05 & -0.05 & 0.1 \end{bmatrix}$$

encode the following messages:

190. *Behind the second brick.*

191. *The orange cat.*

192. Using the inverse of **A**, decode the following message.

2.9, 6.15, 2.65, 3.5, 6.15, 5.65, 2.5, 2.35, 0.55,
4.1, 2.15, −0.25, 2, 1.75, 3.2.

C·H·A·P·T·E·R 4 Mathematical Functions and Their Use as Models

CONSIDER a situation faced by B. Lever, vice president in charge of production for High-Tech Manufacturing Company:

High-Tech has just placed on the market a new electronic flashlight that market research indicates can be sold in substantial quantities at $3 a unit. The weekly fixed cost of the work center that produces this product is $5,000. Variable costs of production, including raw materials, labor, and overhead, are $1.75 per unit at all levels of production from 1 through 5,000 units of output per week. Mr. Lever wonders how many units must be produced and sold each week in order for the sales revenue generated by the product to exactly cover the costs of production. Or, perhaps more important, what level of production and sales would be required to achieve the maximum possible profit (that is, excess of revenue over cost)?

It seems likely that some cost savings may be achieved by dealing with a new supplier of the raw materials used. The savings would reduce per-unit variable cost for the product by 10 percent. How would such a cost reduction affect profit realized at various levels of production and sales?

Labor contract negotiations currently underway could result in wage increases which would bring per-unit variable cost for the product up to 150 percent of the present level. How would such a cost increase affect profit realized at various production and sales levels?

What sales level would have to be achieved at a per-unit selling price of $4 in order to achieve the same profit as the maximum possible profit with a selling price of $3?

A second work shift, with fixed weekly costs of $3,500 and per-unit variable cost of $2, could be used in the work center to raise the total possible production level to 10,000 units a week. What sales level would have to be achieved in order to justify use of this second shift?

We shall see that B. Lever will be able to construct a fairly simple and straightforward "mathematical replica" of this production situation and use that replica to answer these, and many other, questions.

Mathematics may be approached as a purely theoretical, highly abstract, science; or, mathematics may be viewed very pragmatically, as an efficient and effective way to describe everyday, real-world relationships.

The essence of mathematics is its formal, rigorous structure of relationships among numbers, its precise rules for expressing these relationships, and its logical and efficient techniques for investigating the existing relationships. Practical situations which we might wish to study using mathematical methods are not always characterized by this same degree of precision and exactness of structure. Nonetheless, when certain properties of the objects under investigation can be related by semantical rules to certain properties underlying the number series, the analyst may stand to gain a great deal through the use of general mathematical methods.

The procedure of using numbers and mathematical statements to represent underlying structures in a situation, real or hypothesized, is termed MATHEMATICAL MODEL BUILDING. The purpose of this chapter is to introduce the reader to the idea of describing situations in mathematical terms and of using rules governing the manipulation of such mathematical statements to extract useful and meaningful information about the situation.

4.1 THE IDEA OF A MATHEMATICAL FUNCTION

In mathematical model building we are especially concerned with deriving a procedure by which we can pair elements from one SET with appropriate elements from another set. In a typical decision situation, one of the sets might consist of alternative courses of action open to the decision maker, while the other set consisted of the expected consequences of those actions. Such relations exist for many business phenomena. For instance, there presumedly exists some correspondence between total number of units of product sold and total sales revenue realized from the sale of the product, between the per-unit selling price of a commodity and the number of units of that commodity which can be sold, between outlay for advertising and the number of units of product sold, between number of units of product carried in inventory and total inventory-holding cost, between total number of items manufactured and average production cost, and so on and on. In order to make intelligent decisions, we must somehow determine which action results in which consequence. That is, we must "pair" elements from the two sets.

This brings us to a definition of a mathematical function.

> **A function** consists of two sets of elements and a rule which can be used to pair with each element of one set *one and only one* element from the second set.

When the sets, or groups, of objects can be represented by real numbers—such as *number* of units sold, or *number* of dollars of revenue realized—the function is a REAL-VALUED FUNCTION. The objects or concepts with which these sets of numbers are associated are called VARIABLES.

To describe in mathematical terms the relationship between members of the set "number of units sold each day" and members of the set "total daily sales revenue," we might use a symbol such as x to represent the number of units of the product that are sold. The set of numerical values associated with this variable might be comprised of the integers equal to or greater than zero and be denoted $\{0, 1, 2, 3, \ldots\}$. This set would act as a measure of the quantity of product sold on various days.

Another variable might be symbolized y and used to represent total daily sales revenue (in dollars) realized from the sale of the product. The set of numerical values associated with y might be made up of the quantities $\{0, \$5, \$10, \$15, \ldots\}$. We would expect specific numbers from one set to be paired with specific numbers from the other set. That is, on each day when quantity sold is zero, we would expect the sales revenue to be zero. On each day when quantity sold is one, we would expect the revenue realized to be a certain amount, say \$5. Zero from the x set is consistently paired with zero from the y set, one from the x set is consistently paired with \$5 from the y set, and so on. Each element in the "quantity sold" set would be consistently paired with one, and only one, element from the "sales revenue" set.

This is the basic idea of a mathematical function; nothing could appear to be more simple and straightforward. However, we shall see that functions come in a delightful variety, are tantalizing and elusive to construct, can provide useful and intriguing information about the variables, but must always be handled with great care and some skepticism.

4.2 WAYS OF SYMBOLIZING THE FUNCTIONAL RELATIONSHIP

There are a number of acceptable ways of representing the mathematical relationship between two variables. In general, if any two variables x and y are so related that when the value of x is given there is only one value of y, then y is said to be a function of x, and this relationship is expressed symbolically as

$$y = f(x)$$

read "y is a function of x" or simply "y equals f-of-x." The term FUNCTIONAL NOTATION is used to refer to this system of expressing the correspondence between the variables.

This functional notation simply indicates that a variable y has its value determined by the value of another variable x. Generally, it does not matter whether we call the function f or g or h or C, and so on, as long as we do not use the same symbol for two different functional relationships in one and the same model. If the variable is labeled something other than x, we might write $g(p)$, or $r(q)$, ... The functional symbol is one symbol and does not in itself denote any definite mathematical operation. Specifically, it does not mean that f is multiplied times x, or g multiplied times p.

To show that total revenue realized is functionally related to the number of units of product sold, we might write

$$revenue = f(units\ sold)$$

which is read "revenue is a function of units sold." The notation generally can be extended to show the mathematical rule of correspondence. For instance, the typical relationship between revenue and number of units sold is

$$revenue = (selling\ price\ per\ unit) \times (number\ of\ units\ sold)$$

and could be expressed symbolically as

$$r = f(q) = pq$$

where

$r = f(q) =$ revenue, in dollars
$q =$ number of units sold (or quantity sold)
$p =$ selling price per unit, in dolalrs

This notation indicates that the variable r has a value which is determined by the value of another variable q and by a mathematical rule of correspondence between the two. Furthermore, the rule is that the value of r is the value of q multiplied by the value of p.

Even the general formulation of a functional relationship can be made more specific by stating precisely what the rule of correspondence happens to be in a particular case. Assume that the per-unit selling price for a product is $5. Then the function relating number of units sold q and total revenue r could be expressed as

$$r = f(q) = 5q$$

This notation states not only that the value of r is determined by the value of q and a rule that associates the two, but it tells us, as well, that the rule in this particular case is to "multiply the value of q by the constant 5, which is price."

Often the notation "$r =$" is dropped and the specific relationship simply expressed

$$f(q) = 5q$$

An alternative method of representing the relationship between the two variables might use a TABULAR FORMAT. For instance, quantity sold and corresponding revenue realized could be depicted as follows:

Number of units sold, q	0	1	2	3	. . .
	↓	↓	↓	↓	
Revenue, in dollars, $r = f(q)$	0	5	10	15	. . .

Or, the paired numbers might be enclosed in parentheses and listed in ROSTER FORM as $\{(q = 0, r = 0), (q = 1, r = 5), (q = 2, r = 10), (q = 3, r = 15), \ldots\}$. Typically the notations "$q =$" and "$r =$" are omitted and the corresponding values simply listed $\{(0,0), (1,5), (2,10), (3,15), \ldots\}$. Remember that these are ORDERED PAIRS and it would be understood that all first elements come from the set of values associated with number of units sold and all second elements come from the set of values associated with revenue realized.

SET-BUILDER NOTATION provides still another method for describing a function, a method which emphasizes the idea that the function is a set of ordered pairs. For example, the notation

$$f = \{(q,r) \mid r = 5q\} \qquad \text{or} \qquad f = \{(q, f(q)) \mid f(q) = 5q\}$$

is read "f is the set of ordered pairs (q,r) such that r is five times the value of q." The expression $r = 5q$ is an OPEN SENTENCE in two variables that describes the function; it is the RULE of the function and tells what to do with the value of one variable in order to obtain the corresponding value of the other variable. Set-builder notation has a "mathematical" look about it and upon first encounter perhaps seems slightly intimidating. But once we understand how it is read, we can appreciate its convenience. The student should develop a facility for reading and working with this method, as well as with the tabular and roster methods, of denoting the functional relationship.

EXERCISES

Express each of the following relationships in functional notation.

1.

x	0	1	2	3	4
	↓	↓	↓	↓	↓
y	3	5	7	9	11

2. $f = \{(1,1),(2,4),(3,9),(4,16),(5,25),(6,36)\}$
3. $f = \{(x,y) | y = 3x + 9\}$
4.

x	-2	-1	0	1	2
	↓	↓	↓	↓	↓
$f(x)$	-8	-1	0	1	8

5. $f = \{(1/4,1),(2/4,2),(3/4,3)\}$
6.

q	0	1	2	3	4	5
	↓	↓	↓	↓	↓	↓
$C(q)$	500	600	700	800	900	1000

7. $g = \{(0,-1),(1,0),(2,1),(3,2)\}$
8. $f = \{(x\,f(x)), \,|\, f(x) = x^2 + x + 1\}$

Describe each of the following functions in set-builder notation, using an open sentence in two variables x *and* y *to state the rule.*

9. The value of y is half the value of x.
10. The value of y is 10 less than the value of 5 times x.
11. The value of y is 100 plus the value of x squared.
12. The value of y is obtained by subtracting 2 from the value of x and squaring the difference.
13. The value of y is the positive root of x.

A woodworker in a furniture factory is paid a piecework bonus on each unit of furniture completed each week as outlined in the table below:

Number of units completed, x	0	1	2	3	4	5	6
Bonus received, y	0	\$20	\$40	\$60	\$80	\$100	\$120

14. Express the relationship between number of units of furniture completed during a week and the bonus received (in dollars) in roster form.
15. Use functional notation to express the relationship.

4.3 DERIVING SPECIFIC FUNCTIONS

MATHEMATICAL MODEL BUILDING is the process of describing a situation in terms of the functional relationships between the relevant variables. The model builder must have not only a working knowledge of mathematical operations but also a thorough understanding of the situation he wishes to depict. A few examples will help to illustrate the process.

Example 1 A retail clothing store determined the selling price of an item by adding to the original cost of the item an amount called markup. The markup amount is always a fixed percentage of original cost. Write a statement expressing the general functional relationship between selling price and cost.

The function is

$$y = f(x) = x + ax = x(1 + a)$$

where

$y = f(x) =$ selling price, in dollars
$x =$ original cost, in dollars
$a =$ markup percentage

This statement is a mathematical model of the relationship between cost and selling price of an item. []

Example 2 A bank charges each depositor a fixed fee of $3 a month plus an additional 15 cents for each check written during the month. Express in functional notation the fact that the customer's total monthly service charge is determined by the number of checks he writes during the month.

Total service charge may be determined by

$$S = h(n) = 3 + 0.15n$$

where

$S = h(n) =$ total monthly service charge, in dollars
$n =$ number of checks written during the month

This mathematical model can be used to determine a customer's total monthly service charge, given information about the number of checks written. []

Example 3

A rectangular area is to be enclosed by 1,000 yards of fence. Express the area A as a function of the length L of one side of the area.

The area enclosed may be determined by multiplying the length of one side times the width. The combined distance of two sides (one length and one width) equals half the total amount of fencing (or 500 yards). If one side is of length L, the other must be of length $500 - L$. Thus, the model is

$$A = g(L) = L(500 - L) = 500L - L^2$$

where

$A = g(L) =$ area enclosed, in square yards
$L =$ length of one side, in yards []

Example 4

Write a function for converting temperatures on the Fahrenheit scale to temperatures on the Celsius scale. Recall that 0° Celsius equals 32° Fahrenheit. Also, the interval between the freezing and boiling temperatures is divided into 100 equal intervals on the Celsius scale and into $212 - 32 = 180$ equal intervals on the Fahrenheit scale. This means that a change of d degrees Fahrenheit corresponds to a change of $(100/180)d = (5/9)d$ degrees Celsius.

The mathematical function which converts temperatures Fahrenheit to temperatures Celsius is

$$C = f(F) = (5/9)(F - 32)$$

where

$C = f(F) =$ degrees on Celsius scale
$F =$ degrees on Fahrenheit scale []

Functions Involving More than Two Variables

A variable can be, and quite often is, a function of more than one other variable. The amount of interest earned on a savings account depends upon the principal amount and the length of time involved as well as the rate of interest. The transportation charge on a parcel is based on its weight, its freight classification, and the distance it is sent. The volume of a rectangular container is a function of its length, its height, and its width.

A function f of two variables, x and y, associates with each possible combination of values of the variables (x,y) one and only one value for the other variable $z = f(x,y)$. Or a function g of three variables, x, y, and z, associates with each triplet of values for the variables (x,y,z) a value for the other variable $w = g(x,y,z)$. And so on for any number of variables.

Example 5

The direct cost of manufacturing an item, C, is a function of the number of units of raw material used, M, and the number of units of labor used, L. Each unit of raw material used costs \$6 and each unit of labor costs \$9. Use functional notation to express the relationship between cost, raw material used, and labor used.

The function is

$$C = f(M,L) = 6M + 9L$$

where

$C = f(M,L)$ = total direct cost of producing an item, in dollars
M = number of units of raw material used
L = number of units of labor used □

EXERCISES

16. Apples cost 69 cents a pound. Write a mathematical function to describe the relationship between total cost and quantity purchased. Define each variable used.
17. A salesman receives a base salary of \$800 a month plus a commission of 10 percent of the dollar amount of his monthly sales. Use functional notation to specify the relationship between his total monthly earnings and the dollar amount of goods that he sells. Define each variable used.
18. An amount P is deposited in a savings account which draws 8 percent interest, compounded annually. Write a statement showing the total amount that will be in the account after one year if the interest is not withdrawn. Define each variable used.
19. A storage bin has a square base with each side of the base being x feet long. The height of the bin is 2.5 times the length of one side of the base. Write a function which shows how the storage volume is related to the base length. Define each variable used.
20. The cost of renting a car for a day is \$19.50 plus 15 cents for each mile driven. Express the cost as a mathematical function of the number of miles. Define each variable used.
21. A factory worker makes \$7.50 an hour. Deductions for taxes, insurance, union dues, and so on, amount to 30 percent of gross pay. State the relationship between the worker's take-home pay and the number of hours worked each week. Define each variable used.
22. Salespersons for the Abbott Company are paid a travel allowance which consists of 15 cents a mile for each mile traveled during a week plus \$30 for each night which must be spent "on the road" away from the home office location. Using functional notation, express the relationship between total travel allowance, miles traveled, and number of nights on the road.
23. Employees at the Benson Plant are paid \$8.50 an hour for each hour worked. No employee is allowed to work more than 40 hours a week, but many employees work fewer than 40 hours. In addition

to their hourly wage, employees are paid a "production incentive" of $1.95 for each unit of output completed. Use functional notation to express the relationship between an employee's weekly earnings, the number of hours worked, and the number of units of output completed during the week.

4.4 FINDING THE VALUE OF THE FUNCTION

The mathematical rule—the equation—defining a function is often described as a "black box" whose inputs are the various values of one of the variables and whose outputs are the corresponding values of the other variable. (See Figure 4.1.)

Figure 4.1

The Function as a "Black Box"

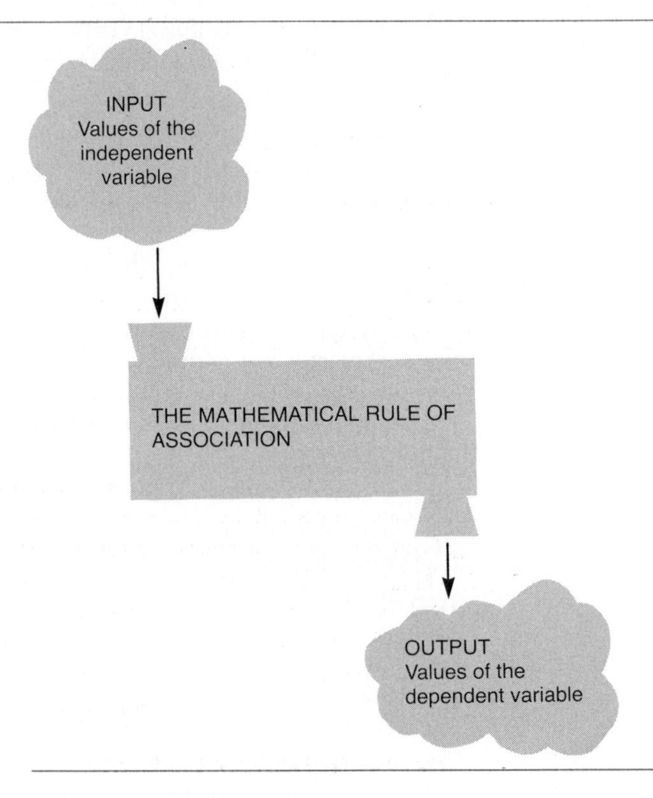

If specific values for an ordered pair are desired, a value of x is selected and substituted into the formula—used to activate the black box—thereby producing the FUNCTION VALUE or the VALUE OF THE FUNCTION for that value of x. Given a function

$$f(x) = 4x + 1$$

we may find the value of the function when $x = 3$ by substituting this value of x into the equation, as

$$f(3) = (4)(3) + 1 = 13$$

The symbol $f(3)$ is read "f-of-3" and means the value of the second variable, usually denoted by y, when $x = 3$.

We could compute, in a similar manner,

$$f(4) = (4)(4) + 1 = 17$$

or

$$f(5) = (4)(5) + 1 = 21$$

and so on.

The input variable is conventionally labeled the INDEPENDENT VARIABLE and the output variable the DEPENDENT VARIABLE. The input, or independent, variable is also known as the ARGUMENT and the dependent variable the VALUE of the function.

Example 6 An automobile-towing service charges a fixed fee of \$30, plus \$1.25 a mile for the distance towed. The function showing how the total charge is related to the number of miles towed is

$$C = f(M) = 30 + 1.25M$$

where

$C = f(M) =$ total charge, in dollars
$M =$ distance towed, in miles

"Distance towed" is the independent variable; "total charge" is the dependent variable.

Once we have modeled the relationship between total charge and distance towed, it becomes easy to answer such queries as: (1) What would be the total charge if the automobile were towed 10 miles? (2) What would be the total charge if the distance were 20 miles? (3) If the total charge were \$67.50, how many miles was the automobile towed?

To determine the total charge for towing an automobile 10 miles, we compute

$$f(10) = 30 + (1.25)(10) = \$42.50$$

The total charge for 20 miles is

$$f(20) = 30 + (1.25)(20) = \$55$$

We can determine the value of the independent variable, given the corresonding value of the dependent variable, by setting $C = \$67.50 = f(M)$ and solving for M. We get

$$\begin{aligned} 67.50 &= 30 + 1.25M \\ M &= (67.50 - 30)/1.25 \\ &= 30 \text{ miles} \end{aligned}$$

[]

EXERCISES

A firm has determined that the total revenue R, in dollars, from the sale of Q units of a product is

$$R = f(Q) = 12Q$$

24. What will be the total revenue generated by the sale of 800 units of the product?
25. How many units must be sold in order to generate $24,000 in revenue?

A salesperson receives a base salary of $500 a month plus a commission of 5 percent of the gross amount of sales (in dollars).

26. Write the function showing how total salary is related to gross dollar amount of sales. Which is the independent variable? Which is the dependent variable?
27. Determine what the salary will be when sales are $12,000, when sales are $18,000.
28. What would sales have to be in order to generate a salary of $20,000? a salary of $25,000?

Storage cost of an item is given by the function

$$S = g(C) = 0.05C + 2.50$$

where

C is the cost of the item

29. Find the cost of storing an item which costs $12.
30. Find the cost of an item whose storage cost is $3.

A firm has determined that the volume V of sales (in units) of a certain product is inversely proportional to the quantity (6 + P) where P is the per-unit selling price of the product.

31. State the function relating sales volume to selling price. Which is the independent variable? the dependent variable?
32. If 5,000 units are sold when the selling price is $10, what is the specific function relating volume to price?
33. How many units would be sold at a price of $9?

4.5 THE DOMAIN AND RANGE OF A FUNCTION

The set of permissible values for the independent variable is called the DOMAIN of the function, while the set of permissible values for the dependent variable is called the RANGE of the function.

A given functional relationship may be appropriate only for specified values of the independent variable. For instance, it may be possible to sell any number of units of a product *up to a maximum of 100 units* for $5 each. If there are any such restrictions imposed upon the value set of the independent variable, these restrictions must be clearly stated along with the rule of association.

If the relationship $f(x) = 5x$ is appropriate only for nonnegative, integer values of x which are less than 101, this restriction on the independent variable is usually specified along with the definition of the function, as

$$f(x) = 5x \qquad \text{where } x = 0, 1, 2, \ldots, 100$$

Or, using set-builder notation, we might specify the restrictions imposed upon the domain of the function by writing

$$f = \{(x, f(x)) \mid f(x) = 5x; x = 0, 1, 2, \ldots, 100\}$$

This particular notation also lets us know that x is a DISCRETE VARIABLE; that is, a variable that can assume only specified values within some interval, those values being separated from one another on a scale by some nonzero space.

In contrast to a discrete variable, a CONTINUOUS VARIABLE can assume any value within an interval, such as all points along a line segment or within an area plane. The points are not separated by any definable space.

If a rule of association $f(x) = 3x + 5$, is appropriate for a continuous independent variable x over the interval of values from 0 to 10, inclusive, the definition of the function and the domain of f might be written

$$f(x) = 3x + 5 \qquad 0 \le x \le 10$$

or

$$f = \{(x, f(x)) \mid f(x) = 3x + 5; \ 0 \le x \le 10\}$$

Sometimes one rule of correspondence is appropriate over a specified set of values for the independent variable and then another rule is appropriate for another set of values, as

$$f(x) = \begin{cases} 2x & 0 \le x \le 5 \\ 3x - 5 & 5 < x \le 10 \end{cases}$$

If no statement specifying the domain of the function is given, the domain shall be considered to be the set of all real numbers, or at least the largest set of real numbers for which the definition of the function is valid or logical. For instance, in many situations negative values for the independent variable would be illogical—such as quantity of product manufactured, interest earned on investment, or dollars spent on advertising. In other instances, noninteger values for the variable might not be allowed—such as number of workers employed full-time, number of automobiles purchased during the year, or number of retail outlets belonging to a trade association.

Also, restrictions on the domain of a function are frequently imposed by the conventions and definitions of mathematics. Values excluded from the domain will include (1) any value of the independent variable which would result in the denominator in a quotient function assuming a zero value, as well as (2) any value of the independent variable which would result in a situation where there are no real roots of a radical.

For instance, the function

$$f(x) = 1/(3x - 9)$$

is not defined when $x = 3$ since division by zero is not a meaningful mathematical operation.

The relationship

$$f(x) = \sqrt{x}$$

is defined only for nonnegative values of x, since the square root of a negative number is not a real number.

Restrictions imposed by the conventions of mathematics are not generally stated explicitly as are the restrictions imposed by some nonmathematical aspect of the situation. Also, the permissible values of the dependent variable are usually not explicitly stated if they can be inferred from the value set of the independent variable.

EXERCISES

For each of the following, state whether the variable is discrete or continuous.

34. Number of miles a salesman travels during a week.
35. Number of units of product sold during a week.
36. Number of conventioneers visiting a city during a year.
37. Number of gallons of gasoline consumed in driving from Boston to San Francisco.
38. Number of employees belonging to a credit union.

Determine the domain of each of the following functions.

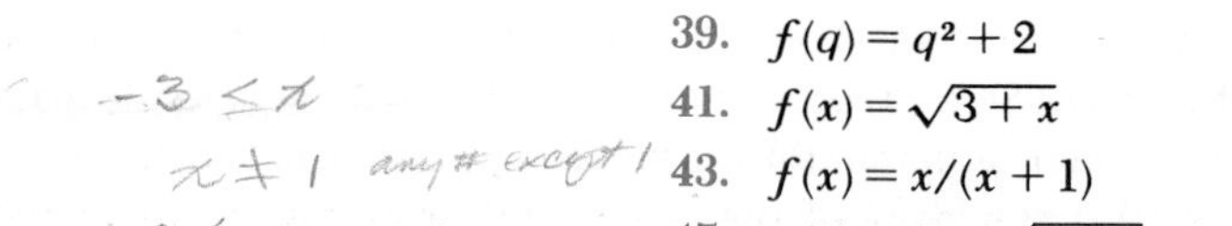

39. $f(q) = q^2 + 2$
40. $g(w) = w/2$
41. $f(x) = \sqrt{3 + x}$
42. $h(t) = 1/(t - 4)$
43. $f(x) = x/(x + 1)$
44. $f(x) = 4/(x^2 - x - 2)$
45. $g(x) = 1 - \sqrt{x - 2}$

A salesperson receives a base salary of \$600 plus 6 percent commission on gross sales exceeding \$5,000 a month.

46. Write a function showing how total salary is related to gross sales amount. State the domain of the function.

47. What will be the total salary when sales are $4,000?

48. What will be the total salary when sales are $6,000?

A manufacturer orders raw material from a supplier who offers a quantity discount based on the schedule given below.

When quantity ordered (Q) is:	Price per unit (P) is:
Fewer than 500 units	$1.00
500 to 1000 units	$0.90
More than 1000 units	$0.825

49. Formulate a function that gives total cost (C) as related to the quantity ordered (Q). Define carefully the domain of the function.

50. What will be the total cost when quantity ordered is 300 units? 600 units? 1,200 units?

4.6 GRAPHICAL REPRESENTATION OF A FUNCTION

We quite often can better understand the functional relationship between variables if we have a picture of this association. The construction of such pictures is an important adjunct to model building.

The relationship between two variables—between two sets of numbers—can be easily pictured using a geometric device called the CARTESIAN, or RECTANGULAR, COORDINATE SYSTEM. Two number lines, one HORIZONTAL and one VERTICAL, intersecting at right angles, are constructed in a plane, thus dividing the plane into four regions. These regions are called QUADRANTS and are numbered counterclockwise I, II, III, and IV (as shown in Figure 4.2). Each point in this two-dimensional space represents an ordered pair (x,y) of numbers, and each possible ordered pair of real numbers represents a point.

The two intersecting lines are called COORDINATE AXES, or simply AXES. Conventionally, the values of the independent variable are shown on the horizontal axis and the values of the dependent variable on the vertical axis. Thus, assuming the use of x for the independent variable and y for the dependent variable, the horizontal number line is often termed the x-AXIS while the vertical number line is called the $f(x)$-AXIS or y-AXIS. The point at which these axes intersect is called the ORIGIN. The origin represents the ordered pair (0,0) and provides a point of central reference for the coordinate system. Positive values of x are measured to the right of the origin and negative values to the left. Positive values of y are measured upward of the origin and negative values downward.

A specific pair of values for the variables is represented geometrically by a POINT in the rectangular coordinate system. The point is located in the plane by measuring its perpendicular direction and distance from each of the axes. These two distances, with signs to indicate their direc-

Figure 4.2

Cartesian, or Rectangular, Coordinate System

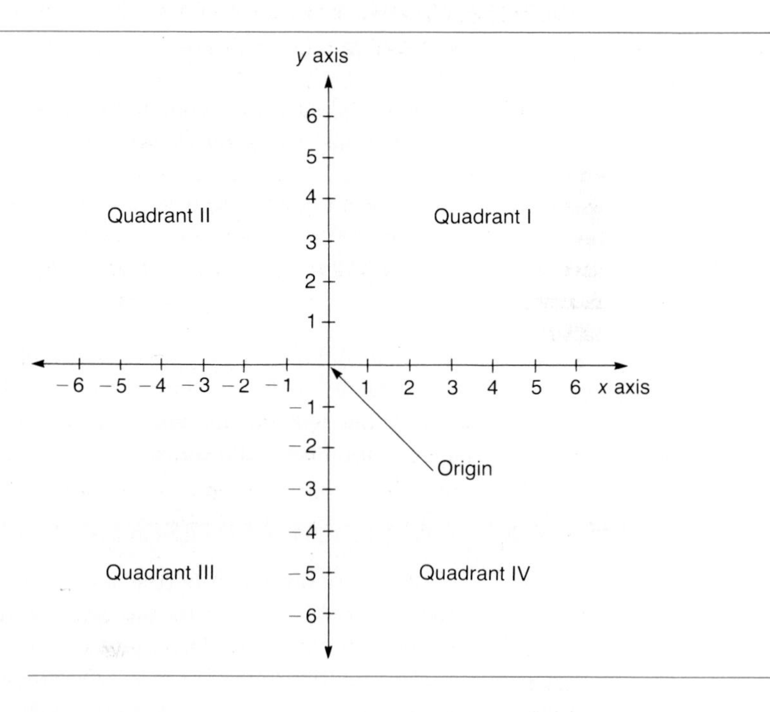

Figure 4.3

Locating Points in the Rectangular Coordinate System

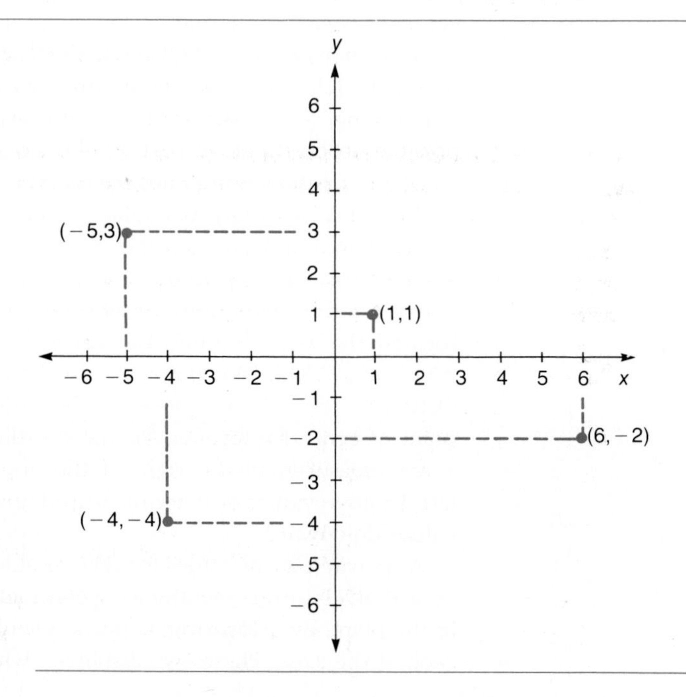

Figure 4.4

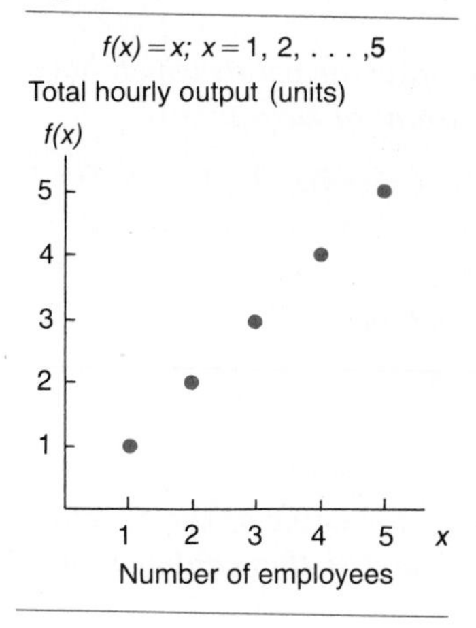

Figure 4.5

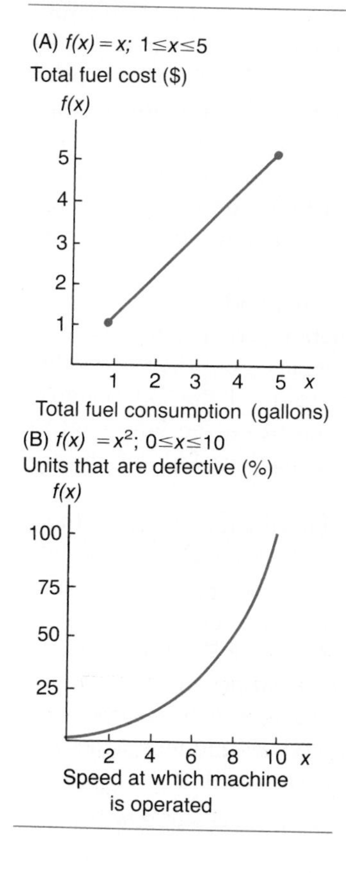

tions, are the COORDINATES of the point. The x-COORDINATE of a point, called the ABSCISSA of the point, indicates the direction and distance of the point relative to (to the right or left of) the y-axis. The y-COORDINATE of a point, called the ORDINATE of the point, indicates the direction and distance of the point relative to (above or below) the x-axis. To give geometric meaning to an ordered pair of numbers, we project the x value vertically from its x-axis and the y value horizontally from its y-axis. The intersection of these projections defines the graph of the ordered pair. The position of a point is indicated by writing its coordinates, in x,y order, enclosed in parentheses, as: (1,1), (6,−2), (−4,−4), or (−5,3). The ordered pair is often referred to as the LOCATION ADDRESS of the point. (See Figure 4.3.)

The rectangular coordinate system can be used to represent any geometric figure that corresponds to an equation in two variables. The geometric figure corresponding to an equation consists of all points, and only those points, whose coordinates form solutions of the equation. If a function is to be plotted on a rectangular coordinate system, its ordered pairs should be computed. These ordered pairs are then used to locate the point in the coordinate plane.

If the independent variable is restricted to a limited number of discrete values, its graph may consist of a series of points, as shown in Figure 4.4. In principle, these points should not be connected to form a smooth line, since to do so would imply that noninteger values of the independent variable were permissible. In practice, however, discrete functions are often depicted by smooth lines simply for the sake of convenience.

When picturing the function of a continuous independent variable, values of the independent variable are selected somewhat arbitrarily and used in plotting selected points. When a sufficient number of points have been plotted, they are then connected with a smooth line so as to show the continuous nature of the domain of the function. (See Figure 4.5.) While we can obtain a pretty good idea of the shape of the graph of a function by plotting a sufficient number of points and connecting these, there are much easier methods of obtaining the graph. We shall study these techniques in the chapters that follow.

EXERCISES

Set up a rectangular coordinate system and locate the points which have the following coordinates. Identify the quadrant of each point.

51. (1,4),(−1,4),(4,1),(4,−1)
52. (−1,−4),(−4,−1),(0,4),(0,−4)
53. (−4,0),(4,0),(4,1/4),(1/4,1/4)

Sketch a graph of each of the following functions.

54. $f=\{(x,f(x))|f(x)=2x;\ x=0, 1, 2, \ldots, 6\}$
55. $g=\{(z,g(z))|g(z)=z/4;\ z=0, 4, 8, 12\}$

For each of the following functions, compute the value of the function for selected values of the independent variable. Use these ordered pairs to sketch the graph of the function.

56. $f(x)=5-x$
57. $g(p)=p^2$
58. $f(x)=x+7$

4.7 MORE ABOUT THE DEFINITION OF A FUNCTION

Not all mathematical equations are considered to be functions. Any rule which associates one or more values of the dependent variable with each value of the independent variable defines a set of ordered pairs termed a mathematical RELATION. The term *function* is reserved for that special type of mathematical association between variables wherein *for each real number in the domain there exists only one corresponding number from the range of the function.* No two ordered pairs in the set of ordered pairs comprising the function can have the same first element.

Since functions pair only one value of the dependent variable with each value of the independent variable, functions are a special type of mathematical relation. Other relations, which pair more than one value of the dependent variable with at least one value of the independent variable, are not functions. Thus, relations constitute the larger set of which functions are a subset. All functions are relations, but not all relations are functions.

Two types of correspondence are included in the definition of a function. When each allowable value of the dependent variable is matched with exactly one value of the independent variable and each value of the independent variable is matched with exactly one value of the dependent variable, there is said to exist a ONE-TO-ONE CORRESPONDENCE. The function $f(x) = 2x + 9$ is an example of a one-to-one function, as shown in Figure 4.6A. Each value of x is associated with only one value of $f(x)$, and each value of $f(x)$ is uniquely associated with only one value of x. For example, for $f(x) = 2x + 9$,

$f(0) = 9$, and no other value of x generates an $f(x)$ value of 9

$f(1) = 11$, and no other value of x generates an $f(x)$ value of 11 and so on.

If each value of the independent variable is associated with only one value of the dependent variable, but more than one value of the independent variable yields the same value of the dependent variable, there is said to be a MANY-TO-ONE CORRESPONDENCE, as shown in Figure 4.6B. Each value of x is associated with only one value of $f(x)$ but the same value for $f(x)$ may be generated by more than one value of x. The function $f(x) = x^2$ is an example of such a correspondence. Notice that

$$f(-1) = 1 \quad \text{and} \quad f(1) = 1$$
$$f(-2) = 4 \quad \text{and} \quad f(2) = 4$$

Figure 4.6

(A) A one-to-one correspondence for a function: $f(x) = 2x + 9$

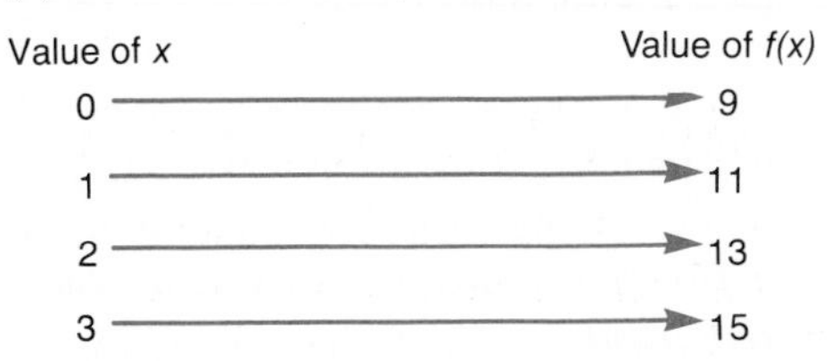

(B) A many-to-one correspondence for a function: $f(x) = x^2$

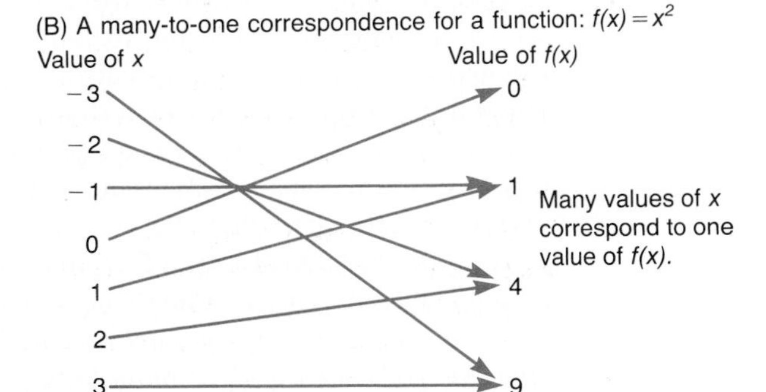

(C) A mathematical relation, but not a function: $y^2 = x$

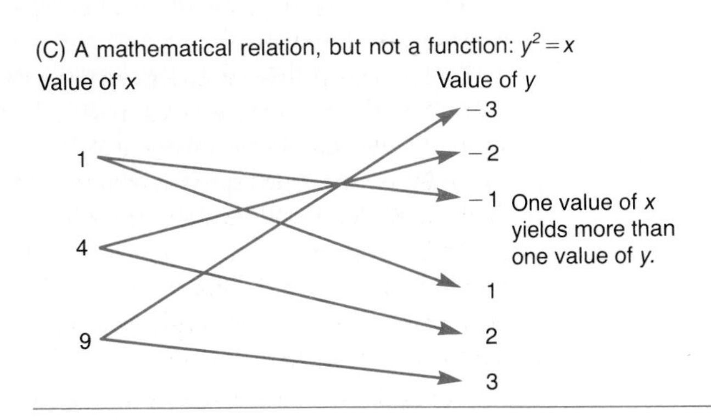

In either case—the one-to-one correspondence or the many-to-one correspondence—for a specific value of x there exists only one value of $f(x)$.

There are many mathematical relationships in which a single value of the independent variable leads to *more than one* value of the dependent variable. An example of such a relationship is

$$y^2 = x$$

which is equivalent to

$$y = \pm\sqrt{x}$$

For each value of x(except for $x = 0$), there are two values of y. For instance, if $x = 1$, $y = \pm 1$; if $x = 4$, $y = \pm 2$; or if $x = 9$, $y = \pm 3$. This is in violation of the requirement that there exist only one value of the dependent variable for each value of the independent variable of a function. Consequently, $y = \pm\sqrt{x}$ is not a function; it is a relation. (See Figure 4.6C.)

The Vertical-Line and the Horizontal-Line Tests

Given the graph of an equation, there is a VERTICAL-LINE TEST which can be easily used to determine whether or not the equation meets the special definition for a mathematical function. *A graph depicts a function provided no vertical line intersects the graph at more than one point.*

Note in Figure 4.7A, where the graph of a function is pictured, that no vertical line erected in the rectangular coordinate system will cross the line representing the graph of the function at more than one point. However, in Figure 4.7B, which shows the graph of a relation which is not a function, we see that certain vertical lines cross the line representing the graph at two points. This means that two values of y are associated with that particular value of x.

A HORIZONTAL-LINE TEST may be used to determine whether a function is a one-to-one function or a many-to-one function. *A function is of a one-to-one correspondence provided it is the case that no horizontal line intersects the graph at more than one point.*

In Figure 4.8A, note that each value of x is paired with only one value of y (the vertical-line test) and also that each value of y is paired with only one value of x (the horizontal-line test). However, in Figure 4.8B we see that, while each value of x is paired with only one value of y, certain y values are paired with more than one x value. The graph in Figure 4.8A represents a one-to-one function; the graph in Figure 4.8B represents a many-to-one function.

Figure 4.7

Vertical-Line Test for a Function

(A) A function

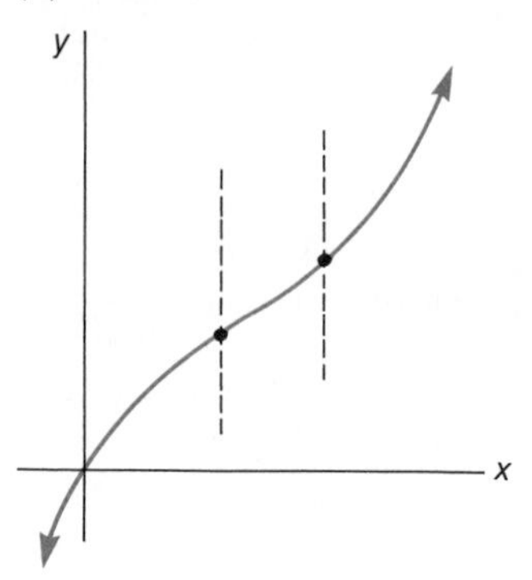

(B) A relation, not a function

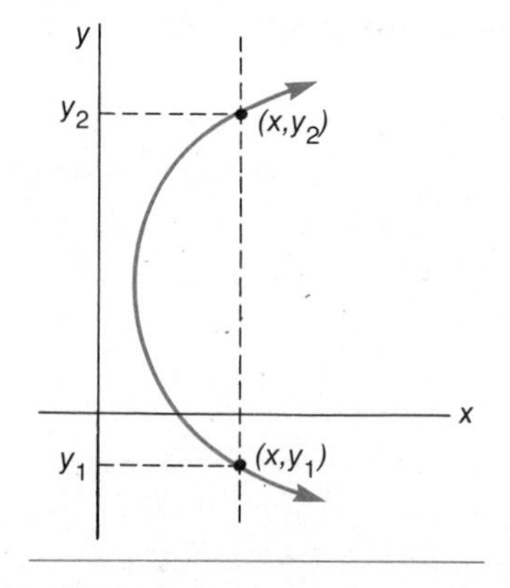

Figure 4.8

Horizontal-Line Test

(A) A function—one-to-one correspondence

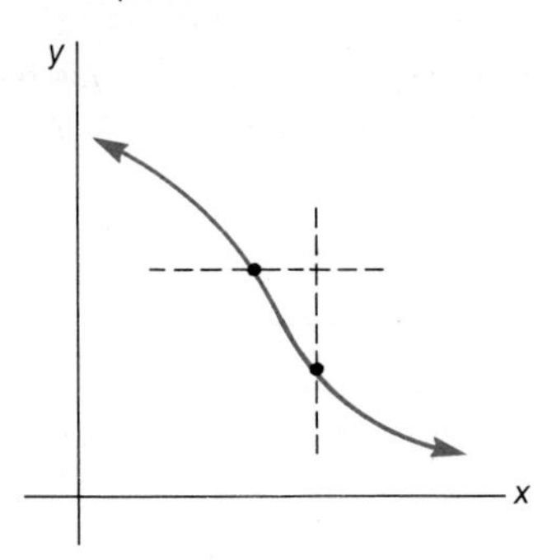

(B) A function—many-to-one correspondence

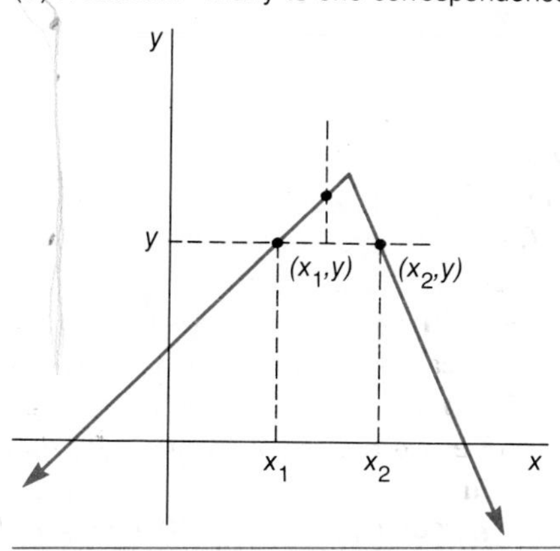

EXERCISES

Indicate which of the following are functions and which are not. For each that is a function, indicate the type of correspondence—one to one or many to one.

59. $\{(0,1),(2,5),(3,9)\}$
60. $\{(1,-1),(2,-1),(3,-5),(4,-5)\}$
61. $\{(-1,1),(-1,2),(-5,3),(-5,4)\}$
62. $y = \{(x,y) \mid y = x^2 + 1\}$
63. $y = \{(x,y) \mid y = 3x + 4;\ 0 \le x \le 100\}$
64. $g = 6w - 9$
65. $h(p) = (1/3)p$
66.

x	0	1	2	3	4
	↓	↓	↓	↓	↓
$h(x)$	−6	−4	−2	0	2

Use the vertical-line test to determine which of the following graphs represent functions. For those that are functions, use the horizontal-line test to see if they are one-to-one functions.

67. function 1 to 1

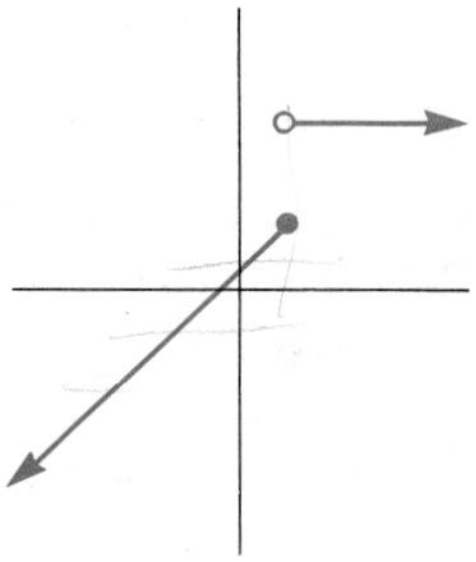

68. function 1 to 1

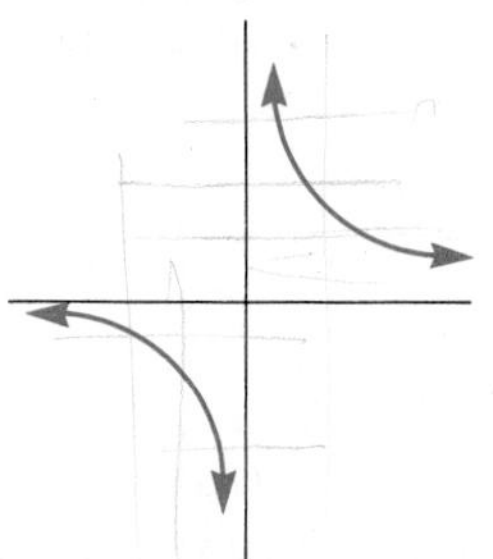

69. relation many to 1

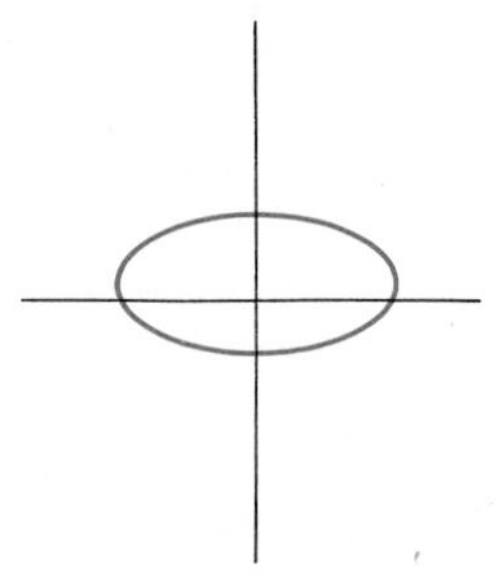

70. function 1 to 1

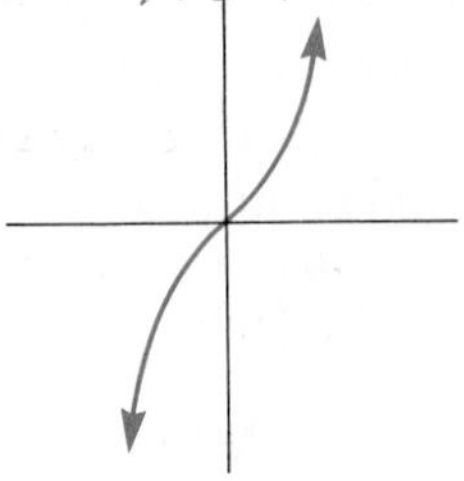

71. function 1 to 1

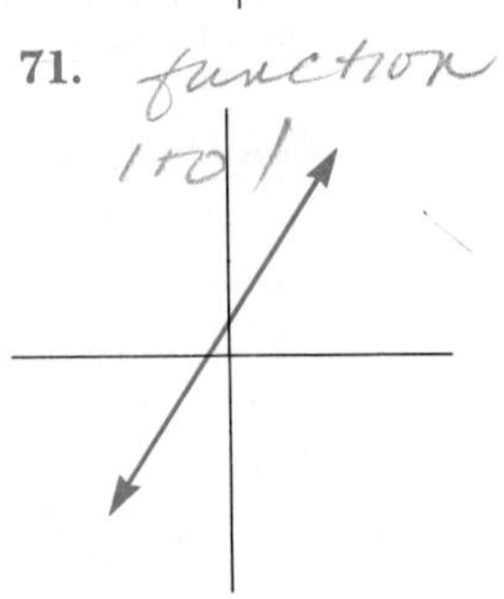

72. function many to 1

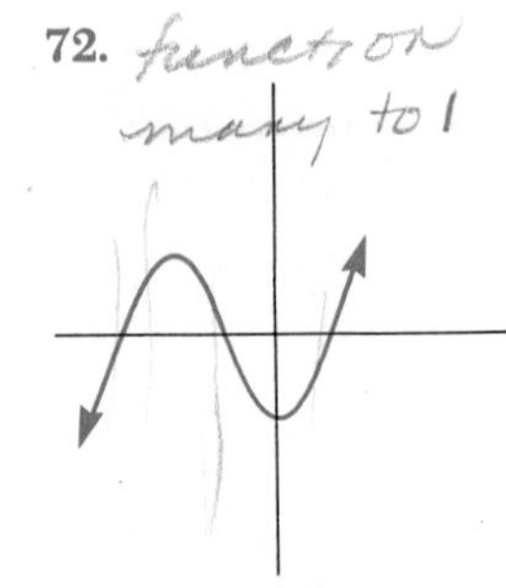

73. func 1 to 1

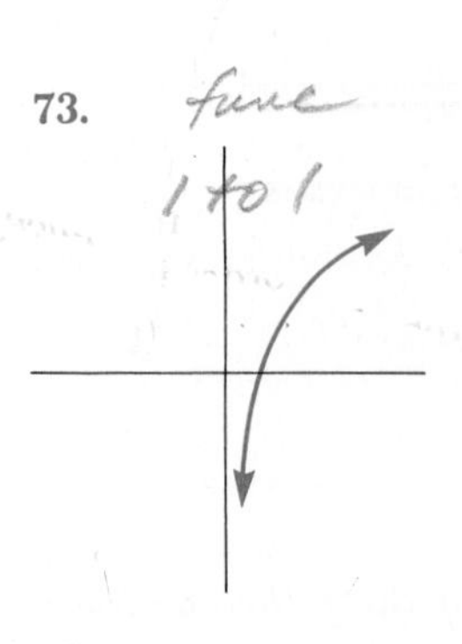

74. func many to 1

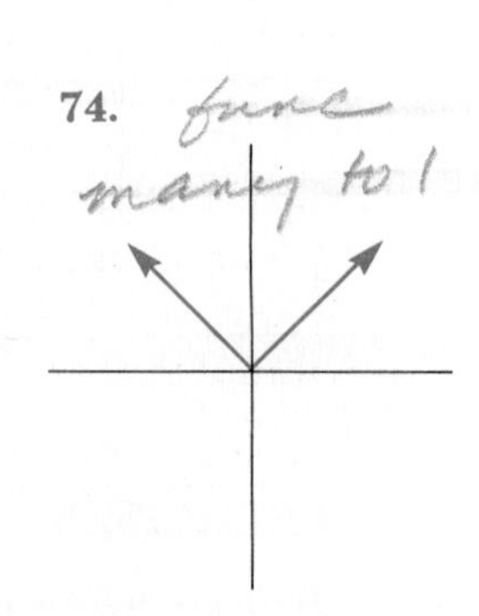

75. Relation

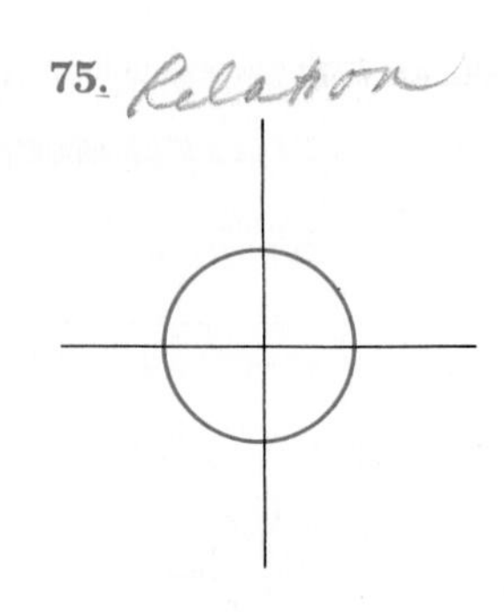

76.

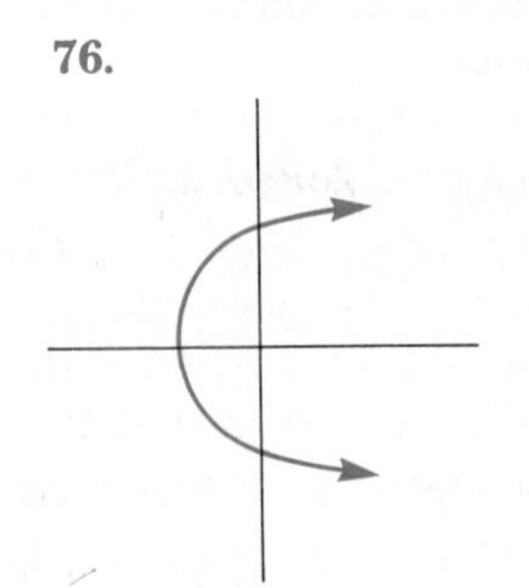

77.

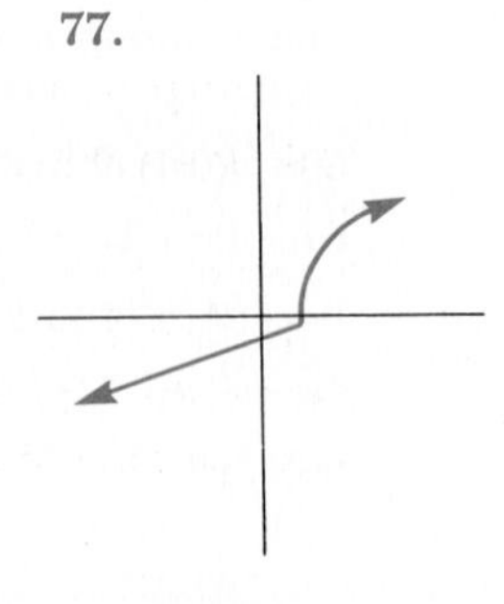

78.

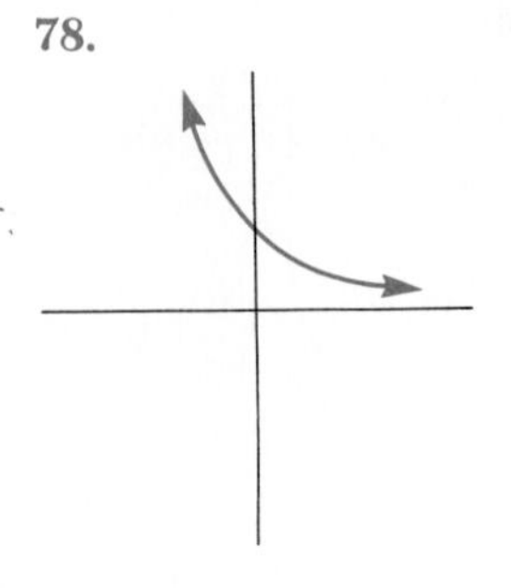

4.8 IMPLICIT, EXPLICIT, AND INVERSE FUNCTIONS

A particular rule relating two variables can be expressed in several formats without actually changing the rule.

An equation may be written in such a way as to imply a mutual relationship between the variables. That is, an equation may be given in the format $h(x,y) = 0$, as, for example

$$y - 3x + 1 = 0$$

Here y is said to be an IMPLICIT FUNCTION OF x, and x is said to be an IMPLICIT FUNCTION OF y. Given a value of y, the equation implicitly defines x; or given a value of x, the equation implicitly defines y.

Alternately, a rule may be expressed in the format $y = f(x)$, as, in the above case,

$$y = f(x) = 3x - 1$$

This statement depicts y as an EXPLICIT FUNCTION OF x. That is, the equation expresses y directly in terms of x and clearly indicates that the value of y is determined by the value of x (although a cause-and-effect situation should not automatically be assumed to exist).

We can also rewrite the original equation with x treated as though it were the dependent variable, as

$$x = g(y) = (y + 1)/3$$

In this expression, x is shown as an EXPLICIT FUNCTION OF y. Thus, the same association between the variables x and y has been expressed in three different ways: as an implicit function of both x and y, as an explicit function of x, and as an explicit function of y.

The following table of ordered pairs shows that the domain and range of $f(x)$ are interchanged for $g(y)$ in the two explicit functions given above.

Explicit function of x $f(x) = 3x - 1$								Explicit function of y $g(y) = (y + 1)/3$	
x ↓	Domain	−1	0	1	2	3	4	$g(y)$ ↑	Range
$f(x)$	Range	−4	−1	2	5	8	11	y	Domain

In Figure 4.9A we see that these two functions yield the same line if both x and $g(y)$ are plotted on the horizontal axis and both y and $f(x)$ are plotted on the vertical axis. This follows, since they are both forms of the same implicit equation.

However, if we adhere to the convention of using the symbol x to denote the independent variable and the symbol y to denote the dependent variable of a function, we would rewrite the explicit function

$$x = g(y) = (y + 1)/3$$

Figure 4.9

Explicit and Inverse Functions

(A) Graph of y as an explicit function of x—$y = f(x) = 3x\text{-}1$— and of x as an explicit function of y—$x = g(y) = (y+1)/3$

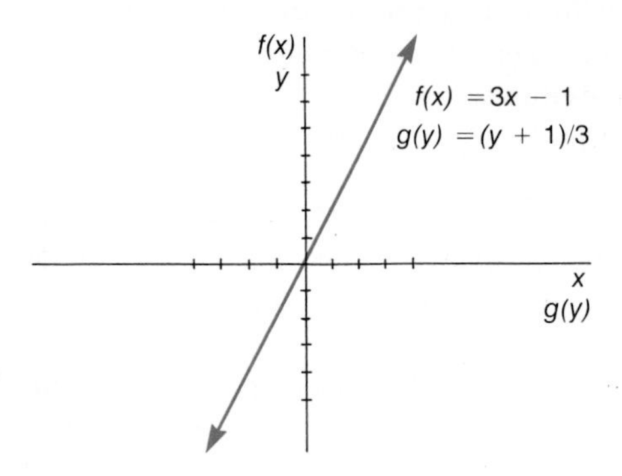

(B) Graph of the function $f(x) = 3x\text{-}1$ and its inverse function $f^1 1(x) = (x+1)/3$

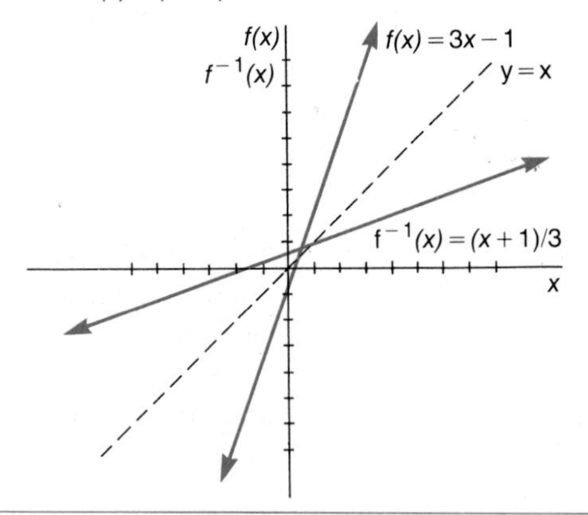

as

$$f^{-1}(x) = (x + 1)/3$$

where the notation f^{-1} identifies the new function as the INVERSE of the original function $f(x) = 3x - 1$. (The superscript, in this case, is not an exponent!)

Geometrically, a function f and its inverse f^{-1} present a mirror image of each other about the diagonal line $y = x$ (See Figure 4.9B.)

The relationships among explicit, implicit, and inverse functions are summarized in Figure 4.10.

Figure 4.10

Explicit, Implicit, and Inverse Functions

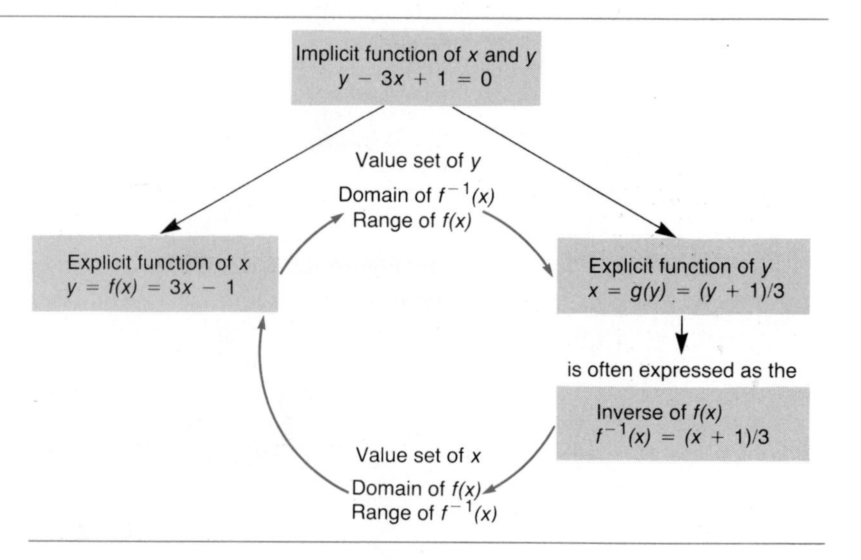

Example 7 Given $y = f(x) = x^3$, find its inverse function $f^{-1}(x)$.

The procedure for finding the inverse of a function $y = f(x)$ involves, basically, solving the equation for x in terms of y and then replacing the symbol y by x and the symbol x by y.

Solving

$$y = x^3$$

for x, we raise each side of the equation to the 1/3 power to determine

$$y^{1/3} = x$$

Then we interchange symbols used for independent and dependent variables to obtain

$$f^{-1}(x) = x^{1/3}$$

The graphs of $f(x) = x^3$ and its inverse $f^{-1}(x) = x^{1/3}$ are shown in Figure 4.11. [ ]

Figure 4.11

The Function $y = f(x) = x^3$ and Its Inverse Function $f^{-1}(x) = x^{1/3}$

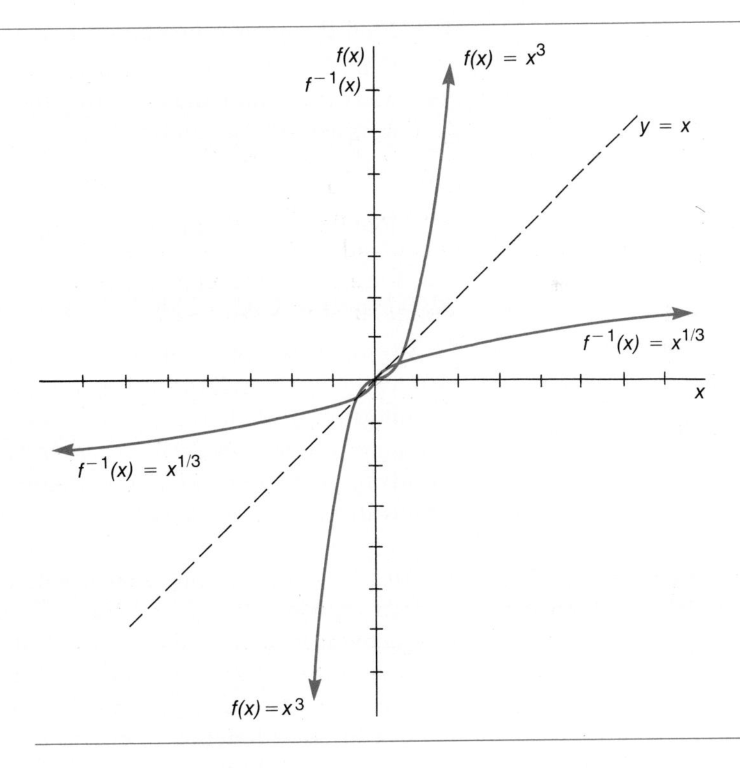

Not all functions have inverses that are themselves functions. Given a function such as

$$y = f(x) = x^2$$

we find that the inverse

$$y = \pm\sqrt{x}$$

is a relation, but not a function. The inverse relation is a function if and only if the original function is a one-to-one function.

EXERCISES

Find the inverse of each of the following functions. Graph f(x) *and its inverse* $f^{-1}(x)$. *Determine whether the inverse itself is a function.*

79. $f(x) = 2x$
80. $f(x) = x + 1$
81. $f(x) = 5 - 2x$
82. $f(x) = 4x - 10$
83. $f(x) = x^2 - 2$
84. $f(x) = 3x^2 + 1$
85. $y = 4/x$
86. $y = 10^x$
87. If C is the number on the Celsius scale for a temperature, then the same temperature on the Fahrenheit scale F can be found by

$$F = (9/5)C + 32$$

Find the inverse function which converts degrees Fahrenheit into degrees Celsius.

88. Market research indicates that the quantity Q of a good demanded at price $\$P$ is given by

$$Q = 125 - 0.25P$$

Find the inverse function and give it an interpretation.

4.9 MATHEMATICAL MANIPULATION OF FUNCTIONS

A primary advantage of the use of mathematical models in describing real-world situations is that the mathematical representation can be easily manipulated to show the effects of changes in the environment while the same type of manipulation cannot always be carried out on the real-world objects or concepts. Let us, therefore, consider some of the computations that can be made with, and on, mathematical functions.

Combining Functions for Model Development

Functions can be subjected to essentially the same algebraic manipulations as can single numbers. That is, functions can be added together, subtracted from one another, multiplied, and divided (provided the divisor is not equal to zero, of course). This property of functions makes it possible for us to begin with simple functions of one variable, combine these with other functions, perhaps of other variables, and thus derive more-complicated functions of perhaps many variables.

Given that f and g are functions of an independent variable x, and k is a constant, it is often meaningful to write

$$kf(x) \qquad (f+g)(x) \qquad (f-g)(x) \qquad (f \cdot g)(x) \qquad (f/g)(x)$$

A somewhat more-complex operation involves COMPOSITE FUNCTIONS such as

$$f(g(x)) \qquad \text{or} \qquad f(g(h(x)))$$

read "f-of-g-of-x" or "f-of-g-of-h-of-x," and states that one function is a function of another function.

Multiplication of a function by a constant A function f may be multiplied by a constant. For instance,

$$f(x) = 2x^2 + 5x + 1$$

may be multiplied by the constant 3 to obtain

$$3f(x) = 3(2x^2 + 5x + 1) = 6x^2 + 15x + 3$$

Addition and subtraction of functions Two functions of the same independent variable, as $f(x)$ and $g(x)$, may be added together to obtain a new function, $h(x) = f(x) + g(x)$, of the same independent variable. Or two functions of different independent variables, as $f(x)$ and $g(y)$, may be added together to obtain a third function, $r(x,y) = f(x) + g(y)$, of two independent variables.

Example 8 The variable x represents the number of units of item X produced. Total direct labor cost, L, is a function of the number of units of the item produced; that is, $L = f(x)$. If the per-unit direct labor cost is a constant $10, the relationship can be symbolized as

$$L = f(x) = 10x$$

Total direct material cost, M, is also a function of the number of units of item X produced. If the per-unit direct material cost is a constant $12, the relationship between total direct material cost and the number of units produced is

$$M = f(x) = 12x$$

Total direct cost of item X can be obtained by summing total direct labor cost and total direct material cost, as

$$L + M = h(x) = f(x) + g(x) = 10x + 12x = 22x$$ []

Example 9 A firm's product line consists of two different items, product X and product Y. Let x represent the number of units of product X sold and $f(x)$ the total revenue from sales of product X. If each unit of product X sells for $10, the relationship is

$$f(x) = 10x$$

Now let y represent the number of units of product Y sold and $g(y)$ the total revenue from sales of product Y. If each unit of Y sells for $12, the relationship is

$$g(y) = 12y$$

Sales of both products would yield a total revenue that is given by

$$r(x,y) = f(x) + g(y) = 10x + 12y \qquad []$$

Multiplication of functions Two, or more, functions may be multiplied or divided.

Example 10

Let us consider a situation wherein we have reference to the maxim

$$\text{Revenue} = \text{Price} \times \text{Quantity}$$

Sam has 100 tickets on the 50-yard line for this year's All-Star game. He can sell all 100 tickets if he sets the price at $10 each. He could sell some of the tickets at a higher price, but he believes that for each $1 increase in asking price, the number of tickets he will be able to sell will decrease by five. All tickets must be sold at the same price.

The price per ticket then can be represented as

$$f(x) = 10 + x$$

where x represents the number of $1 price increases (above the base $10 price).

The number of tickets sold is also a function of the number of $1 price increases and can be symbolized as

$$g(x) = 100 - 5x$$

where x is defined the same as above.

Now, total revenue from the sale of the tickets is a function of both the price per ticket and the number of tickets sold, as

$$h(x) = f(x) \cdot g(x) = (10 + x) \cdot (100 - 5x) = 1000 + 50x - 5x^2 \qquad []$$

Combination of functions by the process of composition Sometimes one function is treated as a function of another function.

Example 11

It is believed that the quantity of product A sold is a function of the amount spent on advertising. The relationship is appropriately expressed as

$$y = g(x) = x + 100$$

where

x = advertising expenditures a week in dollars
y = the number of units of product A sold during that week

Furthermore, the revenue realized on the sale of the product is a function of the number of units sold. The units sell for $5 each so that

$$r = f(y) = 5y$$

where

y = the number of units of product A sold
r = total revenue

Thus, revenue is a function of quantity sold, which is, in turn, a function of advertising expenditures. The relationship between revenue and advertising expenditures can be developed as follows:

$$r = f(y) = f(g(x)) = 5g(x) = 5(x + 100) = 5x + 500$$

It is then possible to write revenue as a function of advertising expenditures, x, alone, as

$$r = h(x) = 5x + 500$$ []

A function that is expressed as a function of another function is known as a COMPOSITE FUNCTION. Note that, in the preceding example, we started with an independent variable that was controllable, advertising expenditures, and depicted its effect upon the dependent variable that was not directly controllable, quantity sold. Quantity sold was, in turn, treated as an independent variable and the association between it and the dependent variable of primary interest, revenue realized, was expressed. The composite function eliminated the intermediate variable, quantity sold, from the relationship modeled and treated revenue as directly dependent upon the controllable variable, advertising expenditures.

Shifting Functions Vertically and Horizontally

Changes in the problem environment may necessitate adjustments in the model describing the functional relationship between two variables. One such adjustment involves shifting the x- and y-axes to some new location such that the new axes are parallel to the original ones. Such a shift has the effect of moving all points that comprise a given function in such a way that their relative positions are maintained. Mathematically, it is a simple matter to shift a function vertically, either upward or downward, on the y-axis, or horizontally, either to the left or to the right, along the x-axis. The computational procedures are outlined below.

Vertical shifts To shift the graph of a function upward b units, replace $f(x)$ by $[f(x) + b]$ wherever $f(x)$ occurs in the function rule. To shift the graph of a function downward b units, replace $f(x)$ by $[f(x) - b]$ wherever $f(x)$ occurs in the function rule. The new rule generated by this procedure defines a new function having the desired location.

Given the function $f(x) = 2x + 4$, we shift the function vertically (on the y-axis) as follows:

To shift upward by 3 units:

$$\begin{aligned} g(x) &= f(x) + 3 \\ &= (2x + 4) + 3 \\ &= 2x + 7 \end{aligned}$$

To shift downward by 3 units:

$$\begin{aligned} h(x) &= f(x) - 3 \\ &= (2x + 4) - 3 \\ &= 2x + 1 \end{aligned}$$

Remember, we wanted to shift the function on the vertical axis, on the $f(x)$-axis, so the adjustment was made to $f(x)$. The graph of the original function and the new functions is shown in Figure 4.12.

Example 12 The monthly manufacturing cost of a product is given by the function

$$f(x) = 10x + 500$$

where

$x =$ the number of units produced
$500 =$ the fixed cost each month in dollars
$10 =$ the unit variable cost in dollars

Figure 4.12

Shifting a Function Vertically

(A) A vertical shift—upward

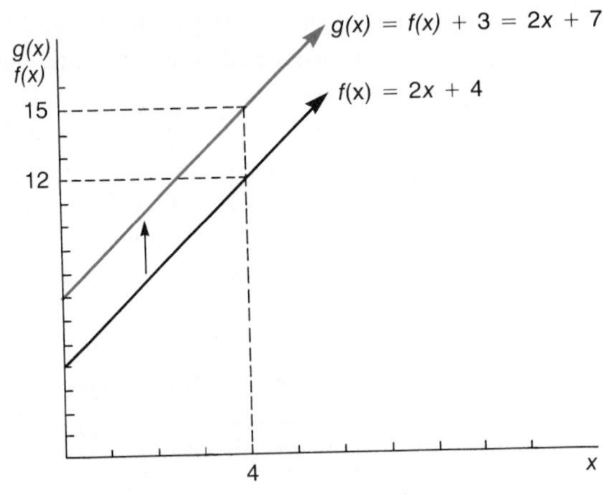

(B) A vertical shift—downward

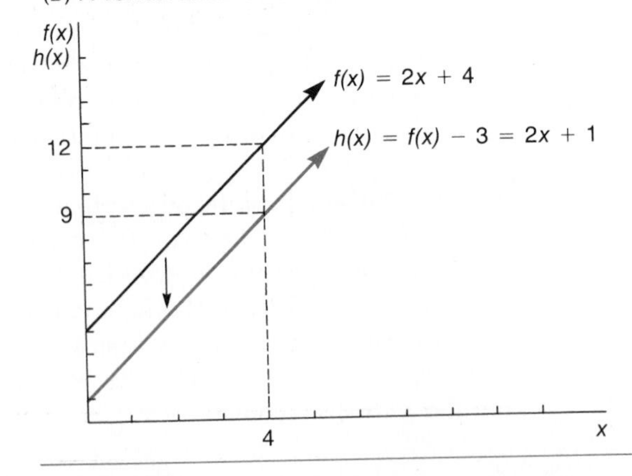

Technological changes are made so that the fixed cost is reduced by \$100 a month. What is the new cost function?

The old cost function shifts vertically, downward, by 100 units on the $f(x)$-axis. Thus, the new cost function is given by

$$\begin{aligned} g(x) &= f(x) - 100 \\ &= (10x + 500) - 100 \\ &= 10x + 400 \end{aligned}$$

[]

Horizontal shifts To shift the graph of a function f by a units to the left, replace x by $(x + a)$ wherever x occurs in the function rule. To shift the graph of a function f by a units to the right, replace x by $(x - a)$ wherever x appears in the function rule.

Given the function $f(x) = 2x + 4$, we shift the function horizontally (on the x-axis) as follows:

Figure 4.13

Shifting a Function Horizontally

(A) A horizontal shift—to the left

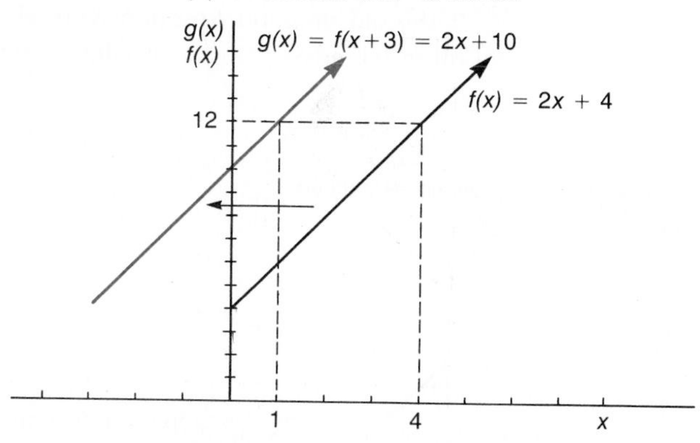

(B) A horizontal shift—to the right

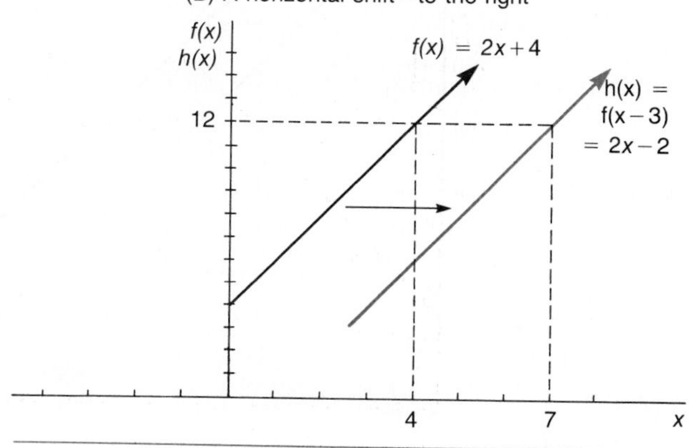

To shift left by 3 units:

$$g(x) = f(x+3) = 2(x+3)+4 = 2x+10$$

To shift right by 3 units:

$$h(x) = f(x-3) = 2(x-3)+4 = 2x-2$$

The graph of the original function and the new functions is shown in Figure 4.13.

Example 13 The DEMAND FUNCTION for a product can be defined as the various quantities of the product purchasers will take off the market at various price levels. Suppose the demand function for a product has been determined to be

$$f(x) = 100 - 10x$$

where

x = the price at which the commodity is offered for sale
$f(x)$ = the quantity demanded in the marketplace

The firm selling the product believes that an intensive advertising campaign would shift the demand function two places to the right. What would the new demand function be?

If the old demand function were shifted two places to the right by an intensive advertising campaign, the new demand function would be

Figure 4.14

A Demand Function Shifts Right because of an Intensive Advertising Campaign

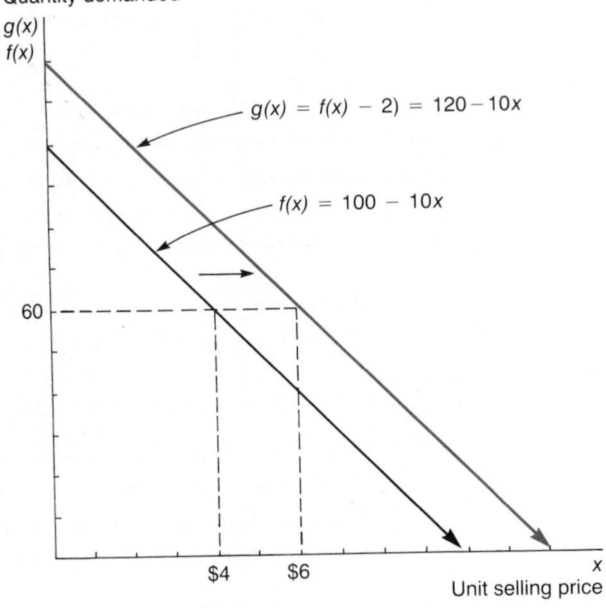

$$\begin{aligned} g(x) &= f(x-2) \\ &= 100 - 10(x-2) \\ &= 120 - 10x \end{aligned}$$

Both the old demand function and the new demand function are pictured in Figure 4.14. Note that with the new demand function the same quantity will be demanded at a unit price of $6 as was previously, before the intensive advertising campaign, demanded at a unit price of $4.

[]

EXERCISES

Given $f(x) = x^2$ *and* $g(x) = x + 2$, *find*

89. $(f+g)(x)$
90. $(f \cdot g)(x)$
91. $(f/g)(x)$

Given $f(x) = x^2 + 2x + 5$ *and* $g(x) = 3x - 7$, *find*

92. $4f(x)$
93. $3f(x) + 2g(x)$
94. $(f+g)(x)$
95. $(f \cdot g)(x)$
96. $(f/g)(x)$
97. $(g/f)(x)$

Given $f(y) = 2y - 3$ *and* $y = g(x) = 4x^2 - 1$, *find*

98. $f(g(x))$
99. $f(g(2))$
100. $f(g(1/3))$
101. $f(g(5))$
102. The number of units x of a product that can be sold at a price of p dollars per unit is given by

$$x = q(p) = 50 - (p/2)$$

Determine a function describing total revenue received from the sale of the product. (Hint: Total revenue equals number of units sold times selling price per unit.)

103. The total cost of manufacturing x units of a product is given by

$$C(x) = x^2 + 100x + 1{,}000$$

Determine the average cost function, where average cost is total cost divided by the number of units produced.

Given $f(x) = 2 - x$, *construct a graph of the function.*

104. Shift the function three points to the right, and construct a graph on the same coordinate system used above.
105. Shift the function two points to the left, and construct a graph on the same coordinate system used above.

Given $f(x) = 0.1x^2$, *construct a graph of the function.*

106. Shift the function upward five points, and construct a graph on the same coordinate system used above.

107. Shift the function downward by four points, and construct a graph on the same coordinate system used above.

108. If the cost of sales is $C(x) = (x/3) + 1$ and the revenue from sales is $R(x) = (x/2) - 4$, determine the profit function. (Hint: Profit = revenue − cost.)

A manufacturing firm has determined that the total number of units of output of a certain product per day is a function of the number of employees working on that day. If Q *represents the number of units of output and* E *represents the number of employees, then*

$$Q = f(E) = 20E - E^2/20$$

Furthermore, total revenue received from selling Q *units of the product is*

$$R = g(Q) = 250Q$$

109. Express output as a function of the number of employees.

110. Express revenue as a function of the number of units of output.

111. Express revenue as a function of the number of employees.

4.10 AN ILLUSTRATION OF MODEL BUILDING AND USE

Many seemingly complex mathematical models are built, step by step, from very simple relationships between variables. These models can then be manipulated so that various possible "situations" can be analyzed. The models can be adjusted to reflect changes that may take place in the decision environment. We hope that the following example will begin to illustrate these procedures.

The BREAK-EVEN MODEL is a tool widely used by management in the analysis of cost-volume-profit relationships. A BREAK-EVEN CHART typically accompanies the mathematical model and pictorially compares the relationships between revenue, cost, and profit, and units sold or units produced. TOTAL COST in most instances is made up of two components, FIXED COST and TOTAL VARIABLE COST. FIXED COST is that component of total cost that does not vary with the number of units of product produced and sold (like rent or insurance on the plant). VARIABLE COST per unit is a cost component which changes directly with the number of units produced and sold (such as labor, material costs, and sales commissions). TOTAL VARIABLE COST would, then, be per-unit variable cost times the number of units produced and sold. PER-UNIT REVENUE is often a constant; that is, it is the NET SELLING PRICE of the item. TOTAL REVENUE is the product of per-unit revenue and the number of units sold. PROFIT is generally defined as total revenue minus total cost.

Recall the production situation faced by B. Lever of High-Tech Manufacturing Company, which was mentioned at the opening of this chapter. High-Tech has just placed on the market an electronic flashlight which it can sell in substantial quantities for $3 a unit. If R represents total revenue and x represents the number of units of the product that are sold, then

$$R = f(x) = 3x$$

For the work center that produces this product, the firm has fixed costs of $5,000 a week. Variable costs are $1.75 per unit at all levels of production within the range from 0 to 5,000 units of output. The function describing fixed cost is

$$F = 5{,}000$$

while the function describing total variable cost is

$$V = 1.75x$$

If C represents total weekly manufacturing costs, both fixed and variable, then

$$C = g(x) = F + V = 5{,}000 + 1.75x \qquad 0 \le x \le 5{,}000$$

The firm's total revenue and total cost functions are depicted in Figure 4.15, which is a typical break-even chart.

One of the first questions that B. Lever has posed concerning the new activity is: At what value of x will the activity break even; that is, how many units of product must be produced and sold in order that the cost of the activity exactly equals revenue generated by the activity? This will be the level of production and sales where cost equals revenue and profit (defined as revenue minus cost) is zero; this level of activity is referred to as the BREAK-EVEN POINT (BEP).

For this activity, the BEP occurs where

$$R - C = P = 0$$

where P represents the profit for the activity. The profit function can be obtained by

$$P = h(x) = f(x) - g(x) = (3x) - (5{,}000 + 1.75x) = 1.25x - 5{,}000$$

The BEP is the level of output where profit equals zero; thus, for High-Tech

$$1.25x - 5{,}000 = 0$$
$$x = 4{,}000 \text{ units } = \text{BEP}$$

If 4,000 units of the product are produced and sold, total costs for the activity will exactly equal revenue generated by the activity, and the

Figure 4.15

Break-Even Chart for High-Tech Manufacturing Company

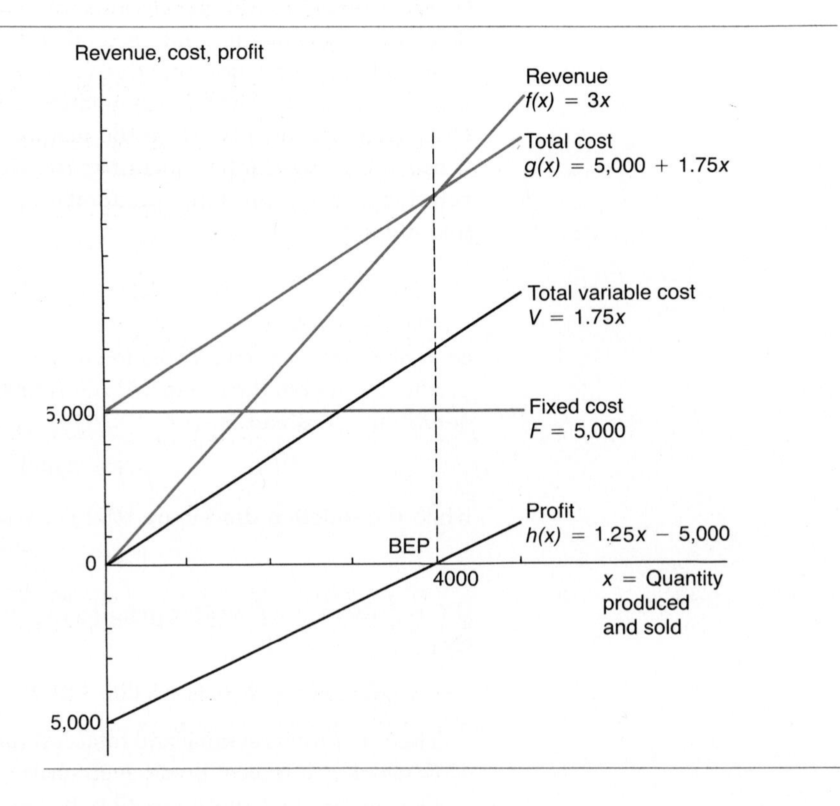

firm will break even. If fewer than 4,000 units are produced and sold, costs will exceed revenue, and the activity will operate at a loss; if more than 4,000 units are produced and sold, revenue will exceed cost, and the activity will operate at a profit.

Other questions typical of those which might be asked by the activity manager are listed below.

1. At what value of x will profit be maximized?

If both the total revenue and total cost functions are linear (that is, are straight lines), as they are here, and if the slope of the revenue function is steeper than that of the cost function (that is, if unit selling price is greater than unit variable cost), the value of x should be set as high as possible. (In general, the maximum or minimum of any linear function will occur on a boundary of the domain of that function.) In this activity, the maximum number of units that can be produced is 5,000. This would be the production level that would maximize profit. At this production level, profits would be

$$h(5{,}000) = 1.25(5{,}000) - 5{,}000 = \$1{,}250$$

2. What would be the profit from 4,500 units?

We may compute

$$h(4{,}500) = 1.25(4{,}500) - 5{,}000 = \$625$$

3. Suppose fixed costs could be reduced by $1,000. What would be the new profit function? Where would BEP occur under these conditions?

If fixed costs could be reduced by $1,000, the total cost function would be shifted downward by 1,000 units, as

$$G(x) = [g(x) - 1{,}000] = (5{,}000 + 1.75x) - 1{,}000 = 4{,}000 + 1.75x$$

The profit function would become

$$H(x) = 3x - (4{,}000 + 1.75x) = 1.25x - 4{,}000$$

and the new BEP would occur when the production and sales level was

$$1.25x - 4{,}000 = 0$$
$$x = 3{,}200 \text{ units}$$

4. Suppose fixed costs were increased by $1,000. What would be the new profit function? Where would BEP occur under these circumstances?

If fixed costs were increased by $1,000, the total cost function would be shifted upward by 1,000 units, as

$$G(x) = [g(x) + 1{,}000] = (5{,}000 + 1.75x) + 1{,}000 = 6{,}000 + 1.75x$$

Then the new profit function would be

$$H(x) = 3x - (6{,}000 + 1.75x) = 1.25x - 6{,}000$$

and BEP would be

$$1.25x - 6{,}000 = 0$$
$$x = 4{,}800 \text{ units}$$

Figure 4.16 depicts the effect changes in fixed cost would have on the break-even level of production.

5. Suppose variable costs are decreased by 10 percent per unit. What would be the new total cost function? What would be the new profit function? Where would BEP occur?

The new unit variable cost is calculated as

$$1.75 - 0.1(1.75) = 1.575$$

Therefore

$$g(x) = 5{,}000 + 1.575x$$
$$h(x) = 3x - (5{,}000 + 1.575x) = 1.425x - 5{,}000$$

Figure 4.16

Effect of Changes in Fixed Cost on Break-Even Point

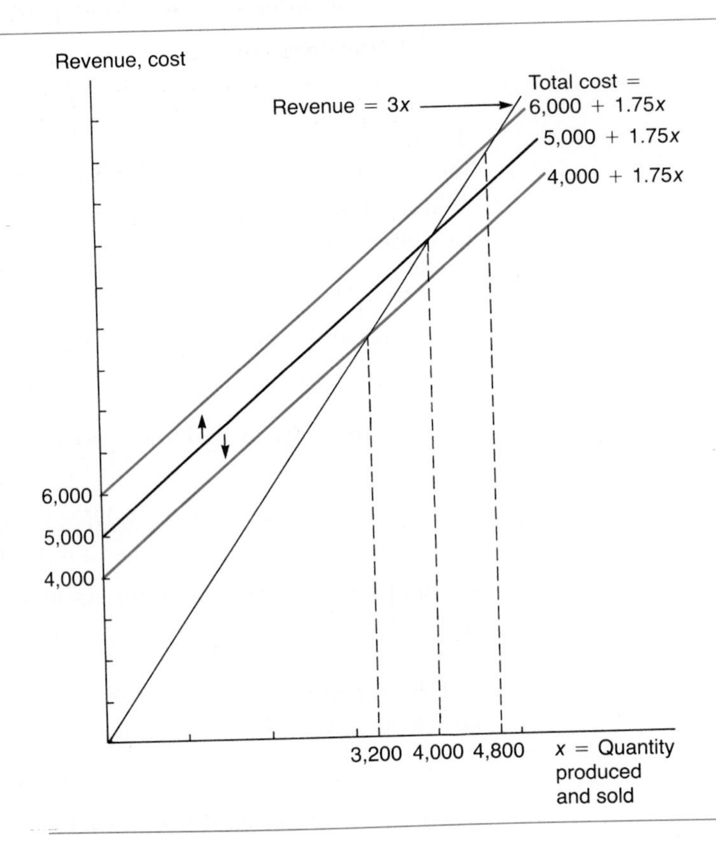

And BEP is calculated as

$$1.425x - 5{,}000 = 0$$
$$x = 3{,}509 \text{ units}$$

6. Suppose variable costs were increased to 150 percent of their previous level. What would be the new variable cost per unit? What would be the new total cost function? What would be the new profit function? Where would BEP occur?

The new unit variable cost is calculated as

$$1.50(1.75) = 2.625$$

Therefore

$$g(x) = 5{,}000 + 2.625x$$
$$h(x) = 3x - (5{,}000 + 2.625x) = 0.375x - 5{,}000$$

And BEP is calculated as

$$0.375x - 5{,}000 = 0$$
$$x = 13{,}333 \text{ units}$$

which is far in excess of the maximum possible production level.

7. If both variable cost per unit and selling price per unit increased by 20 percent, where would BEP occur?

The unit variable cost is calculated as

$$1.75 + 0.2(1.75) = 2.10$$

The unit selling price is calculated as

$$3 + 0.2(3) = 3.60$$

Therefore

$$h(x) = 3.6x - (5{,}000 + 2.1x) = 1.5x - 5{,}000$$

And BEP is calculated as

$$\begin{aligned} 1.5x - 5{,}000 &= 0 \\ x &= 3{,}333 \text{ units} \end{aligned}$$

8. Given the original costs ($5,000 fixed and 1.75 unit variable), where would BEP occur with various possible selling prices?

If unit selling price were $5, the profit function would become

$$h(x) = 5x - (5{,}000 + 1.75x) = 3.25x - 5{,}000$$

and BEP would be calculated as

$$\begin{aligned} 3.25x - 5{,}000 &= 0 \\ x &= 1{,}539 \text{ units} \end{aligned}$$

If unit selling price were $4, the profit function would become

$$h(x) = 4x - (5{,}000 + 1.75x) = 2.25x - 5{,}000$$

and BEP would be calculated as

$$\begin{aligned} 2.25x - 5{,}000 &= 0 \\ x &= 2{,}223 \text{ units} \end{aligned}$$

The break-even chart in Figure 4.17 displays the various break-even volumes with different selling prices for the product.

9. If a full second shift were employed, a maximum of another 5,000 units could be produced. Fixed costs, involving supervisory costs, additional depreciation of equipment, utilities, and so on, would increase by $3,500 and, primarily because of the wage differential paid to second-shift workers, unit variable cost would increase by 25 cents. The total cost function per item would then be defined as

$$g(x) = \begin{cases} 1.75x + 5{,}000 & 0 \le x \le 5{,}000 \\ 2(x - 5{,}000) + 1{,}7250 & 5{,}000 < x \le 10{,}000 \end{cases}$$

There are now two break-even points, at $x \doteq 4{,}000$ units and at $x = 7{,}250$ units as shown in Figure 4.18.

Figure 4.17

Break-Even Point at Various Possible Selling Prices

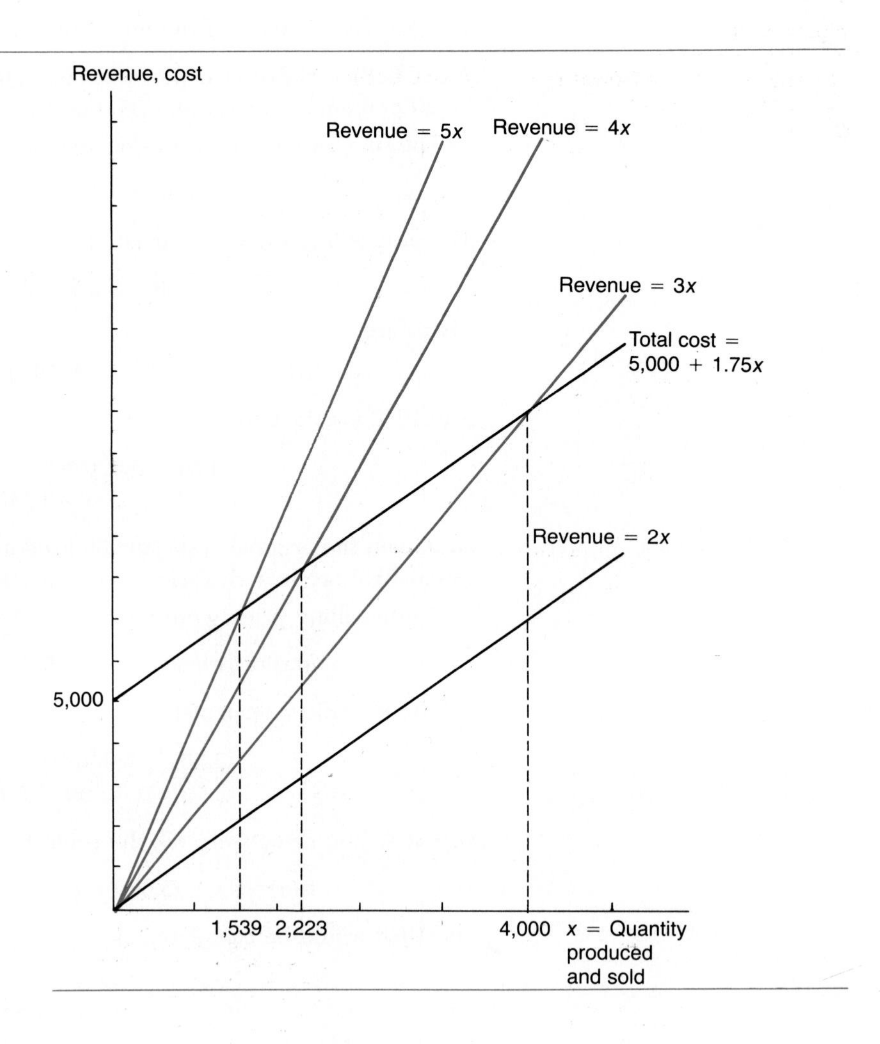

Figure 4.18

"Kinked" Cost Curve Results in Two Break-Even Points

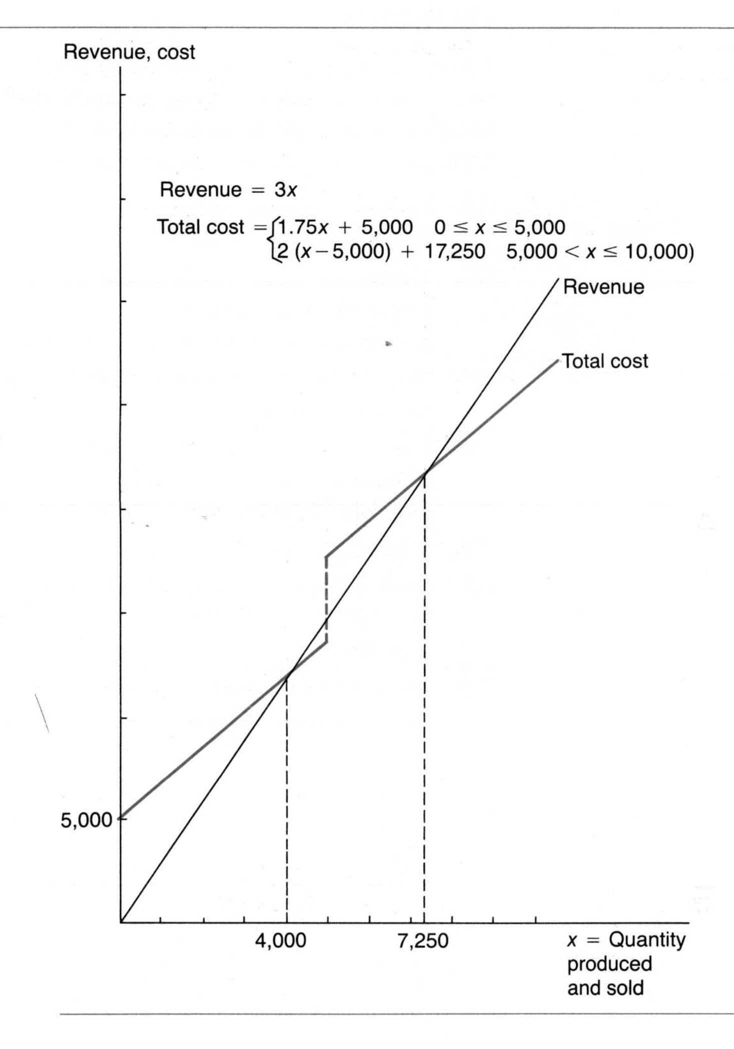

EXERCISES

A firm manufactures and markets a product that sells for $15 per unit. Fixed costs associated with the activity that produces the product total $4,000 a month, while variable cost per unit is $6.50. A maximum of 8,000 units can be produced each month.

112. Derive the total revenue function. What is total revenue when 5,000 units are produced and sold? 6,000 units? 7,000 units?
113. Derive the total cost function where total cost is the sum of fixed cost and total variable cost. What is total cost when 5,000 units are produced? 6,000 units? 7,000 units?
114. Sketch the total revenue function and the total cost function in the same coordinate system.
115. Derive the profit function where profit is defined as total revenue minus total cost. What is total profit when 5,000 units are produced and sold? 6,000 units? 7,000 units?
116. What is the break-even point?
117. Derive the new total cost function given that fixed cost increases by $800 a month. What is the new profit function? Where is the new BEP?
118. Derive the total cost function given that variable cost increases by 20 percent. What is the new profit function? Where is the new BEP?

C·H·A·P·T·E·R 5

Linear Equations, Functions, and Inequalities

A PREVIEW of the types of problems you will be able to work after studying this chapter:

Determine the equation for the straight line going through the points (1,3) and (3,–2).

Fit a straight line through a set of data points using either the freehand method, the method of semiaverages, or the method of least squares.

Sketch a graph of the inequality $3x - 8y \geq -24$.

Sketch a graph of $6x + 9y + 3z = 18$ in three-dimensional space.

Determine the demand function for a product, assuming that the function is linear and given that weekly demand for the product is 100 units when the price is $100 per unit and 500 units when the price is $20.

Determine the equilibrium price and quantity given the demand and the supply functions for a product.

Determine the effect on the supply function of a product when a $5 excise tax is imposed on the manufacturer.

Determine the effect on equilibrium price and quantity when a $5 excise tax is imposed on the manufacturer of a product.

Determine the possible production strategies for a chemical company after it finds that the manufacture of its product produces a sulphur dioxide pollutant. To satisfy federal regulations, this pollutant must be held to a maximum of 600 units a day. Each ton of chemical 1 results in 4 units of sulphur dioxide pollutant and each ton of chemical 2 results in 6 units of pollutant.

The break-even problem of the last chapter was portrayed by a graph consisting entirely of straight lines. Functions which graph as straight lines are called LINEAR. In this chapter we will explain linear functions in general.

A linear equation in n variables is an equation of the form

$$a_1x_1 + a_2x_2 + \ldots + a_nx_n = b \qquad (5.1)$$

where the x_i are the variables, the a_i and b are constants, and where not all $a_i = 0$.

The variables may all be denoted x_i, with the subscript serving to distinguish one variable from another, as $x_1, x_2, \ldots, x_n$; or the variables may each be designated by a different alphabetical character, as x, y, z, and so on.

Each term of a linear equation consists of either a constant or a constant times a variable. The variables appear only to the zero or one power. There are no products of variables within a term, no roots, no logarithms, and no special trigonometric operations such as sines or cosines.

Examples of Linear Equations

$x + 4y = 7$ (Linear equation in two variables)

$x = 6y + z - 1$ (Linear equation in three variables)

$x_1 - 9 = 2x_2 - 3x_3 + x_4$ (Linear equation in four variables)

Examples of Equations That Are Not Linear

$x^2 + 5 = y$ (Variable carried to power higher than one)

$x - \log y = 0$ (Logarithm of the variable)

$5^x = 100$ (Constant carried to a variable power)

$x_1x_2 + 9x_3 = 53$ (Product of two variables within one term)

Linear equations have many important applications. Aside from the fact that they provide exact models for many relationships, they are also frequently used to approximate more-complex kinds of relationships. Because they are easy to handle mathematically, linear functions provide an excellent starting place for a thorough investigation of functional relationships.

5.1 LINEAR FUNCTIONS IN TWO VARIABLES

The STANDARD IMPLICIT FORM for a linear equation in two variables, x_1 and x_2, is

$$a_1x_1 + a_2x_2 = b \tag{5.2}$$

where a_1 and a_2 are not both zero. Or, to avoid the subscript notation, we often write

$$Ax + By = C \tag{5.3}$$

Although they do not always appear in this standard format, all linear functions in two variables, which we shall refer to as SIMPLE LINEAR FUNCTIONS, can be recast in this mold.

Example 1

$y = (5x + 10)/3$ may be rewritten as $3y - 5x = 10$

$6t - 7 = s + 3$ may be rewritten as $6t - s = 10$

$x_1 = x_2 + 5$ may be rewritten as $x_1 - x_2 = 5$ [ ]

Any pair of values for the variables that causes an equation to be a true statement is a SOLUTION to the equation. Consider the linear equation in two variables, $x + 2y = 6$. Obviously, this linear equation is "true" for many different pairs of values for x and y. For example, the equality holds for $x = 0$ and $y = 3$ as well as for $x = 2$ and $y = 2$, or $x = 4$ and $y = 1$. This is generally the case for linear equations; there will be many ordered pairs of values for the variables that satisfy the equality. The set of all such ordered pairs is called the SOLUTION SET for the equation.

EXERCISES

Which of the following equations, if any, are linear equations? For those that are not linear, explain why they are not. Rewrite all linear equations in standard implicit form.

1. $x_2 = 2x_1 + 5$
2. $x + 6y - 12 = 0$
3. $\log(x_1 + 5) = x_2$
4. $7y + 8xy + 12x = 6$
5. $x^2 + x + 5 = 0$
6. $0.5x + 0.8 = y - 0.1$
7. $x - y = 0$
8. $3x = 25 + (1/y)$
9. $(x^3 - 27)/y = 0$
10. $y = 2x + 10$
11. $15^{x-1} = y$
12. $\sqrt{x} - \sqrt{y} = -4$
13. $r = s + 3$
14. $3(x - 5) = 4(y - 1)$
15. $x_1x_2 = 9 + x_3$
16. $(1/2)y - (2/3)x = 5/8$

5.2 GRAPHING A LINEAR FUNCTION

The graph of a linear function in two variables is a STRAIGHT LINE. There is one and only one straight line through any two given points; and all ordered pairs generated by a linear function will fall on the straight line connecting any two points on that line. Thus, if the equation for a simple linear function is given, we can plot its graph by finding any two points which represent solutions and connecting these with a straight line.

Given an equation such as

$$3x - y = -2$$

we may select any permissible value for either of the variables and use this to obtain the other coordinate of one location address. For instance, selecting $x = -1$, we obtain the associated y value, $y = -1$. The location address of this point is $(-1,-1)$. Using any other permissible x value, such as $x = 2$, and solving for the associated y value, we obtain $y = 8$. These two points—$(-1,-1)$ and $(2,8)$—joined by a straight line will completely describe the equation $3x - y = -2$. (See Figure 5.1.)

Using Intercepts to Plot the Line

While any two numbers from the domain of the function may be used to obtain the two ordered pairs needed for plotting the linear function, one easy way to obtain the points is to compute the INTERCEPTS. THE INTERCEPT POINTS of a line representing the graph of a function ARE THE POINTS WHERE THE LINE CROSSES (or touches) THE COORDINATE AXES.

The VERTICAL INTERCEPT, or y-INTERCEPT, is the y-value (if any) where the line representing the graph of the function crosses the y-axis. Its value is determined by setting $x = 0$ (because at the y-axis the value of x is zero) in the equation and solving for y. The HORIZONTAL INTERCEPT, or x-INTERCEPT, of a line is the x-value (if any) where the line crosses the x-axis and can be determined by setting $y = 0$ in the equation and solving for x. Thus, we find the intercepts of a straight line by alternately setting each of the variables equal to zero and solving for the corresponding value of the other variable.

Figure 5.1

Graphing the Equation $3x - y = -2$ by Connecting Two Points on the Graph

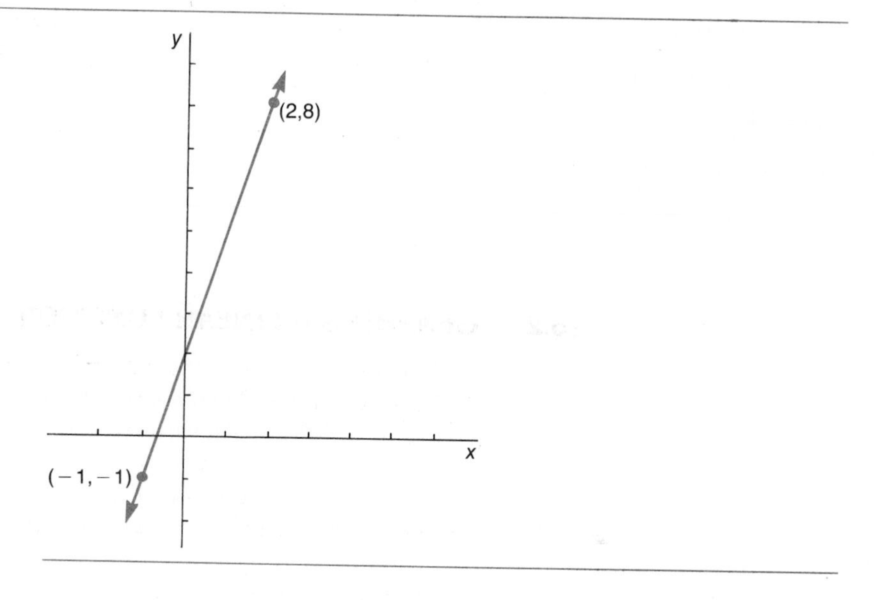

Example 2 Consider the function

$$3x + 2y = 12$$

We compute the intercepts for the graph of the function as follows:

Horizontal intercept ($y = 0$):

$$3x + 2(0) = 12$$
$$x = 4$$

Vertical intercept ($x = 0$):

$$3(0) + 2y = 12$$
$$y = 6$$

The two intercept points are, thus, (4,0) and (0,6). These points have been plotted in the rectangular coordinate system in Figure 5.2 and

connected with a straight line. The graph represents the set of all ordered pairs (x,y) belonging to the solution set for the function $3x + 2y = 12$.
[]

If, when substituting $x = 0$ into an equation to determine the y-intercept (or substituting $y = 0$ to determine the x-intercept), we obtain the point (0,0), the line representing the graph of the function goes through the origin. Then, simply select any other value of a variable and determine a second point.

Figure 5.2

Using Intercepts to Graph the Equation $3x + 2y = 12$

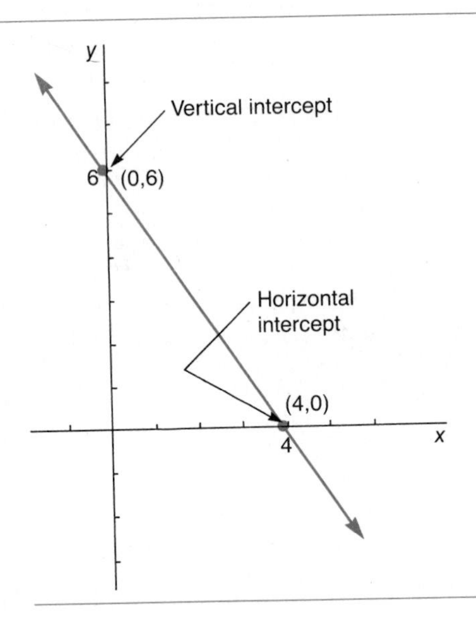

EXERCISES

Sketch a graph of each of the lines described below. Specify the x*-intercept and the* y*-intercept for each.*

17. $4x + 3y - 24 = 0$
18. $x - y + 5 = 0$
19. $5y + 6x - 18 = 0$
20. $2y + 8x = 10$
21. $4x = 10 - 5y$
22. $x_1 = 6 + 3x_2$
23. $3x + y - 12 = 0$
24. $(2/3)x = (5/12) - (3/4)y$
25. $f = \{(x,y) | y = 1 - x\}$
26. $g = \{(m,n) | n = m - 5\}$

5.3 THE SLOPE OF A LINEAR FUNCTION

One important property of a linear relationship in two variables that is highlighted by its graph is the SLOPE of the line. (See Figure 5.3.) The SLOPE of the line is defined as the change taking place along the vertical axis relative to the corresponding change taking place along the horizontal axis, or, THE CHANGE IN THE VALUE OF y RELATIVE TO A ONE-UNIT CHANGE IN THE VALUE OF x.

The SLOPE of the straight line segment joining two points (x_1,y_1) and (x_2,y_2) is given by

$$\text{slope} = \frac{y_2 - y_1}{x_2 - x_1} = \frac{\Delta y}{\Delta x} \tag{5.4}$$

provided

$\Delta x \neq 0$

where the Greek symbol Δ, read "delta," is used to denote the difference between two values.

For any straight line, the slope is constant. The slope between any two points on the line is the same as between any other pair of points on the line and is not dependent upon the value of x. This is, in fact, the basic property of a straight line.

There are several ways to determine the slope of a straight line. One method involves a simple rearrangement of the standard implicit linear

Figure 5.3

The Slope of a Straight Line is $\Delta y/\Delta x$

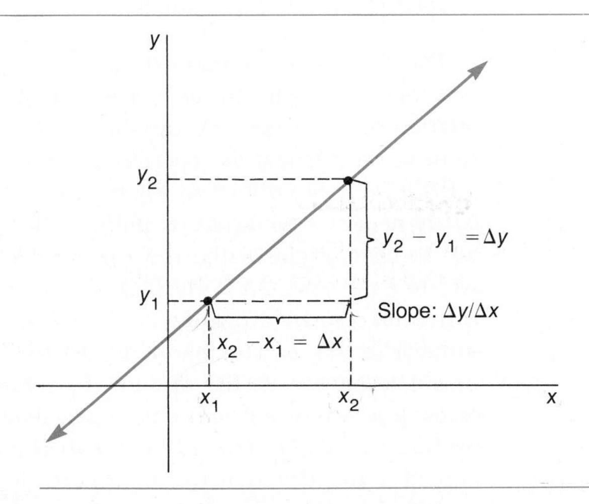

equation. This expression is often rearranged to obtain an equivalent expression but one using functional notation. First, assuming $B \neq 0$,

$$Ax + By = C$$
$$By = -Ax + c$$
$$y = (-A/B)x + (C/B)$$

Then, letting $m = -A/B$ and $b = C/B$, we may write the equation as shown below.

The SLOPE-INTERCEPT FORM of a linear equation is

$$y = mx + b \qquad \text{or} \qquad f(x) = mx + b \qquad (5.5)$$

The coefficient, m, of the x term gives the change in the value of y for each one-unit increase in the value of x; hence, it defines the slope of the line. The constant term, b, is the value of $f(0)$ and is, thus, the vertical intercept. The slope-intercept form is also often seen symbolized $f(x) = ax + b$ or even $f(x) = a + bx$. We mention this, not to confuse our readers, but to caution them that there is no universally recognized "correct" set of symbols and to urge them to be flexible. In the final analysis, it does not matter which letters are used to represent the constants. What is important is that THE COEFFICIENT OF THE INDEPENDENT VARIABLE IS THE SLOPE OF THE LINE AND THE CONSTANT THAT STANDS ALONE IS THE VERTICAL INTERCEPT. That is, the linear equation in slope-intercept form is written

Dependent variable = (Slope × Independent variable) + Intercept

The slope of a line may be positive, negative, or zero. If y increases as x increases, both Δy and Δx will be positive. The line depicting the relationship between the variables will move from lower left to upper right in the rectangular coordinate system. Such lines are said to have a POSITIVE SLOPE. (See Figure 5.4A.) If y decreases as x increases, Δy is negative while Δx is positive. The line will move from upper left to lower right in the rectangular coordinate plane. Such lines are said to have a NEGATIVE SLOPE. (See Figure 5.4B.) If the value of y is not affected by a change in the value of x, the linear equation will simplify to $y = b$. The line depicting this relationship is horizontal and is said to have a ZERO SLOPE. (See Figure 5.4C.) (Indeed, in these cases, y is a constant and not a variable.) If x is a constant unaffected by the value of y, the linear equation reduces to $x = k$. The graph depicting this situation will be a vertical line, and the slope is said to be UNDEFINED. (See Figure 5.4D.)

Figure 5.4

The Slope of a Linear Function May Be (A) Positive, (B) Negative, (C) Zero, or (D) Undefined

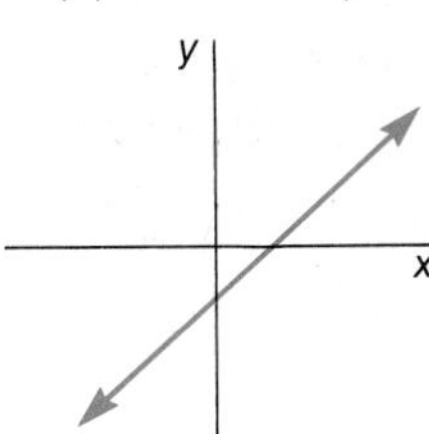

(B) Negative slope

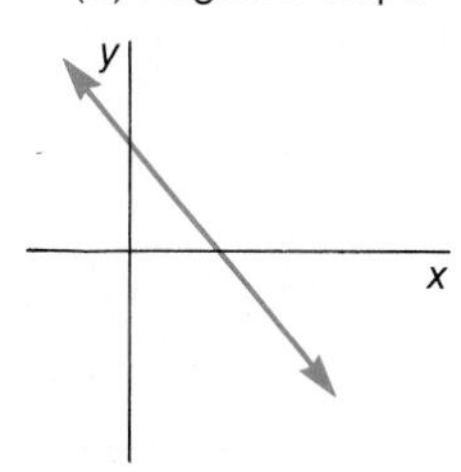

(C) Zero slope

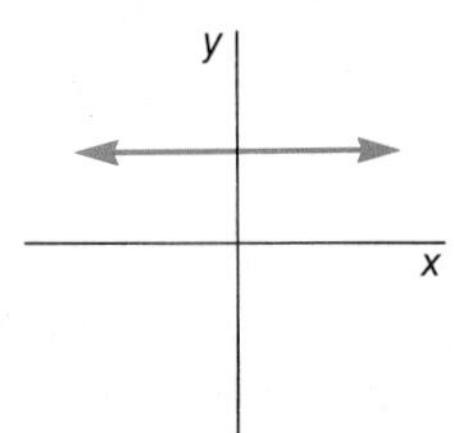

(D) Undefined slope

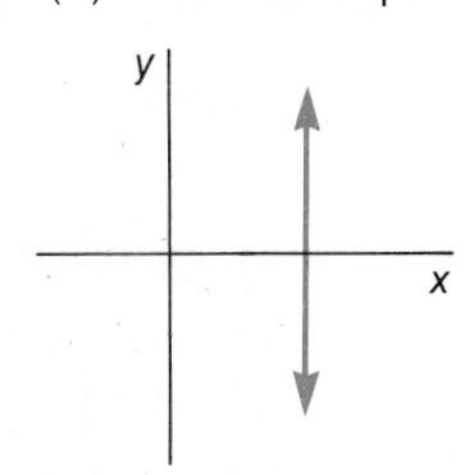

Example 3

The equation

$$3x + 2y = 12$$

may be rewritten in slope-intercept form as

$$y = f(x) = 6 - 1.5x$$

The slope is seen to be -1.5. For each one-unit increase in the value of x, the value of y decreases by 1.5 units. The vertical intercept is $y = 6$. The horizontal intercept is

$$0 = 6 - 1.5x$$
$$x = 6/1.5 = 4$$

The line moves from upper left to lower right in the coordinate plane, as shown in Figure 5.5. []

Figure 5.5

$3x + 2y = 12$

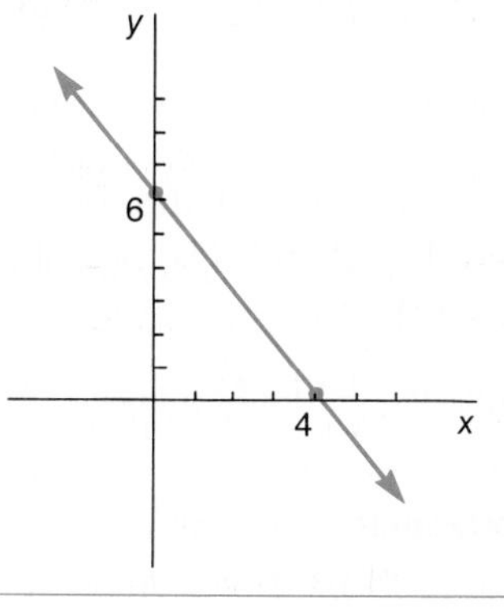

Example 4

Construct a graph of the linear function which passes through the point (1,1) and has slope $+2/3$.

We may begin by plotting a point at the location address (1,1) as shown in Figure 5.6. Then, the slope tells us that y increases by two units while x increases by three units. Thus, we move the equivalent of three places horizontally and to the right, parallel to the x-axis. From there, we move two places vertically and upward, parallel to the y-

Figure 5.6

The Graph of the Linear Function Which Passes through the Point (1,1) and Has Slope +2/3

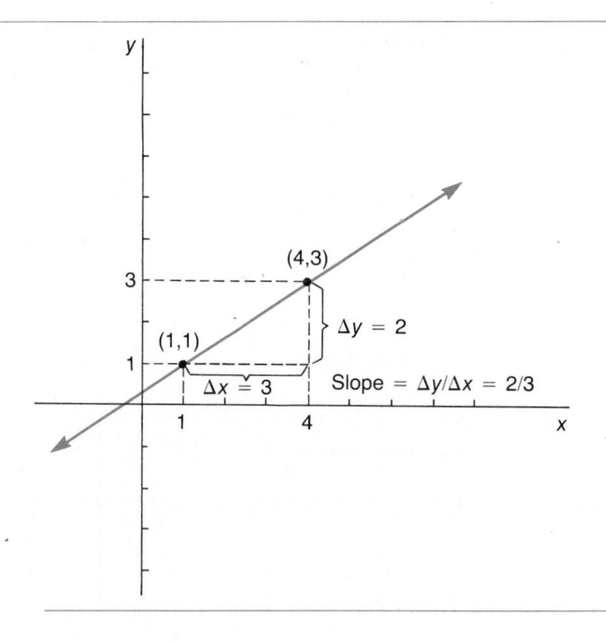

axis. Here we locate a second point on the graph of the function. Connecting these two points with a straight line, we obtain a graph of the linear function which passes through the coordinate (1,1) and has slope +2/3. []

Steepness of the Slope

Compare the slopes of the lines shown in Figure 5.7A and note that the higher the positive value of the slope, the steeper the line. Also, note (in Figure 5.7B) that the more negative the value of the slope, the steeper the line. Thus, THE GREATER THE ABSOLUTE VALUE OF THE SLOPE, THE MORE VERTICAL THE LINE.

In summary, these properties of straight lines are defined by different values for the slope:

1. When SLOPE = 0, the straight line is horizontal.
2. When SLOPE > 0, the straight line has a positive slope where y increases as x increases (that is, the line will slant upward from left to right).
3. When SLOPE < 0, the straight line will have a negative slope where y decreases as x increases (that is, the line will slant downward from left to right).
4. The larger the absolute value of the slope, the steeper, or more vertical, the line.
5. A vertical line, one parallel to the y-axis, has an UNDEFINED SLOPE.

Figure 5.7

Comparison of Steepness of Slope of Linear Functions

(A) Line with higher positive slope is steeper.

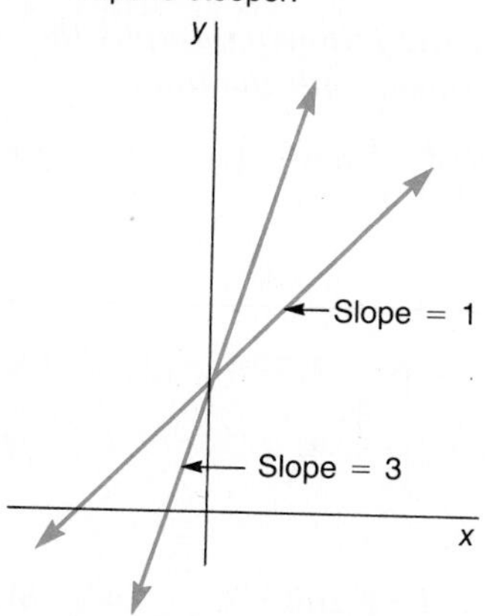

(B) Line with more negative slope is steeper.

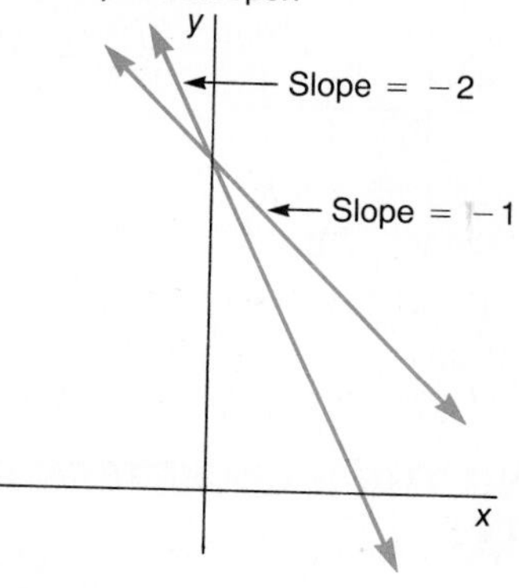

EXERCISES

Rewrite each of the following linear equations in slope-intercept form. Read off the slope and the y-intercept. Determine the x-intercept. Sketch a graph of the function.

27. $4x_1 + x_2 = 8$
28. $2x - (4/3)y - (3/7) = 0$
29. $4y = 12 - 3x$
30. $x = 15 - 3y$
31. $1 - x = y$
32. $6x_1 - 3x_2 - 2 = 0$
33. $3y + 9 = 0$
34. $f = \{(x,y) | 4y = 8 - 3x\}$
35. $3x_1 + 7x_2 = 13$
36. $4x = 20$

Sketch a graph of each of the following linear functions.

37. Graph passes through the point (2,−1) with slope of 3.
38. Graph passes through the point (−3,−2) with slope of −1.
39. Graph passes through the point (2,4) with slope of 0.

40. Graph passes through the point (0,0) with slope of 2.
41. Graph passes through the point (5,0) with undefined slope.

For each of the following graphs the slopes of the two lines are given. Match the slopes with the lines.

42. Slopes of +1 and −1.

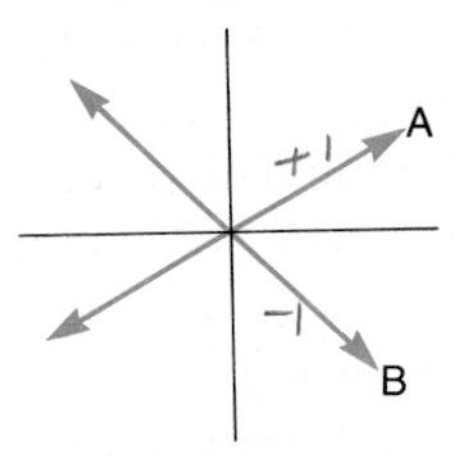

43. Slopes of +2 and +4.

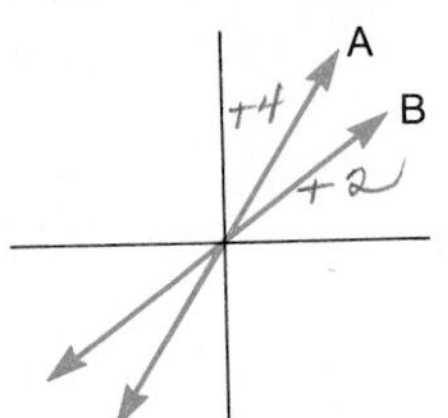

44. Slopes of −2 and −3.

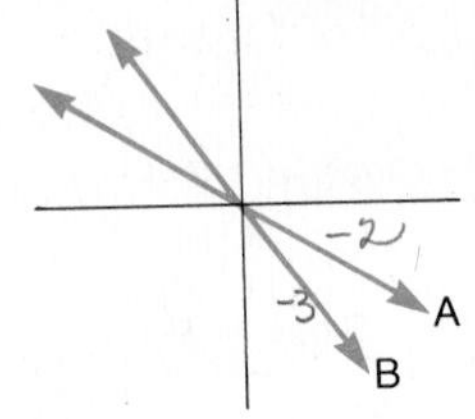

45. Slopes of "zero" and "undefined."

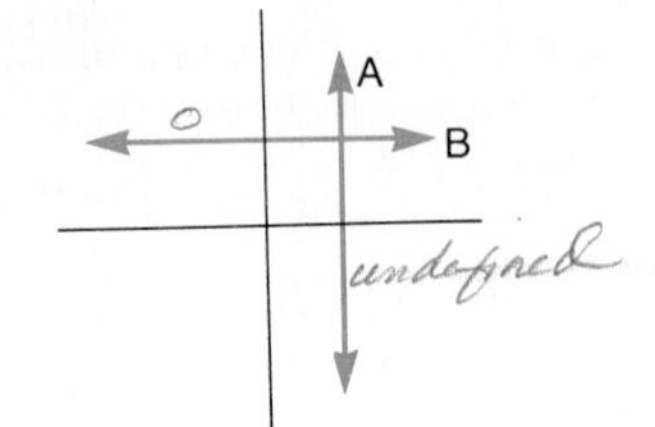

5.4 DERIVING THE PARAMETERS FOR A SPECIFIC LINEAR FUNCTION

The model builder must not only hypothesize the appropriate functional form describing the relationship between variables but must also determine the specific parameters for the function. For example, once we assume that $y = ax + b$ will model a phenomenon, we must then determine appropriate values for a and b. Since we must determine two entities, a and b, we must have at least two pieces of information about the relationship. The following types of situations may occur.

Slope and y-Intercept Given

Often the slope and y-intercept for a specific linear function are obtained directly from the description of the situation we wish to model. When this is the case, we need only to write these values into the slope-intercept equation.

Example 5

A salesman has a fixed base salary of $200 a week. In addition, he receives a sales commission that is 20 percent of his total dollar volume of sales. State in symbolic form the relationship between the salesman's total weekly salary and his sales for the week.

Total salary (let us call this variable y) is dependent upon sales volume, in dollars, (let us call this variable x); that is,

$$\text{salary} = f(\text{sales})$$
$$y = f(x)$$

The information given specifies that the y-intercept is \$200; salary is \$200 when the sales volume is zero. The slope is specified to be +0.2; total salary increases by \$0.20 for each \$1 increase in sales. The linear equation is appropriate because the slope is constant; the 20 percent commission rate is applicable no matter what the volume of sales. Hence, the specific functional relationship in this situation is

$$y = f(x) = 0.2x + 200 \qquad \text{for} \qquad x \geq 0$$ []

Point and y-Intercept Given

The two bits of information about a linear relationship that we are able to extract directly from a description of the situation may be the y-intercept and one point through which the line passes. The value of the y-intercept should be written directly into the linear equation. Then, we can solve for the value of the slope by substituting into the equation the coordinates of the ordered pair given.

Example 6

A salesman has a base salary of \$200 and, in addition, receives a sales commission which is a fixed percentage of his total sales. When his weekly sales are \$1,000, his total salary for the week is \$400. Determine the mathematical relationship between salary and sales.

Let x represent the independent variable, sales volume in dollars, and y, the dependent variable, total salary. The linear functional form is appropriate since the commission rate is constant. The y-intercept is given as \$200. This allows us to write

$$y = f(x) = ax + 200$$

We are also told that when $x = 1{,}000$, $y = 400$. Substituting these values into the equation we have just written, we solve for the value of the slope, as

$$f(1{,}000) = a(1{,}000) + 200 = 400$$
$$a = (400 - 200)/1{,}000 = 0.2$$

The equation modeling the relationship between sales and salary is, thus,

$$y = f(x) = 0.2x + 200 \qquad \text{for} \qquad x \geq 0$$ []

Two Points Given

The two pieces of information available about a linear relationship may be the coordinates of two points through which the line passes. We recall that the slope of a straight line is given by $(y_2 - y_1)/(x_2 - x_1)$. Also, the slope of the line is a constant; it is the same between these two known points as between either one of these points and any other point on the line. Let the general notation (x,y) represent any point whatever on the line and select arbitrarily either one of the known points, (x_1,y_1) or (x_2,y_2). Let us select (x_1,y_1). Then,

$$slope = (y - y_1)/(x - x_1)$$

The two expressions for slope may be set equal to each other, as

$$(y - y_1)/(x - x_1) = (y_2 - y_1)/(x_2 - x_1)$$

and rearranged to obtain a general equation for a straight line through any two given points. This is called the TWO-POINT FORM of the linear equation.

> **Two-point form** of the linear equation:
>
> $$(y - y_1) = \left(\frac{y_2 - y_1}{x_2 - x_1}\right)(x - x_1) \qquad (5.6)$$

Example 7

A salesman has a base salary and, in addition, receives a commission which is a fixed percentage of his sales volume. When his weekly sales are \$1,000, his total salary is \$400. When his weekly sales are \$500, his total salary is \$300. From this information, determine his base salary and his commission percentage and express the relationship between sales and salary in functional notation.

The coordinates of two points on the line that represents the functional relationship between sales and salary are (1,000,400) and (500,300). These two points are substituted into the two-point form of the linear equation, yielding

$$y - 400 = \frac{300 - 400}{500 - 1{,}000}(x - 1{,}000)$$

$$y - 400 = (0.2)(x - 1{,}000)$$

$$y = 0.2x + 200$$

[]

Point and Slope Given

If we rewrite the TWO-POINT FORM of the linear equation by replacing the term $(y_2 - y_1)/(x_2 - x_1)$, which is the slope, by m, a symbol representing the slope, we obtain an expression called the POINT-SLOPE FORM of the equation for a straight line. This expression is conveniently used when the two pieces of information readily available about a linear relationship between two variables are the slope of the line and a point (x_1,y_1) through which the line passes.

> **Point-slope form** of the equation for a straight line:
>
> $$(y - y_1) = m(x - x_1) \qquad (5.7)$$

Example 8 A salesman earns a weekly base salary plus a sales commission of 20 percent of his total sales. When his weekly sales total \$1,000, his total salary for the week is \$400. Derive the formula describing the relationship between total salary and sales.

We begin by noting that the slope of the linear function is +0.2 and that the line goes through the point (1,000,400). Substituting this information in the point-slope formula, we obtain

$$y - 400 = (0.2)(x - 1{,}000)$$
$$y = 0.2x + 200 \qquad []$$

The various ways of deriving a straight-line equation are summarized in Table 5.1.

Table 5.1

Different Forms of the Straight-Line Equation

Form	Format	Used when the information available is:
General implicit form	$Ax + By = C$	Coefficients A and B and Constant C
Slope-intercept form	$y = mx + b$	Slope m and y-intercept b
Point-slope form	$(y - y_1) = m(x - x_1)$	Slope m and point (x_1,y_1)
Two-point form	$(y - y_1 = \left(\dfrac{y_2 - y_1}{x_2 - x_1}\right)(x - x_1)$	Points (x_1,y_1) and (x_2,y_2)

EXERCISES

Find the equation for each of the following linear functions.

46. With slope +1 and going through point (6,–3).
47. With slope –3 and going through point (3,4).
48. Going through the origin and point (1/3,2/3).
49. Going through the points (7,4) and 10,2).
50. Going through the points (1,3) and (3,–2).
51. Going through the point (5,–6) and parallel to the x-axis.
52. Going through the point (1,6) and parallel to the y-axis.
53. Crossing the y-axis at y = 1.25 and going through the point (3.75,–4.5).
54. Crossing the x-axis at x = 2 and having a slope of +4.
55. Crossing the y-axis at y = 3 and the x-axis at x = 4.
56. Passing through the point (3,–2) and having the same slope as $4x + 3y + 1 = 0$.

5.5 FITTING A STRAIGHT LINE TO MANY POINTS

In mathematical model building, we often find that the relationship between variables is approximately—but not precisely—linear. According to the linear function, $y = mx + b$, each unit increase in the value of x is accompanied by a constant change in the value of y, and given the value of x, we can obtain the exact value of y. In the real world, these "perfect" relationships don't always exist. The value of the dependent variable may be affected by many factors other than the value of x. Seldom would it be feasible, or even desirable, to attempt to identify and analyze all the possible explanatory variables. In addition, there is the dilemma of the basic unpredictable randomness of nature. So we settle for a description of the underlying, or approximate, relationship between the variables.

In order to obtain a function that approximates the underlying relationship, many ordered pairs of values for the variables are required. Let us consider the pairs of observations on the two variables *advertising expenditures* (symbolized x) and *number of units of product sold* (symbolized y) given in Table 5.2 and depicted graphically in Figure 5.8.

Table 5.2

Advertising expenditures (x) ($000)	Number of units of product sold (y) (000)
4	7
8	9
6	8
5	8
2	3
3	5
5	6
3	4
36	50

It is evident that it will be impossible to get a smooth line to run through each of the points. But there are a number of procedures we may follow to fit a line to the basic pattern of the points.

Graphic, or Freehand, Method

One of the easiest things to do would be to simply draw, freehand, through the plotted points, the line which, in our best judgment, fairly represents the underlying relationship between the variables. Once the line is placed on the graph, two (x,y) location addresses can be read and used to obtain the parameters for the line using the two-point formula for a straight line.

For example, in Figure 5.9 we have drawn a straight line, freehand, through the data points representing advertising expenditures and number of units of product sold. We read the coordinates of the two starred

Figure 5.8

Ordered Pairs Representing Advertising Expenditures and Sales, Depicted Graphically

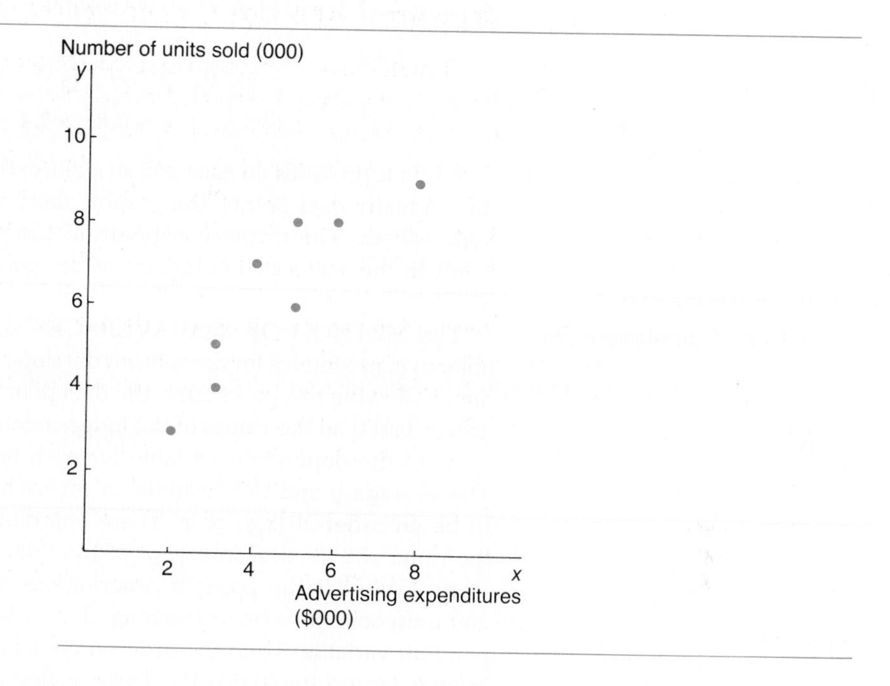

Figure 5.9

A Freehand Line through Data Points

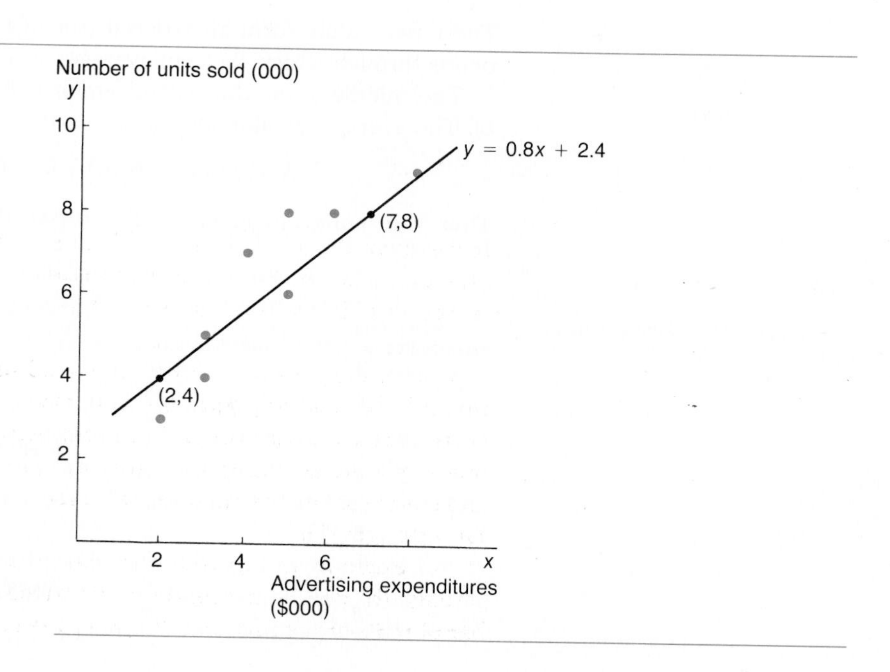

(*) points—(2,4) and (7,8). Then we compute, using the two-point formula,

$$y - 4 = \frac{(8-4)}{(7-2)}(x-2)$$
$$y = 0.8x + 2.4$$

While it provides an easy and straightforward way of fitting a straight line to many data points, the graphic method is not universally held in high esteem. The primary criticism of the procedure is that too much is left to the subjective judgment of the person drawing the line.

Method of Semiaverages

The METHOD OF SEMIAVERAGES also uses a simple, but more objective, procedure for computing the slope and intercept of the straight line. Following this procedure, the data points are divided into two equal groups based on the values of the independent variable. Next the average value of the dependent variable for each of these groups is computed. The average y and the "middle" x for each group are then considered to be an ordered (x,y) pair. These coordinates are used to determine the slope and vertical intercept of the line.

In Table 5.3 the pairs of observations on advertising expenditures and units sold have been rearranged in order of magnitude of the independent variable. Then the ordered pairs have been separated into two groups. Group I contains the four smallest values of x, while Group II contains the four largest values. The middle x value for Group I is $x = 3$. The average y value for this group is computed

$$\text{Average } y = (3+4+5+7)/4 = 4.75$$

These two values form an ordered pair (3,4.75) and give us one of the points through which the semiaverages straight line will pass.

The middle x for the second group is 5.5 (halfway between 5 and 6). The average y value for this group is

$$\text{Average } y = (6+8+8+9)/4 = 7.75$$

Thus, the semiaverages line will also pass through the point (5.5,7.75).

Table 5.3

Computations for Semiaverages Line

Advertising expenditures (x) ($000)	Units of product sold (y) (000)	
2	3	Group I
3	4	Middle x value: 3
3	5	Average y value: (3+4+5+7)/4 = 4.75
4	7	Coordinates of point: (3,4.75)
5	6	Group II
5	8	Middle x value: 5.5
6	8	Average y value: (6+8+8+9)/4 = 7.75
8	9	Coordinates of point: (5.5,7.75)

Using the two-point formula, we find that the calculation for the straight line passing through the points (3,4.75) and 5.5,7.75) is

$$y - 4.75 = \frac{(7.75 - 4.75)}{(5.5 - 3)}(x - 3)$$
$$y = 1.2x + 1.15$$

This semiaverages line has been graphed in Figure 5.10.

Figure 5.10

A Straight-Line Fit by the Semiaverages Method

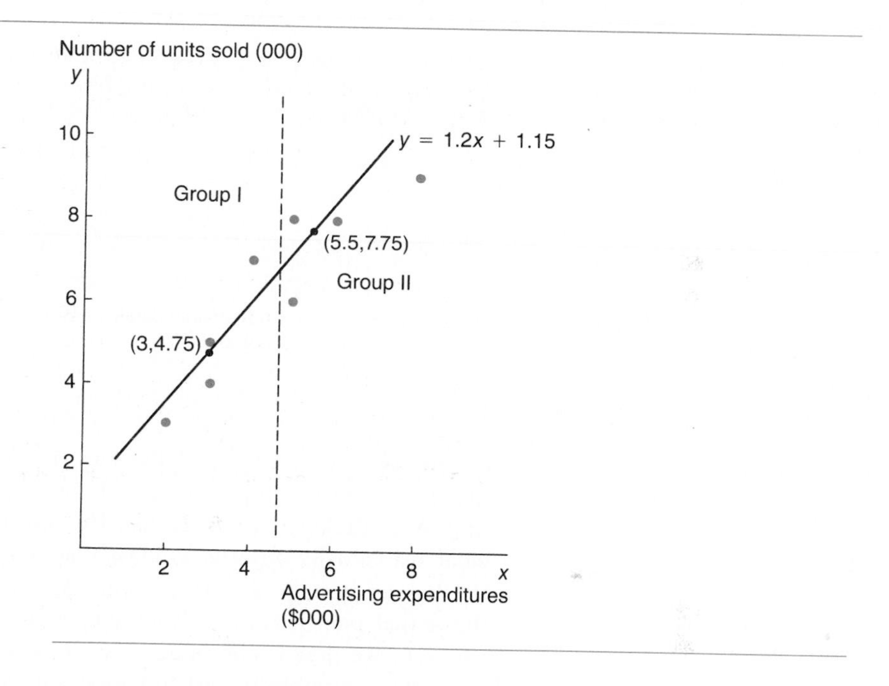

Method of Least Squares

Another method for fitting a line to many data points, and one advocated by statisticians when dealing with statistical inference, is called the METHOD OF LEAST SQUARES. This general procedure can be used to fit not only a straight line but also a curve of any shape to a set of data values on two variables. Indeed, it may be used when there are more than two variables.

The theoretical mathematical properties of this procedure which cause it to be so highly regarded will not be discussed here. It will be sufficient for our purposes to examine the least-squares principle on a strictly intuitive basis. The line drawn in Figure 5.11 is the least-squares linear function relating the set of x and y values given in Table 5.2. Let us label the actual observed values of the dependent variable as y_i and the values of the dependent variable as read from the least-squares line as y_c. Not all y_i coincided with the y_c. But we knew this would

Figure 5.11

A Straight-Line Fit by the Method of Least Squares

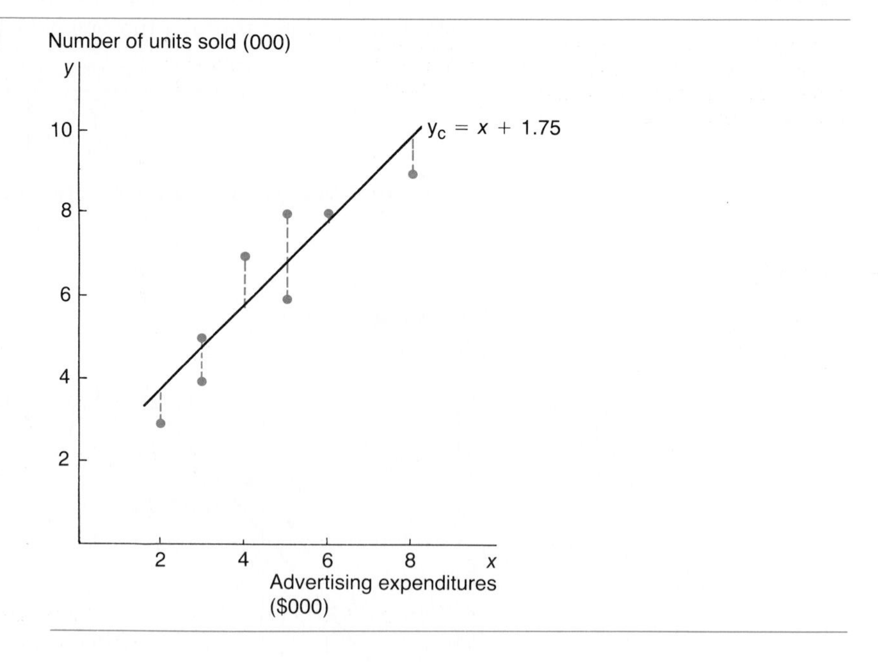

be the case; we recognized the fact that we would derive a function that described the approximate relationship, or the "average" relationship. Nonetheless, we would like the approximation to "approximate" what we actually observe as closely as possible. The criterion defining the line of best fit should, in some sense, minimize the deviations of the actual points (the y_i) from the points on the line (the y_c). If, for every x_i, we denote the differences between the observed value of the dependent variable y_i and the approximated value y_c as $(y_i - y_c)$, the method of least squares will MINIMIZE THE SUM OF THE SQUARED DEVIATIONS. That is, the least-squares criterion is

$$\text{MINIMIZE } \Sigma (y_i - y_c)^2$$

In a later chapter of this text we shall explore the "higher math" needed to move on from here. For the present, we shall simply state that the values for the slope (m) and the vertical intercept (b) for a linear least-squares line are found by solving

$$m = \frac{n\Sigma xy - \Sigma x \Sigma y}{n\Sigma(x^2) - (\Sigma x)^2} \tag{5.8}$$

$$b = \frac{\Sigma y - m\Sigma x}{n} \tag{5.9}$$

where n is the number of sets of ordered (x,y) pairs used in the analysis.

Table 5.4

Computations for Least-Squares Linear Function between Advertising Expenditures (x) and Number of Units Sold (y)

Advertising expenditures ($000) x	Units of product sold (000) y	x^2	xy
4	7	16	28
8	9	64	72
6	8	36	48
5	8	25	40
2	3	4	6
3	5	9	15
5	6	25	30
3	4	9	12
$\Sigma x = 36$	$\Sigma y = 50$	$\Sigma x^2 = 188$	$\Sigma xy = 251$

$$m = \frac{(8)(251) - (36)(50)}{(8)(188) - (36)^2} = 1$$

$$b = \frac{50 - (1)(36)}{8} = 1.75$$

$$y_c = x + 1.75$$

Computations required to determine the least-squares linear function for the relationship between advertising expenditures and number of units sold of the current illustration are given in Table 5.4.

EXERCISES

The following table shows data collected by the fleet manager at Trucks-for-Hire, Inc., relative to number of months a truck has been in use (x) *and the average monthly maintenance and operating costs* (y), *in dollars.*

Age (x)	3	6	9	12	15	18
Cost (y)	215	220	255	278	309	325

57. Use the semiaverages method to fit a straight line to these data which depicts the average underlying relationship between the variables.
58. Use the method of least squares to fit a straight line to these data which depicts the average underlying relationship between the variables.
59. Plot the ordered pairs in a rectangular coordinate system. Draw, freehand, a straight line which in your best judgment depicts the average underlying relationship between the variables. Use the coordinates of two points through which your line passes to determine the parameters of the line.

The following table shows data collected by the sales manager for Bradshaw Company relative to scores (x) *made by newly employed salespersons on an aptitude test and the first-year sales* (y) *in thousands of dollars for these salespersons.*

Score (x)	25	20	15	16	26	14	22	18
Sales (y)	300	200	140	150	325	150	275	190

60. Fit a straight line to these data which depicts the average underlying relationship between the variables using the semiaverages method.
61. Fit a straight line to these data which depicts the average underlying relationship between the variables using the method of least squares.
62. Use the model obtained in Exercise 60 to predict the average sales for salespersons who made a score of 20 on the aptitude test.
63. Use the model obtained in Exercise 61 to predict the average sales for salespersons who made a score of 18 on the aptitude test.

5.6 A MARKET EQUILIBRIUM MODEL

Economic theory of value deals extensively with the relationships between the selling price of a product and the quantities of product demanded and supplied in the marketplace. In essence, the demand for a commodity depends upon the price a buyer has to pay, while the available supply of a good depends upon the price the seller can receive. Of course, in any real-world marketplace, there are many interacting factors, including the number of buyers and the number of sellers and the intensity of competition, the exact nature of the product, taxes and subsidies and other governmental regulations. We shall defer a discussion of the intricate details of the interplay of these and other economic forces in the marketplace to a course in economic theory.

Economists have developed several simplifed models to describe market situations. One such model depicts the market as being characterized by PURE, or PERFECT, COMPETITION, meaning these conditions prevail: (1) the products available are homogeneous, so that buyers have no valid reason to prefer the output of one supplier over that of another supplier; (2) the number of buyers and sellers is so large that the quantity demanded or supplied by a single individual is not large enough to have a significant impact upon the market price; (3) competition takes place freely in a market devoid of artificial barriers; and (4) all economic units have complete knowledge of prices asked, prices purchasers are willing to pay, quantities demanded, and quantities supplied.

Under these assumptions, there will exist at any given point in time a unique schedule of demand for the commodity as well as a unique schedule of supply. These can be represented as mathematical functions.

The Demand Function

The DEMAND FUNCTION for a commodity describes the relationship between the different prices at which the commodity might be offered for sale and the number of units that could be sold at these various prices. In most economic literature, price is treated as the dependent variable (and shown on the vertical axis) while quantity is treated as the independent variable (and shown on the horizontal axis). Demand functions are typically curvilinear functions; however, it may not be too unreasonable in some special cases to assume that the function is linear or approximately linear. We will make that assumption of linearity in this section and consider nonlinear demand functions in later chapters of the text. In any case, because price and quantity demanded are inversely related, the demand function has a negative slope. (There are exceptions, of course, such as luxury or prestige items that become more desirable as their price increases. Or, in rare cases the slope of the demand function may be zero—indicating a constant price regardless of quantity demanded; or the slope may be undefined—indicating a constant demand regardless of price. But we will consider the general case.)

Figure 5.12

A Demand Function: The Mathematical Relationship between Selling Price and Quantity Demanded

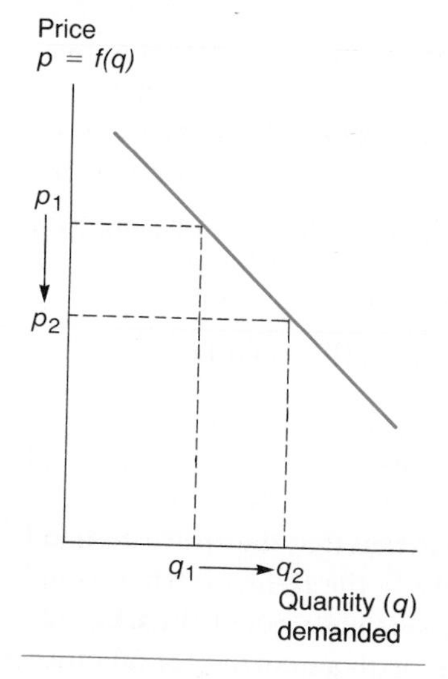

Note that, on the demand function in Figure 5.12, for each level of quantity demanded, there is one and only one price, and for each price there is one and only one level of demand. Economists make a clear distinction between *movement along a specific demand curve* and *a shift in the demand curve for a product.* A change in some underlying market factor can result in a shift of the entire demand curve (see Figure 5.13). If the total demand function shifts to the right (upward), the quantity demanded *at some specific price* will increase. Contrarily, if the total demand function shifts to the left (downward), a lesser quantity will be demanded at the specified price. The shift may be brought about by such forces as an intensive advertising campaign with elaborate promotional gimmicks, by a decrease in consumer esteem for a product because of its consistently low quality, or by changes in the level of consumer income.

Figure 5.13

A Shift in the Demand Function

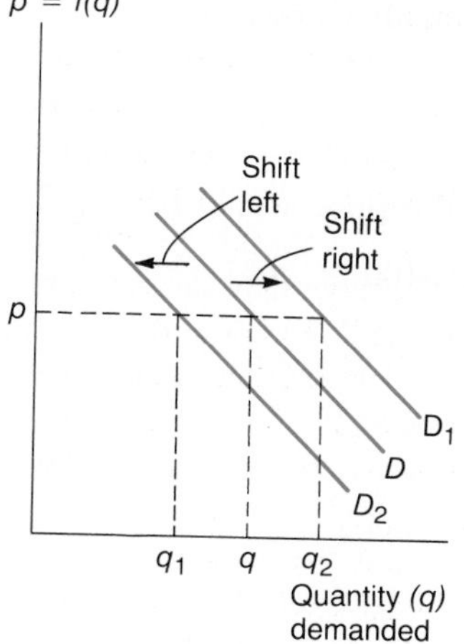

Example 9

Demand for a certain product is linear and is such that 15 units are sold at a selling price of $100 each but only 10 units are sold when the price is $125. (*a*) What is the demand function? (*b*) What would be the price in order to sell 20 units? (*c*) What quantity could be sold at $110?

Figure 5.14

Demand Function: $p = f(q) = 175 - 5q$

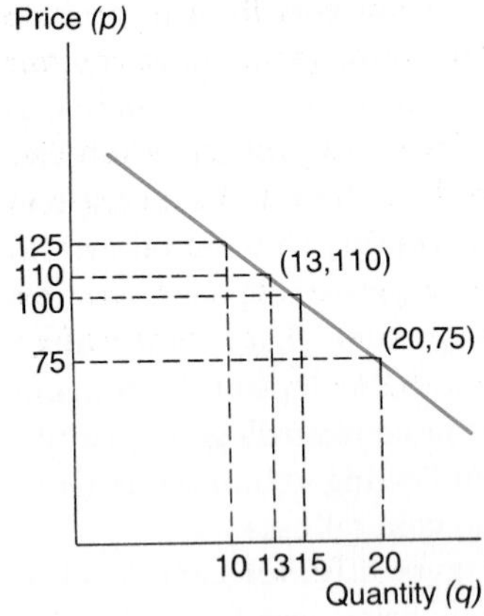

Using the two-point formula and the information (15,100) and (10,125), we compute

$$p - 100 = \frac{(125 - 100)}{(10 - 15)}(q - 15)$$

$$p - 100 = -5(q - 15)$$

$$p = 175 - 5q \qquad \text{or} \qquad f(q) = 175 - 5q$$

(See Figure 5.14.)

The price at which 20 units of the product could be sold would be

$$f(20) = 175 - 5(20) = \$75$$

The number of units that could be sold at $110 would be

$$100 = 175 - 5q$$

$$q = 65/5 = 13 \text{ units}$$

[]

Example 10

If market conditions changed to such an extent that the entire demand function for this commodity shifted five units to the right, (*a*) what would be the new demand function? (*b*) How many units could then be sold at $110? (*c*) What price would be associated with a demand for 20 units?

The new demand function would be

$$h(q) = f(q - 5) = 175 - 5(q - 5)$$

$$= 200 - 5q$$

(See Figure 5.15)

Figure 5.15

Demand Function Shifts to the Right: Quantity Demanded at Price *p* Increases

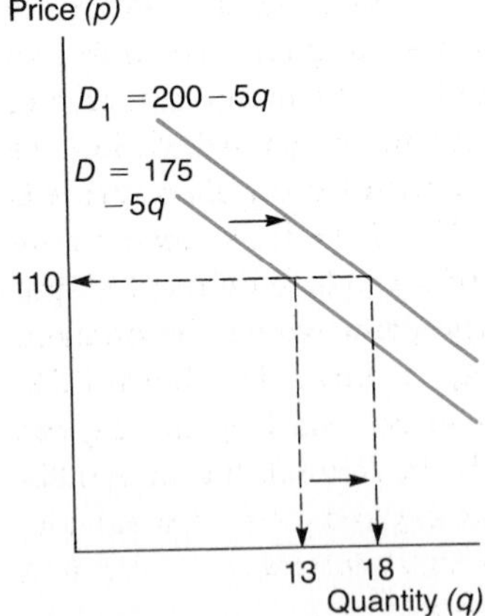

If the price were $110, the quantity demanded would be

$$110 = 200 - 5q$$

$$q = 18 \text{ units}$$

In order for 20 units to be demanded, the price would have to be

$$h(20) = 200 - 5(20) = \$100$$

(See Figure 5.16.)

[]

The Supply Function

Figure 5.16

Demand Function Shifts to the Right: Price at Which Quantity q Can Be Sold Increases

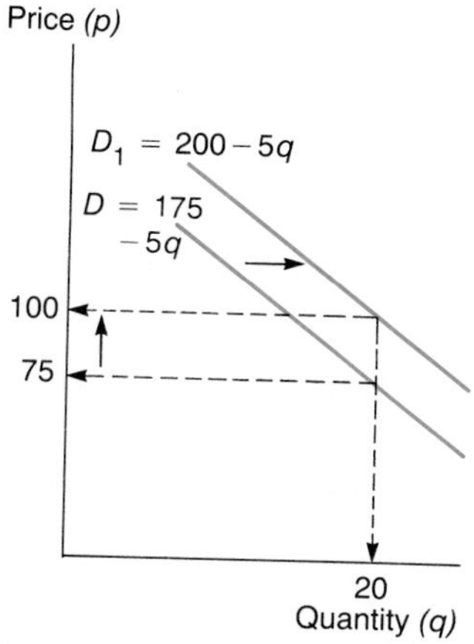

The SUPPLY FUNCTION for a commodity describes the relationship between the prices a supplier is able to secure for its goods and the number of units of product it is willing to offer for sale at those various prices. As before, quantity is conventionally treated as the independent variable and price as the dependent variable. Thus, the typical supply curve has a positive slope; as suppliers are typically willing to provide larger and larger quantities of the commodity as the price moves higher and higher. (See Figure 5.17.)

It is only when the entire supply curve shifts to the right that a supplier is willing to furnish a larger quantity of product without an increase in the price it receives. Shifts in the supply function may be associated with such things as changes in the costs or methods of production, changes in quantities of resources available for production of the commodity, or changes in governmental restrictions and regulations.

Market Equilibrium

Figure 5.17

A Supply Function: The Mathematical Relationship between Selling Price and the Quantity Supplied

Neither the demand function nor the supply function taken alone determines what price will actually prevail in the marketplace. Both must be considered simultaneously. (See Figure 5.18.) At price level p_1 suppliers are willing to supply q_{s_1} units of the commodity, but buyers are willing to purchase only q_{d_1}, resulting in a surplus of $q_{s_1} - q_{d_1}$ units on the market. Price p_1 cannot prevail for long because, as suppliers perceive the accumulation of the surplus, they tend to cut prices under the assumption that with lower prices they can sell the excess goods. Prices drop to level p_2, the price at which consumers are willing to purchase q_{d_2} units. But now at the lower price, suppliers are willing to supply only q_{s_2} units. Demand exceeds supply, and competition among the buyers for the product tends to drive the price upward. As long as the quantity demanded is not equal to the quantity supplied, price is unstable. The price stabilizes, and MARKET EQUILIBRIUM occurs, only at that price which equates the quantity of a product demanded and the quantity supplied. Graphically, this is the point where the demand function and the supply function intersect, as shown in Figure 5.19. Thus, the equilibrium price and quantity can be found by solving the supply and demand functions simultaneously. In general, for an equilibrium to be meaningful, both price and quantity must be nonnegative; hence, the demand and supply functions must intersect in the first quadrant of the rectangular coordinate system.

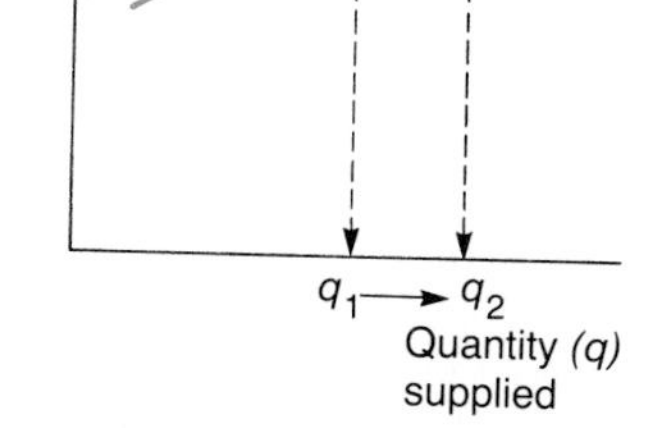

Figure 5.18

Market Equilibrium Is Achieved Only at a Price Such That Quantity Demanded Equals Quantity Supplied

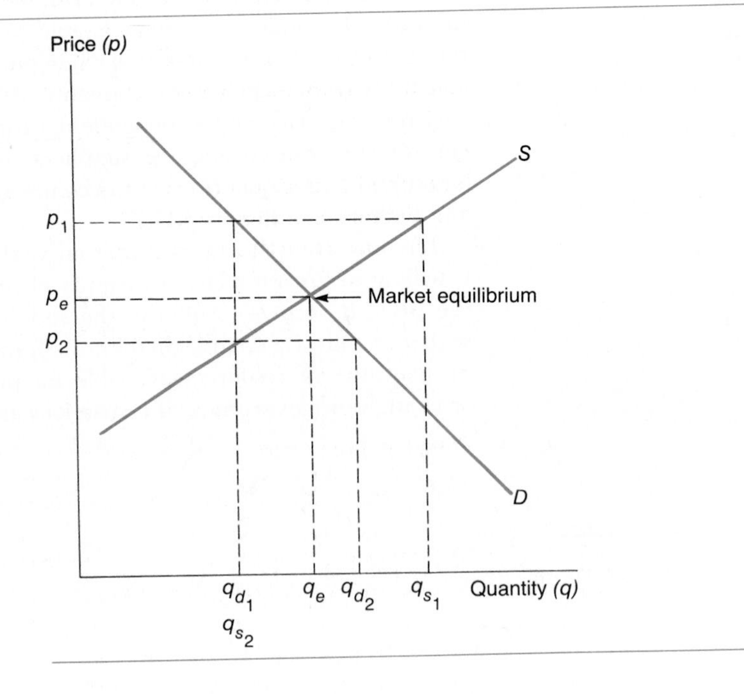

Figure 5.19

Market Equilibrium

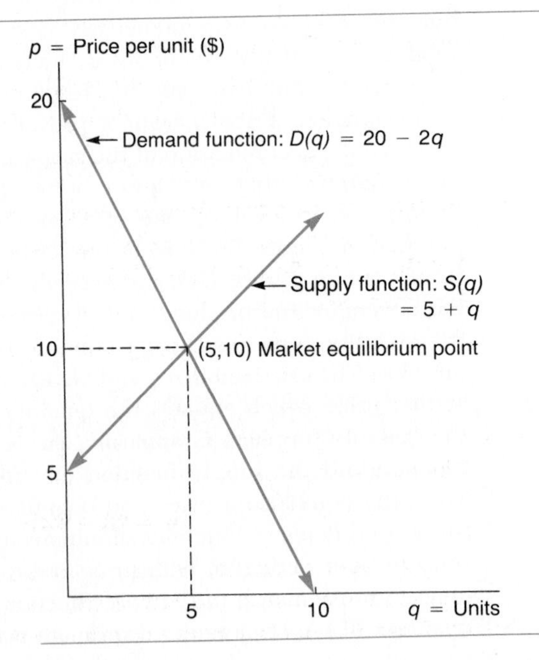

Example 11

The demand function for a certain commodity is given by

$$p = D(q) = 20 - 2q$$

and the supply function by

$$p = S(q) = 5 + q$$

where p is the price per unit for the commodity and q is the number of units demanded, or supplied.

We can determine the market-equilibrium quantity by computing

$$\begin{aligned} \text{quantity demanded} &= \text{quantity supplied} \\ 20 - 2q &= 5 + q \\ q &= 5 \text{ units} \end{aligned}$$

Then, market-equilibrium price is found to be

$$p = 20 - 2(5) = \$10 \text{ per unit}$$

[]

Effects of Changes in the Marketplace

Changes may occur in the marketplace which result in a shift in the demand and/or supply functions for a product.

Example 12

Now suppose that an excise tax of \$3 is imposed on each unit of the product whose demand and supply functions were given in the preceding example. The effect of the tax is to reduce the profit the seller realizes on each unit of the product. The new selling price (after the excise tax) will have to be \$3 more than the old selling price (before the excise tax) in order to be equally attractive to the seller. After the excise tax, the seller will be willing to supply the same quantity q units only at a price p + \$3, where p is the price before the excise tax. All this means that the supply function will shift upward as follows:

$$p_t = (5 + q) + 3 = 8 + q$$

where p_t is the price after the excise tax is levied. (See Figure 5.20.)

The demand function for the commodity should remain unchanged, but the point of market equilibrium will shift. Now, equilibrium quantity becomes

$$\begin{aligned} \text{quantity demanded} &= \text{quantity supplied} \\ 20 - 2q &= 8 + q \\ q &= 4 \text{ units} \end{aligned}$$

The equilibrium price after the excise tax is

$$p_t = 20 - 2(4) = \$12$$

The result of the \$3 excise tax imposed on the seller is, in essence, that consumers must pay \$2 of the tax and the sellers pay the other \$1. Quantity sold in the marketplace drops from five units to four units.

[]

Figure 5.20

An Excise Tax Causes Shift in Supply Function and New Market Equilibrium Point

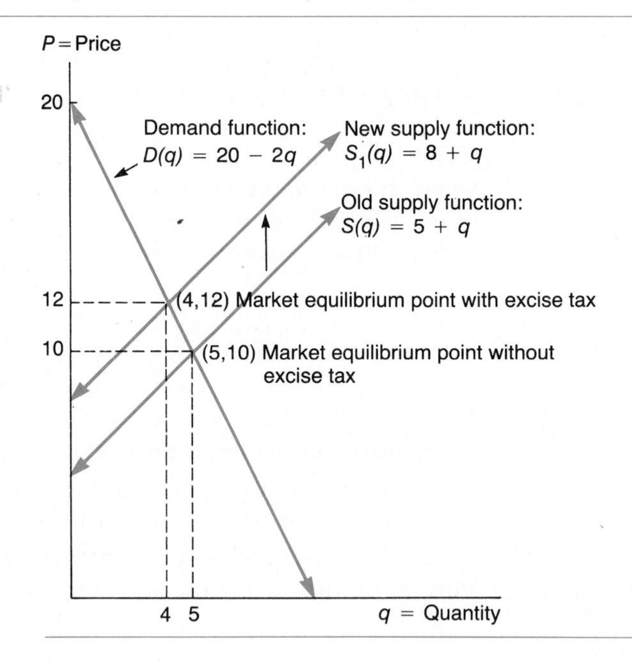

EXERCISES

Determine which of the following are demand equations and which are supply equations. Sketch a graph of each equation.

64. $Q + 5P = 15$
65. $P = 5Q + 11$
66. $2P - 3Q = 12$
67. $P = 16 - 2Q$

Find the appropriate linear demand or supply function in each of the following cases.

68. When the price is \$5 per unit, 100 units can be sold but when the price is \$7, only 50 units can be sold.
69. When the price is \$30, no units can be sold but for each \$2 decrease in price there will be a demand for 40 additional units.
70. When the market price is \$4, producers will not be willing to furnish any units of the product, but for each \$1 increase in price, producers will be willing to supply an additional 12 units.
71. When the selling price is \$25, suppliers will furnish 150 units but at a price of \$40, sellers will be willing to furnish 300 units.

For each pair of demand and supply functions, sketch the appropriate graph. Find the equilibrium price and quantity.

72. D: $5p + 3q = 60$
S: $3p - q = 12$

73. D: $p = 45 - 10q$
S: $p = 5q$

74. D: $p = 10 - 0.5q$
S: $p = 7.5 + 2q$

75. D: $p = 120 - 4q$
S: $p = 8q$

In a certain marketplace the demand and supply functions for a commodity are as follows:

$$\text{D: } p = 100 - 5q$$
$$\text{S: } p = 20 + 4q$$

76. What are the initial equilibrium price and quantity?
77. Assume an imaginative advertising campaign shifts the demand function two places to the right. Sketch the initial demand and supply functions. Sketch the new demand function. Find the new equilibrium price and quantity.
78. Assume that a tax of \$1 per unit is levied on the seller. What will be the effect on the supply function? Depict the situation graphically. Determine the new equilibrium price and quantity.

A manufacturer has the supply function

$$\text{S: } p = 0.6q + 40$$

The weekly demand for the product of this manufacturer is 100 units when the price is \$100 per unit and 500 units when the price is \$20.

79. Determine the demand function, assuming that it is linear.
80. Sketch the demand and the supply functions. Determine the equilibrium price and quantity.

A tax of \$5 per unit is to be imposed on the manufacturer.

81. How will the original supply function be affected? Depict the situation graphically.
82. What will be the new equilibrium price and quantity, assuming demand remains unchanged?
83. How will total revenue for the manufacturer be affected by the tax?

5.7 LINEAR INEQUALITIES AND THEIR GRAPHS

LINEAR INEQUALITIES in two variables describe the conditions

$$Ax + By > C \qquad Ax + By < C$$
$$Ax + By \geq C \qquad Ax + By \leq C$$

See notes for 9-30-83

Such inequalities are frequently encountered in mathematical models and, in particular, in the linear programming model which we shall study in Chapters 7 and 8.

The graph of a linear inequality in two variables is the collection of all points, and only those points, whose coordinates satisfy the inequality. The procedure for graphing a linear inequality is as follows:

1. Begin by graphing the equality $Ax + By = C$. Represent this equality as a *broken line* if the equality is not included in the original inequality (as with the conditions $<$ or $>$). If the equality drawn is included in the original inequality (as with the conditions $\leq$ or $\geq$), represent it as a *solid line.*

   This line divides the rectangular coordinate plane into two HALF-PLANES. Points above the line form the UPPER HALF-PLANE; points below the line form the LOWER HALF-PLANE. (Should the equality plot as a vertical line, we would speak of the LEFT HALF-PLANE and the RIGHT HALF-PLANE.)
2. Select a TEST POINT somewhere in the plane *but not on the line representing the equality.* Substitute the coordinates of the test point into the inequality.
3. If the inequality is satisfied by that test point, the half-plane containing the test point is included in the graph of the original inequality. If the inequality is not satisfied by that test point, the opposite half-plane is included in the graph of the original inequality. The half-plane included as part of the graph should be shaded in some way.

Example 13 Graph the inequality $y - x \geq 1$.

We may graph the equality $y - x = 1$ by plotting its intercepts. These are (0,1) and (−1,0), as shown in Figure 5.21. This is depicted by a solid line. All ordered pairs that satisfy $y - x = 1$ are represented by points on this line.

All ordered pairs that satisfy $y - x > 1$ lie in one of the half-planes formed by the line that has been drawn. To determine which half-plane represents these ordered pairs, we select one point—not on the line itself—and determine whether or not its coordinates satisfy the inequality. The origin (0,0) will be easy to use. We evaluate $y - x > 1$ at (0,0) and find

$$0 - 0 \ngtr 1$$
$$0 \ngtr 1$$

The original inequality is not satisfied by points in the lower half-plane. This means the upper-half plane should be shaded to indicate that it forms a part of the solution set.

(We may verify this solution by selecting any point in the upper half-plane—say (−2,2)—and evaluating $y - x > 1$ at this point. We find

$$2-(-2)>1$$
$$4>1$$

All ordered pairs that satisfy $y - x > 1$ are in the upper half-plane.)

All ordered pairs satisfying the inequality $y - x \geq 1$ are on and above the solid line drawn in Figure 5.21. []

Figure 5.21

Graph of the Inequality $y - x \geq 1$

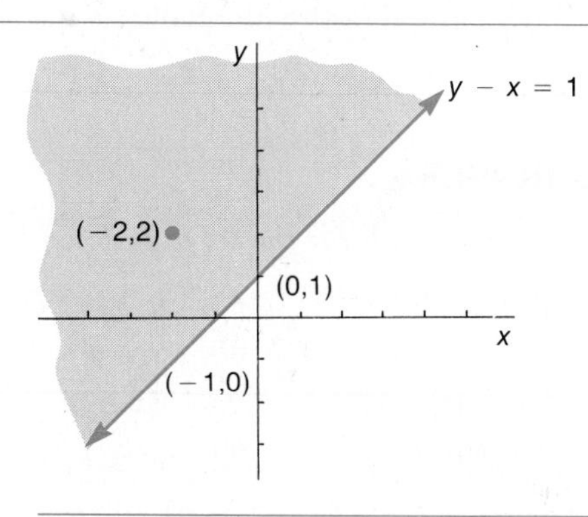

Example 14 Graph the inequality $y < 6 - 2x$.

First we graph the straight line that is $y = 6 - 2x$, but we represent this as a broken line, since points on the line itself do not satisfy the given inequality. (See Figure 5.22.) However, the line does form the two half-planes, one of which contains all the points in the solution set of $y < 6 - 2x$.

Figure 5.22

Graph of the Inequality $y < 6 - 2x$

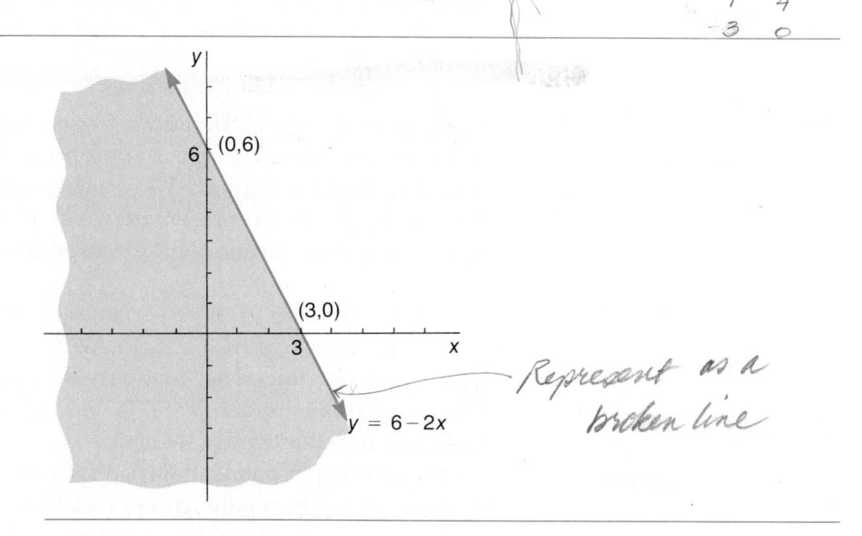

To determine which half-plane contains the appropriate ordered pairs, we select a test point. Again, the origin (0,0) may be used. Then we evaluate $y < 6 - 2x$ at (0,0) to obtain

$$0 < 6 - 2(0)$$
$$0 < 6$$

Thus, (0,0) from the lower half-plane is in the solution set of $y < 6 - 2x$. The graph of the inequality is the shaded region shown in Figure 5.22. []

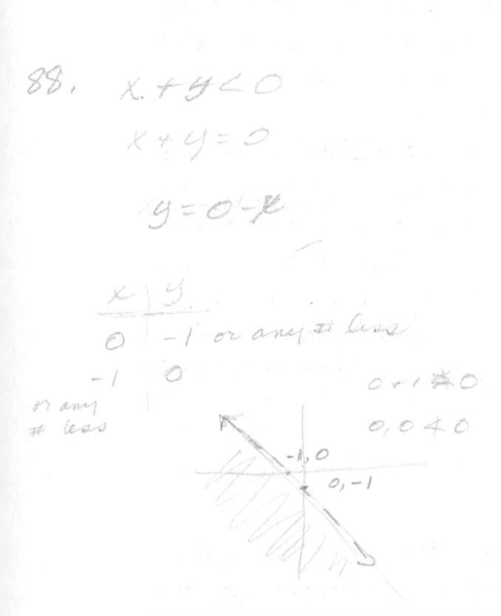

EXERCISES

Construct a graph for the set of ordered pairs satisfying each inequality.

84. $y \geq x + 5$
85. $y > x + 5$
86. $y \leq x + 5$
87. $2x + y > 8$
88. $x + y < 0$
89. $y \leq 3 - 2x$
90. $3x - 4y < 12$
91. $2x \geq y$
92. $y \geq 6$
93. $x - 4 \geq 0$
94. $y < (x/2) - 4$
95. $3x - 8y \geq -24$
96. Jacques Jones has up to $1,500 to spend on summer vacation. Each day spent in Savannah costs $20, while each day spent in Newport costs $50. Determine the possible vacation plans Jacques could make.
97. Flowery Branch Chemical Company finds that in the manufacture of its product a sulphur dioxide pollutant results. To satisfy federal regulations, this pollutant must be held to a maximum of 600 units a day. If each ton of chemical 1 results in four units of sulphur dioxide and each ton of chemical 2 results in six units of sulphur dioxide, determine the possible production strategies.

5.8 PICTURING LINEAR EQUATIONS IN 3-SPACE

Figure 5.23

The Coordinate Axes in 3-Space

z
x,z plane
y,z plane
y
x,y plane
x

Because real-world situations frequently involve linear equations in more than two variables, it is appropriate that we discuss the problem of graphing such equations. We are able to construct on paper somewhat satisfactory pictorial representations of equations of three variables. However, we have no means of graphing equations of more than three variables.

Linear equations in three variables graph as PLANES in three-dimensional space (in 3-SPACE). A PLANE is a flat surface like a sheet of paper, without thickness, that extends indefinitely in all directions in three-dimensional space.

Figure 5.23 illustrates a set of three mutually perpendicular coordinate axes which define points in 3-space. These axes are number lines representing the values of the three variables. We shall think of the linear equation in three variables as being symbolized $Ax + By + Cz = D$ and, accordingly, label the axes x, y, and z, as shown. The *broken-line* extensions of the axes represent the negative portions of the number line. The ORIGIN is the point where the three axes intersect.

The plane formed by the x-axis and the y-axis is termed the X,Y COORDINATE PLANE; the plane formed by the x-axis and the z-axis is termed the X,Z COORDINATE PLANE; and the plane formed by the y-axis and the z-axis is termed the Y,Z COORDINATE PLANE.

Figure 5.24

Locating the Point (3,4,5) in 3-Space

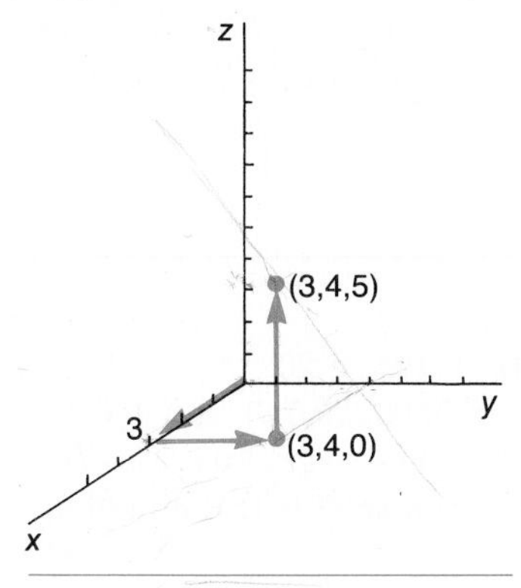

Picturing a three-dimensional configuration on a two-dimensional sheet of paper is, inherently, a somewhat distorting oversimplification. The reader is urged to form a mental image of 3-space by considering the far left corner of the room as the origin. The two adjacent walls form the x,z plane and the y,z plane; the floor forms the x,y plane. The z-axis is a straight line formed by the intersection of the two walls. The x- and y-axes are formed where the walls meet the floor. Of course, we need to visualize these as continuing indefinitely (rather than as being terminated at the next corner or at the ceiling), and we must be able to "see through" the walls and floor to visualize the negative extensions of the axes.

Because the equation is of three variables, an ORDERED TRIPLET of values (x,y,z) will be needed to identify a solution and its corresponding point in 3-space. Figure 5.24 illustrates the plotting of such points. The (3,4,5), representing $x = 3$, $y = 4$, and $z = 5$, is located by moving three units in the positive direction along the x-axis; from that point, moving four units in the direction of the positive y-axis and parallel to the y-axis; and from there moving five units in the direction of the positive z-axis and parallel to the z-axis.

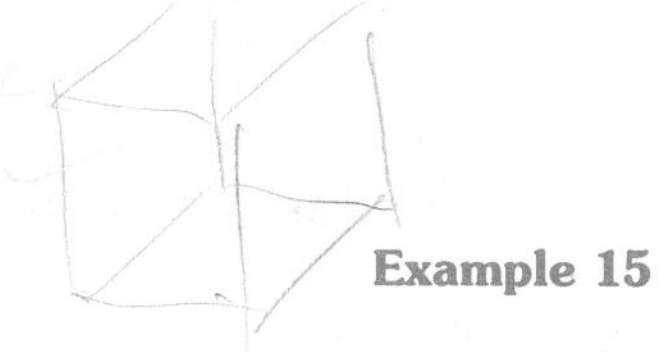

Example 15

Locate the point (3,−3,6) in 3-space.

Figure 5.25 depicts the location of the point. We begin by moving three units in the positive direction along the x-axis; from there we move three units in the direction of the negative y-axis and parallel to the y-axis. Finally we move six units in the direction of the positive z-axis and parallel to the z-axis. This point represents the ordered triplet (3,−3,6) in 3-space. []

Figure 5.25

Locating the Point (3,−3,6) in 3-Space

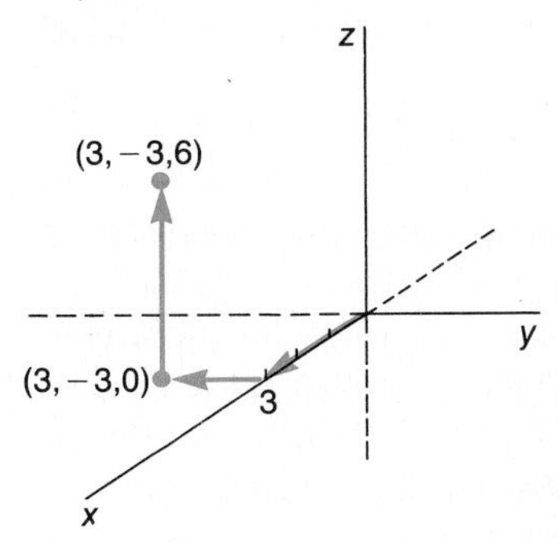

To depict a linear equation in three variables in 3-space, three points which satisfy the equation are required. The three points must be collinear; that is, they must not lie on the same line. Three points which are usually easily identified are the intercept points. These are found by setting two of the variables equal to zero and solving for the remaining variable. These intercepts are joined by line segments to form a plane which represents the graph of the equation. Any point on the plane satisfies the equation, and any point whose coordinates satisfy the equation will be on the plane.

Example 16

Sketch the graph of the equation $2x + 3y + 4z = 12$.

We may begin by finding the intercepts. The x-intercept is $x = 6$; the y-intercept is $y = 4$; and the z-intercept is $z = 3$. Thus, we get the intercept points (6,0,0), (0,4,0), and (0,0,3). These points are plotted and joined by line segments as shown in Figure 5.26.

Recall that a line drawn in the two-dimensional rectangular coordinate system to represent an equation in two variables of unrestricted domain extended indefinitely in each direction. In the same manner, to represent all solution sets for the linear equation in three variables, the plane must extend indefinitely in all directions. So what we are able to picture is only a portion of the solution set, but its position in the 3-space allows us to visualize its extensions. []

Figure 5.26

Plane Representing $2x + 3y + 4z = 12$

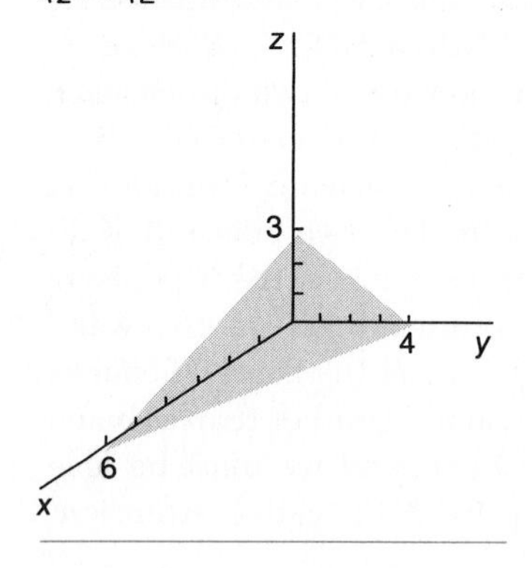

Example 17

Sketch the graph of the equation $2x + 4y + 10z = 20$.

The coordinates of the intercept points are (10,0,0), (0,5,0), and (0,0,2). These are plotted in Figure 5.27 and joined by line segments. The resulting plane represents a portion of the solution set for the equation $2x + 4y + 10z = 20$. []

Figure 5.27

Plane Representing $2x + 4y + 10z = 20$

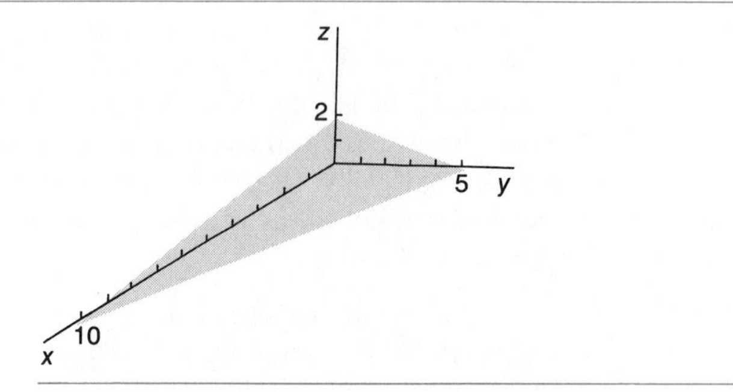

Special planes are obtained when one or more of the variables are missing in the three-dimensional linear equation. In general, if a variable is missing in the equation of a plane, the plane is parallel to the axis of the missing variable. If two variables are missing, the plane is parallel to both their axes.

Example 18

Sketch the plane $x + 0y + 0z = 0$, or simply $x = 0$, in 3-space.

In attempting to find three points which satisfy the equation, we note that so long as $x = 0$, y and z may take on *any* real values at all. Graphically, any point in the y,z plane has an x coordinate of zero, and any point for which x equals zero lies in the y,z plane. Hence $x = 0$ is the equation for the y,z plane itself. This means that the equation $x = 0$ graphs as a plane perpendicular to the x-axis and passing through $x = 0$, as shown in Figure 5.28.

In much the same way, $0x + y + 0z = 0$ (or simply $y = 0$) is an equation for the x,z plane, while $0x + 0y + z = 0$ (or $z = 0$) is an equation for the x,y plane. [ ]

Figure 5.28

A Portion of the Plane $x = 0$ in 3-Space

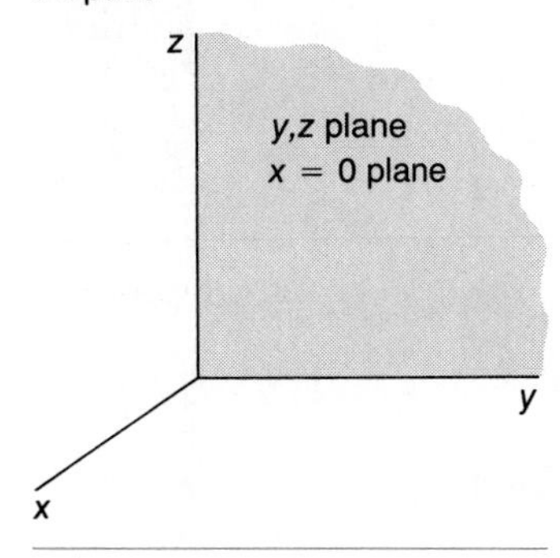

Example 19

Sketch the plane representing $y = 5$ in 3-space.

The variables x and z may assume any values so long as $y = 5$. This means that the plane representing $y = 5$ will be parallel to the x,z plane, perpendicular to the y,z plane, and intersecting the y-axis at the point $y = 5$.

Planes, of course, extend indefinitely; so the illustration in Figure 5.29 represents only a finite portion of the plane $y = 5$. [ ]

Figure 5.29

A Portion of the Plane $y = 5$ in 3-Space

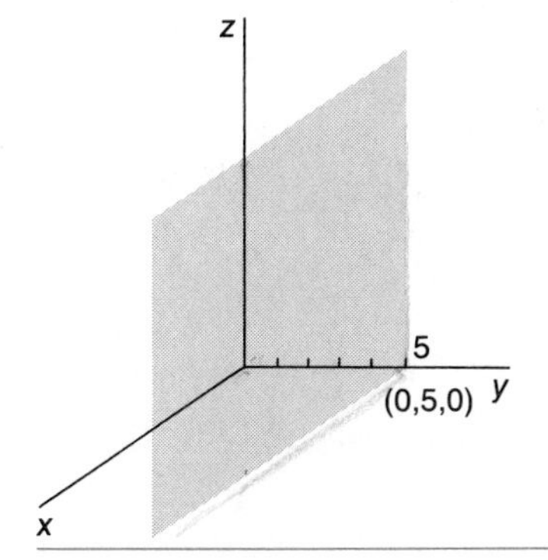

Example 20

Sketch the plane representing $0x + y + 4z = 4$.

The plane intersects the y-axis at (0,4,0) and the z-axis at (0,0,1). It runs parallel to the x-axis and is perpendicular to the y,z plane. (See Figure 5.30.) [ ]

Figure 5.30

Portion of the Plane $y + 4z = 4$ in 3-Space

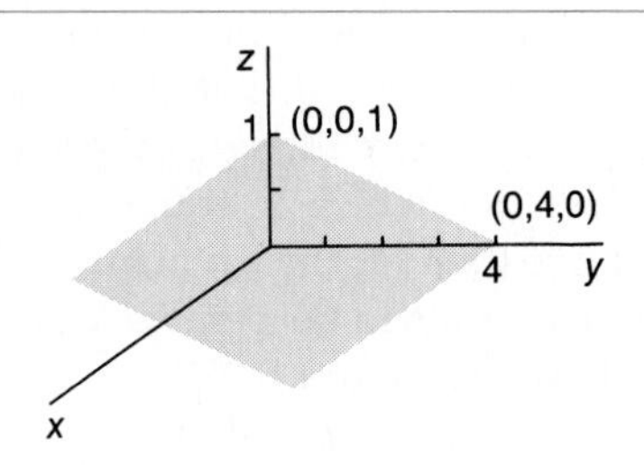

EXERCISES

Locate each of the following points (x,y,z) in 3-space.

98. (2,2,4)
99. (1,6,3)
100. (1,−4,8)
101. (−1,3,2)
102. (2,4,−3)
103. (−2,−4,3)
104. (−2,4,−3)
105. (−4,−5,−6)

Sketch the graphs of each of the following equations in 3-space.

106. $x + y + z = 5$
107. $8x + 2y + 4z = 16$
108. $x + 2y + 7z = 14$
109. $6x + 9y + 3z = 18$
110. $2z - y = 8$
111. $4x + y = 8$
112. $3x + 2z = 6$
113. $x + y = 5$
114. $x = 6$
115. $y = 8$
116. $y = -2$
117. $z = 7$

C·H·A·P·T·E·R 6

Systems of Linear Equations and Inequalities

CONSIDER a decision situation faced by Pepper Ligustrum, president of Smalltown Civic Association, Smalltown, U.S.A. (population 1,056).

Pepper is responsible for organizing Smalltown's 125th annual Fourth of July political rally and picnic. All of the details of the day's festivities seem to have been worked out, except the problem of feeding the crowd. Each year free box lunches are distributed to everyone who attends the rally. Pepper has persuaded A. Leading Citizen to contribute $1,450 to cover the cost of these free lunches. Only one condition has been imposed on the donation: the money must be spent at Small Town Caterers, an establishment operated by A. Leading's wife's mother.

Prices set by the caterer are $1.25 for each southern fried chicken box lunch and $1.50 for each bar-b-que box lunch. Pepper has only to call and let them know how many boxes of each type to prepare.

Pepper realizes, of course, that there are many combinations of the two types of lunches that would result in a cost of exactly $1,450. He could, for example, buy $1,450 ÷ $1.25 = 1,160 chicken lunches and no bar-b-que lunches. But he realizes how much some of the townspeople enjoy the bar-b-que. So, he calculates that he could buy only 1,100 chicken boxes and [1,450 − (1,100 × 1.25)] ÷ 1.50 = 50 bar-b-que boxes. Or, perhaps, 600 bar-b-que boxes and [1,450 − (600 × 1.50)] ÷ 1.25 = 440 chicken boxes. In fact, the possibilities seem to be almost endless.

Then Pepper remembers that each year *everyone in Smalltown* attends the rally and picnic. So he can depend on having to serve exactly 1,056 people. And it would be prudent to have exactly that number of lunches. If fewer were available, some people would not be fed, creating ill will toward the Smalltown Civic Association. Or, if there were an excess of lunches, A. Leading would be sure to notice, decide that his gift had been overly generous, and make an appropriate adjustment next year.

So Pepper returns to his computations. Surely, there are, again, many possibilities regarding the total number of boxes. He could order 1,056 bar-b-que boxes and no chicken boxes, or 1,055 bar-b-que boxes and 1 chicken box, . . . , or 1,056 chicken boxes and no bar-b-que boxes. But 1,056 bar-b-que boxes at $1.50 each would cost more than $1,450. Or, if he ordered 1,056 chicken boxes, he would not spend all of A. Leading's donation—and this would not be wise.

Is there any combination of chicken and bar-b-que lunches that would total exactly 1,056 lunches, at a cost of exactly $1,450?

Embodied in many problem situations are several linear relationships which must be considered simultaneously, requiring the study of a SYSTEM OF LINEAR EQUATIONS and their SIMULTANEOUS SOLUTION.

6.1 AN OLD, FAMILIAR METHOD OF SOLVING SYSTEMS OF EQUATIONS

Given two linear equations

$$\begin{cases} a_{11}x_1 + a_{12}x_2 = b_1 & \textbf{(E1)} \\ a_{21}x_1 + a_{22}x_2 = b_2 & \textbf{(E2)} \end{cases}$$

we often have occasion to ask if there exists a pair of values for the variables which satisfies BOTH of the equations. The two equations are said to form a SYSTEM OF EQUATIONS, and a pair of real numbers (n_1,n_2) is a SOLUTION OF THE SYSTEM if both equations are satisfied when we substitute $x_1 = n_1$ and $x_2 = n_2$ into each of the equations. That is, S is the SOLUTION SET FOR THE SYSTEM OF EQUATIONS where $S = \{(x_1,x_2)|(x_1,x_2)$ is a solution to (E1) and (E2)$\}$.

Just as two individual equations are said to be equivalent if they have the same solution, two separate systems of equations are said to be EQUIVALENT if they have the same SOLUTION SETS. To find solutions to systems of linear equations, we make use of procedures which allow us to move from one system to an equivalent system, but one in simpler form. We continue until we obtain a system from which we can read the solution directly. The operations which may be performed on a system of equations to produce an equivalent system are as follows:

Operations permissible on a system of linear equations

Type 1. An equation in a system of linear equations may be multiplied by a nonzero constant.

Type 2. A multiple of any equation in a system of linear equations may be added to any other equation in the system.

The difficulty with systems of equations stems from the fact that there exists more than one variable. We use the rules listed above to "eliminate" one variable at a time until we are left with one equation in one variable. The following examples illustrate this general procedure.

Example 1 Find the solution to the system of linear equations

$$\begin{cases} 2x + y = 60 & \textbf{(E1)} \\ x + 3y = 105 & \textbf{(E2)} \end{cases}$$

Whether x or y is eliminated from the system first is somewhat immaterial. We will, arbitrarily, select to eliminate x. This we can do by multiplying the second equation (E2) by -2 and adding the result to the first equation (E1).

First, we obtain the system

$$\begin{matrix} \\ -2 \times \text{(E2)} \end{matrix} \quad \begin{cases} 2x + y = 60 & \textbf{(E1)} \\ -2x - 6y = -210 & \textbf{(E3)} \end{cases}$$

Then, adding (E1) and (E3), we obtain

$$\begin{aligned} -5y &= -150 \\ y &= 30 \qquad \textbf{(E4)} \end{aligned}$$

To determine the value of x, we substitute $y = 30$ into *either* (E1) or (E2). We select (E2) and find that

$$x + 3(30) = 105$$
$$x = 15 \quad \text{(E5)}$$

To verify our computations, we can substitute both these values—$x = 15$ and $y = 30$—into each of the original equations to see whether they are, indeed, satisfied. We compute

$$2(15) + 30 = 60 \qquad \text{and} \qquad 15 + 3(30) = 105$$
$$60 = 60 \qquad\qquad\qquad 105 = 105$$

Thus, $x = 15$ and $y = 30$ is the solution to the system of equations. This is the only ordered pair (x,y) that satisfies both equations. We write $S = \{(15,30)\}$.

When a system of linear equations has a unique solution, this solution will correspond to the point of intersection of the two lines representing the equations when these lines are depicted in a rectangular coordinate system. (See Figure 6.1.) [ ]

Figure 6.1

The Solution to the System of Equations

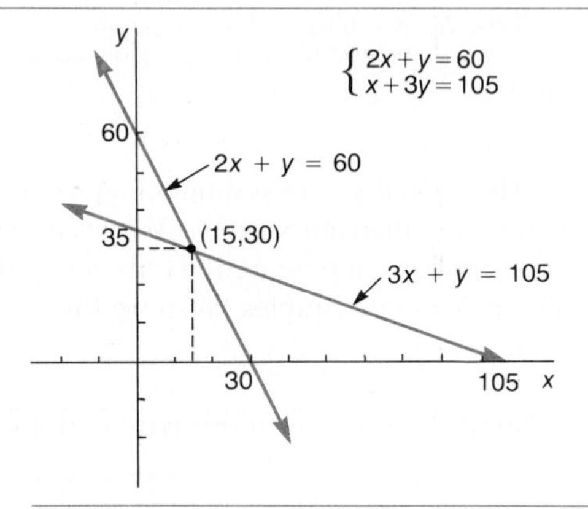

Example 2 Pepper Ligustrum could have found the answer to the dilemma of what kind of food order to place with Smalltown Caterers for the Fourth of July political rally and picnic had he thought to build a mathematical model describing the situation.

Had he used x to represent the number of fried chicken boxes and y to represent the number of bar-b-que boxes, the total number of lunch boxes required could have been described by the statement

$$x + y = 1{,}056 \quad \text{(E1)}$$

And, of course, there are many combinations of values for x and y that would make this a true statement. These are all of the points on the line in Figure 6.2A. (Note that because neither x nor y can logically assume negative values, the graph of the equation is confined to the first quadrant.)

Figure 6.2

The Smalltown Civic Association's Picnic Lunch Box Situation

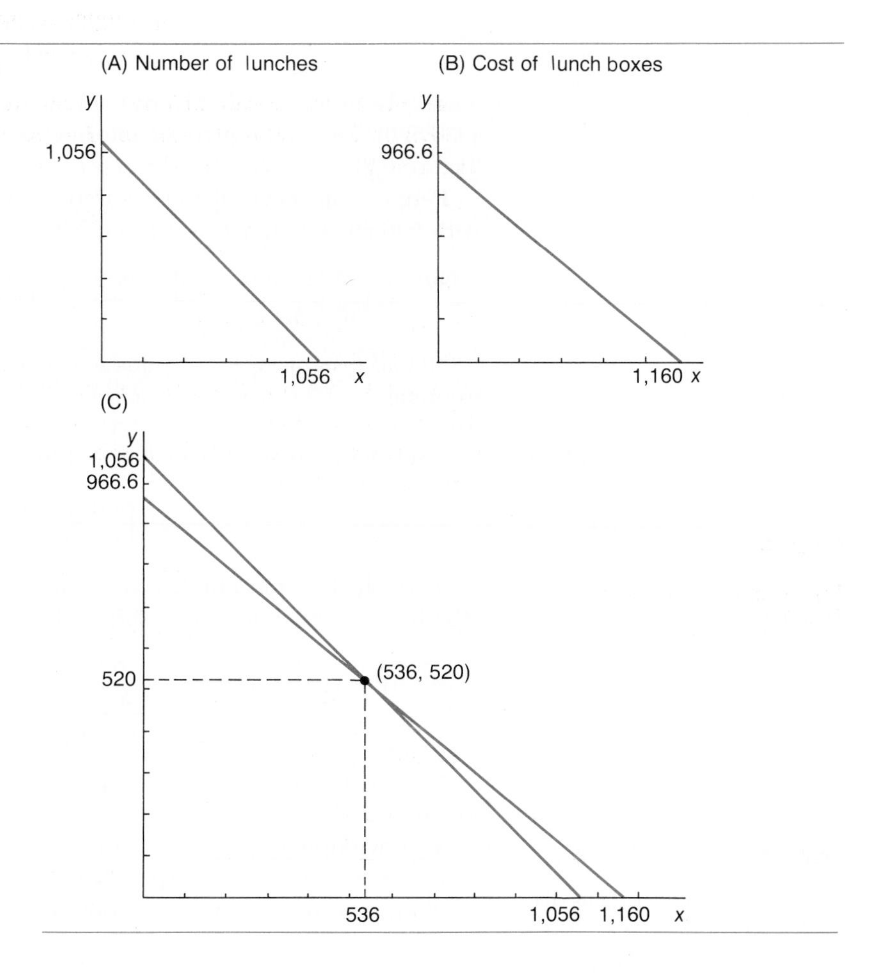

The total cost of the lunch boxes ordered can be modeled by

$$1.25x + 1.50y = 1{,}450 \qquad \textbf{(E2)}$$

Any combination of x and y values that makes this a true statement represents a combination of chicken and bar-b-que boxes that costs exactly $1,450. (These are the points on the line in Figure 6.2B.)

Now to meet both criteria simultaneously, Pepper must solve both equations simultaneously, as follows. Multiply (E1) by -1.25 and combine the result with (E2), as

$$\begin{array}{lrl} -1.25 \times \text{(E1)} & -1.25x - 1.25y = -1{,}320 & \textbf{(E3)} \\ & 1.25x + 1.50y = 1{,}450 & \textbf{(E2)} \\ \hline & 0.25y = 130 & \\ & y = 520 & \textbf{(E4)} \end{array}$$

Thus, $y = 520$ and

$$x + 520 = 1{,}056 \qquad \textbf{(E1)}$$
$$x = 536 \qquad \textbf{(E5)}$$

The only ordered pair of (x,y) values that satisfies *both* equations is (536,520). This is the point of intersection of the two lines, as shown in Figure 6.2C.

Pepper can meet all the restrictions imposed upon the food order with 536 chicken lunch boxes and 520 bar-b-que lunch boxes:

Number of lunches equals 1,056	Cost of lunches equals $1,450
$536 + 520 = 1{,}056$	$536(\$1.25) + 520(\$1.50) = \$1{,}450$ []

Not all systems of linear equations have solutions. Consider the next example.

Example 3 Find the solution to the system of linear equations

$$\begin{cases} x + y = 4 & \textbf{(E1)} \\ x + y = 3 & \textbf{(E2)} \end{cases}$$

We begin by attempting to eliminate x from the system by multiplying (E2) by -1 and adding the result to (E1). We obtain

$$\begin{array}{rl} x + y = 4 & \textbf{(E1)} \\ -x - y = -3 & \textbf{(E3)} \\ \hline 0x + 0y = 1 & \textbf{(E4)} \end{array}$$

Clearly, this represents an impossible situation; there are no values for the variables that will satisfy this condition.

The two original equations in the system are graphed in Figure 6.3. Notice that the two lines are parallel and have no (x,y) values in common.

If we rewrite both (E1) and (E2) in slope-intercept form, as

$$y = 4 - x \qquad \text{and} \qquad y = 3 - x$$

we see that they each have the same slope but different y-intercepts. Straight lines having the same slope but different intercepts can never cross; they are parallel.

This system of equations has no solution. []

Figure 6.3

A System of Linear Equations with No Solution

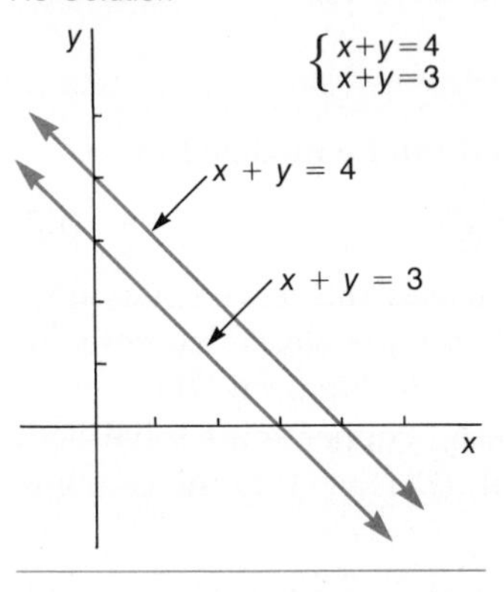

A system of equations that has no solution is said to be INCONSISTENT. If there exists *at least one solution,* the system is said to be CONSISTENT. And, yes, a system of two linear equations in two variables may have more than one solution! What's more, if it does, it will have infinitely many solutions. This happens when the equations are, in effect, equivalent; they represent the same line and have the same graph. Consider the next example.

Example 4 Find the solution to the system of linear equations

$$x + 2y = 4 \qquad \textbf{(E1)}$$
$$2x + 4y = 8 \qquad \textbf{(E2)}$$

We begin by eliminating x. We multiply (E1) by -2 to obtain the system

$$-2x - 4y = -8 \qquad \text{(E3)}$$
$$2x + 4y = 8 \qquad \text{(E2)}$$

Then, adding (E3) and (E2), we obtain

$$0x + 0y = 0 \qquad \text{(E4)}$$

Of course, ***any values*** at all for x and y will satisfy this equation. Hence, there are infinitely many solutions to the system of equations. These are all solutions of (E1) *or* (E2) and are represented by all points on their common graph.

Notice that if, in the original system of equations, we multiply (E1) by 2, we obtain the second equation (E2). The two equations of the

Figure 6.4

Systems of Two Linear Equations in Two Variables

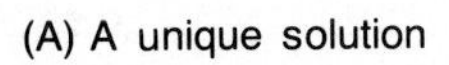
(A) A unique solution

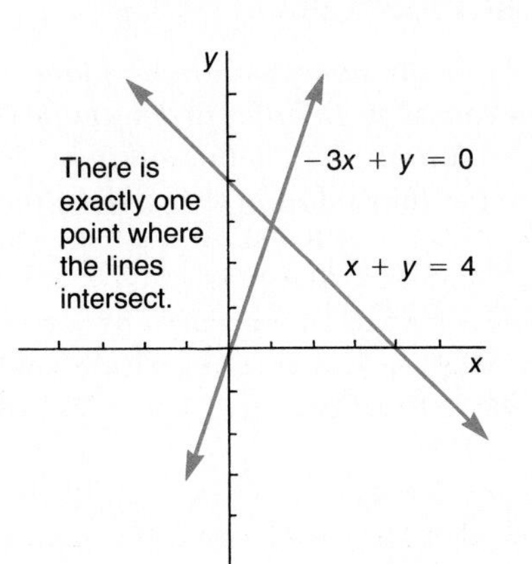
y
There is exactly one point where the lines intersect.
−3x + y = 0
x + y = 4
x

(B) No solution (an inconsistent system)

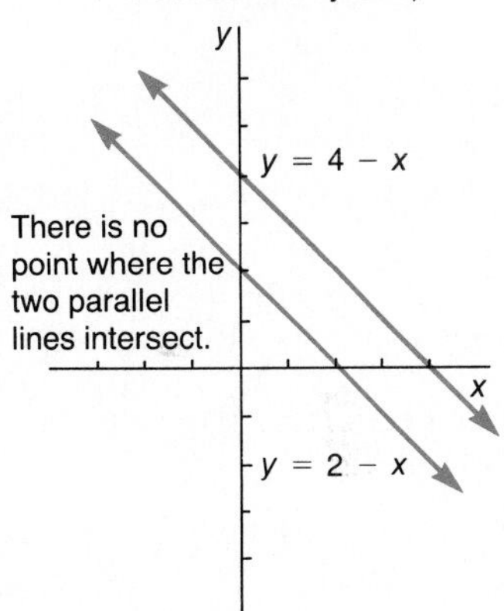
y
y = 4 − x
There is no point where the two parallel lines intersect.
x
y = 2 − x

(C) Infinitely many solutions

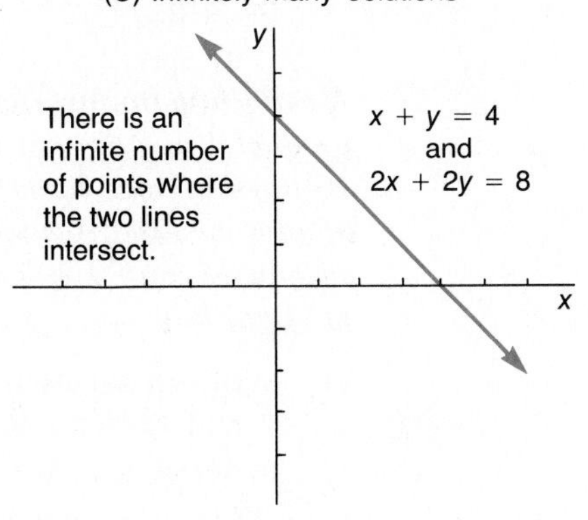
y
There is an infinite number of points where the two lines intersect.
x + y = 4
and
2x + 2y = 8
x

system are, in fact, equivalent, and any solution value (x,y) satisfying one will also satisfy the other. [ ]

To summarize, a system of two linear equations in two variables has either

1. Exactly one solution (the lines intersect at one point).
2. No solution (the lines have the same slope but different y-intercepts and, therefore, are parallel).
3. Infinitely many solutions (the lines have the same slope and the same y-intercept and are, in fact, one and the same line).

Although we have thus far considered only linear systems of two equations in two variables, we shall see that the same results holds for any system of linear equations; that is, EVERY SYSTEM OF LINEAR EQUATIONS HAS EITHER NO SOLUTION, EXACTLY ONE SOLUTION, OR INFINITELY MANY SOLUTIONS. (See Figure 6.4.)

EXERCISES

Graph, in the same coordinate plane, each of the following systems of linear equations in order to determine if they have (a) a unique solution, (b) no solution, or (c) many solutions. If the system has a unique solution, determine that solution using the elimination-of-variable procedure.

1. $2x + 3y = 21$
   $3x + 5y = 34$
2. $2x + 6y = 14$
   $x + 2y = 7$
3. $x_1 + x_2 = 2$
   $3x_1 + 6x_2 = 4$
4. $2y = 1 + x$
   $3x - 2 = y$
5. $x_1 = 3x_2 - 2$
   $x_2 - 3 = 4x_1$
6. $x_2 = 3 + 2x_1$
   $2x_2 = 4x_1 + 4$
7. $2x - 1 = y$
   $4y = 3x - 4$
8. $x_2 = 5 - x_1$
   $3x_1 + 3x_2 = 15$
9. $x_1 + x_2 = 5$
   $3x_1 + 3x_2 = 12$
10. $4x = 12$
    $3x + y - 6 = 0$

A certain manufacturer produces two products, P and Q. Each unit of product P requires in its production 20 units of raw material A and 10 units of raw material B. Each unit of product Q requires 30 units of raw material A and 50 units of raw material B. There is a limited supply of only 1,200 units of raw material A and 950 units of raw material B.

11. Write an equation which expresses the number of units of products P and Q that can be produced using exactly 1,200 units of raw material A.
12. Write an equation which expresses the number of units of P and

Q that can be produced using exactly 950 units of raw material B.

13. Graph the two equations as a means of determining if there is a combination of units of P and Q that would use exactly 1,200 units of raw material A and exactly 950 units of raw material B.
14. If the graph in Exercise 13 indicates that there is a unique solution to the system of equations, use the elimination-of-variable procedure to find that unique solution.

Attendance records indicate that 1,600 people attended the Whitley Chapel–Montrose basketball game on Thanksgiving Day. Total ticket receipts were $2,800. Admission prices were $1.50 for students and $2.50 for others. We would like to determine the number of students and nonstudents who attended the game.

15. Write an equation describing total attendance in terms of number of students and number of nonstudents.
16. Write an equation describing ticket receipts.
17. Use the elimination-of-variable procedure to determine the number of paid admissions of each type, student and nonstudent.

6.2 SOLVING SYSTEMS OF EQUATIONS USING MATRIX METHODS

The elimination-of-variable procedure is generally satisfactory for solving systems of two equations in two variables. However, for systems containing more than two equations and involving more than two variables, this procedure is not as efficient as we would like; so we find ourselves turning to another solution technique.

The alternative solution procedure we shall use employs a MATRIX FORMAT to organize the data.

The two linear equations of the system

$$\begin{cases} a_{11}x_1 + a_{12}x_2 = b_1 \\ a_{21}x_1 + a_{22}x_2 = b_2 \end{cases}$$

can be represented by the MATRIX EQUATION $\mathbf{AX} = \mathbf{B}$ where

$$\mathbf{A} = \begin{bmatrix} a_{11} & a_{12} \\ a_{21} & a_{22} \end{bmatrix}$$

is called the COEFFICIENT MATRIX,

$$\mathbf{X} = \begin{bmatrix} x_1 \\ x_2 \end{bmatrix}$$

is the SOLUTION VECTOR, or VECTOR OF UNKNOWNS,

$$\mathbf{B} = \begin{bmatrix} b_1 \\ b_2 \end{bmatrix}$$

is called the RIGHT-HAND-SIDE VECTOR, or VECTOR OF CONSTANTS.

The vector $\mathbf{X} = \begin{bmatrix} n_1 \\ n_2 \end{bmatrix}$ represents a solution of the system if both equations are satisfied when we substitute $x_1 = n_1$ and $x_2 = n_2$.

The Gaussian Method

One "matrix" procedure that can be used to solve systems of linear equations is known as the GAUSSIAN METHOD. This procedure was developed by the mathematician Karl F. Gauss (1777–1855).

We begin with the augmented matrix representation $[\mathbf{A} \mid \mathbf{B}]$ of the original system of equations. Then we systematically reduce the augmented matrix to equivalent augmented matrices *in simpler form*. We continue until we have reached the simplest possible augmented matrix representation, one from which the solution of the system can be obtained by inspection. Indeed, we seek to transform the augmented matrix $[\mathbf{A} \mid \mathbf{B}]$ into the augmented matrix $[\mathbf{I} \mid \mathbf{X}]$, where $\mathbf{I}$ is the identity matrix and $\mathbf{X}$ is the solution vector.

The operations that may be performed on the augmented matrices to convert them to simpler form are the ELEMENTARY ROW OPERATIONS (EROs) which we studied in Chapter 3. We list these again for your convenience.

> Operations permitted on the rows of a matrix are called ELEMENTARY ROW OPERATIONS. They are as follows:
>
> Type I. Any pair of rows in a matrix may be interchanged. (*EXCHANGE operation*)
>
> Type II. A row can be multiplied by any nonzero real number. (*MULTIPLY operation*)
>
> Type III. A multiple of any row can be added to any other row. (*ADD-A-MULTIPLE operation*)

Again, it is recommended that we follow the same step-by-step procedure for obtaining the identity matrix format that we used when calculating an inverse. That is, working one column at a time, obtain a one in the correct cell using ERO Type II. Then, obtain zeros in all other cells of the column using ERO Type III and multiples of the row which has the one.

Example 5

Using the Gaussian procedure, find the solution to the system of equations

$$\begin{aligned} 2x + y &= 60 \\ x + 3y &= 105 \end{aligned}$$

(This is the system of equations considered in Example 1.)

We set up the augmented matrix tableau

$$\left[\begin{array}{cc|c} 2 & 1 & 60 \\ 1 & 3 & 105 \end{array}\right]$$

To obtain a 1 in the row one–column one cell, we multiply row one by ½, to obtain

$$\left[\begin{array}{cc|c} 1 & \frac{1}{2} & 30 \\ 1 & 3 & 105 \end{array}\right] \begin{array}{l} R_1 \leftarrow (\frac{1}{2})R_1 \\ \\ \end{array}$$

Now, a zero is required in the row two–column one cell. Employing ERO Type III, we subtract row one from row two, to obtain

$$\left[\begin{array}{cc|c} 1 & \frac{1}{2} & 30 \\ 0 & \frac{5}{2} & 75 \end{array}\right] \begin{array}{l} \\ R_2 \leftarrow R_2 - R_1 \end{array}$$

Column one is now in the identity matrix format, and we turn our attention to column two. To obtain a 1 in the row two–column two cell, we multiply row two by ⅖, as

$$\left[\begin{array}{cc|c} 1 & \frac{1}{2} & 30 \\ 0 & 1 & 30 \end{array}\right] \begin{array}{l} \\ R_2 \leftarrow (\frac{2}{5})R_2 \end{array}$$

Finally, to obtain a 0 in the row one–column two cell, we subtract ½ times row two from row one, as

$$\left[\begin{array}{cc|c} 1 & 0 & 15 \\ 0 & 1 & 30 \end{array}\right] \begin{array}{l} R_1 \leftarrow R_1 - (\frac{1}{2})R_2 \\ \\ \end{array}$$

The solution vector is

$$\mathbf{X} = \begin{bmatrix} 15 \\ 30 \end{bmatrix}$$

which is interpreted as $x = 15$, $y = 30$. [ ]

Example 6 Use the Gaussian method to solve the system of equations

$$\begin{cases} 2x + 6y - z = 18 \\ y + 3z = 9 \\ 3x - 5y + 8z = 4 \end{cases}$$

We begin by writing the augmented matrix of the system

$$\left[\begin{array}{ccc|c} 2 & 6 & -1 & 18 \\ 0 & 1 & 3 & 9 \\ 3 & -5 & 8 & 4 \end{array}\right]$$

We want to obtain a 3×3 identity matrix in the first three columns of the augmented matrix. When we do this, we will develop the solution vector in the last column.

To obtain a 1 in the row one–column one cell, we multiply row one by ½,

$$\left[\begin{array}{ccc|c} 1 & 3 & -\frac{1}{2} & 9 \\ 0 & 1 & 3 & 9 \\ 3 & -5 & 8 & 4 \end{array}\right] \begin{array}{l} R_1 \leftarrow (\frac{1}{2})R_1 \\ \\ \\ \end{array}$$

We already have a 0 in the first cell of row two. To change the first element in row three to 0, we multiply row one by -3 and add the result to row three.

$$\left[\begin{array}{ccc|c} 1 & 3 & -\frac{1}{2} & 9 \\ 0 & 1 & 3 & 9 \\ 0 & -14 & \frac{19}{2} & -23 \end{array}\right] \begin{array}{l} \\ \\ R_3 \leftarrow R_3 - 3R_1 \end{array}$$

This transforms the first column into the identity matrix format. We transform the second and third columns in a similar way.

$$\left[\begin{array}{ccc|c} 1 & 0 & -\frac{19}{2} & -18 \\ 0 & 1 & 3 & 9 \\ 0 & 0 & \frac{103}{2} & 103 \end{array}\right] \begin{array}{ll} R_1 \leftarrow R_1 - 3R_2 & ② \\ R_2 & ① \\ R_3 \leftarrow R_3 + 14R_2 & ③ \end{array}$$

$$\left[\begin{array}{ccc|c} 1 & 0 & 0 & 1 \\ 0 & 1 & 0 & 3 \\ 0 & 0 & 1 & 2 \end{array}\right] \begin{array}{ll} R_1 \leftarrow R_1 + (\frac{19}{2})R_3 & ② \\ R_2 \leftarrow R_2 - 3R_3 & ③ \\ R_3 \leftarrow (\frac{2}{103})R_3 & ① \end{array}$$

We are able to read from the final augmented matrix $x = 1$, $y = 3$, and $z = 2$. Hence, the solution set for the system is $S = \{(1, 3, 2)\}$. []

EXERCISES

Use the Gaussian method to determine the solution set for each of the following systems of linear equations.

18. $x_1 + x_2 = 3$
    $2x_1 + x_2 = 4$
19. $2x - 5y = 16$
    $x + 3y = -3$
20. $x + y = 1$
    $4x + 8y = 7$
21. $x - y = 1$
    $4x - 6y = 10$
22. $2x + 3y = 28$
    $x + 4y = 29$
23. $s = t + 1$
    $t = 2s - 5$
24. $2x + y = 10$
    $2y = 5x + 20$
25. $2x - 3y - 5 = 0$
    $x - 4y - 5 = 0$
26. $x - y + 2z = 13$
    $2x - y - 2z = -1$
    $3x - 2y - z = 8$
27. $x_1 + 5x_2 - 2x_3 = 5$
    $3x_1 - x_2 + x_3 = 4$
    $4x_1 + x_2 - x_3 = 3$
28. $2x + 3y + z = 10$
    $x - 4y - z = -24$
    $x + 6y - 2z = 25$
29. $2x + 3y + 4z = 7$
    $x - y = 1$
    $-x - y + 2z = -1$
30. $4x - 2y + z = 50$
    $3x + 4y + z = 21$
    $5x + y - 2z = 52$
31. $x + 3y - 6z = 7$
    $x + y + 2z = -1$
    $2x - y + 2z = 0$

32. J. P. Shagg Carpet Company has in inventory 1,500 square yards of wool and 1,800 square yards of nylon for the manufacture of carpeting. Two grades of carpeting are produced. Each roll of superior-grade carpeting requires 20 square yards of wool and 40 square yards of nylon. Each roll of quality-grade carpeting requires 30 square yards of wool and 30 square yards of nylon. If J. P. Shagg would like to use all the material in inventory, how many rolls of superior and how many rolls of quality carpeting should be manufactured?
33. A wholesaler is able to supply three different types of wine: Burgundy, Sauterne, and vin rosé. A buyer wishes to purchase a total of 600 bottles of wine such that (*a*) the number of bottles of Burgundy purchased equals the sum of the number of bottles of Sauterne and vin rosé, and (*b*) the number of bottles of Sauterne is twice the number of bottles of vin rosé. Determine how many bottles of each type she should buy.
34. A manufacturer is closing out one product line which consists of three different models, A, B, and C. These models are assembled from three types of parts, 1, 2, and 3. The manufacturer would like, in a final production run, to produce such quantities of the three models as to completely deplete the inventory of parts on hand. Each model A uses one unit of part 1, three units of part 2, and two units of part 3. Each unit of model B uses one unit of part 1, two units of part 2, and one unit of part 3. Each model C uses two units of part 1, and three units of part 3. Inventory records show that there are on hand 1,500 units of part 1 and 1,900 units each of parts 2 and 3. How many of each model should the manufacturer plan to produce?

6.3 MULTIPLE B-VECTORS

In some situations we are involved over and over again with a system where the coefficient matrix **A** remains the same while the right-hand-side vector **B** changes. In such situations, if the **B**-vectors are all presently known, the Gaussian procedure may be used on an augmented matrix.

$$[\mathbf{A} \mid \mathbf{B}_1 \mid \mathbf{B}_2 \mid \ldots \mid \mathbf{B}_k]$$

to obtain the k individual solution vectors for each of the k systems.

Example 7 Perelli and White, Inc., manufactures electric law mowers and garden tillers, among other products. Each of these two primary products must be processed in each of two work centers where processing time is severely limited and, indeed, varies from day to day. These two work centers are the wiring department and the final-assembly department.

Each lawn mower requires one hour of processing in the wiring department and two hours of processing in the final-assembly department. Each garden tiller requires three hours of processing in the wiring department and two hours of processing in the final-assembly department. If, then, the firm manufactures x_1 lawn mowers and x_2 garden tillers, it will require

$$x_1 + 3x_2$$

hours of processing time in the wiring department and

$$2x_1 + 2x_2$$

hours of processing time in the final-assembly department.

We may write the two equations

$$\begin{cases} x_1 + 3x_2 = b_1 \\ 2x_1 + 2x_2 = b_2 \end{cases}$$

where b_1 represents the number of hours of processing time available in the wiring department and b_2 represents the number of hours of processing time available in the final-assembly department.

The production manager knows that the processing time available in each of these departments over the next four days will be as follows:

	1st day	2nd day	3rd day	4th day
Finishing department	16	18	21	9
Final-assembly department	16	16	22	10

Given this information, he would like to determine the number of units of each product he should schedule for production on each of the four days in order to use all of the processing time that will be available.

We form right-hand-side vectors for each of the four days, as

$$\mathbf{B}_1 = \begin{bmatrix} 16 \\ 16 \end{bmatrix}, \qquad \mathbf{B}_2 = \begin{bmatrix} 18 \\ 16 \end{bmatrix}, \qquad \mathbf{B}_3 = \begin{bmatrix} 21 \\ 22 \end{bmatrix}, \qquad \text{and} \qquad \mathbf{B}_4 = \begin{bmatrix} 9 \\ 10 \end{bmatrix}$$

Then we set up an augmented matrix

$$[\mathbf{A} \mid \mathbf{B}_1 \mid \mathbf{B}_2 \mid \mathbf{B}_3 \mid \mathbf{B}_4]$$

and use elementary row operations to transform **A** into identity matrix format. Four solution vectors will be generated, as

$$[\mathbf{I} \mid \mathbf{X}_1 \mid \mathbf{X}_2 \mid \mathbf{X}_3 \mid \mathbf{X}_4]$$

From the initial augmented matrix

$$\left[\begin{array}{cc|c|c|c|c} 1 & 3 & 16 & 18 & 21 & 9 \\ 2 & 2 & 16 & 16 & 22 & 10 \end{array}\right]$$

we proceed as follows:

$$\left[\begin{array}{cc|c|c|c|c} 1 & 3 & 16 & 18 & 21 & 9 \\ 0 & -4 & -16 & -20 & -20 & -8 \end{array}\right] \mathrm{R}_2 \leftarrow \mathrm{R}_2 - 2\mathrm{R}_1$$

$$\left[\begin{array}{cc|c|c|c|c} 1 & 0 & 4 & 3 & 6 & 3 \\ 0 & 1 & 4 & 5 & 5 & 2 \end{array}\right] \begin{array}{l} R_1 \leftarrow R_1 - 3R_2 \\ R_2 \leftarrow (-1/4)R_2 \end{array}$$

The four solution vectors for the four days are:

$$\mathbf{X}_1 = \begin{bmatrix} 4 \\ 4 \end{bmatrix}, \quad \mathbf{X}_2 = \begin{bmatrix} 3 \\ 5 \end{bmatrix}, \quad \mathbf{X}_3 = \begin{bmatrix} 6 \\ 5 \end{bmatrix}, \quad \text{and} \quad \mathbf{X}_4 = \begin{bmatrix} 3 \\ 2 \end{bmatrix}$$

Thus, on the first day four lawn mowers and four garden tillers should be scheduled for production; their production would require exactly 16 hours of processing in the wiring department and 16 hours of processing in the final-assembly department. On the second day, three lawn mowers and five garden tillers should be scheduled; their production would require all of the available wiring department and final-assembly department time on that day. On the third day six lawn mowers and five garden tillers should be scheduled, and on the fourth day the plant should produce three lawn mowers and two garden tillers. []

EXERCISES

Solve the following sets of simultaneous equations, each of which has multiple right-hand-side values.

		(a)	(b)	(c)
35.	$x_1 + 2x_2 =$	3	4	8
	$-2x_1 + 3x_2 =$	1	−1	5

		(a)	(b)	(c)	(d)
36.	$5x - 2y =$	14	28	31	40
	$2x + 5y =$	52	46	53	45

		(a)	(b)	(c)
37.	$x_1 + x_2 + x_3 =$	3	5	4
	$x_1 - x_2 + 2x_3 =$	2	10	0
	$3x_1 - x_2 + 4x_3 =$	6	22	4

		(a)	(b)	(c)	(d)
38.	$2x - y + z =$	11	2	3	0
	$x + 2y - z =$	7	2	−1	0
	$x + y + z =$	12	3	0	0

39. The supervisor for Dependable Home Improvement and Repair Service schedules jobs according to the number and type of workers—carpenters, plumbers, and electricians—that will be available each day. Job type A requires four hours of carpenter labor, two hours of plumbing labor, and two hours of electrical labor. Job type B requires three hours of carpenter labor and six hours of electrical labor. Job type C requires two hours of carpenter, three hours of plumbing, and two hours of electrical labor. If the supervisor knows that on the next three days the following kinds and quantities of

labor will be available, which jobs should be scheduled for each day?

	Monday	Tuesday	Wednesday
Carpenter labor (hours)	19	20	24
Plumbing labor (hours)	12	18	15
Electrical labor (hours)	16	14	24

6.4 USING THE INVERSE TO SOLVE A SYSTEM OF EQUATIONS

Given an $n \times n$ system of linear equations, represented in matrix notation as $\mathbf{AX} = \mathbf{B}$, we may be able to find the solution vector using the inverse of $\mathbf{A}$. If the coefficient matrix $\mathbf{A}$ is invertible, the system has a unique solution which is given by

$$\mathbf{X} = \mathbf{A}^{-1}\mathbf{B}$$

This result stems from the fact that, if $\mathbf{A}$ has an inverse, both sides of the matrix equation can be multiplied by that inverse, as

$$\begin{aligned} \mathbf{AX} &= \mathbf{B} \\ \mathbf{A}^{-1}\mathbf{AX} &= \mathbf{A}^{-1}\mathbf{B} \\ (\mathbf{A}^{-1}\mathbf{A})\mathbf{X} &= \mathbf{A}^{-1}\mathbf{B} \\ \mathbf{IX} &= \mathbf{A}^{-1}\mathbf{B} \\ \mathbf{X} &= \mathbf{A}^{-1}\mathbf{B} \end{aligned}$$

Example 8 Let us reconsider the Perelli and White decision situation of Example 7. This firm manufacturers lawn mowers (x_1) and garden tillers (x_2), each of which requires processing in both the wiring department and the final-assembly department. The production manager would like to develop a model that could be used to determine how many units of each product should be scheduled each day to use exactly the time available in each department that day.

Two linear equations describe the resources required by the two products

$$\begin{aligned} x_1 + 3x_2 &= b_1 \qquad \text{(Wiring department)} \\ 2x_1 + 2x_2 &= b_2 \qquad \text{(Final-assembly department)} \end{aligned}$$

where b_1 represents the number of hours of wiring department time available and b_2 represents the number of hours of final-assembly department time available.

The inverse of the coefficient matrix

$$\mathbf{A} = \begin{bmatrix} 1 & 3 \\ 2 & 2 \end{bmatrix}$$

is found by using elementary row operations to convert the augmented matrix $[\mathbf{A} \mid \mathbf{I}]$ into the format $[\mathbf{I} \mid \mathbf{A}^{-1}]$. This inverse is

$$\mathbf{A}^{-1} = \begin{bmatrix} -\frac{1}{2} & \frac{3}{4} \\ \frac{1}{2} & -\frac{1}{4} \end{bmatrix}$$

Given the number of available processing hours in each department, this inverse can be used to determine the appropriate production plan for that day. If, for example, 16 hours are available in each department, the number of units of lawn mowers and garden tillers that should be scheduled is obtained by

$$\begin{bmatrix} -\frac{1}{2} & \frac{3}{4} \\ \frac{1}{2} & -\frac{1}{4} \end{bmatrix} \cdot \begin{bmatrix} 16 \\ 16 \end{bmatrix} = \begin{bmatrix} x_1 \\ x_2 \end{bmatrix}$$

$$\begin{bmatrix} (-\frac{1}{2} \times 16) + (\frac{3}{4} \times 16) \\ (\frac{1}{2} \times 16) + (-\frac{1}{4} \times 16) \end{bmatrix} = \begin{bmatrix} x_1 \\ x_2 \end{bmatrix}$$

$$\begin{bmatrix} x_1 \\ x_2 \end{bmatrix} = \begin{bmatrix} 4 \\ 4 \end{bmatrix}$$

When the information becomes available that 18 hours of processing time in the wiring department and 16 hours of processing time in the final-assembly department can be used on the second day, the production schedule for this day can be determined by

$$\begin{bmatrix} -\frac{1}{2} & \frac{3}{4} \\ \frac{1}{2} & -\frac{1}{4} \end{bmatrix} \cdot \begin{bmatrix} 18 \\ 16 \end{bmatrix} = \begin{bmatrix} x_1 \\ x_2 \end{bmatrix}$$

$$\begin{bmatrix} x_1 \\ x_2 \end{bmatrix} = \begin{bmatrix} 3 \\ 5 \end{bmatrix}$$

The inverse could be used in this manner to determine each day's appropriate production schedule. []

Example 9 Consider the system of linear equations

$$\begin{aligned} x_1 + 2x_2 + 3x_3 &= 1 \\ 2x_1 + 3x_2 + 4x_3 &= 3 \\ x_1 + 2x_2 + x_3 &= 3 \end{aligned}$$

The coefficient matrix **A** and its inverse $\mathbf{A}^{-1}$ are

$$\mathbf{A} = \begin{bmatrix} 1 & 2 & 3 \\ 2 & 3 & 4 \\ 1 & 2 & 1 \end{bmatrix} \qquad \text{and} \qquad \mathbf{A}^{-1} = \begin{bmatrix} -\frac{5}{2} & 2 & -\frac{1}{2} \\ 1 & -1 & 1 \\ \frac{1}{2} & 0 & -\frac{1}{2} \end{bmatrix}$$

(The reader can easily verify that $\mathbf{AA}^{-1} = \mathbf{A}^{-1}\mathbf{A} = \mathbf{I}$.)

Multiplying $\mathbf{A}^{-1}\mathbf{B}$ gives the solution vector **X**, as

$$\begin{bmatrix} -\frac{5}{2} & 2 & -\frac{1}{2} \\ 1 & -1 & 1 \\ \frac{1}{2} & 0 & -\frac{1}{2} \end{bmatrix} \cdot \begin{bmatrix} 1 \\ 3 \\ 3 \end{bmatrix} = \begin{bmatrix} x_1 \\ x_2 \\ x_3 \end{bmatrix}$$

$$\begin{bmatrix} 2 \\ 1 \\ -1 \end{bmatrix} = \begin{bmatrix} x_1 \\ x_2 \\ x_3 \end{bmatrix}$$

The unique solution to the system is $x_1 = 2$, $x_2 = 1$, and $x_3 = -1$. []

Thus, if we are given a system of n linear equations in n variables with an invertible coefficient matrix $\mathbf{A}$, we can solve the system by either the Gaussian method or by inverting $\mathbf{A}$ and computing $\mathbf{X} = \mathbf{A}^{-1}\mathbf{B}$.

EXERCISES

Solve each of the following systems of equations by the inverse method.

40. $x_1 + 6x_2 = 34$
    $x_1 + x_2 = 9$

41. $3x_2 = 6 + x_1$
    $2x_1 - x_2 = 3$

42. $2x_1 + x_2 + 3x_3 = 3$
    $4x_1 + 3x_2 + x_3 = 1$
    $x_1 + x_3 = 4$

43. $3x_1 = x_2$
    $-14 - 2x_1 = 4x_3$
    $x_1 + x_2 + x_3 = 0$

44. $x_1 + 3x_2 + 3x_3 = 15$
    $-2x_2 - 3x_3 = 12$
    $-x_1 - 2x_2 - x_3 = 10$

45. $x_1 + 2x_2 = 5 - x_3$
    $4x_2 = 10 + 3x_1 - 5x_3$
    $2x_2 + 6x_3 = 8 + 4x_1$

Solve the following systems of equations, which differ only in their right-hand-side values. Use the inverse method.

46.

	(a)	(b)	(c)
$x_1 + x_2 + x_3 =$	$6/7$	$4/5$	1
$2x_1 + 3x_2 - x_3 =$	$5/7$	$1\frac{1}{5}$	$1\frac{1}{3}$
$4x_2 + 3x_3 =$	$2\frac{2}{7}$	$1\frac{2}{5}$	$2\frac{1}{3}$

47.

	(a)	(b)	(c)
$2x_1 + x_2 - x_3 =$	2	1.1	0
$4x_1 + x_3 =$	2.5	1.8	−8
$x_2 - 5x_3 =$	−2	0.1	18

48. Jason invested a total of $10,000 in three different savings accounts. The accounts paid simple interest at an annual rate of 8 percent, 9 percent and 7.5 percent, respectively. Total interest earned for the year was $845. The amount in the 9 percent account was twice the amount invested in the 7.5 percent account. How much did Jason invest in each account?

49. Mason's Brick Works manufacturers three types of brick—a face brick, a common brick, and a firebrick. The bricks are made chiefly from three types of materials—clay, shale, and concrete. Each batch of face brick processed requires five units of clay, two units of shale, and one unit of concrete. Each batch of common brick processed requires two units of clay, two units of shale, and three units of concrete. Each batch of firebrick processed requires one unit of clay, two units of shale, and four units of concrete.

    The clay, shale, and concrete used at Mason's is unique and available only in limited quantities. Each week's production is planned depending on the quantities of each of these materials that can be secured for the week. If during the coming week 40 units of clay, 36 units of shale, and 53 units of concrete will be available, how many batches of face brick, common brick, and firebrick should be scheduled for production? (Use the inverse method.)

If the production manager subsequently learns that during the following week he will be able to obtain 50 units of clay, 44 units of shale, and 64 units of concrete, what production should he plan for that week? (HINT: Use the inverse obtained earlier in the problem.)

6.5 INPUT-OUTPUT MODELS AND OTHER MATRIX MODELS

Many decision situations are described by a system of linear equations. Matrix methods can be used very conveniently in these situations.

Input-Output Models

INPUT-OUTPUT MODELS were first developed by W. W. Leontief as a means of tracing the flow of production in an industrial economy where the outputs of various goods and services are highly interdependent. In such economies, an increase in the output of one good, say automobiles, requires an increase in the output of many other goods needed to make automobiles. This *induced* increase in the output of other goods may, in turn, result in a further increase in demand for automobiles. Leontief proposed that, in order to calculate the levels of output in various industries which would be required by particular levels of demand for final goods, the production of all industries in an economy should be treated as a system of linear equations.

Assume an economy that consists solely of n industries. Each industry produces a single homogeneous commodity. The industries are related in the sense that each must use some of the others' product in order to operate. In addition, each industry produces some finished products for final demand. Transactions in such an economy can be described by an INPUT-OUTPUT TABLE such as the one shown in Table 6.1. The b_{ij} are the dollar amounts of the products of industry i used by industry j; d_i is the final demand for the product of industry i; and $x_i = \sum_{j=1}^{n} b_{ij} + d_i$ is the total output of industry i.

Table 6.1

Input-Output Table for an Economy

	User					
Producer	1	2	...	n	Final demand	Total output
1	b_{11}	b_{12}	...	b_{1n}	d_1	x_1
2	b_{21}	b_{22}	...	b_{2n}	d_2	x_2
⋮			⋮			
n	b_{n1}	b_{n2}	...	b_{nn}	d_n	x_n

A TECHNOLOGICAL COEFFICIENT MATRIX is used to describe the structure of the economy. Denoted

$$\mathbf{A} = [a_{ij}]$$

the $a_{ij} = (b_{ij}/x_j)$ and give the dollar value of the output of industry i that industry j must purchase *to produce $1 worth of its own products.*

The ith industry must produce outputs

$$a_{i1}x_1 + a_{i2}x_2 + \ldots + a_{in}x_n \qquad \text{for } i = 1, 2, \ldots, n$$

in order to supply the needs of all industries. Thus interindustry demands can be represented by **AX** where **A** is the technological coefficient matrix and **X** is the total output vector.

Production of an industry must fulfill both interindustry needs and final demands. If the final demand is represented by the column vector **D**, total output can be expressed, in matrix notation, as

$$\mathbf{X} = \mathbf{AX} + \mathbf{D}$$

We solve for **X** using the methods of matrix algebra

$$\mathbf{X} - \mathbf{AX} = \mathbf{D}$$
$$(\mathbf{I} - \mathbf{A})\mathbf{X} = \mathbf{D}$$
$$\mathbf{X} = (\mathbf{I} - \mathbf{A})^{-1}\mathbf{D}$$

Example 10 For simplicity, assume an economy consisting of only two industries, agriculture and manufacturing. Transactions of the economy during the last period are appropriately described by the input-output table:

	User		Final	
Producer	Agriculture	Manufacturing	consumer demand	Total output
Agriculture	24	80	136	240
Manufacturing	72	40	88	200

Entries in the input-output table are in (billions of) dollars. The total output column shows that, during the period covered by the table, total output for agriculture was $240 (billion) and for manufacturing was $200 (billion). Entries in the top row of the table show that of the total $240 output of agriculture, $24 was consumed by agriculture in its own production process and $80 was consumed by manufacturing, leaving only $136 to meet final consumer demand. Of the $200 total output of manufacturing, $40 was consumed internally by manufacturing, $72 was consumed by agriculture, and $88 was used to meet final consumer demand.

Entries in the first column of the input-output table show that the agricultural industry, in producing a total output of $240, consumed $24 in agricultural goods and $72 in manufactured goods. Thus, we see that in order *to produce $1 worth of its own output,* agriculture requires

24/240 = $0.10 of its own output
72/240 = $0.30 of manufacturing output

Similarly, in order to produce $1 worth of its own output, manufacturing requires

$$80/200 = \$0.40 \text{ of agricultural output}$$
$$40/200 = \$0.20 \text{ of its own output}$$

The technological coefficient matrix is, thus,

$$\mathbf{A} = \begin{bmatrix} 0.1 & 0.4 \\ 0.3 & 0.2 \end{bmatrix}$$

The input-output model is built on the assumption that production in each industry is characterized by fixed proportions and constant returns to scale. Hence, these technological coefficients are applicable from period to period, no matter what the level of output.

Having described the underlying structure of the economy, we can now pose the important forecasting question: If final consumer demand for the output of the two industries during the coming period is predicted to be

$$\mathbf{D} = \begin{bmatrix} 150 \\ 100 \end{bmatrix}$$

what will be the total output required by agriculture? by manufacturing? Total output for the product of an industry must be sufficient not only to meet final consumer demand but also to meet interindustry demand. Thus, planners must take into account not only changes in final consumer demand but also changes in the output of related industries induced by the changes in final demand.

Denoting total output of agriculture and manufacturing as x_1 and x_2 respectively, we can describe the output for the economy by the system of equations

$$0.1x_1 + 0.4x_2 + 150 = x_1$$
$$0.3x_1 + 0.2x_2 + 100 = x_2$$

The system, in matrix notation, is $\mathbf{AX} + \mathbf{D} = \mathbf{X}$, which can be solved by $(\mathbf{I} - \mathbf{A})^{-1}\mathbf{D} = \mathbf{X}$.
First,

$$\mathbf{I} - \mathbf{A} = \begin{bmatrix} 1 & 0 \\ 0 & 1 \end{bmatrix} - \begin{bmatrix} 0.1 & 0.4 \\ 0.3 & 0.2 \end{bmatrix} = \begin{bmatrix} 0.9 & -0.4 \\ -0.3 & 0.8 \end{bmatrix}$$

Then, by the Gaussian method, we compute

$$(\mathbf{I} - \mathbf{A})^{-1} = \begin{bmatrix} 4/3 & 2/3 \\ 1/2 & 3/2 \end{bmatrix}$$

We are now ready to determine the final output vector, as

$$\mathbf{X} = (\mathbf{I} - \mathbf{A})^{-1}\mathbf{D} = \begin{bmatrix} 4/3 & 2/3 \\ 1/2 & 3/2 \end{bmatrix} \cdot \begin{bmatrix} 150 \\ 100 \end{bmatrix} = \begin{bmatrix} 266.\bar{6} \\ 225 \end{bmatrix}$$

Thus, a total output of $266.\overline{6}$ will be required by agriculture to meet both interindustry demands and final consumer demands. A total output of 225 will be required by manufacturing. [ ]

The Parts-Requirements Listing Problem

Input-output type situations occur in many different contexts, as exemplified by the PARTS-REQUIREMENTS LISTING PROBLEM.

Example 11

A factory manufactures several types of parts, subassemblies, and assemblies that are sold as finished goods to other manufacturers. The parts and subassemblies are also required in the production of the assemblies. In all, there are n different parts, subassemblies, and assemblies manufactured by the firm. These we list in *technological order* and denote $a_1, a_2, \ldots, a_n$. (Items $a_1, a_2, \ldots, a_n$ are listed in "technological order" if an item a_i does not appear in the list until all parts and subassemblies required in its production have already appeared on the list.)

We define the QUANTITY MATRIX **Q** where q_{ij} is the number of units of item a_i DIRECTLY needed to assemble one item a_j. When we say that certain items a_i are "directly" needed in the assembly of another item a_j, we mean that if we had these a_i items on hand, we could immediately assemble one unit of the item a_j and have no parts left over. That is, if we had on hand q_{1j} units of a_1, q_{2j} units of $a_2, \ldots,$ and q_{nj} units of a_n, we could immediately construct one unit of a_j. From this definition it should be obvious that all $q_{ii} = 0$. Hence, all primary diagonal entries in **Q** are zero. Also, because the items are listed in technological order, all cells below the primary diagonal contain zeros.

Let us assume a factory which produces $n = 6$ different items and whose quantity matrix **Q** is as follows:

$$\mathbf{Q} = \begin{array}{c} \begin{array}{cccccc} a_1 & a_2 & a_3 & a_4 & a_5 & a_6 \end{array} \\ \left[\begin{array}{cccccc} 0 & 2 & 0 & 2 & 0 & 3 \\ 0 & 0 & 3 & 1 & 0 & 1 \\ 0 & 0 & 0 & 4 & 2 & 0 \\ 0 & 0 & 0 & 0 & 1 & 6 \\ 0 & 0 & 0 & 0 & 0 & 5 \\ 0 & 0 & 0 & 0 & 0 & 0 \end{array}\right] \end{array} \begin{array}{c} \\ a_1 \\ a_2 \\ a_3 \\ a_4 \\ a_5 \\ a_6 \end{array}$$

We read from the third column, for example, that item a_3 requires directly only three units of a_2 for its production. However, from column two, we read that each unit of item a_2 requires two units of a_1. Hence, a_3 requires, *indirectly*, six units of a_1.

The factory receives an order for d_1 units of a_1; d_2 units of $a_2, \ldots$; and d_n units of a_n. We shall list this information in an OUTSIDE DEMAND column vector as

$$\mathbf{D} = \begin{bmatrix} d_1 \\ d_2 \\ \vdots \\ d_n \end{bmatrix}$$

The question is: How many units of each of the items must be produced to fill the order?

Let us define a TOTAL PRODUCTION column vector

$$\mathbf{X} = \begin{bmatrix} x_1 \\ x_2 \\ \vdots \\ x_n \end{bmatrix}$$

where

x_1 = the total number of units of item a_1 to be produced
x_2 = the total number of units of item a_2 to be produced, . . . ,
x_n = the total number of units of item a_n to be produced.

Now, in order to produce x_1 units of item a_1, we need to produce $q_{i1}x_1$ units of a_i; to produce x_2 units of a_2, we need to produce $q_{i2}x_2$ units of a_i; and so on. Summing these

$$q_{i1}x_1 + q_{i2}x_2 + \ldots + q_{in}x_n$$

we get the total number of units of a_i that must be produced and used internally by the factory in order to manufacture the production vector **X**. Clearly, the above computation is simply the ith component of the column vector **QX**.

Thus, **QX** gives the internal production requirements for the factory when it produces the quantities given in the production vector. In order to supply both internal consumption requirements **QX** and the outside demand **D**, the factory must have a total production of

$$\mathbf{X} = \mathbf{QX} + \mathbf{D}$$

We are now faced with the purely mathematical problem of solving for **X**. We proceed by

$$\mathbf{X} - \mathbf{QX} = \mathbf{D}$$
$$(\mathbf{I} - \mathbf{Q})\mathbf{X} = \mathbf{D}$$
$$\mathbf{X} = (\mathbf{I} - \mathbf{Q})^{-1}\mathbf{D}$$

In the example at hand, we compute

$$\mathbf{I} - \mathbf{Q} = \begin{bmatrix} 1 & 0 & 0 & 0 & 0 & 0 \\ 0 & 1 & 0 & 0 & 0 & 0 \\ 0 & 0 & 1 & 0 & 0 & 0 \\ 0 & 0 & 0 & 1 & 0 & 0 \\ 0 & 0 & 0 & 0 & 1 & 0 \\ 0 & 0 & 0 & 0 & 0 & 1 \end{bmatrix} - \begin{bmatrix} 0 & 2 & 0 & 2 & 0 & 3 \\ 0 & 0 & 3 & 1 & 0 & 1 \\ 0 & 0 & 0 & 4 & 2 & 0 \\ 0 & 0 & 0 & 0 & 1 & 6 \\ 0 & 0 & 0 & 0 & 0 & 5 \\ 0 & 0 & 0 & 0 & 0 & 0 \end{bmatrix}$$

$$= \begin{bmatrix} 1 & -2 & 0 & -2 & 0 & -3 \\ 0 & 1 & -3 & -1 & 0 & -1 \\ 0 & 0 & 1 & -4 & -2 & 0 \\ 0 & 0 & 0 & 1 & -1 & -6 \\ 0 & 0 & 0 & 0 & 1 & -5 \\ 0 & 0 & 0 & 0 & 0 & 1 \end{bmatrix}$$

Using the Gaussian method, we compute

$$(\mathbf{I}-\mathbf{Q})^{-1}=\begin{bmatrix} 1 & 2 & 6 & 28 & 40 & 373 \\ 0 & 1 & 3 & 13 & 19 & 174 \\ 0 & 0 & 1 & 4 & 6 & 54 \\ 0 & 0 & 0 & 1 & 1 & 11 \\ 0 & 0 & 0 & 0 & 1 & 5 \\ 0 & 0 & 0 & 0 & 0 & 1 \end{bmatrix}$$

Actually, the entries above the primary diagonal in $(\mathbf{I}-\mathbf{Q})^{-1}$ give the total number of units of item a_i required to produce one unit of item a_j. For example, we read from cell (1,2) that two units of item a_1 are required to produce one unit of item a_2. The six in cell (1,3) indicates that six units of a_1 are required to produce one a_3. The 28 in cell (1,4) indicates that 28 units of a_1 must be produced before one unit of a_4 can be assembled.

Assume the factory has received an order for five units of item a_1, two units of a_2, one unit of a_3, three units of a_5, and four units of a_6. The total production vector may be obtained by computing

$$\mathbf{X}=(\mathbf{I}-\mathbf{Q})^{-1}\mathbf{D}=\begin{bmatrix} 1 & 2 & 6 & 28 & 40 & 373 \\ 0 & 1 & 3 & 13 & 19 & 174 \\ 0 & 0 & 1 & 4 & 6 & 54 \\ 0 & 0 & 0 & 1 & 1 & 11 \\ 0 & 0 & 0 & 0 & 1 & 5 \\ 0 & 0 & 0 & 0 & 0 & 1 \end{bmatrix}\cdot\begin{bmatrix} 5 \\ 2 \\ 1 \\ 0 \\ 3 \\ 4 \end{bmatrix}=\begin{bmatrix} 1627 \\ 758 \\ 235 \\ 47 \\ 23 \\ 4 \end{bmatrix}$$

Thus, the factory must produce a total of 1,627 units of item a_1. Five of these will be sold as finished goods; the others will be used to assemble the other products. To be precise, $758 \times 2 = 1{,}516$ units of a_1 will be assembled into the a_2s, $47 \times 2 = 94$ will be assembled into the a_4s, and $4 \times 3 = 12$ will be assembled in the a_6s, making a total of $1{,}516 + 94 + 12 = 1{,}622$ units of a_1 to be used for internal production.

Other entries in the **X** vector can be interpreted in a similar manner. []

EXERCISES

50. A fast-paced economy is based on just two industries: energy and entertainment. The following input-output table describes the output of the economy during the last period.

	User		Final consumer demand	Total output
Producer	Energy	Entertainment		
Energy	25	60	415	500
Entertainment	50	120	430	600

In the next period, demand for energy is predicted to double while demand for entertainment triples. What total output vector satisfies these demands?

51. The following input-output table describes the two-industry economy of a country.

	User			
Producer	Industry X	Industry Y	Final consumer demand	Total output
Industry X	40	30	130	200
Industry Y	20	90	190	300

If demand for the output of industry X is predicted to increase by 25 percent while demand for the output of industry Y decreases by one third, determine the total output vector that will satisfy these demands.

52. An economy is composed of three basic industries: energy, steel, and agriculture. Production of 1 unit of energy requires ⅕ unit of energy, ⅛ unit of steel, and $\frac{1}{10}$ unit of agriculture. Production of 1 unit of steel requires ¼ unit of energy, $\frac{1}{12}$ unit of steel, and ⅛ unit of agriculture. Production of 1 unit of agriculture requires ⅙ of energy, $\frac{1}{10}$ unit of steel, and $\frac{1}{12}$ unit of agriculture.

Find the total production necessary to provide for final consumer demand of 500 units of energy, 800 units of steel, and 1,000 units of agriculture.

53. Given an economy whose input-output table is as follows (in billions of dollars)

	User			
Producer	Agriculture	Industry	Transportation	Final consumer demand
Agriculture	12	12	5	31
Industry	8	24	10	38
Transportation	5	16	8	11

determine the required level of output for each industry if consumer demand for agricultural commodities is forecast to increase to 35, for industrial commodities to increase to 40, and for transportation services to 12.

54. A firm manufactures four products, A1, A2, A3, and A4. The products are sold as finished goods to other manufacturers and, in addition, certain of the products are used in the assembly of other of the products. Each unit of A2 requires in its assembly two units of A1. Each unit of A3 requires in its assembly three units of A1 and four units of A2. Each unit of A4 requires in its assembly one unit of A1 and five units of A3.

The firm has just received an order for 10 units of A1, 15 units of A2, 20 units of A3, and 5 units of A4. What total production should it plan?

55. A firm manufactures five products, A, B, C, D, and E. The products are sold as finished goods to other manufacturers and are also used in the assembly of the other products. Each unit of product A requires in its manufacture two units of B, one unit of C, and one unit of E. Each unit of product B requires in its manufacture three units of C and two units of D. Each unit of product C requires in its manufacture one unit of D and four units of E. Each unit of product D requires in its manufacture five units of E.

    The firm has just received an order for 20 units of product A, 40 units of product B, 10 units of product C, 15 units of product D, and 25 units of product E. How many units of each product should the firm manufacture to fill the order?

6.6 SYSTEMS WHICH DO NOT HAVE A UNIQUE SOLUTION

We have seen that not all systems of linear equations have a unique solution. Some systems have no solution at all; others may have an infinite number of solutions. Let us see how these latter conditions are detected when the Gaussian method is used.

Example 12 Use the Gaussian method to find the solution to the following system of linear equations.

$$\begin{cases} 2x_1 + x_2 + x_3 = 9 \\ 3x_1 + 2x_2 + (3/2)x_3 = 14 \\ x_1 + 4x_2 + (1/2)x_3 = 8 \end{cases}$$

The elementary row operations performed on the augmented matrix are shown below.

$$\left[\begin{array}{ccc|c} 2 & 1 & 1 & 9 \\ 3 & 2 & 3/2 & 14 \\ 1 & 4 & 1/2 & 8 \end{array}\right] \rightarrow \left[\begin{array}{ccc|c} 1 & 1/2 & 1/2 & 9/2 \\ 0 & 1/2 & 0 & 1/2 \\ 0 & 7/2 & 0 & 7/2 \end{array}\right] \begin{array}{ll} R_1 \leftarrow (1/2)R & ① \\ R_2 \leftarrow R_2 - 3R_1 & ② \\ R_3 \leftarrow R_3 - R_1 & ③ \end{array}$$

$$\left[\begin{array}{ccc|c} 1 & 0 & 1/2 & 4 \\ 0 & 1 & 0 & 1 \\ 0 & 0 & 0 & 0 \end{array}\right] \begin{array}{ll} R_1 \leftarrow R_1 - (1/2)R_2 & ② \\ R_2 \leftarrow 2R_2 & ① \\ R_3 \leftarrow R_3 - (7/2)R_2 & ③ \end{array}$$

Because of the row of zeros at the bottom of the matrix, it is impossible to continue with the row operations. *Whenever, in the application of the Gaussian method, an entire row of zeros is created, this is a signal that the equation whose coefficients appeared in that row was REDUNDANT.* A REDUNDANT EQUATION in a system excludes no points in the coordinate plane as solutions for the system which were not excluded by previous equations in the system.

For this system of equations, the third equation was redundant. It has no solutions that are not already a part of the solution set for the system containing only the first two equations. Because the value of x_3 is not fixed by the third equation, we are permitted to arbitrarily assign any real number value to this variable. We call such a variable a FREE VARIABLE.

The value of x_2 is fixed by the second equation (from the final matrix display) $x_2 = 1$. The value of x_1, by the first equation, is $x_1 = 4 - (1/2)x_3$; or, because we will assign x_3 the value a, $x_1 = 4 = (1/2)a$ where a is any real number.

We denote the solution set for the system as

$$S = \{(x_1, x_2, x_3) | x_1 = 4 - (1/2)a; x_2 = 1; x_3 = a \text{ where } a \text{ is any real number}\}$$

This system has an infinite number of solutions. []

Example 13 Use the Gaussian method to find the solution to the following system of linear equations.

$$\begin{cases} x + 4y - z = 4 \\ 3x - 2y + z = 2 \\ x + y - (1/7)z = 5 \end{cases}$$

We set up the initial augmented matrix, as

$$\left[\begin{array}{ccc|c} 1 & 1 & -1 & 4 \\ 3 & -2 & 1 & 2 \\ 1 & 1 & -1/7 & 5 \end{array}\right]$$

and apply the elementary row operations in the usual way.

$$\left[\begin{array}{ccc|c} 1 & 1 & -1 & 4 \\ 0 & -14 & 4 & -10 \\ 0 & -3 & 6/7 & 1 \end{array}\right] \begin{array}{l} R_1 \\ R_2 \leftarrow R_2 - 3R_1 \\ R_3 \leftarrow R_3 - R_1 \end{array} \left[\begin{array}{ccc|c} 1 & 0 & 1/7 & 8/7 \\ 0 & 1 & -2/7 & 5/7 \\ 0 & 0 & 0 & 22/7 \end{array}\right] \begin{array}{l} R_1 \leftarrow R_1 - 4R_2 \\ R_2 \leftarrow -(1/14)R_2 \\ R_3 \leftarrow R_3 + 3R_2 \end{array}$$

Obviously, we cannot continue. The bottom row, however, does not consist solely of zeros; the B-vector value is nonzero. If we convert the augmented matrix back into the linear equation format, the third equation would read

$$0x + 0y + 0z = 22/7$$

Clearly, no solution exists.

Whenever, in the application of the Gaussian method, we arrive at a row that is all zeros except for a nonzero value in the B-vector, we can conclude that the system is inconsistent; it has no solution. []

EXERCISES

Use the Gaussian method to find the solution set for each of the following systems of equations.

56. $x + y + 15 = 0$
    $x = 4 - y$

57. $2x - 3y = 4$
    $6x - 12 = 9y$

58. $2x - 3y - 4 = 0$
    $2x - 3y + 5 = 0$

59. $3x + 2y = 7$
    $9x + 6y = 21$

60. $2x_1 - x_2 - x_3 = 12$
    $x_1 + 3x_2 + x_3 = 28$

61. $x_1 + x_2 + x_3 = 10$
    $3x_1 + x_2 - 2x_3 = 8$

62. $2x_1 + 4x_2 - 2x_3 = 20$
    $x_1 + x_2 + x_3 = 6$
    $x_1 + 2x_2 - x_3 = 10$

63. $x + y + z = 4$
    $2x + z = y$
    $y + 3z = 9 - 4x$

64. $3x = 45 - 6z$
    $y = 10 - x$
    $2y + 2z = 0$

65. $x + y = 5$
    $x = 7 - y$
    $3x + 12z = 42$

66. $2x + 4y - 2z = 5$
    $x + y + z = 6$
    $x + 2y - z = 10$

67. $x_1 + x_2 = 9$
    $-2x_1 + 3x_2 - 12 = 0$
    $3x_1 + 3x_2 = 27$

68. $2x_1 = x_2 - x_3 + 11$
    $x_1 - 9 = x_2 + 2x_3$

69. $2x + y + 6z + 3 = 0$
    $x + 3z = 9 + y$

70. $x_1 + x_2 = 0$
    $4x_1 + 2x_2 = 2$
    $3x_1 + 2x_2 = 0$

71. $x_1 = 10 - x_2 - x_3$
    $2x_1 + 3x_2 = 15$
    $x_3 = 5 + x_1$
    $2x_1 + 2x_2 = 20 - 2x_3$

6.7 SYSTEMS OF LINEAR INEQUALITIES

Two or more linear inequalities, when considered as a unit, form a SYSTEM OF LINEAR INEQUALITIES. Given a two-variable system, any ordered pair of values for the variables that satisfies *all* inequalities in the system is said to be a SOLUTION OF THE SYSTEM. A NONSOLUTION fails to satisfy one or more of the inequalities of the system.

Graphs are particularly useful in defining the solution set for systems of linear inequalities in two variables. In fact, once you have graphed the system, you have solved it. The following examples will illustrate this procedure.

Example 14 Graph the solution set for the system of linear inequalities

$$\begin{cases} x + 3y \le 75 & \textbf{(I1)} \\ x + y \le 50 & \textbf{(I2)} \end{cases}$$

The graph of this system consists of all points that belong to the intersection (or the OVERLAPPING REGION) of the graphs of the individual inequalities making up the system, since these points would satisfy both inequalities in the system.

We learned in Chapter 5 how to determine the solution set for an individual inequality. We put this knowledge to use and sketch separate graphs of $x + 3y \leq 75$ (see Figure 6.5A) and $x + y \leq 50$ (see Figure 6.5B). The graph of the system of inequalities is obtained by sketching both inequalities in the same coordinate system (see Figure 6.5C).

The solution set for the system consists of all points in the double-shaded region and along the boundary lines of this region, including the point where the boundary lines intersect. Among the solutions are the points (0,0), (15,−15), (0,25) and (37.5,12.5). Nonsolutions include the point (50,50)—which violates both inequalities—and the point (−10,40)—which satisfies one, but not the other, of the inequalities.

The intersection points of the boundary lines are called CORNER POINTS or EXTREME POINTS. This solution set has one corner point, with coordinates (37.5,12.5), which are obtained by solving simultaneously the equations for the lines that intersect at this point. [ ]

Figure 6.5

Solution Set for System of Linear Inequalities

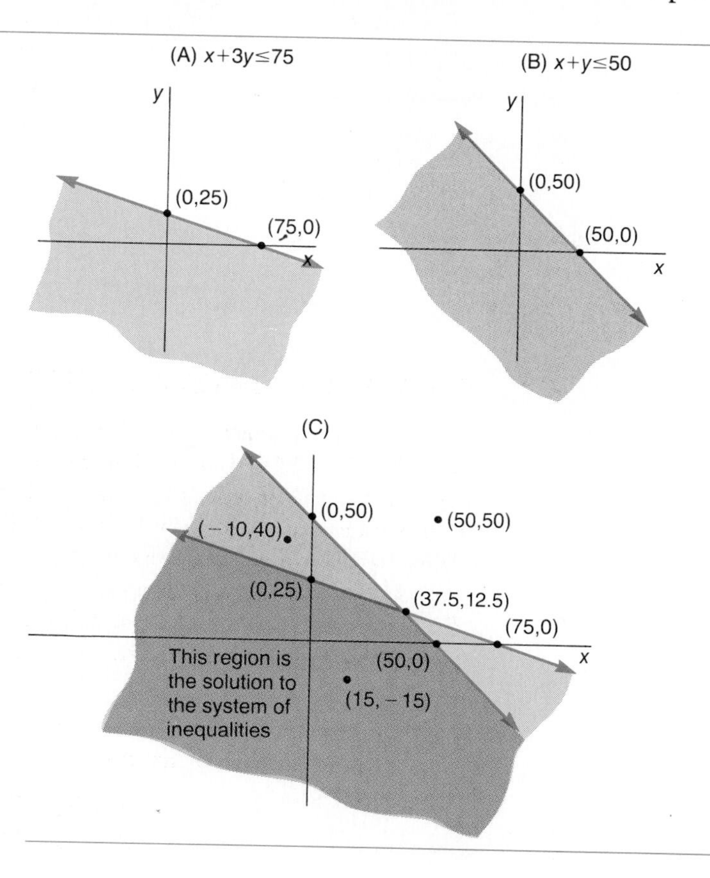

Example 15 Graph the solution set for the system of linear inequalities

$$\begin{cases} 2x + y \le 30 & \text{(IL)} \\ x - 2y \ge 20 & \text{(I2)} \\ -4x + y \le 39 & \text{(I3)} \end{cases}$$

The graphs of the individual inequalities are shown in Figures 6.6A, 6.6B, and 6.6C. The graph of the system is shown in Figure 6.6D.

Notice that the graph of the system is enclosed on all sides. Such a graph is said to be BOUNDED, while a graph which is not enclosed on all sides is said to be UNBOUNDED.

The coordinates of the corner points of the solution region are found

Figure 6.6

Solution Set for System of Linear Inequalities

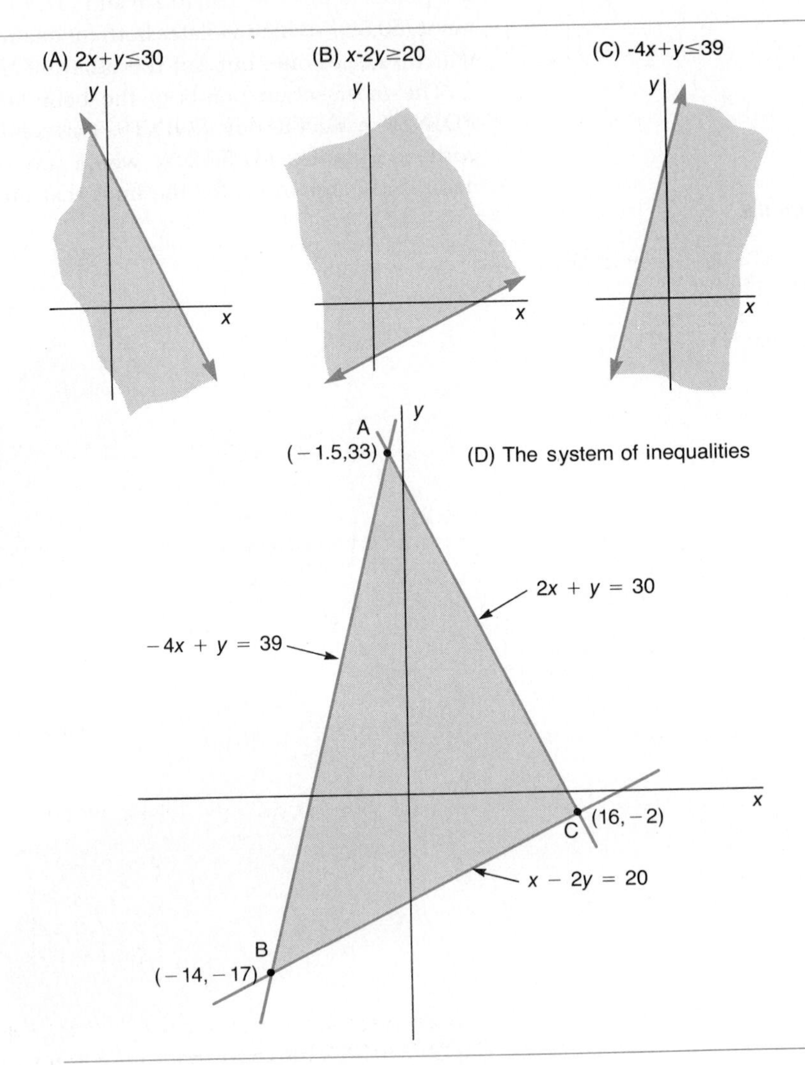

by solving simultaneously the equations for the two lines that intersect at a particular corner point. For example, the coordinates of corner point A are obtained by solving as a system of equations $2x + y = 30$ and $-4y + y = 39$. The coordinates of corner point B are obtained by solving as a system the equations $x - 2y = 20$ and $-4x + y = 39$. The two lines intersecting at corner point C represent the equations $2x + y = 30$ and $x - 2y = 20$. Thus, the coordinates at this corner point are obtained by solving this system of equations. []

Example 16 Graph the solution set for the system of linear inequalities

$$\begin{cases} x + y \geq 15 & \text{(I1)} \\ x + y \leq 12 & \text{(I2)} \end{cases}$$

Since the two lines representing the equalities have the same slope (but different intercepts), they are parallel. The first inequality is satisfied by points *above* the line $x + y = 15$, while the second inequality is satisfied by points *below* the line $x + y = 12$. Thus, the two half-planes have no overlapping region. The system has no solution. (See Figure 6.7.) []

Figure 6.7

A System of Linear Inequalities Which Has No Solution

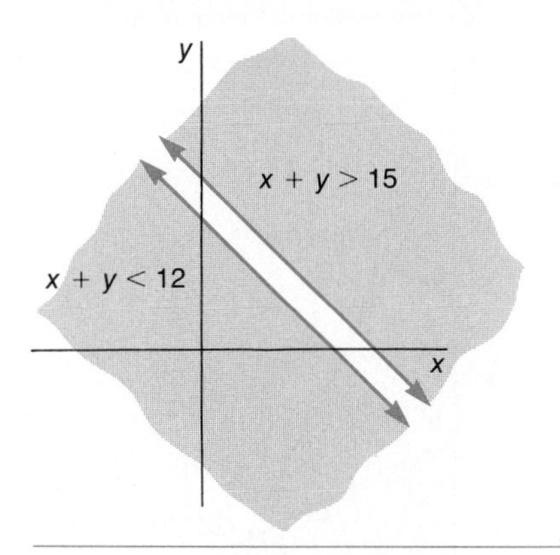

Example 17 Graph the solution set for the system of linear inequalities

$$\begin{cases} x \geq 0 & \text{(I1)} \\ y \geq 0 & \text{(I2)} \\ x + 2y \leq 45 & \text{(I3)} \\ 2x + y \leq 60 & \text{(I4)} \end{cases}$$

The inequalities $x \geq 0$ and $y \geq 0$ are called NONNEGATIVITY REQUIREMENTS for the variables. Geometrically, they restrict the graph to the first quadrant of the rectangular coordinate system. (This nonnegativity requirement appears in many real-world models.)

Then the graph of the system consists of the overlapping region of the graph of the inequalities $x + 2y \leq 45$ and $2x + y \leq 60$ *restricted to the first quadrant.* This is the shaded area in Figure 6.8(C).

The region representing the solution set has four corner points, with coordinates (0,0), (0,22.5), (25,10), and (30,0). []

Figure 6.8

Solution Set for System of Linear Inequalities Which Includes $x \geq 0$ and $y \geq 0$

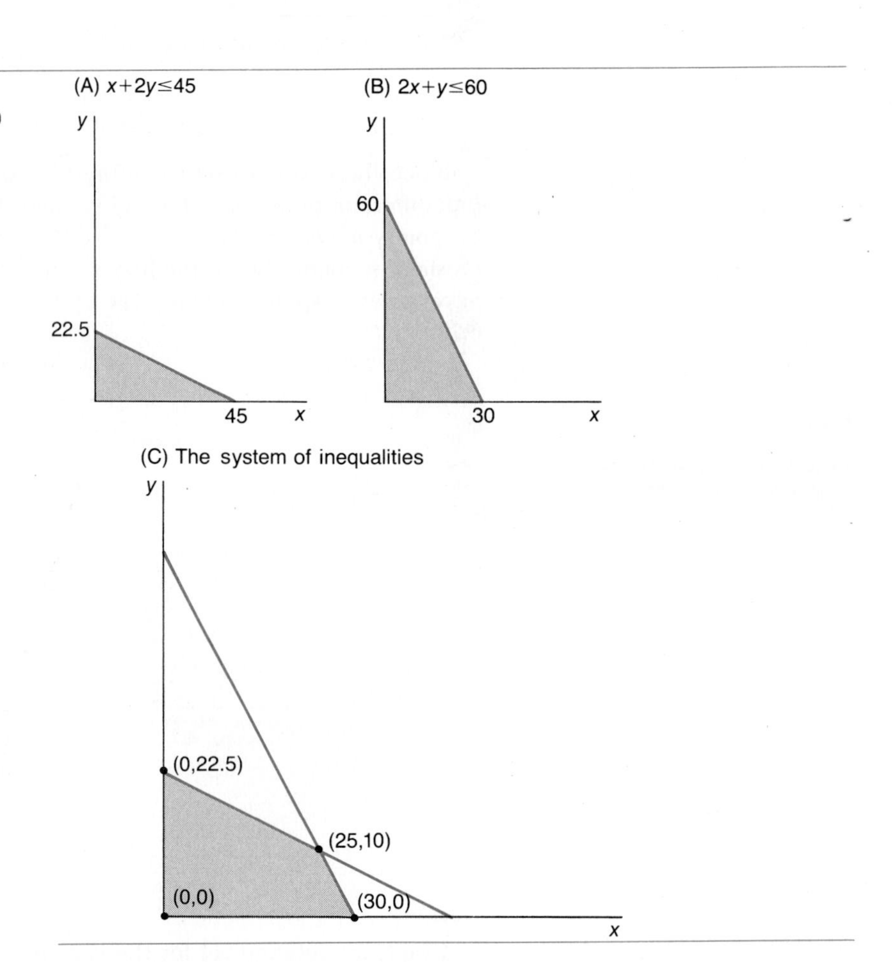

EXERCISES

Graph each of the following systems of linear inequalities as a means of defining the solution set for the system. Find the corner points of the region representing the solution set.

72. $2x - 3y \ge 6$
$x + 5y \le 10$

73. $x + 2y \le 40$
$2x + 3y \ge 90$

74. $x + y \le 10$
$2x + 3y \le 24$

75. $2x + 3y \le 12$
$2x + y \ge 8$

76. $2x + y \le 50$
$2x + y \ge 30$

77. $2x + y \ge 60$
$2x + y \le 40$

78. $3x - 5y \ge 15$
$x + y \le 6$
$y \le 8 + 8x$

79. $x + 2y \le 20$
$x + y \le 12$
$3x + 2y \le 30$

80. $3x + 5y \le 30$
$x + y \le 8$
$x \ge 0$
$y \ge 0$

81. $x + y \ge 12$
$3x + 4y \ge 42$
$x \ge 0$
$y \ge 0$

82. $3x + 2y \le 18$
$2x + 3y \ge 18$
$x \ge 0$
$y \ge 0$

83. $x + 2y \ge 30$
$x + y \ge 20$
$x \le 5$
$y \ge 0$

84. $x \ge 0$
$y \ge 0$
$x + y \le 15$
$x + y \le 12$

85. $x \ge 0$
$y \ge 0$
$3x + y \ge 30$
$2x + 3y \ge 60$

86. $x \ge 0$
$y \ge 0$
$4y \le 24 - 3x$
$x - 2y \ge -2$

87. $x + y \ge 10$
$-x + y \ge 8$
$x - y \ge 6$

88. $y \ge 6 - x$
$3y + 2x \le 24$
$4x \le 32 - 2y$
$x \ge 0$
$y \ge 0$

89. $x \ge 0$
$y \ge 0$
$y \le 10$
$x \le 15$
$y \le 8 + x$
$x \ge 18 - y$

C·H·A·P·T·E·R 7 Linear Programming: An Introduction to the Model

TYPICAL DECISION SITUATIONS WHERE THE LINEAR PROGRAMMING MODEL MIGHT BE USED

Linear programming is one of the most flexible of all mathematical models, having meaningful applications in many disciplines—marketing, production, finance—and in many decision-making contexts. The model has been successfully applied to such problems as determining the maximum-profit combination of products to manufacture, the minimum-cost blend of livestock feeds, the maximum-"reach" allocation of advertising to various media, the minimum-risk investment portfolio, and so on.

A Production Strategy Decision

A manufacturer produces several different types of items, using common production facilities—machines, labor, and raw materials—which are available only in limited quantities. The products differ in their usage rates of these resources: one product requires more machine-hours of processing than do the others, another product requires more hours of labor, and still another product requires more units of the raw material.

The selling prices of the products differ as well, but not in direct proportion to their resource-usage rates. The manufacturer would like to know, given the restricted supply of resources, how many units of each type of product should be manufactured in order to achieve the maximum possible profit.

A Diet Problem

A dietitian at a health spa, working with a few basic types of food, wants to develop a meal plan that satisfies nutritional requirements while providing a minimum intake of calories. Each food item being considered contains different quantities of the nutrients and of calories. The dietitian wants to produce a well-balanced diet but at a minimum cost.

An Investment Portfolio Problem

An investment company has a large amount of cash to invest. Analysts for the company have narrowed the investment possibilities down to just a few. Each of these investment opportunities differs as to growth and income potentialities and as to the risk involved. The current returns from each of the choices are also different. The company wants to maximize the long-term yield on its investment but also feels that it must diversify its portfolio. Because of the risk element involved, the investment in common stock is not to exceed the combined total investment in municipal bonds and savings certificates. Total investment in real estate must be at least as large as investment in mutual funds. Also, management of the firm wishes to restrict its investment in mutual funds to a level not exceeding that of savings certificates. Management of the company would like to determine how the funds should be invested, subject to the specified constraints, so that the maximum return is experienced.

Actually, enumeration of all of the possible areas of application of the linear programming model would be a major undertaking. These few illustrations certainly do not begin to do justice to the versatility of the model.

One of the most prevalent—and frustrating—of all problems faced by a decision maker is that of the optimal allocation of scarce resources to various alternative uses when these different uses contribute in different degrees to the achievement of some objective. LINEAR PROGRAMMING provides a powerful and efficient method for dealing with a special class of allocation problems, namely those in which both the objective to be optimized and all interrelating restrictions and constraints can be described mathematically by *a system of linear equations.*

7.1 CASE 1. THE BENTWOOD PROBLEM

Bentwood Manufacturing Company produces a bentwood rocking chair and a bentwood coffee table, each of superior styling and quality. The net selling prices minus the variable manufacturing costs are such that the company realizes a profit contribution of $40 on each chair and $32 on each table that it manufactures and sells. Because of the strong market demand for these items, Bentwood can sell at the prevailing prices either all the chairs or all the tables, or any combination of chairs and tables that it can produce.

Unfortunately, the production capabilities of the Bentwood plant are severely restricted in several respects. First, a special rosewood which is used as the primary raw material in both the chairs and the tables is available only in very limited quantities. Then, labor of a highly skilled nature is required in the woodworking process and also in the finishing process. Both types of labor are in scarce supply. Burdened by these constraints, the company can manufacture only limited quantities of the chairs and tables and would like to determine what combination of items would be most profitable.

Each chair requires in its construction 40 units of rosewood, while each table requires 20 units in its construction. The only known supplier of rosewood can furnish Bentwood with a maximum of 600 units of wood each day.

Only 100 man-hours of woodworking labor are available to Bentwood each day. Each chair requires 4 hours of woodworking labor, and each table requires 10 hours of this labor.

The final step in the production process, the finishing work, is also performed by highly skilled employees, and a maximum of 38 man-hours of this finishing labor is available each day. Each chair requires two man-hours and each table three man-hours of finishing time.

Clearly, the scarcity of resources places a restriction on the production strategies that are feasible. The products compete with each other for the scarce resources. If the resources are used primarily for the production of chairs, few tables can be manufactured. Or, if the resources are used primarily for the production of tables, few chairs can be made. The problem is, thus, one of allocating the scarce resources to the competing activities in such a way as to best achieve the objective of the firm, which is the maximization of the total profit contribution realized.

Organizing the relevant information about the problem situation is a most important phase of the model-building process. These data for the Bentwood problem are summarized in Table 7.1

Table 7.1

Resource Requirements per Unit of Product, Bentwood Case

Scarce resources	Product		Maximum supply available each day
	Chair (C)	Table (T)	
Rosewood	40 units	20 units	600 units
Woodworking labor	4 man-hours	10 man-hours	100 man-hours
Finishing labor	2 man-hours	3 man-hours	38 man-hours
Profit contribution	$40 per item	$32 per item	

The Decision Variables

Of course, there must be available to the decision maker several alternative courses of action. Unless there is a choice of alternative strategies, there is no decision involved. From these alternative courses of action will evolve the DECISION VARIABLES of the model. These decision variables may represent, depending upon the particular problem situation, such diverse things as the number of units of different products to manufacture, the number of dollars to invest in various projects, the number of acres of land to plant in various crops, or the number of ads to place with different advertising media. Because the decision maker has freedom of choice among these actions, these decision variables are *controllable variables.*

Here, as with any mathematical model, it is imperative that the variables be explicitly defined. Let us approach the construction of the model in the Bentwood case by defining the decision variables as

$C =$ the number of chairs to produce each day
$T =$ the number of tables to produce each day

Because of the favorable market conditions, all of the items produced by Bentwood can be sold at the stated profit contributions.

These variables are related in several different ways that are important to the decision at hand. They are related in the manner in which they contribute to the objective, in the manner in which they use the scarce raw material rosewood, in the manner in which they use the scarce resource woodworking labor, and in the manner in which they use the scarce resource finishing time. Mathematical expressions for each of these different relationships must be developed.

The Constraints

The alternative courses of action available to the decision maker must be interrelated through restrictions imposed by the problem environment. Used in this context, a restriction is anything that defines or limits the feasibility of a proposed course of action. These restrictions form the CONSTRAINTS of the model.

A typical restriction embodies scarce resources: limited labor supply, limited supplies of raw material, limited production capacity, or limited amounts of working capital. This type of restriction imposes a maximum bound on the extent to which a course of action can be pursued. Other restrictions may impose a minimum bound, such as a contractual commitment for the production of a specified minimum quantity of a product, or legal restrictions on such things as minimum liquid reserves for a banking institution.

Because many of the alternative courses of action are governed by the same restrictions, the extent to which a single course of action can be pursued is affected not only by these environmental restrictions but also by the extent to which other courses of action are followed. Hence, the quantity of product A that can be produced is limited not only by the scarcity of raw material but also by the quantity of product B that is produced, since product B is made from the same scarce raw material. These products, in a sense, compete for the available units of scarce resources. Thus, while the decision variables are controllable variables, they are controllable subject to certain constraints.

In the Bentwood case, one constraint is imposed because of the limited supply of rosewood. Only 600 units of the rosewood are available each day. Each chair manufactured requires 40 units of rosewood, and each table manufactured requires 20 units of rosewood. The total quantity of rosewood used each day is, thus, $40C + 20T$. Now, while it is impossible to use more than the 600 units of rosewood that are available each day, there is no requirement that all of this resource be used in this production process. Fewer than 600 units might be used. Hence, the rosewood constraint is expressed, not as an equation, but, rather, as a LESS-THAN-OR-EQUAL-TO inequality. The constraint is, thus, expressed as

$$40C + 20T \leq 600 \qquad \text{(Rosewood constraint)}$$

(We should note that most of the constraints in any linear programming model are inequalities rather than equalities.)

The chairs and tables are also related in a significant way through their usage rates of woodworking labor. Each chair produced requires four hours of woodworking labor; each table requires ten hours of this labor. A maximum of 100 hours of woodworking labor is available each day. The labor actually used will be $4C + 10T$. While this quantity cannot exceed the available supply of 100 units, it might be less than 100 units. So, the constraint is expressed as

$$4C + 10T \leq 100 \qquad \text{(Woodworking labor constraint)}$$

Each chair produced requires two hours of finishing labor, and each table produced requires three hours of finishing labor. A maximum of 38 hours of finishing labor is available each day. Hence, the constraint is expressed as

$$2C + 3T \leq 38 \qquad \text{(Finishing labor constraint)}$$

These are the factors that limit Bentwood's ability to produce untold quantities of the products.

The Objective Function

In the linear programming situation, as in all decision-making situations, there must be an explicitly stated objective. The objective must be refined to the point where it can be expressed in quantitative terms. The contribution of each course of action to the objective must be measurable. The objective, thus, provides a clear-cut criterion by which the relative merits of each of the various courses of action can be evaluated. The strategy chosen must be not only feasible in light of the restrictions imposed but also optimal in light of the objective. This optimality is epitomized by a maximum or minimum in the OBJECTIVE FUNCTION of the model. Frequently, this involves maximization of profit, but there are many possibilities—minimization of cost, maximization of the quantity of product sold, and so on.

In the Bentwood case, the objective function must describe the profit contribution of each type of product. Each chair manufactured and sold by Bentwood contributes $40 to profit; C represents the number of chairs produced and sold; thus $40C$ is the total profit contribution realized from the chairs.

Each table manufactured and sold contributes $32 to profit; T represents the number of tables produced and sold; thus $32T$ is the total profit contribution from the tables. The objective of the firm is to maximize the total profit contribution realized from the items it manufactures. Hence, the objective function for the Bentwood linear programming model is expressed

$$\text{MAXIMIZE } f(C,T) = 40C + 32T = \text{total profit contribution, in dollars}$$

The Nonnegativity Requirement

One additional condition must be met. In general, each variable of a linear programming problem must assume a nonnegative value. Thus,

$$C \geq 0 \qquad \text{and} \qquad T \geq 0$$

are relationships that must exist. This nonnegativity requirement is, in fact, an essential part of all linear programming models and pertains to all decision variables for the model.

The Complete Model in Mathematical Form

The significant mathematical aspects of the Bentwood production strategy problem can be summarized as

$$\text{MAXIMIZE } f(C,T) = 40C + 32T = \text{total profit contribution, in dollars}$$

where

C = the number of chairs to produce each day
T = the number of tables to produce each day

subject to these limiting constraints:

$$\begin{aligned} 40C + 20T &\leq 600 && \text{(Rosewood constraint)} \\ 4C + 10T &\leq 100 && \text{(Woodworking labor constraint)} \\ 2C + 3T &\leq 38 && \text{(Finishing labor constraint)} \end{aligned}$$

with the further requirement that

$$C \geq 0 \qquad \text{and} \qquad T \geq 0$$

In general, we refer to the type of problem exemplified by the Bentwood case as the MAXIMIZING PROBLEM of linear programming.

7.2 A GENERAL STATEMENT OF THE LINEAR PROGRAMMING MAXIMIZATION MODEL

The model summarized above is the specific model for the Bentwood problem. A general statement for all linear programming maximization models can be expressed symbolically as

$$\text{MAXIMIZE } z = c_1x_1 + c_2x_2 + \cdots + c_nx_n$$

subject to:

$$\begin{aligned} a_{11}x_1 + a_{12}x_2 + \cdots + a_{1n}x_n &\leq b_1 \\ a_{21}x_1 + a_{22}x_2 + \cdots + a_{2n}x_n &\leq b_2 \\ &\vdots \\ a_{m1}x_1 + a_{m2}x_2 + \cdots + a_{mn}x_n &\leq b_m \end{aligned}$$

and

$$x_j \geq 0 \text{ for all } j$$

Please note these things about the format of this linear programming model:

1. The objective function is to be MAXIMIZED.
2. All constraints of the model are expressions requiring the variables, or linear combinations of the variables, to be LESS THAN OR EQUAL TO (≤) a NONNEGATIVE CONSTANT. (Note that we are not referring to the nonnegativity requirement as a "constraint.")

We shall see that other linear programming models may assume a slightly different structure. However, this is the structure required for what we shall term the LINEAR PROGRAMMING MAXIMIZATION MODEL.

Two important mathematical properties of the linear programming model should be mentioned before we continue with the Bentwood case.

Mathematical Linearity and Additivity

A primary mathematical assumption of the linear programming model is that of linearity. All functional relationships between the variables, both in the objective function and in the constraints, must be linear in form.

Linearity implies additivity. The alternatives must be additive relative to both their total contribution to objective and total utilization of scarce resources. This means that the total contribution to objective must be the sum of the contributions of each individual alternative; the total amount of resources used must be the sum of the amounts of resources used by each of the individual alternatives.

Complete Divisibility

The linear programming model presupposes that each decision variable is a continuous variable, able to assume any value within a relevant range. This implies that all activity levels and all resource usages are completely divisible; therefore, the decision variables are allowed by the model to take on fractional values. The results may well be an optimal solution which indicates something like the manufacture of 67.5 units of a product, or the usage of 75.003 units of a raw material, or the purchase of ⅔ units of an item.

In many situations it is entirely feasible and appropriate to have fractional values in the levels of activity and resource utilization. However, in many other situations fractional values are neither permissible nor practical. For instance, a firm might conceivably implement a production strategy which calls for the use of 673.5 gallons of a liquid solvent. But, on the other hand, an equipment-purchase strategy involving 1⅛ supersonic jets and ⅝ prop jets might not be so easily executed.

INTEGER PROGRAMMING is an advanced mathematical programming technique used for obtaining the optimal solution to a linear programming problem where integer values for the decision variables are required. While integer programming builds upon the foundation of linear programming, a discussion of this technique is beyond the scope of our presentation.

In the final analysis, we must always remember that any mathematical model is only an abstract representation of the real-world problem. How good a representation it is depends to a large extent upon how closely the actual conditions in the real-world situation correspond to the mathematical assumptions of the model. The closer these assumptions come to being satisfied, the closer the model will come to being an appropriate representation of the real-world problem. Only that model builder who is thoroughly versed in the assumptions of the model and their implications will be able to ascertain that there is, or is not, a correspondence between the structure of the model and the structure of the problem environment.

7.3 A GRAPHICAL APPROACH TO DETERMINING THE OPTIMAL STRATEGY

Graphical methods provide a simple means of solving two-variable linear programming problems. Unfortunately, graphs are restricted to two (or, at the very most, three) dimensions, and most real-world linear programming models involve many more variables. Nevertheless, graphs can be used very effectively in simplified situations, like the Bentwood model, for providing a conceptual understanding of the linear programming procedure.

Picturing the Region of Feasible Solutions

Each of the inequalities representing the constraints in the Bentwood model is displayed graphically in Figure 7.1. The variable C, representing the number of chairs produced each day, labels the horizontal axis, while the variable T, representing the number of tables produced each day, labels the vertical axis.

The nonnegativity requirement of the model—$C \geq 0$ and $T \geq 0$—restricts the region under consideration to the first quadrant of the rectangular coordinate system. (Because the nonnegativity requirement is an integral part of every linear programming model, only the first quadrant is of interest in any such model.)

The rosewood constraint The rosewood constraint is shown in Figure 7.1A. The equality portion of the constraint

$$40C + 20T = 600$$

graphs as a straight line through the points (15,0) and (0,30). The "less than" portion of the constraint

$$40C + 20T < 600$$

encompasses all points to the left and below the line. The complete constraint

$$40C + 20T \leq 600$$

Figure 7.1

The Constraints of the Bentwood Model (restricted to the first quadrant because of the nonnegativity requirement of the model)

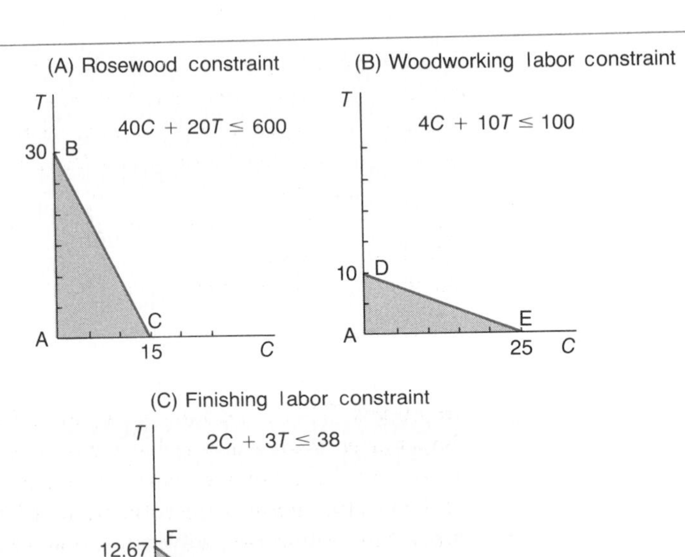

thus includes all points on the BC line in Figure 7.1A, to the left and below the BC line.

Then the "feasible" region defined by the rosewood constraint, together with the nonnegativity requirement, is that region defined by ABC. Points on the BC constraint line itself represent combinations of chairs and tables whose manufacture would require exactly 600 units of rosewood. All points to the left of the BC line, on or above the horizontal axis and on or to the right of the vertical axis, represent combinations of chairs and tables that would require something less than 600 units of rosewood. All of these points, then, represent "feasible" combinations of chairs and tables insofar as the rosewood supply is concerned.

Points to the right of the BC line represent combinations of chairs and tables which cannot be produced because they would require quantities of rosewood in excess of the 600 units that are available each day.

The woodworking labor constraint The woodworking labor constraint is pictured in Figure 7.1B. The equality part of the constraint

$$4C + 10T = 100$$

is represented by the straight line through the points (25,0) and (0,10). The inequality part of the constraint

$$4C + 10T < 100$$

is represented by all points to the left and below the line. The complete constraint

$$4C + 10T \leq 100$$

thus, includes all points on the DE line, to the left and below the line.

The nonnegativity requirement for the values of the variables restricts these to the area above the horizontal axis and to the right of the vertical axis. Therefore, all points along the boundaries ADE and within these boundaries represent feasible combinations of chairs and tables insofar as woodworking labor is concerned, but all points to the right of the DE line represent production combinations that would not be possible since they would require more than the available supply of woodworking labor.

The finishing labor constraint The finishing labor constraint is shown in Figure 7.1C. All points along the AFG boundary and within these boundaries represent feasible combinations of chairs and tables insofar as the finishing labor resource is concerned. But points to the right of the FG line represent combinations of products that are not possible since they would require in their manufacture more than the available supply of finishing labor.

The region of feasible solutions The constraints have been depicted as separate entities when, as a matter of fact, they must be considered simultaneously. All of the scarce resources are required to some extent in the manufacture of either a chair or a table. Hence, only those combinations of chairs and tables that will not violate *any* of the constraints are truly feasible production strategies.

All three constraints are pictured in one graph in Figure 7.2. The shaded region, representing feasible strategies when all constraints, along with the nonnegativity requirements, are considered, is termed the REGION OF FEASIBLE SOLUTIONS for the linear programming model. All points within the ADHJC region or along its boundary lines represent combinations of chairs and tables that could be produced by Bentwood; these points represent FEASIBLE SOLUTIONS to the Bentwood linear programming model.

Figure 7.2

The Region of Feasible Solutions, Bentwood Model

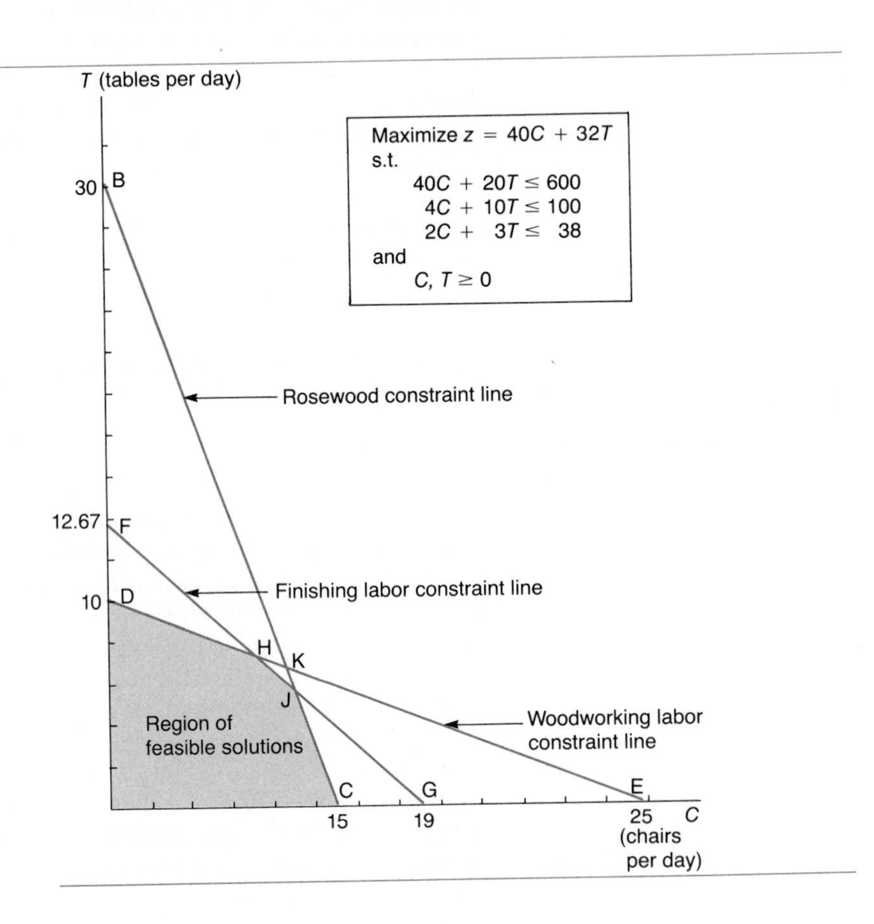

Using Iso-Profit Lines to Determine the Optimal Solution

While there are many, many feasible strategies in the Bentwood problem, the objective is to find that strategy which yields the maximum total profit contribution. Thus, let us now direct our attention to the matter of determining which of the feasible strategies is also optimal. The OPTIMAL SOLUTION for any linear programming problem is that feasible solution which provides the best possible value of the objective function.

The optimal strategy in a linear programming situation is dependent upon more than just the constraints and the exchange rates they represent. The best course of action depends, as well, on the relative contributions to objective of each of the possible courses of action. Thus, one method of graphically locating the optimal solution to a linear programming maximization problem involves the use of ISO-PROFIT LINES, where *an iso-profit line is a line each of whose points designates a solution with the same objective-function value.* A "family" of iso-profit lines is a set of parallel profit lines, each having the same slope but with different intercepts.

The region of feasible solutions in the Bentwood problem is pictured again in Figure 7.3. A series of iso-profit lines has been superimposed over the region. The slope of the profit line is a constant, determined by the ratio of the contribution coefficients to be −1.25. The exact position of the profit line relative to the origin on the graph depends upon the value of the objective function. For example, if we arbitrarily select "profit equals $480," we write

$$40C + 32T = 480$$

and find that the line passes through the points (12,0) and (0,15). Or, if we set

$$40C + 32T = 560$$

Figure 7.3

Using Iso-Profit Lines to Locate the Optimal Solution

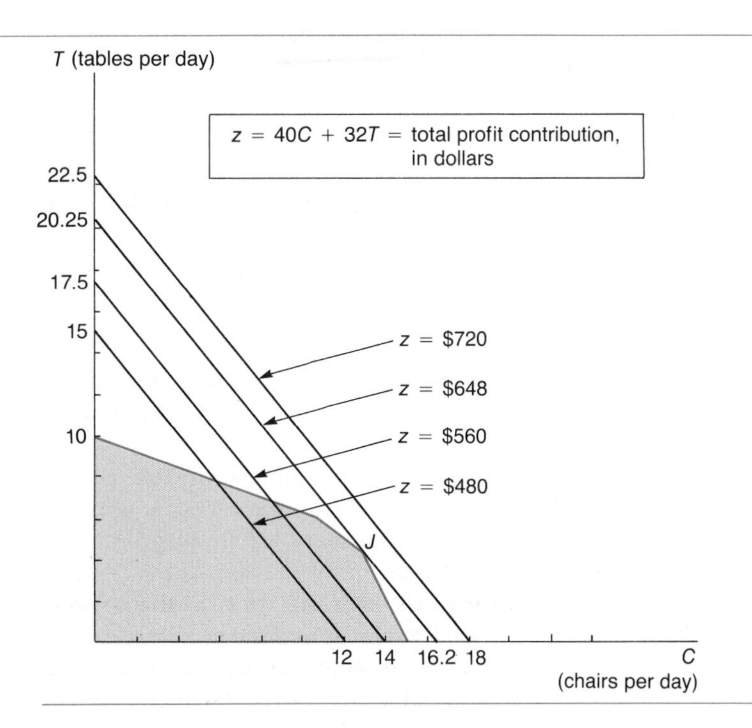

we find that the line moves through the points (14,0) and (0,17.5).

The larger the total profit contribution, the farther the profit line is moved, in a parallel fashion, out from the origin. Thus, with a goal of profit maximization, we want a profit line as far out from the origin as possible. Still, the strategy must be a feasible one; so there must be a point on the profit line that also lies within the region of feasible solutions—THAT POINT WILL ALWAYS BE A CORNER POINT OF THE REGION OF FEASIBLE SOLUTIONS. We see in Figure 7.3 that the point J represents the maximum-profit strategy for Bentwood, yielding a profit contribution of $648.

The lines representing the rosewood constraint and the finishing labor constraint intersect at point J. Solving, simultaneously, the equations that these lines represent, we find the coordinates of this point are (13,4). Thus, the profit-maximizing strategy for Bentwood Manufacturing Company involves the production of 13 chairs and 4 tables each day, with a resulting total profit contribution of $648. No other feasible combination of chairs and tables will yield a larger total profit contribution.

Evaluating Corner Points of the Feasible Region

Our analysis of the region of feasible solutions and of the slope of the objective function of the linear programming model leads to the following theorem, which is of fundamental importance in linear programming.

> **Corner-point theorem.** In a linear programming problem, the optimal solution, if one exists, will always include a CORNER POINT of the region of feasible solutions.

Because the optimal solution (if one exists) to a linear programming model always occurs at a corner point of the region of feasible solutions, the solution procedure for any linear programming problem involving just two decision variables may be summarized as follows.

Corner-point evaluation solution procedure

1. Graph the constraints, thus outlining the region of feasible solutions for the problem.
2. Determine the coordinates of all the corner points of the region of feasible solutions. (This is accomplished by solving simultaneously the equations which the lines intersecting at the corner point represent.)
3. Calculate the value of the objective function at each corner point by substituting the coordinates of the corner point into the objective function.

4. If the objective is to maximize the value of the objective function, the optimal strategy will be given by the coordinates of that corner point yielding the largest value for the objective function. (If the objective is to minimize the value of the objective function, the optimal strategy will be given by the coordinates of that corner point yielding the smallest value for the objective function.)

Evaluating the corner points in the Bentwood problem The feasible region for the Bentwood model is pictured once again in Figure 7.4. The coordinates of each corner point are given. These coordinates were obtained in this way:

Corner point	Intersecting constraint lines	Equations which lines represent	Coordinates
A	Nonnegativity requirements for C and T	$C=0$ $T=0$	(0,0)
D	Nonnegativity requirement for C and woodworking constraint	$C=0$ $4C+10T=100$	(0,10)
H	Woodworking constraint and finishing labor constraint	$4C+10T=100$ $2C+3T=38$	(10,6)
J	Finishing labor constraint and rosewood constraint	$2C+3T=38$ $40C+20T=600$	(13,4)
C	Rosewood constraint and nonnegativity requirement for T	$40C+20T=600$ $T=0$	(15,0)

Once the coordinates for all corner points have been determined, the objective function of the model is evaluated at each of these points. These computations are required in the Bentwood problem:

Corner point	Coordinates	Value of objective function
A	(0,0)	$40(0)+32(0)=0$
D	(0,10)	$40(0)+32(10)=\$320$
H	(10,6)	$40(10)+32(6)=\$592$
J	(13,4)	$40(13)+32(4)=\$648$ ← ← OPTIMAL STRATEGY
C	(15,0)	$40(15)+32(0)=\$600$

The maximum value of the objective function is associated with corner point J, having coordinates (13,4) and representing the production of 13 chairs and 4 tables each day. This production plan would yield a total daily profit contribution of $648, the maximum possible under the prevailing set of constraints at Bentwood.

Figure 7.4

Evaluating the Corner Points of the Region of Feasible Solutions

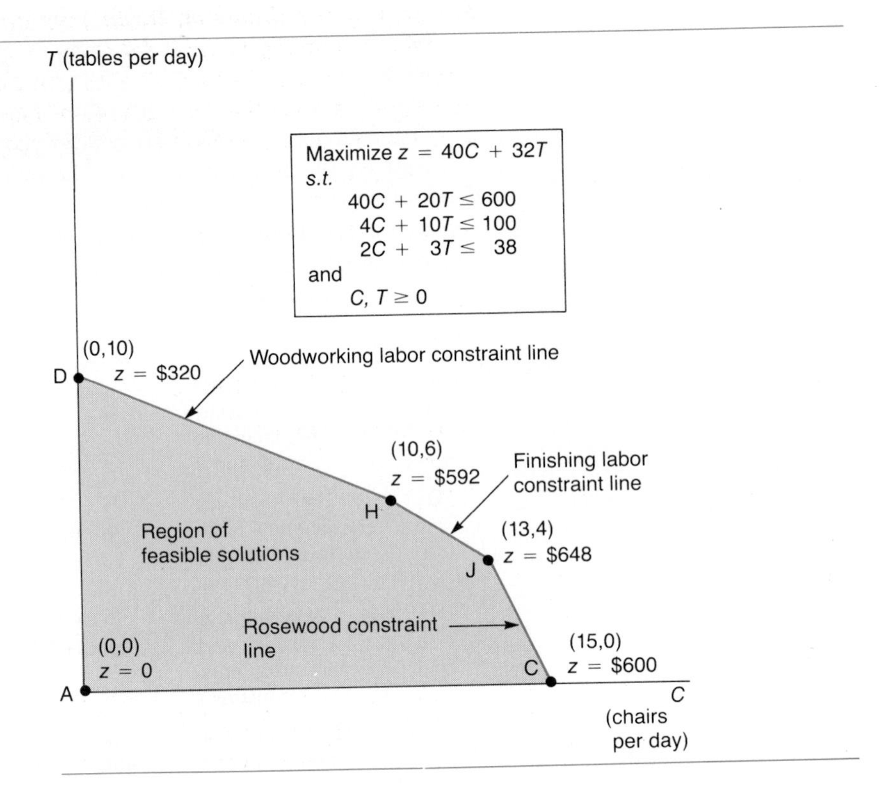

EXERCISES

For each of the following linear programming models (1) picture the region of feasible solutions, (2) use iso-profit lines to determine the optimal corner point, and (3) determine the coordinates of the optimal corner point and the value of the objective function at that point.

1. MAXIMIZE $z = 2x + 3y$

   subject to: $x + y \leq 100$
   $4x + 8y \leq 500$

   and $x, y \geq 0$

2. MAXIMIZE $z = 4x + 2y$

   subject to: $x + y \leq 10$
   $2x + 4y \leq 24$
   $y \leq 5$

   and $x, y \geq 0$

3. MAXIMIZE $z = 4x + 5y$

   subject to: $x + 3y \leq 15$
   $2x + y \leq 10$

   and $x, y \geq 0$

4. MAXIMIZE $z = 0.25x + 0.4y$

   subject to: $x + 3y \leq 15$
   $2x + y \leq 15$
   $x + 1.5y \leq 9$

   and $x, y \geq 0$

For each of the following linear programming models (1) graph the constraints, thereby picturing the region of feasible solutions; (2) determine the coordinates of each corner point of the region of feasible solutions; and (3) find the optimal corner point by evaluating the objective function at each corner point.

5. MAXIMIZE $z = 9x + 6y$

   subject to: $x + y \leq 25$
   $y \geq 4$
   $x \geq 8$

   and $x, y \geq 0$

6. MAXIMIZE $z = 3x + 4y$

   subject to: $5x + 6y \leq 60$
   $x + 2y \leq 16$

   and $x, y \geq 0$

7. MAXIMIZE $z = 8x + 9y$

   subject to: $4x + 5y \leq 70$
   $x + y \leq 16$

   and $x, y \geq 0$

8. MAXIMIZE $z = 6x + 5y$

   subject to: $3x + 2y \leq 6$
   $x + 2y \leq 4$
   $x \geq 0.5$

   and $x, y \geq 0$

9. MAXIMIZE $z = 16x + 9y$

   subject to: $x \leq 5$
   $y \leq 12$
   $2x + y \leq 20$

   and $x, y \geq 0$

10. MAXIMIZE $z = 2x + 4y$

    subject to: $x + 2y \leq 12$
    $3x + y \leq 12$

    and $x, y \geq 0$

7.4 CASE 2. MOODY'S LAWN-CARE SERVICES PROBLEM: A COST MINIMIZATION SITUATION

Frequently the problem situation is of such a nature that the decision maker seeks a strategy which will minimize a cost rather than one which will directly maximize a benefit. To illustrate such a situation, let us consider the case of Moody's Lawn-Care Services.

Moody's Lawn-Care Services contracts with homeowners and business firms for a package of services including planting, fertilizing, weed control, and maintenance of grass lawns. Burr Moody, owner of Moody's Lawn-Care Services and a horticulturist with many years' experience in the field, mixes his own lawn-treatment formulas to meet the special needs of the individual account. Currently, a treatment solution is needed which contains at least 14 measures of chemical A, at least 5 measures of chemical B, and at least 12 measures of chemical C. Two preparations containing these chemicals are sold commercially. Each canister of Solu-X contains four measures of chemical A, one measure of chemical B, and two measures of chemical C. Each canister of Phos-Pho-Gen contains two measures of chemical A, one measure of chemical B, and three measures of chemical C. Each canister of Solu-X costs \$4, while each canister of Phos-Pho-Gen costs \$3.

The total revenue Moody's receives from an account is fixed by contract; thus, the company would like to perform the required services at a minimum cost. How, then, should Mr. Moody combine the two products, Solu-X and Phos-Pho-Gen, to obtain a lawn treatment containing the required quantities of each chemical?

We begin by summarizing the relevant data, as shown in Table 7.2.

Table 7.2

Chemical Requirements and Availabilities, Moody's Lawn-Care Services Case

Chemicals required	Quantities of chemical per container (measures)		Minimum amount required in treatment (measures)
	Solu-X	Phos-Pho-Gen	
Chemical A	4	2	14
Chemical B	1	1	5
Chemical C	2	3	12
Cost, per container	\$4	\$3	

The decision variables are then defined

S = number of canisters of Solu-X to use
P = number of canisters of Phos-Pho-Gen to use

Because Moody's wants to obtain the required mixture at the lowest possible cost, the objective function for the linear programming model is

$$\text{MINIMIZE } z = 4S + 3P = \text{total cost, in dollars}$$

At least 14 measures of chemical A are required for the treatment. Each canister of Solu-X provides four units of chemical A, while each canister of Phos-Pho-Gen provides two units of this chemical. Thus, one specification which must be met by the treatment mixture can be expressed mathematically as

$$4S + 2P \geq 14 \qquad \text{(Minimum requirement for chemical A)}$$

Note that the inequality is of the GREATER-THAN-OR-EQUAL-TO variety, thereby allowing the mixture to exceed this minimum of 14 units of chemical A but not to fall below the minimum required amount.

The requirements for chemical B and chemical C may be expressed

$$S + P \geq 5 \qquad \text{(Minimum requirement for chemical B)}$$

and

$$2S + 3P \geq 12 \qquad \text{(Minimum requirement for chemical C)}$$

And again we have the physically reasonable requirement that both S and P must be nonnegative; that is,

$S, P \geq 0$ (Nonnegativity requirement)

We write the complete model for the Moody's Lawn-Care Services problem as

$$\text{MINIMIZE } z = 4S + 3P = \text{total cost, in dollars}$$

where

S = number of canisters of Solu-X to use
P = number of canisters of Phos-Pho-Gen to use

subject to these requirements

$4S + 2P \geq 14$ (Minimum requirement for chemical A)
$S + P \geq 5$ (Minimum requirement for chemical B)
$2S + 3P \geq 12$ (Minimum requirement for chemical C)

with the further requirement that

$S, P \geq 0$

Notice that this model differs from the Bentwood model in two important respects: the objective function here is to be minimized rather than maximized, and the constraint inequalities are of the greater-than-or-equal-to, rather than the less-than-or-equal-to, variety.

A General Statement of the Linear Programming Minimization Model

The linear programming MINIMIZATION model in its "pure" form can be expressed symbolically as

$$\text{MINIMIZE } z = c_1x_1 + c_2x_2 + \cdots + c_nx_n$$

subject to:

$$\begin{aligned} a_{11}x_1 + a_{12}x_2 + \cdots + a_{1n}x_n &\geq b_1 \\ a_{21}x_1 + a_{22}x_2 + \cdots + a_{2n}x_n &\geq b_2 \\ &\vdots \\ a_{m1}x_1 + a_{m2}x_2 + \cdots + a_{mn}x_n &\geq b_m \end{aligned}$$

and

$$x_j \geq 0 \text{ for all } j$$

These properties characterize the linear programming MINIMIZATION model in its pure form:

1. The objective function is to be MINIMIZED.
2. All constraints of the model are expressions requiring the variables, or linear combinations of the variables, to be GREATER THAN OR EQUAL TO ($\geq$) a NONNEGATIVE CONSTANT.

The Region of Feasible Solutions in the Lawn-Care Model

The graphical procedures used to depict the maximization problems can also be used to depict minimization problems, when these problems have only two decision variables. The constraints in the Lawn-Care Services model are pictured, individually, in Figure 7.5. Here, too, the nonnegativity requirements of the model restrict the region of consideration to the first quadrant of the rectangular coordinate system.

We may begin with the constraint for the minimum quantity of chemical A and graph the equality portion

$$4S + 2P = 14$$

This gives us a line through the points (3.5,0) and (0,7). Any combination of inputs represented by a point on this line—which is the BC line in Figure 7.5A—would provide exactly 14 units of chemical A.

The region described by the inequality portion of the chemical A constraint

$$4S + 2P > 14$$

includes all points above and to the right of the BC line. Such points represent combinations of inputs that would furnish more than 14 mea-

Figure 7.5

The Constraints of the Moody's Lawn-Care Services Model (restricted to the first quadrant by the nonnegativity requirements)

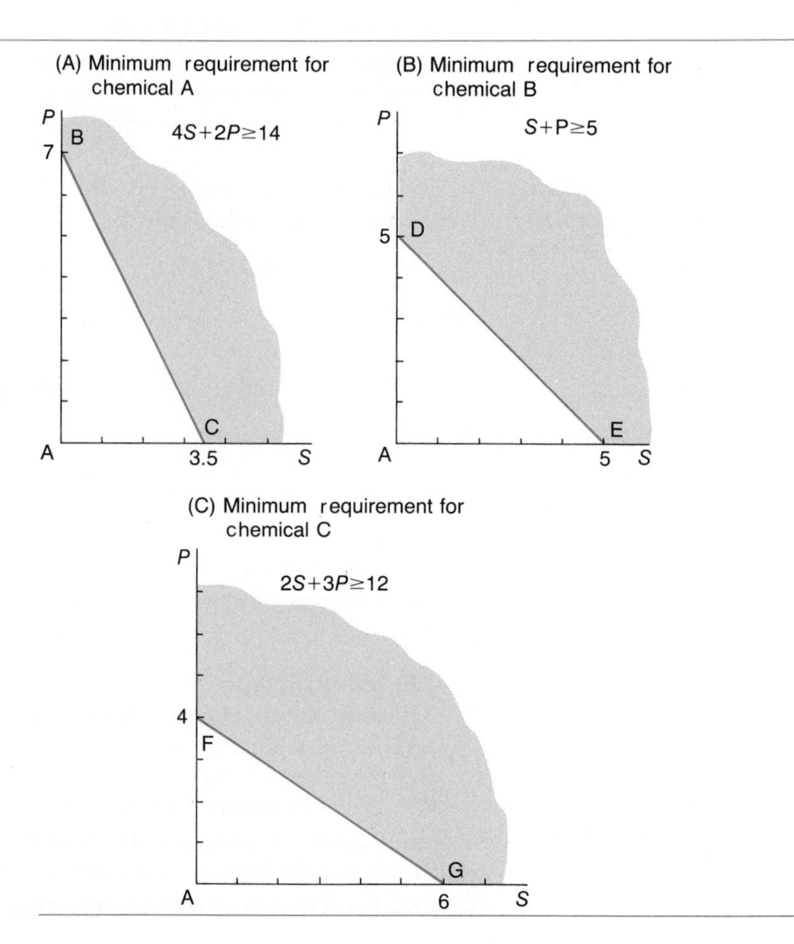

sures of chemical A, but this is a permissible solution since the requirement for chemical A was that the solution should contain *at least* 14 measures. Combinations of inputs represented by points in the first quadrant but below and to the left of the BC line would not provide the minimum amount of chemical A.

The regions specified by the other two constraints—the minimum requirement for chemical B and the minimum requirement for chemical A—are shown in Figure 7.5B and 7.5C.

Finally, in Figure 7.6, the region common to all inequalities of the model, including the nonnegativity requirement, is shown. This is the REGION OF FEASIBLE SOLUTIONS for the Lawn-Care Services model. Notice that this region is unbounded.

Figure 7.6

Region of Feasible Solutions, Moody's Lawn-Care Services Model

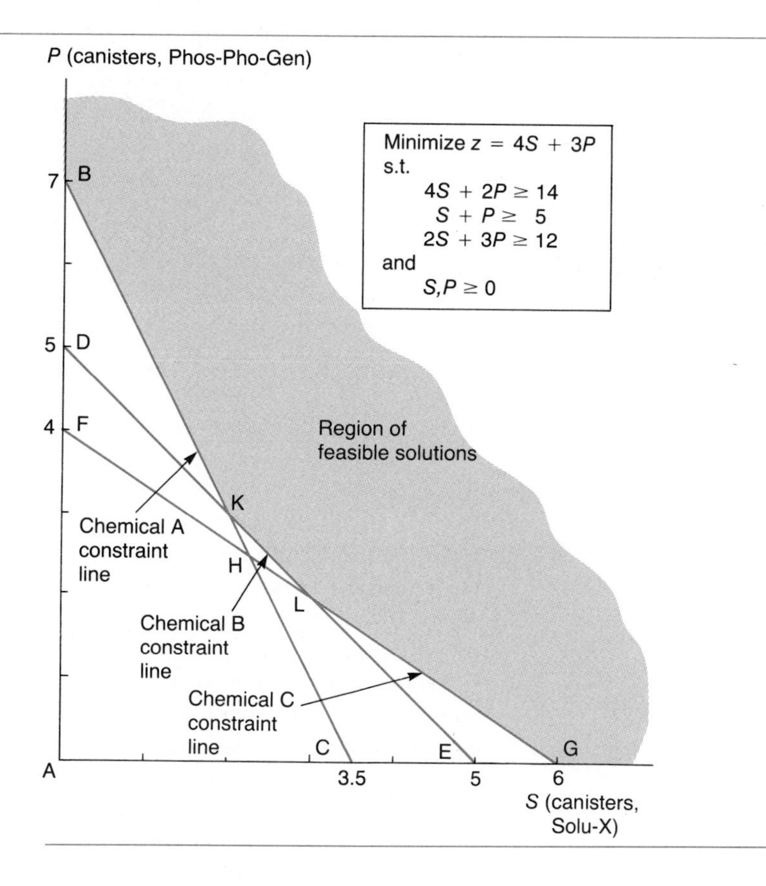

Using Iso-Cost Lines to Determine the Minimum-Cost Solution

ISO-COST LINES can be used to determine the minimum-cost strategy in a linear programming minimization model in much the same way that iso-profit lines can be used to determine the maximum-profit strategy in a linear programming maximization model.

Figure 7.7

Using Iso-Cost Lines to Locate the Minimum-Cost Solution

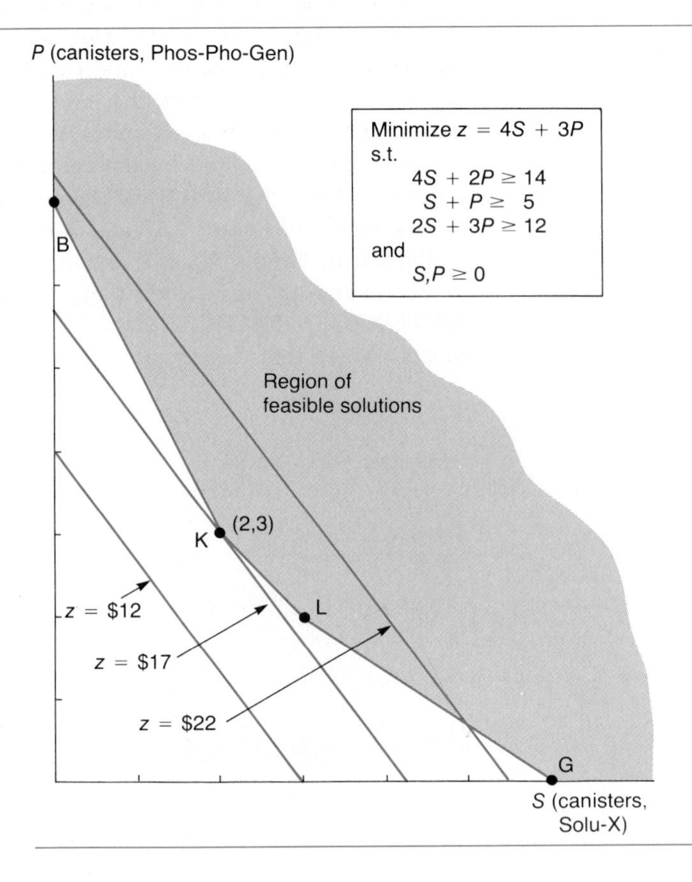

In Figure 7.7 the region of feasible solutions for the Lawn-Care Services model has been pictured. A family of parallel lines—cost $= z = 4S + 3P$—corresponding to various values for total cost has been superimposed over the region. Each line has slope of $-4/3$.

Iso-cost lines closer to the origin represent a smaller cost than do iso-cost lines further out from the origin. Hence, the minimum-cost strategy is given by the iso-cost line which is nearest the origin but still intersecting the region of feasible solutions. We see in Figure 7.7 that the minimum-cost solution to the Lawn-Care Services problem occurs at point K, where the chemical A constraint intersects the chemical B constraint line. The coordinates of this point are found by solving simultaneously the two equations represented by the lines that cross here. These coordinates are (2,3). Thus, the minimum-cost strategy involves the use of two canisters of Solu-X and three canisters of Phos-Pho-Gen, for a total cost of \$17.

Evaluating the Corner Points in the Lawn-Care Services Problem

It is a property of all linear programming models, whether maximization or minimization, that the optimal solution (if one exists) occurs at a corner point of the region of feasible solutions. Thus, an alternative to the use of iso-cost lines to locate the optimal solution in the Lawn-Care Services problem would be to evaluate the objective function at each corner point of the feasible region.

The corner points and their coordinates, shown in Figure 7.8, are as follows:

Corner point	Intersecting constraint lines	Equations which lines represent	Coordinates
B	Nonnegativity requirement for S and chemical A constraint line	$S = 0$ $4S + 2P = 14$	(0,7)
K	Chemical A constraint and chemical B constraint lines	$4S + 2P = 14$ $S + P = 5$	(2,3)
L	Chemical B constraint and chemical C constraint lines	$S + P = 5$ $2S + 3P = 12$	(3,2)
G	Chemical C constraint line and nonnegativity requirement for P	$2S + 3P = 12$ $P = 0$	(6,0)

These computations are required to evaluate the objective function at each of the corner points.

Corner point	Coordinates	Value of objective function	
B	(0,7)	$4(0) + 3(7) = \$21$	
K	(2,3)	$4(2) + 3(3) = \$17$	←— OPTIMAL STRATEGY
L	(3,2)	$4(3) + 3(2) = \$18$	
G	(6,0)	$4(6) + 3(0) = \$24$	

The minimum value of the objective function is associated with corner point K, having coordinates (2,3) and representing the use of two canisters of Solu-X and three canisters of Phos-Pho-Gen. This strategy would produce the required lawn treatment for the total cost of $17, the minimum possible under the prevailing conditions.

Figure 7.8

Locating the Optimal Solution by Evaluating Corner Points of the Region of Feasible Solutions

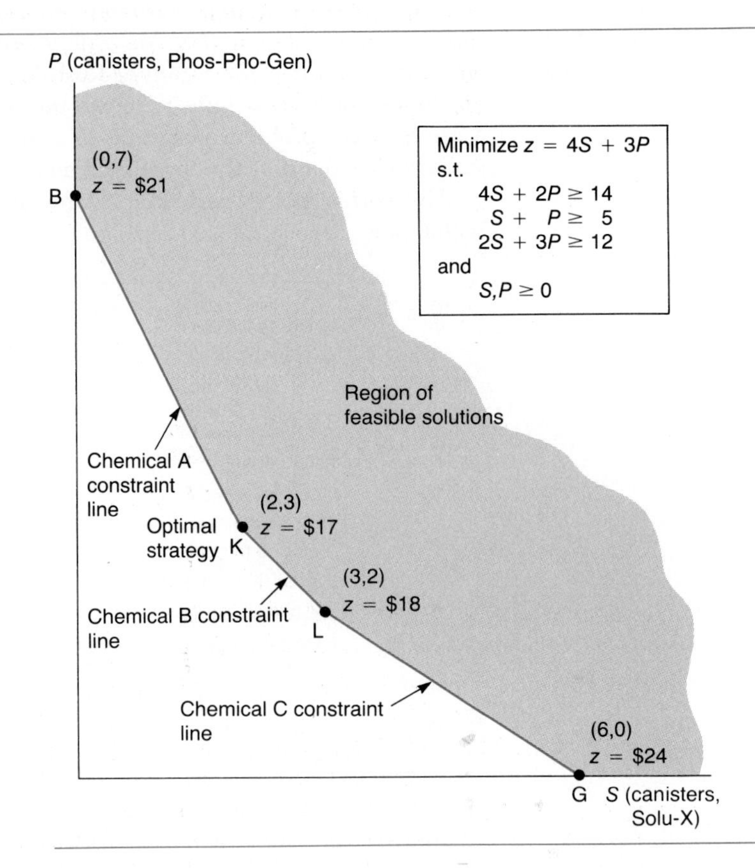

EXERCISES

For each of the following linear programming models: (1) Picture the region of feasible solutions; (2) use iso-cost lines to determine the optimal strategy; (3) determine the coordinates of the optimal corner point and the value of the objective function at that point.

11. MINIMIZE $z = x + y$

    subject to: $3x + 2y \geq 60$
    $x + 2y \geq 40$

    and $x, y \geq 0$

12. MINIMIZE $z = 2x + 4y$

    subject to: $x + 2y \geq 12$
    $3x + y \geq 12$

    and $x, y \geq 0$

13. MINIMIZE $z = 4x + 5y$

    subject to: $x + y \geq 3$
    $x + 2y \geq 4$

    and $x, y \geq 0$

14. MINIMIZE $z = 3x + 2y$

    subject to: $2x + y \geq 10$
    $x + 3y \geq 15$
    $x + y \geq 8$

    and $x, y \geq 0$

For each of the following linear programming problems: (1) graph the constraints; (2) find the coordinates of each corner point of the region of feasible solutions, (3) determine the optimal strategy by evaluating the objective function at each corner point.

15. MINIMIZE $z = 3x + 4y$

    subject to: $1.5x + y \geq 60$
    $2x + 3y \geq 150$
    $x + 2y \geq 90$

    and $x, y \geq 0$

16. MINIMIZE $z = 3x + 2y$

    subject to: $x + y \geq 20$
    $x + 2y \geq 30$
    $y \geq 5$

    and $x, y \geq 0$

17. MINIMIZE $z = 2x + 4y$

    subject to: $4x + y \geq 20$
    $3x + 2y \geq 60$
    $x + 4y \geq 24$

    and $x, y \geq 0$

18. MINIMIZE $z = x + 5y$

    subject to: $x + 2y \geq 8$
    $x + 4y \geq 10$
    $x \leq 14$

    and $x, y \geq 0$

7.5 A FEW ILLUSTRATIVE CASES

Case 3. The Poultry-Plus Feed Mill Problem: A "Diet" Problem

Most linear programming models are neither "pure" maximization problems with all less-than-or-equal-to constraints nor "pure" minimization problems with all greater-than-or-equal-to constraints. Most, instead, are composed of a combination of constraint types, as exemplified by the following "diet" problem.

One of the most important products produced at the Poultry-Plus Feed Mill is Gro-Chick Mix. This is a feed designed to stimulate fast growth in baby chicks. Two basic ingredients and a fiber filler are used in the mix. While these may be blended together in various proportions, there are certain standards that must be met. The feed mix must not contain more than 30 percent filler. Each pound of the mix must contain not more than 45 units of nutrient A nor fewer than 16 units of nutrient B. Each pound of basic ingredient 1 contains 90 units of nutrient A and 40 units of nutrient B. Each pound of basic ingredient 2 contains 50 units of nutrient A and 20 units of nutrient B. The fiber filler contains no nutrients.

Each pound of the feed mix sells for the same price regardless of its actual composition as long as it meets these established standards. This price remains fairly stable throughout the season. Ingredients 1 and 2 cost \$0.25 and \$0.20 per pound, respectively, while each pound of the filler costs \$0.05.

The production manager at Poultry-Plus is Fred Foul. Mr. Foul needs to know in what proportions he should combine the ingredients in order to meet the standards at a minimum cost. The data relevent to this decision are summarized in Table 7.3.

Table 7.3

Nutritional Requirements for Gro-Chick Mix, Poultry-Plus Feed Mill Case

Quantities, per pound of ingredient	Ingredients: Basic ingredient 1	Basic ingredient 2	Fiber filler	Required quantities in pound of mix
Nutrient A	90 units	50 units	—	Not more than 45 units
Nutrient B	40 units	20 units	—	Not fewer than 16 units
Cost, per pound	$0.25	$0.20	$0.05	

Because the decision involves the quantity of each ingredient to use in the feed mix, let us define the decision variables, not as the quantity of product to produce, but rather as

x_1 = pounds of ingredient 1 to use in each pound of Gro-Chick Mix
x_2 = pounds of ingredient 2 to use in each pound of Gro-Chick Mix
x_3 = pounds of fiber filler to use in each pound of Gro-Chick Mix

The model will be used to determine the minimum-cost blend of ingredients for *one pound* of the mix; therefore,

$$x_1 + x_2 + x_3 = 1 \qquad \text{and} \qquad x_3 = 1 - x_1 - x_2$$

This means that the model can be stated mathematically as a two-decision-variable model.

The objective function of the model, is, then

$$\text{MINIMIZE } z = 25x_1 + 20x_2 + 5x_3 = 25x_1 + 20x_2 + 5(1 - x_1 - x_2)$$
$$= 20x_1 + 15x_2 + 5$$
$$\text{(Cost, in cents, per pound)}$$

The restrictions that are imposed on the possible strategies that the feed mill might adopt become the constraints of the model. We notice a greater diversity of mathematical forms for the constraints in this model than in either the Bentwood model or the Lawn-Care Services model. Let us discuss these one at a time.

One requirement is that the filler must not constitute more than 30 percent of the total mix. Because the filler is defined in terms of the other two ingredients, this constraint is expressed as

$$1 - x_1 - x_2 \leq 0.3 \qquad \text{(Maximum proportion of filler)}$$

This expression is rewritten as

$$-x_1 - x_2 \leq -0.7$$
$$x_1 + x_2 \geq 0.7$$

This constraint is pictured in Figure 7.9A.

Figure 7.9

Constraints in the Poultry-Plus Feed Mill problem

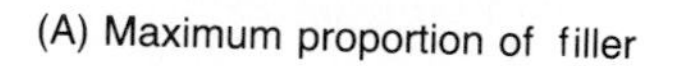

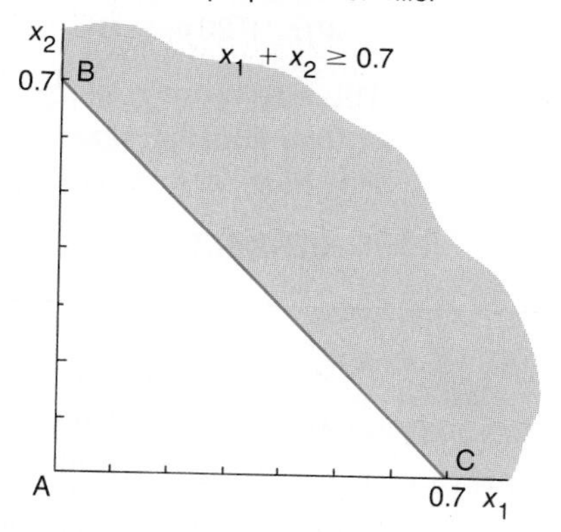

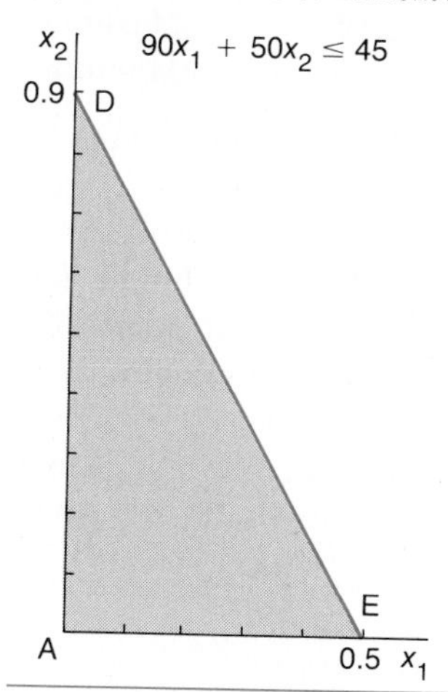

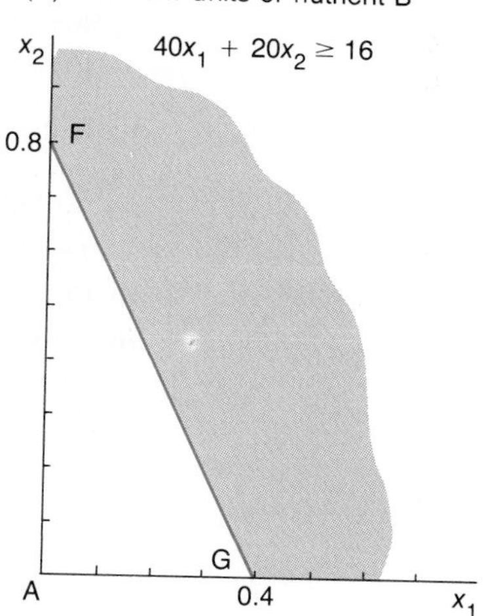

Another constraint is required to ensure that the appropriate number of units of nutrient A is contained in each pound of the feed mix. This number must not be more than 45. Each pound of basic ingredient 1 contains 90 units of nutrient A, and each pound of basic ingredient 2 contains 50 units of the nutrient. The filler contains no units of nutrient A. This constraint inequality is written

$$90x_1 + 50x_2 \leq 45 \qquad \text{(Maximum number of units of nutrient A)}$$

This constraint is pictured in Figure 7.9B.

Finally, each pound of the Gro-Chick mix must contain not fewer than 16 units of nutrient B. Each pound of ingredient 1 contains 40 units of nutrient B, and each pound of ingredient 2 contains 20 units of the nutrient. The filler contains no units of nutrient B. This constraint can be expressed

$$40x_1 + 20x_2 \geq 16 \qquad \text{(Minimum number of units of nutrient B)}$$

This constraint is pictured in Figure 7.9C.

The linear programming model for the Poultry-Plus Feed Mill problem may be summarized

$$\text{MINIMIZE } z = 20x_1 + 15x_2 + 5 \qquad \text{(Cost, in cents, per pound)}$$

where

x_1 = pounds of ingredient 1 to use in each pound of Gro-Chick Mix
x_2 = pounds of ingredient 2 to use in each pound of Gro-Chick Mix
$1 - x_1 - x_2$ = pounds of filler to use in each pound of Gro-Chick Mix

subject to these constraints:

$$x_1 + x_2 \geq 0.7 \qquad \text{(Maximum amount of filler)}$$
$$90x_1 + 50x_2 \leq 45 \qquad \text{(Maximum number of units of nutrient A)}$$
$$40x_1 + 20x_2 \geq 16 \qquad \text{(Minimum number of units of nutrient B)}$$

with the further requirement that

$$x_1, x_2 \geq 0$$

The region of feasible solutions for the Poultry-Plus model is shown in Figure 7.10. The corner points of this feasible region are evaluated in light of the objective function, as follows:

Corner point	Lines intersecting at this point	Equations which lines represent	Coordinates	Value of objective function
D	Nonnegativity requirement for x_1 and nutrient A constraint line	$x_1 = 0$ $90x_1 + 50x_2 = 45$	(0,0.9)	18.5
F	Nonnegativity requirement for x_1 and nutrient B constraint line	$x_1 = 0$ $40x_1 + 20x_2 = 16$	(0,0.8)	17
H	Nutrient B constraint line and filler constraint line	$40x_1 + 20x_2 = 16$ $x_1 + x_2 = 0.7$	(0.1,0.6)	16← ← OPTIMAL STRATEGY
J	Filler constraint line and nutrient A constraint line	$x_1 + x_2 = 0.7$ $90x_1 + 50x_2 = 45$	(0.25,0.55)	18.25

The minimum-cost strategy dictates that each pound of Gro-Chick Mix should contain 0.1 pounds of basic ingredient 1, 0.6 pounds of basic ingredient 2, and 0.3 pounds of fiber filler. The cost will be 16 cents per pound of Gro-Chick Mix.

Figure 7.10

Region of Feasible Solutions, Poultry-Plus Feed Mill Problem

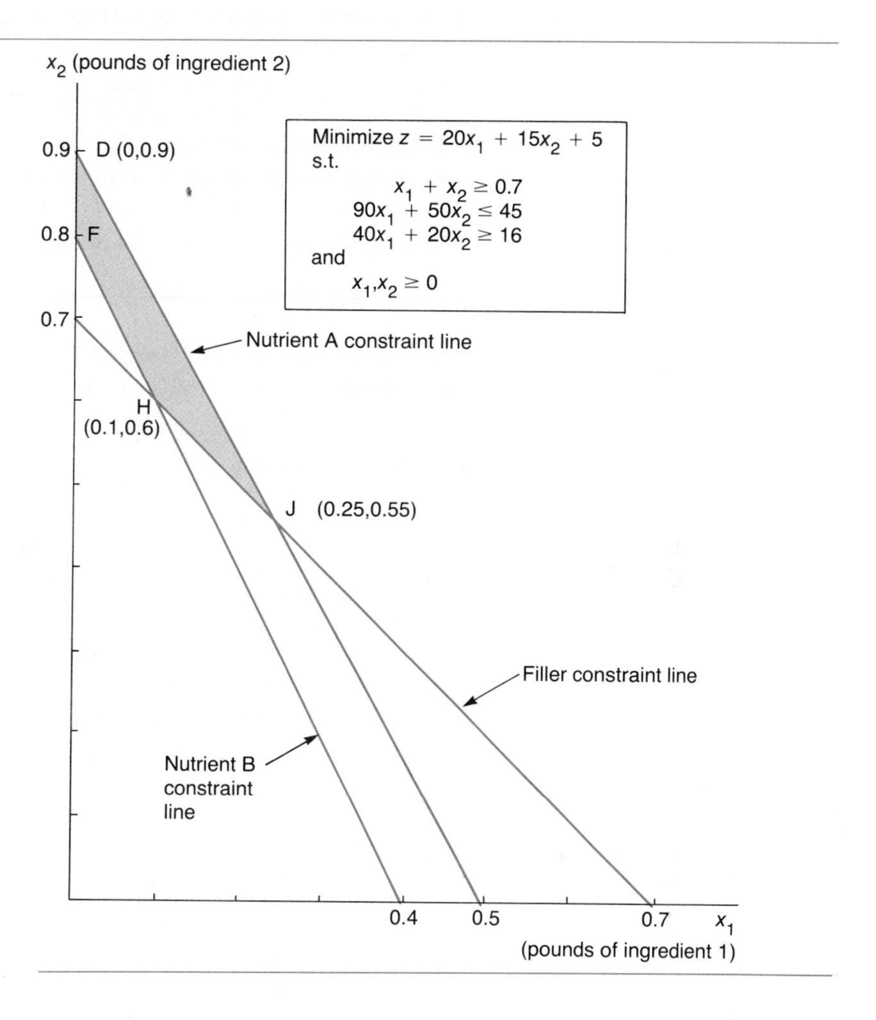

Case 4. The Haynes-Webster, Inc., Problem: A Portfolio-Selection Situation

Portfolio managers—of mutual funds, credit unions, insurance companies, banks, and other financial institutions—are responsible for selecting specific investments for available funds from among the possible investment alternatives. In these portfolio-selection decisions, the objective is generally either to maximize expected return on investment or to minimize the risk involved. Constraints may be imposed not only by the existence or nonexistence of attractive investment opportunities but also by policies of the firm or by governmental regulations.

Haynes-Webster, Inc., manages investment portfolios for its clients. One client has up to $100,000 for immediate investment and has specified that the money be invested either in Montana Mining Company stock or in Osage Oil Corporation stock. The Montana Mining stock is a medium-risk stock with an expected annual return of 15 percent, while the Osage Oil stock is a high-risk stock with an expected annual yield of 25 percent. Because of the risk factor, the client has specified that the amount invested in oil stock should not exceed one third of the total amount invested in both stocks. The client also requires that at least $25,000 be invested in mining stock. Under the specified conditions, how should the funds be invested so as to maximize the investment income?

Let the decision variables be defined

x = amount (in thousands of dollars) invested in Montana Mining Company stock
y = amount (in thousands of dollars) invested in Osage Oil Corporation stock

The objective is to maximize the investment-income function; that is, the objective function of the linear programming model is

MAXIMIZE $z = 0.15x + 0.25y$
$= \text{total investment income (in thousands of dollars)}$

The inequality

$$x + y \leq 100 \qquad \text{(Funds available for investment)}$$

expresses the condition that the client has up to $100,000 he wishes to invest. This constraint is pictured in Figure 7.11A.

The inequality

$$y \leq (1/3)(x + y) \qquad \text{(Maximum proportion of oil stock)}$$

expresses the client's requirement that the investment in oil stock should not exceed one third of the total amount invested in both stocks. This inequality can be equivalently written as

$$(2/3)y - (1/3)x \leq 0$$

or

$$-x + 2y \leq 0$$

This constraint is pictured in Figure 7.11B.

Finally, the client's desire that at least $25,000 be invested in mining stock is expressed by the inequality

$$x \geq 25 \qquad \text{(Minimum investment in mining stock)}$$

This constraint is pictured in Figure 7.11C.

The complete linear programming model is

MAXIMIZE $z = 0.15x + 0.25y$
$= \text{total investment income, in thousands of dollars}$

where

x = amount, in thousands of dollars, invested in Montana Mining Company stock
y = amount, in thousands of dollars, invested in Osage Oil Corporation stock

subject to these constraints

$$x + y \leq 100 \qquad \text{(Funds available for investment)}$$
$$-x + 2y \leq 0 \qquad \text{(Maximum proportion of oil stock)}$$
$$x \geq 25 \qquad \text{(Minimum investment in mining stock)}$$

with the further requirement that

$$x, y \geq 0$$

Figure 7.11

The Constraints in the Haynes-Webster Model

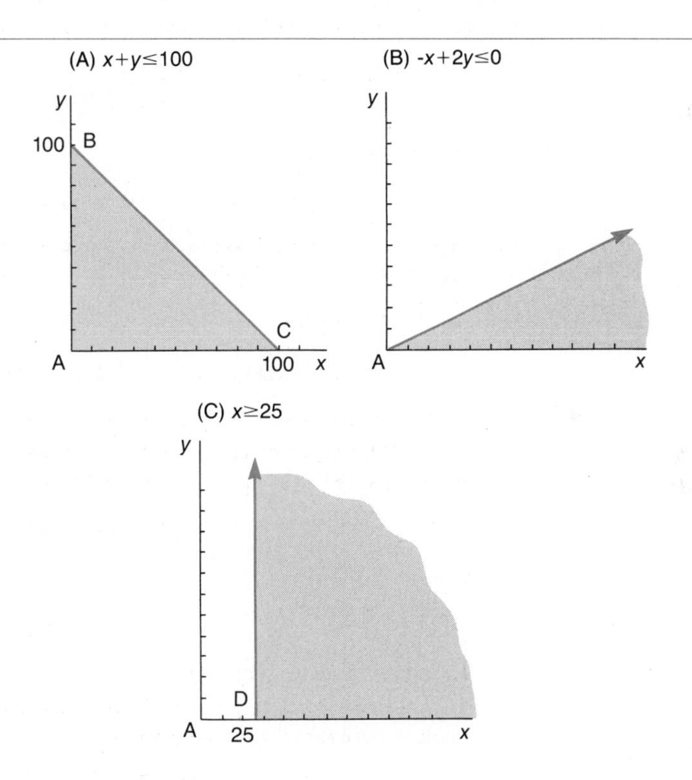

The region of feasible solutions for this model is depicted in Figure 7.12. The points D, E, F, and C are corner points of the feasible region. The optimal solution—the strategy yielding the maximum possible investment income—is represented by the coordinates of one of these four corner points.

The corner points are evaluated as follows:

Corner point	Lines intersecting at this point	Equations which lines represent	Coordinates	Value of objective function
D	Nonnegativity requirement for y and minimum amount of mining stock constraint line	$y = 0$ $x = 25$	(25,0)	\$3.75 thousand
E	Minimum amount of mining stock and maximum proportion of oil stock constraint lines	$x = 25$ $-x + 2y = 0$	(25,12.5)	\$6.875 thousand
F	Maximum proportion of oil stock and funds available for investment constraint lines	$-x + 2y = 0$ $x + y = 100$	($66\frac{2}{3}$, $33\frac{1}{3}$)	\$$18\frac{1}{3}$ thousand ← ← OPTIMAL STRATEGY
C	Funds available for investment constraint line and nonnegativity requirement for y	$x + y = 100$ $y = 0$	(100,0)	\$15 thousand

Figure 7.12

Region of Feasible Solutions, Haynes-Webster Model

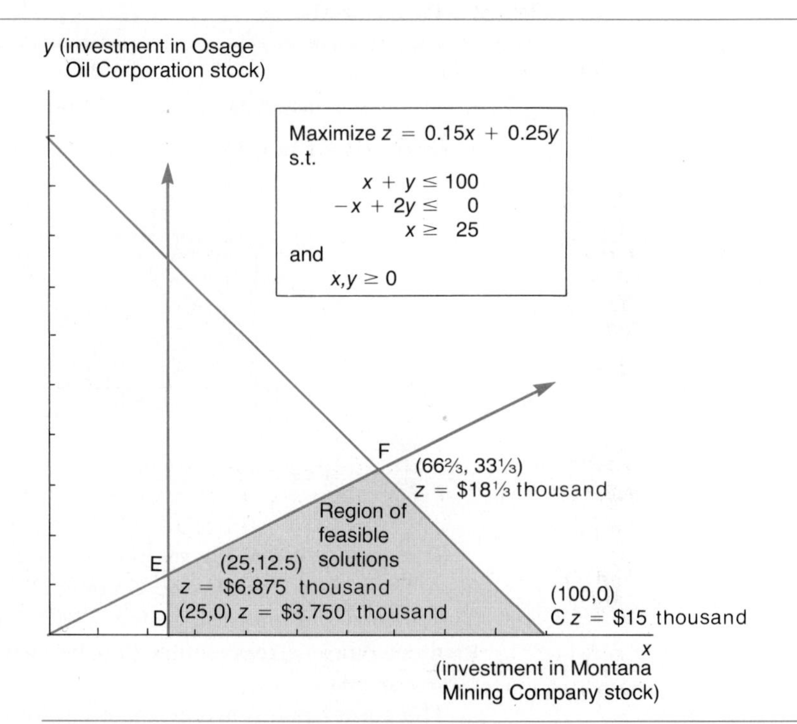

Corner point F represents the optimal strategy. This strategy involves an investment of $66,666 in Montana Mining Company stock along with an investment of $33,333 in Osage Oil Corporation stock. The expected annual income from such an investment plan is $18,333.

Case 5. The Mid-States Cooperative Problem: A Transportation Problem

TRANSPORTATION PROBLEMS commonly have the following structural features. Quantities of a homogeneous product are available at a number of different SOURCES (such as factories, warehouses, or distribution centers). Specified quantities of the product are demanded at various DESTINATIONS (such as sales outlets, or customers). The cost of transporting the product from the different sources to the different destinations varies depending upon the shipping route. The objective is to determine the minimum-cost shipping plan.

The Mid-States Cooperative has grain warehouses in Concordia and North Platte. Grain is shipped from these warehouses to customers throughout the region. The cooperative has just received orders for immediate delivery of barley in the following quantities: 1,500 units to Scotts Bluff, 500 units to Plainview, and 2,000 units to Fremont. There are available 1,800 units of barley at the Concordia warehouse and 2,200 units at the North Platte warehouse.

The cost of shipping one unit of barley from Concordia to Scotts Bluff is $3.20, from Concordia to Plainview is $2.40, and from Concordia to Fremont is $1.10. The costs of shipping one unit of barley from the North Platte warehouse to Scotts Bluff, Plainview, and Fremont are $1.90, $2.25, and $3.80, respectively. How, then, should the grain be shipped so as to minimize the total shipping cost? (The cost data relevant to the problem are summarized in Table 7.4.)

Table 7.4

Schedule of Shipping Costs, per Unit of Barley, Mid-States Cooperative Case

	To			
From	Scotts Bluff	Plainview	Fremont	Amount available
Concordia	$3.20	$2.40	$1.10	1,800 units
North Platte	$1.90	$2.25	$3.80	2,200 units
Amount Required	1,500 units	500 units	2,000 units	4,000 units

Because the amount of barley available (4,000 units) equals the amount of barley required by the customers, the situation can be formulated as a two-decision-variable model (see Table 7.5). Let

x = number of units of barley shipped from Concordia to Scotts Bluff
y = number of units of barley shipped from Concordia to Plainview

The remaining barley at the Concordia warehouse will be shipped to Fremont. Thus,

Table 7.5

Definition of Decision Variables in Mid-States Cooperative Case

Number of units shipped from	Number of units shipped to			Total number of units available
	Scotts Bluff	Plainview	Fremont	
Concordia	x	y	$1{,}800 - (x + y)$	1,800
North Platte	$1{,}500 - x$	$500 - y$	$2{,}000 - [1{,}800 - (x + y)]$	2,200
Total Number of Units Required	1,500	500	2,000	4,000

$1{,}800 - x - y =$ number of units of barley shipped from Concordia to Fremont

The difference between the total quantity ordered by the customer in Scotts Bluff and the quantity shipped to this customer from Concordia will be shipped from the warehouse in North Platte; that is,

$1{,}500 - x =$ number of units of barley shipped from North Platte to Scotts Bluff

In the same manner,

$500 - y =$ number of units of barley shipped from North Platte to Plainview

and

$2{,}000 - (1{,}800 - x - y) = 200 + x + y =$ number of units of barley shipped from North Platte to Fremont

The total cost C of shipping these quantities over these routes will be

$$
\begin{aligned}
z &= 3.20x + 2.40y + 1.10(1{,}800 - x - y) + 1.90(1{,}500 - x) \\
&\quad + 2.25(500 - y) + 3.80(200 + x + y) \\
&= 4x + 2.85y + 6{,}715
\end{aligned}
$$

Now, the amount sent from each source to each destination must be constrained to be nonnegative, as

$$
\begin{array}{lcr}
1{,}800 - x - y \geq 0 & \text{or} & x + y \leq 1{,}800 \\
1{,}500 - x \geq 0 & \text{or} & x \leq 1{,}500 \\
500 - y \geq 0 & \text{or} & y \leq 500 \\
200 + x + y \geq 0 & \text{or} & x + y \geq -200
\end{array}
$$

The complete model is

MINIMIZE $z = 4x + 2.85y + 6{,}715 =$ total cost, in dollars

where

x = number of units of barley shipped from Concordia to Scotts Bluff
y = number of units of barley shipped from Concordia to Plainview

subject to these constraints

$$x + y \leq 1{,}800$$
$$x \leq 1{,}500$$
$$y \leq 500$$
$$x + y \geq -200$$

with the further requirement that

$$x,y \geq 0$$

The region of feasible solutions for the model is shown in Figure 7.13.

Figure 7.13

Region of Feasible Solutions and Optimal Corner-Point, Mid-States Cooperative Case

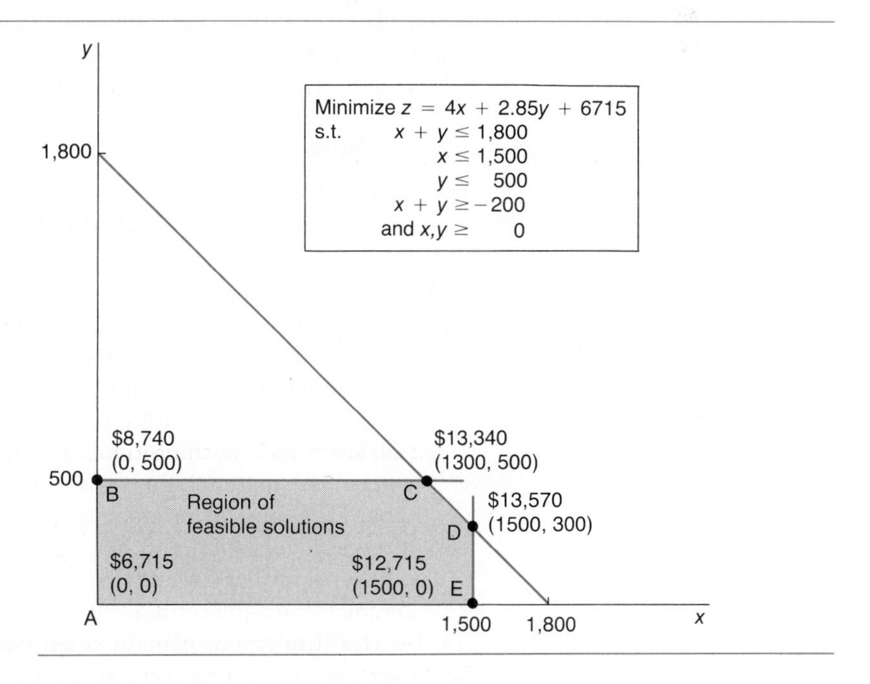

Notice that the fourth constraint ($x + y \geq -200$) is redundant. The nonnegativity requirement restricts both x and y to nonnegative values; hence, their sum would have to be greater than −200. This redundant constraint is not pictured.

The corner points of the feasible region are evaluated as follows:

Corner point	Equations representing lines which intersect at this point	Coordinates	Value of objective function
A	$x=0$ $y=0$	(0,0)	$6,715 ← ← OPTIMAL STRATEGY
B	$x=0$ $y=500$	(0,500)	$8,740
C	$y=500$ $x+y=1,800$	(1300,500)	$13,340
D	$x+y=1,800$ $x=1,500$	(1500,300)	$13,570
E	$x=1,500$ $y=0$	(1500,0)	$12,715

Thus, the minimum-cost shipping plan will involve a total cost of $6,715. This is achieved by shipping no barley from Concordia to Scotts Bluff and no barley from Concordia to Plainview. This means that 1,800 units must be shipped from Concordia to Fremont, 1,500 units from North Platte to Scotts Bluff, 500 units from North Platte to Plainview, and 200 units from North Platte to Fremont.

EXERCISES

19. Lewis Manufacturing Company manufactures two different products. The demand for both products is strong enough so that the firm can sell as many units of either product, or of both, as it can produce and at such a price as to realize a per-unit profit contribution of $16 on product 1 and $10 on product 2. Unfortunately, the production capacity of the company's plant is severely limited. This limitation stems from the fact that the manufacture of the products involves the utilization of three scarce resources: raw material, labor, and machine time. Each unit of product 1 requires four units of raw material, three units of labor, and two units of machine time. Each unit of product 2 requires two units of raw material, three units of labor, and five units of machine time. The firm has a daily supply of 24 units of raw material, 21 units of labor, and 30 units of machine time.

    The firm's management team would like to determine, given these circumstances, how many units of each product should be manufactured in order to maximize total profit contribution realized on the items produced.

    Formulate a linear programming model that can aid in this decision-making process. Graph the constraints of the model. Determine the coordinates of the corner points of the region of feasible solutions. Find the optimal strategy by evaluating the objective function at each corner point. Explain the optimal strategy and the optimal value of the objective function.

20. The White Knight Company produces two types of citizens-band radios—model B and model C. Each radio must be processed on each of two assembly lines. Processing times required are as follows:

	Model B	Model C
Assembly line 1	0.5 hours	0.4 hours
Assembly line 2	0.25 hours	0.6 hours

Assembly line 1 will be available for 40 hours each week but, because of maintenance requirements, assembly line 2 will be available for only 36 hours. Model B radio yields a profit contribution of $80 per unit sold, while model C yields a profit contribution of $60 per unit sold. Demand for the radios far exceeds the production capacity of the plant. The question is: How many units of each model should the company produce in order to maximize profit contribution?

Formulate a linear programming model that can aid in this decision-making process. Graph the constraints of the model. Determine the coordinates of the corner points of the region of feasible solutions. Find the optimal strategy by evaluating the objective function at each corner point. Explain the optimal strategy and the optimal value of the objective function.

21. The dietician at the athletic dining hall at State University wants to provide at least four units of vitamin A and five units of vitamin B per serving of food. There are two foods, F1 and F2, which may be served. These contain vitamins A and B in the amounts shown below:

Food	Vitamin A	Vitamin B
F1	1 unit/oz.	2 units/oz.
F2	2 units/oz.	1 unit/oz.

Each ounce of F1 costs $0.40 and each ounce of F2 costs $0.50.

Formulate an appropriate linear programming model. Graph the constraints of the model. Find the optimal strategy by evaluating the objective function at each corner point of the feasible region. Explain the optimal strategy and the optimal value for the objective function.

22. An investment banker has $8,000 to invest. She can purchase type A bonds, which yield 8 percent on the amount invested, or she can purchase type B bonds, which yield 15 percent return on the amount invested, or she can purchase some combination of each. But she can purchase no more than $7,500 in type A bonds and no more than $2,500 in type B bonds.

Formulate a linear programming model that could be used to help the banker determine how much should be invested in each type of bond in order to maximize the return on her investment.

Find the optimal strategy by evaluating the objective function at each corner point of the feasible region.

23. Pierre Carso, a struggling young tailor, has just opened his own shop and is urgently concerned with incoming revenues. He presently has 20 square yards of cotton material, 60 square yards of woolen material, and 12 square yards of silk material on hand. He can use these materials to produce either "standard" or "fashion-wise" suits, or both. The standard suit is a three-piece suit consisting of jacket, vest, and trousers and requires 2 square yards of cotton material, 7.5 square yards of woolen material, and 1 square yard of silk material. A fashion-wise suit consists of jacket, vest, and two pairs of trousers and requires 2.5 square yards of cotton, 6 square yards of wool, and 2 square yards of silk material. The standard suit sells for $160, while the fashion-wise suit sells for $200.

    Formulate a linear programming model that could help Pierre determine how many suits of each type he should make from the material on hand in order to maximize the revenue. Graph the constraints of the model. Determine the optimal strategy by evaluating the objective function at each corner point of the feasible region. Explain the optimal strategy.

24. McGraw Metals has a contract to provide 8 tons of copper, 20 tons of iron, and 7.5 tons of nickel. Two types of ore, A and B, are available from which the metals can be extracted. Type A ore costs $14 per ton and contains 4 percent copper, 15 percent iron, and 4 percent nickel. Type B ore costs $20 per ton and contains 6 percent copper, 12 percent iron, and 5 percent nickel. The management of McGraw Metals would like to know how many tons of each type of ore should be bought to minimize the cost of filling the contract.

    Formulate an appropriate linear programming model. Graph the constraints of the model. Determine the optimal strategy by evaluating the objective function at each corner point of the feasible region. Explain the optimal strategy.

25. The Cole Chemical Company blends coal for the purpose of obtaining certain by-products. Coal is purchased from two different sources. Analysis of the coal from source 1 indicates that each ton of this coal yields 1,500 pounds of coke, 400 pounds of tar, and 140 pounds of other tar derivatives, when processed. This coal costs $16 per ton. Coal from source 2 costs $12 per ton, and analysis indicates that each ton of this coal yields 750 pounds of coke, 600 pounds of tar, and 100 pounds of other tar derivatives.

    The company needs to mix coal from the two sources in such a manner that the resulting blend will contain not less than 6,000 pounds of coke, 2,400 pounds of tar, and 700 pounds of other tar derivatives.

Set up a linear programming model that will help the company decide what quantities of each type of coal should be blended together in order to obtain the required by-products at the minimum cost. Graph the constraints of the model. Determine the optimal strategy by evaluating the objective function at each corner point of the feasible region. Explain the optimal strategy.

26. Abraham Manufacturing Company manufactures two different products. The two products are processed on two different machines. Product P1 requires four hours of processing on machine M1 and one hour of processing on Machine M2. Product P2 requires two hours on Machine M1 and three hours on Machine M2. There are eight hours of available capacity on Machine M1 and nine hours on Machine M2 each day. Each unit of P1 produces a profit contribution of $20, while each unit of P2 produces a profit contribution of $15.

    Formulate a linear programming model that can aid in determining the maximum-profit-contribution strategy. Graph the constraints of the model. Determine the coordinates of the corner points of the area of feasible solutions. Find the optimal strategy by evaluating the objective function at each corner point. What is the maximum possible profit contribution Abraham can expect?

27. Sam Moskowitz, a truck farmer, grows tomatoes and cauliflower on his 80-acre farm. It costs at the beginning of the growing season—for soil preparation, seed, and so on—$40 for each acre of tomatoes and $20 for each acre of cauliflower. Sam has on hand only $2,500 to meet these expenses. In addition, during the growing season, 10 man-hours of labor are required for each acre of tomatoes and 30 man-hours for each acre of cauliflower grown. At most, 1,800 man-hours of labor will be available.

    If profits from an acre of tomatoes are $290 and from an acre of cauliflower are $200, how many acres of each should Sam plan to grow in order to maximize total profit for the growing season? What is the maximum possible profit?

28. The Sandusky Nuts and Bolts Company manufactures hex nuts and bolts. Output is restricted because of three scarce resources. The relevant data are summarized below.

Resource	Resource requirements (in units) per pound of output		Units available each day
	Hex nuts	Bolts	
Man-hours	1	3	150
Lathe-hours	1	1.5	90
Grinder-hours	1	0	50
Profit contribution	$3	$2	

How many units of each item should be produced each day to achieve maximum profit contribution? What is the maximum possible profit contribution?

29. Nature Snacks, Inc., has an inventory of 20 pounds of banana chips, 15 pounds of almonds, 18 pounds of raisins, and 18 pounds of carob-covered peanuts. These ingredients are mixed and sold in four-ounce packages. Mixture I, which is labeled Tahitian Tasties, contains 1.5 ounces of banana chips, 1.5 ounces of raisins, and 1 ounce of carob-covered peanuts and sells for $2.25. Mixture II, labeled Tasty Treat, contains 1.5 ounces of almonds, 1.5 ounces of carob-covered peanuts, and 1 ounce of banana chips and sells for $2.95.

    How many packages of each mixture should be made in order to obtain the maximum revenue? What is the maximum possible revenue?

30. The Bartow County Arts Alliance is comparing the costs and benefits of newspaper and radio advertising. Each dollar's worth of advertising in newspapers is believed to reach 50 readers in the "under $30,000 income" bracket and 100 readers in the "over $30,000 income" bracket. Each dollar's worth of radio advertising is believed to reach 60 persons in the "under $30,000" bracket and 25 in the "over $30,000" group. The Alliance wants to reach at least 120,000 persons in the "under $30,000" group and at least 150,000 in the "over $30,000" group but at a minimum advertising cost.

    How should the advertising be allocated between newspapers and radio to achieve this goal? What is the minimum advertising cost?

31. The Delacorte Manufacturing Corporation manufactures two grades of plastic. Both grades of plastic are made from the same basic ingredient but are processed differently. All of the basic ingredient goes first into process I, whose capacity is 200 input units per day. There is a 10 percent shrinkage loss in this process.

    Material destined to become grade A plastic goes from process I to process II; material to become grade B plastic goes from process I to process III. In process II, which has a capacity of 100 input units per day, there is a further shrinkage loss of 20 percent. In process III, which has a capacity of 75 input units per day, there is a shrinkage loss of 5 percent.

    The basic ingredient costs $4 per unit. The processing costs are $2 per input unit for process I, $1.25 per input unit for process II, and $1 per input unit for process III. Grade A plastic sells for $21 per unit; Grade B plastic sells for $15 per unit. Delacorte has a contract with the labor union that requires that the production level not fall below 75 output units per day.

    The management team at Delacorte would like to determine that combination of products that would maximize profit contribution. Formulate a linear programming model that could aid in this decision. Graph the constraints of the model. Find the optimal strategy by evaluating the objective function at each corner point of the feasible region. Explain the optimal strategy and the optimal value of the objective function.

32. The Leisure Living House has two warehouses, I and II, and three sales outlets. The sales outlets are located in Claremont, Rochester, and Burlington. The Claremont store requires 60 wrought-iron dining sets, the Rochester store 40 such sets, and the Burlington store 60 sets, for a current sales promotion. There are 70 of these dining sets in stock in warehouse I and 90 of the sets in stock in warehouse II. The costs to ship one set from each of the two warehouses to each of the sales outlets are as follows:

    From warehouse I to the Claremont store, $5

    From warehouse I to the Rochester store, $6

    From warehouse I to the Burlington store, $4

    From warehouse II to the Claremont store, $7

    From warehouse II to the Rochester store, $7

    From warehouse II to the Burlington store, $3

    Use an appropriate linear programming model to determine the minimum-cost shipping strategy.

7.6 A FEW UNUSUAL CONDITIONS AND THEIR GRAPHICAL COUNTERPARTS

From a routine computational point of view, most linear programming models behave very nicely. Occasionally, however, irregularities do arise. We should be able to recognize these anomalies when they occur and to deal with them effectively.

The Case of Many Optimal Solutions

If the slope of the objective function is identical to the slope of one of the linear constraints, there may be infinitely many optimal solutions to the problem, each resulting in the same value for the objective function. Such a situation is depicted in Figure 7.14. The two corner points H and J and all points on the HJ line segment represent optimal solutions; each differs from the others in the strategy represented, but all yield the same value for the objective function.

Thus, a general rule of linear programming is that, if there is one optimal solution to the problem, that solution will occur at one of the finite number of corner points of the region of feasible solutions. If the optimal solution is not a unique one (that is, if there are many strategies yielding this same optimal value), then two of the optimal solutions will be represented by adjacent corner points and the others by the points on the line segment connecting these two corner points. Hence, the

Figure 7.14

The Case of Many Optimal Solutions

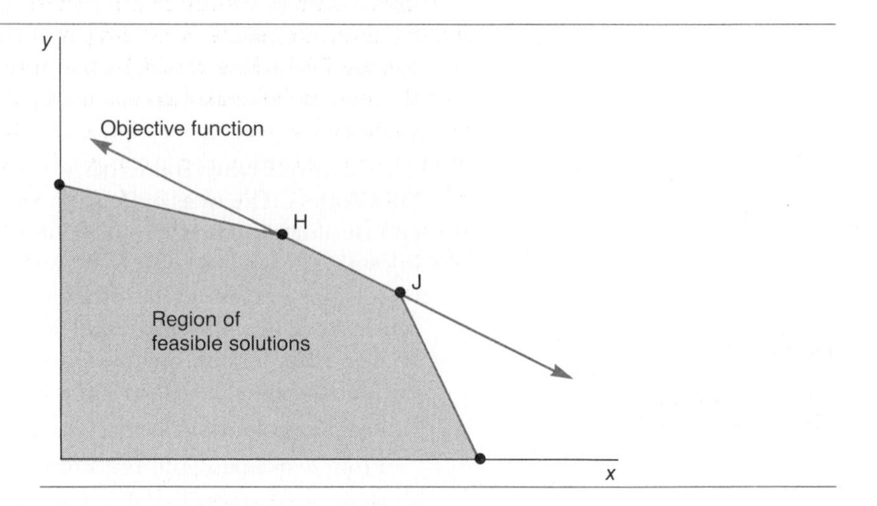

corner points will always include the optimal solution, although this same solution value may, on occasion, be represented by other points as well

Problems Having No Feasible Solution

An important irregularity is that situation where the model has no feasible region, meaning that no feasible solution can be achieved. (See Figure 7.15.) This situation may be attributable to an error in problem formulation, or it may result from conflicting management policies. Whatever the origin of the difficulty, the mathematical solution procedure we shall use will signal that some inconsistency exists.

Figure 7.15

A Problem Having No Feasible Region

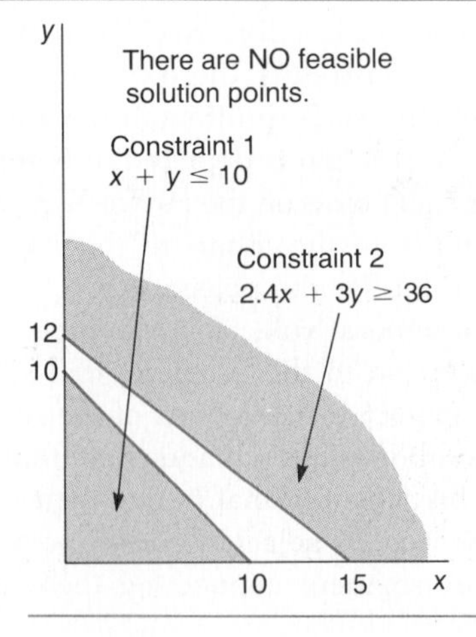

The Case of an Unbounded Solution

If there is no boundary to the region of feasible solutions *in the optimizing direction,* there is no limit on values the variables can assume. (See Figure 7.16.) This would be the case if the objective function of a model were to be maximized and all of the constraints were of the greater-than-or-equal-to variety. A variable could, thus, be introduced in unlimited quantity; the objective-function value could be made infinitely large. Clearly, in any real-world situation, it is not possible to increase the value indefinitely, and this condition indicates that something has gone wrong in defining the problem.

Figure 7.16

An Unbounded Solution for a Maximization Problem

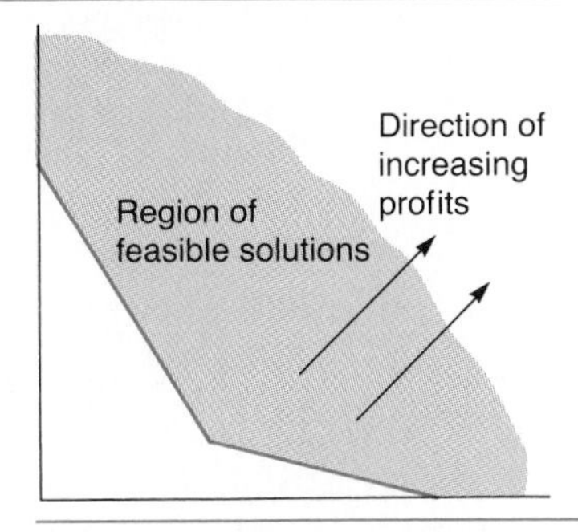

Redundant Constraints

Consider the set of constraints

$$(1)\quad 3x + 6y \leq 60 \qquad \text{and} \qquad (2)\quad 2x + 4y \leq 30$$

as pictured in Figure 7.17. All nonnegative values for (x,y) which satisfy the second constraint satisfy the first constraint as well. The reverse,

Figure 7.17

Redundant Constraints

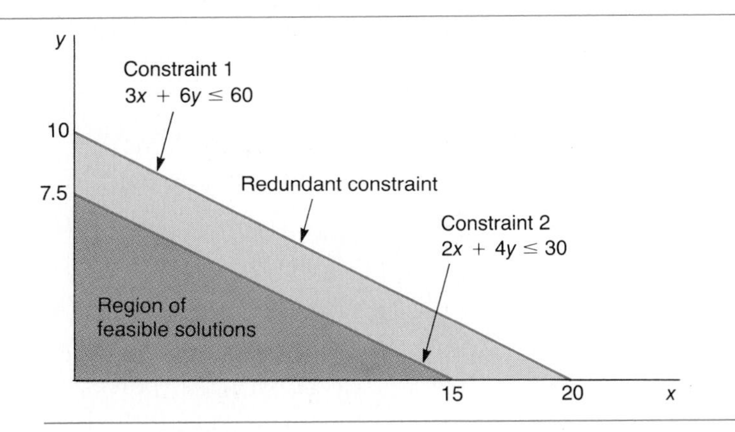

however, is not true. The second constraint is clearly the more restrictive of the two; the first constraint really plays no role in defining the region of feasible solutions. Any constraint such as the first one in this set is said to be a REDUNDANT CONSTRAINT.

Redundant constraints are superfluous and may be omitted altogether from the linear programming model. Their effect, if they are not eliminated, will be to compound the mathematical manipulations necessary to arrive at the optimal solution.

C·H·A·P·T·E·R 8

Linear Programming: The Simplex Algorithm

Few real-world linear programming models can be depicted graphically and the solution determined by the methods described in the preceding chapter. Nonetheless, the logical search for the optimal solution must involve a systematic evaluation of the corner points of the region of feasible solutions. The SIMPLEX ALGORITHM is such a mathematical search procedure. This procedure was developed in the late 1940s by Professor George B. Dantzig and has been further refined since that time by many others. The Simplex method is a formalized mathematical routine for locating the optimal strategy in a linear programming model. It searches from corner point to corner point of the feasible region, evaluating the strategy at each corner point in light of the objective and determining whether or not a move away from that point would be profitable. When it has found a corner point from which it is not desirable to move onward, it stops. This point represents an optimal solution for the model.

8.1 THE FIRST STEP: CONVERTING INEQUALITIES TO EQUATIONS

The optimal solution to a linear programming model, if one exists, always occurs at a corner point of the region of feasible solutions. Corner points occur where lines cross, and lines are representations of equations rather than inequalities. Thus, *the first step in preparing the model for manipulation by the Simplex algorithm is to convert the constraint statements that are inequalities into statements that are equalities.*

Let us continue our study by returning to the Bentwood model from the last chapter. So that it will be conveniently available, we restate the model.

MAXIMIZE $z = 40C + 32T$ = total profit contribution, in dollars

where

C = the number of chairs to produce each day
T = the number of tables to produce each day

subject to these limiting constraints:

$40C + 20T \leq 600$ (Rosewood constraint)
$4C + 10T \leq 100$ (Woodworking labor constraint)
$2C + 3T \leq 38$ (Finishing labor constraint)

with the further requirement that

$C, T \geq 0$

We can begin with the rosewood constraint. We must have an expression that is an equality rather than an inequality, but because we do not wish to change the basic nature of the relationship—because we do not wish to force the use of all available rosewood—we cannot simply replace the $\leq$ with $=$. We instead introduce a variable to take up the "slack" and write

$$40C + 20T + s_1 = 600$$

where s_1 represents the unused units of rosewood. Clearly, the units of rosewood used—$40C + 20T$—plus the units of rosewood not used—s_1—will equal the total number of units of rosewood available for use.

Following this same line of reasoning, we convert the woodworking labor constraint inequality into an equation by adding another slack variable, as

$$4C + 10T + s_2 = 100$$

where s_2 represents the unused units of woodworking labor.

Similarly, the finishing labor constraint inequality becomes

$$2C + 3T + s_3 = 38$$

where s_3 represents the unused units of finishing labor.

Variables such as s_1, s_2, and s_3 which are added to less-than-or-equal-to inequalities in order to transform them into equalities are called SLACK VARIABLES. The nonnegativity requirement applies to slack variables just as it does to other linear programming variables. Thus, while s_1, s_2, and s_3 may take on zero values, they may not take on negative values. We are, in this way, assured that their use will not cause us to violate any constraint.

The Bentwood Model Restated as a System of Equations

We can now restate the Bentwood model, this time as a system of equations.

$$\text{MAXIMIZE } z = 40C + 32T + 0s_1 + 0s_2 + 0s_3$$
$$= \text{total profit contribution, in dollars}$$

subject to these limiting constraints:

$$40C + 20T + s_1 + 0s_2 + 0s_3 = 600 \quad \text{(Rosewood constraint)}$$
$$4C + 10T + 0s_1 + s_2 + 0s_3 = 100 \quad \text{(Woodworking labor constraint)}$$
$$2C + 3T + 0s_1 + 0s_2 + s_3 = 38 \quad \text{(Finishing labor constraint)}$$

with the further requirement that

$$C, T, s_1, s_2, s_3 \geq 0$$

where

C = number of chairs to produce each day
T = number of tables to produce each day
s_1 = unused units of rosewood
s_2 = unused units of woodworking labor
s_3 = unused units of finishing labor

The Simplex method requires that each inequality in the linear programming model be rewritten as an equation, with new variables being used to take up the allowable slack as symbolized by the inequality portion of the constraints. When this has been done, the model is said to be in STANDARD FORM.

EXERCISES

Restate each of the following linear programming models in standard form.

1. MAXIMIZE $z = 3x + 4y$

   subject to: $2x + y \leq 20$
   $x + y \leq 15$

   and $x,\ y \geq 0$

2. MAXIMIZE $z = x + 4y$

   subject to: $x + y \leq 16$
   $x + 2y \leq 18$
   $x + 3y \leq 24$

   and $x,\ y \geq 0$

3. MAXIMIZE $x = x_1 + 4x_2 + 2x_3$

   subject to: $4x_1 + x_2 \leq 2$
   $-10x_1 + x_2 + 3x_3 \leq 1$

   and $x_1,\ x_2,\ x_3 \geq 0$

4. MAXIMIZE $z = x_1 + 2x_2 + 3x_3$

   subject to: $x_1 + 4x_2 + x_3 \leq 6$
   $2x_1 + 3x_2 + x_3 \leq 4$
   $x_3 \leq 3$

   and $x_1,\ x_2,\ x_3 \geq 0$

Restate each of the following models in standard form, defining explicitly each new variable that is used.

5. The Lewis Manufacturing Company model, Exercise 19, Chapter 7.
6. The White Knight Company model, Exercise 20, Chapter 7.
7. The investment banker model, Exercise 22, Chapter 7.
8. The Pierre Carso model, Exercise 23, Chapter 7.
9. The Abraham Manufacturing Company model, Exercise 26, Chapter 7.
10. The Sam Moskowitz model, Exercise 27, Chapter 7.

11. The Sandusky Nuts and Bolts Company model, Exercise 28, Chapter 7.
12. The Nature Snacks model, Exercise 29, Chapter 7.

8.2 SOME IMPORTANT TERMINOLOGY

Notice that there are five variables but only three equations in the set of mathematical statements describing the restrictions imposed in the Bentwood problem. In the standard format, the linear programming model will always consist of a set of m linear equations (for the m constraints) having $n + m$ variables (the n decison variables plus the m slack variables). Whenever there are more variables than equations in a system, that system cannot have a unique solution. Thus, a linear programming problem will always have either many solutions or no solution.

Given a system of m equations with $n + m$ variables, however, a unique solution may exist if n of the variables are set equal to zero. In the terminology of linear programming, a BASIC SOLUTION is a solution to the system of m equations in $n + m$ variables that is obtained by setting n variables equal to zero and solving the m equations for the values of the m remaining variables. The n variables which are assigned zero values are called NONBASIC VARIABLES, or VARIABLES NOT IN SOLUTION, while the m remaining variables are termed BASIC VARIABLES, or VARIABLES IN SOLUTION. These m basic variables form a BASIC SOLUTION or simply a BASIS. There will always be a finite number of basic solutions to a linear programming problem.*

See page 241

In the Bentwood model there are 10 basic solutions. These are represented by the points A, D, F, B, K, H, J, C, G, and E in Figure 8.1. These basic solutions occur wherever lines cross, both inside and outside the feasible region, including the points where the constraint lines intersect the coordinate axes. We can show that at each of these intersections, $n = 2$ of the variables have zero values and $m = 3$ of the variables have nonzero values. For example, consider the following points.

Point A	Point D	Point B	Point K
$C = 0$	$C = 0$	$C = 0$	$C = 12.5$
$T = 0$	$T = 10$	$T = 30$	$T = 5$
$s_1 = 600$	$s_1 = 400$	$s_1 = 0$	$s_1 = 0$
$s_2 = 100$	$s_2 = 0$	$s_2 = -200$	$s_2 = 0$
$s_3 = 38$	$s_3 = 8$	$s_3 = -52$	$s_3 = -2$

(The values for the decision variables are obtained from the coordinates of the point. The values for the slack variables are obtained by substituting the coordinates into each of the constraint equations and solving.)

* If there are m equations and $n + m$ variables, there cannot be more than ${}_{n+m}C_m = (n + m)!/m!n!$ basic solutions. ${}_{n+m}C_m$ is the number of different groups of m items possible from $n + m$ items, and the symbolism $m!$ is read "m factorial" and means $m! = (m) \cdot (m - 1) \cdot (m - 2) \cdot \ldots \cdot (1)$.

Figure 8.1

Region of Feasible Solutions, Bentwood Model

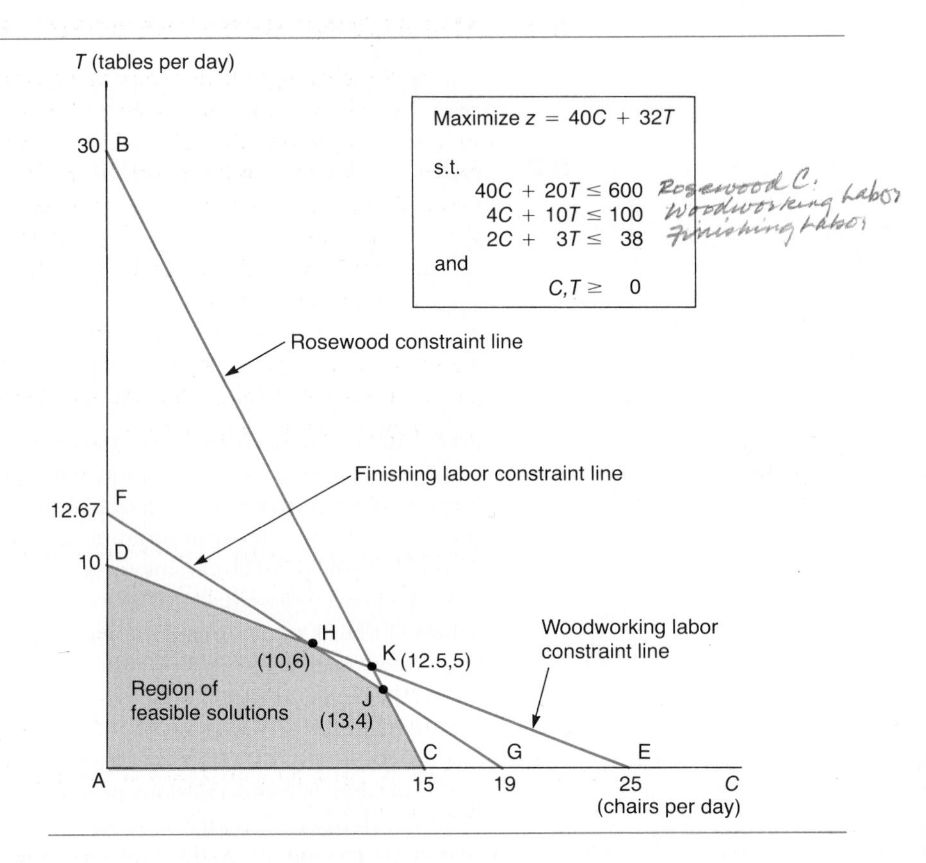

Because the nonzero values of the basic variables may be either positive or negative (see points B and K above), *not all basic solutions are feasible.* A FEASIBLE SOLUTION is a set of values for the $n + m$ variables that satisfies all the restrictions of the model, including the nonnegativity requirement. The region outlined by the constraints of the model is the collection of feasible solutions. There may be, then, an infinite number of feasible solutions.

A BASIC FEASIBLE SOLUTION is a feasible solution that is also a basic solution. Basic feasible solutions always occur at a corner point of the feasible region. There will be only a finite number of basic feasible solutions. For the Bentwood problem these are represented by the points A, D, H, J and C in Figure 8.1.

An OPTIMAL SOLUTION is a feasible solution that optimizes the value of the objective function. Only those feasible solutions that occur at the corner points of the region of feasible solutions qualify as candidates for the optimal solution. Hence, an optimal solution, if one exists, will always be found among the set of basic feasible solutions, and it is upon these solutions that we focus our attention.

8.3 THE SECOND STEP: THE INITIAL SIMPLEX TABLEAU

The Simplex algorithm provides a systematic search process wherein basic feasible solutions are evaluated in an orderly progression of steps, each step improving the value of the objective function, and which stops when the optimal solution has been determined. Because the model describing the situation has been reduced to a system of linear equations, the mathematics of the search procedure resemble very closely those of the matrix method used in Chapter 6 for solving simultaneously systems of linear equations.

An initial matrix—here the matrices are commonly called TABLEAUS—is set up by copying the extended coefficient matrix **A** and the **B** vector (containing the right-hand-side values from the constraint equations) in an orderly array, as

$$\left[\begin{array}{c|c} \mathbf{A} & \mathbf{B} \\ \hline -\mathbf{c} & \end{array}\right]$$

The values on the bottom row of the initial tableau are obtained by rewriting the objective function

$$z = c_1x_1 + c_2x_2 + \ldots + c_nx_n$$

as

$$-c_1x_1 - c_2x_2 - \ldots - c_nx_n + z = 0$$

The constraints and objective function in the Bentwood model form the set of equations

$$\begin{aligned} 40C + 20T + s_1 + 0s_2 + 0s_3 &= 600 \\ 4C + 10T + 0s_1 + s_2 + 0s_3 &= 100 \\ 2C + 3T + 0s_1 + 0s_2 + s_3 &= 38 \\ -40C - 32T + 0s_1 + 0s_2 + 0s_3 + z &= 0 \end{aligned}$$

and yield the initial tableau shown in Figure 8.2.

Each row in the main body of a Simplex tableau represents one constraint equation. Each of these rows is labeled with the label of the variable whose value is shown in the **B**-vector position of that row; these are the basic variables. *The basic variables and their values may always be read directly from the row labels and the* ***B****-vector.* Variables not labeling rows are nonbasic and have zero values. The initial tableau

Figure 8.2

Initial Tableau, the Bentwood Problem

Basis	C	T	s_1	s_2	s_3	z	b
s_1	40	20	1	0	0	0	600
s_2	4	10	0	1	0	0	100
s_3	2	3	0	0	1	0	38
z	−40	−32	0	0	0	1	0

always represents the "do nothing" strategy, so that the decision variables are initially nonbasic. (Pictorially, this strategy is represented by the origin on a graph.) In the Bentwood case, the initial strategy is to make zero chairs and zero tables. All available units of the scarce resources are, as yet, unused. Thus, the initially basic variables are s_1, s_2, and s_3, with values $s_1 = 600$, $s_2 = 100$, and $s_3 = 38$. The nonbasic variables are C and T with values $C = 0$ and $T = 0$.

As we proceed with the Simplex method, various changes will be made in the tableau. As each stage of the process is completed, the resulting tableau represents a basic feasible solution to the linear programming problem. As other strategies are considered, other variables will become basic. Thus, the numbers in the body of the tableau and the row labels will change as the model moves from one corner point to another in its search for the optimal solution.

The values on the bottom row of the initial tableau are the objective-function coefficients. The symbol z is commonly used to represent the dependent variable in the objective function and as such serves as the label for this row of the tableau. Although the values on this row will be altered from one tableau to the next, this row label is not changed. The numbers on this row are sometimes referred to as INDICATORS (or SHADOW PRICES) and the row as the INDICATOR ROW.

The beginning value of the objective function, which is usually, but not always, zero, is entered in the rightmost column of the indicator row of the initial tableau. This bottom right-hand-side cell of the tableau will always contain the current value of the objective function.

Each column of the tableau, except the far right-hand-side column, contains the coefficients of one specific variable and is labeled with the symbol used to represent that variable. The numbers in these columns represent exchange rates, or marginal rates of substitution, of the variable labeling the row for the variable labeling the column. The columns for the basic variables will always have the number +1 in that row which is labeled by this same variable, and all other entries in the column will be zeros. The column labels do not change as we progress through our mathematical manipulations, but the numbers in the columns will be altered to reflect different marginal rates of substitution at the various corner points.

A column labeled z is included in the tableau since z appears in the objective function equation with a coefficient of +1. All other entries in this z column are zeros. We shall see that the numbers in this column have no real effect on the Simplex procedure. In fact, the numbers in the z column never change. For this reason, the column will be carried through the first few examples but after that it will be dropped completely.

Before continuing with the mathematical manipulation of the tableau, let us take a closer look at the interpretation of the numbers it contains.

Interpreting the Numbers in the Simplex Tableau

The columns for the basic variables always are in the same format, with a +1 on the row that carries that same variable label and zeros in all other cells.

The entries in the columns for the nonbasic variables always represent marginal rates of substitution of the row variables for the column variable. Refer to the C column in Figure 8.2. From the "do-nothing" strategy of the initial tableau, in order to gain one unit of C (to produce one chair), Bentwood must give up all of the things listed in the C column. This includes 40 units of s_1 (unused rosewood), 4 units of s_2 (unused woodworking labor), and 2 units of s_3 (unused finishing time), as well as MINUS $40 profit contribution. In other words, if Bentwood wants to alter the present strategy of making zero chairs and zero tables by making one chair, they must use 40 units of previously unused rosewood, use 4 units of woodworking labor, and use 2 units of finishing labor, and, as a result of this action, profit contribution would be INCREASED by $40. (Note that *giving up* MINUS $40 is equivalent to *gaining* PLUS $40!)

The T column is read: To gain one unit of T (a table), Bentwood must give up 20 units of s_1 (unused rosewood), 10 units of s_2 (unused woodworking labor), and 3 units of s_3 (unused finishing time), and MINUS $32 profit. Or, to alter the present strategy of making zero chairs and zero tables by making one table, Bentwood must use 20 units of raw material, 10 units of woodworking labor, and 3 units of finishing labor, and this action would increase the value of the objective function by $32.

Every column, except the **B**-vector, in every tableau is read in this same way: TO GAIN A UNIT OF THE COLUMN VARIABLE, IT IS NECESSARY TO GIVE UP WHATEVER IS LISTED IN THAT COLUMN.

The marginal rates of substitution shown in the initial tableau represent the exchange rates that prevail at this initial strategy only. The search procedure will generate new tableaus, each representing a different strategy. Each tableau will display the marginal rates of substitution that would be applicable to changes made in the strategy depicted by that tableau. Nonetheless, the columns of every tableau may always be read in this same way.

As we move from one tableau to the next, the indicator-row numbers will continue to represent the net change affected in the value of the objective function by a one-unit increase in the value of the column variable. Again, note carefully how the algebraic signs are read. A negative indicator represents an increase in the value of the objective function when the value of the column variable is increased; a positive indicator represents a decrease in the value of the objective function when the value of the column variable is increased.

8.4 MOVING TO ANOTHER BASIC FEASIBLE SOLUTION

In order to evaluate another basic feasible solution, the tableau must be transformed so that it represents the conditions prevailing at an ADJACENT CORNER POINT of the region of feasible solutions. Two corner points are said to be adjacent if they have in common $(m - 1)$ variables, where m is the number of basic variables needed to define a corner point. Hence, one of the variables that is not now in solution will be brought into solution, and one variable that is presently in solution will have to go out of solution. But only one nonbasic variable is brought into solution at a time. Pictorially, we travel along one boundary of the region of feasible solutions from one corner to the next without skipping over corner points.

The tableau must be manipulated so that one of the basic variables becomes nonbasic (assumes a zero value) and one of the nonbasic variables becomes basic (assumes a nonzero value). The various constants and coefficients must be changed accordingly, so that the appropriate marginal rates of substitution will be displayed in the tableau. This process of moving from one basic feasible solution to another is called an ITERATION, or a PIVOTING, of the tableau.

Deciding Which Nonbasic Variable Will Become Basic

The shadow prices on the indicator row of the tableau indicate the change affected in the value of the objective function by a one-unit increase in the value of the column variable. When the shadow price is negative, the value of the objective function will be increased if the value of the column variable is increased. The column variable with the most negative shadow price (the negative number farthest removed from zero in the negative direction) is the variable that could make the greatest per-unit contribution to the objective. It is this variable that is customarily chosen as the variable to be brought into solution. Note, nonetheless, that the introduction of any variable which has a negative shadow price would result in an increase in the value of the objective function. If there are no negative shadow prices in the tableau, the value of the objective function cannot be further improved; hence, the current tableau represents the optimal solution.

Refer to Figure 8.3 where the initial tableau for the Bentwood model is reproduced. Both of the nonbasic variables C and T have negative shadow prices; either would, if brought into solution, result in an increase in total profit contribution. The variable C (representing chairs with a $40 per-unit contribution to profit) would make a greater per-unit contribution to profit than would T (representing tables and a $32 per-unit contribution). Thus, it seems reasonable to select C as the INCOMING VARIABLE.

The column corresponding to the nonbasic variable that is coming into solution is known as the PIVOT COLUMN. Thus, the C column is the pivot column for this pivoting of the tableau.

Figure 8.3

Selecting the Incoming and Outgoing Variables: Initial Tableau

Basis	C	T	s_1	s_2	s_3	z	b	b_i/a_{ik}	
s_1	(40)	20	1	0	0	0	600 ←	600/40 = 15	—PIVOT ROW
s_2	4	10	0	1	0	0	100	100/4 = 25	(smallest
s_3	2	3	0	0	1	0	38	38/2 = 19	ratio)
z	−40	−32	0	0	0	1	0		

PIVOT ELEMENT — PIVOT COLUMN (most-negative indicator)

Deciding Which Basic Variable Will Become Nonbasic

The nonbasic variable which is to become basic is C. We will introduce into solution as many units of this variable as possible at the current exchange rates (so as to travel all the way to a corner point of the region of feasible solutions rather than stopping somewhere along a boundary line between two corner points). To determine the maximum number of units of a variable that can be brought into solution at one time, we establish ratios between the **B**-vector values and the POSITIVE, NONZERO, pivot-column elements. The **B**-vector values give the quantities of the row variables that are presently in solution; the pivot-column values indicate the quantities of each row variable that would have to be exchanged for one unit of the column variable. The limiting constraint is the one that is most restrictive. Thus, the SMALLEST b_i/a_{ik} ratio (where $a_{ik} > 0$ and k is the pivot column) represents the maximum number of units of the incoming variable that can be introduced into solution on this iteration.

This smallest b_i/a_{ik} ratio represents the quantity of the pivot column variable that would fully deplete the most restrictive constraint. The variable labeling this row will take on a zero value, changing from a basic to a nonbasic variable. This variable is often referred to as the OUTGOING VARIABLE. The row occupied by the basic variable that is going out of solution is known as the PIVOT ROW. It is a rule that *the row yielding the smallest b_i/a_{ik} ratio MUST BE SELECTED as the pivot row.* To select any other row would result in the violation of some constraint or of the nonnegativity requirement of the model. The number at the intersection of the pivot row and the pivot column is known as the PIVOT ELEMENT.

In a maximizing linear programming model, when determining which row will become the pivot row, ratios are established only for rows which have positive, nonzero, values in the pivot column. A zero in the pivot column would denote that the introduction of this column variable would have no effect on the value of the row variable, meaning that insofar as this row variable is concerned an infinite quantity of the column variable could be brought into solution. A negative number in the pivot column would denote that with each unit of the column variable introduced into solution the value of the row variable would actually increase and, again, an unlimited quantity of the column variable could be introduced. Because the MOST RESTRICTIVE CONSTRAINT must be identified, ratios are established only for the rows which have positive, nonzero, entries in the pivot column cells. Should it ever become impossible to select a pivot row because appropriate ratios cannot be established, the algorithm is signaling that some inconsistency exists, possibly an unbounded solution set.

These b_i/a_{ik} ratios for the first iteration in the Bentwood model are shown in Figure 8.3. They can be interpreted in this way:

For the s_1 row: To gain one unit of C, Bentwood must give up 40 units of rosewood; there are at this time 600 unused units of rosewood; hence, $600/40 = 15$ units is the maximum number of units of C that could be gained *if* rosewood were the only constraint.

For the s_2 row: To gain one unit of C, Bentwood must give up 4 units of woodworking labor; there are available 100 units of unused woodworking labor; hence, $100/4 = 25$ units is the maximum number of units of C that could be gained *if* woodworking labor were the only constraint.

For the s_3 row: To gain one unit of C, Bentwood must give up 2 units of finishing labor; there are 38 unused units of finishing labor; hence, $38/2 = 19$ is the maximum number of units of C that could be gained *if* finishing labor were the only constraint.

On this first iteration of the Bentwood model, the smallest ratio is 15. The maximum number of chairs that can be brought into solution is 15. If this action is taken, all available units of rosewood will be used and s_1 will assume a zero value, thereby becoming nonbasic. The s_1 row is the pivot row; 40 is the pivot element.

(At this time, please refer again to the region of feasible solutions for the Bentwood model, as shown in Figure 8.1 and note how the b_i/a_{ik} ratios relate to the lines on this graph.)

The Second Bentwood Tableau

As soon as we have specified the incoming and the outgoing variable, the next tableau is partially determined. The new basic variables are C, s_2, and s_3, so that the new tableau will have to adhere to the format

Basis	C	T	s_1	s_2	s_3	z	b
C	1	—	—	0	0	0	—
s_2	0	—	—	1	0	0	—
s_3	0	—	—	0	1	0	—
z	0	—	—	0	0	1	—

where the —s indicate that these entries have yet to be determined.

Columns s_2, s_3, and z are already in the required format, but the pivot column, column C, will have to be transformed

	C (old)		C (new)
	40		1
from	4	to	0
	2		0
	−40		0

and in the process the values in the **B**-vector will be altered. This transformation can be accomplished using ELEMENTARY ROW OPERATIONS for matrices. Specifically,

1. A row of a matrix can be multiplied by any nonzero real constant.
2. A multiple of any row in a matrix can be added to any other row in that matrix.

We transform the C column from the old to the required new format by these computations.

1. *Calculate the new pivot row entries.* First, the pivot row is multiplied by the reciprocal of the pivot element as

$$(1/40) \times [40 \quad 20 \quad 1 \quad 0 \quad 0 \quad 0 \quad 600]$$
$$= [1 \quad 0.5 \quad 0.025 \quad 0 \quad 0 \quad 0 \quad 15]$$

   This new row vector is entered in a new tableau in the row position formerly occupied by the outgoing variable and is labeled with the label of the incoming variable, as shown in Figure 8.4A.

2. *Transfer other rows into the new tableau.* Next, suitable multiples of the pivot row, as it appears in the new tableau, are added to the other rows, so that the remaining entries in the pivot column become zeros.

   a. The new second row is computed as follows: $R_2 \leftarrow R_2 - 4R_1$

$$[4 \quad 10 \quad 0 \quad 1 \quad 0 \quad 0 \quad 100] - 4[1 \quad 0.5 \quad 0.025 \quad 0 \quad 0 \quad 0 \quad 15]$$
$$= [0 \quad 8 \quad -0.1 \quad 1 \quad 0 \quad 0 \quad 40]$$

   This vector is entered in the new tableau in the second-row position. The label of the row is not changed. (See Figure 8.4B).

   b. The new third row is computed as follows: $R_3 \leftarrow R_3 - 2R_1$

$$[2 \quad 3 \quad 0 \quad 0 \quad 1 \quad 0 \quad 38] - 2[1 \quad 0.5 \quad 0.025 \quad 0 \quad 0 \quad 0 \quad 15]$$
$$= [0 \quad 2 \quad -0.05 \quad 0 \quad 1 \quad 0 \quad 8]$$

   This vector is entered in the row-three position of the new tableau; its label is not changed. (See Figure 8.4C.)

   c. The new indicator row is computed as follows: $R_4 \leftarrow R_4 + 40R_1$

$$[-40 \quad -32 \quad 0 \quad 0 \quad 0 \quad 1 \quad 0] + 40[1 \quad 0.5 \quad 0.025 \quad 0 \quad 0 \quad 0 \quad 15]$$
$$= [0 \quad -12 \quad 1 \quad 0 \quad 0 \quad 1 \quad 600]$$

This vector is entered in the indicator row position of the new tableau. (See Figure 8.4.D.)

The pivoting is not completed. Note that this second tableau represents the strategy associated with corner point C of the region of feasible solutions in Figure 8.1.

Figure 8.4

Second Tableau for the Bentwood Model

(A) Partial Second Tableau: The New Pivot-Row Entries

Basis	C	T	s_1	s_2	s_3	z	b
C	1	0.5	0.025	0	0	0	15
	—	—	—	—	—	—	—

(B) Partial Second Tableau; The New s_2-Row Entries Added

Basis	C	T	s_1	s_2	s_3	z	b
C	1	0.5	0.025	0	0	0	15
s_2	0	8	−0.1	1	0	0	40
	—	—	—	—	—	—	—

(C) Partial Second Tableau: The New s_3-Row Entries added

Basis	C	T	s_1	s_2	s_3	z	b
C	1	0.5	0.025	0	0	0	15
s_2	0	8	−0.1	1	0	0	40
s_3	0	2	−0.05	0	1	0	8
	—	—	—	—	—	—	—

(D) Completed Second Tableau: The New Indicator-Row Entries Added

Basis	C	T	s_1	s_2	s_3	z	b
C	1	0.5	0.025	0	0	0	15
s_2	0	8	−0.1	1	0	0	40
s_3	0	2	−0.05	0	1	0	8
z	0	−12	1	0	0	1	600

8.5 SUMMARY OF PROCEDURE TO FOLLOW IN PIVOTING THE TABLEAU

The steps to be taken in pivoting the Simplex tableau may be summarized as follows.

1. *Determine which nonbasic variable will become basic.* (This is the INCOMING VARIABLE.) Determine which nonbasic variables will, when their value is increased, result in an increase in the value of the objective function. The contributions, or shadow prices, of the column variables are shown as the indicator row entries of the tableau. Any variable with a negative shadow price will, when introduced into solution, cause an increase in the value of the objective function. The greatest per-unit contribution to the objective will be made by the variable with the most-negative shadow price. It is customary to select this variable as the incoming variable. This column (let us designate it k) becomes the PIVOT COLUMN.

2. *Determine which basic variable will become nonbasic.* (This is the OUTGOING VARIABLE.) Determine which basic variable will reach a zero value first as the value of the incoming variable is increased. This can be done by establishing ratios between the **B**-vector values and the positive, nonzero values in the pivot column. For each a_{ik} in the pivot column k *which is greater than zero,* set up a ratio b_i/a_{ik}. The variable labeling the row that yields the SMALLEST RATIO must be the outgoing variable. This row is the PIVOT ROW. The tableau element at the intersection of the pivot row and the pivot column is labeled the PIVOT ELEMENT.
3. *Compute the new basic feasible solution and the new marginal rates of substitution.* (These are the entries in the new tableau.) Use ELEMENTARY ROW OPERATIONS to obtain the following:
   a. A "+1" in the pivot element cell. (This can be done by multiplying the old pivot row by the reciprocal of the pivot element.)
   b. Zeros in all other cells of the pivot column, including the indicator row cell. (This should be done by adding appropriate multiples of the new pivot row to the other rows in the tableau.)

   The label of the incoming variable will become the label of the pivot row. Labels for the other rows will not change. Column labels do not change.
4. *Determine whether or not this solution is an optimal solution.* Check the indicator row entries of the newly completed tableau to see whether or not there are any negative shadow prices. If there is a negative shadow price, return to Step 1 and repeat the process outlined above. If there are no negative shadow prices, this tableau represents an optimal solution.

8.6 ANOTHER PIVOTING OF THE TABLEAU

The second Bentwood tableau (see Figure 8.4) represents a strategy calling for the production of 15 chairs and no tables, for a total profit contribution of $600. The next question is: Can we improve upon this strategy? The negative shadow price in the T (tables) column indicates that the second tableau does not represent an optimal solution. Units of T introduced into solution would result in an increase in the total profit contribution.

Thus, the second tableau must be pivoted to obtain still another basic feasible solution. The only variable with a negative shadow price is T; this will be the incoming variable. The T column will be the pivot column. The smallest b_i/a_{ik} ratio is associated with the s_3 row (see Figure 8.5). This s_3 row will be the pivot row, and s_3 will be the outgoing variable. Computations are made according to the steps summarized.

Because there are no negative shadow prices in the third tableau, this tableau represents the optimal solution to the problem modeled.

Figure 8.5

Second and Third Tableaus, the Bentwood Model

(A) Second Tableau

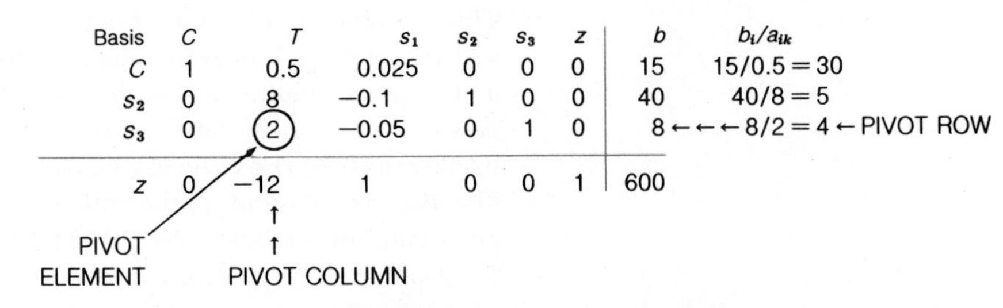

Basis	C	T	s_1	s_2	s_3	z	b	b_i/a_{ik}
C	1	0.5	0.025	0	0	0	15	15/0.5 = 30
s_2	0	8	−0.1	1	0	0	40	40/8 = 5
s_3	0	(2)	−0.05	0	1	0	8	← ← ← 8/2 = 4 ← PIVOT ROW
z	0	−12	1	0	0	1	600	

Step 1. *Calculate the new pivot row entries.* The pivot row from the second tableau is multiplied by the reciprocal of the pivot element. The resulting row vector is entered in the third tableau in the row position formerly occupied by the outgoing variable. This row is now labeled with the label of the incoming variable, as shown below.

(B) Partial Third Tableau: The New Pivot-Row Entries

Basis	C	T	s_1	s_2	s_3	z	b
—	—	—	—	—	—	—	—
T	0	1	−0.025	0	0.5	0	4
—	—	—	—	—	—	—	—

Step 2. The new C row is computed by $R_1 \leftarrow R_1 - 0.5R_3$. This vector is entered in the third tableau in the top-row position; the row label is not changed, as shown below.

(C) Partial Third Tableau: The New C-Row Entries Added

Basis	C	T	s_1	s_2	s_3	z	b
C	1	0	0.0375	0	−0.25	0	13
—	—	—	—	—	—	—	—
T	0	1	−0.025	0	0.5	0	4

Step 3. The new s_2 row is computed $R_2 \leftarrow R_2 - 8R_3$. The resulting vector is entered in the second-row position of the third tableau; the row label is not changed.

(D) Partial Third Tableau: The New s_2-Row Entries Added

Basis	C	T	s_1	s_2	s_3	z	b
C	1	0	0.0375	0	−0.25	0	13
s_2	0	0	0.1	1	−4	0	8
T	0	1	−0.025	0	0.5	0	4
—	—	—	—	—	—	—	—

Step 4. The new indicator row is computed $R_4 \leftarrow R_4 + 12R_3$. This vector is entered in the bottom row of the third tableau. This pivoting is now completed.

(E) Completed Third Tableau: New Indictor-Row Entries Added

Basis	C	T	s_1	s_2	s_3	z	b
C	1	0	0.0375	0	−0.25	0	13
s_2	0	0	0.1	1	−4	0	8
T	0	1	−0.025	0	0.5	0	4
z	0	0	0.7	0	6	1	648

8.7 INTERPRETING THE FINAL TABLEAU

From the row labels in the final tableau and the far right-hand-side vector, we read

$$C = 13 \qquad s_2 = 8 \qquad \text{and} \qquad T = 4$$

The production plan that will yield the maximum possible profit contribution under the conditions prevailing in the Bentwood case calls for the manufacture of 13 chairs and 4 tables each day. This combination of products will yield a total daily profit contribution of $648.

All available units of two of the scarce resources will be used in the optimal production plan. There are no unused units of rosewood and no unused units of finishing time. (Both s_1 and s_3 are nonbasic variables with zero values in the final tableau.) The other resource, woodworking labor, proved not to be a limiting constraint after all, as there remain eight unused units of this labor.

8.8 A MODEL WITH THREE DECISION VARIABLES

Let us go through one more example to be sure that we understand the steps that are involved in the Simplex procedure. The linear programming model we shall use is

$$\text{MAXIMIZE } z = 2x_1 + 3x_2 + x_3$$

$$\text{subject to:} \quad x_1 + x_2 + x_3 \le 10$$
$$2x_1 + 3x_2 + 4x_3 \le 24$$

$$\text{and } x_1, x_2, x_3 \ge 0$$

Slack variables are added to each constraint inequality, and the objective function is rewritten to yield the set of equations

$$x_1 + x_2 + x_3 + s_1 + 0s_2 = 10$$
$$2x_1 + 3x_2 + 4x_3 + 0s_1 + s_2 = 24$$
$$-2x_1 - 3x_2 - x_3 + 0s_1 + 0s_2 + z = 0$$

These coefficients and the right-hand-side values are transferred into the initial tableau as follows:

Basis	x_1	x_2	x_3	s_1	s_2	z	b	b_i/a_{ik}
s_1	1	1	1	1	0	0	10	10/1 = 10
s_2	2	(3)	4	0	1	0	24 ←	24/3 = 8 ← PIVOT ROW (smallest ratio)
z	−2	−3	−1	0	0	1	0	
		↑ PIVOT COLUMN (most-negative indicator)						

PIVOT ELEMENT (points to the circled 3)

The pivot column, column x_2, is selected because it has the most-negative indicator. The pivot row, row s_2, is selected because it yields the smallest ratio. Thus, x_2 is the INCOMING VARIABLE; s_2 is the OUTGOING VARIABLE.

The second tableau is computed by this sequence of steps, which is designed to change the format of the pivot column from that of a nonbasic variable to that of a basic variable. That is, the x_2 column must be changed

	x_2 (old)		x_2 (new)
	1		0
from	3	to	1
	−3		0

The easiest procedure to follow is to obtain the "+1" first in the pivot column by multiplying the old pivot row by the reciprocal of the pivot element. Then, use appropriate multiples of this new pivot row to obtain zeros in all other cells of the pivot column. This sequence of steps can be followed:

Step 1. Calculate the new pivot row entries: $R_2 \leftarrow (1/3)R_2$. Enter these results in the row position of the outgoing variable, and change the label of this row to that of the incoming variable. The new partial tableau is as follows:

Partial Second Tableau

Basis	x_1	x_2	x_3	s_1	s_2	z	b
	—	—	—	—	—	—	—
x_2	2/3	1	4/3	0	1	0	8
	—	—	—	—	—	—	—

Step 2. Transfer row one (the s_1 row) into the new tableau: $R_1 \leftarrow R_1 - R_2$. When these results are entered in the new partial tableau, it reads as follows

Partial Second Tableau

Basis	x_1	x_2	x_3	s_1	s_2	z	b
s_1	1/3	0	−1/3	1	−1	0	2
x_2	2/3	1	4/3	0	1	0	8
	—	—	—	—	—	—	—

Step 3. Transfer the indicator row into the new tableau: $R_3 \leftarrow R_3 - 3R_2$. When these results are entered in the new tableau, the second pivoting is completed.

Completed Second Tableau

Basis	x_1	x_2	x_3	s_1	s_2	z	b
s_1	1/3	0	−1/3	1	−1	0	2
x_2	2/3	1	4/3	0	1	0	8
z	0	0	3	0	1	1	24

There are no negative indicators in the second tableau; hence, this tableau represents the optimal solution. From the row labels and the right-hand-side vector we read the basic variables and their values: s_1

$= 2$ and $x_2 = 8$. The nonbasic variables are those that do not label rows. Here they are x_1 and s_2 with values $x_1 = 0$ and $s_2 = 0$. The maximum value of z is 24.

Again we note that the column z does not affect any of the Simplex operations. Hence, this column will be omitted from the linear programming tableaus that follow.

EXERCISES

Use the Simplex method to find the optimal solution to each of the following linear programming models.

13. MAXIMIZE $z = 2x + 3y$

    subject to: $x + y \leq 100$
    $x + 2y \leq 250$

    and $x, y \geq 0$

14. MAXIMIZE $z = 4x + 5y$

    subject to: $4x + 2y \leq 10$
    $x + 3y \leq 18$

    and $x, y \geq 0$

15. MAXIMIZE $z = 2x + y$

    subject to: $x + y \leq 10$
    $x + 2y \leq 12$
    $y \leq 5$

    and $x, y \geq 0$

16. MAXIMIZE $z = 5x + 8y$

    subject to: $2x + y \leq 15$
    $x + y \leq 15$
    $2x + 3y \leq 18$

    and $x, y \geq 0$

17. MAXIMIZE $z = 3x + 2y$

    subject to: $x + y \leq 25$
    $y \leq 20$
    $x \leq 15$

    and $x, y \geq 0$

18. MAXIMIZE $z = 3x + 4y$

    subject to: $5x + 6y \leq 60$
    $x + 3y \leq 18$

    and $x, y \geq 0$

19. MAXIMIZE $z = 8x_1 + 9x_2 + 6x_3$

subject to:
$$4x_1 + 5x_3 \leq 70$$
$$x_1 + x_2 \leq 16$$
$$x_1 \leq 15$$

and $x_1, x_2, x_3 \geq 0$

20. MAXIMIZE $z = x_1 + 2x_2 + 3x_3$

subject to:
$$x_1 + 4x_2 + x_3 \leq 60$$
$$2x_1 + 3x_2 + x_3 \leq 40$$

and $x_1, x_2, x_3 \geq 0$

Using the Simplex algorithm, determine the optimal solution to each of the following linear programming problems. Interpret this solution.

21. Lewis Manufacturing Company problem, Exercise 19, Chapter 7.
22. White Knight Company problem, Exercise 20, Chapter 7.
23. Investment Banker Problem, Exercise 22, Chapter 7.
24. Pierre Carso problem, Exercise 23, Chapter 7.
25. Abraham Manufacturing Company problem, Exercise 26, Chapter 7.
26. Sam Moskowitz problem, Exercise 27, Chapter 7.
27. Sandusky Nuts and Bolts Company problem, Exercise 28, Chapter 7.
28. Nature Snacks problem, Exercise 29, Chapter 7.
29. Otto Quotta, sales manager for Unique Office Machines Company, is trying to determine how he should allocate his salespeople to the company's three sales territories during the next six weeks. During this time period, various salespersons will be taking their annual vacations, leaving only 14 salespersons available for duty each week. In territory 1, each salesperson can sell, on the average, merchandise yielding a profit contribution of $500. For territory 2 and for territory 3, the profit contribution per salesperson assigned averages $450 and $400, respectively.

    Mr. Quotta believes that this sales level can be maintained in territory 1 only if not more than five salespersons are assigned to that territory. He also feels that the number of salespeople assigned to territory 2 should not be more than twice the number assigned to territory 3.

    The selling expense budget for these weeks may also pose a problem. A budget of only $900 a week has been approved. Selling expenses are $100 per week for each salesperson assigned to territory 1, are $75 per week for each salesperson assigned to territory 2, and are $60 a week for each salesperson in territory 3.

    Formulate a linear programming model to help determine how

many salespeople should be assigned to each territory in order to maximize profit contribution. Use the Simplex method to obtain the optimal solution.

30. The Triad Manufacturing Corporation produces three products, each of which is manufactured from three raw materials. Vital production information is as follows:

	Each unit of product requires		
	Raw material A	Raw material B	Raw material C
Product 1	4 units	3 units	5 units
Product 2	6 units	4 units	2 units
Product 3	1 unit	3 units	2 units
Maximum quantity available each day	300 units	210 units	400 units
Cost, per unit	$1	$1.50	$2

Products 1, 2, and 3 sell for $40, $20, and $15 per unit, respectively. Triad can sell, at these prices, as many units of any, or all, of these products as it can produce. Triad would like to produce that combination of items which would yield the maximum total profit contribution where profit contribution is defined as selling price minus cost of raw material.

Determine the profit contribution of each of the three products. Formulate a linear programming model that could be used by Triad to help decide which combination of products would maximize profit contribution. Use the Simplex method to obtain the optimal solution to the model.

31. Sandy Land, a farmer, has for cultivation during the coming season 200 acres of highly productive tillable land and another 100 acres of less-fertile land. Crops which can be grown in this soil and for which there is a high demand are peas and corn. Relevant information on yields and market prices are outlined below.

	Yield per acre (in units)		
Crop	High-grade soil	Lower-grade soil	Selling price (per unit)
Peas	100	75	$6
Corn	65	45	$9

Each acre, whether it is the high-grade soil or the lower-grade, planted in peas will require 15 units of labor during the season and 3.0 units of fertilizer. Each acre planted in corn will require 10 units of labor and 0.5 units of fertilizer. There will be available during the season only 3,000 units of labor. Labor costs $5 per unit. Fertilizer costs $30 per unit. There will be other variable costs of $125 per acre for each acre of peas and $30 per acre for each acre of corn.

Sandy would like to be able to determine the crop-planting program that would maximize contribution to profits. Formulate an appropriate linear programming model. Use the Simplex method to determine the optimal solution to the problem modeled. Interpret the solution.

32. A. LaSaey, Ltd., is a roaster of fine coffee. The firm blends three types of coffee—Premium, Superior, and Special—each with its own individual taste characteristics. Three types of coffee beans are used in the blends: Brazilian, Colombian, and Guatemalan. The composition of each blend is as follows:

	Type of coffee bean (percent)		
Cofee	Brazilian	Colombian	Guatemalan
Premium	40%	35%	25%
Superior	30	30	40
Special	35	50	15

The roaster realizes a profit contribution of 40 cents on each pound of Premium, 46 cents on each pound of Superior, and 32 cents on each pound of Special. The maximum quantity of each type of bean available each week is: Brazilian, 25,000 pounds; Colombian, 20,000 pounds; and Guatemalan, 32,000 pounds. The roasting plant has a maximum capacity of 65,000 pounds of coffee each week. The Premium and Special Coffees can be sold in whatever quantities are processed, but not more than 6,000 pounds of the Superior coffee can be sold weekly.

Prepare an appropriate linear programming model which the roaster might use as an aid in the decision process. Use the Simplex method to find the optimal solution to the problem modeled. Interpret this solution.

8.9 THE SIMPLEX METHOD AND MODELS WITH MIXED CONSTRAINTS

Thus far we have developed the Simplex method for maximization problems with less-than-or-equal-to constraints only, where all elements in the **B** vector are positive. The procedure can be extended so that constraints may be of any form—greater-than-or-equal-to, less-than-or-equal-to, or exactly-equal to. The objective function may be either one which is to be maximized or one that is to be minimized.

Let us consider again the model designed to aid Poultry-Plus Feed Mill determine the optimal blend of ingredients for its Gro-Chick Mix (see Case 3, Chapter 7). The model is

$$\text{MINIMIZE } z = 20x_1 + 15x_2 + 5 = \text{cost, in cents, per pound}$$

where

x_1 = pounds of ingredient 1 to use in each pound of Gro-Chick Mix
x_2 = pounds of ingredient 2 to use in each pound of Gro-Chick Mix
$1 - x_1 - x_2$ = pounds of fiber filler to use in each pound of Gro-Chick Mix

subject to these constraints:

$x_1 + x_2 \geq 0.7$ (Maximum amount of filler)
$90x_1 + 50x_2 \leq 45$ (Maximum number of units of nutrient A)
$40x_1 + 20x_2 \geq 16$ (Minimum number of units of nutrient B)

and

$x_1, x_2 \geq 0$

Preparing the Model for the Simplex Procedure

The Simplex procedure requires that the objective function for the model be stated in a "maximizing" format. Any linear programming objective function that is to be minimized can be transformed into one that should be maximized simply by multiplying the function by −1. Hence, we take the objective function of the Poultry-Plus model

$$\text{MINIMIZE } z = 20x_1 + 15x_2 + 5 = \text{cost, in cents, per pound}$$

and transform it into an equivalent linear programming objective function

$$\text{MAXIMIZE } z' = -20x_1 - 15x_2 - 5$$

where z' will be cost, affixed with a negative sign. Note that when the objective function is rewritten in the format needed for entering it on the bottom row of the initial tableau, it becomes

$$+20x_1 + 15x_2 + z = -5$$

Thus, the indicators on the bottom row of the initial tableau will be positive. A "−5" will be entered in the rightmost cell of the bottom row as the initial value of the objective function.

The Simplex method also requires that each constraint be stated first as a LESS-THAN-OR-EQUAL-TO inequality and then converted into an equality. One of the Poultry-Plus constraints—the restriction on the number of units of nutrient A—is already in this format; so we need only to use a slack variable to convert this inequality to an equality. We write

$90x_1 + 50x_2 + s_1 = 45$ (Maximum number of units of nutrient A)

The other two constraint inequalities of the model are now expressed as greater-than-or-equal-to inequalities. They must first be converted to less-than-or-equal-to inequalities. Recall that multiplying an inequality by −1 reverses the direction of the inequality. Thus, an alternative way of expressing these two constraints is

$$x_1 + x_2 \geq 0.7 \qquad 40x_1 + 20x_2 \geq 16$$
$$-x_1 - x_2 \leq -0.7 \qquad -40x_1 - 20x_2 \leq -16$$

Once the constraints are in the less-than-or-equal-to format, they are converted to equations by the addition of another variable. Thus, we write

$$-x_1 - x_2 + s_2 = -0.7 \qquad \text{(Amount of filler)}$$

and

$$-40x_1 - 20x_2 + s_3 = -16 \qquad \text{(Units of nutrient B)}$$

Please note, because this is most important, that *first* the greater-than-or-equal-to inequality is converted to a less-than-or-equal-to inequality (by multiplying by −1); *then* the inequality is converted to an equation (by use of the additional variable). Variables such as s_2 and s_3, introduced to change what was originally a greater-than-or-equal-to constraint inequality into an equation, are called SURPLUS VARIABLES. A surplus variable represents the amount by which a minimum, or "floor," is exceeded, whereas the slack variable represents the amount by which a strategy falls below a maximum, or "ceiling."

We are now prepared to restate the Poultry-Plus Feed Mill model, this time as the following system of equations.

$$90x_1 + 50x_2 + s_1 + 0s_2 + 0s_3 = 45$$
$$-x_1 - x_2 + 0s_1 + s_2 + 0s_3 = -0.7$$
$$-40x_1 - 20x_2 + 0s_1 + 0s_2 + s_3 = -16$$

and

$$20x_1 + 15x_2 + 0s_1 + 0s_2 + 0s_3 + z' = -5$$

The Simplex Tableaus

The initial tableau for the Poultry-Plus model is shown in Figure 8.6. Note that there are no negative shadow prices in the initial tableau. This is often the case in a model designed to determine the minimum-

Figure 8.6

Initial Simplex Tableau, Poultry-Plus Feed Mill Model:

Basis	x_1	x_2	s_1	s_2	s_3	b	
s_1	90	50	1	0	0	45	
s_2	−1	−1	0	1	0	−0.7	
s_3	−40	−20	0	0	1	−16	← ← ← PIVOT ROW (negative value in B vector)
z'	20	15	0	0	0	−5	
	↑						

z_j/a_{rj} 20/−40 = −0.5 ←PIVOT COLUMN (largest ratio)
15/−20 = −0.75

cost strategy because the minimum-cost strategy will ordinarily be a "do-nothing" strategy. Any activity undertaken increases costs. It is only because of the greater-than-or-equal-to constraints which establish minimum levels of activity that this do-nothing strategy is not feasible, and thus not the optimal strategy. Minimization models will always be characterized by at least one greater-than-or-equal-to constraint.

How, given that there are no negative shadow prices, do we choose a pivot row and column? Recall the nonnegativity requirement of the linear programming model and note the negative values in the right-hand-side vector of the initial tableau. Two of the initially basic variables have negative values, in violation of the nonnegativity requirement. This, of course, means that the initial tableau represents a strategy that is not feasible. The region of feasible solutions for the Poultry-Plus model is shown again in Figure 8.7. Note that the origin (which is represented

Figure 8.7

Region of Feasible Solutions, Poultry-Plus Feed Mill Problem

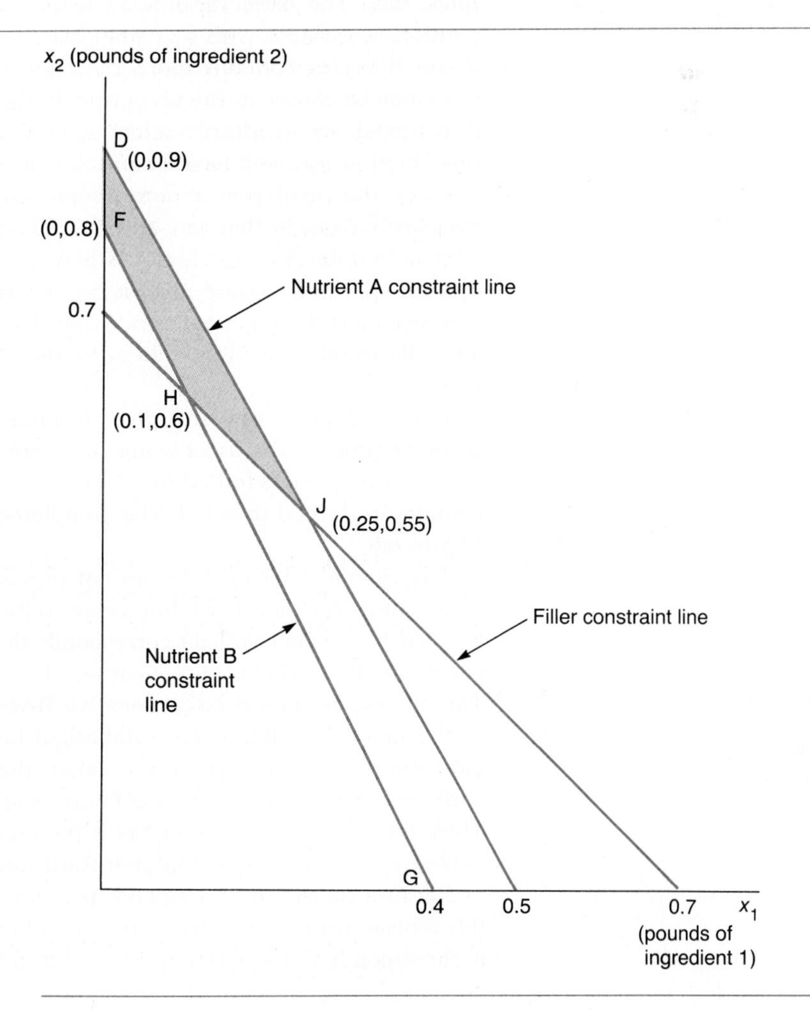

by the initial tableau) is not a corner point for the feasible region. At each corner of the feasible region, all variables have nonnegative values. Hence, all **B**-vector values must become nonnegative before the tableau will represent a feasible solution.

Whenever there are negative values in the **B** vector of the initial tableau, the Simplex algorithm is made up of two phases. PHASE I constitutes a search for a basic feasible solution—for a beginning corner point of the feasible region. All values in the **B** vector must become nonnegative before this search is completed. PHASE II continues with the search around the boundary of the feasible region for the optimal corner point.

Thus, whenever there are negative **B**-vector values, whether the situation be a minimizing or a maximizing one, these negative values must be removed before any other factors or goals can be considered. *A row with a negative* **B**-*vector value, if any exist, must be selected as the pivot row.* The basic variable with the negative value will go out of solution; a variable with a nonnegative value will come into solution. If the **B** vector contains more than one negative value, any of these rows may be chosen as the pivot row. In the first pivoting of the Poultry-Plus model, we arbitrarily select s_3 as the pivot row, although the s_2 row could just as well have been selected.

Once the pivot row is determined, ratios are established between *negative* entries in that row and the values in the indicator row of the tableau to determine the least-costly way to meet the requirement. The column with the *largest* z_j/a_{rj} ratio (where r is the pivot row and $a_{rj} < 0$) becomes the pivot column. (Note that because in the first pivoting all of these ratios are less than zero, the "largest" is the one closest to zero.)

Once the pivot row and pivot column have been determined, the pivoting proceeds as usual using the elementary row operations. First, the pivot element is forced to a "+1"; then all other entries in the pivot column are forced to zeros. The completed second tableau is shown in Figure 8.8.

This second tableau still does not represent a feasible solution since it has a negative right-hand-side value on the second row. (Note in Figure 8.7 that the second tableau corresponds to the strategy represented by corner point G, which is not a corner of the region of feasible solutions.) The s_2 row, because it has a negative **B**-vector value, must be selected as the pivot row. Ratios are established between the numbers on the indicator row and the pivot row values that are negative. The column with the largest ratio, the x_2 column, is selected as the pivot column. Thus, for the second pivoting, x_2 is the incoming variable and s_2 is the outgoing variable. The completed third tableau is shown in Figure 8.8.

Because there are no negative **B**-vector values in the third tableau, this tableau represents a basic feasible solution. (Note in Figure 8.7 that it corresponds to the strategy represented by corner point H, which is

Figure 8.8

The Tableaus in the Poultry-Plus Feed Mill Problem

Initial Tableau

Basis	x_1	x_2	s_1	s_2	s_3	b	
s_1	90	50	1	0	0	45	
s_2	−1	−1	0	1	0	−0.7	
s_3	(−40)	−20	0	0	1	−16	← PIVOT ROW
z'	20	15	0	0	0	−5	
	↑ PIVOT COLUMN						

Second Tableau

Basis	x_1	x_2	s_1	s_2	s_3	b	
s_1	0	5	1	0	−2.25	0.9	
s_2	0	(−0.5)	0	1	−0.025	−0.3	← ← PIVOT ROW
x_1	1	0.5	0	0	−0.025	0.4	
z'	0	−5	0	0	0.5	−13	
		↑					
−5/−0.5 = 10		↑ PIVOT COLUMN			0.5/−0.025 = −20		

Third Tableau

Basis	x_1	x_2	s_1	s_2	s_3	b
s_1	0	0	1	1	−2.275	0.6
x_2	0	1	0	−2	0.05	0.6
x_1	1	0	0	1	−0.05	0.1
z'	0	0	0	10	0.75	−16

a corner point of the feasible region.) Because there are no negative indicators in the third tableau, this tableau represents the optimal solution.

The optimal strategy then is to use 0.1 pounds of basic ingredient 1 and 0.6 pounds of basic ingredient 2 in each pound of Gro-Chick Mix. This means that 0.3 pounds of fiber filler will be used in each pound of the mix. The cost will be 16 cents a pound. (The minus sign attached to the objective-function value in the tableaus stems from the fact that z = cost in the minimizing objective function became z' = *minus* cost in the maximizing objective function. Remember.)

In this particular model, the first feasible solution was also the optimal solution. Nevertheless, in those problems where both negative **B**-vector values and negative shadow prices exist, all negative **B**-vector values must be cleared away *first.* Then pivoting on negative shadow prices takes place. The optimal solution is represented by the first tableau that has neither negative **B**-vector values nor negative shadow prices.

8.10 THE SIMPLEX METHOD, IN SUMMARY

We may summarize the Simplex procedure as follows.

PREPARATION

Step 1. State the objective function in the MAXIMIZING format. An objective function which is to be minimized can be converted into one which should be maximized by multiplying through by −1.

Step 2. Write the objective function in implicit form as

$$-c_1x_1 - c_2x_2 - \ldots + z = 0$$

Step 3. State all constraints as inequalities of the LESS-THAN-OR-EQUAL-TO (≤) type.

a. Multiply each greater-than-or-equal-to inequality by −1 to reverse the direction of the inequality.

b. Replace each exactly-equal-to constraint by two constraints, one a less-than-or-equal-to inequality and the other a greater-than-or-equal-to inequality. Then multiply the latter by −1 to reverse the direction of the inequality.

Step 4. Convert all constraint inequalities into equations by adding slack and/or surplus variables.

Step 5. Copy the coefficients and the right-hand-side values of both the constraint equations and the objective function equation into a matrix tableau of the format

$$\left[\begin{array}{c|c} \mathbf{A} & \mathbf{B} \\ \hline \multicolumn{2}{c}{-\mathbf{c}} \end{array}\right]$$

PROCESSING PHASE I—FINDING AN INITIAL BASIC FEASIBLE SOLUTION

Step 6. Are all entries in the **B** vector (except possibly the indicator-row entry) nonnegative?

a. If yes, go to Phase II, Step 8.

b. If no, select as the PIVOT ROW r the row with the most negative **B**-vector value.

Step 7. Are all entries on the pivot row nonnegative?

a. If yes, stop; no feasible solution exists.

b. If no, set up ratios z_j/a_{rj} for all columns with negative entries in the pivot row. Select the column with the largest ratio as the PIVOT COLUMN.

GO TO PIVOTING, Step 10.

PROCESSING PHASE II—FINDING THE OPTIMAL SOLUTION

Step 8. Are all entries on the indicator row of the tableau nonnegative?

a. If yes, stop; current solution is optimal.
b. If no, select column with most-negative indicator as PIVOT COLUMN k.

Step 9. Are all entries in the pivot column negative?

a. If yes, stop; the solution is unbounded.
b. If no, set up ratios b_i/a_{ik} for all positive entries in the pivot column. Select row with smallest ratio as PIVOT ROW.

GO TO PIVOTING, STEP 10.

PIVOTING

Step 10. Use elementary row operations on entire tableau to obtain the following:

a. A "+1" at the intersection of the pivot row and pivot column. Relabel pivot row with label of pivot column.
b. Zeros in all other cells of the pivot column.

RETURN TO PROCESSING PHASE I, Step 6.

Note in Steps 7*a* and 8*a* that the Simplex procedure sends out signals when certain special conditions exist in the model. Two other special conditions are worthy of our attention.

1. A zero shadow price for a nonbasic variable means that alternate optimal solutions exist. The strategy at the adjacent alternate corner point can be obtained by pivoting on the nonbasic column with the zero value in the indicator row of the tableau.
2. A zero **B**-vector value for a basic variable signals a condition that is known as DEGENERACY. Pictorially this means that more than two constraint lines intersect at the same point. Ordinarily this condition presents no real problem and simply leaves us with fewer than m basic variables with positive values. We know that degeneracy is going to develop if we have a "tie" when selecting the pivot row.

EXERCISES

Use the Simplex algorithm to obtain the optimal solution to the problem modeled in each case stated below. For the two-decision-variable models, sketch the region of feasible solutions and follow the tableaus through this sketch.

33. MINIMIZE $z = 3x + y$

subject to: $x + y \geq 2$
$2x + y \geq 3$

and $x, y \geq 0$

34. MINIMIZE $z = x + y$

subject to: $3x + 2y \geq 60$
$x + 2y \geq 40$

and $x, y \geq 0$

35. MINIMIZE $z = x + y$

subject to: $x + 2y \leq 25$
$y \leq 10$
$x + y \geq 8$

and $x, y \geq 0$

36. MINIMIZE $z = x_1 + x_2 + x_3$

subject to: $2x_1 + x_2 \geq 8$
$2x_2 + x_3 \geq 6$

and $x_1, x_2, x_3 \geq 0$

37. MAXIMIZE $z = x + y$

subject to: $x + 2y \leq 25$
$y \leq 10$
$x + y \geq 8$

and $x, y \geq 0$

38. MAXIMIZE
$z = 2x_1 + 2x_2 + 3x_3$

subject to: $x_1 + x_2 + x_3 \leq 18$
$x_1 + 2x_2 + x_3 \geq 6$
$x_1 \geq 3$

and $x_1, x_2, x_3 \geq 0$

39. MINIMIZE $z = x + 2y$

subject to: $2x + y \leq 10$
$x \leq 3$
$x + y \geq 6$

and $x, y \geq 0$

40. MAXIMIZE $z = x + y$

subject to: $3x + y \leq 18$
$x + 2y \geq 8$
$x \geq 4$

and $x, y \geq 0$

Use the Simplex algorithm to find the optimal solution to each of the following problems. In each case, interpret the optimal solution.

41. Dietician at athletic dining hall, Exercise 21, Chapter 7.
42. McGraw Metals Company, Exercise 24, Chapter 7.
43. Cole Chemical Company, Exercise 25, Chapter 7.
44. Bartow County Arts Alliance, Exercise 30, Chapter 7.
45. Delacorte Manufacturing Company, Exercise 31, Chapter 7.
46. Leisure Living House, Exercise 32, Chapter 7.
47. Moody's Lawn-Care Services, Case 2, Chapter 7.
48. Haynes-Webster, Inc., Case 4, Chapter 7.
49. Mid-States Cooperative, Case 5, Chapter 7.
50. Purefoy Metals, Inc., has received an order for an aluminum alloy that meets these specifications:

Content	Specification
Copper	Between 25% and 35%, inclusive
Zinc	Not more than 18%
Silicon	Not less than 12%
Aluminum	Balance

The following inputs can be used to produce the alloy, with aluminum being used to bring the total up to 100 percent:

	Composition			
Input	Copper	Zinc	Silicon	Cost, per ton
A	0.7	0.2	0.1	$80
B	0.5	0.1	0.4	$40
C	0.3	0.6	0.1	$60

Aluminum costs $50 per ton.

Purefoy would like to determine that blend of inputs which would minimize the cost of the alloy. Formulate a linear programming model which could be used as an aid in this decision. Use the Simplex method to obtain the optimal solution to the problem modeled.

51. Krumble, Inc., a firm selling fine cookies and candies, is designing a new assortment of cookies which will be sold in two-pound tins. Three different kinds of cookies—sugar cookies, malted milk biscuits, and gingersnaps—will be used in the assortment. Sugar cookies cost $0.90 a pound; malted milk biscuits cost $1.25 a pound, and gingersnaps cost $1.50 a pound. The new assortment will sell for $5.50 for a two-pound container.

Krumble would like to determine the minimum-cost assortment which meets the following quality standards:

a. Gingersnaps must comprise at least 20 percent of the package.
b. Sugar cookies must not comprise more than 60 percent of the package.
c. The proportion of malted milk biscuits must not be less than 50 percent of the proportion of sugar cookies.

Formulate a linear programming model that Krumble might use to determine the maximum-profit assortment. Use the Simplex algorithm to determine the optimal solution to the problem modeled.

C·H·A·P·T·E·R 9

Nonlinear Functions

In this chapter you will become friends with many different types of mathematical functions, including QUADRATIC FUNCTIONS, CUBIC FUNCTIONS, HIGHER-ORDER POLYNOMIALS, RATIONAL FUNCTIONS, POWER FUNCTIONS, ABSOLUTE-VALUE FUNCTIONS, EXPONENTIAL FUNCTIONS, AND LOGARITHMIC FUNCTIONS.

You will be able to "match" the following functions with their graphs

Functions

1. $f(x) = 1/a^x$
2. $f(x) = ax^2 + bx + c$ for $a > 0$
3. $f(x) = x^{2/3}$
4. $f(x) = 1/x$
5. $f(x) = |x|$
6. $f(x) = ce^{kt}$
7. $f(x) = ce^{-kt}$
8. $f(x) = A - ce^{-kt}$
9. $f(x) = ax^4 + bx^3 + cx^2 + d$
10. $f(x) = 1/x^2$
11. $f(x) = x^3$
12. $f(x) = a^x$ for $a > 1$
13. $f(x) = \dfrac{A}{1 + ce^{-kt}}$
14. $f(x) = log_a\, x$ for $a > 1$

Graphs

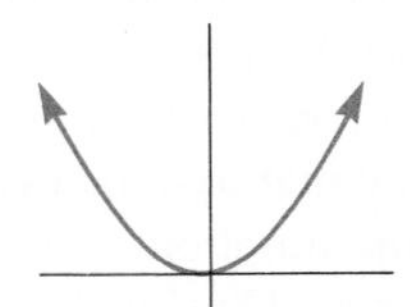
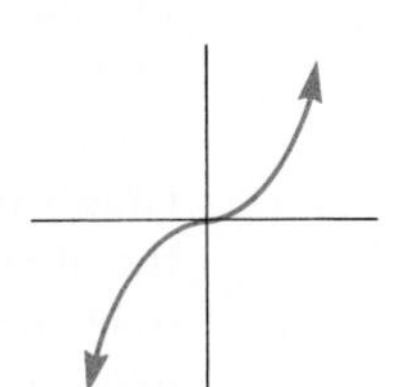
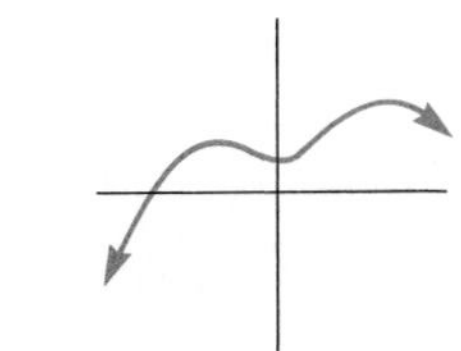
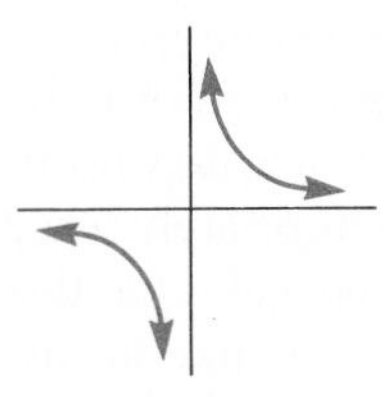
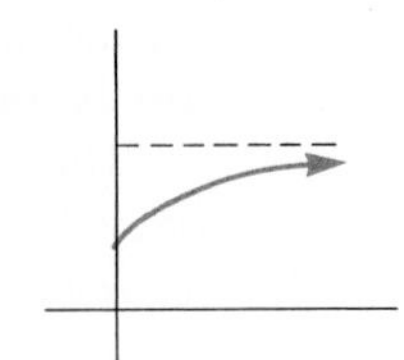
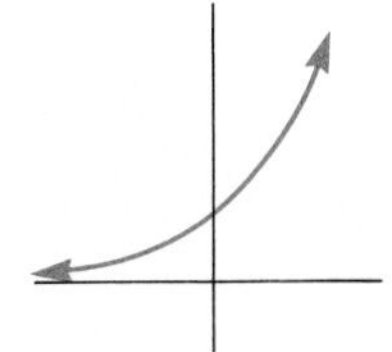
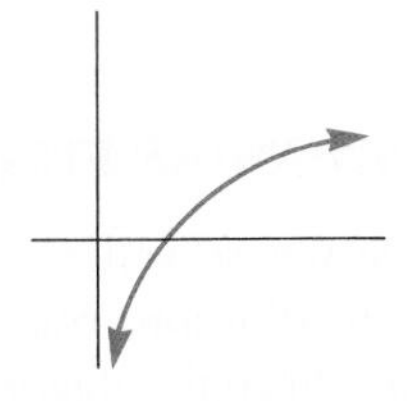
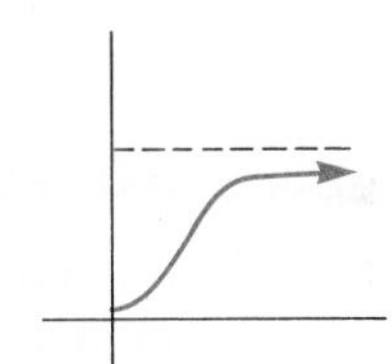
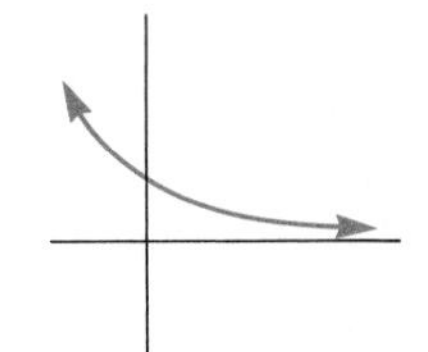
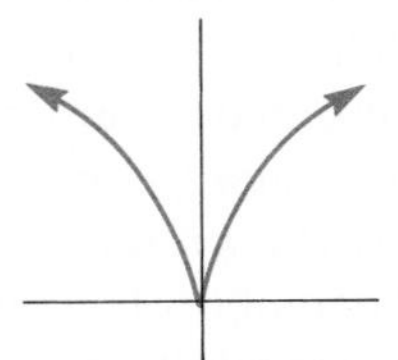
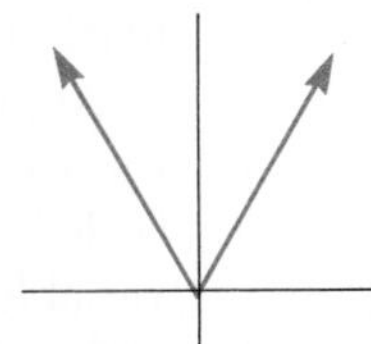
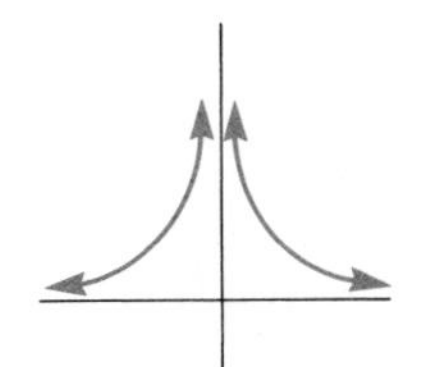

And you will learn how various functions are appropriately used in a variety of types of models, including

Market equilibrium.
Profit maximization.
Cost minimization.
Cost analysis.
Analysis of sales patterns.
Decreasing purchasing power.
Population growth.
Carbon-14 dating.
"Learning" processes.
Demand for a product.

We have studied linear functions and a few of the many important linear models. These functions are appropriately used when the relationship between the variables is such that each one-unit change in the value of the independent variable brings about a constant amount of change in the value of the dependent variable. Many relationships, however, are not linear by nature and cannot be adequately approximated by linear functions. For instance, the per-unit variable cost of producing a product is not necessarily constant. It may change as the number of units produced increases because of various possible economies, or diseconomies, of scale (such as quantity discounts received on large purchases of raw material, or overtime pay differentials). If this is the case, the total cost function will be a nonlinear function of the number of units produced. Or, functions which are linear may, when combined with other functions in the model-building process, yield nonlinear functions. Whenever, for example, the demand function for a product is linear, the revenue function for that product, defined as quantity sold times selling price per unit, is nonlinear.

There exist a wide variety of nonlinear (or curvilinear) functions which may be appropriately used in those situations where each one-unit change in the value of the independent variable is not accompanied by a constant change in the value of the dependent variable. We begin our study in this area with the POLYNOMIAL FUNCTIONS.

9.1 THE POLYNOMIAL FUNCTIONS

A function whose value is determined through addition, subtraction, multiplication, division, raising a variable to a power, or extracting a root (that is, through *algebraic operations*) is called an ALGEBRAIC FUNCTION. Functions that are not algebraic are TRANSCENDENTAL FUNCTIONS.

Algebraic functions are composed of MONOMIALS, or TERMS, in the variables, where a monomial in certain variables, such as x and y, is defined as the product of a nonzero constant times powers of the variables. Examples of monomials in one variable, x, are: $3x^2$, $0.5x^3$, and $-x^4$. Monomials in two variables, x and y, are: $3xy$, x^2y, and $-4x^2y^3$.

The value of the power to which a variable is raised is the DEGREE OF THE VARIABLE; the DEGREE OF A TERM is the sum of the degrees of the variables appearing in that term; the DEGREE OF AN EQUATION is the degree of its term to the highest degree. Thus, $x^2 - 4 = 0$ and $x^2 - xy = 0$ are both *second-degree equations;* $8x^3 - 64 = 0$ and $3x^2y + x - 14 = 0$ are both *third-degree equations;* while $5x^2 + 4x^3y^3 = 0$ is a *sixth-degree equation.*

One very important group of algebraic functions is the group of POLYNOMIALS. A POLYNOMIAL FUNCTION in x is the sum of any

Figure 9.1

Categories of Mathematical Functions

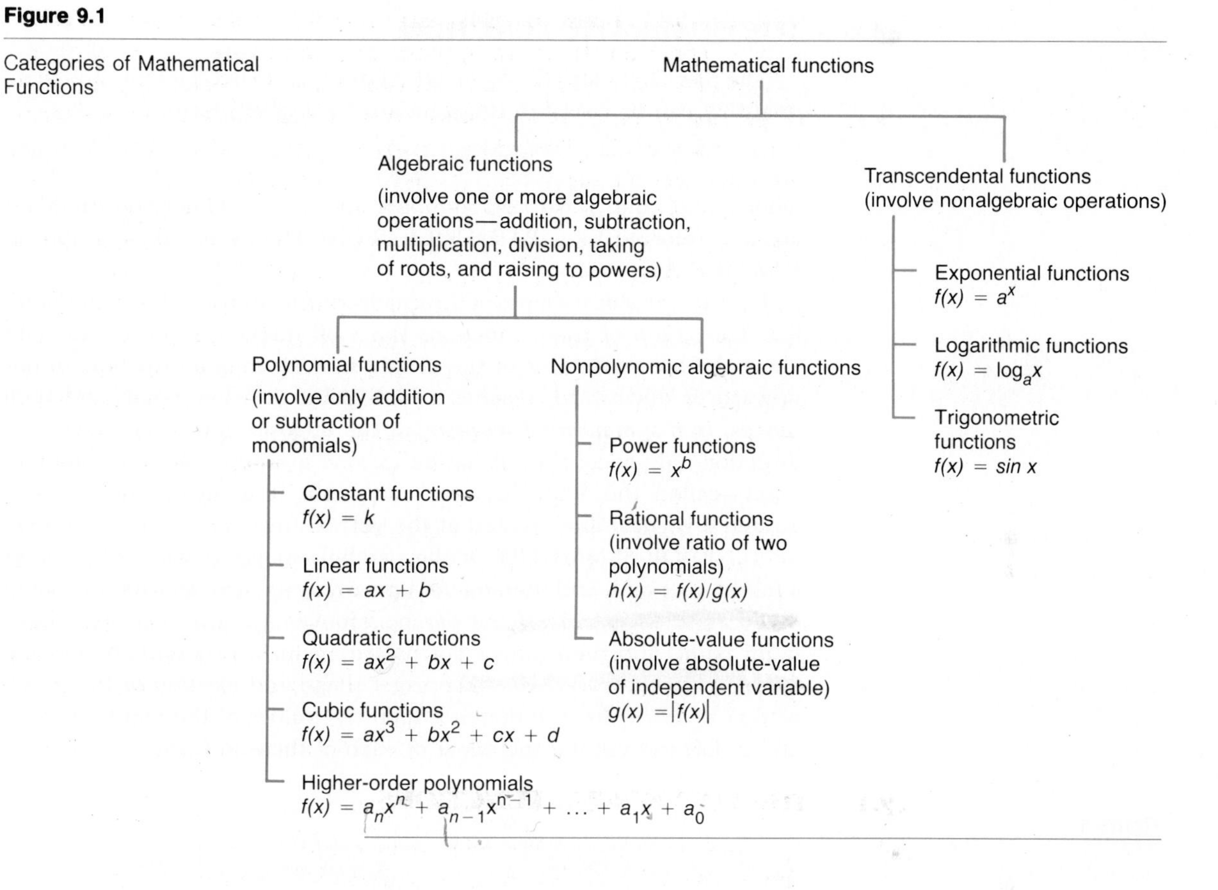

number of monomials in x in which powers of the variable are nonnegative integers.

> **A function f** defined by
>
> $$f(x) = a_0 + a_1x + a_2x^2 + \cdots + a_{n-1}x^{n-1} + a_nx^n \qquad (9.1)$$
>
> where n is a nonnegative integer, $a_n \neq 0$, and the a_n are real numbers, is called a POLYNOMIAL FUNCTION of degree n.

A polynomial function of *degree zero,* as given by $f(x) = a_0$ (or $f(x) = a$) is called a CONSTANT FUNCTION. Its graph is a horizontal line. A polynomial function of *degree one,* as given by $f(x) = a_1x + a_0$ (or $f(x) = ax + b$) where $a_1 \neq 0$, is called a LINEAR FUNCTION. We have previously studied each of these functions. This chapter continues with polynomials of a higher degree.

9.2 THE QUADRATIC FUNCTION

The SECOND-DEGREE, or QUADRATIC, FUNCTION is a polynomial that can be recast in the standard (or explicit) form

$$y = f(x) = ax^2 + bx + c \tag{9.2}$$

where a, b, and c are real numbers and $a \neq 0$. Quadratic functions are also referred to as PARABOLIC FUNCTIONS and their graphs as PARABOLAS.

Possible graphical shapes of the quadratic function are shown in Figure 9.2. The graph of these functions starts off (moving from left to right along the horizontal axis of the graph) moving in one direction, either upward or downward, reaches a high point or a low point, and then moves, *in a symmetrical manner,* in the opposite direction. It changes direction only once; thus, it has either one maximum or one minimum point—called the VERTEX—but not both. It is always symmetrical around a vertical line erected at the vertex; this line is called the AXIS (or LINE) OF SYMMETRY. If the parabola moves downward, reaches a minimum point, and then moves upward again, it is described as being CONCAVE UPWARD. If the parabola moves upward, reaches a maximum point, and then moves downward again, it is described as being CONCAVE DOWNWARD. The exact shape and location of the graph of $f(x) = ax^2 + bx + c$ depend upon the values of the constants a, b, and c. Let us examine the effect of each of these in turn.

Figure 9.2

Quadratic Functions (or Second-Order Polynomials)

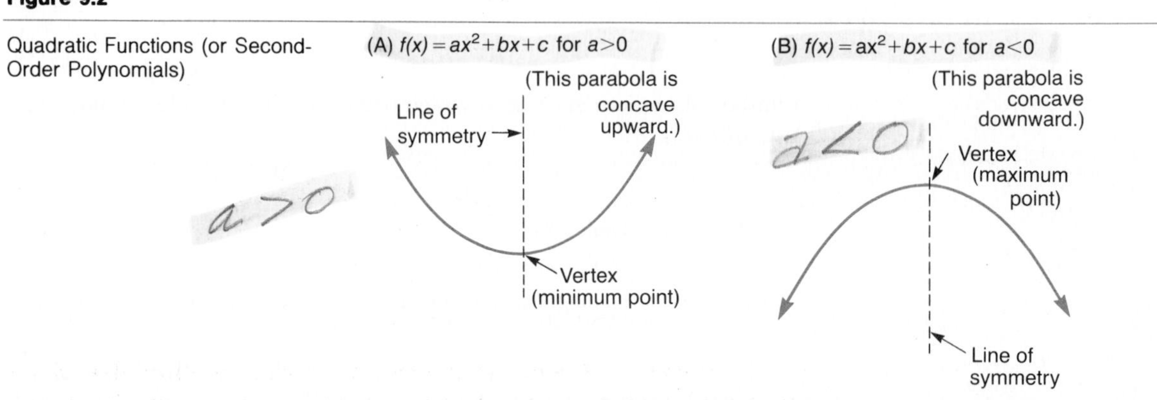

The Role of *a*

The coefficient of the x^2 term determines the way the parabola opens. All functions of the form $f(x) = ax^2 + bx + c$ for $a > 0$ are similar in shape. Each opens upward, or is CONCAVE UPWARD, and the vertex is the minimum point on the curve (see Figure 9.2A). The graph of $f(x) = ax^2 + bx + c$ for $a < 0$ opens downward and is described as being CONCAVE DOWNWARD. The vertex of this function is a maximum point on the curve (see Figure 9.2B).

The coefficient a also determines the "spread" of the parabola. The larger the absolute value of a, the "skinnier" the graph (as shown in Figure 9.3). Thus, a, the coefficient in the x^2 term, determines not only the direction taken by the graph (open upward or open downward) but also the rate of ascent or descent.

Figure 9.3

The Absolute Value of a Determines the Rate of Ascent or Descent of $f(x) = ax^2$

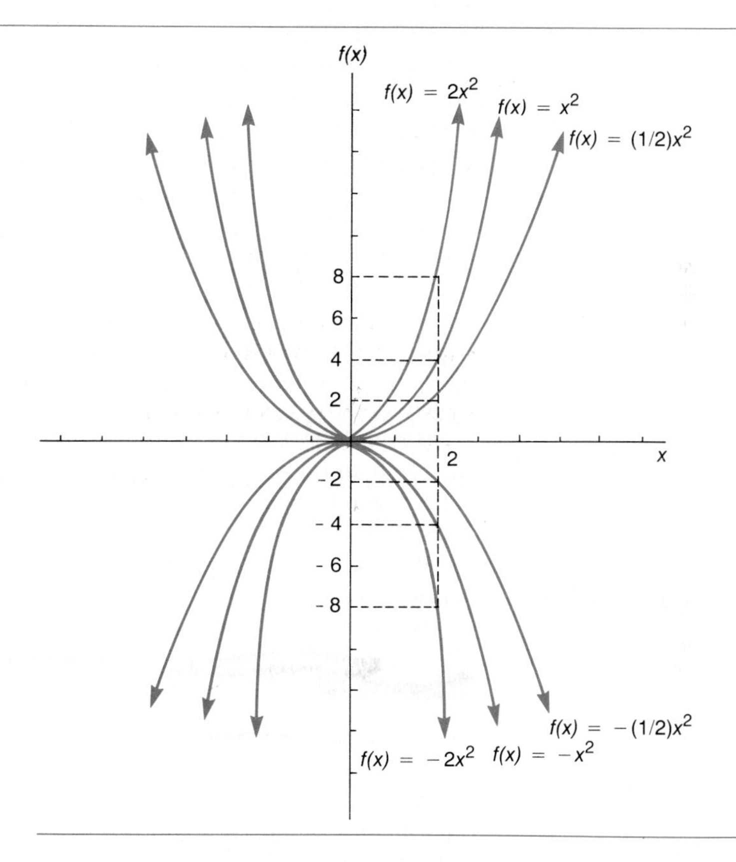

The Roles of b and c

The coefficient b and the constant c, working together, determine the vertical and horizontal displacement of the curve. In $f(x) = ax^2$, both $b = 0$ and $c = 0$. The graphs of $f(x) = ax^2$ (whether $a > 0$ or $a < 0$) are all symmetrical with respect to the y-axis and have the point (0,0) as a vertex.

Because for $f(x) = ax^2 + bx + c$, $f(0) = c$, it is the constant c that is the y-intercept and, consequently the VERTICAL DISPLACEMENT of the curve on the rectangular coordinate system. The graph of $f(x) = ax^2 + c$, where $c \neq 0$, is obtained by shifting the graph of

Figure 9.4

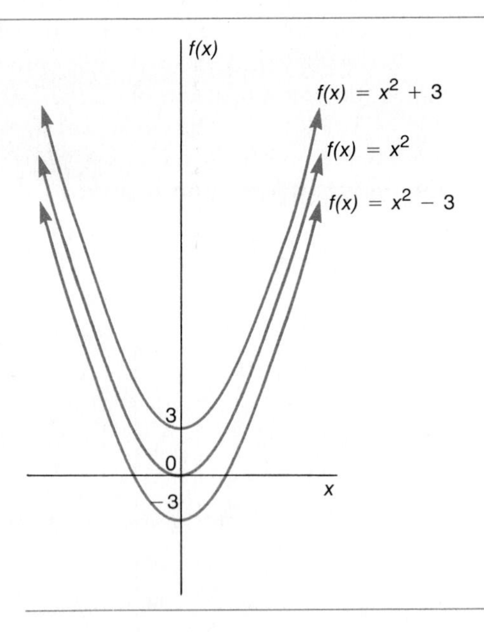

The Algebraic Value of the Constant c Determines the Vertical Displacement of $f(x) = ax^2 + c$

$f(x) = ax^2$ vertically by c units. The graph is shifted upward $|c|$ units if $c > 0$ and downward $|c|$ units if $c < 0$ (as illustrated in Figure 9.4). These graphs remain symmetrical about the y-axis, but the vertex of each is $(0,c)$.

When $b \neq 0$, the parabola experiences both a horizontal and a vertical displacement (see Figure 9.5).

We shall be able to show, after we have studied the calculus in Chapter 12, that the coordinates for the vertex of any second-degree polynomial are given by

$$x = -b/2a \qquad \text{and} \qquad y = (4ac - b^2)/4a \tag{9.3}$$

Example 1 Given the quadratic equation $f(x) = x^2 + 5x + 12$, determine the coordinates of the vertex. Describe, in general terms, the graph of the function.

The vertex occurs where

$$x = -b/2a = -5/2(1) = -2.5$$

and

$$y = (4ac - b^2)/4a = [4(1)(12) - (5)^2]/4(1) = 23/4 = 5.75$$

Thus, the coordinates of the vertex of the function are $(-2.5, 5.75)$.

Because $(a = 1) > 0$, the parabola is concave upward. The vertex represents a minimum point. The line of symmetry is a vertical line erected at $x = -2.5$. The y-intercept is $(0,12)$. (See Figure 9.6.) [ ]

Figure 9.5

The Effect of $b \neq 0$ on $f(x) = ax^2 + bx + c$

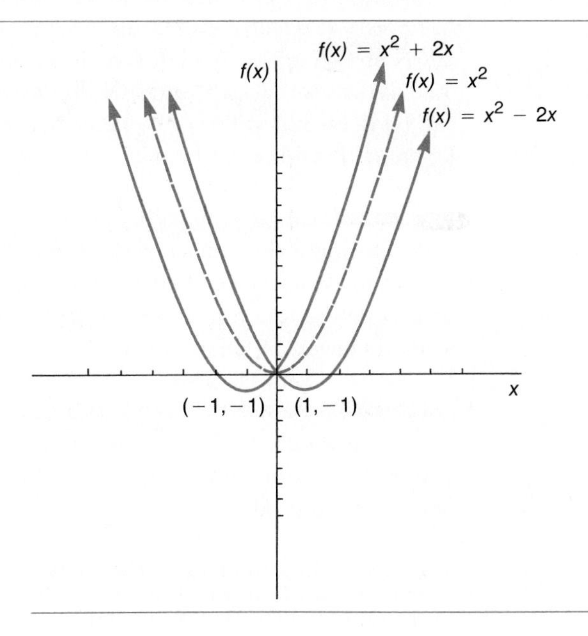

Figure 9.6

$f(x) = x^2 + 5x + 12$

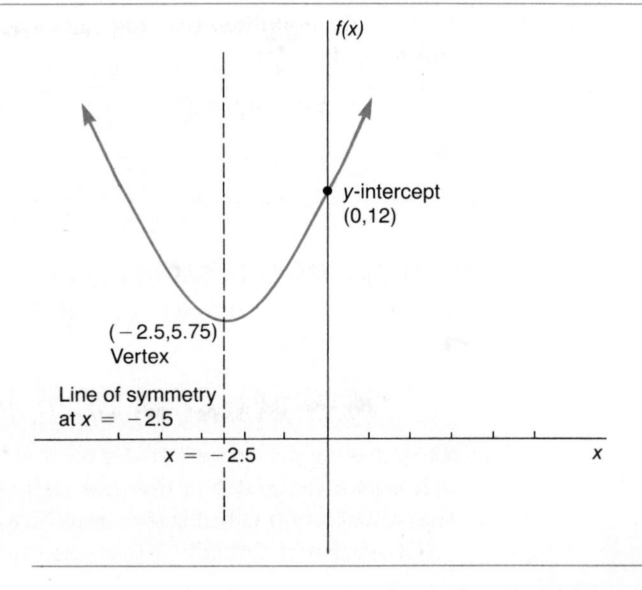

Finding the x-Intercepts

Because of the symmetry of all parabolas, each will either (1) touch the x-axis at exactly one point, which will be the vertex, (2) cross the x-axis at two points, which are equidistant from the line of symmetry, or (3) be entirely above, or below, the x-axis. The x-intercepts, or ROOTS, can be determined either by factoring or by use of the Quadratic Roots Formula (Formula 9.4 below).

Finding the roots by factoring If we can find the factors of a function and determine the values of the independent variable that cause those factors to equal zero, then we have also found values of the variable that cause the function itself to equal zero. (The procedures of factoring were reviewed in Chapter 1.)

Using the Quadratic Roots Formula Functions are not always easy, or even possible, to factor, and other methods of finding roots are required. If the function is a quadratic function, the Quadratic Roots Formula may be used.

> Quadratic Roots Formula. The roots of a quadratic function written in the standard form $ax^2 + bx + c = 0$, if any exist, may be found using the QUADRATIC ROOTS FORMULA:
>
> $$x = \frac{-b \pm \sqrt{b^2 - 4ac}}{2a} \qquad (9.4)$$

Example 2 Find the roots of the function $f(x) = 3x^2 - x - 4$.

We are looking for the values of x which satisfy the equality $3x^2 - x - 4 = 0$. In preparing to use the Quadratic Roots Formula, we note that, for this function, $a = 3$, $b = -1$, and $c = -4$. Then

$$x = \frac{-(-1) + \sqrt{(-1)^2 - 4(3)(-4)}}{2(3)} = \frac{1 + \sqrt{49}}{6} = \frac{1+7}{6} = 1\ 1/3$$

and

$$x = \frac{-(-1) - \sqrt{(-1)^2 - 4(3)(-4)}}{2(3)} = \frac{1 - \sqrt{49}}{6} = \frac{1-7}{6} = -1$$

These values of x—$x = 1\frac{1}{3}$ and $x = -1$—are the roots of the equation. These points—$(1\frac{1}{3}, 0)$ and $(-1, 0)$—are the points where the curve crosses the horizontal axis. []

Of course, not all parabolas cross the x-axis. If the graph of a quadratic function lies entirely above, or entirely below, the x-axis, it has no roots; there are no $x = a$ such that $f(a) = 0$. The other possible situation is that where the graph of the quadratic function touches the horizontal axis at just one point, but does not cross that axis.

Evaluation of the DISCRIMINANT— which is $b^2 - 4ac$—enables us to quickly discover the nature of the roots of the quadratic equation. If $(b^2 - 4ac) > 0$, the function intersects the x-axis at two points. If $(b^2 - 4ac) = 0$, the function touches the x-axis at only one point. If $(b^2 - 4ac) < 0$, the function does not touch the x-axis.

The Graph of the Quadratic Function, in Summary

The graph of $f(x) = ax^2 + bx + c$ is called a PARABOLA and has these properties:

1. It changes direction exactly once and thus has either a maximum point or a minimum point but not both.
2. It is concave upward and has a minimum point if $a > 0$ or is concave downward and has a maximum point if $a < 0$.
3. It is symmetrical around a vertical line called the LINE OF SYMMETRY erected at $x = -b/2a$.
4. It has exactly one maximum or minimum point called the VERTEX, with coordinates $x = -b/2a$ and $f(x) = (4ac - b^2)/4a$.
5. Its y-intercept occurs at $f(0) = c$.
6. Its x-intercepts, if any, occur at $x = a$ such that $f(a) = 0$ and can be determined either by factoring or by using the Quadratic Roots Formula:

$$x = \frac{-b \pm \sqrt{b^2 - 4ac}}{2a}$$

EXERCISES

For each of the following functions (a) *determine if the parabola is concave upward or concave downward,* (b) *determine the y-intercept,* (c) *determine the location of the line of symmetry,* (d) *determine the coordinates of the vertex,* (e) *determine the x-intercepts, if any, and* (f) *sketch the graph of the function.*

1. $f(x) = 3x^2$
2. $f(x) = -3x^2$
3. $f(x) = 3x^2 + 4$
4. $f(x) = -3x^2 + 4$
5. $f(x) = 3x^2 - 5$
6. $f(x) = -3x^2 - 5$
7. $f(x) = 3x^2 + 2x$
8. $f(x) = -3x^2 + 2x$
9. $f(x) = 3x^2 - 4x$
10. $f(x) = -3x^2 - 4x$
11. $f(x) = 3x^2 + 2x + 4$
12. $f(x) = -3x^2 + 2x + 4$
13. $f(x) = 3x^2 - 4x + 4$
14. $f(x) = -3x^2 - 4x + 4$
15. $f(x) = 3x^2 - 4x - 5$
16. $f(x) = -3x^2 - 4x - 5$
17. $f(x) = -x^2 - 3x + 6$
18. $f(x) = (x - 3)^2 - 64$
19. $y = x^2 + 9x + 18$
20. $y = x^2 - 25$

9.3 APPLICATIONS OF QUADRATIC MODELS

Many relationships between variables are appropriately described by a quadratic function.

Example 3 **Area of a rectangle** A dairy farmer has 1,000 feet of fencing material that he wishes to use to enclose a rectangular holding area for his cows. A tall stone wall is located in an appropriate place so that it can serve as one side of the area. The other three sides will have to be fenced, as shown.

Find the dimensions of the rectangle that will maximize the size (in square feet) of the holding area.

Let x represent the width of the area. Then, $1{,}000 - 2x$ will be the length. We recall that

$$\begin{aligned} \text{Area} &= \text{Width} \times \text{Length} \\ &= x(1{,}000 - 2x) \\ &= 1{,}000x - 2x^2 \end{aligned}$$

This is a parabola with vertex at

$$x = -b/2a = -1{,}000/2(-2) = 250$$

The vertex represents a maximum since $(a = -2) < 0$, so that the parabola is concave downward.

Dimensions of the area are: width = 250 feet, length = 500 feet, and total enclosed area = 125,000 square feet. []

Example 4 **Market equilibrium** The demand for a certain product seems to be appropriately described by the function

$$p = D(x) = -0.005x^2 + 800$$

where x is the quantity demanded at price p per unit. (See Figure 9.7.)

The supply function for the same product is given by

$$p = S(x) = 0.0078x^2$$

Market equilibrium occurs at that price where the quantity demanded is equal to the quantity supplied, that is,

$$\begin{aligned} D(x) &= S(x) \\ -0.005x^2 + 800 &= 0.0078x^2 \\ 0.0128x^2 &= 800 \\ x &= 250 \end{aligned}$$

This quantity is associated with a price of

$$p = S(250) = 0.0078(250)^2 = \$487.50$$

[]

Figure 9.7

Market Equilibrium Occurs Where Demand Function and Supply Function Intersect

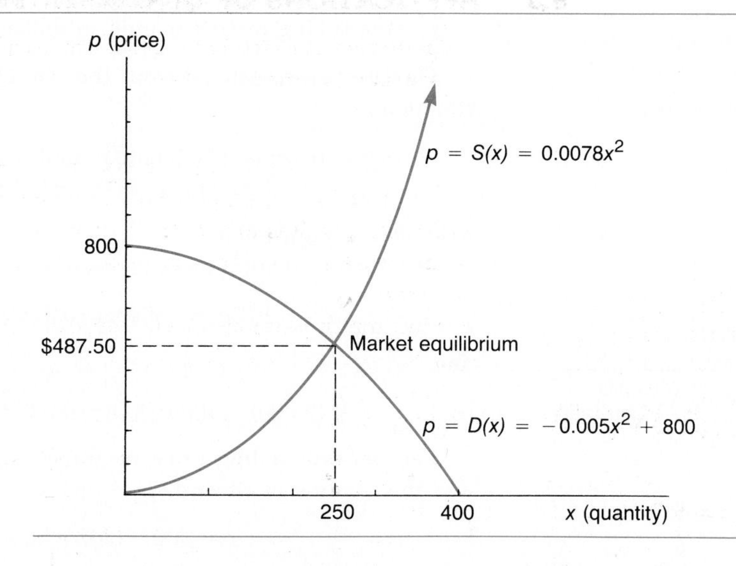

Quadratic Revenue Functions

Whenever the demand function for a product is linear with a negative slope, the revenue function for the sale of the product will be quadratic. We define

$$\text{Revenue} = (\text{Quantity sold}) \times (\text{Selling price per unit})$$

If the demand function is linear with a negative slope, it can be represented by a function of the general form

$$p = D(x) = ax + b$$

where x is the quantity demanded and p is the selling price per unit and $a < 0$ to yield a downward-sloping demand function. If we represent the revenue function $R(x)$, we have

$$R(x) = (x)(ax + b) = ax^2 + bx$$

which is a concave-downward quadratic function. Because of the nature of the situation, only that portion of the function where both $p > 0$ and $x > 0$ has a logical interpretation.

Example 5 **Quadratic revenue function** The Carey-Coleman Products Company has determined that the quantity demanded of its product depends upon the price at which the product is offered for sale. Market research indicates that the relationship is

$$p = f(x) = 300 - 0.025x$$

where

Figure 9.8

Quadratic Revenue Function
$R(x) = 300x - 0.025x^2$

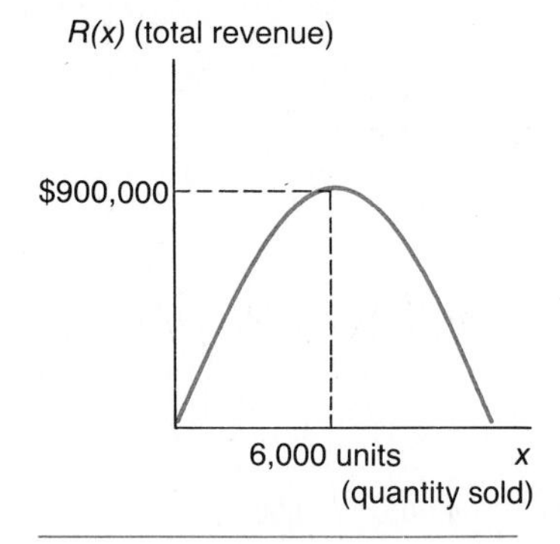

x = the quantity demanded, in units
p = the selling price per unit, in dollars

Total revenue realized from the sale of the product is given by the function

$$\text{Total revenue} = (\text{Quantity sold}) \times (\text{Selling price per unit})$$
$$R(x) = (x) \times (300 - 0.025x) = 300x - 0.025x^2$$

as shown in Figure 9.8.

The total revenue function reaches its maximum point at

$$x = -b/2a = -300/2(-0.025) = 6{,}000 \text{ units sold}$$

and

$$f(6{,}000) = 300(6{,}000) - 0.025(6{,}000)^2 = \$900{,}000$$

The per-unit selling price required to generate this volume of sales and this volume of revenue is

$$p = 300 - 0.025(6{,}000) = \$150$$ []

Example 6 **Quadratic revenue and linear cost functions** The demand function for a product is given by

$$p = D(x) = 90 - x$$

where x units are demanded at a price of p dollars per unit.

Total revenue, expressed as a function of the selling price of the product, is

$$R(x) = (x)(90 - x) = 90x - x^2$$

This revenue function is pictured in Figure 9.9.

The fixed cost of producing the product is \$1,375, while variable cost, per unit, is \$10. Hence,

$$\begin{aligned} \text{Total cost} &= \text{Fixed cost} + \text{Total variable cost} \\ &= \text{Fixed cost} + (\text{variable cost per unit}) \cdot \\ &\quad (\text{Number of units produced}) \\ C(x) &= 1{,}375 + 10x \end{aligned}$$

The total cost function is linear and is also pictured in Figure 9.9.

The total cost and total revenue functions intersect at the points $x =$ 25 and = 55. These points of intersection can be determined by setting the two functions equal to each other and solving, as

$$\begin{aligned} R(x) &= C(x) \\ 90x - x^2 &= 1{,}375 + 10x \\ -x^2 + 80x - 1{,}375 &= 0 \end{aligned}$$

This is a quadratic equation in one variable, which can be solved using the Quadratic Roots Formula or by factoring.

Figure 9.9

Total Revenue, Total Cost, and Total Profit Functions for a Product

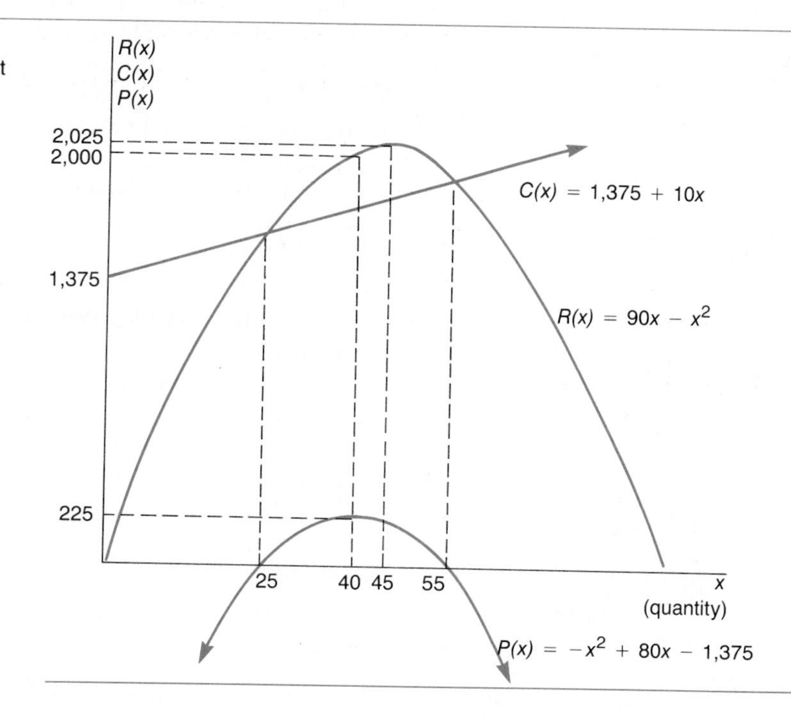

For production levels below $x = 25$ and above $x = 55$, total cost exceeds total revenue, and the firm operates at a loss. For production levels of $x = 25$ and $x = 55$, total revenue exactly equals total cost, and the firm breaks even. For production levels $25 < x < 55$, total revenue exceeds total cost, and the firm realizes a profit.

The total profit function for the firm is given by

$$\begin{aligned} \text{Total profit} &= \text{Total revenue} - \text{Total cost} \\ P(x) &= R(x) - C(x) \\ &= (90x - x^2) - (1{,}375 + 10x) \\ &= -x^2 + 80x - 1{,}375 \end{aligned}$$

The graph of this function is superimposed over those of the total revenue and the total cost in Figure 9.9. The profit function yields a concave-downward parabola with a maximum point at $x = -80/2(-1) = 40$. The profit associated with this level of production is $P(40) = -(40)^2 + 80(40) - 1{,}375 = \$225$.

Notice that the production level yielding the maximum profit is not the same production level as that yielding maximum revenue. Maximum occurs when $x = -90/2(-1) = 45$ for a total revenue $R(45) = 90(45) - (45)^2 = 2{,}025$, and a total profit of $P(45) = -(45)^2 + 80(45) - 1{,}375 = 200$. This is \$25 less than the maximum possible profit. []

Quadratic Cost Functions

The per-unit variable cost of producing a product is not necessarily a constant. It may change as the number of units produced increases because of various possible economies, or diseconomies, of scale. If this is the case and if per-unit variable cost can be appropriately depicted by a linear function, then total cost will be a quadratic function.

Example 7

Quadratic cost function The Raiford Lighting Company manufactures table lamps at a total variable cost of $6x^2 - 300x$, where x is the weekly production in hundreds of units. Weekly fixed costs of production are \$12,000. What is the total weekly costs function?

Total costs are total variable costs plus fixed costs, or, for Raiford Lighting Company,

$$C(x) = 6x^2 - 300x + 12{,}000$$

Figure 9.10

The Quadratic Cost Function $C(x) = 6x^2 - 300x + 12{,}000$

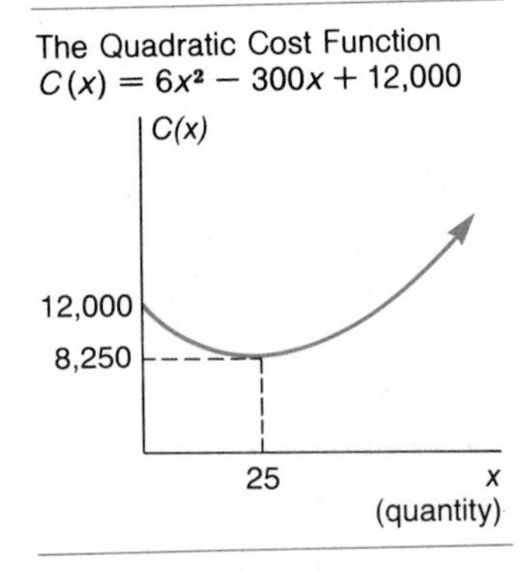

The function is pictured in Figure 9.10.

Total costs are at a minimum when production level is

$$x = -(-300)/(2)(6) = 25 \text{ units}$$

and production costs are

$$C(25) = 6(25)^2 - 300(25) + 12{,}000 = \$8{,}250$$

[]

Example 8

Quadratic revenue and quadratic cost functions Price for a certain product is given by $p = D(x) = 100 - 0.1x$ where x is the quantity demanded, in units, at a price of p dollars per unit. Thus, the total revenue function for sale of the product is $R(x) = x(100 - 0.1x) = 100x - 0.1x^2$.

For the same product, variable cost per unit follows the linear function $v(x) = 6.7 + 0.033x$. Fixed costs are \$8,000. Thus, total costs are $C(x) = 8{,}000 + x(6.7 + 0.033x) = 8{,}000 + 6.7x + 0.033x^2$.

The total revenue and the total cost functions are shown in Figure 9.11. To determine the break-even points, we compute

$$\begin{aligned} \text{Total revenue} &= \text{Total cost} \\ 100x - 0.1x^2 &= 8{,}000 + 6.7x + 0.033x^2 \\ -0.133x^2 + 93.3x - 8{,}000 &= 0 \end{aligned}$$

The roots are found, using the Quadratic Roots Formula, to be $x = 100$ and $x = 601.5$.

The profit function for the product is given by

$$\begin{aligned} \text{Total profit} &= \text{Total revenue} - \text{Total cost} \\ P(x) &= (100x - 0.1x^2) - (8{,}000 + 6.7x + 0.033x^2) \\ &= -0.133x^2 + 93.3x - 8{,}000 \end{aligned}$$

This is a concave-downward quadratic function with its maximum point at $x = -93.3/(2)(-0.133) = 350.75$ units. Total profit at this level of production is P(350.75) = \$8,362.58. []

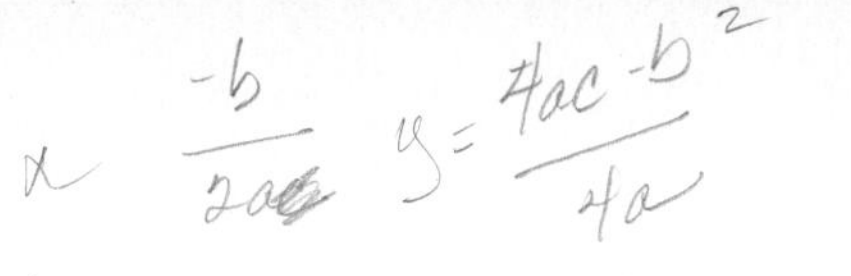

Figure 9.11

Quadratic Revenue and Quadratic Cost Functions

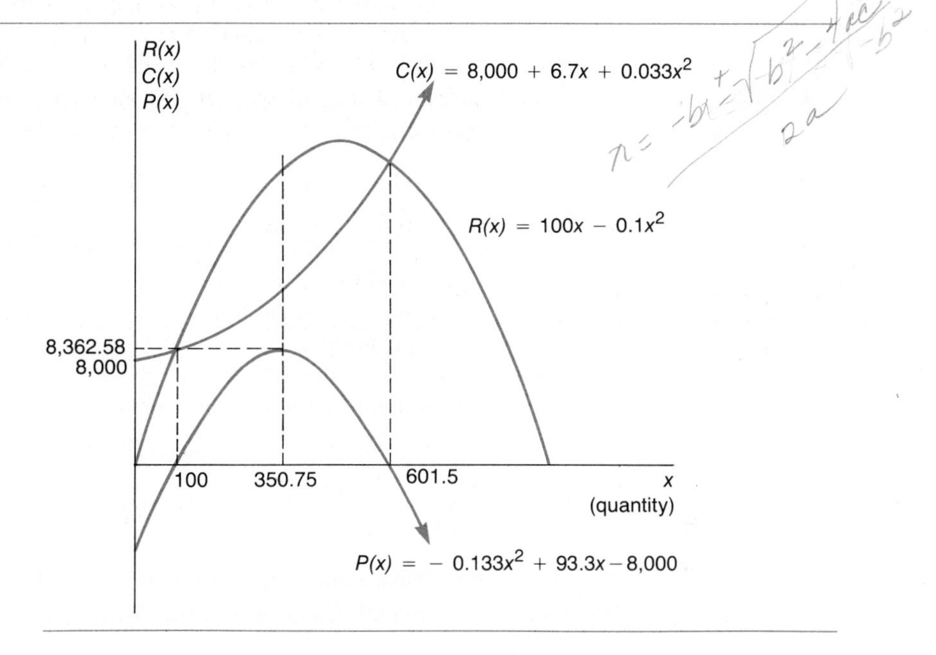

EXERCISES

The owner of Pinetree Apartments has 100 apartments available for rent. A market survey indicates that all 100 apartments can be rented at \$400 a month, but that for each \$10 increase in rent, one fewer apartment will be rented.

21. Determine the revenue function for the apartments.
22. What rent should be charged each month in order to maximize the revenue from apartment rental?
23. How many apartments will be rented at this monthly rate?
24. What will be the total revenue realized at this monthly rate?

A craftsman can sell, each month, 25 handmade hammocks at \$75 per hammock. For each \$5 increase in price, he will be able to sell one fewer hammock each month.

25. Determine the revenue function from the sale of hammocks for the craftsman.
26. What price should he charge in order to maximize his revenue from the sale of hammocks?
27. How many hammocks will he sell at this price?
28. What will be the total revenue realized from the hammocks?

The supply S(x) *and demand* D(x) *functions for a certain product are*

$$p = S(x) = 0.02x^2 \qquad \text{and} \qquad p = D(x) = -0.005x^2 + 90$$

where x *represents quantity, in thousands of units, at a price of* p *dollars per unit.*

29. What is the demand for the product at a selling price of \$50? \$75? \$85?
30. What must the price of the product be in order to generate a supply of 50 units? 75 units? 100 units?
31. What is the equilibrium price? What quantity will be supplied at this price?

The demand function for a certain product is

$$p = D(x) = -0.04x^2 + 400$$

where x *is the quantity demanded, in units, at a price of* p *dollars per unit.*

32. How many units will be demanded at a price of \$100? \$150? \$170?
33. Sketch the graph of the demand function.

The supply function for this same product is

$$p = S(x) = 0.025x^2$$

34. How many units will be supplied at a price of \$100? \$150? \$170?
35. Sketch the graph of the supply function on the same coordinate axes as the demand function.
36. What are the equilibrium price and quantity?

The market-research department for a company estimates that the market for a certain product the company produces is appropriately described by

$$x = f(p) = 5{,}000 - 25p$$

where x *is the number of units of the product that consumers are likely to buy each month at a price of* p *dollars per unit.*

The cost control department of the same firm estimates that the costs of producing the product follow the function

$$C(x) = 75{,}000 + 60x$$

where \$75,000 is the fixed cost and \$60 is the variable cost per unit.

37. What is the revenue function for the product?
38. What per-unit price generates maximum revenue?
39. What is the maximum revenue?
40. Sketch the graphs of the revenue function and the cost function in the same coordinate system.

41. What is the profit function for the product, given that profit is defined as total revenue minus total cost?
42. What price generates maximum profit?
43. What is the maximum profit?
44. Sketch the profit function on the same coordinate system as the revenue and cost functions.
45. What are the break-even points?
46. How much profit would be lost if the firm priced its product so as to maximize revenue rather than to maximize profit?

A firm finds that the market for its product is such that the selling price and quantity demanded of the product are related by the function

$$p = D(x) = 6 - 0.001x$$

where x *is the number of units demanded at a price of* p *dollars per unit.*

Monthly production costs are given by the cost function

$$C(x) = 3{,}000 + 2x$$

47. Find x such that the firm maximizes profit.
48. What will be the per-unit selling price at this production level?
49. What is the dollar amount of maximum profit?

A firm has revenue and cost functions defined by

$$R(x) = 600x - 5x^2 \qquad \text{and} \qquad C(x) = 2x^2 + 40x + 8{,}000$$

where x *is the quantity of product manufactured and sold.*

50. Sketch the graphs of the revenue and cost functions on the same coordinate system.
51. Determine the break-even output levels for the firm.
52. Determine the profit function for the firm and the output level at which profit is maximized.

A refinery has revenue and cost functions defined by

$$R(x) = 520x - 2x^2 \qquad \text{and} \qquad C(x) = 0.5x^2 + 20x + 3{,}500$$

where x *is units of weekly output and* R(x) *and* C(x) *are total revenue and total cost, in dollars, respectively.*

53. For what level of output does the firm's total revenue equal its total cost?
54. Sketch the graph of total revenue and total cost on the same coordinate system.
55. Determine the total profit function and sketch this function on the same coordinate system as the revenue and cost functions.
56. What level of output maximizes profit?

57. What is the maximum profit?
58. For what level of output does the firm realize maximum revenue? What is the profit for this level of output?

9.4 POLYNOMIALS OF A STILL HIGHER ORDER

Third-degree and fourth-degree and, on occasion, even higher-degree polynomial functions are encountered in many mathematical models. Examples of such polynomial functions are:

$f(x) = 3x^3 - 4x^2 + 5.5x + 7$ (Third-degree, or CUBIC, function)

$f(x) = 0.6x^4 - 15$ (Fourth-degree, or QUARTIC, function)

$f(x) = x^5 - x^4 + x^3 - x^2 + x - 1$ (Fifth-degree polynomial)

Graphs of these and other higher-degree polynomial functions will always be smooth curves, having perhaps several ups and downs but with no breaks, holes, or sharp corners.

Figure 9.12

$f(x) = x^n$ Where n Is a Positive Even Integer

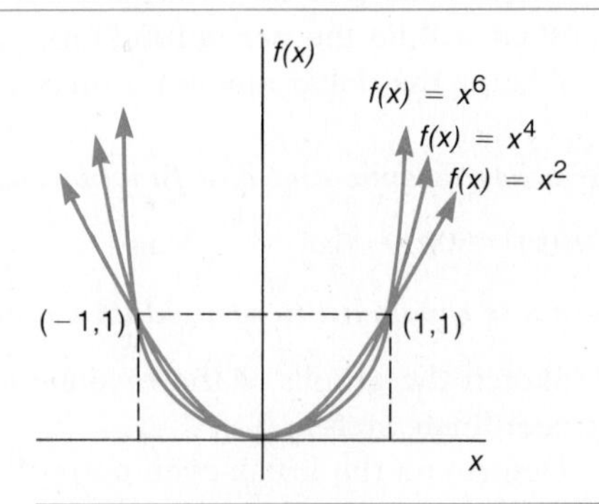

If the polynomial takes on the form $f(x) = x^n$, where n is a positive *even* integer, its graph will have the general shape illustrated in Figure 9.12. All of these functions intersect at the point $(-1,1)$ and again at $(1,1)$. As n increases, the graph flattens in the interval $-1 < x < 1$ and grows steeper for $x < -1$ and $x > +1$. Each of these functions is SYMMETRICAL WITH RESPECT TO THE y-AXIS and, accordingly, termed an EVEN FUNCTION. (See Figure 9.13.)

Figure 9.13

A Function Has SYMMETRY WITH RESPECT TO THE y-AXIS if $f(-x) = f(x)$ for All x

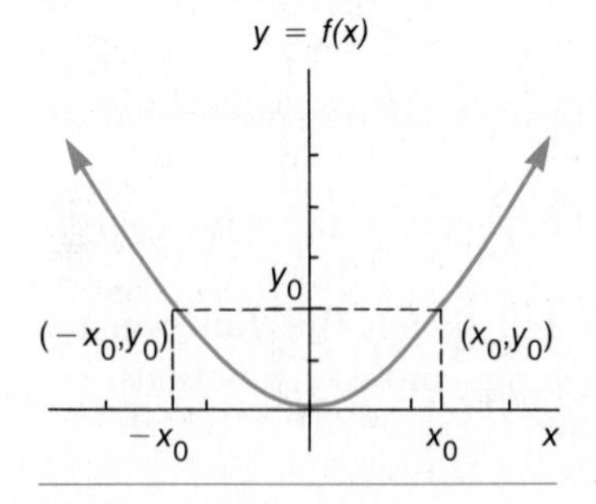

A function is termed an EVEN FUNCTION (and has SYMMETRY WITH RESPECT TO THE y-AXIS) if $f(-x) = f(x)$ for all x and $-x$ in the domain of f.

Figure 9.14

$f(x) = x^n$ Where n Is a Positive Odd Integer

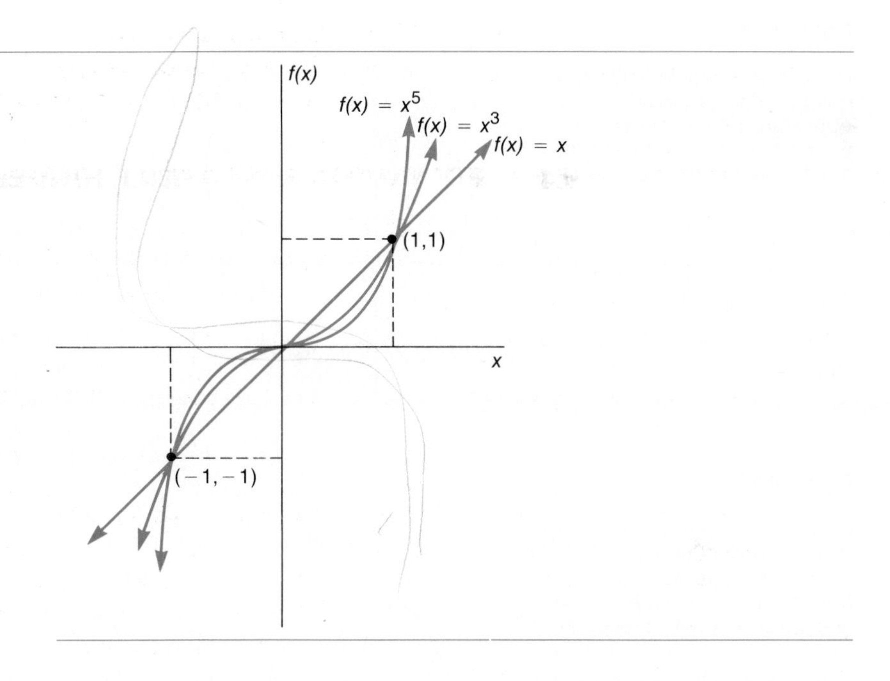

Where n is a positive *odd* integer, the graphs of the functions $f(x) = x^n$ have the general shape shown in Figure 9.14. All of these functions intersect at the points $(-1,-1)$ and again at $(1,1)$. As n increases the graphs are successively flatter near the origin (0,0) and become successively steeper as they move away from the origin. Each of these functions is SYMMETRICAL WITH RESPECT TO THE ORIGIN and, accordingly, termed an ODD FUNCTION. (See Figure 9.15.)

Figure 9.15

A Function Has SYMMETRY WITH RESPECT TO THE ORIGIN if $f(-x) = -f(x)$ for All x and $-x$ in the Domain of f

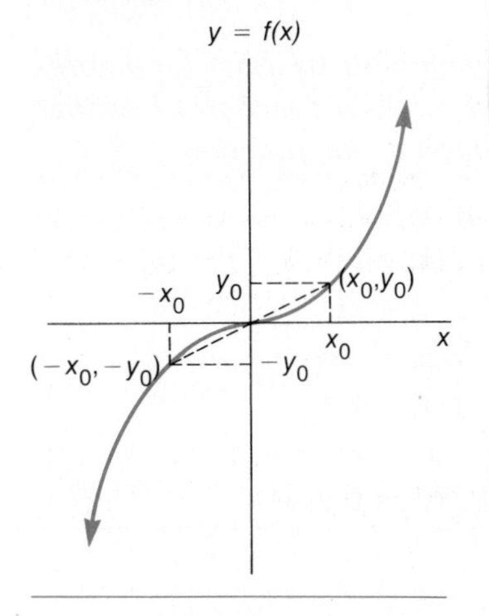

> **A function is termed an ODD FUNCTION** (and has SYMMETRY WITH RESPECT TO THE ORIGIN) if $f(-x) = -f(x)$ for all x and $-x$ in the domain of f.

When more terms are added to the polynomials, the graphs take on more-elaborate shapes. The graphs of these polynomials have a roller-coaster appearance, with, generally, one less turning point than the highest exponent in the function. If the degree of the function is an even number, the extreme ends of the graph will arrive and leave the graph from the same direction. (See Figure 9.16.) If the degree is an odd number, the extreme ends will arrive from one direction and depart to the opposite direction. (See Figure 9.17.)

The task of plotting the graph of a higher-order polynomial function is greatly simplified by a little knowledge of the calculus. This topic is

Figure 9.16

Many Higher-Degree Polynomials Have a "Roller-Coaster" Appearance (even-degree polynomials enter and leave the graph from the same direction)

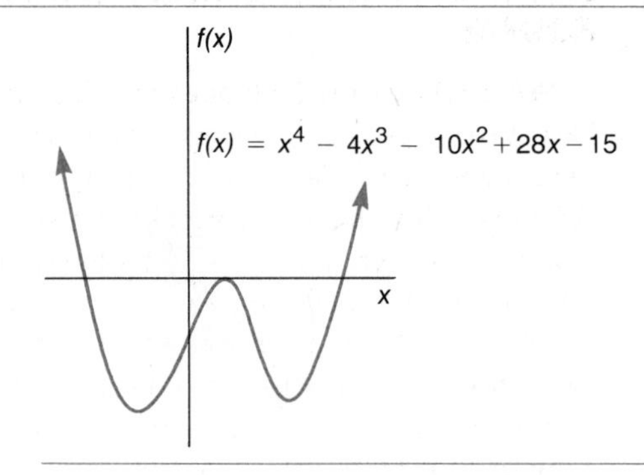

Figure 9.17

Many Higher-Degree Polynomials Have a "Roller-Coaster" Appearance (odd-degree polynomials enter and leave the graph from opposite directions)

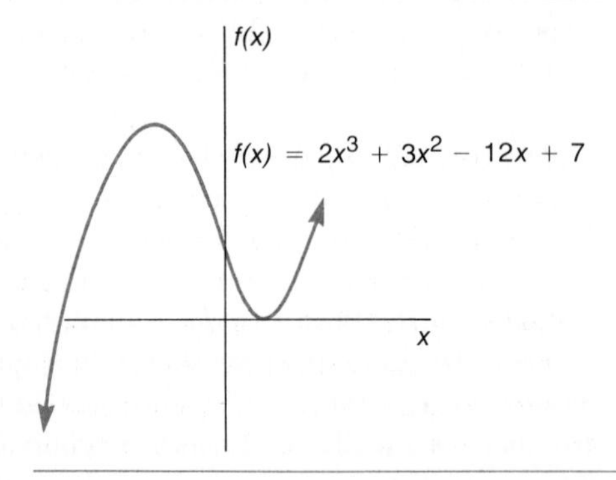

covered in later chapters of this text; so we will wait until then to discuss graphing a polynomial of third or higher degree.

EXERCISES

For each of the following functions, (a) *determine its order* (*or degree*), (b) *determine if it is an odd function or an even function or neither, and* (c) *describe, in general terms, the graph of the function.*

59. $f(x) = x^2$
60. $f(x) = x^3$
61. $f(x) = -x^3$
62. $f(x) = x^3 + 1$
63. $f(x) = x^3 + x^2$
64. $f(x) = 4x^3$
65. $f(x) = x^4$
66. $f(x) = -x^4$
67. $f(x) = x^4 + 1$
68. $f(x) = 4x^4$
69. $f(x) = x^4 - x^2$
70. $f(x) = x^5$
71. $f(x) = x^6$
72. $f(x) = x^6 + x^5$
73. $f(x) = 6 + 7x + x^2 + x^3$

9.5 FITTING A POLYNOMIAL FUNCTION TO REFERENCE POINTS

When a functional relationship inherent in a problem situation is linear by nature, it is not uncommon to have the parameters of the equation determined directly from a description of the situation. The story is different when the association between variables is nonlinear. Nonlinear functions in many cases result from combining linear functions. In many other cases, bits and pieces of information must be extracted about the situation, used to hypothesize the general functional form of the relationship, and then applied to estimate the specific parameters.

In some instances, the polynomial must be "fitted" through a limited number of (x,y) coordinates. For a polynomial of degree k, $(k + 1)$ data points are required. Thus, three coordinates are necessary for fitting a quadratic equation, four coordinates are needed for fitting a third-order polynomial, and so on. The procedure is to establish the $(k + 1)$ data points into the general form of the equation and solve the resulting equations simultaneously for the values of the parameters.

Example 9

A team of analysts visualize the relationship between total cost of production, $f(x)$, and the number of units produced, x, as shown in Figure 9.18. That is, they believe the fixed costs of production are \$100. As production quantity increases, the total cost of production will increase rather slowly at first and then more rapidly as various diseconomies of large-scale operations are experienced. Clearly, the relationship is not linear. In fact, the team of production engineers and costs analysts agrees that a second-degree polynomial can be reasonably used to approximate the quantity-cost relationship and would like to find the parameters for the specific quadratic function that describes this particular quantity-cost relationship.

Figure 9.18

A Quadratic Function, Fit to Reference Points

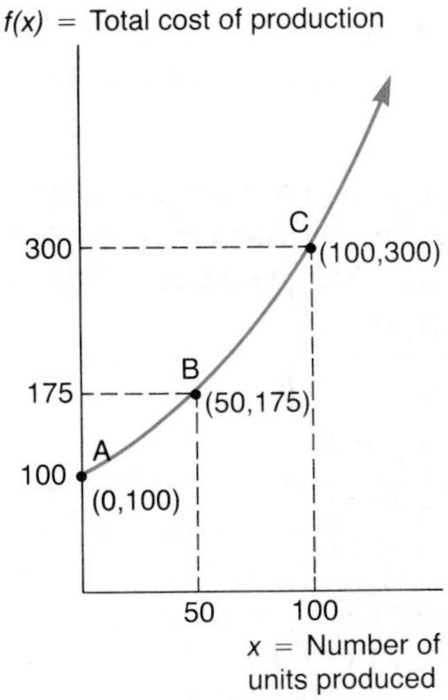

Because the quadratic is a second-degree equation, in order to "fit" a special function to their visualization of the production situation the analysts will need three specific points of reference. One point of reference can be the "zero volume, \$100 fixed cost" point. They also know—based on past experience, their intimate knowledge of the production-cost situation, careful analysis of the relationships, and so on—that when 50 units are produced, total production costs are around \$175, and costs increase to \$300 when 100 units are produced.

These three ordered pairs—(0,100), (50,175), and (100,300)—are used to set up three equations of the quadratic form.

$$f(x) = ax^2 + bx + c$$

Point A: $f(100) = (100)^2a + 100b + c = 10{,}000a + 100b + c = 300$

Point B: $f(50) = (50)^2a + 50b + c = 2{,}500a + 50b + c = 175$

Point C: $f(0) = (0)^2a + 0b + c = c = 100$

The three resulting equations are solved simultaneously for the parameters a, b, and c.

The parameters for the production-cost function are $a = 0.01$, $b = 1$, and $c = 100$. The functional relationship pictured by the analysts can be described mathematically by $f(x) = 0.01x^2 + x + 100$. []

Polynomials of a higher order can be fitted to reference points in this same manner.

EXERCISES

Determine the parameters defining the quadratic functions anchored by each of the following sets of reference points.

74. (1,0),(2,–1),(3,0)
75. (0,5),(2,15),(3,40)
76. (1,0),(5,32),(0,–3)
77. (2,4),(4,16),(5,25)
78. (1,–3),(–1,–3),(3,5)
79. (0,–6),(–2,–20),(1.2)
80. The supply function for a product is quadratic. Three points which lie on the function are ($25,110), ($30,200), and ($50,800). Determine the equation for the supply function. What quantity will be supplied at a price of $40? $60?
81. The demand function for a product is quadratic. Three points which lie on the function are ($5,2000), ($10,1600), and ($20,900). Determine the equation for the demand function. What quantity will be demanded at a price of $25? $30?

9.6 OTHER NONLINEAR ALGEBRAIC FUNCTIONS

In addition to the polynomials, there are many nonpolynomic functions which are important in model building.

Rational Functions

A RATIONAL FUNCTION is the ratio of two polynomial functions.

> If g and h are polynomial functions, then a function f defined by
>
> $$f(x) = g(x)/h(x) \tag{9.5}$$
>
> where $h(x) \neq 0$ is called a RATIONAL FUNCTION.

The definition of a rational function includes all of the polynomial functions. To see this, simply set $h(x) = 1$ (which would be the constant function). The resulting rational function is $f(x) = g(x)/1$ where $g(x)$ is a polynomial. Certainly, however, not all rational functions are polynomials.

Examples of rational functions which are not polynomials:

$f(x) = 5/x$ $f(x) = (2x + 3)/(x - 4)$
$f(x) = 2x/(x + 1)$ $f(x) = (x^2 + 3x + 1)/(x - 1)$
$f(x) = 1/(x^2 + 1)$ $f(x) = x^4/(x^2 - 7)$

Because division by zero is not permitted, any input value of the independent variable which would produce a zero in the denominator of a rational function must be excluded from the domain of that function. Such exclusion of certain real numbers from the domain of a rational function may result in natural "breaks" or "holes" in the graph of the function.

One of the most important of the rational functions is defined as $f(x) = 1/x$ and is pictured in Figure 9.19. Note that the function is not defined at $x = 0$. But look at what happens to the value of the function as x moves closer and closer to zero. First, as x moves nearer and nearer to zero from the left, through numbers that are smaller than zero, the function assumes negative values farther and farther from zero. But then, as x moves closer and closer to zero from the right, through numbers that are larger than zero, the value of the function becomes larger and larger and larger. Because, as the value of x gets closer and closer to zero—but never quite equals zero—the $|f(x)|$ gets larger and larger, the graph has a VERTICAL ASYMPTOTE at $x = 0$.

A line that the graph of a function *approaches* —gets closer and closer to but never quite reaches—is known as an ASYMPTOTE. Asymptotes may be vertical lines, horizontal lines, or diagonal lines. The graphs of rational functions commonly have asymptotes. Look again at the graph

Figure 9.19

Graph of the Function $f(x) = 1/x$

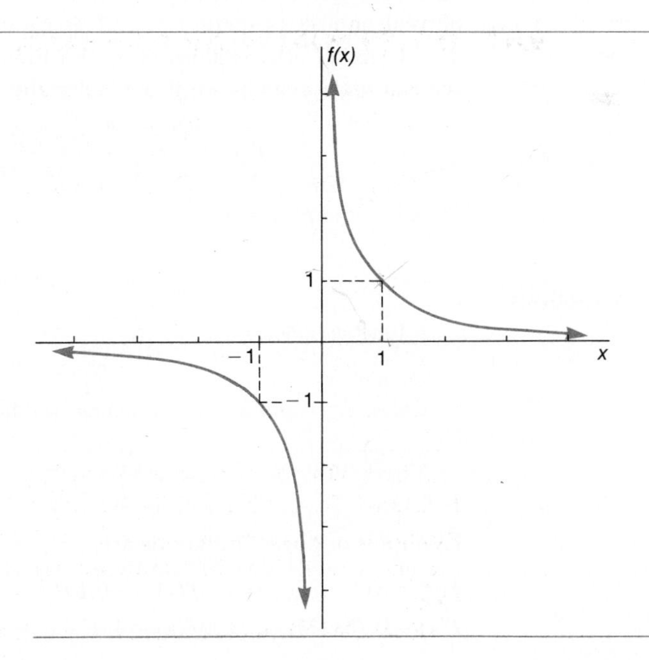

Figure 9.20

The General Shape of Graphs of $f(x) = c/x^n$ Where c Is a Positive-Constant and n Is an ODD Positive Integer

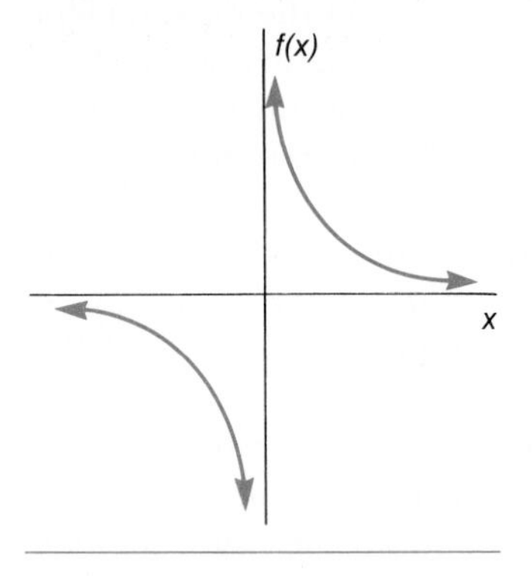

of $f(x) = 1/x$. As $|x|$ gets larger and larger, the value of $f(x)$ moves closer and closer to zero—but never quite reaches zero. Thus, the graph has a HORIZONTAL ASYMPTOTE at $y = 0$. (We shall study asymptotes of graphs in more depth in Chapter 10.)

A configuration such as that seen in Figure 9.19 is called a HYPERBOLA. A hyperbola consists of two separated branches, one the mirror image of the other, and having two intersecting asymptotes. Each branch of the hyperbola approaches each of the asymptotes.

Graphs of the functions $f(x) = c/x^n$ where c is a positive constant and n is an *odd* positive integer have the general shape shown in Figure 9.20. These are ODD FUNCTIONS having symmetry with respect to the origin. Graphs of the functions $f(x) = c/x^n$ where c is a positive constant and n is an *even* positive integer have the general shape shown in Figure 9.21. These are EVEN FUNCTIONS having symmetry with respect to the y-axis. In each case, the vertical asymptote is the y-axis and the horizontal asymptote is the x-axis.

Graphs of the functions $f(x) = 1/(x - h)^n$ have the same shape as graphs of $f(x) = 1/x^n$ but are translated $|h|$ units to the right or left. Thus, $x = h$ serves as the vertical asymptote for both the right and left branches of the graph. Graphs of the functions $f(x) = a + 1/x^n$ will have the same shape as $f(x) = 1/x^n$ but will be translated $|a|$ units up or down, with $y = a$ as the horizontal asymptote.

Figure 9.21

The General Shape of Graphs of $f(x) = c/x^n$ Where c Is a Positive Constant and n Is An EVEN Positive Integer

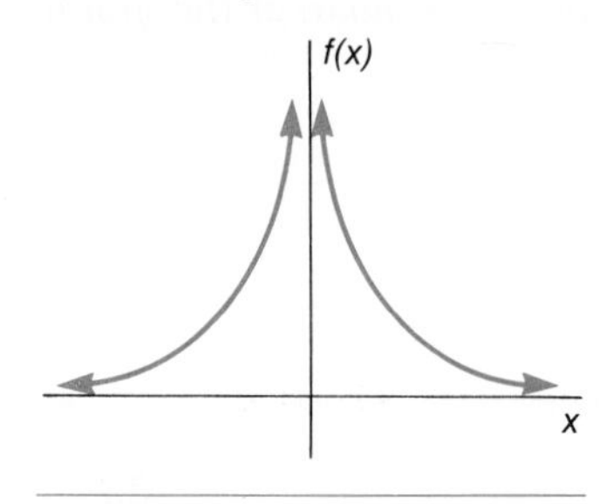

Not all rational functions graph as hyperbolas. Consider

$$f(x) = (x^2 - 1)/(x - 1)$$

and its graph as illustrated in Figure 9.22. The function is defined for all real numbers except $x = 1$. This results in a "hole" in the graph at $x = 1$. At all other values of x, the function is a linear function. This we can more readily see if we factor the function, as

$$f(x) = \frac{x^2 - 1}{x - 1} = \frac{\cancel{(x-1)}(x + 1)}{\cancel{(x-1)}} = x + 1$$

for all x except $x = 1$.

Power Functions

> A function defined by
>
> $$f(x) = cx^b \tag{9.6}$$
>
> where $x > 0$ and b is a real number is called a POWER FUNCTION.

Examples of power functions are:

$f(x) = 6x^3$	$f(x) = -0.1x^{1/3}$	$f(x) = 3x^{-1}$
$f(x) = 0.75x^{-0.5}$	$f(x) = 4x^{1.75}$	$f(x) = 3x^{0.25} + 2x^{0.25}$

Figure 9.22

$f(x) = (x^2 - 1)/(x - 1)$

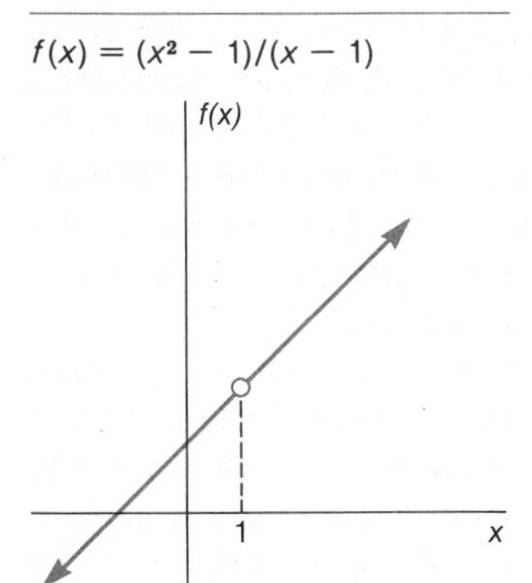

Notice that if b is a positive integer, then the power function is a polynomial function. However, the definition of the power function does not restrict b to positive integer values; therefore, not all power functions are polynomials.

Graphs of power functions cannot be neatly classified by type. However, the graph of one interesting power function—$f(x) = x^{2/3}$—is shown in Figure 9.23.

Figure 9.23

A Graph of the Power Function $f(x) = x^{2/3}$

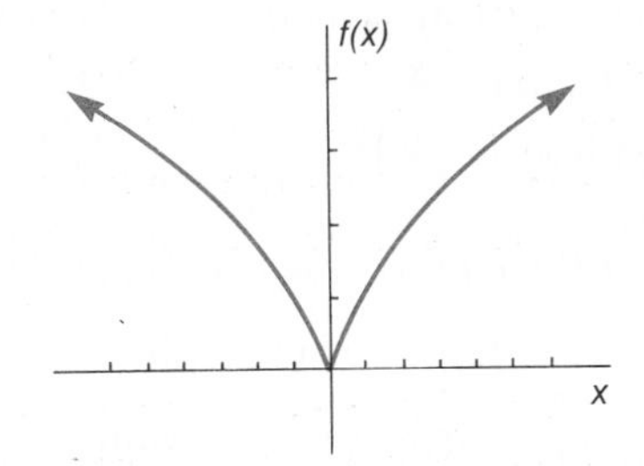

The Absolute-Value Function

The ABSOLUTE-VALUE FUNCTION is the function defined as

$$f(x) = |x| = \begin{cases} x & \text{if } x \geq 0 \\ -x & \text{if } x < 0 \end{cases} \tag{9.7}$$

Figure 9.24

A Graph of the Absolute-Value Function $f(x) = |x|$

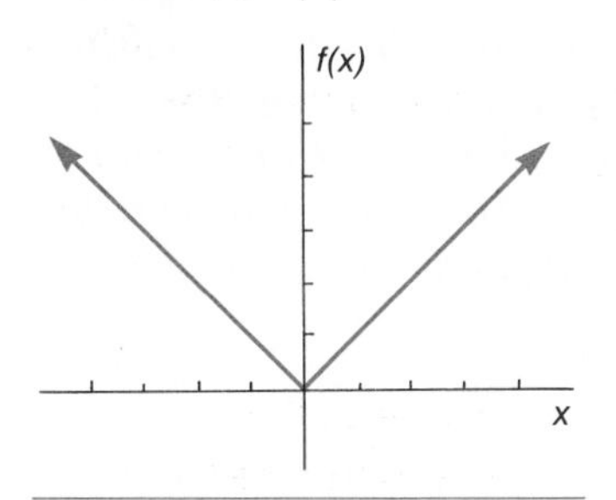

If x is positive or zero, the absolute-value function leaves it unchanged. If, however, x is negative, the absolute-value function makes it positive. The graph of this function is distinctive because of the sharp "corner" it produces. (See Figure 9.24.)

EXERCISES

Which of the following are algebraic functions? Which are polynomic? Identify algebraic functions by type. Identify polynomic functions by order, or degree.

82. $f(x) = 4$
83. $f(x) = x^4 + 6$
84. $5y = 3x + 7$
85. $f(x) = 5x^3 + 3$
86. $f(x) = \log_3(8 - x)$
87. $f(x) = (x + 6)/5$
88. $g(x) = \cos x$
89. $h(m) = m/4$
90. $f(x) = 1/\sqrt{x}$
91. $y = x^2 2^x$
92. $f(x) = 1/x^2$
93. $f(x) = -11$
94. $f(x) = 1/(x - 2)$
95. $f(x) = -(1/2)x^3$
96. $g(y) = -0.58y^3$
97. $g(x) = 1/(x^2 - 1)$
98. $f(x) = (x + 9)^4$
99. $f(x) = \log_{10}(2x + 5)$
100. $10 - x^2 - y = 0$
101. $f(x) = 3x + 13$
102. $f(x) = x^2 + 2x + 1$
103. $f(x) = x^3 + x^2$
104. $f(x) = (x^2 + 4)/(x - 5)$
105. $g(y) = y + 0.875$
106. $y = \log_{10}(x - 8)$
107. $f(x) = 3^x - 4$
108. $f(x) = x^2/(x^2 + 1)$
109. $f(x) = x/(2x + 3)$

Given that the total cost function for a firm is

$$C(x) = 3{,}500 + 1.5x \qquad 0 \le x \le 300$$

where C(x) *is the total cost, in dollars, of producing* x *units of product, the AVERAGE COST FUNCTION is given by*

$$\overline{C}(x) = [C(x)]/x = [3{,}500 + 1.5x]/x$$

110. What is the domain of the average cost function?
111. What is the total cost of producing 100 units of product?
112. What is the average cost of a unit of product when 100 units are produced?
113. What is the average cost of a unit of product when 200 units are produced?

A chemical manufacturing company has found that the cost of removing most of the environmental pollutants from its waste products is fairly reasonable; but the cost of removing the very last, small proportion of the pollutant rapidly becomes very expensive. The cost of removing the pollutant as a function of the percent of pollutant removed is given by

$$f(x) = \frac{15x}{137.5 - x} \qquad 0 \le x \le 100$$

where f(x) *is the cost, in thousands of dollars, and* x *is the percent of pollutant removed.*

114. What is the cost of removing 50 percent of the pollutant?
115. What is the cost of removing 75 percent of the pollutant?
116. By what percentage do costs increase as the percent of pollutant removed increases from 75 to 85?
117. By what percentage do costs increase as the percent of pollutant removed increases from 75 to 95?
118. What is the cost of removing 100 percent of the pollutant?
119. Sketch a graph of the cost function.

A rectangular lot, bordered on one side by an existing wall, is to be fenced on the other three sides. The area of the lot is to be 4,800 square yards.

120. Express the length of fencing required as a function of the length x of the unfenced side (the side bordering the existing wall).
121. Fencing costs $18 per linear yard. Express the cost of the fence as a function of the length x of the unfenced side.
122. Sketch a graph of the cost function.
123. What happens to the cost as the length of the unfenced side gets very small? Why?
124. What happens to the cost as the length of the unfenced side gets very large? Why?

9.7 EXPONENTIAL FUNCTIONS

Functions that are not algebraic are transcendental. One of the more important of these functions is the EXPONENTIAL FUNCTION.

> A function of the form
>
> $$f(x) = ca^{g(x)} + b \qquad (9.8)$$
>
> where x is the independent variable, $g(x)$ is some functional expression of the independent variable, a, b, and c are real constants, with $a > 0$ and $a \neq 1$, is called an EXPONENTIAL FUNCTION with base a.

Each of the following is an exponential function:

$f(x) = 3^x$	$f(x) = (0.5)^{x+1}$	$f(x) = (1/3)^x$
$f(x) = 4^{1/x}$	$f(x) = (10)^{-5x}$	$f(x) = 2^x + 7$
$f(x) = e^x$	$f(x) = e^{x+1}$	$f(x) = e^{x^2}$

Figure 9.25

Exponential Functions

(A) $f(x) = a^x$ $a > 1$
(a monotonically increasing function)

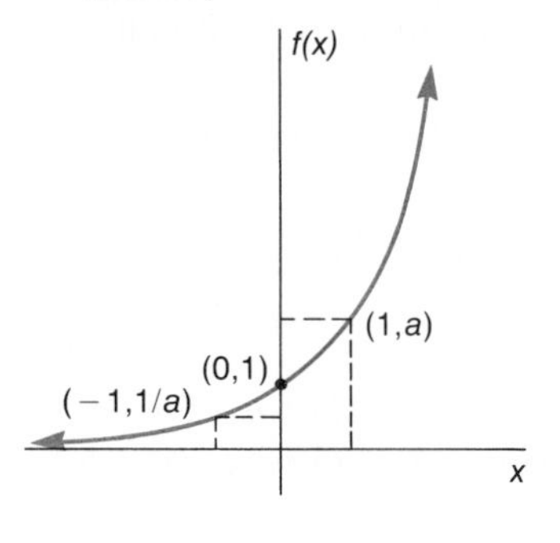

(B) $f(x) = a^x$ $0 < a < 1$
(a monotonically decreasing function)

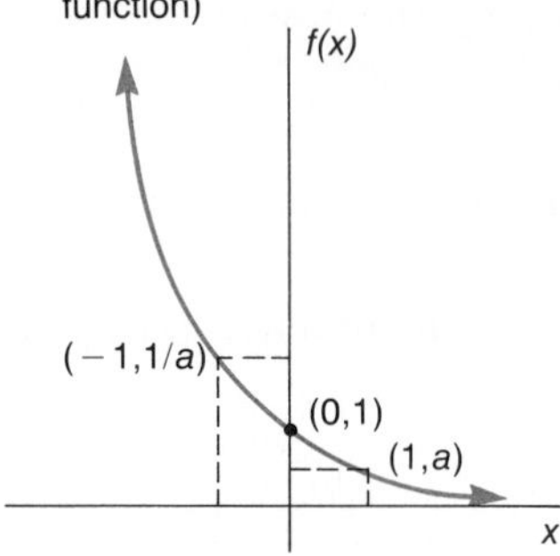

The reader is cautioned not to confuse the exponential function $f(x) = a^x$ with the power function $f(x) = x^a$. In the exponential function, the constant base a is raised to a variable power x. In the power function, the base x is a variable and this variable is raised to the constant power a. Hence, $f(x) = x^3$ is a power function, but $g(x) = 3^x$ is an exponential function.

The base a of an exponential function may be any real number greater than zero but not equal to one. For $a = 1$, the function would simply be a horizontal line, since $f(x) = 1^x = 1$ for all x. For $a < 0$, $f(x) = a^x$ would be defined in the real number system only for integer values of x. The number probably most widely used as a base for an exponential expression is the irrational number $e = 2.71828\ldots$. This constant occurs in many types of economic analyses and in various models involving compound interest, population growth, and so on.

The domain of the function $f(x) = a^x$ is the set of all real numbers. The range of the function is the set of all positive real numbers.

In its simplest form—as $f(x) = a^x$—the exponential function has two basic shapes, depending on the value of a. If $a > 1$, $f(x) = a^x$ is a MONOTONICALLY INCREASING FUNCTION; that is, for each $x_i < x_j$, $f(x_i) < f(x_j)$ for all i and j. Conversely, for $0 < a < 1$, $f(x) = a^x$ is a MONOTONICALLY DECREASING FUNCTION; that is, for each $x_i < x_j$, $f(x_i) > f(x_j)$. Each of these types of functions is displayed in Figure 9.25.

Notice in Figure 9.25A that, as x continues to decrease through the negative values, the value of the dependent variable moves closer and closer to zero; that is, the graph of the function approaches the x-axis as a horizontal asymptote. Then, as x increases without bound through its positive values, the value of $f(x)$ increases steadily and rapidly. The value of $f(x)$ is positive for all x. These properties are typical of the exponential function whose base is larger than one. The value of a determines the exact curvature of the function. The rise and the fall of the curve are sharper for larger values of the base a and are more gradual for values of a near one, as illustrated in Figure 9.26A.

If $0 < a < 1$, $f(x) = a^x$ is a steadily decreasing function. As x increases indefinitely in value, $f(x)$ moves closer and closer to zero but is always positive. Thus, the graph of the function approaches the x-axis as a horizontal asymptote. On the other hand, for increasingly negative values of x, the dependent variable takes on larger and larger and larger positive values. Again, while these properties are typical of all curves $f(x) = a^x$ for $0 < a < 1$, the exact curvature of the graph depends on the exact value of a. For values of a closer to one (but less than one), the graph rises and falls gradually; for values of a closer to zero (but greater than zero), the graph rises and falls more rapidly, as shown in Figure 9.26B.

By the rules of exponents

$$(1/a)^x = a^{-x}$$

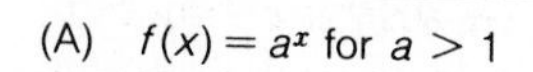

Figure 9.26

(A) $f(x) = a^x$ for $a > 1$

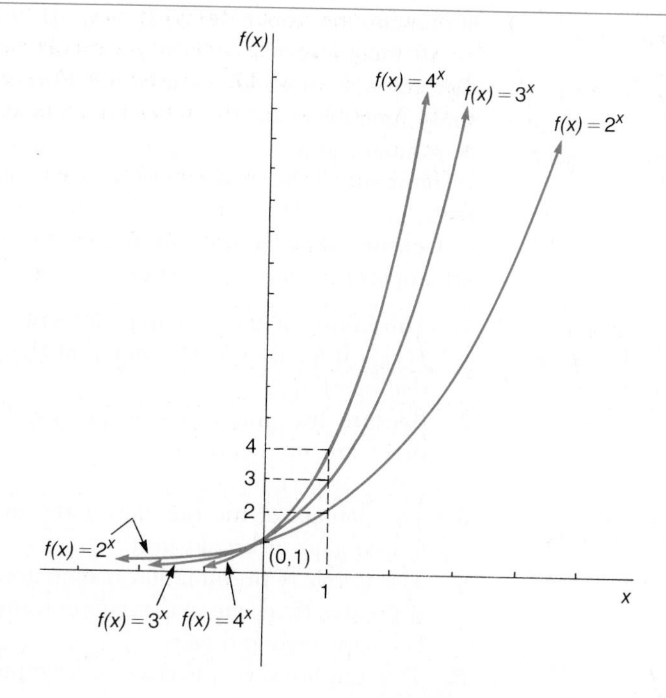

(B) $f(x) = a^x$ for $0 < a < 1$

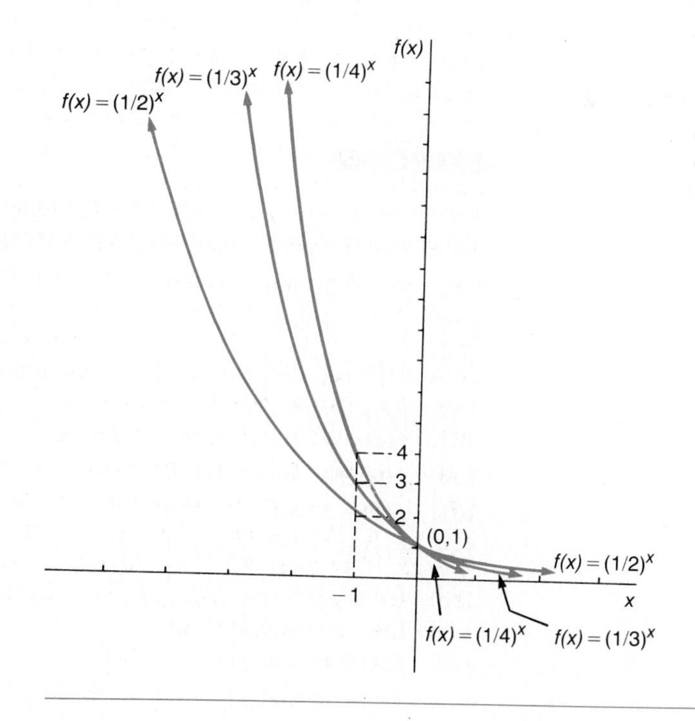

Figure 9.27

$f(x) = 2^x$ and $f(x) = (1/2)^x$

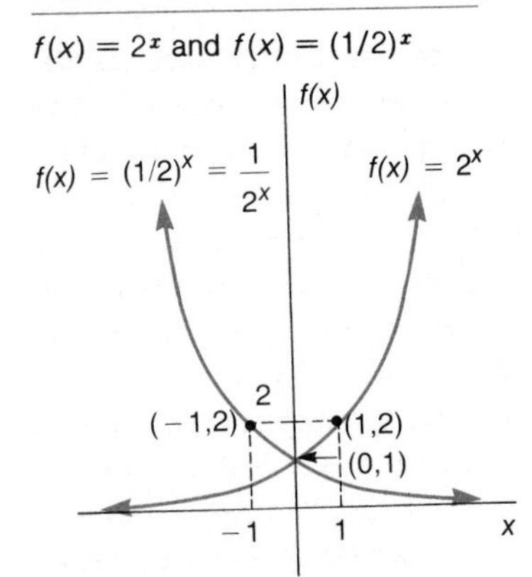

For this reason, the graphs of $f(x) = 2^x$ and $f(x) = (1/2)^x$ are reflections of one another about the positive y-axis. (See Figure 9.27.)

All functions of the form $f(x) = a^x$, whether $a > 1$ or $0 < a < 1$, pass through three key points: $(-1,1/a)$, $(0,1)$, and $(1,a)$. The graphs of these functions can, then, be determined essentially by plotting these coordinates and connecting them with a smooth curve (although it is recommended that several other coordinates also be calculated and plotted).

Certain other basic properties of the exponential function $f(x) = a^x$ are apparent from the preceding graphs.

1. The graph of the function lies entirely above the x-axis because $f(x) > 0$ for $a > 0$. The range of the function is the set of positive real numbers.
2. Because the graph of the function lies entirely above the x-axis, there are no x-intercepts. The y-intercept is always $y = 1$ since $f(0) = a^0 = 1$ for all a.
3. The domain of the function is the set of real numbers. The graph has no gaps or breaks in it.
4. The graph of the function is monotonically increasing if the base a is greater than one and monotonically decreasing if the base a lies between zero and one.
5. The graph has as a horizontal asymptote the x-axis.

EXERCISES

Sketch each pair of functions on the same coordinate system and state the conclusions you can draw from a comparison of the graphs.

125. $f(x) = 3^x$ and $f(x) = 3^{-x}$
126. $f(x) = 3^x$ and $f(x) = (1/3)^x$
127. $f(x) = 3^x$ and $f(x) = -(3^x)$
128. $f(x) = 3^x$ and $f(x) = 3^{2x}$
129. $f(x) = 3^x$ and $f(x) = 3^{-2x}$
130. $f(x) = 3^x$ and $f(x) = 3^{x+1}$
131. $f(x) = 3^x$ and $f(x) = 3^{x-1}$
132. $f(x) = 3^x$ and $f(x) = 1 + 3^x$
133. $f(x) = 3^x$ and $f(x) = 1 - 3^x$
134. $f(x) = e^x$ and $f(x) = e^{-x}$
135. $f(x) = e^x$ and $f(x) = (1/2)e^x$
136. $f(x) = e^x$ and $f(x) = e^{x/2}$

9.8 SELECTED EXPONENTIAL MODELS

The basic exponential function and modifications of this function describe many types of growth processes. In addition to the compounding of interest, some of the many relationships in business and economics which are appropriately represented by growth curves are: sales of a product as a function of the amount spent on advertising, sales of a product as a function of the length of time the product has been on the market, maintenance cost of a machine as a function of the length of time the machine has been in use. In other areas, the growth function is often used to describe population growth or wealth and capital growth.

The fundamental property of these growth processes is that they are monotonically increasing. They may or may not have an upper asymptote. With some processes the growth begins rapidly and levels off as it approaches an upper asymptote; other processes may start slowly, increase more rapidly, and eventually approach an upper asymptote.

The following examples include some of the more widely encountered of these models.

Constant-Growth-Rate Model

One of the most important properties of the basic type of exponential functions $f(x) = a^x$ is the fact that, as the independent variable increases or decreases by a constant amount, the dependent variable changes by a constant percentage. Notice that for $f(x) = 2^x$, as x increases by one unit, from 1 to 2, or from 2 to 3, or 3 to 4, $f(x)$ always doubles in value, as from 2 to 4, from 4 to 8, or from 8 to 16. Thus, as x increases as an ARITHMETIC PROGRESSION, $f(x)$ increases as a GEOMETRIC PROGRESSION.

In general, given the base a, the change in $f(x)$ will be a $(a - 1)100$ percent change over the previous value for a one-unit increase in x. Thus, for base $a = 3$, with each one-unit change in x, $f(x)$ will change by $(3 - 1)100$ percent $= 200$ percent. For base $a = 0.3$, with each one-unit change in x, $f(x)$ will change by $(0.3 - 1)100$ percent $= (-0.7)$ 100 percent (or will decrease by 70 percent).

If, beginning with some initial quantity, growth takes place at a constant rate per time period, the process may be described by the model

$$f(t) = A(1 + r)^t \tag{9.9}$$

where

$A =$ the initial amount
$r =$ the growth rate per time period
$t =$ the time period

A negative rate, r, would indicate that a decrease, rather than in increase, is taking place.

Example 10 **Increasing costs** McCabe Electronic Equipment Sales has determined that the cost of training a new salesperson is presently $15,000 but is increasing at an average rate of 9 percent a year. Determine the estimated cost in six years.

The functional relationship is modeled by

$$f(t) = A(1 + r)^t$$

where A, the amount at time $t = 0$, is $15,000; r, the constant percentage rate of growth, is 0.09. Thus, the cost of training a new salesperson, as a function of time t, is

$$f(t) = 15{,}000(1 + 0.09)^t$$

And the cost six years hence is estimated to be

$$f(6) = 15{,}000(1.09)^6$$

Now, the immediate obstacle is the evaluation of $(1.09)^6$. If a hand-held calculator with a y^x function key is available, we can easily use this to determine $(1.09)^6 = 1.6771$. Then we proceed by

$$f(6) = 15{,}000(1.6771) = \$25{,}156.50.$$

If we do not have access to such a calculator, the function may be evaluated using logarithmic transformation. (Calculations using logarithms are reviewed in Chapter 1.) The procedure, using logarithms, is as follows: Take the log of each side of the equation, as

$$ln\ y = ln\ 15{,}000 + 6\ ln\ 1.09$$

Then, using the ln function on a hand-held calculator, or a Table of Natural Logarithms, we write

$$\begin{aligned} ln\ y &= 9.6158055 + 6(0.0861777) \\ ln\ y &= 10.132872 \\ y &= e^{10.132872} = \$25{,}156.50 \end{aligned}$$

If the 9 percent rate of increase annually holds over the next six years, training costs at that time will have reached $25,156.50. []

Example 11 **Decreasing purchasing power** Due to inflation, the purchasing power of the dollar is constantly decreasing. Assuming a 6.5 percent annual rate of inflation, what will be the purchasing power of a $20,000 annual pension in 10 years?

The model

$$f(t) = A(1 + r)^t$$

is used with the initial amount being $20,000 and the annual rate being *minus* 0.065. Thus,

$$\begin{aligned} f(t) &= 20{,}000(1 - 0.065)^t \\ f(10) &= 20{,}000(0.935)^{10} \\ &= \$10{,}212.83 \end{aligned}$$

Thus, if the present rate of inflation continues, 10 years from now the \$20,000 annual pension will have the purchasing power of only approximately \$10,212. It will have lost \$9,787 in purchasing power due to inflation. []

Unlimited-Growth Model

If the rate of increase of a quantity takes place continuously and is proportional to the amount present, the phenomenon can be described by

$$f(t) = ce^{kt} \qquad k > 0 \tag{9.10}$$

where

$c =$ the amount present at time $t = 0$
$t =$ the time period
$k =$ the growth rate

This model is appropriately used to depict such processes as the growth of money when interest is compounded *continuously* or processes like short-term population growth. (See Figure 9.28.)

Example 12 The population of a certain community grew continuously from 8,000 in 1970 to 12,000 in 1980. What will be the population of this community in 1990 if this growth rate continues?

First, we must determine the growth rate; this is called the CONSTANT OF PROPORTIONALITY. Let $t =$ length of time, in years, with $t = 0$ in 1970. Initially, at $t = 0$, population is 8,000; hence, $c =$ 8,000. The appropriate model is

$$f(t) = 8{,}000e^{kt}$$

At $t = 10$ (in 1980), population is 12,000; hence, $f(10) = 12{,}000$. We write

$$12{,}000 = 8{,}000e^{10k}$$
$$e^{10k} = 1.5$$

To proceed, we use a logarithmic transformation

$$\ln e^{10k} = \ln 1.5$$
$$10k = 0.405465$$
$$k = 0.0405465$$

Therefore, the model that can be used to predict population in this community at any future time t is

$$f(t) = 8{,}000e^{0.0405465\,t}$$

Specifically, in 1990, when $t = 20$,

$$f(20) = 8{,}000e^{(0.0405465)(20)}$$
$$= 8{,}000e^{0.81093}$$
$$= 8{,}000(2.25)$$
$$= 18{,}000$$

Figure 9.28

Unlimited-Growth Model
($f(t) = ce^{kt}$ $\quad k > 0$)

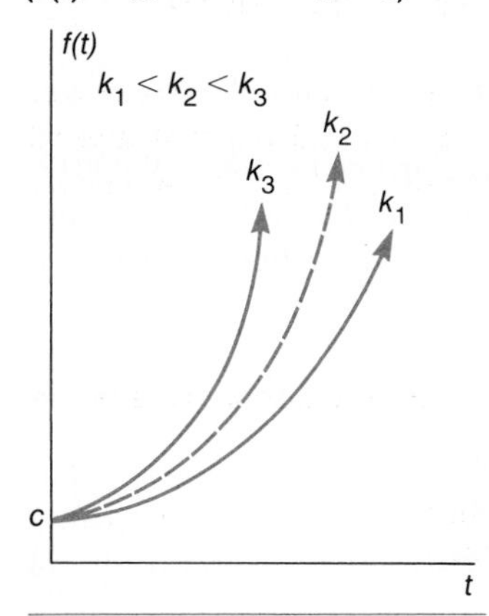

Hence, if the population growth rate continues, in 1990 the population of this community should be approximately 18,000. []

Unlimited-Decay Model

Figure 9.29

Unlimited-Decay Model
$f(t) = ce^{-kt}$

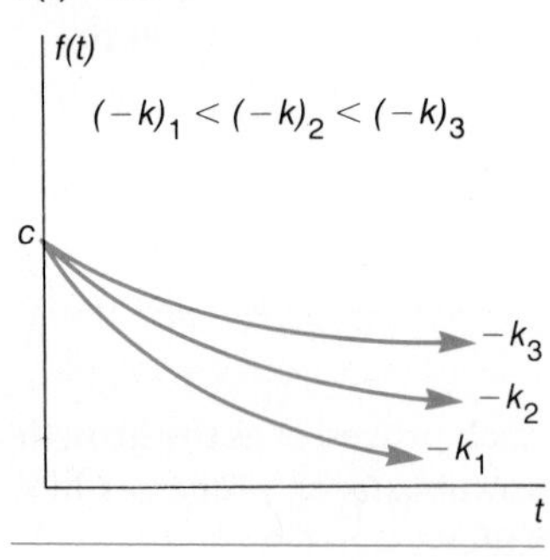

> If the continuous rate of change of a quantity is proportional to the quantity present but is a decrease rather than an increase, the process may be modeled by the EXPONENTIAL DECAY MODEL,
>
> $$f(t) = ce^{-kt} \qquad (9.11)$$
>
> where
>
> t = the time period
> k = the rate of decay
> c = the amount present at time $t = 0$

These models are often used to describe such phenomena as radioactive decay, the decay of advertising effect, or the amount of annual depreciation charges. (See Figure 9.29.)

Example 13 **Carbon 14 dating and exponential decay** Many radioactive materials are known to decay at a rate that is proportional to the amount of the substance that has not yet decayed and can, thus, be modeled by

$$f(t) = ce^{-kt}$$

This property of radioactive substances has lead to a procedure for estimating the ages of objects that is known as *carbon 14 dating.* Radioactive carbon 14 decays exponentially and has a half-life of approximately 5,760 years; that is, about 5,760 years are required for a given quantity of radioactive carbon 14 to decay to one half its original mass. Thus, where c is the original amount,

$$\begin{aligned}(1/2)c &= ce^{-k(5760)}\\ 0.5 &= e^{-k(5760)}\\ ln\ 0.5 &= ln\ e^{-k(5760)}\\ -0.6931 &= -k(5760)\\ k &= 0.00012\end{aligned}$$

Thus, the *decay constant* for the radioactive carbon 14 is $k = 0.00012$.

Recently an archaeological expedition discovered an artifact containing 62 percent of its original amount of radioactive carbon 14. The age of the object would be estimated to be

$$\begin{aligned}0.62c &= ce^{-0.00012t}\\ 0.62 &= e^{-0.00012t}\\ ln\ 0.62 &= ln\ e^{-0.00012t}\\ -0.4780 &= -0.00012t\\ t &= 3{,}983.6 \text{ years}\end{aligned}$$

[]

Limited-Growth Model: The Learning Curve

Many processes experience rapid growth as they approach some maximum possible level. Here the constant rate of change is appropriately applied to the difference between a fixed constant and the quantity present. The model is

$$f(t) = A - ce^{-kt} \tag{9.12}$$

where

t = the time period
k = the rate of change
A = the upper limit
$A - c$ = the initial amount

Figure 9.30

Limited-Growth Model: The Learning Curve ($f(t) = A - ce^{-kt}$)

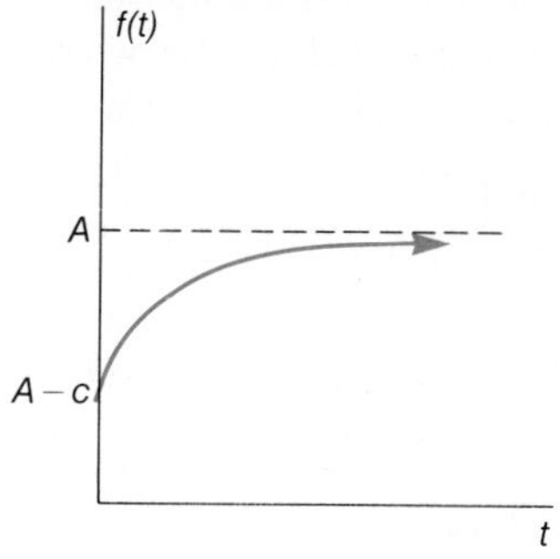

f(t)
A
A − c
t

Figure 9.30 shows that the graph of such a function begins at a level $A - c$, rises steeply at the beginning, and then flattens out and eventually approaches a horizontal asymptote, A.

The model is based on the assumption that individuals tend to learn from experience. Consequently, in many production operations, the time required by a worker to perform a task will lessen as the worker performs the task over and over again. The more often a person repeats an operation, the more efficient he or she becomes in performing that operation. There is a limit to any worker's efficiency, however, and as this limit is approached the amount of time required to perform the task will level off. If the rate of improvement is regular, the LEARNING CURVE can be used to predict future time requirements for tasks and, thus, become a valuable tool for planning and control.

Example 14

It has been determined by careful time and motion studies that the average number of items assembled each day by a worker t days after he or she is first put on this particular task is given by

$$f(t) = 50 - 50e^{-0.25t}$$

What is the optimum (maximum) performance to be expected of the worker? What percentage of this optimum performance will be achieved after 5 days? after 10 days? How soon can a person be expected to achieve, say, 99 percent of the optimum performance level?

The optimum performance level is given by the constant A of the model; here this level is 50 units of output.

To determine the level of production that will be achieved after $t = 5$ days, we compute

$$\begin{aligned} f(5) &= 50 - 50e^{-0.25(5)} \\ &= 50(1 - e^{-1.25}) \\ &= 50(1 - 0.287) \\ &= 35.65 \text{ units of output} \end{aligned}$$

which is 71.3 percent of the optimum.

After $t = 10$ days, the level of production that will have been reached is

$$\begin{aligned} f(10) &= 50(1 - e^{-0.25(10)}) \\ &= 50(1 - e^{-2.5}) \\ &= 50(1 - 0.082) \\ &= 45.9 \text{ units of output} \end{aligned}$$

which is 91.8 percent of the maximum production level attainable.

We can determine how many days must elapse before the worker will be at 99 percent of maximum performance level by computing

$$\begin{aligned} 99 \text{ percent of } 50 &= 49.5 \text{ units} \\ 49.5 &= 50(1 - e^{-0.25t}) \\ 0.99 &= 1 - e^{-0.25t} \\ e^{-0.25t} &= 0.01 \\ ln\, e^{-0.25t} &= ln\ 0.01 \\ -0.25t &= -4.605 \\ t &= 18.42 \text{ days} \end{aligned}$$

The worker should reach approximately 99 percent of maximum production in 18 or 19 days. []

Example 15 **A modified growth model for sales** Annual sales of a new wood-burning heater are expected to increase rapidly when the product is first placed on the market and then to level off as sales approach a maximum expected level. Market researchers believe sales, in units, will follow the function

$$f(t) = 9{,}000 - 7{,}000e^{-0.4t}$$

where t is the number of years the product has been on the market.

According to the model, annual sales for the heater are predicted to be as follows over the first 10 years:

Number of years on market	Annual sales, in units	Increase over previous year
1	4,308	4,308
2	5,855	1,547
3	6,892	1,037
4	7,587	695
5	8,053	466
6	8,365	312
7	8,574	209
8	8,715	141
9	8,808	93
10	8,872	64

[]

Limited-Growth Model: The Logistics Curve

Many processes experience an initial period of gradual growth, then show a more-rapid increase before leveling off. The model appropriately used in these cases may be

$$f(t) = \frac{A}{1 + ce^{-kt}} \tag{9.13}$$

where

A = the upper limit approached by the value of the function
k = the rate of growth
t = the time period
$A/(1 + c)$ = the amount present at time $t = 0$

Figure 9.31

Limited-Growth Model: The Logistics Curve $\left(f(t) = \frac{A}{1 + ce^{-kt}}\right)$

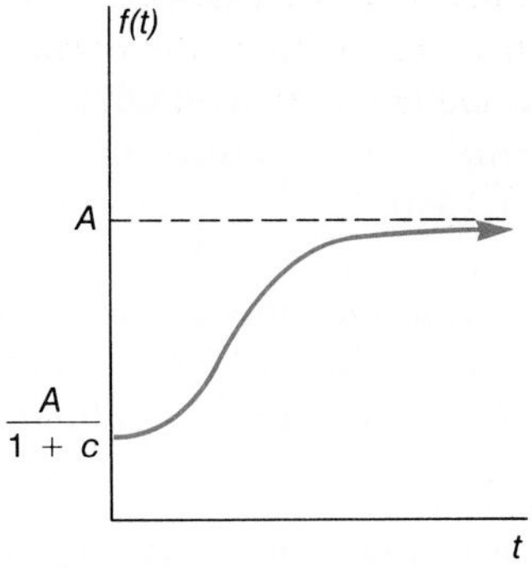

The logistics curve is illustrated by Figure 9.31. Phenomena displaying this type of growth pattern may be sales of a new product, growth patterns in industries, or long-term population growth.

Example 16 **A logistics curve** In many cases the sales of a product increase rather slowly when the product is first introduced on the market. Then sales increase more rapidly, and finally they level off. Such a situation is often modeled by a logistics curve.

Sales of a new electronic-calculator wristwatch are predicted to follow the pattern

$$f(t) = \frac{25{,}000}{1 + 100e^{-0.8t}}$$

Then, sales in time $t = 1$ are predicted to be

$$\begin{aligned} f(1) &= \frac{25{,}000}{1 + 100e^{-0.8(1)}} \\ &= 25{,}000/(1 + 100e^{-0.8}) \\ &= 25{,}000/45.9 \\ &= 545 \text{ units} \end{aligned}$$

In time $t = 5$, sales are predicted to have reached the level

$$\begin{aligned} f(5) &= 25{,}000/(1 + 100e^{-0.8(5)}) \\ &= 25{,}000/2.83 \\ &= 8{,}834 \text{ units} \end{aligned}$$

How soon are sales expected to reach the 15,000-unit level?

$$15{,}000 = 25{,}000/(1 + 100e^{-0.8t})$$
$$1 + 100e^{-0.8t} = 25{,}000/15{,}000 = 1.6667$$

$$100e^{-0.8t} = 0.6667$$
$$e^{-0.8t} = 0.006667$$
$$ln\ e^{-0.8t} = ln\ 0.006667$$
$$-0.8t = -5.0106$$
$$t = 6.26 \text{ years}$$

[]

EXERCISES

137. Sales of Acme-Burlington Products Company's Item #XZ28 are $8,000 in 1981. Assuming a 12 percent annual increase in sales of this item, project sales for 1986; for 1990.

Union County Utility Company records indicate that the demand for electricity in Union County is growing at an annual rate of 12.75 percent. In 1981 the records indicate that 8 billion kilowatt-hours of electricity were used.

138. Assuming this usage rate continues, how much electrical power will be needed in 1987?
139. If the maximum capacity of the utility's power plant is 500 billion kilowatt-hours, when will this maximum capacity be reached?

Sara Schuster plans to retire in 10 years with a $35,000 annual pension. Inflation causes the purchasing power of a dollar to decrease.

140. Assuming a 5 percent annual rate of inflation over the next 10 years, what will be the purchasing power of the $35,000 pension?
141. How much purchasing power will be lost due to inflation?

Jason Jacobowitz will retire in 1985 with a fixed annual pension of $15,000.

142. Assuming a 5 percent annual rate of inflation, what will be the purchasing of $15,000 in 1990? in 1995?
143. How long will it take for the purchasing power of the pension to be reduced by half?

The population of Tinytown is growing continuously at an annual rate of 2.5 percent.

144. If there are 1,500 residents in Tinytown on January 1, 1980, what will be the population on January 1, 1985? on January 1, 1990?
145. When will the population of Tinytown reach 2,500?
146. How long will it take the population to double?

The world population is now estimated to be growing continuously at an annual rate of 1.9 percent.

147. If it continues to grow at this rate, how long will it take the world's population to double?
148. If the world population in 1975 was approximately 4 billion, estimate the population in 1985; in 1995; in 2000.

A demographer is studying the pattern of population growth in Chatham County. In 1970, population was 10,000 and by 1975 it had reached 10,400.

149. Assuming that the exponential growth model is appropriate, estimate the annual rate of growth.
150. Estimate the population of Chatham County in 1990; in 2000.

The number of bacteria present in a certain culture follows the exponential growth law. There were 5,000 bacteria present initially; after six hours, 11,000 bacteria were present.

151. Determine an appropriate model for describing the growth of bacteria in the culture.
152. Estimate the number of bacteria present after 9 hours; after 12 hours.
153. An artifact from an ancient tomb was discovered and was determined to have 25 percent of the original amount of radioactive carbon 14 present. Estimate the age of the artifact. (Note: $f(t) = ce^{-0.00012t}$.)

The salvage value of a certain piece of heavy-duty construction equipment after t *years is estimated to be given by*

$$S(t) = 400{,}000e^{-0.15t}$$

154. What will be the salvage value after the fifth year of use? after the eighth year of use?
155. When will the salvage value reach $100,000?

A manufacturer of a new type of alkaline battery has found that the proportion still operative after t *years of use is described by*

$$f(t) = e^{-0.45t}$$

156. What proportion of the batteries last for at least one year?
157. What proportion of the batteries last for at least two years?

The number of units of a product manufactured each day t *days after the start of a production run is given by*

$$f(t) = 500 - 500e^{-0.25t}$$

158. What is the maximum level of daily output possible under these circumstances?

159. What percentage of this maximum level has been achieved after 5 days? after 10 days?
160. When will 50 percent of the maximum output level be achieved? 75 percent? 90 percent?

Sales of a new model popcorn popper are approximated by

$$S(t) = 600 - 500e^{-t}$$

where S(t) *represents sales in thousands of units and* t *represents the number of years the item has been on the market.*

161. Estimate sales in the first year the product is on the market.
162. Estimate sales in the second year the product is on the market.
163. Sketch a graph of $S(t)$.

The proportion of potential customers responding to the advertisement of a new product after it has been on the market t *weeks is given by*

$$f(t) = 1 - e^{-0.15t}$$

The market area contains 500,000 potential customers. Each response to the advertisement results in profit—exclusive of advertising—to the company of $380. The fixed cost of preparing an advertising campaign is $40,000, with a variable cost of $5,000 for each week the campaign runs.

164. What proportion of potential customers respond after the first 5 weeks of advertising? after 10 weeks? after 15 weeks?
165. Sketch the sales function.
166. What is the advertising cost function? Sketch this function.
167. What is the net profit function after advertising costs?

The number of units a worker can assemble per day, t *days after first beginning the task, is given by the function*

$$f(t) = 300 - 300e^{-0.2t}$$

168. What is the maximum performance level that can be expected from a worker?
169. What is the performance level that should be achieved after 10 days?
170. What proportion of the maximum performance level does this represent?
171. When should a worker reach 95 percent of the maximum performance level?

Sales of a new product are expected to follow the function

$$f(t) = 9{,}000/(1 + 150e^{-t})$$

where f(t) *represents total sales in units and* t *is the number of months the product has been on the market.*

172. Predict sales at time $t = 2$; $t = 4$; $t = 6$. Sketch this curve.
173. When will sales reach the 7,500-unit level?
174. When will sales reach the 8,500-unit level?

9.9 LOGARITHMIC FUNCTIONS

A logarithmic function with base a is the inverse of an exponential function with base a.

> A function of the form
>
> $$f(x) = log_a\, g(x) \qquad (9.14)$$
>
> where x is the independent variable, $g(x)$ is some functional expression of x with $g(x) > 0$, and a is a real constant greater than zero but not equal to one, is a LOGARITHMIC FUNCTION with base a.

Figure 9.32

Logarithmic Functions

(A) $f(x) = log_a\, x$ for $a > 1$

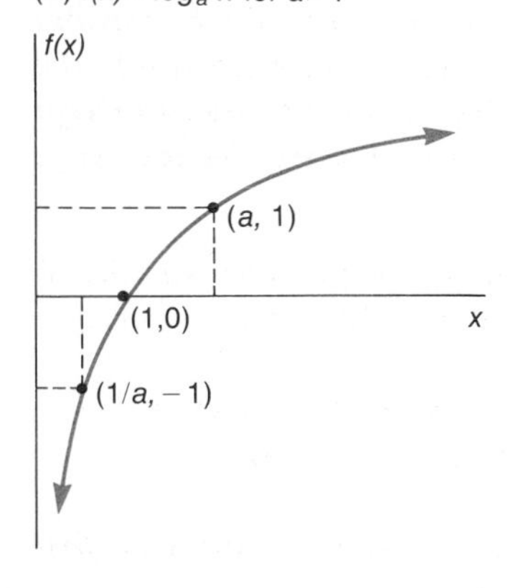

(B) $f(x) = log_a\, x$ for $0 < a < 1$

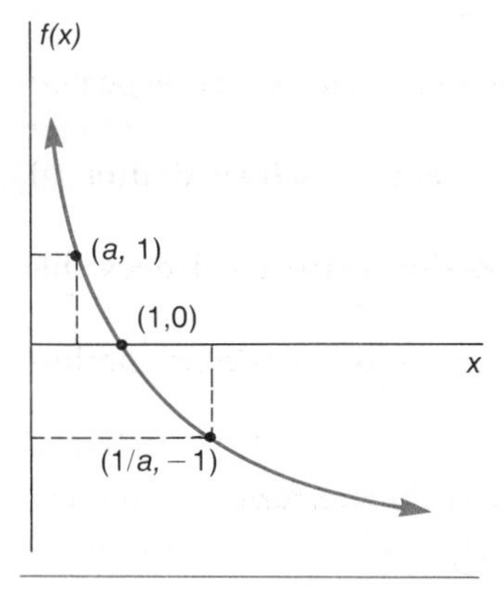

Each of the following is a logarithmic function:

$f(x) = log\ x$	$f(x) = log_2\ x$	$f(x) = log_8(5x)$
$f(x) = ln\ x$	$f(x) = ln(x + 1)$	$f(x) = log\ x^2$

We shall begin with the basic logarithmic function

$$f(x) = log_a\ x$$

The domain of such a function is the set of positive real numbers, while the range is the set of all real numbers. The basic shape of the graph of the function is determined by the value of the base a. For $a > 1$, $f(x)$ is a monotonically increasing function. (See Figure 9.32A.) The graph passes through the point (1,0). Between $x = 0$ and $x = 1$, the value of $f(x)$ is negative, approaching the y-axis (which is a vertical asymptote). For x values greater than 1, $f(x)$ is positive, increasing but then leveling off somewhat as x increases indefinitely. (The graph does not, however, have a horizontal asymptote.) The curve increases less rapidly for a larger base a than for a smaller base a. (See Figure 9.33A.)

For $0 < a < 1$, $f(x) = log_a\ x$ is a monotonically decreasing function, as shown in Figure 9.32B. The graph passes through the point (1,0). For x greater than zero but less than one, the value of $f(x)$ is positive, approaching the y-axis as a vertical asymptote as x approaches zero from the right. For x greater than one, the value of $f(x)$ is negative, decreasing, but beginning to level off as x increases indefinitely. (It does not, however, have a horizontal asymptote.) The curve becomes steeper for values of a closer to one (with $0 < a < 1$), as shown in Figure 9.33B.

Figure 9.33

Logarithmic Functions: The Effect of the Magnitude of the Base

(A) For Base $a>0$

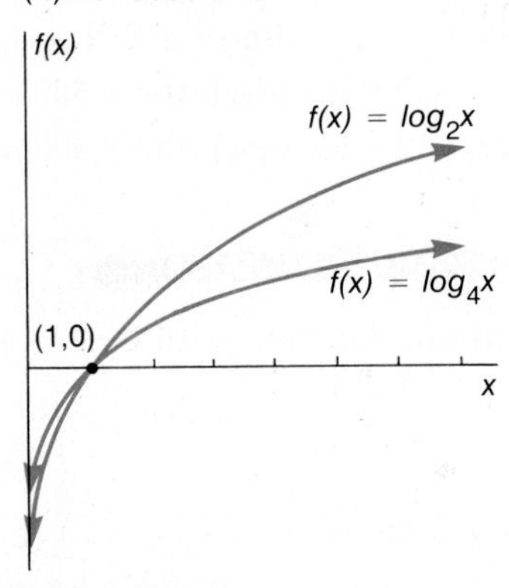

(B) For Base a with $0<a<1$

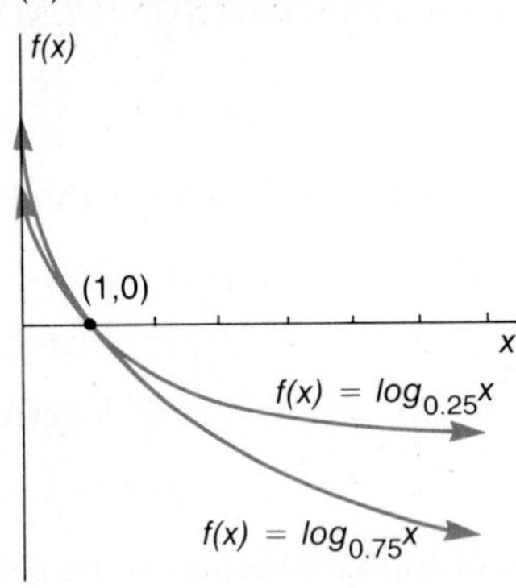

For any base $a > 0$ and $a \neq 1$, the graph of the function $f(x) = log_a\ x$ lies entirely in the first and fourth quadrants of the coordinate system. It passes through three key points. $(1/a, -1)$, (1,0), and $(a, 1)$.

Relationship between Exponential and Logarithmic Functions

Recall that the *log* base a of x is the power to which base a would have to be raised to yield x; that is

$$log_a\ x = y \text{ means } a^{log_a x} = x \text{ or } a^y = x$$

Hence, the exponential and logarithmic functions are inverse functions of each other. This relationship can be seen in Figure 9.34. The domain of an exponential function includes all real numbers, and its range is the set of positive real numbers. Thus, the domain of a logarithmic function is the set of all positive real numbers, and its range is the set of all real numbers.

Figure 9.34

Comparison of Exponential and Logarithmic Functions

(A) For $a>1$

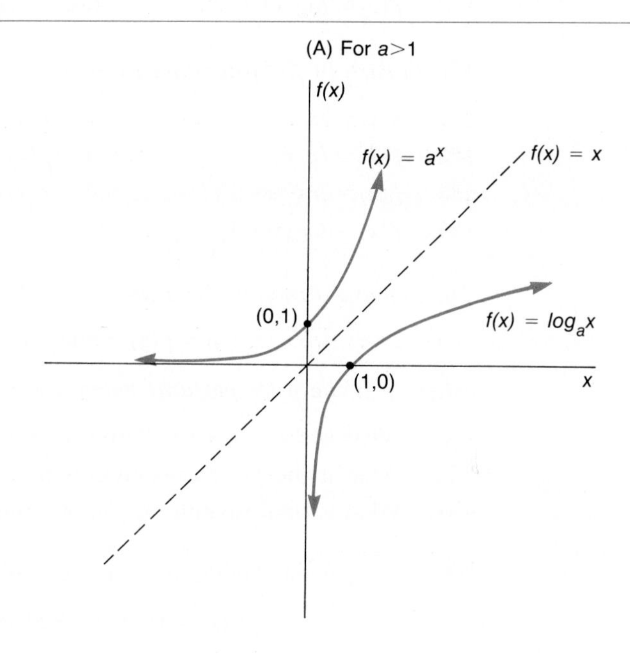

(B) For $0<a<1$

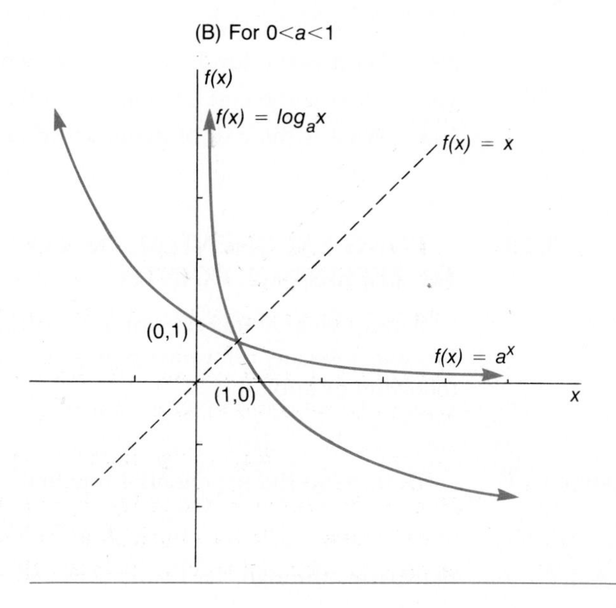

EXERCISES

For what values of x are the following functions defined?

175. $f(x) = log\ x$
176. $f(x) = log_2(x^2 + 1)$
177. $f(x) = ln(x - 5)$
178. $f(x) = ln(x + 1)$
179. $f(x) = log(x^3 + 27)$
180. $f(x) = ln(x^4 - 16)$

Sketch each of the following functions.

181. $f(x) = log_2\ x$
182. $f(x) = log\ x$
183. $f(x) = ln\ x$
184. $f(x) = -ln\ x$
185. $f(x) = ln(-x)$
186. $f(x) = ln(1 + x)$
187. $f(x) = log(x + 1)$
188. $f(x) = log\ x^2$

The demand function for a product is given by

$$p = f(x) = 500/ln\ (x + 15)$$

where p *is the price per unit when* x *units are demanded.*

189. What is the price associated with $x = 20$ units demanded?
190. What is the price associated with $x = 25$ units demanded?
191. What is total revenue when 30 units are demanded and sold?

The cost function for the manufacture of a certain product is given by

$$C(x) = 15{,}000 + 800\ ln(2x^2 + 3)$$

where x *is the number of units manufactured.*

192. What is the fixed cost (the cost when $x = 0$)?
193. What is the cost of manufacturing 100 units?
194. What is the cost of manufacturing 500 units?

9.10 FITTING EXPONENTIAL OR LOGARITHMIC FUNCTIONS TO REFERENCE POINTS

An exponential function, or a logarithmic function, can be established through relevant reference points. The procedure is illustrated by the following examples.

Example 17 Determine the exponential function of the general form

$$f(x) = ce^{kx}$$

that passes through the points (2,5) and (8,10). (See Figure 9.35.)

Substituting the two data points gives the two equations

$$5 = ce^{2k} \qquad \text{and} \qquad 10 = ce^{8k}$$

Now, writing each of these expressions in terms of logarithms, we obtain

$$\begin{aligned} \ln 5 &= \ln ce^{2k} & \ln 10 &= \ln ce^{8k} \\ \ln 5 &= \ln c + \ln e^{2k} & \ln 10 &= \ln c + \ln e^{8k} \\ \ln 5 &= \ln c + 2k & \ln 10 &= \ln c + 8k \end{aligned}$$

Figure 9.35

Fitting an Exponential Function to Reference Points

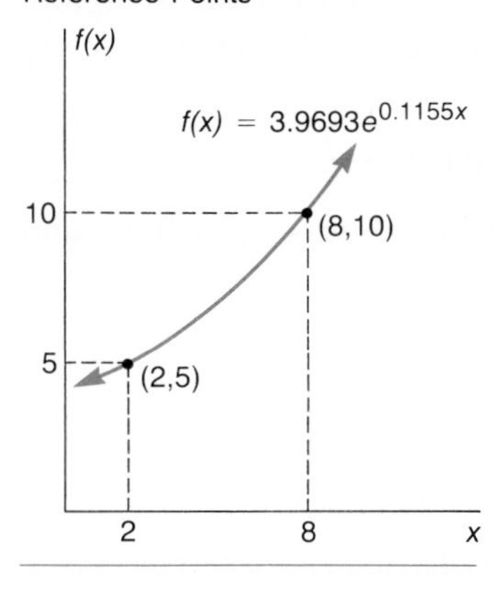

Because $\ln 5 = 1.6094$ and $\ln 10 = 2.3026$, we can write

$$\begin{aligned} \ln c + 2k &= 1.6094 \\ \ln c + 8k &= 2.3026 \end{aligned}$$

Solving the two equations simultaneously yields $k = 0.1155$ and $\ln c = 1.3786$. The antilog of $\ln c$ is $c = 3.9693$. The exponential function is thus

$$f(x) = 3.9693e^{0.1155x}$$

Example 18

Fit a logarithmic function of the general format

$$f(x) = \ln(a + bx)$$

through the reference points (2,1) and (10,3).

The logarithmic function is fit to the reference points in much the same way as is an exponential function. First, we write

$$1 = \ln(a + 2b) \qquad \text{and} \qquad 3 = \ln(a + 10b)$$

Figure 9.36

Fitting a Logarithmic Curve to Reference Points

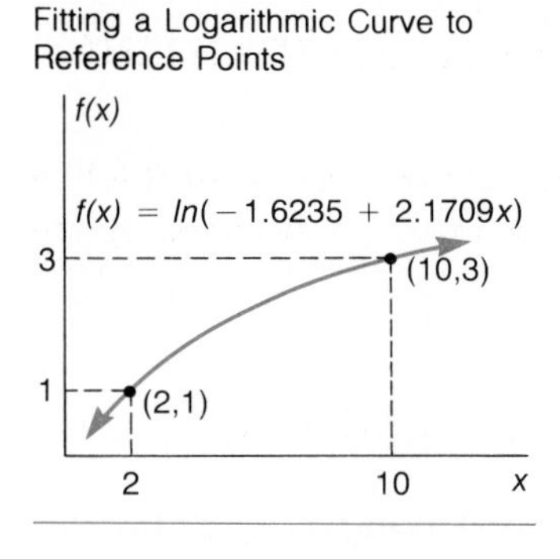

Using the exponential transformation, we rewrite the expressions as

$$\begin{aligned} e^1 &= e^{\ln(a+2b)} & \text{and} \qquad e^3 &= e^{\ln(a+10b)} \\ e^1 &= a + 2b & e^3 &= a + 10b \end{aligned}$$

Now, because $e^1 = 2.7183$ and $e^3 = 20.0855$, the statements can be expressed

$$\begin{aligned} 2.7183 &= a + 2b \\ 20.0855 &= a + 10b \end{aligned}$$

When we solve these equations simultaneously, we obtain $a = -1.6235$ and $b = 2.1709$. Then, the logarithmic function fit to the two reference points given is

$$f(x) = \ln(-1.6235 + 2.1709x)$$

[]

Example 19 If the situation described by a learning curve can be pictured and the limit which the curve is approaching and one other reference point determined, as shown in Figure 9.37, the specific function can be determined as follows.

In the general learning curve functional form

$$f(x) = A - ce^{-kx}$$

where A is the limit that the curve is approaching and c is the difference between the initial amount and the limit, A. The horizontal asymptote is shown in the graph to be 1,000,000 and the initial amount to be 0. Thus, we can begin by writing

$$f(x) = 1{,}000{,}000 - 1{,}000{,}000e^{-kx}$$

Now we substitute in the one reference point given—(2,750000)—and obtain

$$750{,}000 = 1{,}000{,}000 - 1{,}000{,}000e^{2k}$$
$$1{,}000{,}000e^{2k} = 250{,}000$$

Next we rearrange the terms and write

$$e^{2k} = 0.25$$
$$2k = ln\ 0.25 = -1.386$$
$$k = -0.693$$

Thus, the specific equation is

$$f(x) = 1{,}000{,}000 - 1{,}000{,}000e^{-0.693x}$$ []

Figure 9.37

Fitting a Learning Curve to Reference Points

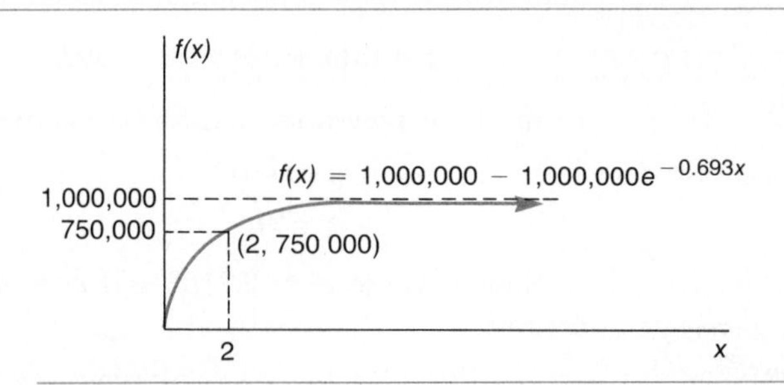

EXERCISES

195. Fit an exponential function of the general form $f(x) = ce^{kx}$ through the reference points (3,6) and (9,16).

196. Fit an exponential function of the general form $f(x) = ce^{kx}$ through the points (4,9) and (10,16).

197. Fit a logarithmic function of the general form $f(x) = ln(a + bx)$ through the reference points (3,2) and (13,4).

198. Fit a learning curve of the general form $f(x) = A - ce^{-kx}$ to the information given. The horizontal asymptote is 50,000 and the initial amount is 10,000. The curve also passes through the point (3,40000).

199. Fit a logarithmic function of the general form $f(x) = log(a + bx)$ through the reference points (2,0.95) and (5,1.18).

200. Determine the parameters for a learning curve of the general form $f(x) = A - ce^{-kx}$ whose horizontal asymptote is 100,000 and whose initial amount is 10,000. The curve also passes through the point (2,92600).

C·H·A·P·T·E·R 10

Limits and Continuity of a Function

THE PARADOXES OF ZENO

With this chapter we begin our study of THE CALCULUS. The calculus is the study of changing quantities—of dynamic, rather than static, situations.

Ironically, among the most intriguing paradoxes of mathematical theory are those of the Eleatic philosopher Zeno (ca. 450 B.C.) by which he tried to prove the absurdity of believing that reality is made up of changing things.

Different schools of mathematical reasoning were developing among the early Greek mathematicians and philosophers. One school of thought was based on the assumption that a magnitude is infinitely divisible; another school of thought assumed that a magnitude is made up of a very large number of small, indivisible atomic parts. Some of the logical difficulties encountered in either assumption were underscored by the paradoxes of Zeno. He reasoned that, under either assumption, motion is impossible.

The Dichotomy This argument maintains that there is no motion because that which is moved must arrive at the middle of its course before it arrives at the end.

In order to traverse a line segment it is necessary first to reach the midpoint of that segment; and in order to do this it is necessary to reach the midpoint of the first half segment; and in order to do this it is necessary to reach the midpoint of the first quarter segment; and so on *ad infinitum.* To traverse a line segment one must pass through an infinite number of spaces—which is impossible. Indeed, if a line segment is infinitely divisible, not only can motion not take place, motion can never even begin.

The Arrow If everything is either at rest or moving when it occupies a space equal (to itself) while the object moved is always in the instant (in the *now*), the moving arrow is unmoved.

If time (and space) is considered as made up of indivisible instants (or parts), the nature of time itself renders movement impossible. It is strictly impossible that the arrow can move in the indivisible instant, for if it changed position, the instant would at once be divided. Hence a moving arrow is always at rest, for at any instant it is in a fixed position. Since this is true for every instant, the arrow never moves.

10.1 LIMITS

A LIMIT is a rigorously defined mathematical concept, fundamental to quantitative models used in physics, chemistry, psychology, statistical research, economics, and many other fields of study. The approach taken here will be more intuitive.

Basically, we are concerned with what happens to the value of the dependent variable $f(x)$ as the value of the independent variable x approaches some constant a. Consider, for example, the function f defined by $f(x) = x + 2$ and notice what happens to the value of $f(x)$ as the value of x moves closer and closer to 2.

Let us set up a table of x and corresponding $f(x)$ values, as

x	$1.9 \to 1.99 \to 1.999 \to 1.9999$	$2.0001 \leftarrow 2.001 \leftarrow 2.01 \leftarrow 2.1$
$f(x) = x + 2$	$3.9 \to 3.99 \to 3.999 \to 3.9999$	$4.0001 \leftarrow 4.001 \leftarrow 4.01 \leftarrow 4.1$

Clearly, as x moves closer and closer to 2, $f(x)$ moves closer and closer to 4. We call the real number approached by $f(x)$ as x approaches some specified constant a the LIMIT OF $f(x)$ AS x APPROACHES a.

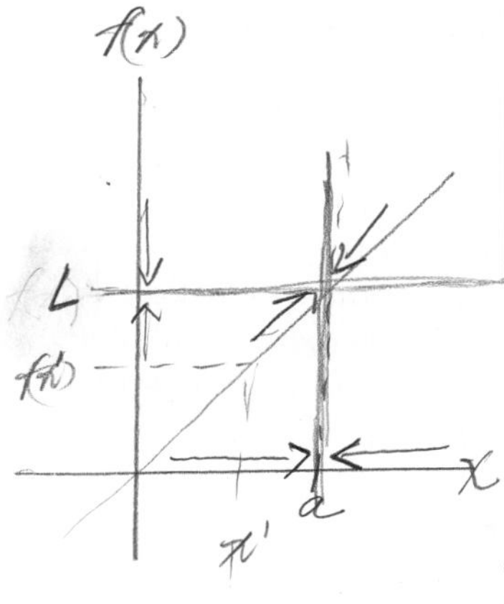

> **A function f has a limit L** (where L is some real number) as x approaches a if the values of the dependent variable $f(x)$ differ arbitrarily little from L for all values of x which lie very close to a. The limit as x approaches a is symbolized
>
> $$\lim_{x \to a} f(x) = L$$

Several important points regarding the limit of a function must be emphasized. First, *the concept of the limit of a function as* x *approaches* a *should not be confused with the concept of the value of the function at* x = a. The limit as x approaches a may exist, and the function may, or may not, be defined at a. The function may be defined at a, and the limit may, or may not, exist. The limit as x approaches a may exist, and the function may be defined at a, and their values may, or may not, be the same.

Second, it should be noted that, generally, x *can approach* a *from either of two directions,* through values that are less than a or through values that are greater than a.

And, *the limit* L *must be a finite number.*

Let us give each of these points further consideration.

Left-Hand and Right-Hand Limits

When the variable x approaches the value $x = a$ but always remains less than a, we say x is approaching a from the left. If the value of $f(x)$ gets closer and closer to a real number L^- as x approaches a from the left, we call L^- the LEFT-HAND LIMIT and employ the symbolism

$$\lim_{x \to a^-} f(x) = L^-$$

Only those values of x to the left of a are used to compute the left-hand limit. For example, to determine

$$\lim_{x \to 3^-} (x - 1)$$

we might compute $f(2.9) = 1.9$, $f(2.99) = 1.99, \ldots$, $f(2.999\ldots) = 1.999\ldots$. Clearly, the values of $f(x)$ are getting closer and closer to the number 2, and we write

$$L^- = \lim_{x \to 3^-} (x - 1) = 2$$

Similarly, the RIGHT-HAND LIMIT of a function f as x approaches a is the value L^+, which $f(x)$ converges on as x approaches the point

Figure 10.1

$(L^- = 2) = (L^+ = 2) = L$ So That $\lim_{x \to 3} (x - 1) = 2$

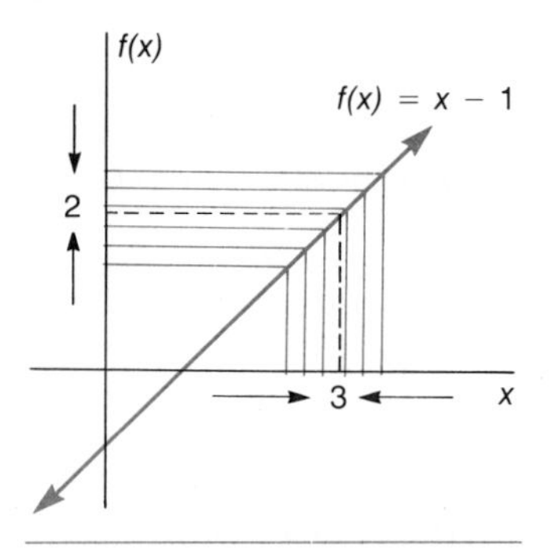

$x = a$ from the right, but always remains greater than a. We symbolize this limit as

$$\lim_{x \to a^+} f(x) = L^+$$

Only those values of x to the right of a on the number line are used to obtain the right-hand limit. We might compute the limit

$$L^+ = \lim_{x \to 3^+} (x - 1) = 2$$

by evaluating $f(3.1) = 2.1$, $f(3.01) = 2.01, \ldots, f(3.000001) = 2.000001$.

A function f has a limit L as x approaches a only if the left-hand and right-hand limits are equal. Their common value is the limit of f as x approaches a. Here $(L^- = 2) = (L^+ = 2) = L$, and we write

$$\lim_{x \to 3} (x - 1) = 2$$

This equation is graphed in Figure 10.1.

$L^- \neq L^+$

We naturally ask: Do the values of $f(x)$ always get closer and closer to a single real number L as x approaches a from both the left and the right? The answer is *NO!* To see this, let us consider the function

$$f(x) = \frac{|x|}{x}$$

and its limit as x approaches zero.

We set up a table of x and $f(x)$ values, as

x	$-0.1 \rightarrow$	$-0.01 \rightarrow$	-0.001	0.001	$\leftarrow 0.01$	$\leftarrow 0.1$
$f(x) = \frac{\lvert x \rvert}{x}$	-1	-1	-1	1	1	1

As x moves closer and closer to zero from the left and from the right, the values of $f(x)$ are not converging on a *single* real number. Because $(L^- = -1) \neq (L^+ = 1)$, we write

$$\lim_{x \to 0} \frac{|x|}{x} \text{ does not exist}$$

This formula is graphed in Figure 10.2.

Figure 10.2

$\lim_{x \to 0} \frac{|x|}{x}$ Does Not Exist

Because $L^- = -1$ and $L^+ = 1$

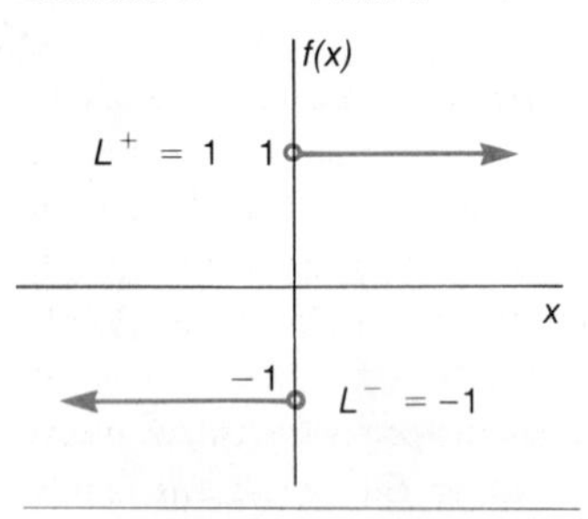

Or, consider the function

$$f(x) = \begin{cases} 2x & 0 \leq x \leq 5 \\ 6x - 10 & x > 5 \end{cases}$$

(as shown in Figure 10.3), and its limit as x approaches 5. As x approaches 5 from the left (through numbers that are less than 5), the value of $f(x)$ approaches 10 as a limit. Thus, $L^- = 10$. However, as x approaches 5 from the right (through numbers that are greater than 5), the value of $f(x)$ approaches 20, not 10, as a limit. Thus, $L^+ = 20$; and $L^- \neq L^+$.

Figure 10.3

$L^- \neq L^+$; Hence L Does Not Exist

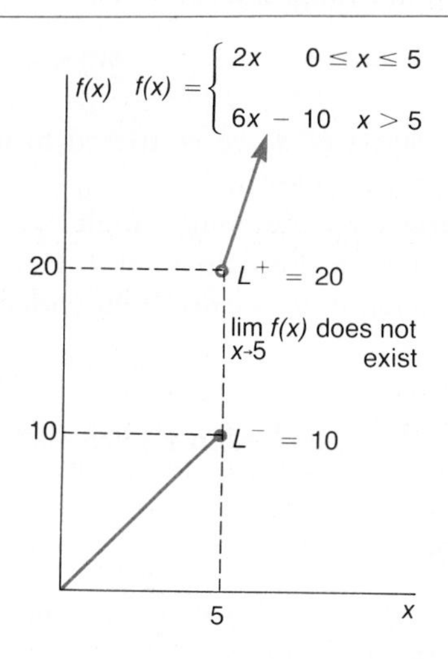

Because the left-hand and the right-hand limits are not equal, the function has no limit as x approaches 5.

Graphically, whenever there is a vertical gap in the graph of f at $x = a$, the limit as x approaches a fails to exist.

$f(a)$ Not Defined

The existence of a limit as x approaches a does not require that a actually be in the domain of f. The concept of a limit requires only that the function be defined as x moves closer and closer to a. It is quite possible that f may not be defined at the point $x = a$ itself.

Consider the function pictured in Figure 10.4 and defined as

$$f(x) = \begin{cases} 4 - x & \text{for } x < 4 \\ x - 4 & \text{for } x > 4 \end{cases}$$

Figure 10.4

$f(a)$ Not Defined; $\lim_{x \to a} f(x)$ Does Exist

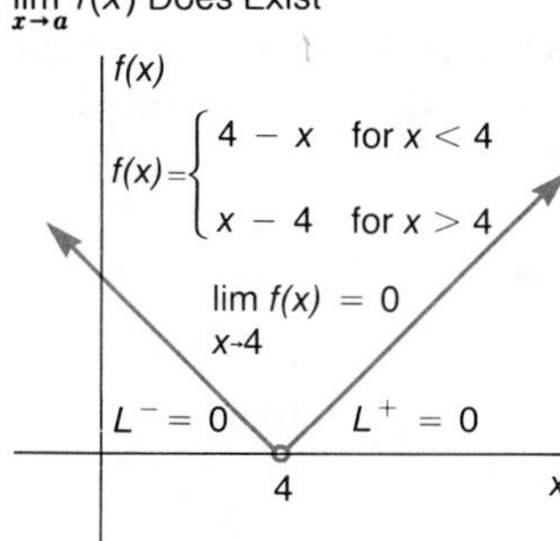

The point $x = 4$ is not in the domain of the function. Still, both

$$L^- = \lim_{x \to 4^-} f(x) = 0 \qquad \text{and} \qquad L^+ = \lim_{x \to 4^+} f(x) = 0$$

so that

$$\lim_{x \to 4} f(x) = 0$$

Whenever $f(a)$ is undefined, yet f is defined for all other values of x in some interval about a, there will be a point missing in the graph of the function. Nonetheless, the limit does exist if that part of the graph to the left of a vertical line erected at $x = a$ and that part of the graph to the right of such a line run into the vertical line at the same point.

"a" at an Endpoint

If the number a which x is approaching is an endpoint for the domain of f, either one or the other of the right-hand or the left-hand limits

Figure 10.5

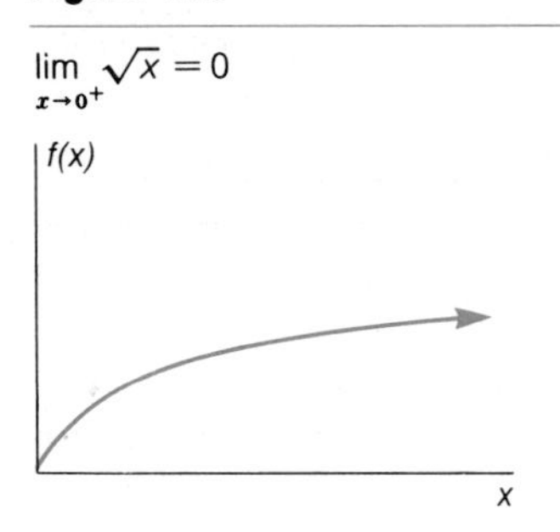

will not be meaningful. Consider

$$\lim_{x \to 0} \sqrt{x}$$

Since the values of x are restricted to nonnegative numbers, the left-hand limit is not relevant.

For purposes of defining a limit, we are interested in convergence on a from within the domain of f. Hence, we must conclude that the limit of a function may exist at an endpoint of the function. Then

$$\lim_{x \to 0^+} \sqrt{x} = 0$$

This equation is graphed in Figure 10.5.

EXERCISES

Use a table of x *and* f(x) *values to find the indicated limit for each of the following functions.*

1. $\lim_{x \to 3} (3x + 1)$
2. $\lim_{x \to 0} 5$
3. $\lim_{x \to 0} x$
4. $\lim_{x \to 2} (x^2 - 1)$
5. $\lim_{x \to 2} -2x^3$
6. $\lim_{x \to 0} |x|$
7. $\lim_{x \to 2} (2/x)$
8. $\lim_{x \to 0} \dfrac{x^2 + 3x}{x}$
9. $\lim_{x \to 4} 2\sqrt{x}$
10. $\lim_{x \to 3} e^x$
11. $\lim_{x \to 5} \ln x$
12. $\lim_{x \to 3} 2^x$

Use the following graphs to find the indicated limits, if they exist.

13.

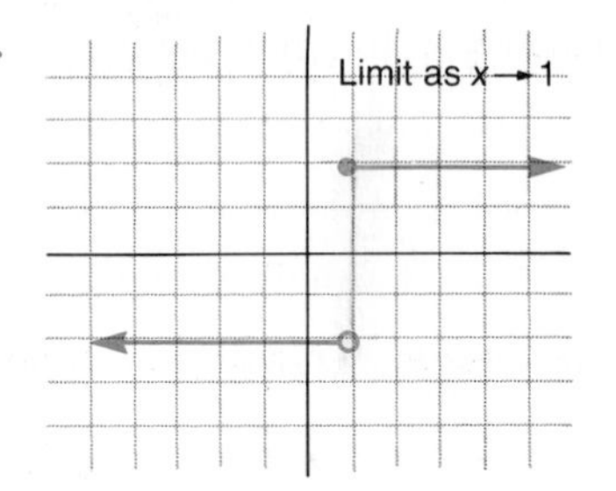

14.

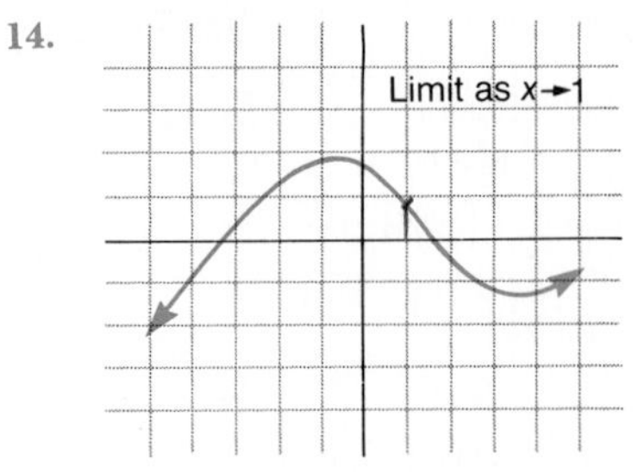

15.

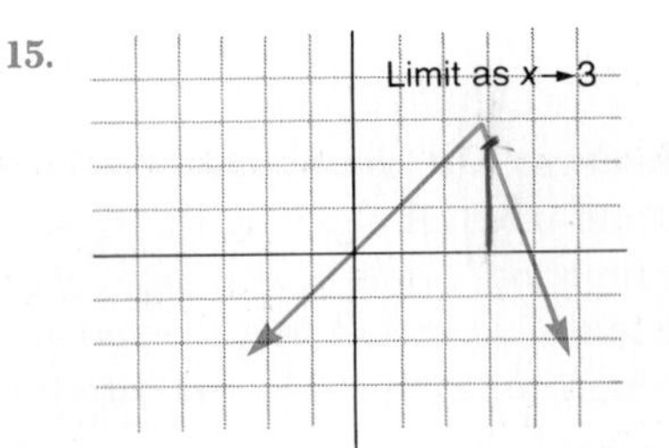

16.

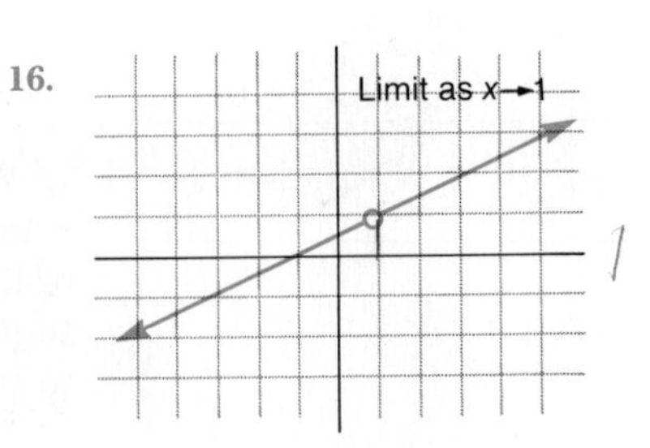

17.

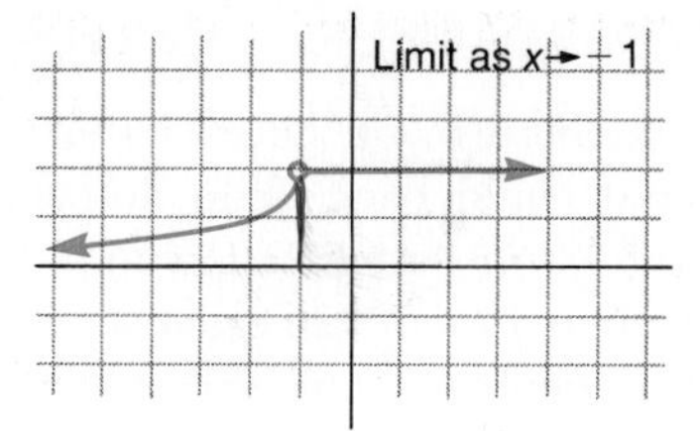

18.

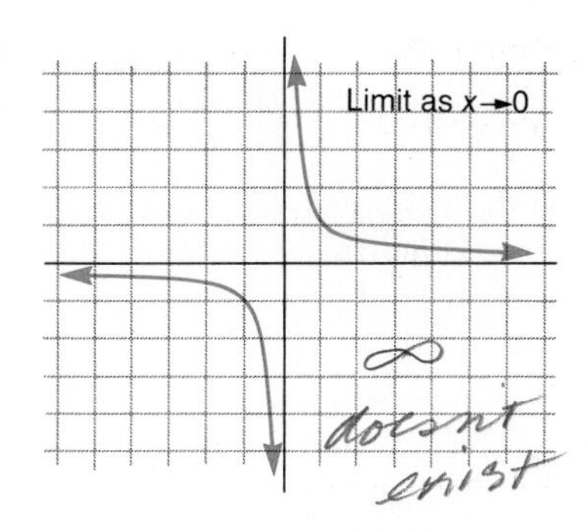

Sketch a graph of each of the following functions. Use the graph to determine the indicated limit, if it exists.

19. Given $f(x) = \begin{cases} 3x+2 & \text{for } x \leq 2 \\ 10-x & \text{for } x > 2 \end{cases}$

Find (*a*) $\lim_{x\to 2^-} f(x)$

(*b*) $\lim_{x\to 2^+} f(x)$

(*c*) $\lim_{x\to 2} f(x)$

20. Given $f(x) = \begin{cases} x-1 & \text{for } x < 5 \\ x+1 & \text{for } x > 5 \end{cases}$

Find (*a*) $\lim_{x\to 5^-} f(x)$

(*b*) $\lim_{x\to 5^+} f(x)$

(*c*) $\lim_{x\to 5} f(x)$

21. Given $f(x) = \begin{cases} x^2 & \text{for } x \leq 0 \\ x+4 & \text{for } x > 0 \end{cases}$

Find (*a*) $\lim_{x\to 0^-} f(x)$

(*b*) $\lim_{x\to 0^+} f(x)$

(*c*) $\lim_{x\to 0} f(x)$

22. Use a table of x and $f(x)$ values to evaluate

$$\lim_{x\to 0} (1+x)^{1/x}$$

What is the special significance of this limit?

23. The cost $C(x)$ of producing x units of a product is given by

$$C(x) = \begin{cases} 3x + 2{,}500 & \text{for } 0 \leq x \leq 5{,}000 \\ 19{,}000 + 3.5x & x > 5{,}000 \end{cases}$$

Find $\lim_{x\to 5{,}000} C(x)$.

10.2 THEOREMS USEFUL IN THE EVALUATION OF LIMITS

Undoubtedly you noticed that in many of the preceding examples the limit could have been determined simply by evaluating $f(a)$. In general, *if the function is either a polynomial, power, exponential, or logarithmic function or is constructed from such functions via the operations of function sum, difference, multiplication, or composition, and if* f *is defined at* a, *then* $\lim_{x \to a} f(x)$ *does equal* f(a). And in these cases, we can determine the limit simply by computing $f(a)$. However, at times $f(a)$ is not defined; or there is a "gap" in the function at $x = a$. Then we must use more-elaborate methods for determining the limit.

We have seen that we may evaluate the limit by setting up tables of x and $f(x)$ values, or by constructing a graph of the function. These can sometimes be cumbersome processes, however, and other methods may be more useful.

The following theorems are most useful in the evaluation of limits.

For any real number a, assuming that $\lim_{x \to a} f(x)$ and $\lim_{x \to a} g(x)$ exist:

1. For any real constant k, $\lim_{x \to a} k = k$.
2. For any real number n, $\lim_{x \to a} x^n = a^n$.
3. $\lim_{x \to a} \sqrt[n]{f(x)} = \sqrt[n]{\lim_{x \to a} f(x)}$ if the root is defined.
4. $\lim_{x \to a} kf(x) = k \lim_{x \to a} f(x)$
5. $\lim_{x \to a} [f(x) \pm g(x)] = \lim_{x \to a} f(x) \pm \lim_{x \to a} g(x)$
6. $\lim_{x \to a} f(x) \cdot g(x) = \lim_{x \to a} f(x) \cdot \lim_{x \to a} g(x)$
7. $\lim_{x \to a} [f(x)/g(x)] = \lim_{x \to a} f(x)/\lim_{x \to a} g(x)$ if $\lim_{x \to a} g(x) \neq 0$

The following examples illustrate the use of these theorems.

Example 1

$$\lim_{x \to 2} (3x^2 - x + 6) = 3 \lim_{x \to 2} x^2 - \lim_{x \to 2} x + \lim_{x \to 2} 6 = (3)(2)^2 - 2 + 6 = 16$$

[]

Example 2

$$\begin{aligned} \lim_{x \to 3} (2x^2 + 1)(3x - 4) &= \lim_{x \to 3} (2x^2 + 1) \cdot \lim_{x \to 3} (3x - 4) \\ &= [(2 \lim_{x \to 3} x^2) + (\lim_{x \to 3} 1)] \cdot [(3 \lim_{x \to 3} x) - (\lim_{x \to 3} 4)] \\ &= [2(3)^2 + 1] \cdot [3(3) - 4] \\ &= (19) \cdot (5) \\ &= 95 \end{aligned}$$

[]

Example 3

$$\lim_{x\to 2} (3x-4)/(x+1) = \frac{\lim_{x\to 2} (3x-4)}{\lim_{x\to 2} (x+1)}$$

$$= \frac{3 \lim_{x\to 2} x - \lim_{x\to 2} 4}{\lim_{x\to 2} x + \lim_{x\to 2} 1}$$

$$= \frac{3(2)-4}{2+1}$$

$$= 2/3$$

[]

Example 4

$$\lim_{x\to 3} \sqrt{2x+3} = \sqrt{\lim_{x\to 3} (2x+3)}$$

$$= \sqrt{2 \lim_{x\to 3} x + \lim_{x\to 3} 3}$$

$$= \sqrt{2(3)+3}$$

$$= 3$$

[]

When the Denominator Limit Equals Zero

But now consider

$$\lim_{x\to 2} \frac{x^2-4}{x-2}$$

When we attempt to use Limit Theorem 7 to evaluate this limit, we obtain $0/0$, an *indeterminate form.* (Do *not* interpret $0/0$ as equal to zero!) The fact that the numerator limit and the denominator limit appear, initially, to equal zero does not necessarily mean that the limit does not exist. Recall that the limit is concerned with the situation *as* x *approaches* a, *not* with $x = a$. Thus, if we can derive an alternate function F, equivalent to the function f stated above at all x except $x = a$, we may be able to circumvent the problem posed by the limit of zero in the denominator. We can often approach the new function by factoring the original function, as follows.

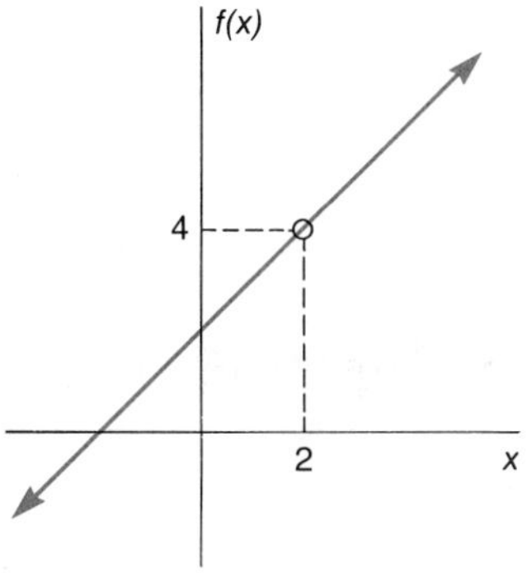

$$f(x) = \frac{x^2-4}{x-2} \quad \text{is transformed into} \quad F(x) = \frac{(x+2)\cancel{(x-2)}}{\cancel{(x-2)}} = x+2$$

and $F(x)$ is exactly equivalent to $f(x)$ except at $x = 2$. Now, we continue

$$\lim_{x\to 2} \frac{x^2-4}{x-2} = \lim_{x\to 2} \frac{(x+2)\cancel{(x-2)}}{\cancel{(x-2)}} = \lim_{x\to 2} (x+2)$$

$$= \lim_{x\to 2} x + \lim_{x\to 2} 2$$

$$= 2+2$$

$$= 4$$

This formulation is graphed in Figure 10.6.

If the function whose limit we are attempting to evaluate is one which cannot be approached by factoring, other algebraic manipulations may

Figure 10.6

$$\lim_{x\to 2} \frac{x^2-4}{x-2} = 4$$

f(x)
4
2
x

yield a tractable function. Or, as a last resort, we may be able to determine the limit by substituting values of x which move closer and closer to a (but are not equal to a).

Example 5

$$\begin{aligned}\lim_{x\to 2}\frac{x^2-3x+2}{x-2} &= \lim_{x\to 2}\frac{(x-1)(x-2)}{(x-2)}\\ &= \lim_{x\to 2}(x-1)\\ &= \lim_{x\to 2}x-\lim_{x\to 2}1\\ &= 2-1\\ &= 1\end{aligned}$$

[]

Example 6

$$\begin{aligned}\lim_{x\to 0}\frac{[1/(x+4)-(1/4)]}{x} &= \lim_{x\to 0}\frac{\dfrac{4}{4(x+4)}-\dfrac{(x+4)}{4(x+4)}}{x}\\ &= \lim_{x\to 0}\frac{4-(x+4)}{4(x+4)}\cdot\frac{1}{x}\\ &= \lim_{x\to 0}\frac{-x}{x[4(x+4)]}\\ &= \lim_{x\to 0}\frac{-1}{4(x+4)}\\ &= \frac{\lim_{x\to 0}-1}{(\lim_{x\to 0}4)(\lim_{x\to 0}x+\lim_{x\to 0}4)}\\ &= -1/[4(0+4)]\\ &= -1/16\end{aligned}$$

[]

EXERCISES

Use the limit theorems to evaluate each of the following limits.

24. $\lim_{x\to 3} 1$
25. $\lim_{x\to -6} 1$
26. $\lim_{x\to 1} 5$
27. $\lim_{x\to 0} 2/3$
28. $\lim_{x\to 0} 0.56825$
29. $\lim_{x\to 3} -9$
30. $\lim_{x\to 1} x$
31. $\lim_{x\to -4} x^2$
32. $\lim_{x\to 0.25} x^2$
33. $\lim_{x\to 2} x^4$
34. $\lim_{x\to -3} x^3$
35. $\lim_{x\to 0.1} x^2$
36. $\lim_{x\to 0} 0.15x^5$
37. $\lim_{x\to 1/2} 3x$

38. $\lim_{x \to 0.3} 0.9x^2$

39. $\lim_{x \to 2} 8x^2$

40. $\lim_{x \to -1} (x^2/2)$

41. $\lim_{x \to 3} -4x^3$

42. $\lim_{x \to 2} (x + 2)$

43. $\lim_{x \to 1/2} (4x^2 + 5)$

44. $\lim_{x \to -1} (x^4 - 1)$

45. $\lim_{x \to 0} (x^2 - x + 12)$

46. $\lim_{x \to 3} (x^2 + x + 1)$

47. $\lim_{x \to 0.4} (3x^2 + 2x + 5)$

48. $\lim_{x \to 2} x^2(x^3 - 1)$

49. $\lim_{x \to 2} (2x^2 - 3)(2x + 5)$

50. $\lim_{x \to 3} (x - 1)(x - 3)$

51. $\lim_{x \to -1} (x^2 - x + 1)(x + 2)$

52. $\lim_{x \to 0.2} (x^2 + 1)(1 - x)$

53. $\lim_{x \to 0} (x + 4)(x - 1)$

54. $\lim_{x \to 5} x/(x + 1)$

55. $\lim_{x \to 2} x^2/(x^3 - 1)$

56. $\lim_{x \to -2} 3x/(x^3 + 1)$

57. $\lim_{x \to 4} (1 - x)/(x^2 + 2)$

58. $\lim_{x \to 2} (5x + 7)/(2x + 6)$

59. $\lim_{x \to -1} (x^2 + 3x - 4)/(2x - 3)$

60. $\lim_{x \to 2} \sqrt{(x^2 + 5)/(x - 1)}$

61. $\lim_{x \to 1} \sqrt{x/(x + 3)}$

62. $\lim_{x \to 2} x^{3/2}$

63. $\lim_{x \to 2} \sqrt{(x^3 + 4x + 6)/(2x^2 + 1)}$

64. $\lim_{x \to 0} (x^2 + x)/x$

65. $\lim_{x \to 1} (x^2 - 1)/(x - 1)$

66. $\lim_{x \to 3} (x^2 - 4x + 3)/(x - 3)$

67. $\lim_{x \to -2} (x^2 + x - 2)/(x + 2)$

68. $\lim_{x \to 2} (x^3 - 8)/(x - 2)$

69. $\lim_{x \to 0} [1/(x + 3) - (1/3)]/x$

10.3 LIMITS INVOLVING INFINITY

We have considered limits of f as the independent variable x approaches some constant a. What happens when the variable x is allowed to increase (or decrease) without bound? The expression "x tends to infinity" is used to indicate that x is not approaching any real number but is increasing without bound. The following symbols are used.

$x \to +\infty$ indicates that x increases without bound through positive values.

$x \to -\infty$ indicates that x decreases without bound through negative values.

It is important that we remember that *infinity* is not a real number. *Infinity* represents a concept, not a point on the real-number line!

Let us consider the behavior of

$$f(x) = 1/x^2$$

Figure 10.7

$\lim_{x \to \infty} 1/x^2 = 0$

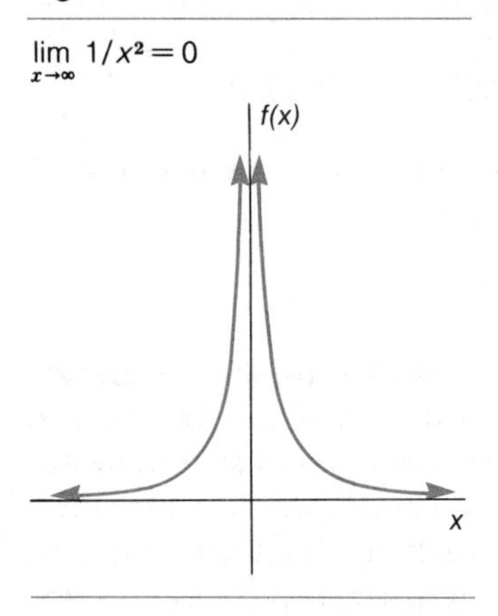

as x increases without bound. If we compile the following table of values

x	± 100	± 1000	± 10000
$f(x) = 1/x^2$	1/10,000	1/1,000,000	1/100,000,000

(which are also plotted in Figure 10.7), we see that as x increases without bound, $f(x)$ approaches zero. This is symbolized

$$\lim_{x \to \infty} 1/x^2 = 0$$

Because x is not approaching a finite value, a new definition for the limit must be formulated.

> If the values of $f(x)$ approach a single real number L as x increases without bound in the positive direction, then L is the limit of the function f as $x \to +\infty$, denoted by
>
> $$\lim_{x \to +\infty} f(x) = L$$
>
> Also, if the values of $f(x)$ approach a single real number L as x decreases without bound in the negative direction, then L is the limit of the function f as $x \to -\infty$, denoted by
>
> $$\lim_{x \to -\infty} f(x) = L$$

Generally, the limit theorems previously given remain unchanged when $x \to a$ is replaced by $x \to +\infty$ or $x \to -\infty$. We do have this additional limit theorem, however.

> **Limit theorem 8.** If n is any positive integer, then
>
> $$(a) \quad \lim_{x \to +\infty} 1/x^n = 0 \qquad (b) \quad \lim_{x \to -\infty} 1/x^n = 0$$

Example 7

$$(a) \quad \lim_{x \to \infty} 1/x = 0 \qquad (b) \quad \lim_{x \to \infty} 3/x^3 = 3 \lim_{x \to \infty} 1/x^3 = (3)(0) = 0$$

In each of these functions, the denominator increases without bound as x increases without bound. Because the numerator is a constant, the value of the ratio moves closer and closer to zero. []

When $f(x)$ Increases without Bound

The value of $f(x)$, like the value of x, may increase without bound. Look again at the function $f(x) = 1/x^2$, pictured in Figure 10.7. Let us this time notice what happens to the values of $f(x)$ as x approaches zero. We may tabulate the following pairs of values:

x	$\pm 1/2$	$\pm 1/4$	$\pm 1/10$	...	$\pm 1/100$	...	$\pm 1/1{,}000$
$f(x)$	4	16	100	...	10,000	...	1,000,000

As x takes on values closer and closer to zero, $f(x)$ increases without bound. We write

$$\lim_{x \to 0} 1/x^2 = \infty$$

However, this notation *does not really mean* that the limit is infinity! The limit must be a real number; infinity is not a real number; hence, the limit cannot be infinity. The notation is useful because it indicates that the limit does not exist and tells why it does not exist. It is important that we distinguish between the condition that "the limit does not exist because $f(x)$ is increasing (or decreasing) without bound" as opposed to the condition that "the limit does not exist because there is a vertical gap in the graph of the function f at $x = a$." This notation serves that purpose. Thus, we list one additional limit theorem.

Limit theorem 9. If n is any positive integer, then

$$\lim_{x \to 0^+} 1/x^n = +\infty$$

and

$$\lim_{x \to 0^-} 1/x^n = \begin{cases} -\infty & \text{if } n \text{ is odd} \\ +\infty & \text{if } n \text{ is even} \end{cases}$$

Whenever the numerator is a nonzero constant and the denominator approaches zero, the fraction will either increase or decrease without bound, and the limit is either plus infinity or minus infinity.

Example 8

(*a*) $\lim_{x \to 0^+} 2/x^3 = +\infty$ (*b*) $\lim_{x \to 0^-} 2/x^3 = -\infty$

(*c*) $\lim_{x \to 0^+} 3/x^4 = +\infty$ (*d*) $\lim_{x \to 0^-} 3/x^4 = +\infty$ []

In some instances both x and $f(x)$ increase without bound. Consider $f(x) = x^2$ as $x \to \infty$. As x gets larger and larger, so does x^2. The limit is symbolized $\lim_{x \to \infty} x^2 = \infty$. Or, in general,

$$\lim_{x \to \infty} x^n = \infty \qquad \textbf{(10.1)}$$

And, because of the importance of the exponential function, we note

$$\lim_{x\to\infty} e^{x}=\infty \qquad \text{and} \qquad \lim_{x\to\infty} e^{-x}=0 \tag{10.2}$$

Example 9 The number of rainbow trout in a lake varies with time according to the function

$$f(t)=\frac{13{,}000}{1+4e^{-t}}$$

Figure 10.8

$$\lim_{t\to\infty}\frac{13{,}000}{1+4e^{-t}}=13{,}000$$

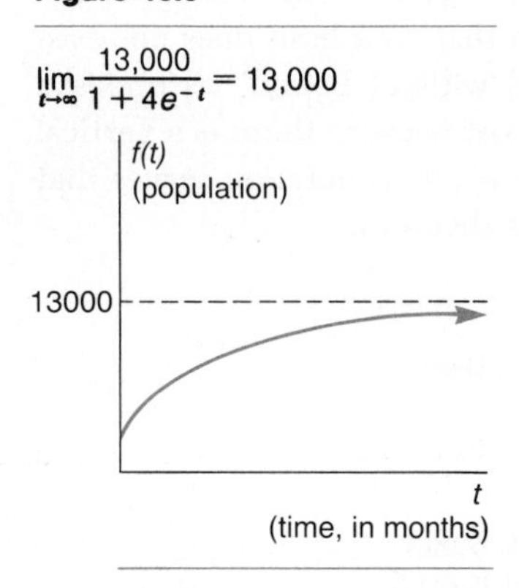

where t is time (in months) elapsed since the lake is stocked. Over an extended period of time, the trout population in the lake will tend to what level?

We compute

$$\lim_{t\to\infty}\frac{13{,}000}{1+4e^{-t}}=\frac{\lim_{t\to\infty} 13{,}000}{\lim_{t\to\infty} 1+4\lim_{t\to\infty} e^{-t}}$$

$$=\frac{13{,}000}{1+(4)(0)}$$

$$=13{,}000$$

Over time the trout population in the lake will tend toward an upper level of 13,000. (See Figure 10.8.) []

A Special Case with Rational Functions

In some instances both the numerator and the denominator of a rational function get larger and larger in the positive sense as x increases without bound. Let us examine the limit

$$\lim_{x\to\infty}\frac{4x}{x+1}$$

One is tempted to say that the value which the function approaches is ∞/∞, but this is a meaningless expression. Mathematical manipulation of the original statement of the function should lead to the correct limit. One technique that will often work with functions of this type is to *divide each term in both the numerator and the denominator of the function by* x^n, *where* n *is the highest exponent of the variable* x *appearing in the function.* Here we may divide both the numerator and the denominator of the function by x (or multiply by $1/x$), to obtain

$$\lim_{x\to\infty} 4x/(x+1)=\lim_{x\to\infty}\frac{(4x)(1/x)}{(x+1)(1/x)}$$

$$=\lim_{x\to\infty}\frac{4}{1+(1/x)}$$

$$= \frac{\lim_{x\to\infty} 4}{\lim_{x\to\infty} 1 + \lim_{x\to\infty} (1/x)}$$

$$= 4/(1+0)$$

$$= 4 \qquad []$$

Example 10

$$\lim_{x\to\infty} \frac{2x^2 - 3x + 4}{x^2 + 4} = \lim_{x\to\infty} \frac{(2x^2/x^2) - (3x/x^2) + (4/x^2)}{(x^2/x^2) + (4/x^2)}$$

$$= \frac{\lim_{x\to\infty} 2 - \lim_{x\to\infty} (3/x) + \lim_{x\to\infty} (4/x^2)}{\lim_{x\to\infty} 1 + \lim_{x\to\infty} (4/x^2)}$$

$$= (2 - 0 + 0)/(1 + 0)$$

$$= 2 \qquad []$$

EXERCISES

Find the indicated limit, if it exists.

70. $\lim_{x\to\infty} 17$
71. $\lim_{x\to\infty} 5x$
72. $\lim_{x\to 0} 5/x$
73. $\lim_{x\to\infty} e^x$
74. $\lim_{x\to 0} 2e^x$
75. $\lim_{x\to\infty} e^{-x}$
76. $\lim_{x\to 0} [1 + (-6/x)]$
77. $\lim_{x\to 0} (1/6x^2)$
78. $\lim_{x\to 0} [(-3/x^3) + 1]$
79. $\lim_{x\to 0} e^x$
80. $\lim_{x\to\infty} (1 - x)$
81. $\lim_{x\to -\infty} (1 - x)$
82. $\lim_{x\to\infty} 2x/(x - 1)$
83. $\lim_{x\to 3} 4/(x - 3)$
84. $\lim_{x\to\infty} (3x^2 + 7)$
85. $\lim_{x\to\infty} 6x^2/(x - 3)$
86. $\lim_{x\to\infty} 19/(x^3 + 12)$
87. $\lim_{x\to\infty} (4x^2 - x + 15)/(3x + 1)$
88. $\lim_{x\to 2} (x + 2)/(x - 4)$
89. $\lim_{x\to\infty} (1 + (1/x))$
90. $\lim_{x\to\infty} (x - 4)/(x + 3)$
91. $\lim_{x\to\infty} (x + 1)/(2x + 1)$
92. $\lim_{x\to\infty} [2 + (5/(x - 1))]$
93. $\lim_{x\to\infty} (5 - x^2)/(2x + 4)$
94. $\lim_{x\to\infty} (7x^2 - 2)/(3x^2 + 10x + 100)$
95. $\lim_{x\to\infty} (x^2 + 6)/(x + 6)$
96. $\lim_{x\to 2} 4x/(2 - x)$
97. $\lim_{x\to\infty} (5 + 3x - 4x^2)/(4 + 5x^2)$
98. $\lim_{x\to 0} (1 + 1/x)$
99. $\lim_{x\to\infty} (2x^2 - x + 9)/(4x^3 + 1)$

100. $\lim_{x \to \infty} x^2/(x + 1)$

101. Daily sales, in units, of a product are a function of the length of time t, in days, that the product has been on the market, as

$$S(t) = 6{,}000 - 4{,}000e^{-t}$$

Over an extended period of time, sales will tend toward what level?

102. The number of trout in a lake varies with time according to the function

$$f(t) = \frac{11{,}000}{1 + 3e^{-t}}$$

where t is time (in months) elapsed since the lake was stocked. To what level will the trout population in the lake tend over an extended period of time?

103. The proportion of workers successfully passing a certain skills test after x hours of training is given by

$$f(x) = 1 - \frac{50}{50 + x}$$

Assuming that training can be continued over an extended period of time, toward what level will the proportion of workers passing the test tend?

104. The total cost C of producing x units of output is given by

$$C(x) = 5{,}000 + 6x$$

Average cost is defined as total cost divided by the number of units of output. What happens to average cost of producing these units as output increases indefinitely?

10.4 CONTINUITY OF A FUNCTION

The important concept of CONTINUITY of a function is developed from the theory of limits.

A function f is continuous at $x = a$ if and only if *all* of these conditions apply to f at a:

1. $f(a)$ is defined (that is, the domain of f includes $x = a$).
2. $\lim_{x \to a} f(x)$ exists.
3. $\lim_{x \to a} f(x) = f(a)$ (whether x approaches a from the left or the right).

Figure 10.9

Continuous and Discontinuous Functions

(A) Continuous

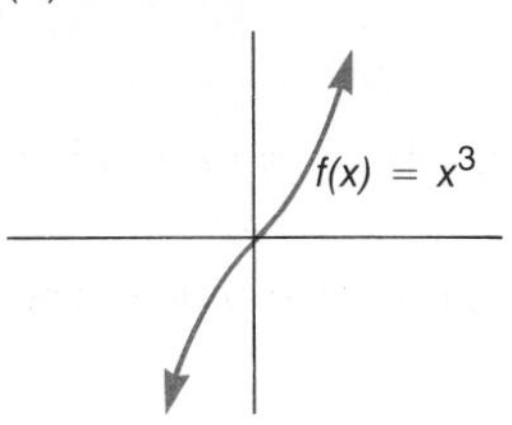

(B) Continuous

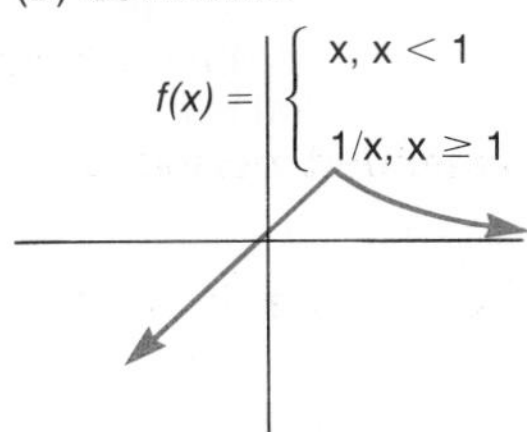

(C) Jump Discontinuity (vertical gap) at $x = 1$

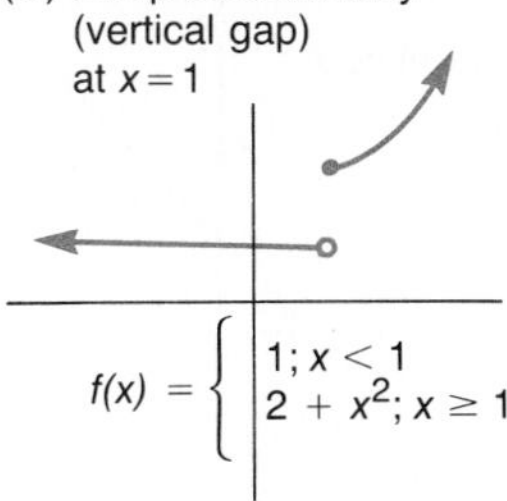

(D) Point-Missing Discontinuity at $x = 4$

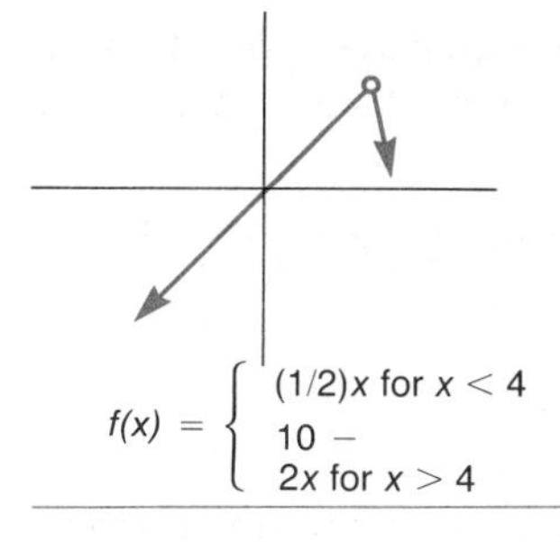

(E) Infinite Discontinuity at $x = 0$

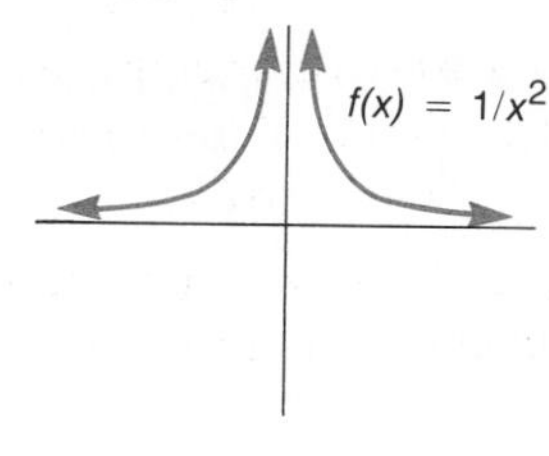

Geometrically, a function f is seen to be continuous at a point $x = a$ when there is neither a hole nor a gap in the graph of f at $x = a$. Figure 10.9 shows both continuous and discontinuous functions.

Four Cases of Discontinuity

If any one of the three requirements specified in the definition for continuity is not satisfied, the function f is said to be DISCONTINUOUS at $x = a$. Different types of discontinuity can be illustrated by examples.

Example 11

A function f may be discontinuous because the value of f at $x = a$ is not defined, although the limit of f as x approaches a exists.

The function

$$f(x) = (x^2 - 4)/(x - 2)$$

is not defined at $x = 2$. Thus Condition 1 of the definition of a continuous function is not satisfied, and the function is discontinuous at this point. This discontinuity is shown by the hole in the graph of f where the point (2,4) has been removed in Figure 10.10A. Still, the limit as x approaches 2 does exist. Also, the function is continuous at all other points in its domain. []

Example 12

A function f may be discontinuous because the limit as x approaches a does not exist. The function f, defined piecewise as

$$f(x) = \begin{cases} 2x & x \le 5 \\ 3x + 5 & x > 5 \end{cases}$$

Figure 10.10

Four Cases of Discontinuity

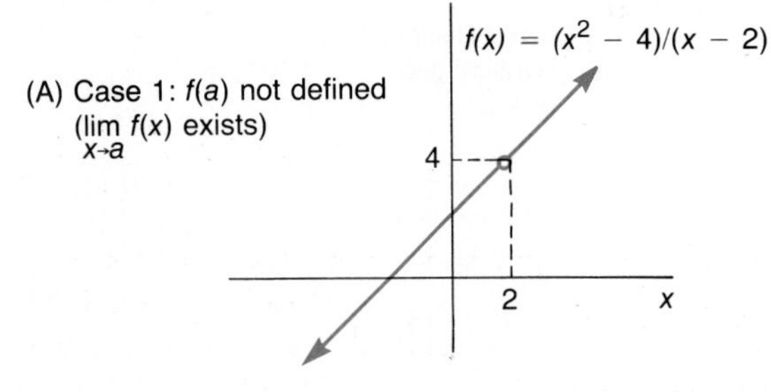

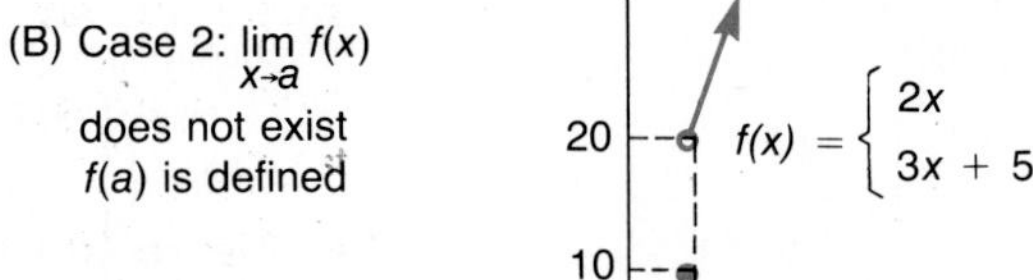

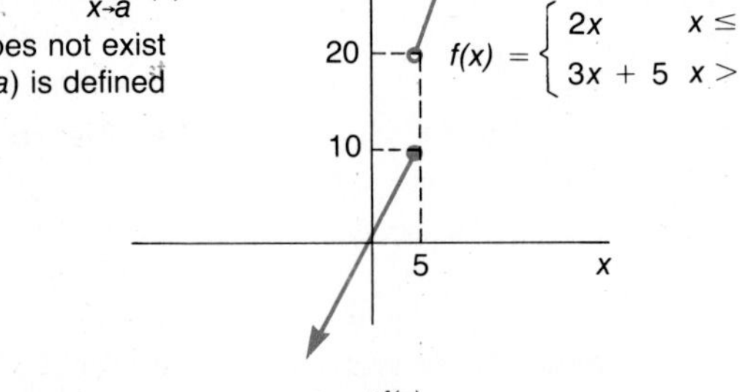

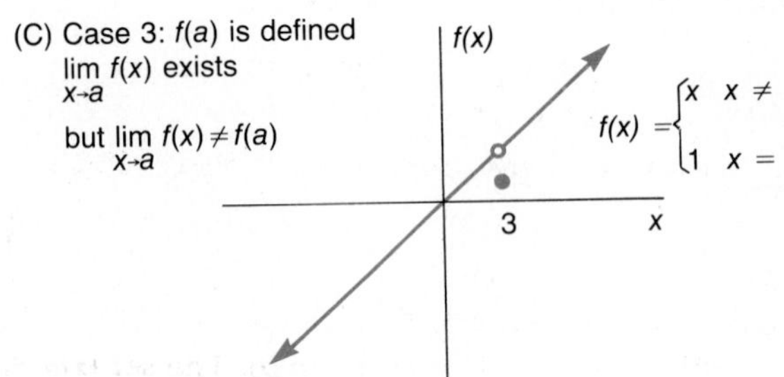

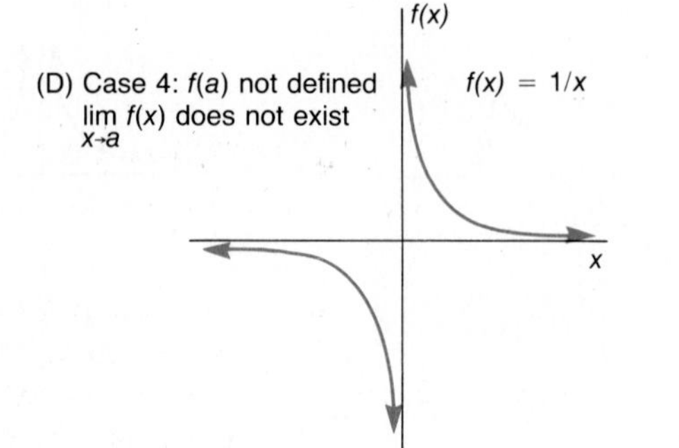

and pictured in Figure 10.10B, has no limit as x approaches 5. As x approaches 5 from the left (where $x < 5$), $f(x)$ moves closer and closer to 10. Thus, the left-hand limit as x approaches 5 is

$$L^- = \lim_{x\to 5^-} f(x) = \lim_{x\to 5^-} 2x = 10$$

As x approaches 5 from the right (where $x > 5$), $f(x)$ moves closer and closer to 20. The right-hand limit is

$$L^+ = \lim_{x\to 5^+} f(x) = \lim_{x\to 5^+} (3x + 5) = 20$$

The limit does not exist as x approaches a if the function has different right-hand and left-hand limits at that point. Consequently, this function does not have a limit as x approaches 5. The graph of the function has a vertical gap at $x = 5$. Such a vertical gap always indicates discontinuity of the function at this point. This type of discontinuity is often called JUMP DISCONTINUITY. []

Example 13 A function may be discontinuous because the limit of f as x approaches a is not the value of f at $x = a$.

The function

$$f(x) = \begin{cases} x & x \neq 3 \\ 1 & x = 3 \end{cases}$$

is discontinuous at $x = 3$ because

$$\lim_{x \to 3} f(x) = 3 \qquad \text{is not equal to} \qquad f(3) = 1$$

(See Figure 10.10C.) []

Example 14 Another type of discontinuity exists. A function is said to have an *infinite discontinuity* at $x = a$ if $f(x)$ becomes an arbitrarily large positive or negative number as $x \to a^-$ or $x \to a^+$. An example of such a discontinuity occurs at $x = 0$ in the function $f(x) = 1/x$. Note that

$$\lim_{x \to 0^-} 1/x = -\infty \qquad \text{and} \qquad \lim_{x \to 0^+} 1/x = +\infty$$

Hence, $\lim_{x \to 0} 1/x$ does not exist, nor is $f(0)$ defined. (See Figure 10.10D.)

[]

Continuity over an Interval

If a and b are real numbers and $a < b$, then the set $\{x|a < x < b\}$ is called an OPEN INTERVAL and is denoted by (a,b). The set $\{x|a \leq x \leq b\}$ is called a CLOSED INTERVAL and denoted by $[a,b]$. The half-open interval $\{x|a \leq x < b\}$ is symbolized $[a,b)$ whereas the half-open interval $\{x|a < x \leq b\}$ is symbolized $(a,b]$. In each case a and b are the ENDPOINTS of the interval, and any x value such that $a < x < b$ is an INTERIOR POINT.

> A function f is *continuous on an open interval* if it is continuous at each number in that interval.
> A function f is *continuous on a closed interval* $[a,b]$ provided the following conditions are satisfied:
>
> 1. f is continuous over the open interval (a,b).
> 2. $f(x) \to f(a)$ as $x \to a$ from within (a,b).
> 3. $f(x) \to f(b)$ as $x \to b$ from within (a,b).

Example 15 Determine whether there are any discontinuities for the function

$$f(x) = 5x/(x^3 - x)$$

The function is not defined and is thus discontinuous for any x such that

$$x^3 - x = x(x - 1)(x + 1) = 0$$

The function is discontinuous at $x = -1$, $x = 0$, and $x = 1$. The function is continuous at all other points. []

If there is no point of discontinuity for a function, that function is described as being a CONTINUOUS FUNCTION; otherwise, it is described as being a DISCONTINUOUS FUNCTION. All polynomials with unrestricted domain are continuous at all points of their domain. Rational functions are continuous except at the points, if any, where the denominator has value zero. In fact, most functions given by algebraic formulas are continuous wherever they are defined. If a function is defined by more than one formula, among the important possible points of disconti nuity of the amalgamated function are those points at which the component domains are adjoined.

EXERCISES

Which of the following graphs represent continuous functions?

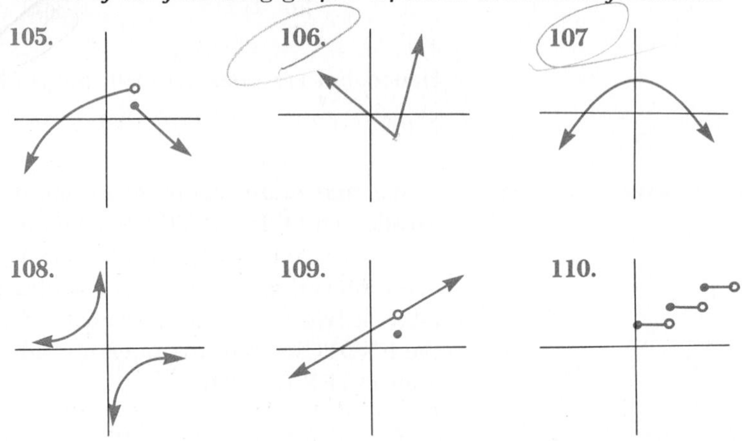

111. A boat-rental facility at a lake resort area charges $25 for the first two hours, or fraction thereof, and an additional $10 for each additional hour, or fraction thereof, for the rental of boats.
Write a function showing total rental charges as related to the number of hours a boat is rented. Graph the function. Is the function continuous?

Sketch each of the following functions. Determine whether or not the three requirements for continuity are met at the points specified.

112. $f(x)=\begin{cases} 10 & \text{for } x \leq 3 \\ -1 & \text{for } x > 3 \end{cases}$ Is f continuous at $x = 3$? Why or why not?

113. $g(x) = 2x + 1$ Is g continuous at $x = 0$? at $x = 1$? Why or why not?

114. $f(x)=\begin{cases} 1/(x+5) & \text{for } x \neq -5 \\ 0 & \text{for } x = -5 \end{cases}$ Is f continuous at $x = 0$? at $x = -5$? Why or why not?

115. $h(x) = x^2 - 1$ Is h continuous at $x = -1$? Why or why not?

116. $f(x)=\begin{cases} 2x+1 & \text{for } x \neq 4 \\ 0 & \text{for } x = 4 \end{cases}$ Is f continuous at $x = 4$? at $x = 0$? Why or why not?

117. $f(x) = \begin{cases} x^2 & \text{for } x \leq 1 \\ 2 - x & \text{for } x > 1 \end{cases}$ Is f continuous at $x = 1$? Why or why not?

118. $f(x) = \begin{cases} -2x^2 & \text{for } x < -1 \\ x - 1 & \text{for } x > -1 \end{cases}$ Is f continuous at $x = -1$? at $x = 1$? Why or why not?

119. $g(p) = p^3 + 4$ Is g continuous at $p = -1$? at $p = 1$? Why or why not?

120. $f(x) = \begin{cases} 1/(x+4)^2 & \text{for } x \neq -4 \\ 5 & \text{for } x = -4 \end{cases}$ Is f continuous at $x = -4$? at $x = 0$? Why or why not?

Determine whether the function is continuous on the interval specified. If discontinuous, specify the point, or points, at which discontinuity occurs.

121. $f(x) = x + 3$ $\quad -10 \leq x \leq 10$

122. $f(x) = x/(x - 4)$ $\quad$ for all x

123. $f(x) = 1/(x^2 + 3)$ $\quad$ for all x

124. $f(x) = |x|/x$ $\quad -100 \leq x \leq 100$

125. $f(x) = \begin{cases} 5x & x < 3 \\ 10 & x = 3 \\ 2x + 4 & x > 3 \end{cases}$ $\quad 0 \leq x \leq 25$

126. $f(x) = \begin{cases} 1/(x - 2) & \text{for } x \neq 2 \\ 8 & \text{for } x = 2 \end{cases}$ $\quad$ for all x

127. $g(x) = (x - 1)/(x^2 - x - 6)$ $\quad$ for all x

128. $f(x) = e^x$ $\quad$ for all x

129. $f(x) = 4/(x^3 - x^2)$ $\quad$ for all x

130. An overnight parcel-delivery service has the following rate structure, based on the weight, in ounces, of the parcel:

$$R(x) = \begin{cases} 4.50 & 0 < x \leq 10 \\ 6.50 & 10 < x \leq 20 \\ 10.50 & x > 20 \end{cases}$$

where x is weight, in ounces. Is the function continuous? If not, identify the points of discontinuity.

10.5 LIMITS AND ASYMPTOTES

A line that the graph of a function approaches—gets closer and closer to but never quite reaches—is known as an ASYMPTOTE. Asymptotes may be vertical lines, horizontal lines, or diagonal lines.

> The line $x = a$ is a VERTICAL ASYMPTOTE of the graph of f if and only if at least one of the following statements is true:
>
> (*a*) $\lim_{x \to a^-} f(x) = +\infty$ (or $-\infty$) $\quad$ or $\quad$ (*b*) $\lim_{x \to a^+} f(x) = +\infty$ (or $-\infty$)

Figure 10.11

Vertical Asymptote Is the y-Axis

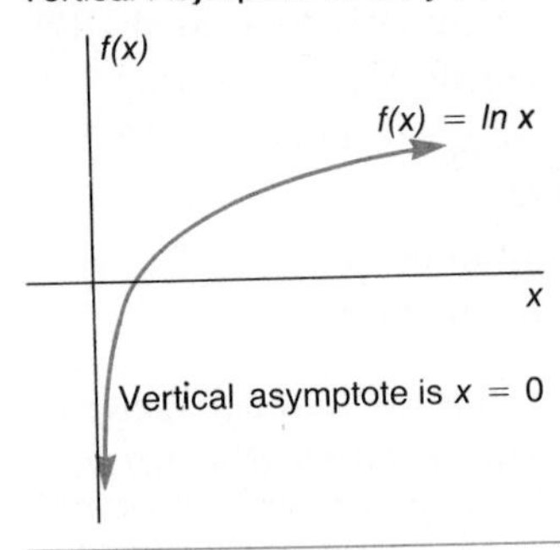

> The line $y = b$ is a HORIZONTAL ASYMPTOTE of the graph of $y = f(x)$ if and only if at least one of the following statements is true:
>
> (a) $\lim_{x \to +\infty} f(x) = b$ or (b) $\lim_{x \to -\infty} f(x) = b$

Consider the function $f(x) = ln\, x$ pictured in Figure 10.11. For this function

$$\lim_{x \to 0^+} ln\, x = -\infty$$

Thus the graph has a vertical asymptote at $x = 0$ (the y-axis).

The graph of $f(x) = e^x$ pictured in Figure 10.12 has a horizontal asymptote at $y = 0$ (the x-axis), as determined by

$$\lim_{x \to -\infty} e^x = 0$$

Example 16

Determine the horizontal and vertical asymptotes for the graph of the function

$$f(x) = \frac{1}{x-2} + 5$$

and sketch a graph of the function.

To test for horizontal asymptotes we find the limits of f as $x \to \infty$ and as $x \to -\infty$.

$$\lim_{x \to \infty} \left[\frac{1}{x-2} + 5\right] = \frac{\lim_{x \to \infty} 1}{\lim_{x \to \infty} x - \lim_{x \to \infty} 2} + \lim_{x \to \infty} 5 = 0 + 5 = 5$$

$$\lim_{x \to -\infty} \left[\frac{1}{x-2} + 5\right] = \frac{\lim_{x \to -\infty} 1}{\lim_{x \to -\infty} x - \lim_{x \to -\infty} 2} + \lim_{x \to -\infty} 5 = 0 + 5 = 5$$

Figure 10.12

Horizontal Asymptote Is the x-Axis

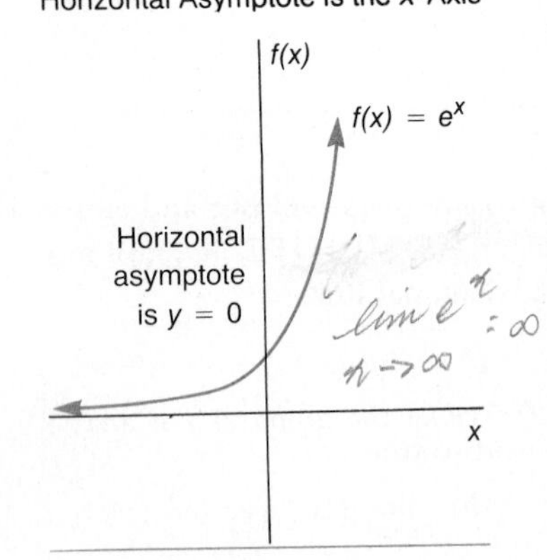

Thus, the graph of f approaches $y = 5$ as a horizontal asymptote as x increases without bound through the positive numbers and as x decreases without bound through the negative numbers.

To sketch the graph, we note that as x approaches $+\infty$, the first term of the function—$[1/(x-2)]$—approaches zero but remains positive. Thus, the value of $f(x)$ would approach 5 but would be slightly greater than 5. Contrarily, as x approaches $-\infty$, the term $[1/(x-2)]$ approaches zero but remains negative. Thus, while the value of $f(x)$ approaches 5, it will always be slightly less than 5.

Now, if $x = a$ is a vertical asymptote, the function cannot be continuous at a; in fact, it will have an infinite discontinuity at a. So, good places to look for vertical asymptotes of a function are those values of x for which the function is undefined—those values of x for which the denominator of a rational function equals zero.

We note that the denominator of $1/(x-2)$ equals zero when $x = 2$. And we test this value for the existence of a vertical asymptote, as

$$\lim_{x \to 2^-} \left[\frac{1}{x-2} + 5\right] = -\infty + 5 = -\infty$$

and

$$\lim_{x \to 2^+} \left[\frac{1}{x-2} + 5\right] = \infty + 5 = \infty$$

We conclude that the value of the function decreases without bound as x approaches $x = 2$ from the left (through numbers that are smaller than 2). Also the value of the function increases without bound as x approaches $x = 2$ from the right (through numbers that are larger than 2).

A sketch of the function is shown in Figure 10.13. []

These facts about functions and asymptotes are useful when we are graphing polynomials or rational functions. *A polynomial function has no asymptotes. A rational function of the form* f(x)/g(x) *has a vertical asymptote in its graph at* x = a *if* g(a) = *0 while* f(a) ≠ 0.

Figure 10.13

$f(x) = \frac{1}{x-2} + 5$

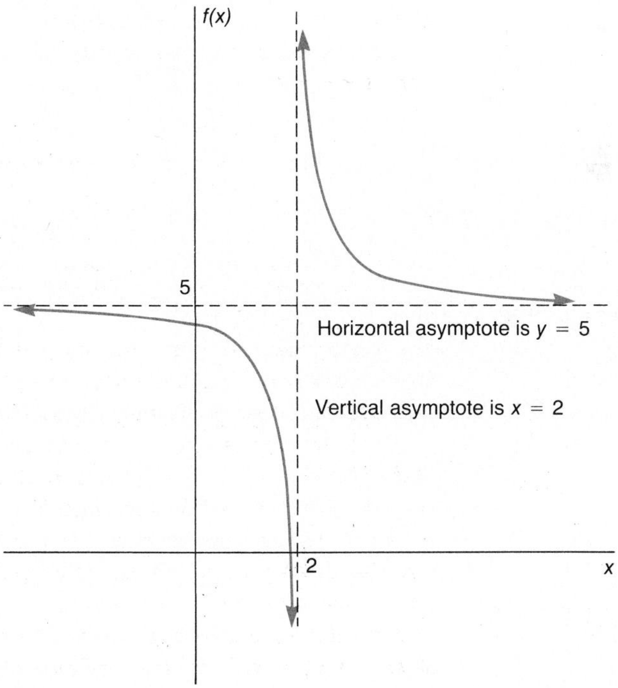

EXERCISES

Determine the vertical and horizontal asymptotes, if any, of the following functions. Sketch a graph of each function.

131. $y = 1/(x + 3)$
132. $y = 3/(x - 3)$
133. $y = \dfrac{1}{x + 3} + 4$
134. $y = \dfrac{2}{x + 3} - 4$
135. $y = x/(x + 3)$
136. $y = 2x/(x + 3)$
137. $y = (x + 1)/(2x + 3)$
138. $y = x/(x^2 - 1)$
139. $y = x/(2x^2 - 1)$
140. $y = (x + 1)/(x^2 - 1)$
141. $y = (2x + 1)/(x^2 - 1)$
142. $y = 1/(x^2 + 3x + 2)$
143. $y = x/(x^2 + 4x - 5)$
144. $y = (x + 1)/(x^2 + 4x - 5)$
145. $y = x^2/(x^2 - 1)$
146. $y = (x^2 + 1)/(x^2 - 1)$
147. $y = (2x^2 - 1)/(3x^2 + 1)$
148. $y = 1/x^3$
149. $y = (x - 1)/x^3$
150. $y = 1/x^4$
151. $y = 1/(x^3 - x)$
152. $y = (x^3 - x)/x$
153. $y = x^2/(x^2 + 1)$
154. $y = (x^2 + 3x + 2)/(x - 1)$

How?

147) $y = \dfrac{(2x^2 - 1)}{(3x^2 + 1)}$ $\lim_{x \to \infty} \dfrac{(2x^2 - 1)}{(3x^2 + 1)} = \dfrac{2}{3}$ H.A. $\lim_{x \to ?} \dfrac{(2x^2 - 1)}{(3x^2 + 1)}$

no V.A.

C·H·A·P·T·E·R 11 The Derivative and Rules of Differentiation

Of primary concern to anyone interested in studying the relationship between variables would be the manner in which the value of one variable (the dependent variable) is changed with given increases or decreases in the value of the other variable (the independent variable). It is through such an analysis of how the value of the dependent variable responds to changes in the value of a controllable independent variable that the decision maker is able to determine the optimal value for that controllable variable.

This need to understand the manner in which the behavior of one variable affects the behavior of another variable leads us to the study of RATES OF CHANGE, and eventually to the study of DERIVATIVES.

The distinguishing characteristic of a linear function is that its slope is a constant; each one-unit change in x, no matter what the value of x, has the same effect on the value of y. It is clear from the graph of any nonlinear function that the slope changes from one value of the independent variable to another (as illustrated in Figure 11.1). Hence, when the relationship between x and y is depicted by a nonlinear function, the effect that a one-unit change in the value of x has upon the value of y is itself a function of x.

Because the slope is not a constant, it is not logical to speak of *the* slope of a nonlinear function as we did for a linear function. We instead may speak of the AVERAGE RATE OF CHANGE BETWEEN TWO POINTS, or, in other instances, the RATE OF CHANGE OF THE FUNCTION AT A POINT (also called the INSTANTANEOUS RATE OF CHANGE).

11.1 AVERAGE RATE OF CHANGE

The AVERAGE RATE OF CHANGE of a function f over an interval x to $x + \Delta x$ is given by the change in $f(x)$ divided by the change in x; that is,

$$\textit{Average rate of change} = \frac{\Delta f(x)}{\Delta x} = \frac{f(x+\Delta x) - f(x)}{\Delta x} \qquad \textbf{(11.1)}$$

This ratio is also called a DIFFERENCE QUOTIENT.

Consider, specifically, the function $f(x) = x^2$. We would like, first, to determine the average rate of change in $f(x)$ as x changes from $x = 2$ to $x = 4$. This average rate of change is computed

$$\frac{\Delta f(x)}{\Delta x} = \frac{f(4) - f(2)}{4-2} = \frac{(4)^2 - (2)^2}{2} = \frac{16-4}{2} = \frac{12}{2} = 6$$

As x increases from $x = 2$ to $x = 4$, $f(x)$ increases by a total of 12 units, or an average increase of 6 units for each 1-unit increase in x.

Figure 11.1

(A) The slope of a linear function is constant

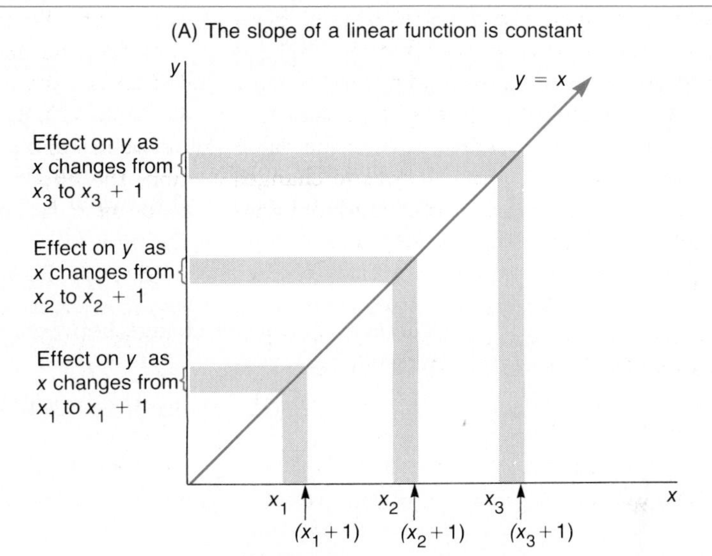

(B) The slope of a nonlinear function is not constant but is, instead, a function of x

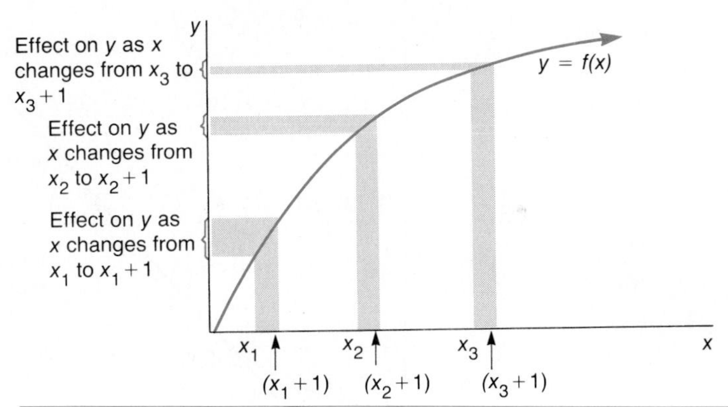

The average rate of change between two points on a curve can be depicted graphically by the slope of a straight line connecting those two points, as shown in Figure 11.2. Such a straight line connecting two points on a curve is called a SECANT LINE.

We can derive a function describing the average rate of change over any interval x to $x + \Delta x$ for $f(x) = x^2$ by computing

$$\frac{\Delta f(x)}{\Delta x} = \frac{f(x + \Delta x) - f(x)}{(x + \Delta x) - x} = \frac{(x + \Delta x)^2 - x^2}{\Delta x}$$

$$= \frac{[x^2 + 2x\Delta x + (\Delta x)^2] - [x^2]}{\Delta x}$$

$$= \frac{2x\Delta x + (\Delta x)^2}{\Delta x}$$

$$= \frac{\Delta x(2x + \Delta x)}{\Delta x}$$

$$= 2x + \Delta x$$

Using this general formulation, we can determine the average rate of change for $f(x) = x^2$ between $x = 2$ and $x + \Delta x = 3$ (so that $\Delta x = 1$) by

$$2x + \Delta x = 2(2) + 1 = 5$$

The average rate of change between $x = 4$ and $x + \Delta x = 4.5$ (so that $\Delta x = 0.5$) is

$$2x + \Delta x = 2(4) + 0.5 = 8.5$$

Figure 11.2

Average Rate of Change of $f(x) = x^2$ between $x = 2$ and $x = 4$

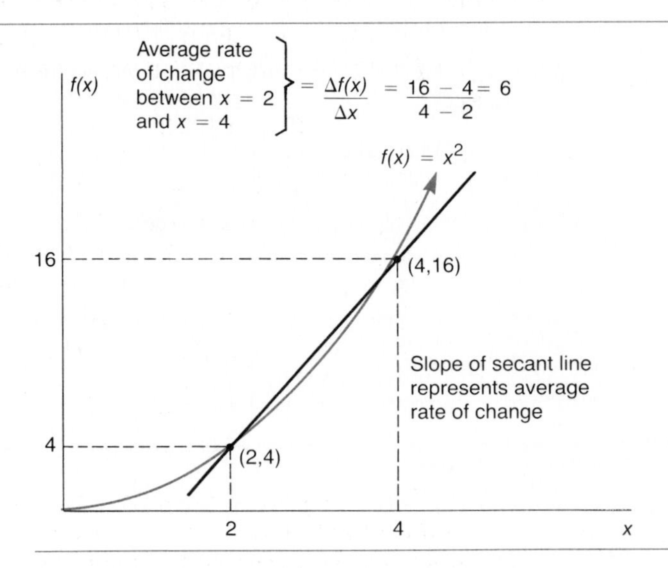

Applications of Average Rate of Change

An analysis of the average rate of change—of profits, of revenue, of cost, of demand, and so on—over a specified interval of values for some controllable independent variable is important in many decision-making situations.

Example 1

A company has determined that the total cost (in dollars) of manufacturing x units of a product is given by

$$C(x) = x^2 + 0.5x + 5{,}000 \qquad 0 \le x \le 150$$

Determine the average rate of change in total cost as the number of units of product manufactured is increased from 60 to 65.

We can determine the average rate of change in total cost between $x = 60$ and $x + \Delta x = 65$ by computing

$$
\begin{aligned}
\frac{\Delta C(x)}{\Delta x} &= \frac{C(65) - C(60)}{65 - 60} \\
&= \frac{[(65)^2 + 0.5(65) + 5{,}000] - [(60)^2 + 0.5(60) + 5{,}000]}{5} \\
&= \frac{9{,}257.5 - 8{,}630}{5} \\
&= 125.5
\end{aligned}
$$

The total cost of production would increase by *an average of* \$125.50 *for each of the 5 additional units* (above 60).

If we wish to analyze several possible changes in production level, it will be advantageous to develop a general function describing the average rate of change in total cost beginning at any specified production level x and involving any specified change in production level Δx. We use the definition of the difference quotient (Equation 11.1) and compute

$$
\begin{aligned}
C(x) &= x + 0.5x + 5{,}000 \\
C(x + \Delta x) &= (x + \Delta x)^2 + 0.5(x + \Delta x) + 5{,}000 \\
&= x^2 + 2x\Delta x + (\Delta x)^2 + 0.5x + 0.5\Delta x + 5{,}000 \\
\Delta C(x) &= C(x + \Delta x) - C(x) \\
&= [x^2 + 2x\Delta x + (\Delta x)^2 + 0.5x + 0.5\Delta x + 5{,}000] \\
&\quad - [x^2 + 0.5x + 5{,}000] \\
&= 2x\Delta x + (\Delta x)^2 + 0.5\Delta x \\
&= \Delta x(2x + \Delta x + 0.5) \\
\frac{\Delta C(x)}{\Delta x} &= \frac{\Delta x(2x + \Delta x + 0.5)}{\Delta x} \\
&= 2x + \Delta x + 0.5
\end{aligned}
$$

Using this general formulation, we can determine the average rate of change in total cost of production if the number of units manufactured were increased from 60 to 70 units by computing

$$2x + \Delta x + 0.5 = 2(60) + 10 + 0.5 = 130.5$$

Total cost would increase by an average of \$130.50 for each of the 10 additional units.

The average rate of change in total cost of production if production level were increased from 60 to 75 units would be

$$2x + \Delta x + 0.5 = 2(60) + 15 + 0.5 = 135.5$$

These results indicate that as management continues to increase the level of production (above 60 units), the average cost of the additional units also increases. □

EXERCISES

Find the equation describing the average rate of change from $x = a$ *to* $x = a + \Delta x$ *for each of the following functions. Evaluate the average rate of change over the intervals specified.*

1. $f(x) = 4x^2 + 7$ (a) from $x = 3$ to $x = 5$; (b) from $x = 6$ to $x = 10$
2. $f(x) = x^2 + 5x + 100$ (a) from $x = 10$ to $x = 14$; (b) from $x = 1$ to $x = 5$
3. $f(x) = -x^2 + 10x - 2$ (a) from $x = 6$ to $x = 9$; (b) from $x = 9$ to $x = 12$
4. $f(x) = x^3$ (a) from $x = 0$ to $x = 2$; (b) from $x = 0$ to $x = 3$
5. $f(x) = x/(x+1)$ (a) from $x = 2$ to $x = 3$; (b) from $x = 3$ to $x = 5$
6. $f(x) = 10 - x - x^3$ (a) from $x = 0$ to $x = 2$; (b) from $x = 1$ to $x = 3$
7. $f(x) = e^x$ (a) from $x = 1$ to $x = 3$; (b) from $x = 1$ to $x = 4$
8. $f(x) = ln(x+1)$ (a) from $x = 1$ to $x = 3$; (b) from $x = 1$ to $x = 4$
9. The Terence Trek family left home Saturday morning at 5:45 A.M. on an automobile trip to Grandmother Trek's house. They arrived at their destination at 11:15 A.M. that same morning. The car's odometer read 24,630 when they began the trip and 24,883 when they arrived. What was their average speed?
10. The formula

$$S = (1/2)gt^2$$

is used in physics to describe the motion of a freely falling body acted upon only by the force due to gravity where

t = time, in seconds
g = acceleration due to gravity, in ft/sec²
S = distance, in feet

If $g = 32$, we have

$$S = (1/2)32t^2 = 16t^2$$

What is the average rate of change of S over the interval $2 \leq t \leq 4$? $4 \leq t \leq 7$?

11. Daily demand for a certain commodity seems to be appropriately described by the function

$$D(x) = 120 - x^2 \qquad 4 \leq x \leq 10$$

where x is selling price per unit, in dollars.

Find the average rate of change in demand for a price change from \$5 to \$7; from \$5 to \$8.

12. A theater has determined that its profit function is given by

$$P(x) = -x^2 + 9x - 11$$

where x is admission ticket price, in dollars.

What would be the average rate of change in profit if the ticket price were increased from \$2 to \$4? from \$4 to \$5? from \$5 to \$7?

Graph the profit function and the secant lines depicting the average rates of change between $x = 2$ and $x = 4$; between $x = 4$ and $x = 5$; between $x = 5$ and $x = 7$.

13. Harben Manufacturing Company has determined that the total cost (in dollars) of manufacturing x units of its product is given by

$$C(x) = (x^2/2) + 20x + 320 \qquad 0 \le x \le 400$$

Determine the average rate of change in total cost as the number of units of product manufactured is increased from 100 to 120; from 120 to 150; from 150 to 175. What conclusions can you draw about further increases in the level of production and the effect on total cost?

14. Newton Novelty Company has determined that its revenue function is given by

$$R(x) = 20x\ log(10{,}000/x)$$

Determine the average rate of change in total cost as the number of units of product x sold increases from 40 to 50; from 50 to 75; from 75 to 100.

11.2 INSTANTANEOUS RATE OF CHANGE

While the average rate of change provides very useful information in many decision situations, there are also many times when we must use another concept, the INSTANTANEOUS RATE OF CHANGE. The instantaneous rate of change of a function is the rate of change, not over an interval of finite length, but *at a point.*

Refer to the secant line PQ in Figure 11.3. Notice that if the point Q were shifted to another location on curve C, another secant line could be formed. In particular, notice what happens to the succession of secant lines as the distance separating P and Q on the x-axis (that is, Δx) continues to decrease. The succession of secant lines approaches a *limiting line,* called the TANGENT LINE at P. The slope of the tangent line at any point x on the x-axis represents the INSTANTANEOUS RATE OF CHANGE of the function f at that point.

The **INSTANTANEOUS RATE OF CHANGE** of f at any point x is the limit of the average rate of change of f over the interval from x to $x + \Delta x$ as Δx approaches zero; that is,

$$\text{Instantaneous rate of change} = \lim_{\Delta x \to 0} \frac{\Delta f(x)}{\Delta x} = \lim_{\Delta x \to 0} \frac{f(x + \Delta x) - f(x)}{\Delta x} \tag{11.2}$$

if this limit exists.

Figure 11.3

Secant Lines to Curve C at P and the Tangent at P

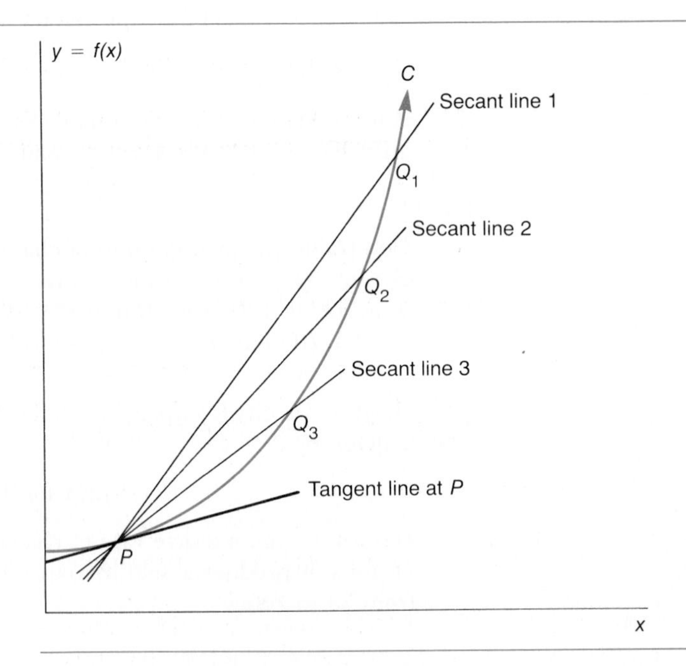

Example 2 The distance (in feet) that a small steel ball dropped from a high tower will fall in t seconds is given by

$$f(t) = 16t^2$$

The average rate of change in distance for an indicated change in time is termed the AVERAGE VELOCITY: that is,

$$\text{Average velocity} = \frac{\text{Distance traveled}}{\text{Time traveling}}$$

We can derive a general expression for the average velocity between any point in time t and any other point in time $t + \Delta t$ as follows:

$$\frac{\Delta f(t)}{\Delta t} = \frac{f(t+\Delta t) - f(t)}{\Delta t} = \frac{[16(t+\Delta t)^2] - [16t^2]}{\Delta t}$$

$$= \frac{16t^2 + 32t\Delta t + 16(\Delta t)^2 - 16t^2}{\Delta t}$$

$$= \frac{\Delta t[32t + 16(\Delta t)]}{\Delta t}$$

$$= 32t + 16(\Delta t)$$

This expression can be used to evaluate the average velocity of the steel ball over any relevant time span. For instance, from $t = 0$ to $t = 3$ (meaning $\Delta t = 3$), the average velocity of the ball is

$$32t + 16(\Delta t) = 32(0) + 16(3) = 48 \text{ feet per second}$$

From the time it is first dropped to the moment when it has been falling three seconds, the ball travels at an average rate of 32 feet per second.

From $t = 3$ to $t = 5$ (meaning $\Delta t = 2$), the average velocity of the ball is

$$32t + 16(\Delta t) = 32(3) + 16(2) = 128 \text{ feet per second}$$

Clearly, the speed at which the ball is traveling is accelerating; its velocity is increasing with time.

Now suppose we are interested in the velocity *at the exact moment* when the steel ball has been falling for t seconds. This is known as the INSTANTANEOUS VELOCITY of the ball and is given by

$$\lim_{\Delta t \to 0} \frac{\Delta f(t)}{\Delta t}$$

Figure 11.4

Slope of Tangent Line at t Gives Instantaneous Rate of Change in $f(t)$ at That Point

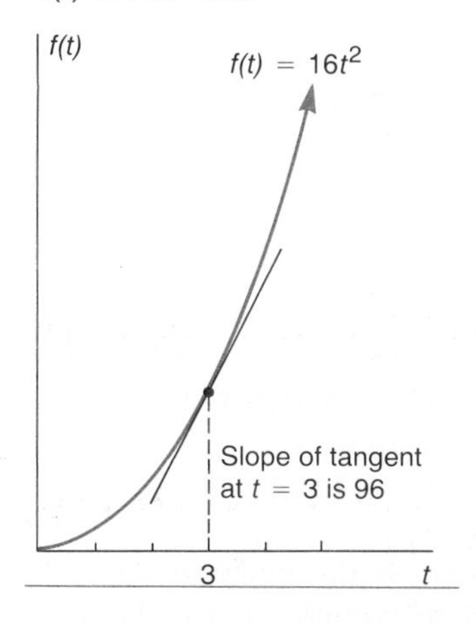

That is, the instantaneous velocity is the limit of the average velocity as the change in time approaches zero.

A general formulation which can be used to find the instantaneous velocity of the steel ball at any moment in time t can be derived by

$$\lim_{\Delta t \to 0} (32t + 16\Delta t) = \lim_{\Delta t \to 0} 32t + 16 \lim_{\Delta t \to 0} \Delta t$$

$$= 32t + 16(0)$$

$$= 32t$$

Thus, the instantaneous velocity of the steel ball at time $t = 3$ is $(32) \cdot (3) = 96$ feet per second, at time $t = 4$, the instantaneous velocity is $(32) \cdot (4) = 128$ feet per second.

The instantaneous rate of change of the function f at the point t is given by the slope of the tangent constructed at that point. (See Figure 11.4.) []

The "Delta Process"

The procedure of deriving the instantaneous-rate-of-change function by direct application of the definition

$$\begin{matrix}\textit{Instantaneous} \\ \textit{rate of change}\end{matrix} = \lim_{\Delta x \to 0} \frac{\Delta f(x)}{\Delta x} = \lim_{\Delta x \to 0} \frac{f(x + \Delta x) - f(x)}{\Delta x}$$

is a four-step process which is generally referred to as the DELTA PROCESS. The steps are:

1. Compute $f(x + \Delta x)$ by substituting $(x + \Delta x)$ for x in the defining rule of the function.
2. Compute the difference $f(x + \Delta x) - f(x)$ and simplify the expression. The result is $\Delta f(x)$.
3. Divide $\Delta f(x)$ by Δx.
4. Take the limit of $\Delta f(x)/\Delta x$ as $\Delta x \to 0$, if this limit exists.

Example 3 Use the delta process to derive a general expression for the instantaneous rate of change of the function

$$f(x) = 3x^2 + 6x - 9$$

at any x.

We make these computations:

Step 1. $f(x + \Delta x) = 3(x + \Delta x)^2 + 6(x + \Delta x) - 9$
$= 3x^2 + 6x\Delta x + 3(\Delta x)^2 + 6x + 6\Delta x - 9$

Step 2. $\Delta f(x) = [3x^2 + 6x\Delta x + 3(\Delta x)^2 + 6x + 6\Delta x - 9] - [3x^2 + 6x - 9]$
$= 6x\Delta x + 3(\Delta x)^2 + 6\Delta x$

Step 3. $\Delta f(x)/\Delta x = [6x\Delta x + 3(\Delta x)^2 + 6\Delta x]/\Delta x$
$= [\Delta x(6x + 3\Delta x + 6)]/\Delta x$
$= 6x + 3\Delta x + 6$

Step 4. $\lim_{\Delta x \to 0} \Delta f(x)/\Delta x = \lim_{\Delta x \to 0} [6x + 3\Delta x + 6)$

$= \lim_{\Delta x \to 0} 6x + 3 \lim_{\Delta x \to 0} \Delta x + \lim_{\Delta x \to 0} 6$

$= 6x + 6$

The instantaneous rate of change of $f(x) = 3x^2 + 6x - 9$ at any point x can be evaluated using the above expression. For example, at $x = 2$, the instantaneous rate of change is $6(2) + 6 = 18$. At $x = -1$ the instantaneous rate of change is $6(-1) + 6 = 0$; at $x = -3$, the instantaneous rate of change is $6(-3) + 6 = -12$. The function f is depicted in Figure 11.5. Tangent lines have been drawn at $x = -3$, $x = -1$, and $x = 2$ to represent the instantaneous rate of change of f at these points. [ ]

Observe carefully in Figure 11.5 the slope of the tangents and the underlying movement of the original function at these points. We may be able to generalize our observations as follows:

1. If the instantaneous rate of change is a positive number, the slope of the tangent is positive, and the graph of f is rising from left to right.

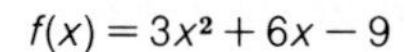

Figure 11.5

$f(x) = 3x^2 + 6x - 9$

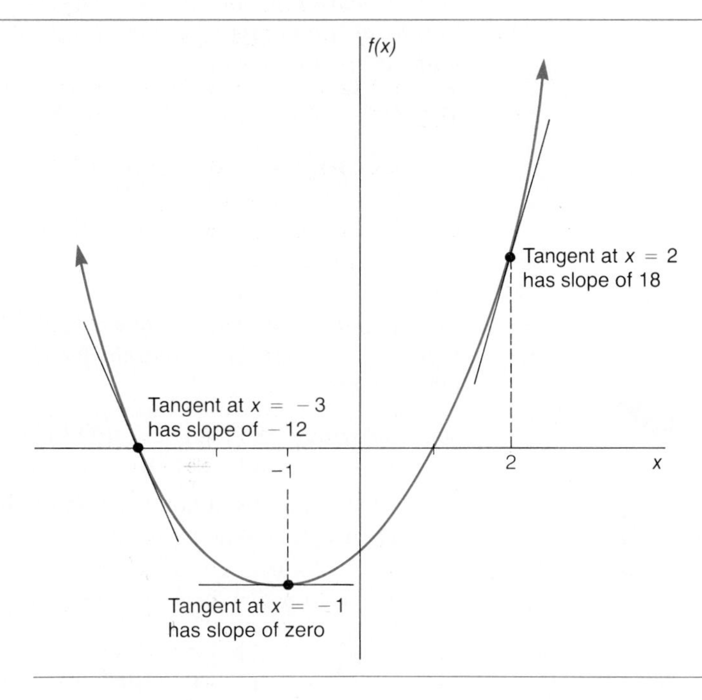

2. If the instantaneous rate of change is a negative number, the slope of the tangent is negative, and the graph of f is falling from left to right.
3. If the instantaneous rate of change is zero, the tangent line is horizontal; the graph of f is neither rising nor falling but is stationary.

As we proceed, we shall have much more to say about the information provided by the instantaneous rate of change of a function!

EXERCISES

Use the delta process to derive the function for the instantaneous rate of change for each of the following functions.

15. $f(x) = 2x^2 + 5$
16. $f(x) = 3x^2$
17. $f(x) = x^2 + 5x + 4$
18. $f(x) = 2x^2 + 3x + 1$
19. $f(x) = 71$
20. $f(x) = 4x$
21. $f(x) = 11 - 7x$
22. $f(x) = x^3 - 1$
23. $f(x) = x^3/3$
24. $f(x) = 3/x$

Graph each of the following functions. Construct tangent lines to the graphs at the points indicated. Find the slope of each of the tangents and interpret its meaning.

25. $f(x) = x^2$ (a) at $x = 3$; (b) at $x = 6$
26. $f(x) = 2x^2 + 1$ (a) at $x = -1$; (b) at $x = 1$
27. $f(x) = 25 - 3x - x^2$ (a) at $x = 2$; (b) at $x = 1/2$
28. $f(x) = x^2 - x$ (a) at $x = 0$; (b) at $x = -5$; (c) at $x = 5$
29. $f(x) = 3x^2 + 2x + 1$ (a) at $x = 1$; (b) at $x = -3$

The proportion of people in an undeveloped country who have been infected by a certain communicable disease prior to time t *is given by* $p(t) = 0.09(t^2 + t)$.

30. Determine the average rate of change in the proportion between time $t = 1$ and $t = 3$; between time $t = 3$ and $t = 5$.
31. What is the instantaneous rate of change of $p(t)$ with respect to t when $t = 1$? When $t = 3$?

A vehicle is at a distance d *(in miles) from its starting point in* t *hours as given by*

$$d(t) = 10t^2 \qquad 0 \le t \le 10$$

32. Find the average rate of change of distance with respect to time from $t = 0$ to $t = 4$. What does this number mean?
33. Derive a function for the instantaneous rate of change. Evaluate this function at $t = 4$. What does this number mean?

A tank is filled by a small inlet pipe. The volume v *(in gallons) in the tank at any time* t *(in hours) after the inlet pipe is opened is given by*

$$v(t) = 400t^2 + 400t \qquad 0 \le t \le 6$$

34. What is the volume in the tank at the end of one hour? at the end of three hours?
35. Derive a function describing the average rate of change in volume. Evaluate this function from $t = 0$ to $t + 3$; from $t = 0$ to $t = 5$.
36. Derive a function describing the instantaneous rate of change in volume. Evaluate this function at $t = 3$; at $t = 4$; at $t = 5$.

An object is dropped from the roof of a 500-foot building. The distance d *(in feet) of the object above the ground is given as a function of* t *(in seconds) by the equation*

$$f(t) = 500 - 16t^2$$

37. Sketch a graph of the function.

38. Derive a function describing the instantaneous rate of change in d.
39. Determine the instantaneous rate of change in d at time $t = 2$. Explain the significance of this measure.

A small rocket is launched from ground level with speed of 900 feet per second. The vertical height S *of the rocket above the ground at time* t *(in seconds) after the launch is given by*

$$S(t) = 900t - 16t^2$$

40. Sketch a graph of the function.
41. Derive a function describing the instantaneous rate of change in S.
42. Determine the instantaneous rate of change in S at time $t = 2$; at $t = 4$. Explain the significance of these measures.

11.3 THE DERIVATIVE

Because of its importance in the analysis of the relationship existent between variables, the instantaneous-rate-of-change function is given a special name; it is called the DERIVED FUNCTION, or the DERIVATIVE.

> **The DERIVATIVE of a function** f is that function, commonly denoted f', whose function value at any x in the domain of f is given by
>
> $$f'(x) = \lim_{\Delta x \to 0} \frac{\Delta f(x)}{x} = \lim_{\Delta x \to 0} \frac{f(x + \Delta x) - f(x)}{\Delta x} \qquad (11.3)$$
>
> if that limit exists.

Each of these methods of denoting the derivative of the function f is equivalent and generally acceptable:

$$f'(x) \qquad D_x f(x) \qquad \frac{df(x)}{dx} \qquad \frac{df}{dx}$$

Or, if $y = f(x)$ is used, the derivative function may be symbolized

$$y' \qquad D_x y \qquad \frac{dy}{dx}$$

We have been using f and x almost systematically, but if other letters are used to denote the function and the independent variable, then the symbolism for the derivative function is adjusted accordingly, as, for example,

$$g'(t) \qquad \text{or} \qquad df(u)/du$$

If the derivative function is to be evaluated at any specific $x = a$, any of the following notations may be used

$$f'(a) \qquad D_x y \Big|_{x=a} \qquad D_x f(x) \Big|_{x=a} \qquad y' \Big|_{x=a} \qquad dy/dx \Big|_{x=a}$$

In the absence of one universally accepted set of symbols, it is desirable that the student should be familiar with each of these alternative notational schemes. Actually, each symbolic form seems most convenient to use under certain circumstances; so all are used in different places in this (and in most other) texts. Most important, the reader must realize that each of these should be viewed as a complete symbol and not a set of symbols related by some arithmetic operation. The symbol dy/dx, for instance, DOES NOT mean that we multiply d by y or divide dy by dx.

Existence of the Derivative of a Function

The process of determining the derivative of a function is termed DIFFERENTIATION. In order for the derivative to exist, the difference quotient $\Delta f(x)/\Delta x$ must have a limit as $\Delta x \to 0$. The limit may exist for some values of x of a function f and fail to exist for others. At each x where this limit does exist, the function is said to be *differentiable*.

For a derivative to exist at a specified value of x, the function must be both *smooth* and *continuous* at that point. The continuity requirement dictates that $\lim_{x \to a} f(x) = f(a)$ and means pictorially that there can

Figure 11.6

To Be Differentiable at the Point $x = a$, a Function Must Be Both Smooth and Continuous at That Point

(A) $f'(a)$ does not exist because f is not continuous at $x = a$

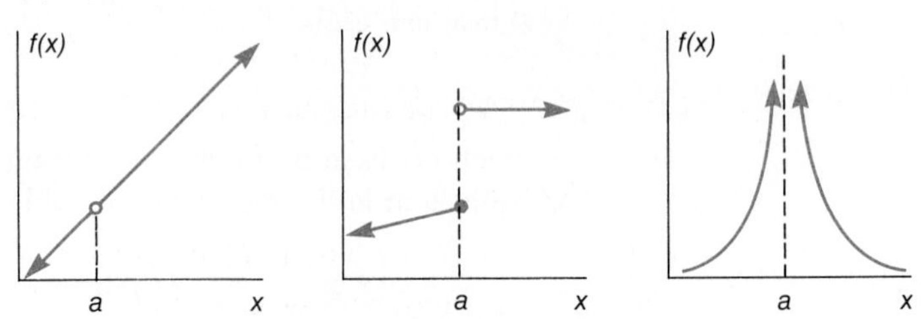

(B) $f'(a)$ does not exist because f is not smooth at $x = a$

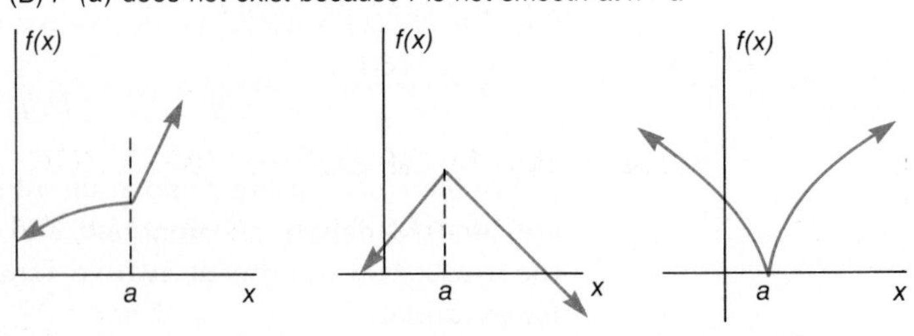

be neither a point missing nor a vertical gap nor an infinite discontinuity in the function at that point. (See Figure 11.6.) IF A FUNCTION IS NOT CONTINUOUS AT A POINT, IT CANNOT HAVE A DERIVATIVE AT THAT POINT.

However, while continuity is a necessary condition for differentiability, it is not a sufficient condition. DIFFERENTIABILITY ALSO REQUIRES THAT THE FUNCTION MUST BE "SMOOTH" AT THE POINT $x = a$. The smoothness requirement dictates that there can be no abrupt change of direction at $x = a$ and no sharp corners at $x = a$. (See Figure 11.6B.)

EXERCISES

Given

$$f(x) = \begin{cases} 10 - x & \text{for } x \leq 2 \\ 2x + 4 & \text{for } x > 2 \end{cases}$$

43. Sketch a graph of f.
44. Is f continuous at $x = 2$?
45. Is f differentiable at $x = 2$?

Given

$$f(x) = \begin{cases} x & \text{for } x \leq 5 \\ x - 2 & \text{for } x > 5 \end{cases}$$

46. Sketch a graph of f.
47. Is f continuous at $x = 5$?
48. Is f differentiable at $x = 5$?
49. Is f continuous at $x = 10$?
50. Is f differentiable at $x = 10$?

Given $f(x) = 1 - x^2$

51. Sketch a graph of the function.
52. Is f continuous at $x = 1$?
53. Is f differentiable at $x = 1$?

11.4 RULES OF DIFFERENTIATION

Differentiation by direct application of the formula

$$f'(x) = \lim_{\Delta x \to 0} \frac{\Delta f(x)}{\Delta x}$$

is sometimes awkward and cumbersome. Fortunately, functions can be classified into certain groups so that the process for finding a derivative when applied to these groups follows a definite pattern. Certain rules of differentiation, each based on the above formulation, have been developed which permit the rapid and efficient differentiation of certain types of functions. A few of the more commonly used rules are listed below.

Derivative of a Constant Function

> **Rule 1** *The derivative of a constant function is zero.*
> If $f(x) = k$, where k is any constant, $f'(x) = 0$.

From a geometrical standpoint, the rule states that any function which is depicted by a horizontal line has a slope of zero.

Example 4

(any constant) (Slope is 0)

(*a*) If $f(x) = 5$, then $f'(x) = 0$	(*b*) $D_x 6 = 0$
(*c*) If $f(x) = -8$, then $f'(x) = 0$	(*d*) $D_x - 16 = 0$
(*e*) If $f(x) = 1/3$, then $f'(x) = 0$	(*f*) $D_x 1/6 = 0$
(*g*) $\frac{d}{dx}(1/9) = 0$	(*h*) $\frac{d}{dx}(23) = 0$ []

Derivative of a Variable Raised to a Constant Power

> **Rule 2** *The derivative of a variable raised to the constant power* n, *where* n *is any real number and* n ≠ 0, *is equal to the power* n *multiplied times the variable raised to the power* n − 1.
> If $f(x) = x^n$, then $f'(x) = nx^{n-1}$

Example 5

a. If $f(x) = x^3$, $f'(x) = 3x^2$	*b.* $D_x x^{101} = 101x^{100}$
c. If $f(x) = x^{-4}$, $f'(x) = -4x^{-5}$	*d.* $D_x x^{9/4} = (9/4)x^{5/4}$
e. If $f(x) = x^{0.2}$, $f'(x) = 0.2x^{-0.8}$	*f.* $\frac{d}{dx}(x^{31}) = 31x^{30}$
g. If $f(x) = x$, $f'(x) = 1x^0 = 1$	*h.* $\frac{d}{dx}(x^{1.25}) = 1.25x^{0.25}$

i. If $f(x) = 1/x = x^{-1}$, $f'(x) = -1x^{-2} = -1/x^2$

j. If $f(x) = \sqrt{x} = x^{1/2}$, $f'(x) = (1/2)x^{-1/2} = \frac{1}{2\sqrt{x}}$ []

EXERCISES

Find each of the following:

54. $f'(x)$ for $f(x) = x^6$
55. y' for $y = x^{10}$
56. $D_x(x^{2/3}) =$
57. $\frac{d}{dx}(x^{-2}) =$

58. $f'(u)$ for $f(u) = u^{3.45}$
59. s' for $s = t^{5/3}$
60. $D_x(1/x^4) =$
61. $f'(1)$ for $f(x) = x^7$
62. $f'(2)$ for $f(x) = x^5$
63. $y'\Big|_{x=3}$ for $y = x^3$
64. $f'(2)$ for $f(x) = 1/x^2$
65. $f'(x)$ for $f(x) = \sqrt[3]{x^2}$

Derivative of a Constant Times a Function

> **Rule 3** *The derivative of a constant times a function is the constant times the derivative of the function, if* $g'(x)$ *exists.*
>
> If $f(x) = kg(x)$, then $f'(x) = k \cdot g'(x)$

Example 6

a. If $f(x) = 5x^3$, then $f'(x) = 5(3x^2) = 15x^2$
b. If $f(x) = -9x^4$, then $f'(x) = (-9)(4x^3) = -36x^3$
c. If $f(x) = 7x^{-2}$, then $f'(x) = (7)(-2x^{-3}) = -14x^{-3} = -14/x^3$
d. If $f(x) = 3x$, then $f'(x) = (3)(1) = 3$ []

EXERCISES

Find each of the following.

66. $f'(x)$ for $f(x) = 15x^6$
67. $f'(1)$ for $f(x) = (1/14)x^{14}$
68. y' for $y = 5x^3$
69. $y'\Big|_{x=2}$ for $y = 9x^2$
70. y' for $y = 3x$
71. $D_x(0.1x^{0.1}) =$
72. $D_x(-8x^4) =$
73. $f'(x)$ for $f(x) = (-1/5)x$
74. $\dfrac{d}{dx}(x^4/4) =$
75. $g'(3)$ for $f(t) = 16t^2$
76. The slope of the tangent to $y = 3x^2$ at (1,3)
77. The slope of the tangent to $f(x) = (1/6)x^3$ at (2,4/3)

Derivative of the Sum or Difference of Functions

> **Rule 4** *The derivative of the sum (or difference) of two functions is the sum (or difference) of their respective derivatives, if these derivatives exist separately.*
>
> If $f(x) = [g(x) \pm h(x)]$, then $f'(x) = [g'(x) \pm h'(x)]$

This rule can be extended to the derivative of any finite number of sums and differences of functions. For instance, if $f(x) = [g(x) + h(x) - p(x) - q(x)]$, then $f'(x) = [g'(x) + h'(x) - p'(x) - q'(x)]$, if all these derivatives exist separately.

Example 7

a. If $f(x) = 6x^4 - 5x^2$, then $f'(x) = 24x^3 - 10x$

b. If $f(x) = 1 + 2x + 3x^2$, then $f'(x) = 2 + 6x$

c. $D_x(5x^3 - x^{1/2}) = 15x^2 - (1/2)x^{(-1/2)} = 15x^2 - \dfrac{1}{2\sqrt{x}}$

d. $D_z(z/5 - 4/z^2) = D_z[(1/5)z - 4z^{-2}] = (1/5) + 8z^{-3} = (1/5) + (8/z^3)$ []

Example 8

A factory is dumping chemical waste into a stream such that the total quantity of waste Q (in tons) having been dumped into the stream after t weeks is given by

$$Q(t) = 0.1t^2 + 1.5t$$

At what *rate* is the amount of waste increasing at the end of three weeks?

The rate at which a total quantity is changing is given by the first derivative of the "total" function. We want, thus, to determine $Q'(t)$ and evaluate this at $t = 3$. We have

$$Q'(t) = 0.2t + 1.5$$

so that

$$Q'(3) = 0.2(3) + 1.5 = 2.1 \text{ tons/week}$$

At the end of the third week, the quantity of waste is increasing at the rate of 2.1 tons a week. []

Example 9

A viral infection spreads through a population so that the total number of people N exposed to the virus in t weeks is given by

$$N(t) = 150t + 25t^2$$

Determine the *rate* at which exposure to the virus is spreading at the end of two weeks; at the end of four weeks.

The rate at which exposure to the virus is changing is given by

$$N'(t) = 150 + 50t$$

To determine the rate at the end of two weeks, we evaluate

$$N'(2) = 150 + 50(2) = 250 \text{ people/week}$$

To determine the rate at the end of four weeks, we evaluate

$$N'(4) = 150 + 50(4) = 350 \text{ people/week}$$ []

EXERCISES

Find each of the following.

78. $f'(x)$ for $f(x) = x^4 - 4x^3 + 5x^2 + 6x + 7$

79. $f'(x)$ for $f(x) = (1/3)x^3 - x$

80. s' for $s = 15t^2 - 6t + 12$
81. y' for $y = (3 + 4x - 5x^2)/6$
82. $D_x(x^5 - 5x^3 + 3x)$
83. $D_x[(1/x^2) - (1/x^3)]$
84. $f'(1)$ for $f(x) = [(x^4/4) + (3/x^3)]$
85. $f'(2)$ for $f(x) = [(1/x^2) + \sqrt{x}\,]$
86. Slope of the tangent to $f(x) = 5x^2 + 3x + 2$ at (1,10)

A vehicle, moving along a straight road, travels a distance D *(in miles) in* t *hours, as given by*

$$D(t) = 0.3t^2 + 0.5t$$

87. What is the total distance traveled at $t = 3$? at $t = 5$?
88. What is the instantaneous velocity of the vehicle at $t = 3$? at $t = 5$?

The population of Cementville, a planned community, is given by

$$P(t) = 4{,}500 + 50t^2$$

where t *is years since the town was incorporated.*

89. Find the total population at $t = 5$.
90. Find the instantaneous rate of change in population (the *growth rate*) at $t = 5$.
91. Find the growth rate at $t = 10$.

A firm estimates that the number N *of units of a product sold after spending* x *dollars on advertising is given by*

$$N(x) = -0.1x^2 + 200x + 60$$

92. How many units will be sold when $800 is spent on advertising?
93. What is the instantaneous rate of change of the number of units sold with respect to the amount spent on advertising?
94. What is the rate of change in sales at $x = 900$? at $x = 1{,}000$? at $x = 1{,}100$? What is the significance of these measures?

An object is hurled upward with an initial velocity of 100 feet per second. After t *seconds, its height above the ground is given by*

$$S(t) = 100t - 16t^2$$

95. Derive a function describing the instantaneous velocity of the object at time t.
96. What is the instantaneous velocity of the object at $t = 1$? at $t = 2$?
97. What is the significance of $t = 25/8$?

Derivative of a Constant to a Functional Power

Rule 5 *If* $f(x) = a^{g(x)}$, *where* $a > 0$ *and* $a \neq 1$, *and* $g(x)$ *is differentiable, then* $f'(x) = g'(x)a^{g(x)}\ln a$.

Because $ln\ e = 1$, if the constant which is carried to the functional power is *e*, the rule for the derivative is simplified.

Rule 5a *If* $f(x) = e^{g(x)}$ *where* $g(x)$ *is differentiable, then* $f'(x) = g'(x)e^{g(x)}$.

Example 10

a. If $f(x) = 5^{x^2}$, then $f'(x) = (2x)(5^{x^2})ln\ 5$

b. If $f(x) = 3^{2x+1}$, then $f'(x) = (2)(3^{2x+1})ln\ 3$

c. If $f(x) = 8^{x^3+x^2+x+1}$, then $f'(x) = (3x^2 + 2x + 1)(8^{x^3+x^2+x+1})ln\ 8$

d. $D_x e^x = (1)e^x = e^x$ (Note that the function and its derivative are exactly the same!)

e. $D_x e^{-x^2} = -2xe^{-x^2}$

f. $D_x e^{(x^2+2x)} = (2x + 2)e^{(x^2+2x)}$ []

EXERCISES

Find the derivative of each of the following functions.

98. $f(x) = e^{2x}$

99. $f(x) = e^{3x+5}$

100. $f(x) = e^{x^2+x+10}$

101. $f(x) = 10^x$

102. $f(x) = 5^{-3x}$

103. $f(x) = 7^{x^2}$

104. $f(x) = 3^{5x+9}$

105. $f(x) = 3e^{-x^3}$

106. $f(x) = e^2 + e^{2x} + e^{x^2}$

107. $f(x) = 2e^{x^3+2x-1} + 5^{4x^4+3}$

108. $f(x) = 20^{x^2+x}$

109. $f(x) = e^{5x} + x^5$

The value of a stock is given by

$$V(t) = 85(1 - e^{-t}) + 25$$

where V *is the value of the stock, in dollars, at time* t, *in months.*

110. What is the value of the stock at time $t = 3$? at time $t = 5$? at time $t = 8$?

111. What is the *rate of change* of the value of the stock at time $t = 3$? at time $t = 5$? at time $t = 8$?

Derivative of the Logarithm of a Function

Rule 6 *If* $f(x) = \log_a[g(x)]$ *where* $g(x)$ *is differentiable, then*

$$f'(x) = \frac{g'(x)}{g(x)} \log_a e$$

or, by the change-of-base rule which states $\log_a b = 1/\log_b a$,

$$f'(x) = \frac{g'(x)}{g(x) \ln a}$$

The derivative rule for the natural logarithm of a function may be written in a simplified form.

Rule 6a *If* $f(x) = \ln[g(x)]$ *where* $g(x)$ *is differentiable, then*

$$f'(x) = \frac{g'(x)}{g(x) \ln e} = \frac{g'(x)}{g(x)}$$

Example 11

a. If $f(x) = ln(5x^2 + 3x + 7)$, then $f'(x) = (10x + 3)/(5x^2 + 3x + 7)$

b. If $f(x) = log(3x + 2)$, then $f'(x) = \dfrac{3}{(3x + 2) ln\ 10}$

c. If $f(x) = log_8(4x^3 + 9x^2)$, then $f'(x) = \dfrac{12x^2 + 18x}{(4x^3 + 9x^2)(ln\ 8)}$

$$= \frac{6(2x + 3)}{x(4x + 9)(ln\ 8)}$$

d. If $f(x) = ln(x - 3x^2)$, then $f'(x) = (1 - 6x)/(x - 3x^2)$ []

Example 12

Differentiate $y = ln(3x + 4)^2$.

First we should simplify the right-hand side of the equation using the properties of logarithms, as

$$y = ln(3x + 4)^2 = 2\ ln(3x + 4)$$

Then,

$$y' = 2[3/(3x + 4)] = 6/(3x + 4)$$ []

EXERCISES

Find the derivative of each of the following functions.

112. $f(x) = log\ x$
113. $f(x) = log_2\ x^2$
114. $f(x) = ln(x + 2)$
115. $f(x) = ln(x^3 + 1)$
116. $f(x) = log\ (2x^3 + 5x + 1)$
117. $f(x) = log(x^{1/2} + x)$
118. $f(x) = ln\ x$
119. $f(x) = ln(3x + 4)^3$

120. $f(x) = ln(x^2 + 3x)$

121. $f(x) = log_8 x^4$

122. $f(x) = 10 \; log \; x$

123. $f(x) = 10^x - log \; x$

A market-research team estimates that the number N *of units of a certain product sold, as a function of the amount* a *(in thousands of dollars) spent on advertising the product, is modeled by*

$$N(a) = 1{,}200 + 150 \; ln \; a \qquad a \geq 1$$

124. Estimate the total number of units sold if \$5,000 is spent on advertising; if \$7,500 is spent on advertising.

125. What is the *rate of change* in the number of units sold when advertising is at the \$5,000 level? when advertising is at the \$7,500 level?

11.5 MARGINAL ANALYSIS

One important use of the derivative in business and economics involves MARGINAL ANALYSIS. Economic theory abounds with the concept of marginalism. In addition to the old standbys MARGINAL REVENUE, MARGINAL COST, and MARGINAL PROFIT, we read of MARGINAL PRODUCT, MARGINAL PRODUCTIVITY, MARGINAL EFFICIENCY OF INVESTMENT, MARGINAL UTILITY, MARGINAL PROPENSITY TO CONSUME, MARGINAL PROPENSITY TO SAVE, and so on.

MARGINAL COST refers to the extra, or incremental, cost of producing an additional unit of output. Although products are most often produced in discrete units, economists, as well as other analysts, customarily use continuous functions to describe the relationship between level of output and cost. When this is the case, the concept of MARGINAL CHANGE becomes analogous to the concept of INSTANTANEOUS RATE OF CHANGE—that is, to the DERIVATIVE. Thus, if

$C(x)$ = Total cost of producing x units of output during some specified period of time

then

$C'(x)$ = Marginal cost at level of production x; that is, rate of change in total cost with respect to an incremental change in the level of output

In the same manner, if

$R(x)$ = Total revenue realized from the sale of x units of product

then

$R'(x)$ = Marginal revenue at level of sales x—that is, the rate of change in total revenue with respect to an incremental change in the number of units sold

Or, if

$P(x) =$ Total profit from the sale of x units of product

then

$P'(x) =$ Marginal profit at level of sales x; that is, the rate of change in total profit with respect to an incremental change in the number of units sold

Example 13 **Marginal Cost** A manufacturer of automobile hubcaps finds that its total cost C of producing x units of product is given by

$$C(x) = 0.001x^3 - 0.21x^2 + 40x + 1{,}000$$

Determine the marginal cost when the level of production is $x = 50$.

The marginal cost function is the derivative of the total cost function. Thus,

$$C'(x) = 0.003x^2 - 0.42x + 40$$

The marginal cost when 50 units are being produced is

$$C'(50) = 0.003(50)^2 - 0.42(50) + 40 = 26.50$$

Roughly speaking, we can say that the 51st unit costs \$26.50 to produce. This statement is not quite accurate, since the level of production will change in unit increments while the derivative is based on the idea that the change in x is approaching zero. Let us determine the change in total cost when the level of production is increased from 50 to 51 units. We find

$$C(50) = 0.001(50)^3 - 0.21(50)^2 + 40(50) + 1{,}000 = \$2{,}600$$

and

$$C(51) = 0.001(51)^3 = 0.21(51)^2 + 40(51) + 1{,}000 = \$2{,}626.41$$

so that

$$C(51) - C(50) = 2{,}626.41 - 2{,}600.00 = \$26.41$$

The cost of the 51st unit is \$26.41 which is *approximately* the value of $C'(50)$.

In general, in the case of a product that has discrete units, marginal cost is only the approximate cost of the $(x + 1)$st item; that is,

$$C'(x) \simeq C(x + 1) - C(x)$$

[]

Example 14 **Marginal Revenue** The revenue R (in dollars) realized by a firm when x units of product are produced and sold is given by

$$R(x) = 15x - 0.01x^2$$

Determine the marginal revenue function. Evaluate marginal revenue when $x = 400$.

Marginal revenue is the increase in total revenue generated by the production and sale of an incremental unit of product. If R is the total revenue function for a firm, its derivative R' is the marginal revenue function. Here

$$R'(x) = 15 - 0.02x$$

When $x = 400$, marginal revenue is

$$R'(400) = 15 - (0.02)(400) = 7$$

Roughly speaking, the revenue generated by the 401st unit of product produced and sold is \$7. []

Example 15 **Marginal Profit** The profit function for a firm is given by

$$P(x) = -0.1x^2 + 80x - 5{,}000$$

where x is the number of units of product produced and sold and $P(x)$ is profit (in dollars). Determine the marginal profit function. What is marginal profit when $x = 300$?

Marginal profit is the increase in total profit associated with an incremental unit of sales. If P is the profit function, its derivative P' is the marginal profit function. Here

$$P'(x) = -0.2x + 80$$

Marginal profit when the level of sales is 300 units is given by

$$P'(300) = -0.2(300) + 80 = 20$$

Thus, the 301st unit of sales generates a profit of \$20. []

EXERCISES

The total cost function for a firm is given by

$$C(x) = 0.001x^3 - 0.25x^2 + 50x + 1{,}200$$

where C(x) *is cost, in dollars, and* x *is the number of units produced.*

126. Determine the marginal cost function.
127. What is marginal cost when $x = 150$? when $x = 160$? when $x = 170$?
128. What is the (approximate) cost of the 101st unit produced?

Revenue R *realized from the sale of* x *units of a product is*

$$R(x) = 600x - x^2$$

129. Determine the marginal revenue function.
130. What is marginal revenue when $x = 250$? when $x = 300$? when $x = 350$? What conclusions can you draw from these numbers?

Profit (in thousands of dollars) from the sale of x *thousand units of product is given by*

$$P(x) = -0.15x^3 - 5x^2 + 6545x - 50{,}000$$

131. Determine the marginal profit function.
132. What is marginal profit when $x = 100$? when $x = 125$?

A manufacturer's cost and revenue functions are

$$C(x) = 3{,}500 + 200x + 0.2x^2 \qquad 0 \le x \le 100$$

and

$$R(x) = 284x - 0.5x^2 \qquad 0 \le x \le 100$$

where C(x) *and* R(x) *are in dollars and* x *is the number of units of product produced and sold.*

133. Compare the marginal cost and marginal revenue when $x = 50$.
134. Compare the marginal cost and marginal revenue when $x = 60$.
135. Compare the marginal cost and marginal revenue when $x = 70$.
136. The cost to produce x units of soybeans is given by

$$C(x) = 4{,}500 + 18x + 8\sqrt{x}$$

Find marginal cost when $x = 25$; when $x = 36$; when $x = 49$.

137. Total profit realized by a publisher from the sale of a certain trade magazine is given by

$$P(t) = 7{,}500 + 6{,}000e^{-0.5t}$$

where t is time, in years, that the publication has been on the market. Determine the marginal profit function.

138. Demand for a product (in thousands of units) is given by

$$D(t) = 2000(8 - 4e^{-0.6t})$$

where t is time, in months, that the product has been on the market. Determine the instantaneous rate of change of this function at $t = 3$. Interpret this measure.

139. Marginal propensity to consume is defined as the instantaneous rate of change in total consumption with respect to a change in income. If the aggregate consumption function for an economy is given by

$$c(x) = 400 + 20\sqrt{x}$$

where $c(x)$ represents total consumption and x represents national income. Determine the marginal-propensity-to-consume function. Evaluate this function at $x = 16$ and interpret the result.

140. Marginal physical productivity is defined as the instantaneous rate

of change in total production with respect to a change in the number of machines or workers. Given the total productivity function

$$p(x) = 12{,}000x - 3x^2$$

where x is the number of workers, determine marginal physical productivity at $x = 1{,}200$; at $x = 1{,}500$; at $x = 1{,}800$.

The demand function for a product is $p = D(x) = 25 - x$, *where* p *is price (in dollars) per unit and* x *is the number of units sold.*

141. Determine the marginal revenue function.
142. What is marginal revenue when 10 units are sold? when 12 units are sold?

11.6 ADDITIONAL DERIVATIVE RULES

Derivative of the Product of Functions

Rule 7 *The derivative of a product of two functions is the first function times the derivative of the second plus the second function times the derivative of the first. That is, if*

$$h(x) = f(x) \cdot g(x)$$

then

$$h'(x) = f(x)g'(x) + g(x)f'(x)$$

(*assuming* $g'(x)$ *and* $f'(x)$ *exist*).

Example 16 If

$$h(x) = (2x^2 - 5)(x + 3)$$

then

$$\begin{aligned} h'(x) &= (2x^2 - 5)(1) + (x + 3)(4x) \\ &= 2x^2 - 5 + 4x^2 + 12x \\ &= 6x^2 + 12x - 5 \end{aligned}$$

[]

Example 17 Given $f(x) = (x^2 - 3)^2$, find $f'(x)$.

We may rewrite the function as $f(x) = (x^2 - 3)^2 = (x^2 - 3)(x^2 - 3)$. Then,

$$\begin{aligned} f'(x) &= (x^2 - 3)(2x) + (x^2 - 3)(2x) \\ &= 2x^3 - 6x + 2x^3 - 6x \\ &= 4x^3 - 12x \\ &= 4x(x^2 - 3) \end{aligned}$$

[]

This product rule can be extended to include the derivative of the product of three or more functions. For example, if $k(x) = f(x) \cdot g(x) \cdot h(x)$, then $k'(x) = f'(x)g(x)h(x) + f(x)g'(x)h(x) + f(x)g(x)h'(x)$.

Example 18

$$D_x(x-1)(x+1)(2x+1) = (1)(x+1)(2x+1) + (x-1)(1)(2x+1) + (x-1)(x+1)(2)$$
$$= 2x^2 + x + 2x + 1 + 2x^2 + x - 2x - 1 + 2x^2 + 2x - 2x - 2$$
$$= 6x^2 + 2x - 2$$
$$= 2(3x^2 + x - 1)$$

[]

It is frequently necessary to use other derivative rules in conjunction with the product rule, as the next example will illustrate.

Example 19

Find $D_x(2x+1)(e^{x^2})$.

We obtain, using the product rule,

$$D_x(2x+1)(e^{x^2}) = (2x+1)(D_x e^{x^2}) + (e^{x^2})[D_x(2x+1)]$$
$$= (2x+1)(2x)(e^{x^2}) + (e^{x^2})(2)$$
$$= 2e^{x^2}[(2x+1)(x) + 1]$$
$$= 2e^{x^2}[2x^2 + x + 1]$$

[]

EXERCISES

Find the derivative of each of the following functions.

143. $y = (x+1)(x+4)$
144. $y = (x+2)(x-2)$
145. $y = x(3x^2 + 6x - 7)$
146. $y = 2x(x^3 - 9x^2 - 11)$
147. $y = (x-9)(3x+2)$
148. $y = (6x^3 + 0.25)(1 - 0.5x^2)$
149. $y = (2x-1)(x^2+3)$
150. $y = (3x+2)(x^3 - 2x^2)$
151. $y = (x^2 - 2x)(5x+2)$
152. $y = (3 - x^3)^2$
153. $y = (x^2 - 3x + 1)(2x^3 + 3x^2 + x)$
154. $y = (2x^4 - 3x^3 + x^2)(x^2 + x - 5)$
155. $y = x[ln(x^2+4)]$
156. $y = (x^2 - x)(e^x + 1)$
157. $y = (2x-7)(3x+1)(x-4)$
158. $y = (x+1)(x+2)(x+3)$
159. $y = (ln\ x)(e^x + x)$
160. $y = (6x+5)(e^{6x+5})$
161. The population of bacteria (in millions) present in a culture at time t is given by

$$N(t) = (t-8)(5t) + 40$$

At what rate is the population changing at time $t = 3$? at time $t = 5$? at time $t = 8$?

162. Demand for a product is modeled by the function

$$p = D(x) = 6(-e^{0.1x})$$

where p is price per unit, in dollars, and x is the number of units sold.

Given that

$$\text{Revenue} = (\text{Number of units sold}) \cdot (\text{Price per unit})$$

determine the marginal revenue function.

163. Average income per capita I (in dollars) in a certain economy at time t (in years) is given by

$$I(t) = 6{,}500 + 490t + 12t^2$$

Population P (in millions) in the economy at time t is given by

$$P(t) = 14 + 0.25t + 0.01t^2$$

Given that gross national product is defined as

$$\text{Gross national product} = (\text{Population size}) \cdot (\text{Per-capital income})$$

Determine the rate of change in gross national product at time $t = 2$; at time $t = 5$.

Derivative of the Quotient of Functions

Rule 8 *The derivative of a quotient of two functions is the denominator times the derivative of the numerator minus the numerator times the derivative of the denominator, all divided by the square of the denominator. That is, if*

$$h(x) = f(x)/g(x)$$

then

$$h'(x) = \frac{g(x)f'(x) - f(x)g'(x)}{[g(x)]^2}$$

(assuming $f'(x)$ *and* $g'(x)$ *exist).*

Example 20

$$D_x[(1-x)/x^2] = \frac{x^2 D_x(1-x) - (1-x)D_x(x^2)}{(x^2)^2}$$

$$= \frac{x^2(-1) - (1-x)(2x)}{(x^2)^2}$$

$$= \frac{-x^2 - 2x + 2x^2}{x^4}$$

$$= \frac{x^2 - 2x}{x^4}$$

$$= \frac{x-2}{x^3}$$

[]

Example 21

$$D_x \frac{(3x^2+4x+5)}{(x-1)} = \frac{(x-1)D_x(3x^2+4x+5) - (3x^2+4x+5)D_x(x-1)}{(x-1)^2}$$

$$= \frac{(x-1)(6x+4) - (3x^2+4x+5)(1)}{(x-1)^2}$$

$$= \frac{6x^2 - 6x + 4x - 4 - 3x^2 - 4x - 5}{(x-1)^2}$$

$$= \frac{3x^2 - 6x - 9}{(x-1)^2}$$

$$= \frac{3(x^2 - 2x - 3)}{(x-1)^2} \qquad []$$

EXERCISES

Find the derivative of each of the following functions.

164. $y = (3x + 1)/(x - 2)$
165. $y = (x^2 - 9)/(x + 3)$
166. $y = (x^2 + 6x - 5)/x$
167. $y = (x + 3)/(x - 4)$
168. $y = (4 - x)/(x^3 - 9)$
169. $y = (x^2 + x + 2)/(2x)$
170. $y = (x + 3)/(x - 3)$
171. $y = (3 - x^3)/(3 + x^3)$
172. $y = (x^4 + 3x^2 + 5)/(x - 1)$
173. $y = x/(x^3 - 3)$
174. $y = (1 - x)/\sqrt{x}$
175. $y = (2x + 3)/(3x + 2)$
176. $y = (e^x + 1)/(e^{x^2+1} + 5)$
177. $y = [ln(5x + 2)]/(x^2 + 3)$
178. $y = ln[(5 + 3x)/(5 - 3x)]$
179. $y = (ln\, x)/x$

The total cost C *(in dollars) of manufacturing* x *units of a product is given by*

$$C(x) = 1{,}200 + 14x^2 + x^3$$

180. Determine the AVERAGE COST function, where average cost is total cost divided by the number of units produced.
181. Determine the MARGINAL AVERAGE COST function.

Derivative of a Composite Function

A COMPOSITE FUNCTION is one in which one function can be considered as the variable in another function. Thus, if y is a function of u and u is a function of x, then y is also a function of x, so that y is a function of a function, or a composite function. The composite function may be symbolized $f(g(x))$.

Example 22 Given $y = f(u) = u^3$ and $u = g(x) = x + 1$, then y may be expressed as a function of x as

$$y = f(u) = f(g(x)) = (x + 1)^3 \qquad []$$

Example 23 Given $y = f(u) = \sqrt{u}$ and $u = g(x) = 2x^2 + 5$, then

$$y = f(g(x)) = \sqrt{2x^2 + 5} \qquad []$$

Example 24 Given $y = f(u) = u^2 + 3u - 8$ and $u = g(x) = x + 3$, then

$$y = f(g(x)) = (x + 3)^2 + 3(x + 3) - 8 \qquad []$$

Rule 9 The chain rule *If* y = f(u) *is a differentiable function of* u *and* u = g(x) *is a differentiable function of* x, *then the derivative of* y *with respect to* x *is the derivative of* y *with respect to* u *times the derivative of* u *with respect to* x; *that is,*

$$\frac{dy}{dx} = \frac{dy}{du} \cdot \frac{du}{dx}$$

Example 25 Given $y = f(u) = 2u^2 - 3u - 4$ and $u = g(x) = x^2 + 1$, find the derivative of y with respect to x.

We compute

$$\frac{dy}{du} = \frac{d}{du}(2u^2 - 3u - 4) = 4u - 3$$

and

$$\frac{du}{dx} = \frac{d}{dx}(x^2 + 1) = 2x$$

Finally,

$$\frac{dy}{dx} = \frac{dy}{du} \cdot \frac{du}{dx} = (4u - 3)(2x) = 8xu - 6x$$

We write the derivative function exclusively in terms of x by replacing u with $u = x^2 + 1$, as

$$\frac{dy}{dx} = 8x(x^2 + 1) - 6x = 8x^3 + 8x - 6x = 8x^3 + 2x = 2x(4x^2 + 1)$$

Note that we might have approached the derivative by expressing y directly as a function of x. This we do by replacing the variable u in $f(u)$ with the function $u = g(x)$. We write

$$\begin{aligned} f(g(x)) &= 2(x^2 + 1)^2 - 3(x^2 + 1) - 4 = 2(x^4 + 2x^2 + 1) - 3x^2 - 3 - 4 \\ &= 2x^4 + 4x^2 + 2 - 3x^2 - 7 \\ f(x) &= 2x^4 + x^2 - 5 \end{aligned}$$

The derivative is

$$f'(x) = 8x^3 + 2x = 2x(4x^2 + 1)$$ []

Example 26 Given $y = f(x) = ln[ln\ x]$, find $f'(x)$.

Using the chain rule, we identify $y = f(u) = ln\ u$ and $u = g(x) = ln\ x$. Then

$$\frac{dy}{du} = \frac{1}{u} \qquad \text{and} \qquad \frac{du}{dx} = \frac{1}{x}$$

Then

$$\frac{dy}{dx} = \frac{1}{u} \cdot \frac{1}{x}$$

and substituting for u, we have

$$\frac{dy}{dx} = \frac{1}{ln\, x} \cdot \frac{1}{x} = \frac{1}{x\, ln\, x}$$ []

The chain rule extends to longer chains. As an example, the chain of functions

$$y = f(u) \qquad u = g(x) \qquad x = h(t)$$

defines the composite function

$$y = f\{g[h(t)]\}$$

and the chain rule for the derivative of y with respect to t is

$$\frac{dy}{dt} = \frac{dy}{du} \cdot \frac{du}{dx} \cdot \frac{dx}{dt}$$

Example 27 Let $y = 2u^2 + 1$; $u = 3x^2$; *and* $x = 2t + t^3$.
Then

$$\frac{dy}{du} = 4u; \quad \frac{du}{dx} = 6x; \quad \frac{dx}{dt} = 2 + 3t^2$$

and

$$\begin{aligned}\frac{dy}{dt} &= (4u)(6x)(2 + 3t^2) \\ &= [4(3x^2)][6(2t + t^3)][2 + 3t^2] \\ &= [4(3)(2t + t^3)^2][6(2t + t^3)][2 + 3t^2] \\ &= 72(2t + t^3)^3(2 + 3t^2) \\ &= 72t^3(2 + t^2)^3(2 + 3t^2)\end{aligned}$$ []

A special case occurs when one of the constituent parts of the composite function is a power function.

> **Rule 10 The extended power rule**
> *If*
>
> $$y = [f(x)]^n$$
>
> *then*
>
> $$y' = n[f(x)]^{n-1} \cdot f'(x)$$

Example 28

a. $D_x(3x^2 + 9)^3 = 3(3x^2 + 9)^2(6x^2) = 18x^2(3x^2 + 9)^2$

b. $D_x(-8\sqrt{7x^2 + 1}) = (-8)D_x(7x^2 + 1)^{1/2} = (-8)(1/2)(7x^2 + 1)^{-1/2}(14x)$
$= \dfrac{-56x}{\sqrt{7x^2 + 1}}$

c. $D_x \dfrac{1}{(4x + 1)^3} = D_x(4x + 1)^{-3} = (-3)(4x + 1)^{-4}(4) = \dfrac{-12}{(4x + 1)^4}$ []

EXERCISES

Differentiate each of the following functions.

182. $y = (x + 3)^4$
183. $y = 3(x^2 + 1)^3 - 5(x^2 + 1)^2 + 4(x^2 + 1) - 9$
184. $y = (x + 1/x)^5$
185. $y = 1/(x^2 + 3x)^2$
186. $y = \sqrt{x^3 + 1}$
187. $y = 4\sqrt{1 - x^3}$
188. $y = (5 - x^2)^6$
189. $y = \ln \ln (x + 1)$
190. $y = 5/(3x^2 + 1)^3$
191. $y = \ln[x/(x + 2)]$
192. $y = (e^x + 1/x^2)^{3/2}$
193. $y = e^{e^x}$
194. $y = \ln \ln x^2$
195. $y = [1/(x + 1)]^5$
196. $y = \dfrac{(x^2 + 1)^2}{1 + (x^2 + 1)^2}$
197. $y = \dfrac{1}{[1 + (1/x)]^2}$
198. $y = \sqrt{(x + 1)/(x - 1)}$
199. $y = (2 \ln x + 5)^4$

Derivative of a Variable to a Variable Power

We have already considered rules for differentiating a variable to a constant power and a constant to a variable power. The following rule allows us to differentiate a variable to a variable power.

Rule 11 *If*

$$f(x) = [g(x)]^{[h(x)]}$$

then

$$f'(x) = g(x)^{h(x)} \left[h'(x) \cdot \ln g(x) + h(x) \cdot \frac{g'(x)}{g(x)} \right]$$

Example 29

(*a*) $D_x x^x = x^x[(1) \ln x + x(1/x)]$

$= x^x[\ln x + 1]$

(*b*) If $f(x) = (2x^3 + 4)^{[x^2+1)}$, then

$$f'(x) = (2x^3 + 4)^{(x^2+1)} \left[(2x) \ln(2x^3 + 4) + (x^2 + 1) \frac{6x^2}{(2x^3 + 4} \right]$$ []

EXERCISES

Differentiate each of the following functions:

200. $y = x^{5x}$
201. $y = x^{x^2}$
202. $y = x^{(x^4+4)}$
203. $y = x^{(3x^2+1}$
204. $y = x^{e^x}$
205. $y = (2x + 1)^x$
206. $y = x^{\sqrt{x}}$
207. $y = x^{(2x+1)}$
208. $y = (x + 5)^{(x+5)}$
209. $y = (2x + 1)^{(x^2+2)}$
210. $y = (\ln x)^{(5x)}$
211. $y = x^{(\ln x)}$

11.7 HIGHER-ORDER DERIVATIVES

If a function f has a derivative for each value of x in some specified interval, then the derivative function f' is defined for that interval. If, in turn, the derivative function itself has a derivative for points in that interval, this new derivative function is called the SECOND DERIVATIVE of the original function, often denoted $f''(x)$. Similarly, if f'' has a derivative, we speak of the THIRD DERIVATIVE of the original function, and note this as $f'''(x)$, and so on.

No new rules of differentiation are needed for HIGHER-ORDER DERIVATIVES. One simply differentiates, and then differentiates again, and again, using the same set of derivative rules.

The only confusing aspect of these higher-order derivatives is the symbolic notation used. Some of the commonly encountered systems of notation are:

Original function	$f(x)$	$f(x)$	$f(x)$	y
First derivative	$f'(x)$	$D_x f(x)$	dy/dx	y'
Second derivative	$f''(x)$	$D_x^2 f(x)$	d_y^2/dx^2	y''
Third derivative	$f'''(x)$	$D_x^3 f(x)$	d_y^3/dx^3	y'''
Fourth derivative	$f^{(4)}(x)$	$D_x^4 f(x)$	d_y^4/dx^4	$y^{(4)}$
⋮	⋮	⋮	⋮	⋮
nth derivative	$f^{(n)}(x)$	$D_x^n f(x)$	d_y^n/dx^n	$y^{(n)}$

The symbolism $f^{(n)}(x)$ and $y^{(n)}$ is adopted to avoid confusion with powers of the functions as represented by $f^n(x)$ and y^n.

Example 30 The first, second, and third derivatives of the function

$$f(x) = -9x^3$$

are as follows:

$$f'(x) = -27x^2$$
$$f''(x) = -54x$$
$$f'''(x) = -54$$

[]

Example 31 Find the first, second, and third derivatives of the function

$$y = 4x^3 - (9/x)$$

These derivatives are

$$y' = 12x^2 + (9/x^2)$$
$$y'' = 24x - (18/x^3)$$
$$y''' = 24 + (54/x^4)$$

[]

Example 32 Find the first, second, third, and fourth derivatives of the function

$$g(x) = (2x + 5)^4$$

These derivatives are

$$g'(x) = 8(2x+5)^3$$
$$g''(x) = 48(2x+5)^2$$
$$g'''(x) = 384x + 960$$
$$g^{(4)}(x) = 384$$

[]

EXERCISES

For each of the following functions, find the first, second, and third derivatives.

213. $y = 5x^3$
214. $y = 3x^3 + 6x + 4$
215. $y = 0.2x^3 - 0.5x^2 + 0.25$
216. $y = (x-1)(x+1)$
217. $y = (x^2 + x + 3)(2x - 1)$
218. $y = x^9 + x^6 - x^3 - 15$
219. $y = x + (1/x)$
220. $y = (3/x) + x^2$
221. $y = 2 + 4x + (3/x^2)$
222. $y = x^2 + 3x + (1/x^2)$
223. $y = 5x^{2/3}$
224. $y = \sqrt{4x}$
225. $y = \sqrt{3x} + 3x + 3$
226. $y = x^{3/5} + (3x/5) - (3/5x)$
227. $y = (x+1)^4$
228. $y = (2x+1)^3$
229. $y = ln\, x$
230. $y = ln(x^2+1)$
231. $y = e^x$
232. $y = e^{(x^2+1)}$
233. $y = x/(x+1)$
234. $y = x^3/(x^2+1)$
235. $y = e^{x^2} + x^2$
236. $y = e^x\, ln\, x$
237. $y = 5x^5(x-1)^4$
238. $y = log(x^3+7)$

C·H·A·P·T·E·R 12

Applications of the Derivative

Derivatives are of particular importance because of the information they provide about the behavior of a function. They give the rate of change in the value of the function relative to infinitesimal changes in the value of the independent variable. They indicate where the function is increasing in value and where it is decreasing. They aid in determining where the function reaches its maximum and/or minimum values.

Among the many, many problem situations where we find a knowledge of derivatives useful are these:

1. Given a functional relationship between profit and level of sales, determining the levels of sales over which profit is increasing, or determining the levels of sales over which profit is decreasing, or determining the level of sales for which profit is at its maximum possible level.
2. Given a functional relationship between level of sales and level of advertising expenditures, determining the levels of advertising expenditures over which sales are increasing at an increasing rate, or determining the levels of advertising expenditures over which sales are increasing at a decreasing rate.
3. Given a functional relationship between per-unit selling price for a product and the total revenue realized by the manufacturer of the product, determining the selling prices over which total revenue is increasing, or determining the selling prices over which total revenue is decreasing, or determining the selling price at which total revenue reaches its maximum possible level.
4. Given a functional relationship between time and the number of persons affected by an epidemic, determining the time period over which the number of persons affected is increasing and determining the point in time at which the number of persons affected begins to decrease.
5. Given a limited quantity of fencing, determining the maximum area that can be enclosed.
6. Determining the dimensions of a cylindrical container that will achieve a required volume at minimum cost.
7. Determining the number of miles an automobile should be driven before replacement in order to minimize the sum of capital cost and average operating cost.

And so on, and on.

12.1 INCREASING AND DECREASING FUNCTIONS

A function is STRICTLY INCREASING over an interval (a,b) if $f(x_1) < f(x_2)$ whenever $a < x_1 < x_2 < b$. Or a function is STRICTLY DECREASING over an interval (a,b) if $f(x_1) > f(x_2)$ whenever $a < x_1 < x_2 < b$. We can determine the intervals on which f is strictly increasing or strictly decreasing by examining the algebraic sign of the first derivative f' over the specified intervals.

The derivative $f'(a)$ gives the slope of the tangent to the curve at point $(a,f(a))$. If the tangent line at point a rises to the right, it has a positive slope. The first derivative $f'(a)$ will have a positive value, and the curve itself will be rising at this point. (See Figure 12.1A.) Contrarily, if a curve is falling at a point b, the tangent line at that point will have a negative slope, and the first derivative $f'(b)$ will have a negative value. (See Figure 12.1B.) Furthermore, the steeper the slope of the tangent,

Figure 12.1

The Algebraic Sign of the First Derivative Indicates Whether the Function f Is Increasing or Decreasing

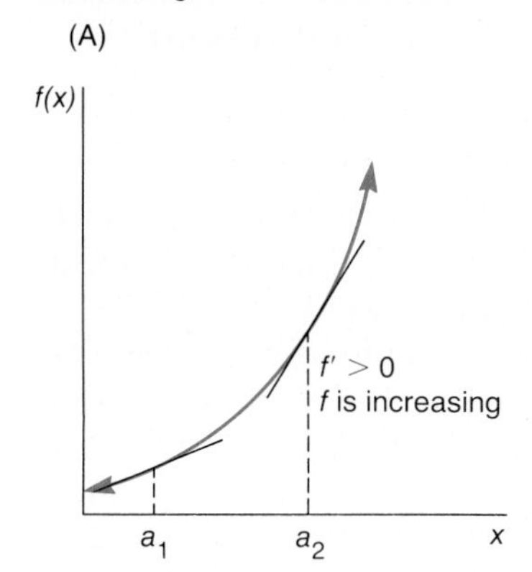

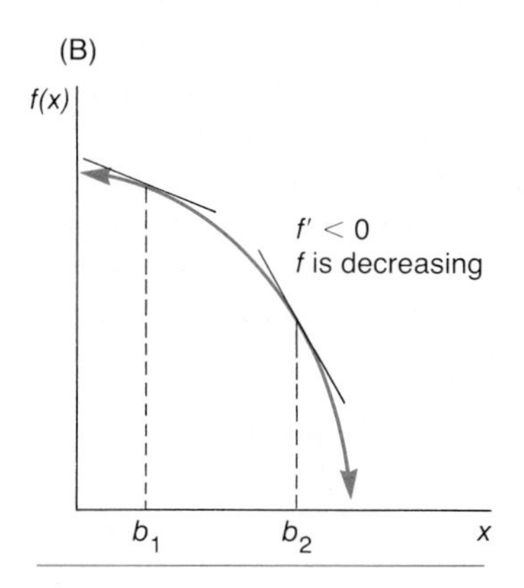

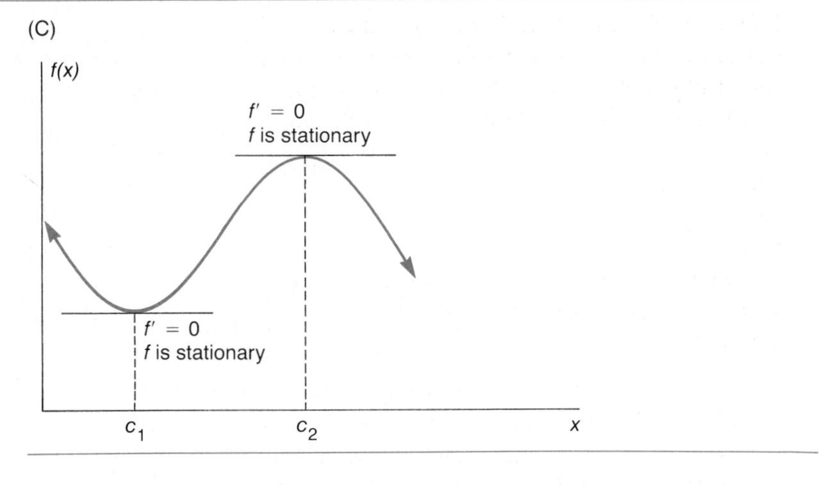

the more rapidly the curve is rising or falling. That is, the higher the positive value of f' or the lower its negative value, the more pronounced is the change in $f(x)$ with a small change in x.

A tangent line that is horizontal has a zero slope; the first derivative at that point has a zero value. At such a point, $f(x)$ is neither increasing nor decreasing. (See Figure 12.1C.) A point $(c,f(c))$ on a curve where the derivative is zero is known as a STATIONARY POINT, or a CRITICAL POINT, and $f(c)$ is called a STATIONARY VALUE of the function.

> **Given a function** f which is continuous on some open interval I and differentiable in that interval
>
> 1. If $f'(x) > 0$ for all x in the interval, then f is strictly increasing over that interval.
> 2. If $f'(x) < 0$ for all x in the interval, then f is strictly decreasing over that interval.
> 3. If $f'(x) = 0$ for all x in the interval, then f is a constant function over that interval.

In general, a continuous function cannot change algebraic sign, either from plus to minus or from minus to plus, without taking on the value zero. (However, a function does not necessarily change algebraic sign as its value crosses over zero!) Hence, it is appropriate, when searching for those intervals over which a function is increasing and those intervals over which a function is decreasing, to ascertain which values of x cause the first derivative function to equal zero. These are the CRITICAL VALUES OF x. These x values should be used to establish open intervals over which the first derivative is evaluated.

A function may also change from positive to negative, or negative to positive, at a point of discontinuity. Thus, all x values at which the function is discontinuous should also be used to establish intervals.

The following examples illustrate these principles.

Example 1 Given $f(x) = x^3 - 12x + 5$, determine the open intervals over which f is increasing and the open intervals over which f is decreasing.

The first derivative

$$f'(x) = 3x^2 - 12 = 3(x^2 - 4)$$

equals zero when $x = -2$, $x = 2$. Because the function is a polynomial, continuous over all x in its domain, there are no x such that the first derivative is undefined. Thus, the set of critical values for x is $x^* = \{-2,2\}$. We use these values of x to establish open intervals as: $(-\infty,-2)$, $(-2,2)$, and $(2,\infty)$. Then we select any interior point from within an interval and evaluate the first derivative at that point to determine whether the algebraic sign of the first derivative over that interval is positive or negative. For instance, for any $x < -2$, $3(x^2 - 4) > 0$; for any $-2 < x < 2$, $3(x^2 - 4) < 0$; and for any $x > 2$, $3(x^2 - 4) > 0$.

We summarize this information.

Interval of x values	Algebraic sign of first derivative $f'(x)$	Direction of movement of graph of function $f(x)$
$(-\infty,-2)$	+	Increasing
-2	0	Stationary
$(-2,2)$	−	Decreasing
2	0	Stationary
$(2,\infty)$	+	Increasing

The graph of $f(x) = x^3 - 12x + 5$ is shown in Figure 12.2. []

Example 2 Given $f(x) = x^3$, determine the intervals over which f is increasing and the intervals over which f is decreasing.

The first derivative is $f'(x) = 3x^2$. Clearly, $f' = 0$ only for $x = 0$ and f' is never undefined. Thus, the set of critical values of x is $x^* = \{0\}$. We establish open intervals using this value of x and evaluate the first derivative over each of these intervals. We summarize this information.

Interval of x values	Algebraic sign of first derivative $f'(x)$	Direction of movement of graph of function $f(x)$
$(-\infty,0)$	+	Increasing
0	0	Stationary
$(0,+\infty)$	+	Increasing

Note that the derivative does not change its algebraic sign as its value crosses over zero. The function $f(x) = x^3$ is a strictly increasing function over all values of x. The graph of $f(x) = x^3$ is sketched in Figure 12.3. []

Figure 12.2

f Is Increasing When *f′* Is Positive; *f* Is Decreasing When *f′* Is Negative

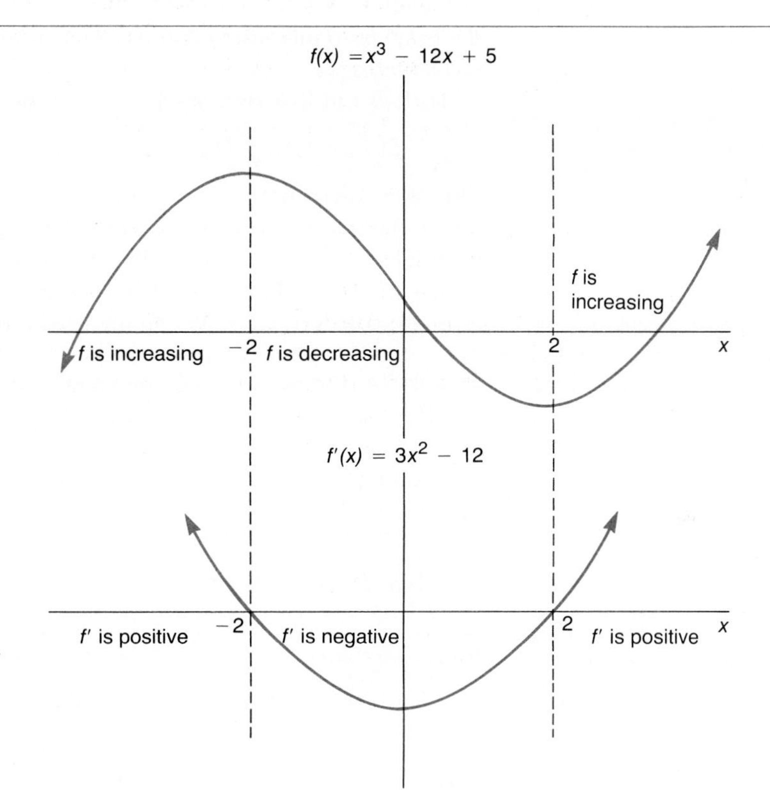

Figure 12.3

When $f'(x) > 0$, $f(x)$ Is Increasing

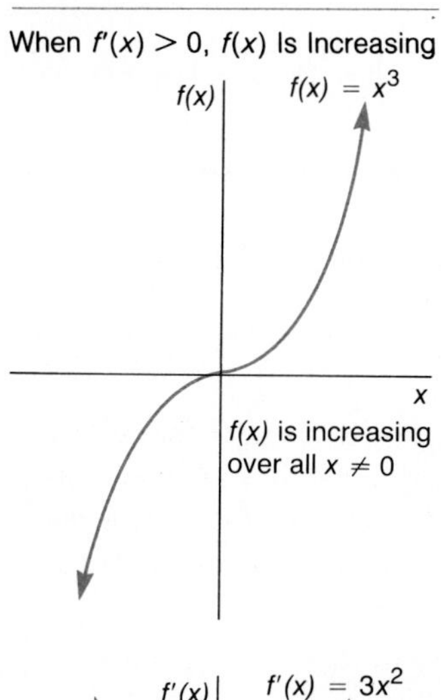

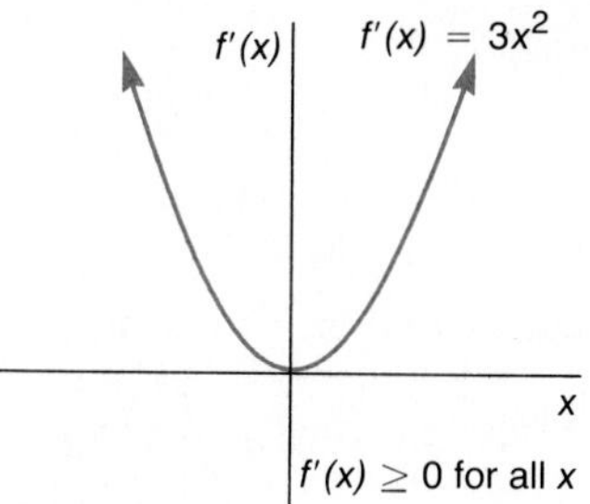

Example 3 Given $f(x) = 1/(x + 1)^2$, determine the open intervals over which the function is increasing and the open intervals over which the function is decreasing.

To find the first derivative, we use the extended power rule. Thus,

$$D_x 1/(x+1)^2 = D_x(x+1)^{-2} = (-2)(x+1)^{-3}(1) = -2/(x+1)^3$$

The derivative function does not take on the value zero since the numerator of this ratio never equals zero. The derivative function, however, is undefined at $x = -1$ (since the denominator equals zero for $x = -1$).

We use this value, $x = -1$, to set up open intervals over which we evaluate the derivative. We summarize the information obtained.

Interval of x values	Algebraic sign of first derivative $f'(x)$	Direction of movement of graph of function $f(x)$
$(-\infty, -1)$	$-/- = +$	Increasing
-1	Undefined	Discontinuous
$(-1, \infty)$	$-/+ = -$	Decreasing

We can verify that $x = -1$ is a vertical asymptote by

$$\lim_{x \to -1^-} [1/(x+1)^2] = \infty \qquad \text{and} \qquad \lim_{x \to -1^+} [1/(x+1)^2] = \infty$$

Note also that

$$\lim_{x \to \infty^-} [1/(x+1)^2] = 0 \qquad \text{and} \qquad \lim_{x \to \infty^+} [1/(x+1)^2] = 0$$

Hence the graph of the function has as a horizontal asymptote the x-axis.

The graph of $f(x) = 1/(x + 1)^2$ is shown in Figure 12.4. []

Figure 12.4

$f(x) = 1/(x + 1)^2$;
$f'(x) = -2/(x + 1)^3$

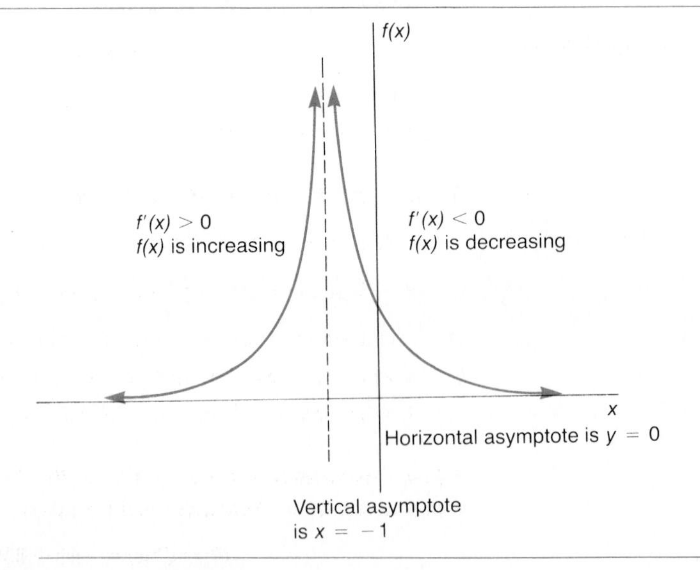

EXERCISES

For each of the following functions, identify the open intervals over which the function is increasing and the open intervals over which the function is decreasing. Identify points where the function is stationary. Use this information to sketch a graph of the function.

1. $f(x) = 6 - 5x$
2. $f(x) = 3x - 5$
3. $f(x) = 4x^2 - 3$
4. $f(x) = 13 - 2x^2$
5. $f(x) = 2x^3 - 12x$
6. $f(x) = x^2 + 6x + 9$
7. $f(x) = 1 - x^3$
8. $f(x) = x^3 - 9x^2 + 15$
9. $f(x) = x^2 + x^4$
10. $f(x) = (x^2 - 1)(x^2 - 4)$
11. $f(x) = 2 + x - 2x^2 + x^3$
12. $f(x) = x^4 + 6x^3 + 4x^2 + 10$
13. $f(x) = 15 - 8x - 5x^2 - x^3$
14. $f(x) = (x^4/4) - (x^3/3)$
15. $f(x) = 1/(x^2 + 2)$
16. $f(x) = x/(x + 2)$
17. $f(x) = e^{x+1}$
18. $f(x) = ln(x + 1)$
19. $f(x) = (x + 1)/(x + 2)$
20. $f(x) = 1/(x + 1)^2$
21. $f(x) = 5x^5 - 3x^3$
22. $f(x) = (x^2 - 4x - 12)^3$
23. $f(x) = 2x/(x + 1)^3$
24. $f(x) = 5^x$
25. $f(x) = (x + 2)(3x^2 + 4)$
26. $f(x) = (x + 1)(x^2 + x - 1)$
27. Lendl and Sons find that profit, in dollars, for their Mulberry Street shop is given by

$$f(x) = 12x - 0.25x^2 \qquad 0 \le x \le 50$$

where x represents sales in units. Is profit increasing at $x = 20$? at $x = 30$? Over what intervals of x is profit increasing? decreasing?

28. Total profit (in thousands of dollars) from the sale of x thousand units of a product is given by

$$P(x) = -x^3 + 9x^2 + 150x - 500$$

Over what level of sales is profit increasing? Over what level of sales is profit decreasing?

A manufacturer has daily cost and revenue functions of

$$C(x) = x^2 + 4x + 180 \qquad \text{and} \qquad R(x) = 104x - x^2$$

where x *is the number of units of product manufactured and sold.*

29. Over what level of production is total cost increasing? decreasing?
30. Over what level of sales is revenue increasing? decreasing?
31. Over what level of sales is profit increasing? decreasing?

A firm estimates that it can sell Q *units of its product with an advertising expenditure of* x *thousand dollars where*

$$Q = Q(x) = -x^2 + 600x + 25$$

32. Over what level of advertising expenditure is the number of units of product sold increasing?
33. Over what level of advertising expenditure is the number of units of product sold decreasing?

The demand function for a product is given by

$$p = D(x) = 100 - 2x$$

where p *is price per unit and* x *is the number of units sold.*

34. Over what interval of x values is total revenue increasing? decreasing?
35. Over what interval of p values (selling price) is total revenue increasing? decreasing?

The number of people in a certain region who are affected by an epidemic of viral pneumonia is given by

$$P(t) = -t^2 + 100t + 15$$

where t *is the number of days after the disease has been detected.*

36. Over what time period will the number of persons affected by the epidemic increase?
37. At what time t will the number of persons affected by the epidemic begin to decrease?

12.2 ADDITIONAL INFORMATION PROVIDED BY THE SECOND DERIVATIVE

Just as f' relates information about the behavior of f, so does f'' relate information about the behavior of f'. It follows, naturally, that f'' also relates information about the behavior of f. In the same way, the third derivative f''' gives information about the second derivative and so on.

We have noted that when the first derivative of a function is positive, the function is increasing; or when the first derivative is negative, the function is decreasing. But we often need information about the *rate of increase or decrease* and for this information we look to the second derivative of the function. The second derivative gives the slope of the first derivative. If $f''(x) > 0$, the first derivative f' is an increasing function at x; if $f''(x) < 0$, the first derivative f' is a decreasing function at x; if $f''(x) = 0$, the first derivative is a constant function at x. Thus, the first and second derivative taken together tell us a great deal more about the function than does the first derivative alone.

For instance, we know that if $f' > 0$ the function f is an increasing function. If $f' > 0$ and $f'' > 0$ as well, the function f is increasing at an increasing rate. The slope of f is positive and is getting steeper. (See

Figure 12.5

The Effect of the Second Derivative on the Slope of the Function

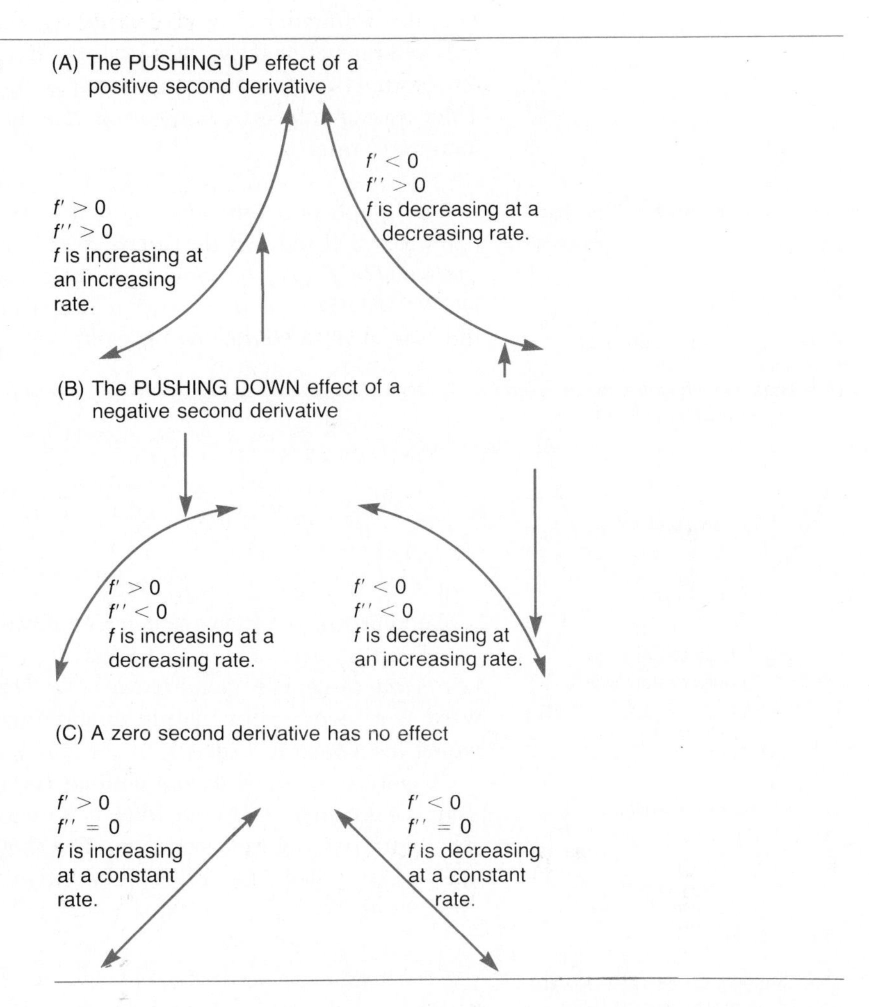
(A) The PUSHING UP effect of a positive second derivative
f' > 0
f'' > 0
f is increasing at an increasing rate.
f' < 0
f'' > 0
f is decreasing at a decreasing rate.
(B) The PUSHING DOWN effect of a negative second derivative
f' > 0
f'' < 0
f is increasing at a decreasing rate.
f' < 0
f'' < 0
f is decreasing at an increasing rate.
(C) A zero second derivative has no effect
f' > 0
f'' = 0
f is increasing at a constant rate.
f' < 0
f'' = 0
f is decreasing at a constant rate.

Figure 12.5.) If $f' > 0$ but $f'' < 0$, the function f is increasing but at a decreasing rate. The slope of f is positive but is becoming less steep. If $f' > 0$ and $f'' = 0$, the function f is increasing at a stationary rate; that is, the function is linear with a positive slope.

Whenever $f' < 0$, the function f has a negative slope; f is a decreasing function. If $f' < 0$ and $f'' < 0$ as well, the slope of f is negative and is becoming steeper. The function f is decreasing at an increasing rate. (See Figure 12.5.) If $f' < 0$ but $f'' > 0$, the slope of f is negative but is becoming less steep. The function f is decreasing at a decreasing rate. If $f' < 0$ while $f'' = 0$, then the function f has a negative slope that is constant; it is a linear function with a negative slope.

One easy way to remember the relationships is to remember that a positive second derivative has a "pushing up" effect on the function f. If the function f were increasing, it becomes increasing at an increasing rate; if the function f were decreasing, it becomes decreasing at a decreasing rate. A negative second derivative has a "pushing down" effect on the function f, so that an increasing function becomes increasing at a decreasing rate or a decreasing function becomes decreasing at an increasing rate.

Concavity and Inflection Points

The graph of a function f is said to be CONCAVE UPWARD at a point $x = a$ if the first derivative f' is an increasing function at a—*as indicated by* $f''(a) > 0$. Conversely, the graph of f is said to be CONCAVE DOWNWARD at a point $x = a$ if the derivative f' is a decreasing function at a—*as indicated by* $f''(a) < 0$.

Figure 12.6

Concavity of the Graph of a Function

(A) Tangents lie below a curve that is concave upward

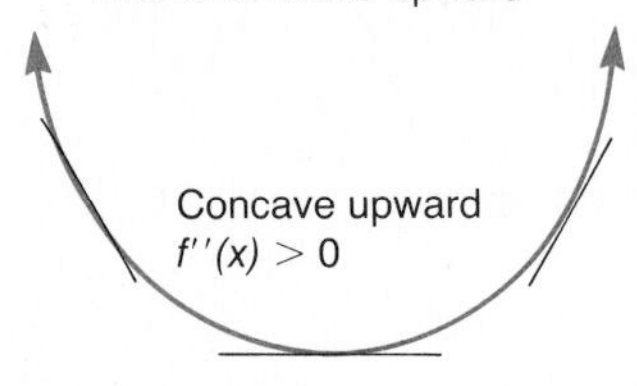

(B) Tangents lie above a curve that is concave downward

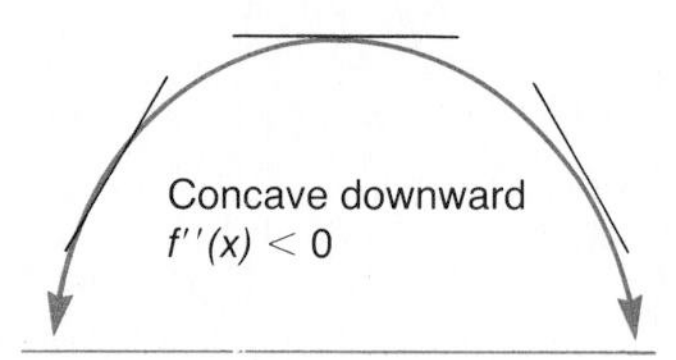

> **If $f''(x) > 0$ for all x** on an interval I, the graph of the function f is CONCAVE UPWARD on that interval.
> **If $f''(x) < 0$ for all x** on an interval I, the graph of the function f is CONCAVE DOWNWARD on that interval.

Geometrically, a curve is concave upward on an interval if at every point on the curve within that interval there exists a tangent line which lies *below* the curve. (See Figure 12.6.) Then, a curve is concave downward if at every point within an interval there exists a tangent line which lies *above* the curve.

A word of caution! do not confuse the ideas of concave upward and concave downward with the ideas of increasing and decreasing functions. These are two different concepts. The sketches in Figure 12.7 illustrate that a decreasing function may be either concave upward or concave

Figure 12.7

The Concepts of Concave Upward and Concave Downward Differ from the Concepts of Increasing Function and Decreasing Function

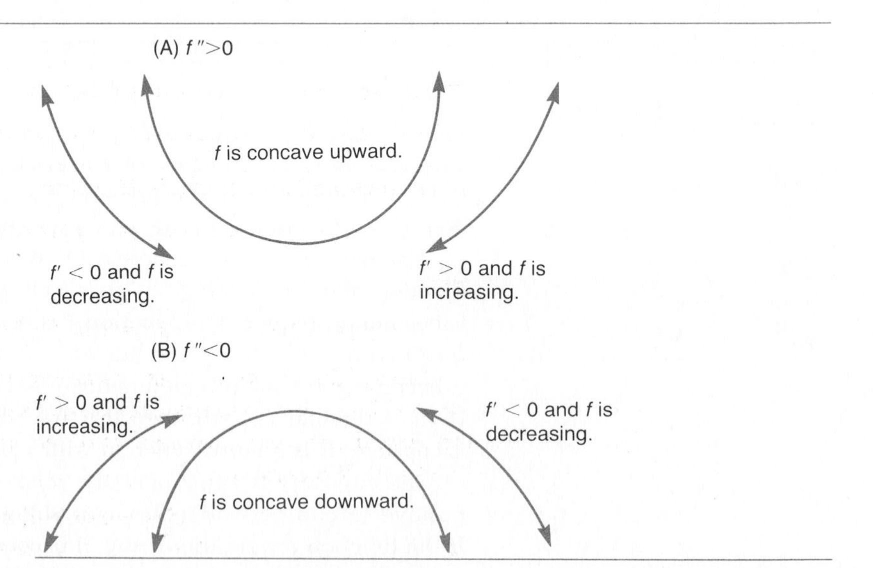

downward, just as an increasing function may be either concave upward or concave downward.

Example 4 Discuss the concavity of the function

$$f(x) = x^2 + 2x + 6$$

The first and second derivatives are

$$f'(x) = 2x + 2$$

and

$$f''(x) = 2$$

Because $[f''(x) = 2] > 0$ for all x, the function f is concave upward for all x. (See Figure 12.8.) []

Figure 12.8

The Graph of the Function $f(x) = x^2 + 2x + 6$ Is Concave Upward for All x

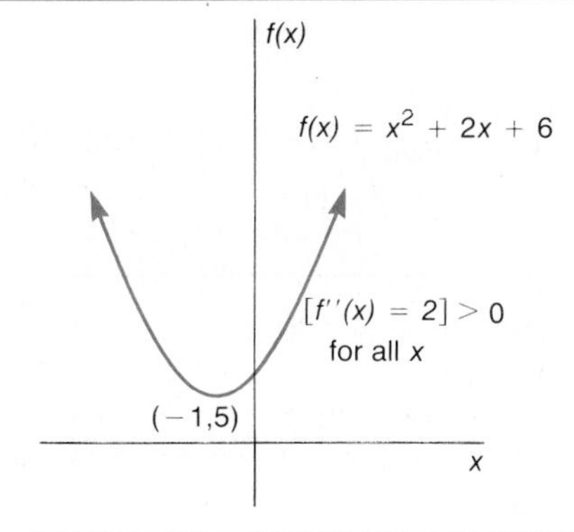

Example 5 Discuss the concavity of the function

$$f(x) = x^4 - 6x^3 + 12x^2 + 10x - 9$$

The first and second derivatives are

$$f'(x) = 4x^3 - 18x^2 + 24x + 10$$

and

$$f''(x) = 12x^2 - 36x + 24 = 12(x^2 - 3x + 2) = 12(x - 1)(x - 2)$$

Clearly, the algebraic sign of the second derivative is not the same for all values of x. To set up appropriate intervals over which to evaluate the second derivative, we need to determine those values of x for which $f''(x) = 0$. We compute

$$12(x - 1)(x - 2) = 0$$
$$x = 1,2$$

We use these values as endpoints of open intervals over which we evaluate the algebraic sign of the second derivative, as shown below.

Interval of x values	Algebraic sign of second derivative $f''(x)$	Type of concavity of graph of function f
$(-\infty,1)$	$(12(-)(-)=+$	Concave upward
$(1,2)$	$(12)(+)(-)=-$	Concave downward
$(2,\infty)$	$(12)(+)(+)=+$	Concave upward

A sketch of the graph of f is shown in Figure 12.9. []

Figure 12.9

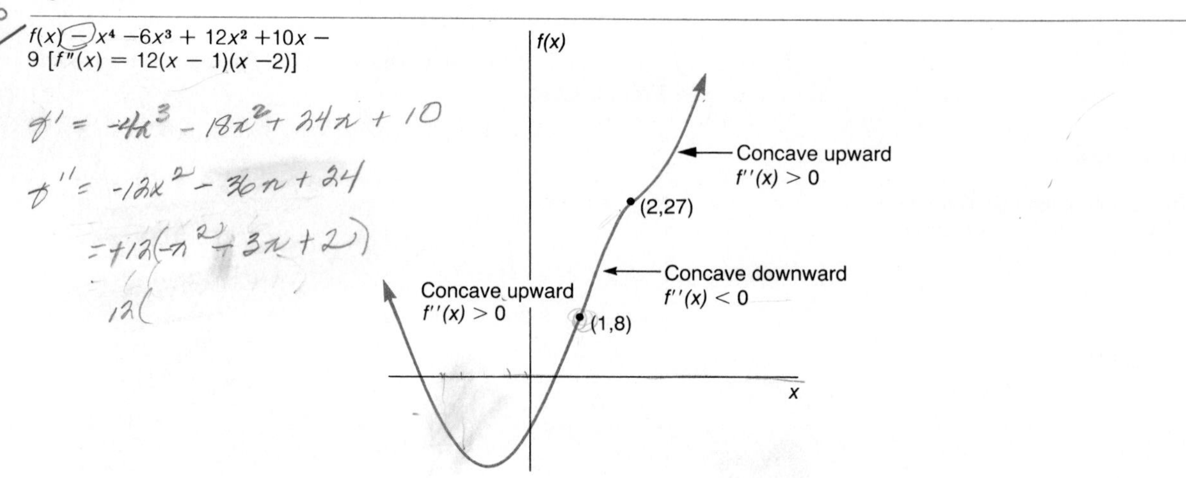

$f(x) = x^4 - 6x^3 + 12x^2 + 10x - 9$ $[f''(x) = 12(x-1)(x-2)]$

A point on a graph where the concavity changes, either from upward to downward or from downward to upward, is called a POINT OF INFLECTION (or an INFLECTION POINT). (See Figure 12.10.) The second derivative is always zero at an inflection point. However, the point where $f''(x) = 0$ may not be an inflection point. That is to say, the condition $f''(x) = 0$ is a necessary but not a sufficient condition for an inflection point.

> **Theorem** *If* (a,f(a)) *is a POINT OF INFLECTION on the graph of a function* f *and if* f'' *is continuous at* a, *then* f''(a) = 0.

The theorem asserts that under the assumption f'' is continuous, the points of inflection, if any, are among those points $(x,f(x))$ for which $f''(x) = 0$. However, although $f''(a) = 0$, $(a, f(a))$ is not necessarily a point of inflection. Also, a point $(a,f(a))$ is a candidate for being a point

Figure 12.10

Points of Inflection on a Function

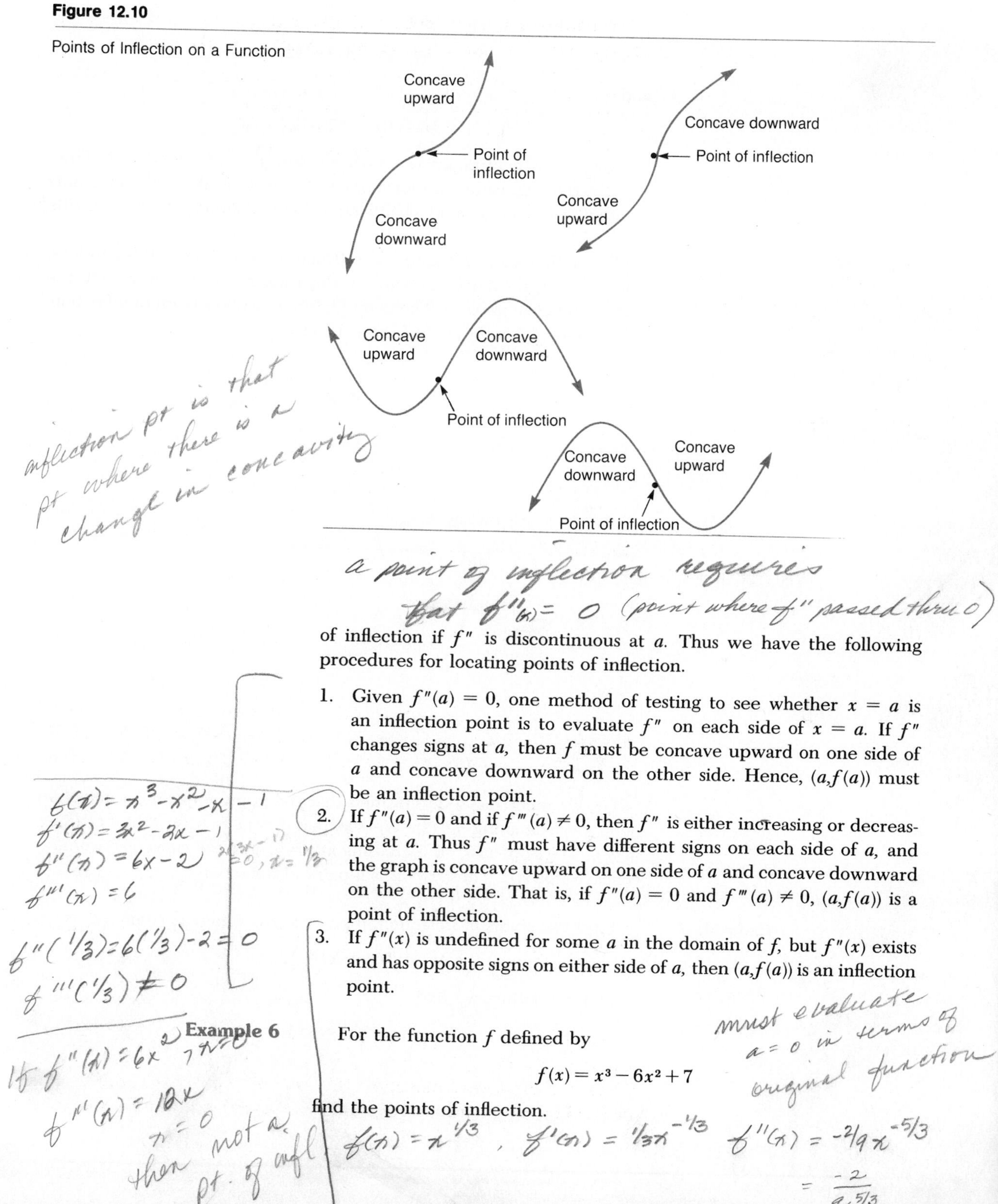

of inflection if f'' is discontinuous at a. Thus we have the following procedures for locating points of inflection.

1. Given $f''(a) = 0$, one method of testing to see whether $x = a$ is an inflection point is to evaluate f'' on each side of $x = a$. If f'' changes signs at a, then f must be concave upward on one side of a and concave downward on the other side. Hence, $(a,f(a))$ must be an inflection point.
2. If $f''(a) = 0$ and if $f'''(a) \neq 0$, then f'' is either increasing or decreasing at a. Thus f'' must have different signs on each side of a, and the graph is concave upward on one side of a and concave downward on the other side. That is, if $f''(a) = 0$ and $f'''(a) \neq 0$, $(a,f(a))$ is a point of inflection.
3. If $f''(x)$ is undefined for some a in the domain of f, but $f''(x)$ exists and has opposite signs on either side of a, then $(a,f(a))$ is an inflection point.

Example 6

For the function f defined by

$$f(x) = x^3 - 6x^2 + 7$$

find the points of inflection.

Differentiating, we obtain

$$f'(x) = 3x^2 - 12x$$

and

$$f''(x) = 6x - 12 = 6(x - 2)$$

Now, f'' is continuous everywhere, and $f'' = 0$ for $x = 2$. Thus, $x = 2$ is the only candidate for a point of inflection. We may determine whether it is such a point of inflection by evaluating f'' on each side of $x = 2$.

We see that if $x < 2$, then $f''(x) < 0$, and if $x > 2$, $f''(x) > 0$. Therefore, when $x < 2$, the curve is concave downward, and when $x > 2$, the curve is concave upward. The point $(2, f(2))$ must be a point of inflection. (The graph of f is shown in Figure 12.11.)

Figure 12.11

$f(x) = x^3 - 6x^2 + 7$
$f''(x) = 6(x - 2)$

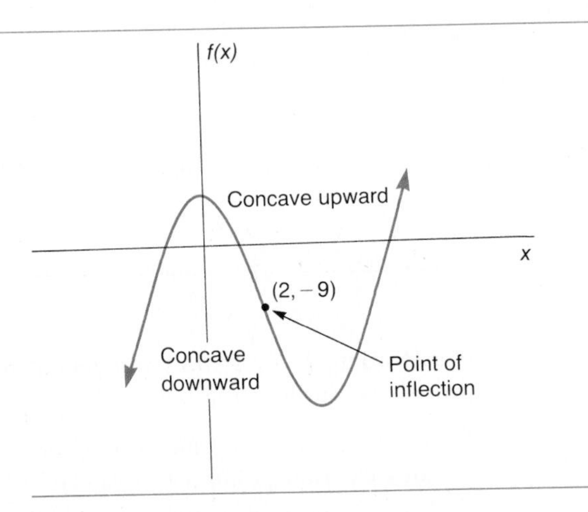

Or, we might have determined that the point $x = 2$ is an inflection point by evaluating $f'''(2)$. We see that $f'''(x) = 6$, so that $[f'''(2) = 6] \neq 0$. The slope of f'' is increasing at $x = 2$, and the curve is becoming concave upward. Thus, $x = 2$ is a point of inflection. [ ]

Example 7 Find the points of inflection, if any, for the function f defined as

$$f(x) = x^4 + 12$$

Differentiating, we find

$$f'(x) = 4x^3$$

and

$$f''(x) = 12x^2$$

Thus, $f''(x) = 0$ when $x = 0$. Because $f''(x)$ is continuous, the point $x = 0$ is the only candidate for a point of inflection.

We check the algebraic signs of f'' on each side of $f''(0)$ and find $f''(x) > 0$ for $x < 0$ and for $x > 0$ as well. Hence, the concavity of the graph of f does not change at $x = 0$. The point $x = 0$ is not a point of inflection. The graph of $f(x) = x^4 + 12$ is shown in Figure 12.12.

Figure 12.12

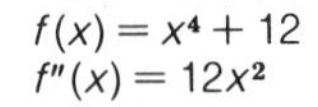

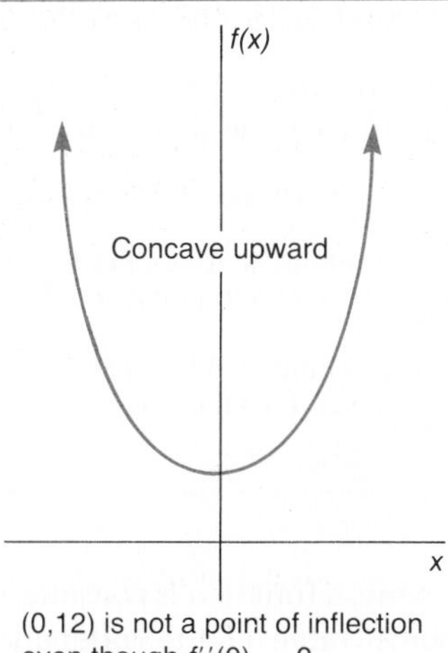

If we evaluated the third derivative at $x = 0$, we would find

$$f'''(x) = 24x$$
$$f'''(0) = 0$$

The second derivative is not changing slope at $x = 0$ (it is stationary at that point). And the function f is not changing concavity at $x = 0$. []

EXERCISES

For each of the following functions: (a) *Find those open intervals over which the function is concave upward and those open intervals over which the function is concave downward;* (b) *find all points of inflection.*

38. $f(x) = x^2 - 3x - 4$
39. $f(x) = 2 + 3x - 2x^2$
40. $f(x) = x^3 - 1$
41. $f(x) = x^3 - x^2 - x - 1$
42. $f(x) = 2x^3 - 3x^2 - 12x + 24$
43. $f(x) = 3x^3 + 2x^2$
44. $f(x) = x^4 + x^2 + 10$
45. $f(x) = x^5 + 5x$
46. $f(x) = 1/(x + 5)$
47. $f(x) = 3/x^2$
48. $f(x) = x + (1/x)$
49. $f(x) = x/(1 + x^2)$

50. $f(x) = x^2/(2 + x)$
51. $f(x) = 3x^4 - 4x^3$
52. $f(x) = x^{2/3}$
53. $f(x) = e^x + 1$
54. $f(x) = 1/x^3$
55. $f(x) = ln(x + 3)$
56. $f(x) = (x - 2)^3$
57. $f(x) = x/\sqrt{x^2 + 1}$
58. $f(x) = (x + 1)^5$
59. $f(x) = (x - 1)^6$

12.3 EXTREME VALUES OF A FUNCTION

Geometrically, the MAXIMUM and/or MINIMUM values of a function are the high and/or low points on the function's graph.

> A function f attains a **RELATIVE MAXIMUM** at $x = a$ if there exists an open interval I containing a such that $f(a) \geq f(x)$ for all x in the interval I.
>
> A function f attains a **RELATIVE MINIMUM** at $x = a$ if there exists an open interval I containing a such that $f(a) \leq f(x)$ for all x in the interval I.

That is to say, a function is considered to achieve a RELATIVE MAXIMUM at some value of the independent variable if the value of the function at that point is at least as large as it is for any other *nearby* point (see Figure 12.13). A function is considered to achieve a RELATIVE MINIMUM at some value of the independent variable if the value of the function at that point is at least as small as it is for any *nearby* point (see Figure 12.14).

Because a relative maximum need not be the highest point for the entire graph but simply a point higher than other points in its immediate

Figure 12.13

The Function f Reaches a Relative Maximum at x_1, x_2, and x_3

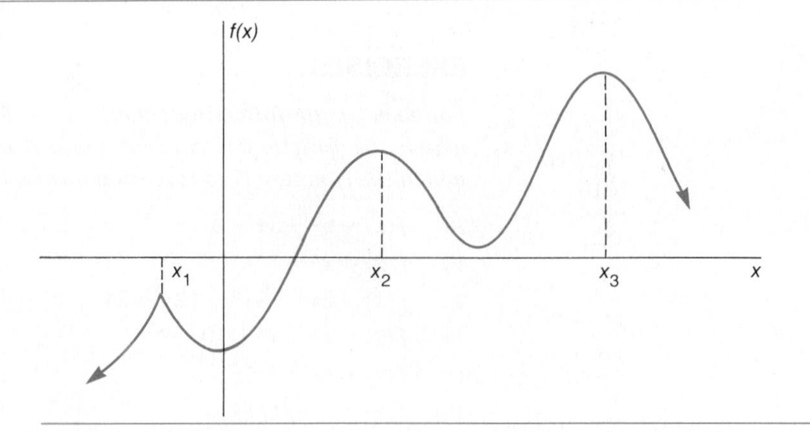

Figure 12.14

The Function f Reaches a Relative Minimum at x_1, x_2, and x_3

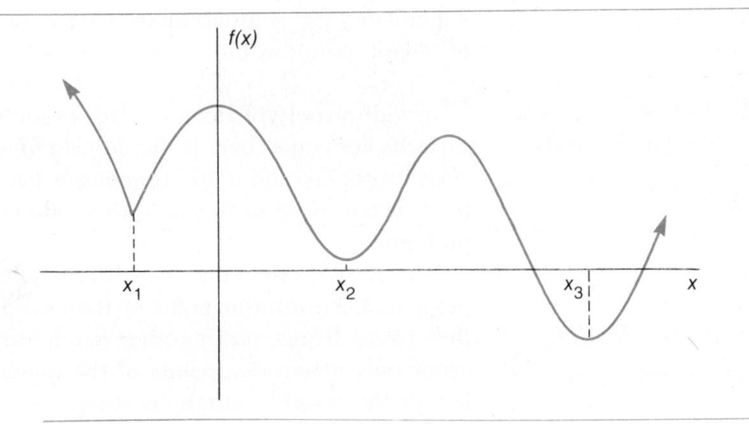

neighborhood, a function may have more than one relative maximum. Similarly, a relative minimum is minimal only in relation to adjacent points, so a function may have more than one relative minimum. Indeed, in Figure 12.15 we see that a relative maximum value of a function may even be less than a relative minimum of that same function.

> A function f attains an **ABSOLUTE MAXIMUM** at $x = a$ if and only if $f(a) \geq f(x)$ for all x in the domain of f.
> A function f attains an **ABSOLUTE MINIMUM** at $x = a$ if and only if $f(a) \leq f(x)$ for all x in the domain of f.

Figure 12.15

Relative Maxima and Minima

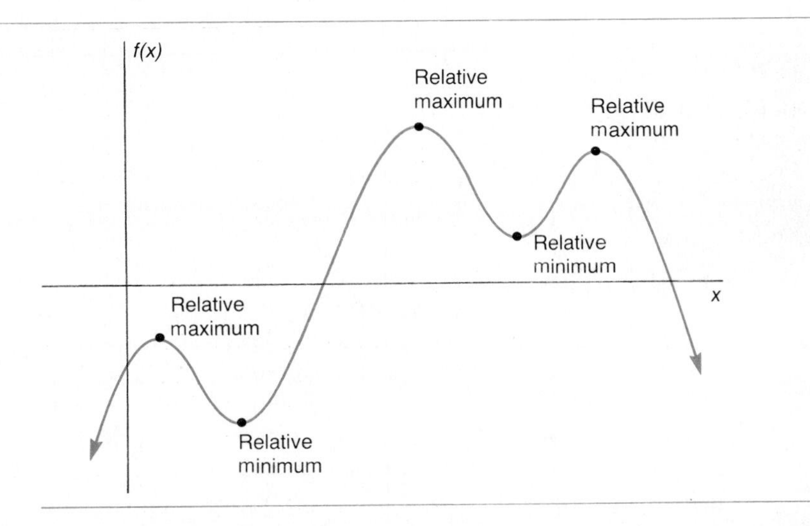

The term ABSOLUTE EXTREMA is used to cover both situations. A point $(a,f(a))$ is an absolute extremum only if it is the very highest, or lowest, point on the entire curve.

When the Domain of f Is a Closed Interval

A wide variety of situations are possible where relative and absolute extrema are concerned. If the domain of a function is restricted to some closed interval, and if the function is continuous on that domain, then the function necessarily has both an absolute maximum and an absolute minimum.

The absolute extrema of such functions may occur either at an end-point or at an interior point of the closed interval, or possibly at both. (Relative extrema, on the other hand, are conventionally considered to occur only at interior points of the domain.) Figure 12.16 illustrates a few of the possible situations when f is a continuous function defined on the closed interval $[a,b]$. In Figure 12.16A both the absolute maximum and the absolute minimum occur at endpoints of the domain, and there are no relative extrema. In Figure 12.16B, both absolute maximum and absolute minimum points are at endpoints of the closed interval over

Figure 12.16

Extrema for a Function Defined over a Closed Interval

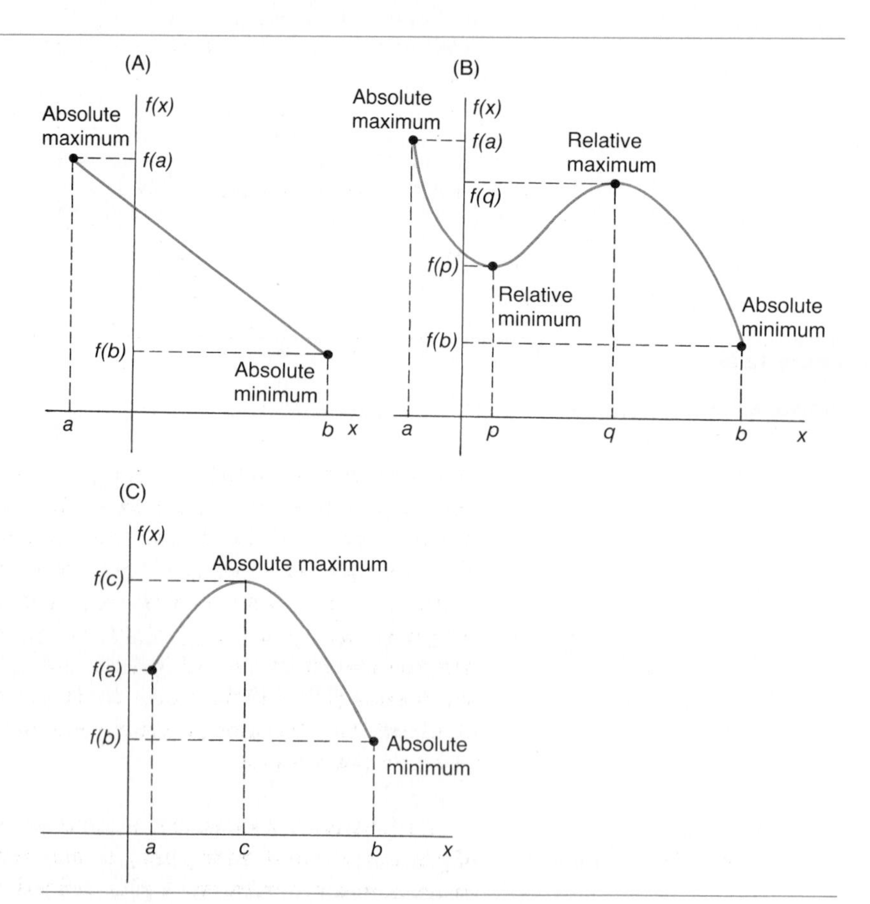

Figure 12.16 (*concluded*)

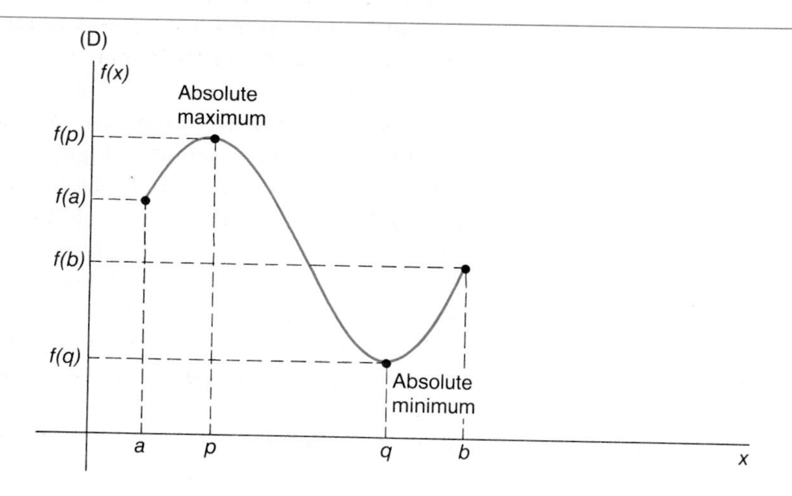

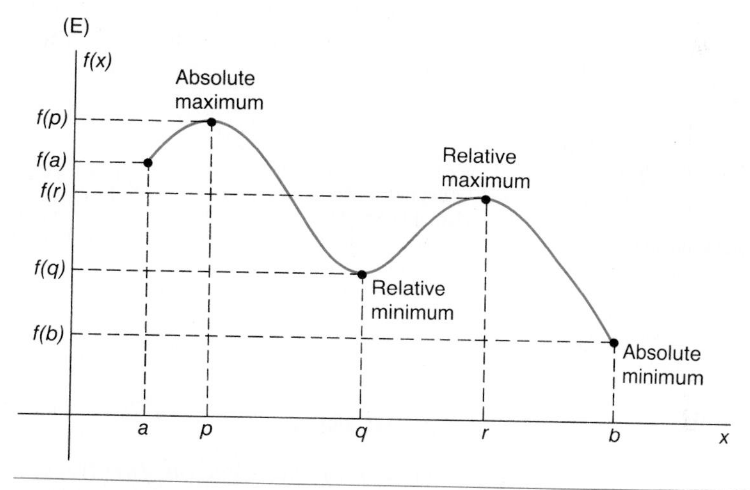

which f is defined. Relative extrema occur at interior points of this interval. In Figure 12.16C the absolute minimum occurs at an endpoint; the other endpoint of the domain does not represent an extremum. The absolute maximum occurs at an interior point of the domain of f. In Figure 12.16D neither endpoint represents an extremum. Both the absolute maximum and the absolute minimum occur at interior points of the domain. In Figure 12.16E the absolute maximum point occurs at an interior point of the interval, while the absolute minimum occurs at an endpoint. Relative maximum and minimum points occur at interior points of the interval.

When the Domain of f Is an Open Interval

If a function f is defined over an open interval (a,b) or if the domain of f is unrestricted, then f may or may not have an absolute maximum or an absolute minimum. If f is defined over the open interval (a,b), then x can only approach a or b, but x cannot equal either a or b. Thus, endpoints are not considered in determining maxima or minima. All extrema for such a function must occur at interior points of its domain.

Figure 12.17 illustrates some of the possible situations concerning extrema for functions defined over an open interval or whose domain is unrestricted. The function in Figure 12.17A has a relative maximum and a relative minimum but no absolute extrema. In Figure 12.17B the function has an absolute minimum but neither a relative nor an absolute maximum. In Figure 12.17C the function has an absolute maximum but no minimum. The function in Figure 12.17D has neither a maximum nor a minimum.

Figure 12.17

Extrema for a Function Whose Domain Is Unrestricted

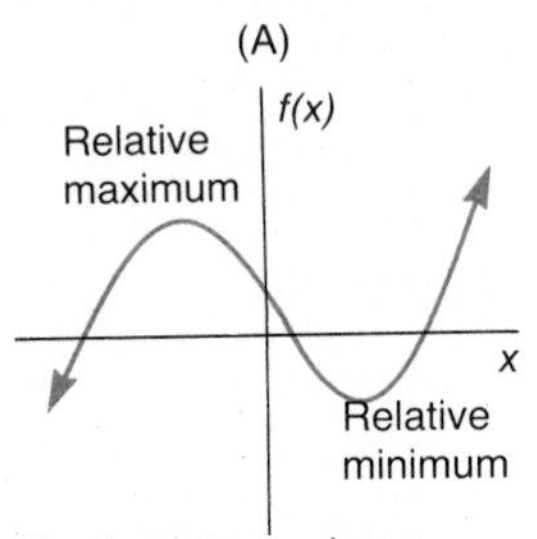

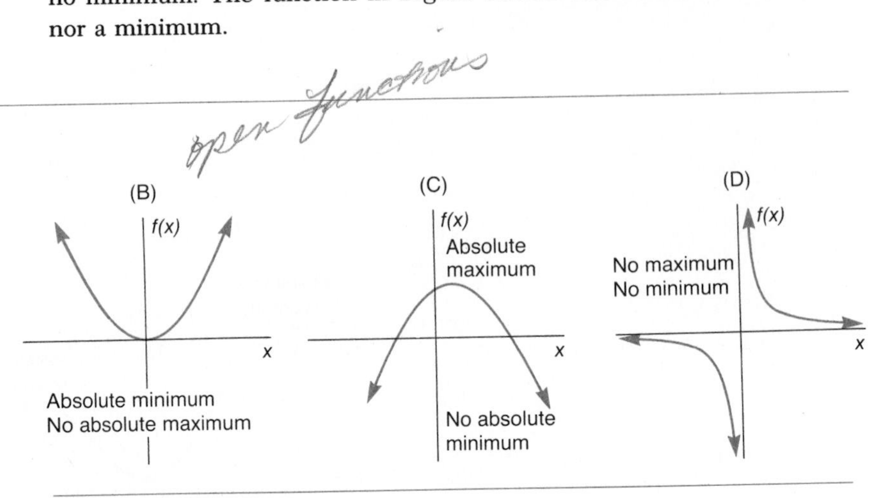

EXERCISES

For each of the following functions, identify relative maxima, relative minima, absolute maxima, absolute minima, intervals over which the function is increasing, intervals over which the function is decreasing, intervals over which the function is concave upward, intervals over which the function is concave downward, and all points of inflection.

60.

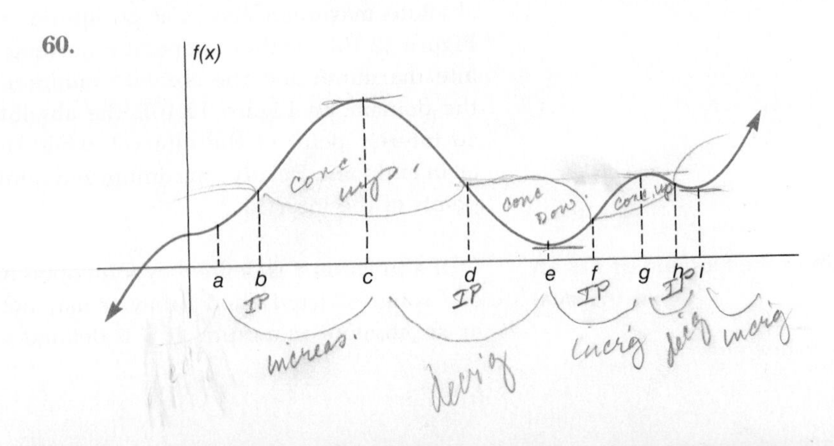

answer in book is reversed

61.

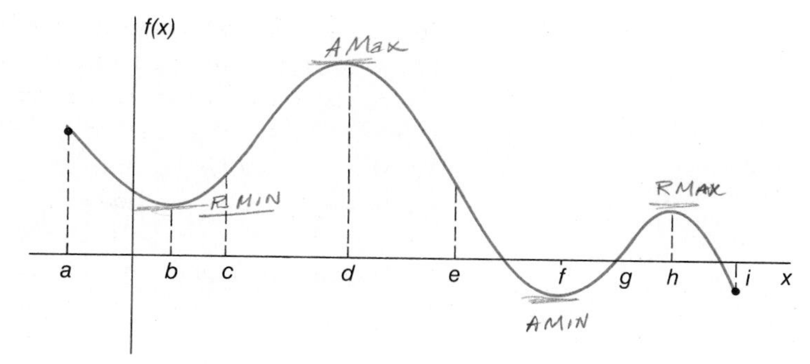

62.

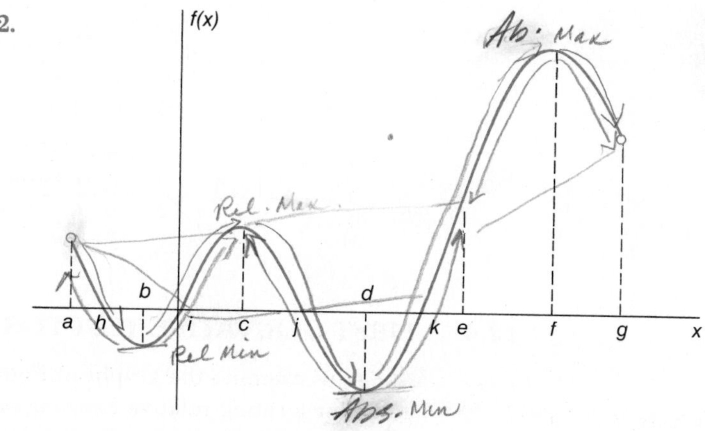

63.

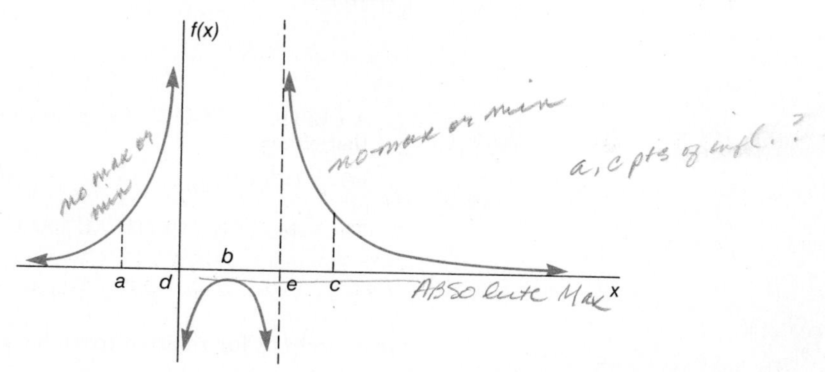

64.

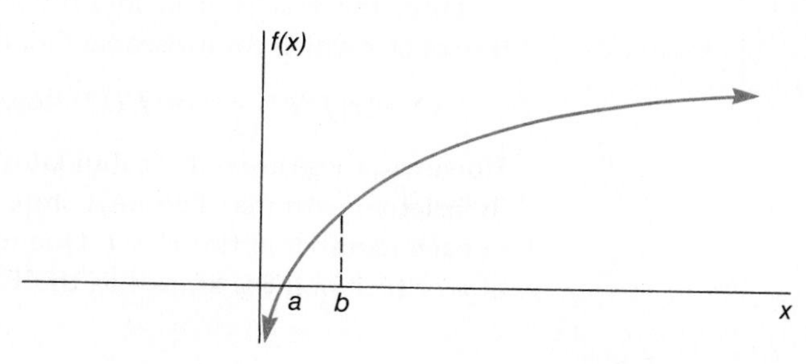

65.

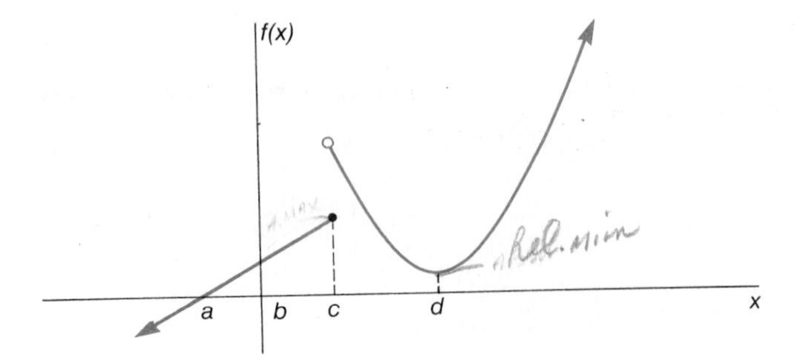

66.

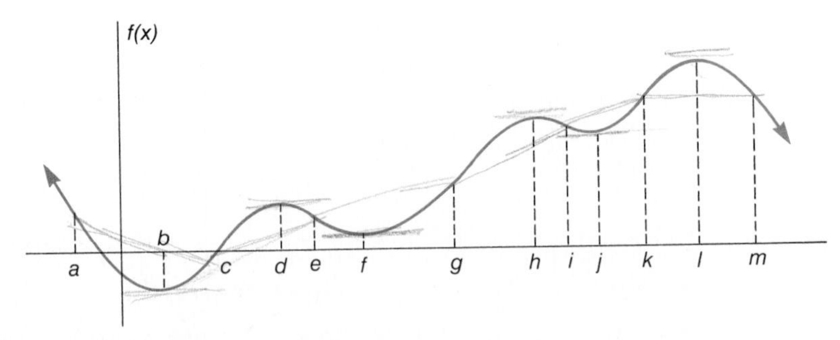

12.4 FIRST DERIVATIVE TEST FOR RELATIVE EXTREMA

Let us examine the graphs in Figure 12.18 and see if we can discern a test for locating relative extrema of a function. First, RELATIVE EXTREMA OCCUR ONLY AT CRITICAL POINTS OF A FUNCTION.

> A CRITICAL POINT of a continuous function f is a point $(x^*, f(x^*))$ such that either
>
> (*a*) $f'(x^*) = 0$ or (*b*) $f'(x^*)$ does not exist but $f(x^*)$ is defined
>
> The value x^* is a CRITICAL VALUE of x.

In searching for relative extrema we need only consider critical values of x.

Thus, the first step in identifying relative extrema is to determine the set of x which comprises candidates for extrema; that is, we determine

$$x^* = \{x \mid f'(x^*) = 0 \text{ or } f'(x^*) \text{ does not exist but } f(x^*) \text{ is defined}\}.$$

However, these are only "candidates" for extrema; they are not necessarily relative extrema. The next step, thus, is to determine the character of each member of the x^* set. One method of determining the character of a critical point is known as the FIRST DERIVATIVE TEST.

Figure 12.18

The First Derivative Test for Locating Relative Extrema

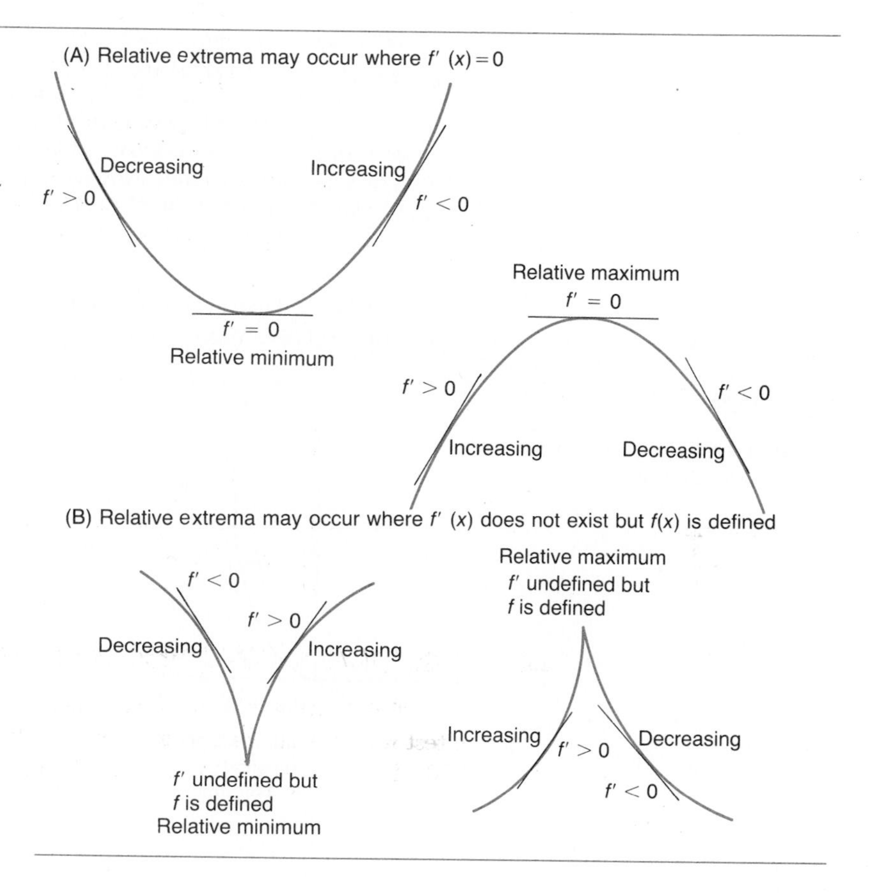

Refer again to Figure 12.18 and note that to the immediate left of a maximum point the function is always increasing (that is, $f' > 0$) and to the immediate right of a maximum point the function is always decreasing (that is, $f' < 0$). Hence, THE ALGEBRAIC SIGN OF THE FIRST DERIVATIVE ALWAYS CHANGES FROM POSITIVE TO NEGATIVE AS THE FUNCTION TRAVERSES A RELATIVE MAXIMUM POINT (always moving from left to right along the horizontal axis).

Similarly, to the immediate left of a minimum point the function is decreasing (that is, $f' < 0$), and to the immediate right of a minimum point the function is always increasing (that is, $f' > 0$). So, THE ALGEBRAIC SIGN OF THE FIRST DERIVATIVE ALWAYS CHANGES FROM NEGATIVE TO POSITIVE AS THE FUNCTION TRAVERSES A RELATIVE MINIMUM POINT.

If the sign of the first derivative does not change as the function traverses a critical point, that point is neither a maximum nor a minimum.

Theorem *If for a continuous function* f, x = x* *is a critical value then*

1. x = x* *is a RELATIVE MAXIMUM of* f *if* f'(x) *changes from positive to negative as* x *changes from just below* x* *to just above* x*.
2. x = x* *is a RELATIVE MINIMUM of* f *if* f'(x) *changes from negative to positive as* x changes from just below x* *to just above* x*.
3. x = x* *is not a relative extrema if* f'(x) *does not change sign as* x *changes from just below* x* *to just above* x*.

Figure 12.19 illustrates a function for which a critical value does not represent a relative extremum.

Figure 12.19

Not All Points Such That $f'(x) = 0$ Represent Extrema

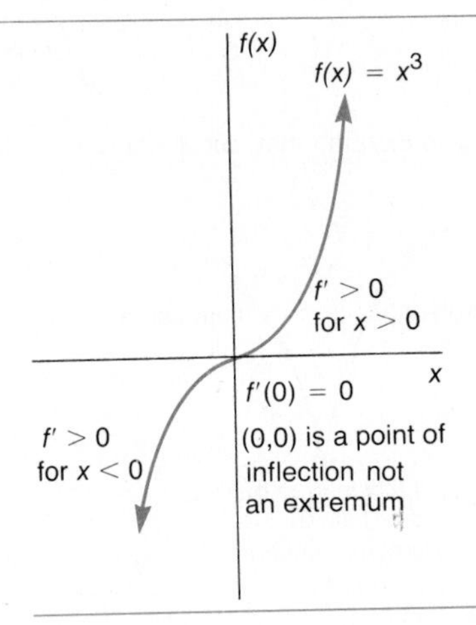

Example 8 Find relative maxima and minima for the function defined by

$$f(x) = x^2 - 6x + 7$$

Using the aids we have discussed, sketch the graph of f.

We locate the candidates for relative extrema by finding the critical values of x. The first derivative is

$$f'(x) = 2x - 6 = 2(x - 3)$$

and equals zero only for $x^* = 3$; f' is never undefined. Thus, $x^* = 3$ is the only candidate for extremum.

We use $f'(x)$ to determine the behavior of f on either side of $x^* = 3$. We conclude

$f'(x) < 0$ for $x < 3$—thus f is decreasing when $x < 3$
$f'(x) > 0$ for $x > 3$—thus f is increasing when $x > 3$

so that at $x = 3$ there is a relative minimum of f. The other coordinate of this extreme point is computed $f(3) = (3)^2 - 6(3) + 7 = -2$.

We use this information to sketch the graph shown in Figure 12.20. Note, incidentally, that the point (3,−2) represents an absolute minimum as well as a relative minimum since there is no other value of f less than $f(3) = -2$. []

Figure 12.20

$f(x) = x^2 - 6x + 7$

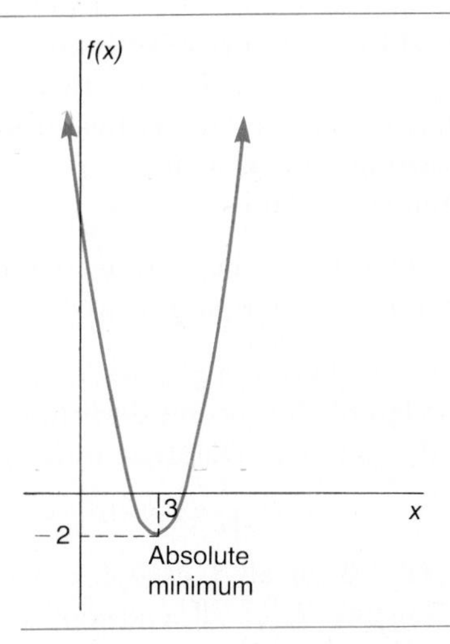

Example 9 Find all relative maxima and minima for the function f defined by

$$f(x) = 2x^3$$

Sketch the graph of f.

We begin by locating the set of critical values of x. Differentiating, we obtain

$$f'(x) = 6x^2$$

Setting $f' = 0$, we find

$$6x^2 = 0$$
$$x^* = 0$$

Because f' is defined on all x, there are no critical values other than $x^* = 0$. This is the only candidate for a local extreme point.

Continuing, we evaluate $f'(x)$ on either side of $x^* = 0$ and find

$f'(x) > 0$ when $x < 0$—thus f is increasing when $x < 0$
$f'(x) > 0$ when $x > 0$—thus f is increasing when $x > 0$

The algebraic sign of f' does not change as we move from left to right across the critical point; therefore, the critical point, $x^* = 0$, is neither a maximum nor a minimum. (It is, instead, a point of inflection.) The function f has neither a maximum nor a minimum point; it is a strictly increasing function. The graph of f appears in Figure 12.21. []

Figure 12.21

$f(x) = 2x^3$

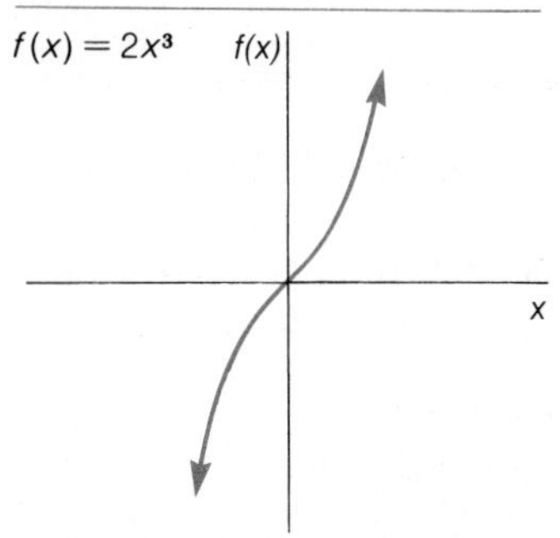

Function has neither a maximum nor a minimum.

Example 10 Find relative maxima and minima for the function defined by

$$f(x) = x^{2/3}$$

Construct a graph of the function.

Differentiating, we obtain

$$f'(x) = (2/3)x^{-1/3} = 2/(3\sqrt[3]{x})$$

We see that the first derivative is not zero for any x. However, observe that f' (0) is not defined. (Nonetheless, $f(0)$ is defined and is $f(0) = (0)^{2/3} = 0$.) Hence, the only critical value and the only candidate for a relative extremum is $x^* = 0$.

Continuing, we find

$f'(x) < 0$ for $x < 0$; thus f is decreasing when $x < 0$

$f'(x) > 0$ for $x > 0$; thus f is increasing when $x > 0$

Thus, at $x = 0$ there is a relative minimum.

Knowledge of the second derivative will aid us in constructing the graph of the function. Differentiating again, we obtain

$$f''(x) = (-2/9)x^{-4/3} = -2/(9\sqrt[3]{x^4})$$

Thus, $f''(x) < 0$ for all x except $x = 0$. The graph of the function is concave downward for all x except $x = 0$. The graph of $f(x) = x^{2/3}$ appears in Figure 12.22. []

Figure 12.22

An Extreme Point Occurs Where $f'(x)$ Is Not Defined

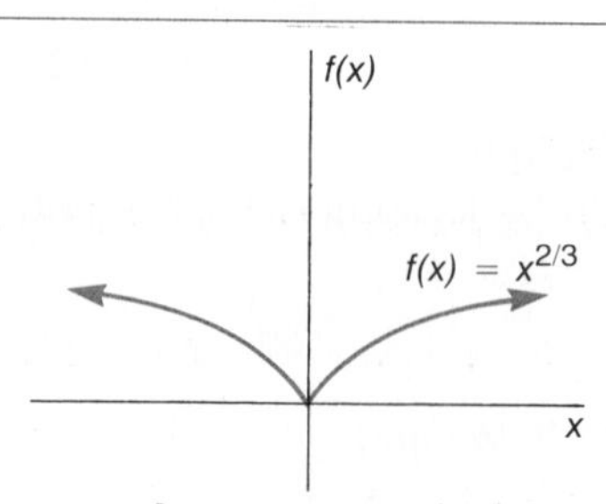

x = 0 represents an absolute minimum.

Example 11 Find relative extrema for the function defined by

$$f(x) = 1/x$$

Differentiating, we find

$$f'(x) = D_x(x^{-1}) = -1x^{-2} = -1/x^2$$

We see that the first derivative is never equal to zero (since the numerator of the ratio never equals zero) but is undefined at $x = 0$. Recall, however, that any x such that $f'(x)$ does not exist is a critical value *only if* $f(x)$ *is defined*. Of course, $f(0) = 1/0$ is not defined. Thus $x = 0$ is not a critical value. Indeed, there are no critical values; the function has no relative extrema (see Figure 12.23). []

Figure 12.23

$f(x) = 1/x$ Has No Relative Extrema

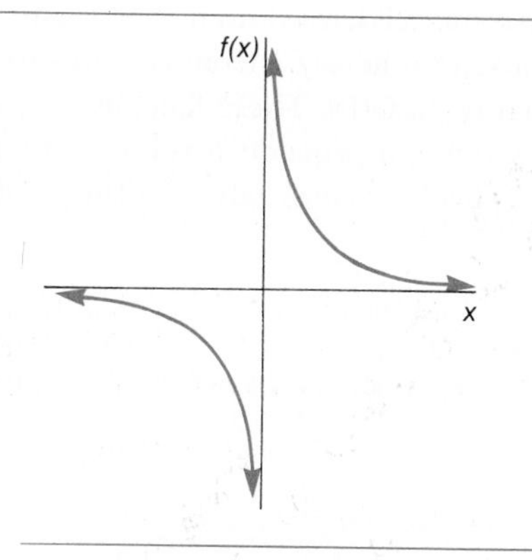

EXERCISES

Find all relative extrema, using the First Derivative Test. Sketch a graph of the function.

67. $f(x) = x^2 - 5x + 6$
68. $f(x) = 3 + 2x - x^2$
69. $f(x) = x^2 + 9$
70. $f(x) = x^3 - 3x^2 - 2x - 4$
71. $f(x) = 1 + x^2 - x^3$
72. $f(x) = 4/x^2$
73. $f(x) = x/(x^2 + 1)$
74. $f(x) = x^5 - 1$
75. $f(x) = (x - 4)^4$
76. $f(x) = (2x - 1)/(x + 2)$
77. $f(x) = (2x - 3)^5$
78. $f(x) = x^4 - 4/x$
79. $f(x) = (x^2 - 9)^2$
80. $f(x) = x^4 - 2x^2 + 2$
81. $f(x) = 1/(x + 1)$
82. $f(x) = x + (1/x)$
83. $f(x) = x^2/(x - 1)$
84. $f(x) = (x^2 + 1)/(x + 1)$
85. $f(x) = e^{2x+1}$
86. $f(x) = 5^{x^3}$
87. $f(x) = ln\ (2x + 7)$
88. $f(x) = \log (x^2 - 1)$
89. $f(x) = x^{4/3}$
90. $f(x) = \begin{cases} x + 1 & \text{for } x < 0 \\ x^2 + 1 & \text{for } x > 0 \end{cases}$

12.5 SECOND DERIVATIVE TEST FOR RELATIVE EXTREMA

We have seen that the second derivative f'' gives information about the rate of change of the first derivative f'. If $f'' > 0$ at some point $x = a$, then f' is an increasing function at that point, and the graph of f is concave upward. Or, if $f'' < 0$ at some point $x = a$, then f' is a decreasing function at that point, and the graph of f is concave downward.

If an extremum occurs at a point x^* where f is concave downward, that extremum is a maximum. (See Figure 12.24A.) If an extremum occurs at a point x^* where f is concave upward, that extremum is a minimum (see Figure 12.24B). These facts lead to a second procedure for locating local extrema, a procedure referred to as the SECOND DERIVATIVE TEST. (This is an alternative to the First Derivative Test.)

Figure 12.24

The Second Derivative Test for Locating Relative Extrema

(A) If $f''(x^*) < 0$, then $f(x^*)$ is a maximum value of f.

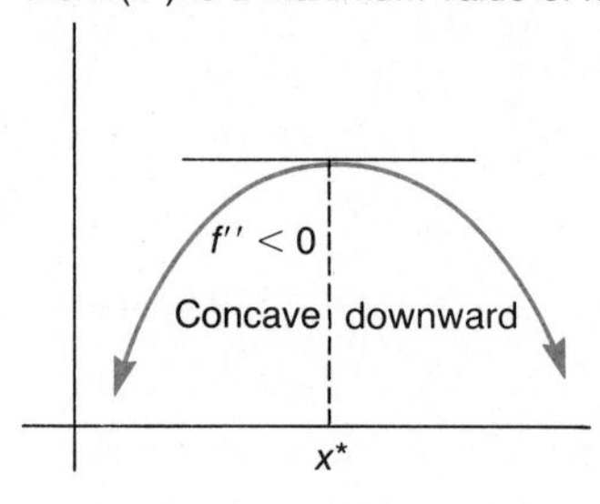

(B) If $f''(x^*) > 0$, then $f(x^*)$ is a minimum value of f.

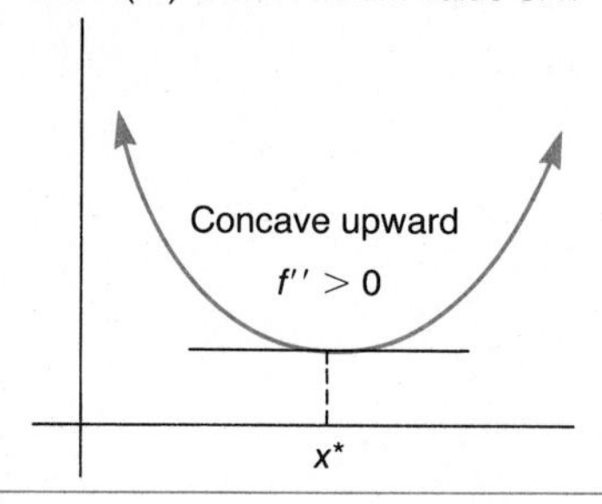

The Second Derivative Test for locating relative extreme proceeds as follows:

1. Find the first derivative f' and the second derivative f'' of the function f.
2. Locate the set of critical values of x.
3. Evaluate the second derivative f'' at each critical value x^*.
   a. If $f''(x^*) < 0$, then $f(x^*)$ is a maximum value of f.
   b. If $f''(x^*) > 0$, then $f(x^*)$ is a minimum value of f.
   c. If $f''(x^*) = 0$, then the test is indeterminate. (In this case, we shall return to the use of the First Derivative Test.)

Example 12 Use the Second Derivative Test to locate the relative extrema for the function f defined by

$$f(x) = x^2 - 4x - 45$$

We begin by finding

$$f'(x) = 2x - 4 \qquad \text{and} \qquad f''(x) = 2$$

Then we solve for the critical values of x by

$$2x - 4 = 0$$
$$x^* = 2$$

The function is never undefined; so this is the only critical value of x.

To ascertain whether this point is a relative maximum, a relative minimum, or neither, we substitute this critical value of x into the second derivative function and evaluate. We obtain

$$f''(2) = 2$$

Because the second derivative at $x^* = 2$ is positive, the function f is concave upward at this point. Hence, the point $(2, f(2))$ is a relative minimum point on the function f. []

Example 13 Use the Second Derivative Test to locate relative extrema on the function f defined by

$$f(x) = (1/3)x^3 + x^2 - 8x + 12$$

To begin, we find the first and second derivatives. These are

$$f'(x) = x^2 + 2x - 8$$
$$f''(x) = 2x + 2$$

The critical values of x are located by setting the first derivative equal to zero and solving

$$x^2 + 2x - 8 = 0$$
$$(x + 4)(x - 2) = 0$$
$$x^* = -4, \; x^* = 2$$

The function f has two stationary points; one at $x^* = -4$ and the other at $x^* = 2$. We must evaluate the second derivative at each of these critical points.

$$f''(-4) = 2(-4) + 2 \qquad f''(2) = 2(2) + 2$$
$$= -6 \qquad\qquad\qquad = 6$$

Because $f''(-4) < 0$, the point $(4, f(4))$ is a maximum point on the graph of the function f. Because $f''(2) > 0$, the point $(2, f(2))$ is a minimum point on the graph of f. []

Example 14 Use the Second Derivative Test to find all relative extrema for the function

$$f(x) = x^4$$

We find

$$f'(x) = 4x^3 \qquad \text{and} \qquad f''(x) = 12x^2$$

Critical values of x are found by

$$4x^3 = 0$$
$$x^* = 0$$

But when substituting $x^* = 0$ into the second derivative, we obtain

$$f''(0) = 0$$

and see that the Second Derivative Test in this case is indeterminate.

We revert to the use of the First Derivative Test and determine that $(0,f(0))$ represents a minimum point of f. []

EXERCISES

Use the Second Derivative Test to locate relative extrema for each of the following functions.

91. $f(x) = x^3 + 2x^2 + 1$
92. $f(x) = 6x - x^2$
93. $f(x) = x^2 - 6x + 7$
94. $f(x) = x^2 - 8x$
95. $f(x) = 4 - x^4$
96. $f(x) = x^2 + 1/x^2$
97. $f(x) = x^5 + 1$
98. $f(x) = x^3 - 9x^2 + 18x - 15$
99. $f(x) = (x^2 + 2x + 1)/(1 + x^2)$
100. $f(x) = e^{-x}$
101. $f(x) = x^{1/3}$
102. $f(x) = x^4 - 8x^3 + 18x^2 - 8$
103. $f(x) = -x^7$
104. $f(x) = (x + 3)^4$
105. $f(x) = x^4 - 6x^2$
106. Show that the quadratic polynomial $f(x) = ax^2 + bx + c$ always has a minimum at $x = -b/2a$ if $a > 0$ and a maximum at $x = -b/2a$ if $a < 0$.

12.6 LOCATING ABSOLUTE MAXIMA AND MINIMA

In locating absolute extrema we must consider whether the function is defined on a closed interval $[a,b]$ or on an open interval (a,b). If the domain is not specified, it is generally considered to be the set of all real numbers, or at least the set of real numbers for which the relationship is possible from a mathematical standpoint.

When the Domain of f Is a Closed Interval

> **Extreme value theorem** *A continuous function* f *defined on a closed interval* [a,b] *must attain both an absolute maximum and an absolute minimum at points in* [a,b].

If f is continuous on $[a,b]$ with critical values $x_1^*, x_2^*, \ldots x_n^*$ at interior points of the interval, the point at which f attains its absolute maximum must be either a critical value of x which represents a relative maximum or at one of the endpoints of the interval a or b. Similarly, f must attain its absolute minimum either at one of the interior-point critical values which represents a relative minimum or at a or b.

Thus, to find the absolute maximum and absolute minimum, we evaluate $f(x_1^*), f(x_2^*), \ldots, f(x_n^*), f(a)$, and $f(b)$. The largest of these is the absolute maximum of f on the interval $[a,b]$; the smallest of these is the absolute minimum of f on the interval $[a,b]$. The other critical values—the other x^*—represent relative extrema. The endpoints a and b are considered as candidates for absolute extrema only, and not as candidates for relative extrema.

Example 15 Find all maxima and minima for the function f defined as

$$f(x) = x^3 - 7.5x^2 + 12x + 10 \qquad 0 \le x \le 5$$

We locate the set of critical values for x by setting the first derivative equal to zero, as

$$f'(x) = 3x^2 - 15x + 12$$

then

$$\begin{aligned} 3x^2 - 15x + 12 &= 0 \\ 3(x^2 - 5x + 4) &= 0 \\ (x-4)(x-1) &= 0 \\ x^* &= 1,4 \end{aligned}$$

We now must evaluate the function at each of these critical values and at each of the two endpoint values, as

$$\begin{aligned} f(0) &= 10 \\ f(1) &= 1^3 - 7.5(1)^2 + 12(1) + 10 = 15.5 \\ f(4) &= 4^3 - 7.5(4)^2 + 12(4) + 10 = 2 \\ f(5) &= 5^3 - 7.5(5)^2 + 12(5) + 10 = 7.5 \end{aligned}$$

We see that the very highest value of the function occurs at (1,15.5); this then is the absolute maximum of f. The very smallest value of the function occurs at (4,2); this is the absolute minimum of f. Note that both of the absolute extrema occurred at interior points of the interval over which f is defined. We do not, in such a case, label the endpoints relative extrema (see Figure 12.25). []

Example 16 Find all maxima and minima for the function f defined as

$$f(x) = x^2 + 2x + 2 \qquad -3 \le x \le 5$$

The first derivative is $f'(x) = 2x + 2$, yielding one critical value $x^* = -1$.

Figure 12.25

Locating Absolute Extrema When the Domain of f Is a Closed Interval $f(x) = x^3 - 7.5x^2 + 12x + 10 \qquad 0 \leq x \leq 5$

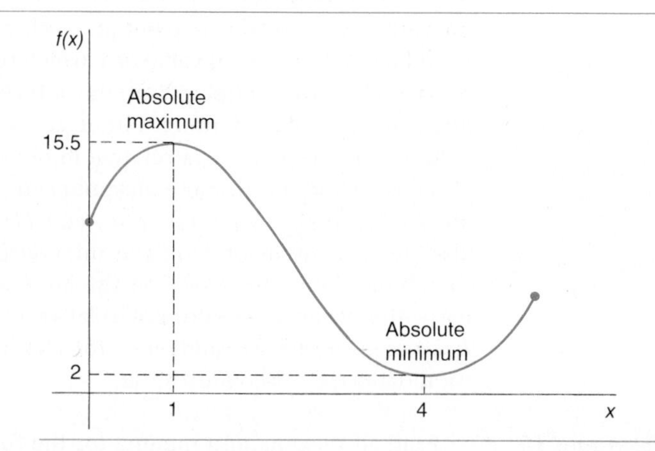

Because the domain of f is a closed interval, endpoints of the interval, along with the critical values of x, must be evaluated. We calculate

$$f(-3) = (-3) + 2(-3) + 2 = 5$$
$$f(-1) = (-1) + 2(-1) + 2 = 1$$
$$f(5) = (5) + 2(5) + 2 = 37$$

We see, thus, that the absolute maximum occurs at (5,37), which is an endpoint of the domain. The absolute minimum occurs at (−1,1), which is an interior point. We do not label the other endpoint a relative extremum. (The graph of the function is shown in Figure 12.26.) [ ]

Figure 12.26

$f(x) = x^2 + 2x + 2$ for $-3 \leq x \leq 5$

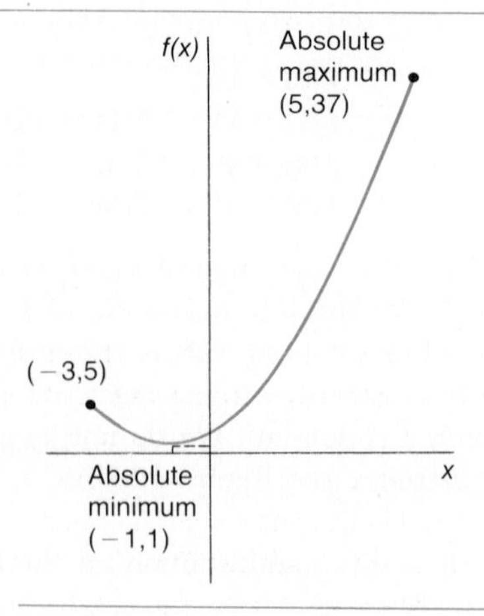

When the Domain of f Is Not a Closed Interval

If the domain of the function is some open interval or if the domain is unrestricted, we have no endpoints to consider and may proceed to locate all interior-point extrema using either the First Derivative Test or the Second Derivative Test. These will be relative extrema; they may or may not be absolute extrema.

In general, if a function f has only one critical point x^* in an open interval (a,b) or $(-\infty, \infty)$, then

1. If x^* represents a maximum, it is the absolute maximum and the function has no minimum.
2. If x^* represents a minimum, it is the absolute minimum and the function has no maximum.

If a function f has more than one critical point x^* in an open interval (a,b) or $(-\infty, \infty)$, then we must evaluate the function at each of the critical values of x as well as determine the value that $f(x)$ approaches as x approaches a and as x approaches b.

EXERCISES

Find all maximum and minimum points of the given function. Indicate for each extremum whether it is absolute or relative. Sketch a graph of the function.

107. $f(x) = x^2 - 6x$ for $0 \le x \le 10$
108. $f(x) = 3x^3 - 15x^2 + 36x$ for $0 \le x \le 10$
109. $f(x) = x^3 + 3x^2 - 9x - 7$ for $-6 < x < 2$
110. $f(x) = 8x - x^2$ for $0 < x < 12$
111. $f(x) = x^3 - 12x$ for $-5 \le x \le 5$
112. $f(x) = 1/x$ for $1 \le x \le 5$
113. $f(x) = x^4$ for $-1 \le x \le 1$
114. $f(x) = 5 - x^5$ for $x \ge 0$
115. $f(x) = (x - 2)^4$ for $0 \le x \le 4$
116. $f(x) = 10x/(x^2 + 2)$ for $0 \le x \le 5$
117. $f(x) = -2/(x + 3)$ for $x \ge 0$

12.7 APPLIED MAXIMA AND MINIMA PROBLEMS

The following examples illustrate only a few of the many common problems which require that some quantity be maximized or minimized.

Example 17

Existing fence

x | A = Enclosed area | x

y

Area enclosed by fence A rancher with 800 feet of fencing wishes to enclose a rectangular holding region for cattle. The region will border on one side of an existing fence, as shown at left.

What is the largest area that can be enclosed? What are the dimensions of this region?

The quantity to be maximized is the area, which we shall denote A. The value of this variable is dependent upon the dimensions of the rectangle, which we shall denote as x = width-by-y = length. Then,

$$A = x \cdot y$$

The dimensions of the lot, x and y, are in turn related to the total amount of fencing available, 800 feet. That is,

$$x + x + y = 800$$
$$2x + y = 800$$
$$y = 800 - 2x$$

Thus, the area A can be expressed

$$A = A(x) = x \cdot y$$
$$= x(800 - 2x)$$
$$= 800x - 2x^2$$

The domain of x is restricted, by the amount of fencing available, to $0 < x < 400$. The problem is to maximize the value of

$$A(x) = 800x - 2x^2 \qquad \text{for } 0 < x < 400$$

We begin by determining the first derivative

$$A'(x) = 800 - 4x$$

and the critical value of x,

$$800 - 4x = 0$$
$$x^* = 200$$

Because $A''(x) = -4$, we see that

$$[A''(200) = -4] < 0$$

and conclude that $x = 200$ yields a maximum on the function A. Because it is the only extremum within the open interval on which the function is defined, it is the absolute maximum (see Figure 12.27).

Figure 12.27

$A(x) = 800x - 2x^2$
$0 < x < 400$

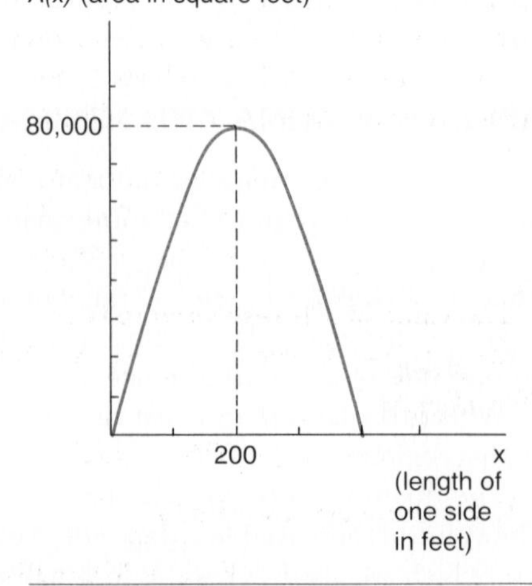

Given that the optimal width is $x = 200$, we find the corresponding length y to be $y = 800 - 2(200) = 400$. The dimensions of the area should be two parallel sides of 200 feet and a third side, parallel to the existing fence, of 400 feet.

The maximum area that can be enclosed with the available fencing is A = (200)(400) = 80,000 square feet. []

Example 18 **Dimensions of a box** An open-top box is to be made from a 24-inch-square sheet of cardboard by cutting smaller squares out of each corner of the sheet and folding up the edges, as shown. What should be the dimensions of the box so as to maximize its volume?

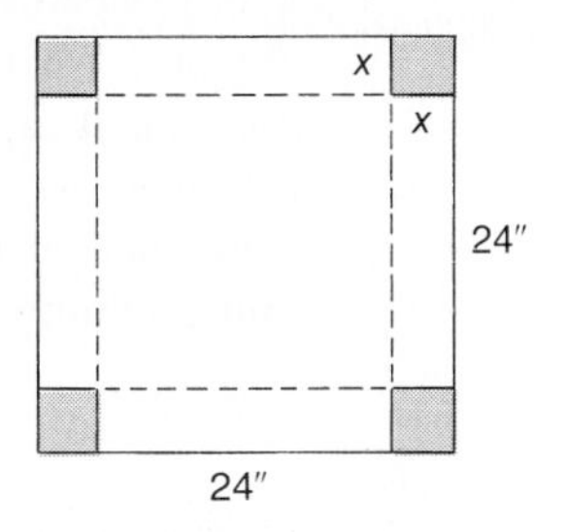

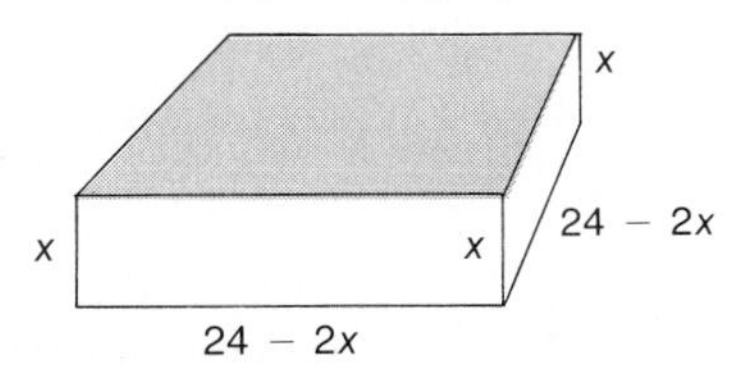

The size of the square to be cut from each corner of the rectangle is the independent variable; the dimensions of the box, and thus its volume, are a function of the size of this square. If the squares to be cut from each corner of the rectangle are of dimensions x by x, the height of the box is x. Its base will be of dimensions $(24 - 2x)$ by $(24 - 2x)$. The volume of the box, which is the quantity to be maximized, and which we shall denote V, is given by

$$\begin{aligned} V &= \text{Length} \times \text{Width} \times \text{Height} \\ &= (24 - 2x) \cdot (24 - 2x) \cdot x \\ &= 576x - 96x^2 + 4x^3 \end{aligned}$$

The value of x is restricted to be $x > 0$ and, because of the dimensions of the cardboard rectangle, $x < 12$. Thus, we want to find the absolute maximum of

$$V = V(x) = 576x - 96x^2 + 4x^3 \qquad 0 < x < 12$$

Differentiating, we obtain

$$V'(x) = 576 - 192x + 12x^2$$

and

$$V''(x) = -192 + 24x$$

Setting $V'(x) = 0$, we obtain the critical values of x, as

$$V'(x) = 576 - 192x + 12x^2 = 0$$
$$12(12 - x)(4 - x) = 0$$
$$x^* = 4 \text{ and } x^* = 12$$

Because $x = 12$ is not within the domain of the function, $x^* = 4$ is the only critical value of importance to the model.

Thus, using the Second Derivative Test, we find that $[V''(4) = -96] < 0$, indicating that a maximum occurs at $x = 4$. Indeed, because it is the only extremum within the open interval on which the function is defined, it is the absolute maximum. Thus, the maximum volume is achieved by cutting squares of dimension four inches by four inches from each corner of the sheet of cardboard (see Figure 12.28). This maximum volume is

$$V(4) \doteq (24 - 2(4))^2(4)$$
$$= 1{,}024 \text{ cubic inches}$$

The resulting box will have dimensions of

$$\text{Height} = 4 \text{ inches}$$
$$\text{Length} = 24 - 2(4) = 16 \text{ inches}$$
$$\text{Width} = 24 - 2(4) = 16 \text{ inches}$$

[]

Figure 12.28

$V(x) = 576x - 96x^2 + 4x^3 \qquad 0 < x < 12$

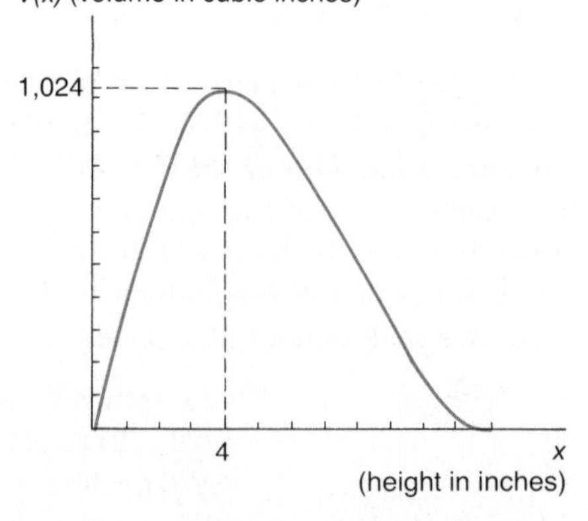

Example 19 **Volume of a cylindrical container** A container-manufacturing company has received an order for a quantity of cylindrical containers, each with a capacity of 800 cubic inches. The only other requirement concerning the dimensions of the container is that the radius must not exceed eight inches.

Material used to make the top and bottom of the container costs \$0.10 per square inch, while material used in the sides of the container costs \$0.06 per square inch. Find the dimensions of the container that will minimize the total cost.

The top, bottom, and lateral sides of the cylindrical container have surfaces as shown:

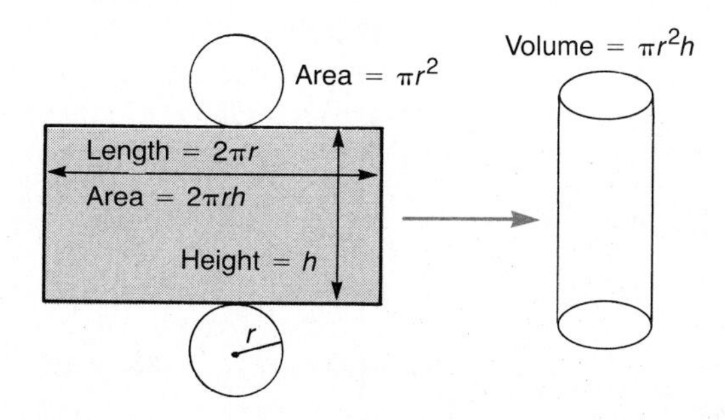

The volume of a cylindrical container is determined by

$$V = \pi r^2 h$$

Now, although both the radius r and the height h of the container to be produced are variables, they are related by the fact that volume must be 800 cubic inches; that is,

$$V = \pi r^2 h = 800$$
$$h = 800/\pi r^2$$

The cost of the material used in making the container is

$$\begin{aligned} \text{Cost} &= \text{Cost of top and bottom} + \text{Cost of lateral sides} \\ &= [(0.10)(2)(\pi r^2)] + [(0.06)(2\pi rh)] \\ &= 0.2\pi r^2 + 0.12\pi r(800/\pi r^2) \end{aligned}$$

or

$$C(r) = 0.2\pi r^2 + 96/r \qquad 0 \leq r \leq 8$$

To find the critical values of r, we first differentiate $C(r)$ with respect to r to obtain

$$C'(r) = 0.4\pi r - 96/r^2$$

Then, setting $C'(r) = 0$ and solving, we find

$$\begin{aligned} C'(r) = 0.4\pi r - 96/r^2 &= 0 \\ (0.4\pi r^3 - 96)/r^2 &= 0 \\ 0.4\pi r^3 - 96 &= 0 \\ 0.4\pi r^3 &= 96 \\ r^3 &= 240/\pi \qquad (\text{Note: } \pi \simeq 3.14159) \\ r^* &\simeq 4.24 \text{ inches} \end{aligned}$$

$r^2($

The second derivative $C''(r) = 0.4\pi + 192/r^3$ is positive for all r in the domain of the function. Hence, the graph of the function is concave upward for all r, and any interior-point r that is an extremum must be a minimum; indeed, it must be the absolute minimum (see Figure 12.29).

Figure 12.29

$C(r) = 0.2\pi r^2 + 96/r$
$0 \le r \le 8$

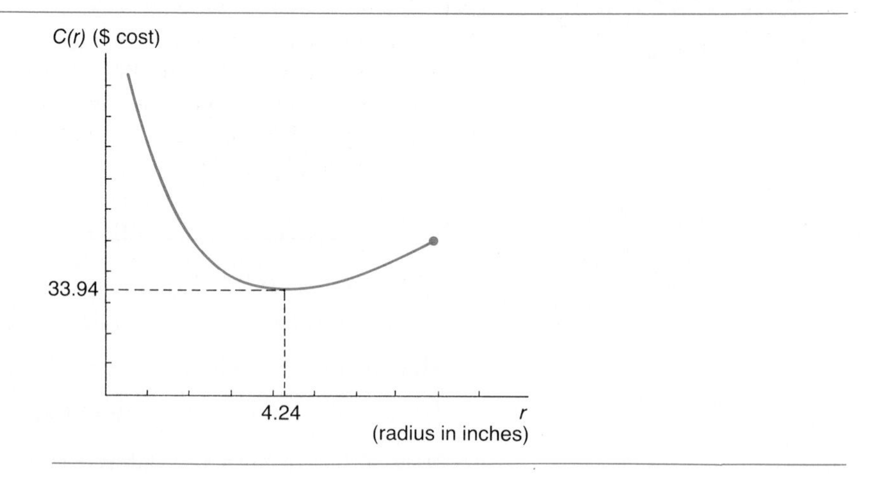

Thus, the radius of the container should be 4.24 inches. Its height will be

$$h = 800/\pi(4.24)^2$$
$$= 14.16 \text{ inches}$$

The cost of the material required to make the container will be

$$C(4.24) = 0.2\pi(4.24)^2 + 96/4.24$$
$$= \$33.94$$

Suppose that it is necessary for the radius to be some increment of 1/4 inch. We then would need to round 4.24 both up and down to the nearest quarter inch and compare the costs

$$C(4) = (0.2)(3.14159)(4)^2 + 96/4 = \$34.05$$

and

$$C(4.25) = (0.2)(3.14159)(4.25)^2 + 96/4.25 = \$33.94$$

The container with radius $r = 4.25$ inches has a cost of \$33.94 (rounded to the nearest cent). This container would have height of

$$h = 800/(3.14159)(4.25)^2 = 14.10 \text{ inches}$$ []

Example 20 **Utility power line cost** An electric utility must run a power line from a generator located at point A (see Figure 12.30) to an offshore oil-drilling

Figure 12.30

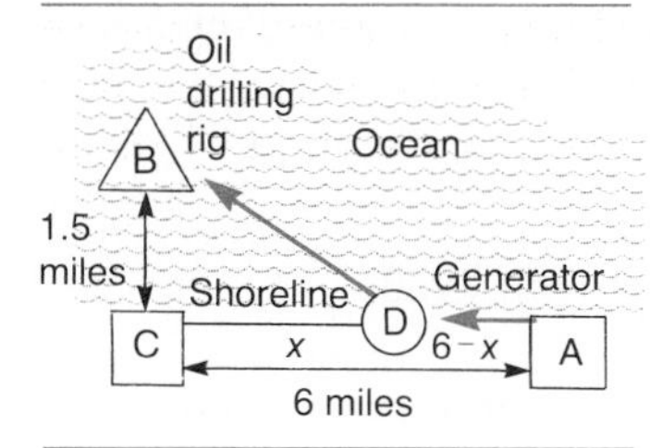

rig located at point B. The offshore rig is 1.5 miles from the nearest shore point, point C. Point C is six miles down the shoreline from A.

The power line is to run straight down the shoreline from A toward C to a point D; from D the power line will go, underwater, straight to B. It costs $2,500 per mile to run a power line underwater and $2,000 per mile to run a line underground along the shoreline. Where should point D be located in order to minimize the total cost of running the line from the generator to the oil rig?

Let us define x as the number of miles from C to D with $0 \leq x \leq 6$. Thus, $6 - x$ = number of miles from A to D. The length of the power line running underwater is the hypotenuse of a right triangle, and by the Pythagorean theorem,

$$c^2 = a^2 + b^2$$
$$c = \sqrt{a^2 + b^2}$$

Here,

$$c = \sqrt{x^2 + (1.5)^2} = \text{Length of the underwater power line}$$

Total cost for the power line is

$$C(x) = \text{Cost of underwater line} + \text{Cost of underground line}$$
$$C(x) = (2{,}500)\sqrt{x^2 + (1.5)^2} + (2{,}000)(6 - x) \qquad 0 \leq x \leq 6$$

We find the minimum point on the cost function by finding the first derivative, as

$$C'(x) = \frac{2{,}500x}{\sqrt{x^2 + 2.25}} - 2{,}000$$

Setting this derivative equal to zero, we obtain the critical values of x

$$\frac{2{,}500x}{\sqrt{x^2 + 2.25}} - 2{,}000 = 0$$
$$\frac{2{,}500x}{\sqrt{x^2 + 2.25}} = 2{,}000$$
$$2{,}000\sqrt{x^2 + 2.25} = 2{,}500x$$
$$\sqrt{x^2 + 2.25} = 1.25x$$
$$x^2 + 2.25 = (1.25x)^2 = 1.5625x^2$$
$$0.5625x^2 = 2.25$$
$$x^2 = 4$$
$$x = 2$$

The second derivative of the cost function is positive for all x within the domain of the function, meaning that the function is always concave upward. Thus, the extremum represented by $x = 2$ is a minimum.

The power line will be run $6 - 2 = 4$ miles along the shoreline from the generator to point D and then will continue underwater from point D to the oil-drilling rig. The distance from D to the rig is given by

$$\sqrt{2^2 + 2.25} = 2.5 \text{ miles}$$

Total cost of the power line is given by

$$C(2) = 2{,}500(2.5) + 2{,}000(4) = \$14{,}250$$ [ ]

Example 21 **Timing of equipment replacement** A question of major concern to many decision makers involves the optimal time to replace capital equipment. Two opposing types of cost must be considered—CAPITAL COST and OPERATING COST.

CAPITAL COST is defined as PURCHASE COST less SALVAGE VALUE. For example, if a piece of equipment initially costs $12,000 and is subsequently sold for $2,000, its capital cost is $10,000. In determining when to replace a piece of equipment, management usually focuses on AVERAGE CAPITAL COST, which tends to decrease with time.

OPERATING COST is the cost of owning, operating, and maintaining the equipment and includes such expenses as insurance, gasoline or electricity, and repairs. AVERAGE OPERATING COST tends to increase with time because the equipment tends to become less efficient as it ages.

Rent-A-Truck maintains a fleet of rental trucks and wants to determine how long it should keep its trucks before replacing them. The trucks are purchased new, and fully equipped, for $14,000. Salvage value of a truck is given by

$$S(x) = 12{,}000 - 0.07x$$

where x is the number of miles the truck is driven.

The average operating cost, in dollars per mile, is

$$f(x) = 0.0000004x + 0.18$$

where x is the number of miles the truck is driven.

Management of Rent-A-Truck would like to determine the number of miles a truck should be driven before replacement in order to minimize the sum of average capital cost and average operating cost. (See Figure 12.31.)

Capital cost is the purchase price minus salvage value and is a function of the number of miles the truck is driven, as

$$g(x) = 14{,}000 - (12{,}000 - 0.07x) = 2{,}000 + 0.07x$$

Average capital cost, per mile driven, then is given by

$$\bar{g}(x) = \frac{2{,}000 + 0.07x}{x}$$

$$= (2{,}000/x) + 0.07 \qquad x > 0$$

Figure 12.31

Timing of Equipment-Replacement Decision

(A) Average capital cost

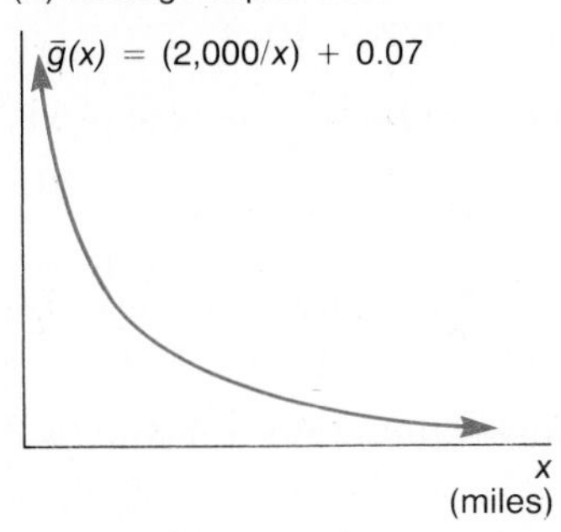

(B) Average operating cost

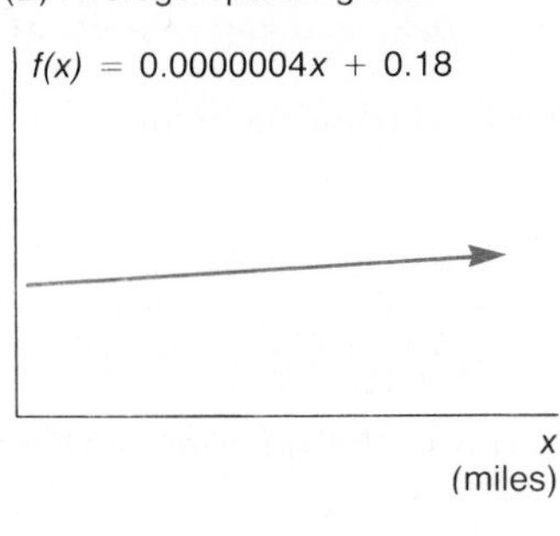

(C) Sum of average capital cost and average operating cost

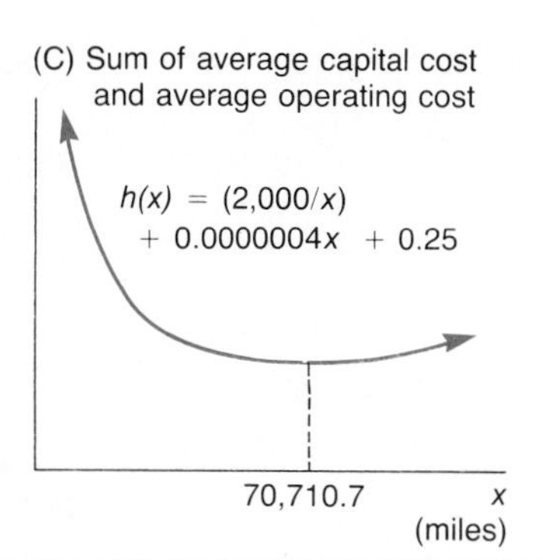

The sum of average capital cost and average operating cost is

$$h(x) = [(2{,}000/x) + 0.07] + [0.0000004x + 0.18]$$
$$= (2{,}000/x) + 0.0000004x + 0.25 \qquad x > 0$$

To find x such that this function is at a minimum point, we set the first derivative

$$h'(x) = (-2{,}000/x^2) + 0.0000004$$

equal to zero and solve for

$$(-2{,}000/x^2) + 0.0000004 = 0$$
$$x^* = 70{,}710.7$$

We verify that this is a minimum by observing that the second derivative

$$h''(x) = 4000/x^3$$

is positive for all x in the domain of the function, indicating that the function is concave upward for all x. The interior point extremum is an absolute minimum.

The sum of the average capital cost and the average operating cost will be a minimum if the truck is driven 70,710.7 miles before replacement. [ ]

Example 22 **Maximum profit** A manufacturer sells x units of a product at a dollar price of $p = p(x) = 6{,}565 - 10x - 0.1x^2$ per unit. The cost of manufacturing the product is

$$C(x) = 0.05x^3 - 5x^2 + 20x + 250{,}000 \qquad \text{for } 0 \le x \le 150$$

How many units should be produced and sold to maximize the resulting profit?

Total revenue R realized from the sale of x units of product will be

$$\text{Total revenue} = (\text{Price per unit}) \cdot (\text{Number of units sold})$$
$$R(x) = (6{,}565 - 10x - 0.1x^2)\ x = 6{,}565x - 10x^2 - 0.1x^3$$

Profit P realized from the manufacture and sale of these x units of product will be

$$\text{Profit} = \text{Revenue} - \text{Cost}$$
$$P(x) = (6{,}565x - 10x^2 - 0.1x^3) - (0.05x^3 - 5x^2 + 20x + 250{,}000)$$
$$= -0.15x^3 - 5x^2 + 6{,}545x - 250{,}000 \qquad 0 \le x \le 150$$

To determine the quantity x that maximizes the profit function, we first find

$$P'(x) = -0.45x^2 - 10x + 6{,}545$$

and, setting this first derivative equal to zero, we solve for critical values of x. Using the Quadratic Roots Formula, we find $x^* = 110$. (The other root is outside the domain of f.)

To ascertain that this critical value represents a maximum, we use the second Derivative Test, as

$$P''(x) = -0.9x - 10 \qquad \text{and} \qquad [P'(110) = -109] < 0$$

indicating that this critical value does indeed represent a maximum. Profit for the firm at a production level of $x = 110$ units will be

$$P(110) = -0.15(110)^3 - 5(110)^2 + 6{,}545(110) - 250{,}000 = \$209{,}800$$

In looking for the absolute maximum, we also evaluate the endpoints of the domain of the function, as

$$P(0) = -250{,}000 \qquad \text{and} \qquad P(150) = \$113{,}000$$

Hence, the production level $x = 110$ units represents the absolute maximum on the profit function. If the firm produces at this level, its profit will be $209,800. []

Example 23 **Optimal price for maximum profit** Owners of Oceanus Oasis, a resort village, have 200 seaside cottages which they rent by the week to summer vacationers. When weekly rent is $400, all units are occupied throughout the summer months. Owners of the village would like to increase the rental fee, but experience has indicated that for *each* $20 per week increase in rent (above $400), five units will be vacant each week. What weekly rental fee should be charged to maximize weekly revenue?

The quantity to be maximized is revenue realized, R, each week and is determined by

$$\text{Revenue} = (\text{Number of cottages rented}) \cdot (\text{Weekly rent per cottage})$$

The controllabe independent variable is the number of \$20 increases in rent put into effect. Let us denote this as

$$x = \text{number of \$20 increases in weekly rental fee put into effect}$$

Because after 40 such increases all cottages would be vacant, the value set of x is restricted to $0 \leq x \leq 40$.

The number of cottages rented will be $(200 - 5x)$, while the weekly rental per cottage is $(400 + 20x)$. Hence, the revenue function is given by

$$R(x) = (200 - 5x) \cdot (400 + 2x) = 80{,}000 + 2{,}000x - 100x^2 \qquad 0 \leq x \leq 40$$

To find the value of x at which this function reaches a maximum, we differentiate to obtain $R'(x) = 2{,}000 - 200x$. Setting the first derivative equal to zero, we find the critical value of x to be

$$2{,}000 - 200x = 0$$
$$x^* = 10$$

Then, by the Second Derivative Test, we see that $R''(x) = -200$ is negative for all x, indicating that this extremum is a maximum point on the graph.

To locate absolute maxima or minima, we also consider the endpoint of the domain of the function. Thus, we evaluate

$$R(0) = 80{,}000 + 200(0) - 100(0)^2 = \$80{,}000$$
$$R(10) = 80{,}000 + 200(10) - 100(10)^2 = \$90{,}000$$
$$R(40) = 80{,}000 + 200(40) = 100(40)^2 = 0$$

Hence, at $x = 10$ there is an absolute maximum.

With 10 increases of \$20 each in the weekly rental fee for a cottage, the rent would increase from \$400 to \$600. The number of cottages rented each week would decrease by (10)(5) = 50, from the present level of 200 down to 150. Revenue realized would be \$600(150) = \$90,000, and would be at a maximum level. []

Example 24 **Optimal price with a cost function** Owners of the resort village Oceanus Oasis (from the preceding example) wonder whether the strategy that yields maximum revenue will also yield maximum profit. Their weekly cost for cleaning and maintenance on a cottage which is rented is \$90, while this cost is only \$10 on a cottage which is vacant.

Given that the number of cottages rented each week is a function of the number x of \$20 increases in weekly rent (above the present \$400),

$$\text{Number of cottages rented} = 200 - 5x$$

and

$$\text{Number of cottages vacant} = 5x$$

Then the cost (of cleaning and maintenance) function is

$$\begin{aligned} C(x) &= 90(200 - 5x) + 10(5x) \\ &= 18{,}000 - 400x \end{aligned}$$

The profit function, thus, is

$$\begin{aligned} \text{Profit} &= \text{Revenue} - \text{Cost} \\ P(x) &= (80{,}000 + 2{,}000x - 100x^2) - (18{,}000 - 400x) \\ &= 62{,}000 + 2{,}400x - 100x^2 \qquad 0 \le x \le 40 \end{aligned}$$

To find the critical values of x for this function, we find the first derivative $P'(x) = 2{,}400 - 200x$, and, setting this equal to zero, solve

$$\begin{aligned} 2{,}400 - 200x &= 0 \\ x^* &= 12 \end{aligned}$$

To locate the absolute maximum, we evaluate $P(x)$ at this critical value as well as at each of the endpoints of the interval of relevant x values, as

$$\begin{aligned} P(0) &= 62{,}000 + 2{,}400(0) - 100(0)^2 = 62{,}000 \\ P(12) &= 62{,}000 + 2{,}400(12) - 100(12)^2 = 76{,}400 \\ P(40) &= 62{,}000 + 2{,}400(40) - 100(40)^2 = -2{,}000 \end{aligned}$$

Thus, at $x = 12$, there is an absolute maximum for the function.

The profit-maximizing strategy—which is not the same as the revenue-maximizing strategy—would be to increase weekly rent on each cottage by (12)($20) = $240, or from $400 to $640. This would mean that (5)(12) = 60 cottages would be vacant each week; only 140 of the 200 cottages would be occupied. Revenue would total ($640)(140) = $89,600. Cost of cleaning and maintenance would be

$$C(12) = 18{,}000 - 400(12) = \$13{,}200$$

leaving a weekly profit of $89,600 − $13,200 = $76,400. [ ]

Example 25 **Optimal level of advertising expenditures** A team of analysts at Tennex Manufacturing Company has been carefully studying the sales-forecasting function of the firm. They began working on the premise that the number of units of a product the company is able to sell during a year is affected not only by the price of the item but also by its quality (which is, in turn, related to the quality-control procedures adopted by the firm along with its endeavors in researching and developing a top-grade product). In addition, sales are affected by the extent to which the company advertises the product. (See Figure 12.23.) After considerable experimentation with the relationship between number of units sold, selling price of the unit, dollars spent on quality control, engineering, research and development, and dollars spent on advertising, the team developed this function:

$$Q = 200 - 15S + 0.2AE + E + 2A$$

where

Q = Number of units sold (in millions)
S = Selling price per unit (in dollars)
A = Advertising expenditures (in millions of dollars)
E = Cost of quality control and of research and development (in millions of dollars)

Total revenue realized from the sale of the product is given by

$$\text{Revenue} = (\text{Selling price per unit}) \times (\text{Number of units sold})$$
$$R = S(200 - 15S + 0.2AE + E + 2A)$$
$$= 200S - 15S^2 + 0.2AES + ES + 2AS$$

where

R = Total revenue (in millions of dollars)

The analysts also investigated the cost pattern of the company and actually had a much easier time of isolating costs and identifying cost relationships than they had in identifying the sales relationships. The total fixed costs, excluding quality control and advertising, were determined to be \$12.5 million. Variable cost was determined to be \$0.50 per unit. Thus, total cost is given by

$$C = 12.5 + 0.5Q + E + A$$

where

C = Total cost (in millions of dollars)

Figure 12.32

Relationships Existent among Variables in Tennex Manufacturing Company Problem

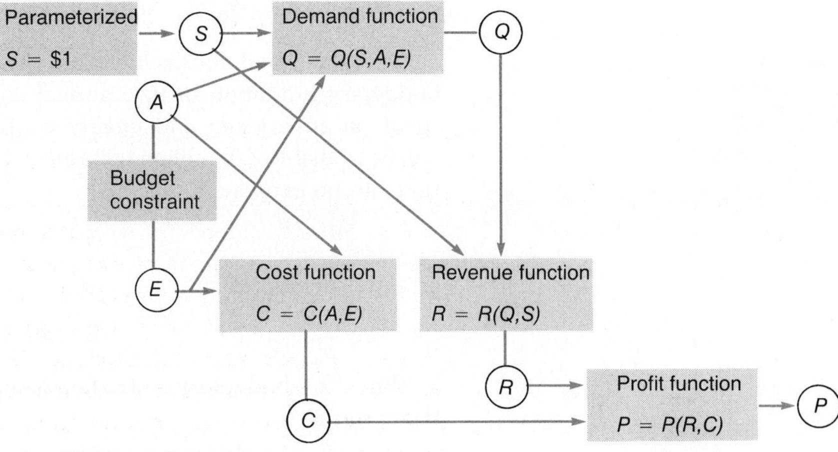

The profit function, based on the revenue and cost functions developed by the analysts, is

$$\begin{aligned}\text{Profit} &= \text{Revenue} - \text{Cost}\\ P &= (200S - 15S^2 + 0.2AES + ES + 2AS - (12.5 + 0.5Q + E + A)\\ &= 200S - 15S^2 + 0.2AES + ES + 2AS - 12.5 - 0.5Q - E - A\end{aligned}$$

where

$P =$ Total profit (in millions of dollars)

The profit function as expressed above contains four independent variables but can be simplified somewhat if Q, quantity sold, is eliminated as a variable. This can be accomplished by substitution, since Q is a function of the other independent variables. Thus

$$\begin{aligned}P &= 200S - 15S^2 + 0.2AES + ES + 2AS - 12.5 - E - A - 0.5(200\\ &\quad - 15S + 0.2AE + E + 2A)\\ &= 207.5S - 15S^2 + 0.2AES + ES + 2AS - 112.5 - 1.5E - 2.5A\\ &\quad - 0.1AE\end{aligned}$$

Total profit can thus be considered to be a function of three independent variables: selling price of the product (S), expenditures for quality control and for research and development (E), and expenditures on advertising (A).

The product has recently been selling for $1 a unit, and the firm would like to determine the best configuration of expenditures on quality control and advertising, assuming that this $1 selling price is maintained. If the selling price is to be considered fixed at $1 per unit, the total profit function can be rewritten as

$$\begin{aligned}P &= 207.5(1) - 15(1)^2 + 0.2AE(1) + E(1) + 2A(1) - 112.5 - 1.5E\\ &\quad - 2.5A - 0.1AE\\ &= 80 + 0.1AE - 0.5E - 0.5A\end{aligned}$$

One additional constraint is imposed upon the situation. There is a budgetary limitation on the annual amount of money available to be spent on advertising and quality control. For the coming year, there will be available $20 million to be allocated to these purposes. This restriction can be expressed as

$$E + A = 20$$

or, equivalently, as

$$E = 20 - A$$

When this budgetary restriction is substituted into the profit function, the result is

$$\begin{aligned}P &= 80 + 0.1A(20 - A) - 0.5(20 - A) - 0.5A\\ &= 70 + 2A - 0.1A^2\end{aligned}$$

The level of advertising that will yield the maximum profit can be determined by finding the first derivative of the profit function, setting the derivative equal to zero, and solving for the critical values of A, as

$$P = f(A) = 70 + 2A - 0.1A^2$$
$$f'(A) = 2 - 0.2A$$

Then

$$f'(A) = 2 - 0.2A = 0$$
$$A^* = 10$$

Continuing with the Second Derivative Test, we compute

$$f''(A) = -0.2$$

and

$$[f''(10) = -0.2] < 0$$

indicating that the critical value does represent a maximum on the graph of the function.

Hence, profit is maximized when advertising expenditures are $10 million. This means that expenditures on quality control and on research and development will also be at the $10 million level. Quantity sold, under these conditions, will be

$$Q = 200 - 15(1) + 0.2(10)(10) + 10 - 2(10)$$
$$= 195 \text{ million units}$$

yielding a total revenue, where the unit selling price is $1, of $195 million.

Total cost will be

$$C = 12.5 + 0.5(195) + 10 + 10$$
$$= \$105 \text{ million}$$

and the maximum attainable profit will be $90 million.

The firm believes that the relationships outlined would not be altered by a minor change in selling price and, consequently, would like to know the effect on total profit of a 5 percent increase and of a 5 percent decrease in selling price. The total amount of money available to be spent on advertising and quality control cannot be increased above the $20 million level.

If selling price were increased by 5 percent, the new selling price would be $1.05. Substituting $S = 1.05$ and $E = 20 - A$ into the profit function, we obtain

$$P = g(A) = 207.5(1.05) - 15(1.05)^2 + 0.2A(20 - A)(1.05) + (20 - A)(1.05)$$
$$+ 2A(1.05) - 112.5 - 1.5(20 - A) - 2.5A - 0.1A(20 - A)$$
$$= 79.8375 + 2.25A - 0.11A^2$$

To locate the maximum point on this function, we compute the first derivative

$$g'(A) = 2.25 - 0.22A$$

and

$$g'(A) = 2.25 - 0.22A = 0$$
$$A^* = 10.227$$

Then

$$[g''(A) = -0.22] < 0$$

for $A = 10.227$, thus indicating that this critical value represents a maximum on the graph of the function.

The maximum profit for this strategy would be

$$P = g(10.227) = 79.8375 + 2.25(10.227) - 0.11(10.227)^2$$
$$= \$91.334 \text{ million}$$

This is $1.344 million greater than the profit attainable with the present selling price of $1.

Similar computations using a selling price of $0.95 indicate that the maximum profit attainable with this selling price would be only $68 million. Thus, of the three prices considered, the optimum would be $1.05. At this price level, advertising expenditures should be $10.227 million and expenditures on quality control should be $9.773 million. The quantity sold would be approximately 234.467 million units. []

EXERCISES

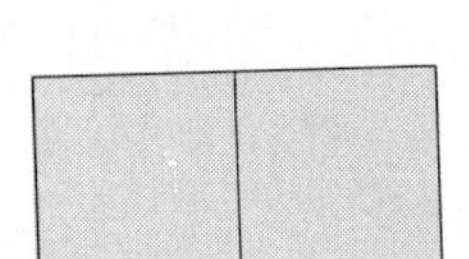

118. A farmer has 600 feet of fencing that he wishes to use to enclose a large rectangular field which he will then divide into two fields with a fence parallel to one of the sides, as shown at left. What is the largest area he can enclose with the fencing he has available? What should be the dimensions of the rectangle?

119. A farmer wishes to enclose with a fence a rectangular field that is adjacent to an existing stone wall so that no fencing will be required along the stone wall. What are the dimensions of the field if 1,200 feet of fencing are available and the area of the field is to be the maximum possible? (The stone wall is 750 feet long, so that the dimensions of the side parallel to the wall cannot exceed 750 feet.)

120. A fence is to be built around an outdoor storage site. The area to be enclosed is 900 square feet. The fence along three sides is to be made of material costing $5 per linear foot, while the material for the fourth side costs $9 per linear foot. Find the dimensions of the rectangle for the most-economical fence.

121. Saad Case, chief design engineer for Cardboard Containers, Inc., has developed a revolutionary new procedure for constructing open-top boxes. The procedure involves removing a square of sides

x inches from each corner of a flat sheet of cardboard and turning up the edges to form the sides of the box. The top is left open.

Given a sheet of cardboard measuring 30 by 36 inches, what should be the size of the cut-out squares so that the resulting box has maximum volume? What is the volume of the resulting box?

Given a sheet of cardboard 48 inches square, what should be the size of the cut-out squares so that the resulting box has maximum volume? What is the volume of the resulting box?

122. A carpenter wishes to construct a rectangular box with square top and bottom and with a 1,800-cubic-inch capacity. Materials used in the top and bottom of the box cost \$0.12 per square inch, while materials used in the sides cost \$0.08 per square inch. What are the dimensions of the least-costly box?

123. A firm wishes to produce a cylindrical container with a capacity of 500 cubic inches. The top and bottom of the container are made from material costing \$0.08 per square inch, while the sides of the container are made from materials costing \$0.06 per square inch. Find the dimensions of the container that will minimize its cost.

(Note: the area of the lateral side is $2\pi rh$, while the area of either the top or the bottom of the container is πr^2, where r is the radius of the top or bottom of the container and h is its height.)

124. A rectangular billboard must have a total area of 1,800 square feet. The sign will have 2-foot margins at top and bottom and 1.5-foot margins at each side. Find the dimensions of the sign so that the center area, used for displaying advertising materials, is a maximum.

125. A printer has contracted to provide one-page advertising circulars which have 120 square inches of printing space. The printing area must have 1-inch margins on each side and 1.5-inch margins at the top and bottom of the page, but no other dimensions are specified. What should the dimensions of the page be so as to minimize cost of paper?

126. It costs $(8 + 0.02x)$ dollars per mile to drive a truck x miles per hour. In addition, a driver receives \$18 per hour. How fast should the truck be driven to minimize the cost per mile?

127. The rate at which hydrocarbons are emitted from an automobile engine is a function of the speed at which the automobile is operating. If the emission rate $T(x)$ in milligrams per minute at speed x (in miles per hour) is

$$R(x) = xe^{-0.016x}$$

at what speed is the emission rate at a maximum?

128. A real estate developer has built a hunting and fishing club on an island situated in a large lake. The club is 2.5 miles from the

nearest point C on shore (as shown in the figure below). The power lines to the club must be connected to a generator located in a straight line 10 miles from point C. The power lines will run underground from the generator to some point D, located between C and the generator, and then underwater from D to the club house. If the cost per foot of running the power line underground is \$11 and the cost per foot of running it underwater is \$18, where should point D be located to minimize the total cost of the power lines?

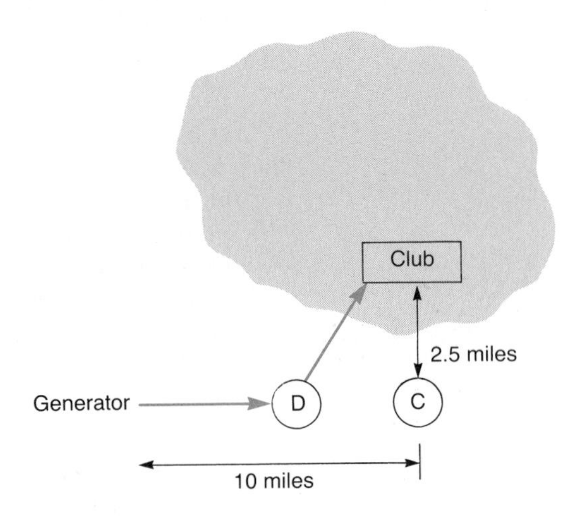

129. A company maintains a fleet of automobiles for use by its salespersons and executives. The cars are purchased new for \$10,500 and have a resale (or salvage) value which is a function of the number x of miles the vehicle is driven, as

$$S(x) = 9{,}000 - 0.085x$$

The average operating cost of an automobile is also a function of the number x of miles driven, as given by

$$f(x) = 0.15 + 0.0000006x$$

Determine the number of miles an automobile should be driven before replacement in order to minimize the sum of average capital cost and average operating cost.

130. A firm has determined that its weekly profit function is given by

$$P(x) = 95x - 0.05x^2 - 5000 \qquad 0 \le x \le 1{,}000$$

where $P(x)$ is profit, in dollars, and x is the number of units of product sold.

For what value of x does profit reach a maximum? What is this maximum profit?

131. Cousin Cora's Chicken Coops are a fast-food chain serving fried chicken and chicken filet sandwiches. Market research indicates that the profits of the chain are related to the number of "coops" open in a region by the function

$$P(x) = 16 - x \ln (x^2/115) \qquad 1 \leq x \leq 20$$

where $P(x)$ is profit, in millions of dollars, and x is the number of coops open in the area.

Determine the optimal number of coops to open in a region. What will be the resulting profit?

132. The demand equation for a certain product is

$$p(x) = 500 - 0.0125x$$

where $p(x)$ is the price per unit, in dollars, and x is the quantity demanded.

For what value of x is revenue a maximum? What is the maximum possible revenue? What is the selling price per unit that maximizes revenue?

133. A manufacturer of electric coffeepots estimates its profit function to be

$$P(x) = 60{,}000 + 20{,}000x - x^2$$

where x is the number of dollars spent on advertising.

What amount spent on advertising will yield maximum profit? What is the maximum profit?

134. A manufacturer estimates that it costs $0.1x^2 + 10x$ dollars to produce x units of a product. If the item is offered for sale at $\$p$ per unit, $1{,}000 - 0.04p$ units will be sold. Where should the product be priced in order to realize the maximum profit?

135. A company finds that the revenue realized from selling x units of product per week is given by

$$R(x) = 10.5x - (x^2/1{,}000) \qquad 0 \leq x \leq 10{,}000$$

The cost function for the firm is

$$C(x) = 7{,}500 + 2.5x \qquad 0 \leq x \leq 10{,}000$$

What output level will result in the maximum profit? What is this maximum profit?

136. A firm manufactures items which it can sell for \$150 each in whatever quantities it can produce. The cost function for the firm is

$$C(x) = (1/3)x^3 - 5x^2 + 170x + 300 \qquad 0 \leq x \leq 8{,}000$$

How many items should the firm manufacture in order to maximize profit?

137. The owner of a mango orchard estimates that if 20 trees are

planted per acre, each tree will yield an average of 60 bushels of fruit each year. For each additional tree planted per acre (up to 10 additional trees, or a total of 30 trees per acre), average yield per tree drops two bushels. How many trees should be planted per acre to maximize the yield per acre? What is the maximum yield per acre?

138. The owner of a peach orchard estimates that if she begins harvesting the peach crop this weekend, she will harvest 1,200 bushels, which she can sell for $5 per bushel. If she waits, she will have a larger harvest, but price per bushel will decrease because of an increased supply of peaches on the market. For each day that she waits (up to a maximum of 30 days), the crop will increase by 50 extra bushels, but the price will decrease by $0.20 per bushel. When should the crop be harvested in order to maximize revenue?

139. Fly-by-Nite Airlines has remodeled one of its planes into a cocktail bar and disco. Each Friday night patrons drink and dance while flying high above downtown Omaha.

The craft will accommodate a maximum of 84 patrons, and Freddie Flaker, owner of Fly-by-Nite, finds that at $100 per person he has a capacity crowd each Friday night. He is considering increasing the fee, but research studies show that, for each $5 increase in fee, the number of patrons would decrease by two. What price should be charged in order to maximize revenue?

140. Mannix Manufacturing Company has developed a deluxe boat, trailer, and outboard-motor rig and is considering marketing the rig before the next boating and fishing season. The firm's market research department, after a thorough study of the potential market, has forecast the following relationship between selling price and demand:

If the selling price is:	The estimated demand (in units) during the first season will be:
$1,000	3,000
$2,000	2,000
$3,000	1,000
$4,000	0

Unit selling prices falling between those specifically listed are possible, and demand would be proportionally adjusted.

After careful analysis of the production and selling process, the cost-accounting department, working with the production-control group, has determined that the fixed costs associated with the manufacture of the boating rig would be $500,000 in the first year. This figure includes tooling and setup, depreciation on the new machinery and equipment that would have to be acquired if the boat were produced, interest charges on the funds that would have to be borrowed, and other items of this nature, as well as fixed overhead for the manufacturing, administrative, and selling functions. Direct labor, direct material, and variable manufacturing expenses would be \$800 per unit, while other variable costs (marketing, sales commissions, and other such costs) would be \$200 per unit.

Mannix would like to set the selling price of the boating rig at such a level that the season's profit on the venture would be as high as possible. What should the selling price be? What is the maximum possible profit?

12.8 CURVE SKETCHING

Sketching an accurate graph of a linear or even a quadratic function is a fairly straightforward process. This is not always the case with more-complex functions. However, we have learned many things about functions that, when taken together, will allow us to sketch a sufficiently accurate graph of even the most complicated function.

In sketching the curve, we will use information about these properties of functional relationships.

1. *Domain and range.* Are there any environmental, or mathematical, restrictions imposed on the value set of the independent variable? of the dependent variable?
2. *Discontinuities.* The function will be discontinuous at any individual x not included in the domain. Discontinuities may take the form of a point missing, a vertical gap, or an infinite discontinuity.
3. *Intercepts.* An INTERCEPT of a graph is a point at which the graph crosses (or touches) one of the coordinate axes. A function will cross the y-axis only once, if at all, but may cross the x-axis several times. The y-intercept is determined by setting $x = 0$ and solving for y. The x-intercepts are determined by setting $y = 0$ and solving for x.
4. *Symmetry.* (*a*) A curve is SYMMETRIC WITH RESPECT TO THE y-AXIS if and only if both of the points (x,y) and $(-x,y)$ are on the curve. (*b*) A curve is SYMMETRIC WITH RESPECT TO THE ORIGIN if and only if both of the points (x,y) and $(-x,-y)$ are on the curve.
5. *Asymptotes.* An ASYMPTOTE is a line, either vertical or horizontal, such that the distance between the line representing the graph of the function and the line which is the asymptote tends to zero.
   (*a*) A VERTICAL ASYMPTOTE is

$$x = a \text{ if } \lim_{x \to a^+} f(x) = \pm\infty \qquad \text{or} \qquad \lim_{x \to a^-} f(x) = \pm\infty$$

   (If $y = f(x)$ is a rational function, check all $x = a$ for which the denominator vanishes but the numerator does not.)

(*b*) A HORIZONTAL ASYMPTOTE is

$$y = b \text{ if } \lim_{x \to -\infty} f(x) = b \qquad \text{or} \qquad \lim_{x \to +\infty} f(x) = b$$

6. *Intervals where function is increasing and intervals where function is decreasing.* The function f is an increasing function wherever $f' > 0$; and f is a decreasing function wherever $f' < 0$.
7. *Maxima and/or minima.* Determine all $x = a$ such that $f'(a) = 0$ or $f'(a)$ is undefined, and test for possible extrema. Use either the First Derivative Test or the Second Derivative Test. If the domain of the function is a closed interval, evaluate its endpoints.
8. *Concavity.* A function f is concave upward wherever $f'' > 0$ and is concave downward wherever $f'' < 0$.
9. *Points of inflection.* POINTS OF INFLECTION occur where the graph changes type of concavity. All $x = a$ such that $f''(a) = 0$ and $f'''(a) \neq 0$ are points of inflection. Also check all x such that f'' is undefined for possible points of inflection by determining the type of concavity on either side of these x values.
10. *Key coordinates.* Determination of both x- and y-coordinates at key points (such as extrema, or points of inflection) on the graph will aid in placing the graph in the coordinate system.
11. *Behavior of function for extreme values of independent variable.* Evaluation of

$$(a) \quad \lim_{x \to -\infty} f(x) \qquad \text{and} \qquad (b) \quad \lim_{x \to +\infty} f(x)$$

will tell us the behavior of the curve for large positive values and large negative values of the independent variable.

Certainly, not all bits of information will be relevant for all functions. Nonetheless, let us illustrate the use of this type of information in curve sketching.

Example 26 Sketch the graph of $f(x) = x^3 + 2x^2 - 6x - 10$. We accumulate this information:

1. *Domain and range.* There are no restrictions on the domain of the function.
2. *Discontinuities.* This is a polynomial; there are no discontinuities.
3. *Intercepts.* The y-intercept is $f(0) = -10$. Because we will use a method of approximation to determine the x-intercepts, we will wait until we are able to construct a rough sketch of the graph to attempt to find these points. The graph will help us make a good estimate of the root as well as letting us know how many roots there are.
4. *Symmetry.* None (Check for (x,y), $(-x,y)$ and $(-x,-y)$).
5. *Asymptotes.* This is a polynomial; there are no asymptotes.

6&7. *Increasing and decreasing intervals and maxima and minima.*

We find $f'(x) = 3x^2 + 4x - 6$. The critical values of x are, then, determined to be

$$\begin{aligned} 3x^2 + 3x - 6 &= 0 \\ 3(x^2 + x - 2) &= 0 \\ 3(x-1)(x+2) &= 0 \\ x^* = -2 \text{ and } x^* &= 1 \end{aligned}$$

Then, we use these points to establish intervals over which we evaluate the movement of the function, as

Interval of x values	Algebraic sign of f'	Direction of movement of f
$(-\infty,-2)$	+	Increasing
-2	0	Stationary—MAXIMUM
$(-2,1)$	−	Decreasing
1	0	Stationary—MINIMUM
$(1,\infty)$	+	Increasing

8&9. *Concavity and points of inflection.* We determine $f''(x) = 6x + 4$ and

$$\begin{aligned} f''(x) = 6x + 4 &= 0 \\ x &= -2/3 \end{aligned}$$

We use this value of x to determine intervals over which the function is concave upward and intervals over which it is concave downward and to check for points of inflection.

Interval of x values	Algebraic sign of f''	Type of concavity of f
$(-\infty,-2/3)$	−	Concave downward
$-2/3$	0	POINT OF INFLECTION
$(-2/3,\infty)$	+	Concave upward

It will be helpful to combine the information provided by the first and second derivatives, as

x		-2		$-2/3$		1	
f''	− −		− − −	|	+ +		+ + +
f'	+ + +	|	− − −		− −	|	+ + +
f	increasing, concave down	|	decreasing, concave down	|	decreasing, concave up	|	increasing, concave up

10. *Coordinates of key points on graph.* We evaluate

Maximum: $f(-2) = (-2)^3 + 2(-2)^2 - 6(-2) - 10 = 2$
Minimum: $f(1) = (1)^3 + 2(1)^2 - 6(1) - 10 = -13$
Point of inflection: $f(-\frac{2}{3}) = (-\frac{2}{3})^3 + 2(-\frac{2}{3})^2 - 6(-\frac{2}{3}) - 10 = -5\frac{11}{27}$

11. *Behavior of graph for large and small values of* x. We evaluate $\lim_{x \to \infty} (x^3 + 2x^2 - 6x - 10) = +\infty$; thus the value of the function continues to increase for larger and larger values of x. (Note: This is *not* an asymptote.)

    We also evaluate $\lim_{x \to -\infty} (x^3 + 2x^2 - 6x - 10) = -\infty$; thus the value of the function continues to decrease for smaller and smaller values of x.

Putting all of this information together, we obtain the sketch of the function shown in Figure 12.33. Using a trial-and-error procedure, we determine the x-intercepts to be $x \approx -2.8775$, $x \approx -1.4763$, and $x \approx 2.3539$. []

Figure 12.33

Sketch of Graph of $f(x) = x^3 + 2x^2 - 6x - 10$

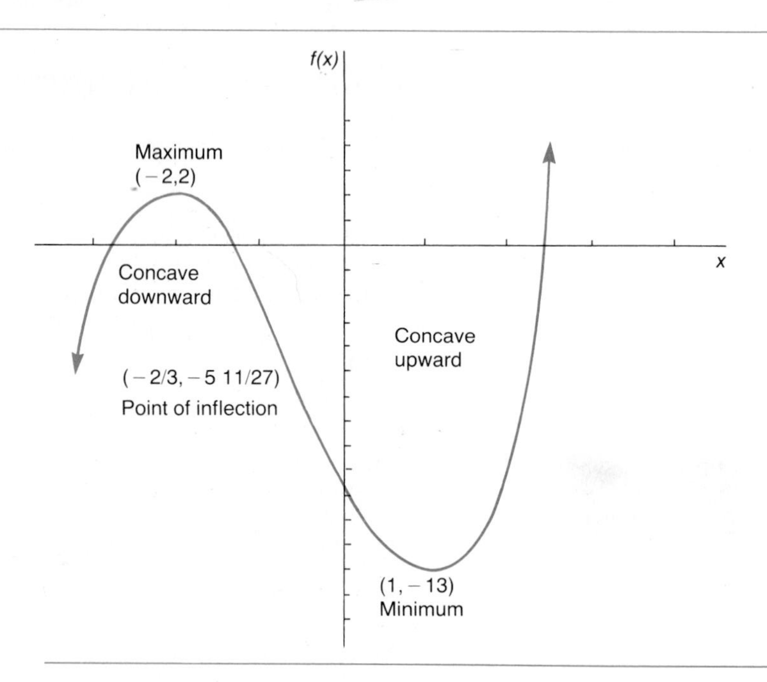

Example 27 Sketch the graph of $f(x) = 4/(x^2 - 1)$. We accumulate this information.

1&2. *Domain and range and discontinuities.* The only real numbers excluded from the domain are ± 1; these values for x would generate a denominator of zero for the rational function.

3. *Intercepts.* $f(0) = 4/-1 = -4$; thus, the y-intercept is $y = -4$. There are no $x = a$ such that $f(a) = 0$; thus, there are no x-intercepts.

4. *Symmetry.* Because

$$\frac{4}{(x)^2 - 1} = \frac{4}{(-x)^2 - 1} \qquad \text{(that is, } (x,y) = (-x,y))$$

the graph is symmetric with respect to the y-axis. However, there is no symmetry with respect to the x-axis.

5. *Asymptotes.* Vertical asymptotes often occur at points of discontinuity. Since this function is discontinuous at $x = +1$ and $x = -1$, we verify

$$\lim_{x \to -1^+} [4/(x^2 - 1)] = -\infty \qquad \text{and} \qquad \lim_{x \to -1^-} [4/(x^2 - 1)] = +\infty$$

$$\lim_{x \to 1^+} [4/(x^2 - 1)] = +\infty \qquad \text{and} \qquad \lim_{x \to 1^-} [4/(x^2 - 1)] = -\infty$$

Hence, vertical asymptotes do occur at $x = -1$ and at $x = +1$; these limits describe the behavior of the graph as it approaches these asymptotes.

Then, searching for horizontal asymptotes, we compute

$$\lim_{x \to -\infty} [4/(x^2 - 1)] = 0 \qquad \text{and} \qquad \lim_{x \to +\infty} [4/(x^2 - 1)] = 0$$

Thus, $y = 0$ is a horizontal asymptote.

6&7. *Intervals of increase and decrease and maxima and minima.* The first derivative is

$$f'(x) = -8x/(x^2 - 1)^2$$

and

$$f'(x) = 0 \text{ only when } -8x = 0$$

This means $x^* = 0$ is the only critical value.

Because the graph of the function may change direction of movement at this critical value of x, or at any point of discontinuity, we use $x = 0$ along with $x = -1$ and $x = 1$ to set up intervals over which we evaluate the algebraic sign of the first derivative.

Interval of x values	Algebraic sign of f''	Direction of movement of f
$(-\infty,-1)$	+	Increasing
-1	Undefined	Discontinuous
$(-1,0)$	+	Increasing
0	0	Stationary—MAXIMUM
$(0,1)$	−	Decreasing
1	Undefined	Discontinuous
$(1,+\infty)$	−	Decreasing

8&9. *Concavity and points of inflection.* The second derivative is

$$f''(x) = \frac{8(3x^2 + 1)}{(x^2 - 1)^3}$$

and could be zero only when $8(3x^2 + 1)$ equals zero. Thus, there are no x values such that $f''(x) = 0$.

However, the type of concavity of a function may change at points of discontinuity, so we set up the following intervals and evaluate the algebraic sign of the second derivative.

Interval of x values	Algebraic sign of f''	Type of concavity of f
$(-\infty,-1)$	$+$	Concave upward
-1	Undefined	—
$(-1,1)$	$-$	Concave downward
1	Undefined	—
$(1,+\infty)$	$+$	concave upward

We combine the information provided by the first and second derivatives, as

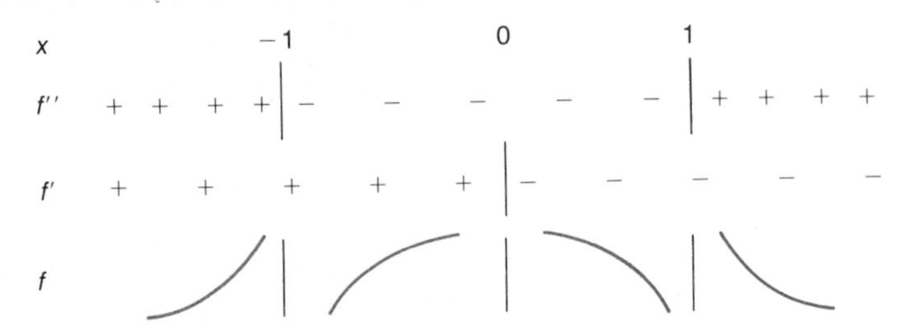

10. *Key coordinates.* Coordinates of key points on the graph are: MAXIMUM: $f(0) = 4/(0^2 - 1) = -4$; thus the coordinates of the maximum point on the graph are $(0,-4)$.

    Two other points selected rather arbitrarily from either side of the points of discontinuity are:

    $$f(-2) = 4/3 \qquad \text{and} \qquad f(2) = 4/3$$

11. *Behavior of function for large and small values of* x. We have previously determined that the graph approaches the x-axis as a horizontal asymptote as x increases without bound and as x decreases without bound.

All of this information is assimilated and the graph of the function pictured in Figure 12.34. []

Figure 12.34

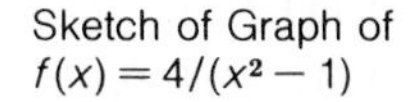

Sketch of Graph of $f(x) = 4/(x^2 - 1)$

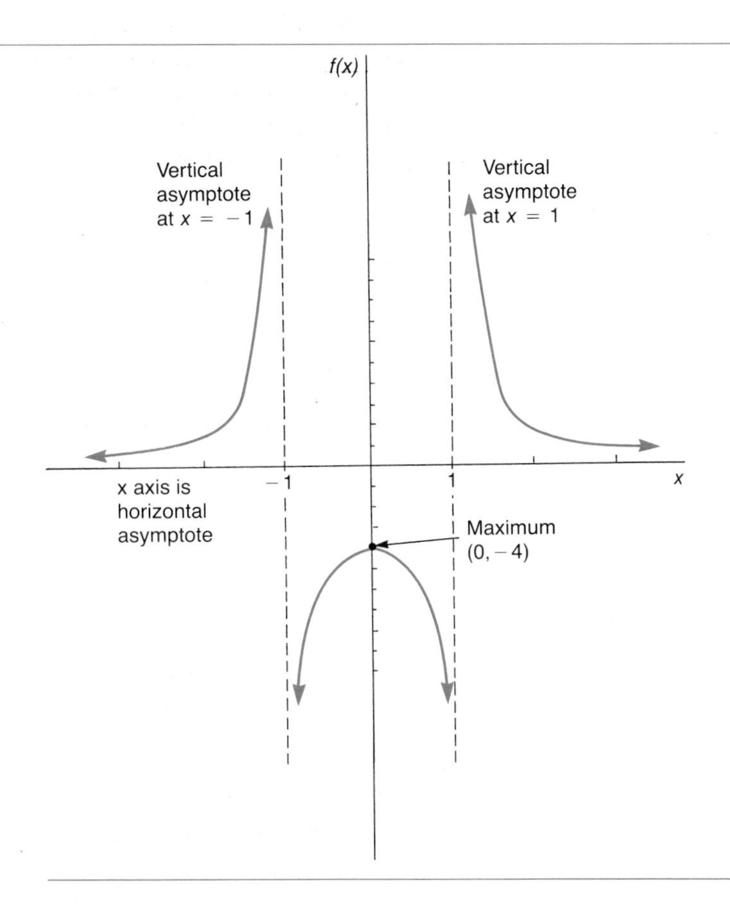

EXERCISES

Follow the curve-sketching procedure outlined in sketching a graph of each of the following functions.

141. $f(x) = -x^3 + 6x^2 - 8x$
142. $f(x) = x^4 - 9x^2$
143. $f(x) = 3x^4 - 4x^3 - 2$
144. $f(x) = (x^2 - 9)/(x + 3)$
145. $f(x) = 1/(x^2 - 9)$
146. $f(x) = x^5 + 1$
147. $f(x) = 2/(x^2 - 1)$
148. $f(x) = (0.1x^2 + 10x + 25)/x$
149. $f(x) = 6/(x^2 + 2)$
150. $f(x) = 3/(x - 1)$
151. $f(x) = 3x/(x^2 + 4)$
152. $f(x) = (x - 3)/(x + 3)$
153. $f(x) = 3x^2/(x^2 - 4)$
154. $f(x) = x^3/(x + 1)$
155. $f(x) = [1/x^2] + [8/x]$
156. $f(x) = (x^2 + x + 2)/(x - 1)$
157. $f(x) = (2x^2 - 1)/x$
158. $f(x) = e^{-x^2}/2$

159. $f(x) = (x^2 - 4)/(x^2 - 1)$

160. $f(x) = 1/(x - 1)^3$

161. $f(x) = (x^2 + 2x)^5$

162. $f(x) = (x^2 - 1)/(x^2 + 1)$

163. $f(x) = x/(x^2 - 9)$

164. $f(x) = 1/(x^2 + 1)$

165. $f(x) = 1/(x^3 - x)$

166. $f(x) = x + (1/x)$

167. $f(x) = x^{3/5} + 1$

C·H·A·P·T·E·R 13 Integral Calculus

Calculus has two main branches: (1) DIFFERENTIAL CALCULUS—which deals with the problems of finding the rate at which a known, but varying, quantity changes; and (2) INTEGRAL CALCULUS—which deals with the problem of finding a quantity, given that we know the rate at which it is changing.

In Chapters 11 and 12 we presented the fundamentals of differential calculus; in this chapter and the next we shall explore the fundamentals of integral calculus.

Although it is the customary procedure in college courses to begin with differential calculus and later consider integral, the ideas of integral calculus developed historically before those of differential calculus. Actually it was the problem of finding certain areas, volumes, and arc lengths that spawned the idea of integration, and it was in the role of a summation device that calculus was first important.

It was some time later that the process of differentiation was created in connection with problems of tangents to curves and with questions concerning maxima and minima of functions.

And still later came the observation that integration and differentiation were related to each other as inverse operations.

Isaac Newton and Gottfried Wilhelm Leibniz are generally credited with having devised the calculus—working separately but at about the same time, toward the end of the 17th century. Because of their extremely important work, mathematics advanced one giant step.

13.1 THE ANTIDERIVATIVE CONCEPT

Most mathematical operations have an "inverse" operation which in effect reverses, or undoes, the previous operation. For instance, the inverse operation for addition is subtraction, and the inverse operation for multiplication is division. If we add a term to a function and then subtract the same term, the result is the initial function. Or, if we multiply a function by a factor, we can return to the original function if we divide by the same factor. We have seen that many matrices have multiplicative inverses and that the inverse of the inverse is the original matrix. We have looked at the procedure for finding the logarithm of a number and for determining the antilog of a logarithm. We have also noted the special inverse relationship which exists between exponential and logarithmic functions.

In this chapter we shall consider reversing the process of differentiation. Given a function $f(x)$, we are acquainted with procedures for finding the derivative function $f'(x)$. There may be many occasions in which we need to reverse this operation. Given the speed at which an object is traveling, we may wish to determine the distance traveled. Given the rate at which a population is growing, we may wish to determine the total population at some time t. We may wish to retrieve a total profit function from a known marginal profit function, or we may wish to discover a total cost function after having discerned the marginal cost relationship. Knowing the derivative function $f'(x)$ then, we wish to determine the function $f(x)$ whose derivative we know. Because this process of finding the original function is the reverse of differentiation, it is called ANTIDIFFERENTIATION, or INTEGRATION. The original function is said to be an ANTIDERIVATIVE of the derivative function.

> A function $F(x)$ is called an ANTIDERIVATIVE of a function $f(x)$ if and only if $f(x) = F'(x)$.

Thus, if $F'(x)$ is the derivative of $F(x)$, then $F(x)$ is an antiderivative of $F'(x)$

Example 1 Consider the function $f(x) = 3$. It is not difficult to conclude that the function $F(x) = 3x$ has a derivative of the form $F'(x) = f(x) = 3$. Hence, $F(x) = 3x$ is an antiderivative of $f(x) = 3$. (That is, given $F'(x) = 3$, we find $F(x) = 3x$.)

But note that the function $F(x) = 3x + 1$ also has a derivative of the form $F'(x) = 3$, and so $F(x) = 3x + 1$ is also an antiderivative of $f(x) = 3$. Similarly, $3x - 9$, $3x + 1/3$, $3x - 0.4$, and, indeed, $3x + C$, for any constant C, are all antiderivatives of $f(x) = 3$. []

Because the derivative of a constant is zero, once an antiderivative of a function is found, another antiderivative of the same function can be formulated simply by adding a nonzero constant to the first antiderivative. Thus, $F(x)$, by itself, is not unique; it is only one member of a *family of antiderivatives* for some specified function, each of these antiderivatives differing from the others only by an arbitrary constant C. Thus, if $F(x)$ is *an* antiderivative of $f(x)$, the general expression $F(x) + C$ is called *the* antiderivative of $f(x)$, or the INDEFINITE INTEGRAL, and is denoted by the symbol $\int f(x)\,dx$.

> Given $F(x)$, an antiderivative of $f(x)$, the INDEFINITE INTEGRAL of $f(x)$ is defined to be
>
> $$F(x) + C = \int f(x)\,dx$$

The notation $\int f(x)\,dx$ is read "the indefinite integral of the function $f(x)$ with respect to the variable x." The symbol $\int$ is called the INTEGRAL SIGN and indicates that the operation of antidifferentiation is to be performed; $f(x)$ is termed the INTEGRAND, and dx indicates that antidifferentiation is to be performed with respect to the variable x. The arbitrary constant, C, in $F(x) + C$, is termed the CONSTANT OF INTEGRATION. (Needless to say, the variable of integration is not always x. We often write $\int g(y)dy = F(y) + C$ or $\int h(t)\,dt = F(t) + C$ and so on.)

Example 2 Find the indefinite integral $\int 2x\,dx$.

Because the derivative of x^2 is $2x$, $F(x) = x^2$ is an antiderivative of $2x$. But other functions also have the derivative $2x$. All these functions will be of the form $F(x) = x^2 + C$, where C is a constant. Thus, we say $F(x) = x^2 + C$ is the antiderivative of $f(x) = 2x$ and symbolize this as

$$\int 2x\,dx = x^2 + C$$

[]

Example 3 Find the indefinite integral of $f(x) = 2x - 7$; that is, evaluate $\int (2x - 7)\,dx$.

Working with each term separately and mentally reversing the rules of differentiation, we conclude that the antiderivative, or indefinite integral, is

$$\int (2x - 7)\,dx = x^2 - 7x + C$$

(We may double-check our work by verifying that $D_x(x^2 - 7x + C) = 2x - 7$).

[]

The constant C can be evaluated, and the precise antiderivative of a particular function determined, when a value of the antiderivative function is known for some value of x. This additional information about the particular antiderivative is called a BOUNDARY CONDITION.

Example 4 Find the function that has a slope of $4x^3$ at each point and that passes through the point (2, 17).

Since $4x^3$ is the derivative of $x^4 + C$, the indefinite integral is

$$\int 4x^3\,dx = x^4 + C$$

All functions belonging to the family $y = x^4 + C$ have the slope $4x^3$ at each x.

To determine which member of this set of functions passes through (2,17), we substitute $x = 2$ and $y = 17$ into the equation to obtain

$$\begin{aligned} y &= x^4 + C \\ 17 &= (2)^4 + C \\ C &= 1 \end{aligned}$$

Thus, the specific function we seek is $y = x^4 + 1$.

[]

EXERCISES

Given the following derivative functions, determine an antiderivative of each.

1. $f(x) = 5.8$
2. $f(x) = -6$
3. $f(x) = 5x^4$
4. $f(t) = 2t$

5. $f(x) = -2x$
6. $f(x) = x/2$
7. $f(x) = 2x + 2$
8. $f(r) = 6r + 6$
9. $f(x) = 1/x$
10. $f(x) = e^x$

Find the following indefinite integrals.

11. $\int 9\,dx$
12. $\int -53\,dx$
13. $\int y\,dy$
14. $\int 4x\,dx$
15. $\int x^3\,dx$
16. $\int x^{-2}\,dx$
17. $\int (x + 5)\,dx$
18. $\int (1/x^2 - 2x)\,dx$
19. $\int 3/8\,dt$
20. $\int u\,du$

Determine F(x) *given* F′(x) *and a point through which* F(x) *passes.*

21. $F'(x) = 2/3$; (3,10)
22. $F'(x) = 6x^5$; (1,12)
23. $F'(x) = 5 + 3x^2$; (2,−2)
24. $F'(x) = -x^{1/2}/2$; (4,5)
25. $F'(x) = 3x^2 + 2x + 2$; (2,0)
26. $F'(x) = 4x^3 - 3x^2/4 + 2x - 5$; (0,10)

13.2 BASIC RULES OF INTEGRAL CALCULUS

Fortunately we are not restricted to determining antiderivatives by the trial-and-error process of mentally reversing the rules of differentiation. Many basic integration "rules" have been formulated from the corresponding theorems concerning derivatives. Thus, if a function has a particular form, a rule may be readily available which allows us to very easily determine its antiderivative. But, as we shall soon see, this will not always be the case. The techniques of antidifferentiation are not as straightforward or as systematic or as comprehensive as are the techniques of differentiation.

The following set of rules facilitates finding the indefinite integral of some common functional forms.

Integration of a Constant

Rule 1 *If* k *is any constant, then*

$$\int k\,dx = kx + C$$

The integral of any constant with reference to a variable is the product of the constant times the variable, plus the constant of integration.

Example 5

a. $\int 3\,dx = 3x + C$ *b.* $\int -4\,dx = -4x + C$

c. $\int 1\,dx = \int dx = x + C$ *d.* $\int 0.06\,du = 0.06u + C$

e. $\int \frac{dt}{5} = \int (1/5)\,dt = t/5 + C$ [ ]

EXERCISES

Evaluate each indefinite integral.

27. $\int 6\,dx$
28. $\int -28\,dx$
29. $\int 4/5\,ds$
30. $\int 0.83\,dy$
31. $\int \sqrt{3}\,dx$
32. $\int \frac{dx}{3}$

Integration of a Variable Raised to a Constant Power

Rule 2 *If* $f(x) = x^n$, *then*

$$\int x^n\,dx = \begin{cases} \dfrac{x^{n+1}}{n+1} + C & \text{if } n \neq -1 \\ \ln|x| + C & \text{if } n = -1 \end{cases}$$

where n *is a constant.*

The notation $|x|$, which means *absolute value* of x, is necessary because the logarithmic function is not defined for negative values of x.

The following examples illustrate this rule.

Example 6

a. $\int x\,dx = x^2/2 + C$

b. $\int x^2\,dx = x^3/3 + C$

c. $\int \sqrt{x}\,dx = \int x^{1/2}\,dx = \frac{x^{3/2}}{3/2} + C = \frac{2x^{3/2}}{3} + C$

d. $\int 1/x^2\,dx = \int x^{-2}\,dx = \frac{x^{-1}}{-1} + C = -1/x + C$

e. $\int 1/\sqrt{x}\,dx = \int x^{-1/2}\,dx = \frac{x^{1/2}}{1/2} + C = 2\sqrt{x} + C$

f. $\int x^{0.2}\,dx = \frac{x^{1.2}}{1.2} + C$ [ ]

EXERCISES

Evaluate each integral.

33. $\int x^4\,dx$
34. $\int 1/x^4\,dx$
35. $\int \sqrt[4]{x}\,dx$
36. $\int 1/\sqrt[4]{x}\,dx$
37. $\int x^{0.4}\,dx$
38. $\int x^{4.1}\,dx$
39. $\int x^{4/5}\,dx$
40. $\int x^{5/4}\,dx$

Integration of a Constant Times a Function

Rule 3 *If* $\int f(x)$ *exists, then*

$$\int kf(x)\,dx = k\int f(x)\,dx$$

where k *is any real-valued constant.*

Whenever a constant can be factored from the integrand, the constant may be taken out of the integral to ease computation. IMPORTANT NOTE: this procedure is appropriate only for constants, not for variables.

Example 7

a. $\int 6x\,dx = 6\int x\,dx = 6(x^2/2) + C = 3x^2 + C$

(The rules of algebra suggest that we should multiply any constant k factored out of the integral by the constant of integration C, as, in this exercise $6[(x^2/2) + C] = 3x^2 + 6C$. However, since the constant of integration represents any real number which could be factored in any manner, this multiplication is unnecessary. Thus, the convention is to simply write C and not some multiple of C.)

b. $\int (4/5)x^3\,dx = (4/5)\int x^3\,dx = (4/5)(x^4/4) + C = x^4/5 + C$

c. $\int x^4/3\,dx = (1/3)\int x^4\,dx = (1/3)(x^5/5) + C = x^5/15 + C$

d. $\int 4/x^2\,dx = 4\int x^{-2}\,dx = (4)(x^{-1}/-1) + C = -4/x + C$

e. $\int -3/x\,dx = (-3)\int 1/x\,dx = (-3)\int x^{-1}\,dx = -3\ ln\ |x| + C$ [ ]

EXERCISES

41. $\int -8x\,dx$
42. $\int 4x\,dx$
43. $\int 6x^3\,dx$
44. $\int 3x^4\,dx$

45. $\int -6/\sqrt{x}\, dx$

46. $\int -x^5/5\, dx$

47. $\int 4x/5\, dx$

48. $\int 3\sqrt{x}\, dx$

49. $\int 4(\sqrt{x})^3\, dx$

50. $\int 1/(3\sqrt{x})\, dx$

51. $\int ex^e\, dx$

52. $\int 3x^{3/2}\, dx$

Integration of the Sum, or Difference, of Functions

Rule 4 *The integral of the sum, or difference, of two, or more, functions is the sum, or difference, of their respective integrals; that is,*

$$\int [f(x) \pm g(x)]\, dx = \int f(x)\, dx \pm \int g(x)\, dx$$

Thus, just as we are able to take derivatives of functions term by term, we can integrate terms of functions separately and then sum these integrals.

Example 8

$$\begin{aligned}\int (2x+5)\, dx &= 2\int x\, dx + \int 5\, dx \\ &= 2(x^2/2 + C_1) + (5x + C_2) \\ &= x^2 + 5x + C \qquad \text{(where } C = C_1 + C_2\text{)}\end{aligned}$$

[]

Although the two integrals technically result in separate constants of integration, these constants are conventionally combined in a single number C.

Example 9

$$\begin{aligned}\int (x^3 + x^2 + x + 1)\, dx &= \int x^3\, dx + \int x^2\, dx + \int x\, dx + \int 1\, dx \\ &= x^4/4 + x^3/3 + x^2/2 + x + C\end{aligned}$$

[]

Example 10 Evaluate $\int (x+2)^2\, dx$.

We can square the function before we integrate, as follows:

$$\begin{aligned}\int (x+2)^2\, dx = \int (x^2 + 4x + 4)\, dx &= \int x^2\, dx + 4\int x\, dx + \int 4\, dx \\ &= x^3/3 + 4(x^2/2) + 4x + C \\ &= x^3/3 + 2x^2 + 4x + C\end{aligned}$$

[]

Example 11 Evaluate

$$\int \frac{2 + 3x^3}{5x^2}\, dx$$

We may rewrite the integrand as

$$\int \frac{2+3x^3}{5x^2}\,dx = \int \frac{2}{5x^2} + \frac{3x^3}{5x^2}\,dx = (2/5)\int x^{-2}\,dx + (3/5)\int x\,dx$$
$$= (2/5)(x^{-1}/-1) + (3/5)(x^2/2) + C$$
$$= -2/5x + 3x^2/10 + C$$

[]

EXERCISES

Evaluate each integral.

53. $\int (3x+6)\,dx$

54. $\int (3x+x^2)\,dx$

55. $\int (x^2+3x+5)\,dx$

56. $\int (4x^2-7x+2)\,dx$

57. $\int (3/x+x/3)\,dx$

58. $\int (x^3+4x)\,dx$

59. $\int (x^4-2x+\sqrt{x})\,dx$

60. $\int (3x^3+1)/x^2\,dx$

61. $\int (x+9)^2\,dx$

62. $\int (5x+4)^2\,dx$

Integration of the Exponential Function

Rule 5 *Where a and k are real constants and the exponent of x equals 1, then*

$$\int a^{kx}\,dx = \frac{1}{k\,ln\,a}\,a^{kx} + C$$
$$\int e^{kx}\,dx = \frac{1}{k}\,e^{kx} + C = \frac{e^{kx}}{k} + C$$

Example 12

a. $\int 5^{2x}\,dx = \dfrac{5^{2x}}{2\,ln\,5} + C$

b. $\int (1/3)^{4x}\,dx = \dfrac{(1/3)^{4x}}{4\,ln\,(1/3)} + C$

c. $\int 3^{1/2x}\,dx = \dfrac{3^{1/2x}}{(1/2)\,ln\,3} + C$

d. $\int e^x\,dx = e^x + C$

e. $\int 9e^{3x}\,dx = 9\int e^{3x}\,dx = 9(e^{3x}/3) + C = 3e^{3x} + C$

f. $\int 1/e^x\,dx = \int e^{-x}\,dx = -e^{-x} + C = -1/e^x + C$

[]

EXERCISES

Evaluate each integral.

63. $\int 3^{2x}\,dx$
64. $\int 8^{0.5x}\,dx$
65. $\int e^{2x}\,dx$
66. $\int 4e^{5x}\,dx$
67. $\int 5e^{-3x}\,dx$
68. $\int (1/2)e^{x/2}\,dx$
69. $\int -0.1e^{-0.1x}\,dx$
70. $\int 9^{x/3}\,dx$
71. $\int (8^t + 5)\,dt$
72. $\int (6^y + y^6)\,dy$

Integration of the Logarithmic Function

Rule 6 *Where* a *and* k *are real constants and the exponent of* x *equals 1, then*

$$\int \log_a (kx)\,dx = x \log_a (kx) - x \log_a e + C$$

$$\int \ln (kx)\,dx = x \ln (kx) - x + C$$

Although this rule does not have an immediately recognizable corollary among the derivative rules we studied, we can verify this rule by determining the derivative of the integrand, as

$$D_x(x \log_a kx - x \log_a e + C) = x(k/kx) \log_a e + \log_a kx - \log_a e = \log_a kx$$

Example 13

a. $\int \ln x\,dx = x \ln x - x + C$

b. $\int \log_{10} x\,dx = x \log_{10} x - x \log_{10} e + C$

c. $\int \ln 4x\,dx = x \ln 4x - x + C$

d. $\int \log_3 2x = x \log_3 2x - x \log_3 e + C$ []

EXERCISES

Evaluate each integral.

73. $\int \log_{10} 5x\,dx$
74. $\int \ln y\,dy$
75. $\int \log_3 4x\,dx$
76. $\int 5 \ln x\,dx$

77. $\int ln\ 6x\ dx$

78. $\int ln\ 13x\ dx$

79. $\int (ln\ 3x + ln\ 2x)\ dx$

80. $\int (e^x + x^e + ln\ x)\ dx$

81. $\int [(ln\ 7x) + 7x]\ dx$

82. $\int (ln\ x)/3\ dx$

13.3 APPLICATIONS OF THE ANTIDERIVATIVE

It is often the case that the rate of change of a function with respect to some variable is known or is relatively easy to ascertain. The problem then may be to find the function whose rate of change is given. For instance, the marginal-revenue or marginal-cost function in a given situation may be known to the decision maker, and it may be necessary for planning purposes to determine the total-revenue or total-cost function. We have seen that an expression for marginal revenue is the derivative of the total-revenue function; hence, the total-revenue function is the antiderivative of the marginal-revenue function. Likewise, the antiderivative of a marginal-cost function is the corresponding total-cost function. The following examples illustrate how the procedure of integration can be used to find total functions given their marginal or rate-of-change counterparts.

Example 14 A company has determined that the marginal-revenue function for one of its products is

$$R'(x) = 80{,}000 - 2x$$

where

$R'(x)$ = marginal revenue, in dollars
x = the number of units of the product that are produced and sold

Given that total revenue is zero when no units are produced and sold, determine that total-revenue function for the product.

Because a marginal-revenue function is the derivative of a total-revenue function, the total-revenue function is the antiderivative of the marginal-revenue function. We determine

$$R(x) = \int R'(x)\ dx = \int (80{,}000 - 2x)\ dx = 80{,}000x - x^2 + C$$

Since $R(0) = 0$, we can determine that

$$0 = 80{,}000(0) - 2(0) + C$$
$$0 = C$$

Thus, total revenue for the company's product is given by

$$R(x) = 80{,}000x - x^2$$ []

Example 15 A company has been able to determine that the marginal cost of producing its product is

$$C'(x) = x + 75$$

where

$C'(x)$ = marginal cost, in dollars

x = the number of units produced

The company has also been able to determine that fixed cost of production is \$50,000. Determine the total-cost function.

To determine the total-cost function given information regarding marginal cost, we must find the antiderivative of the marginal-cost function. Thus, we conclude

$$C(x) = \int C'(x)\,dx = \int (x + 75)\,dx = (x^2/2) + 75x + C = 0.5x^2 + 75x + C$$

Then, given that $C(0) = 50{,}000$, we determine that

$$50{,}000 = 0.5(0)^2 + 75(0) + C$$
$$50{,}000 = C$$

The specific total-cost function for producing the product is

$$C(x) = 0.5x^2 + 75x + 50{,}000$$ []

Example 16 The marginal profit realized from producing and selling a certain product is given by

$$P'(x) = 1{,}200 - 6x$$

where

$P'(x)$ = marginal profit, in dollars
x = the number of units of product produced and sold

When no units are produced and sold, a loss of \$950 is experienced. Determine the total-profit function. What is the amount of profit realized when 200 units are produced and sold?

The total-profit function is the antiderivative of the marginal-profit function; that is,

$$P(x) = \int (1{,}200 - 6x)\,dx = 1{,}200x - 3x^2 + C$$

With the information that $P(0) = -950$, we determine that $C = -950$. Thus, the specific total-profit function for the product is

$$P(x) = -3x^2 + 1{,}200x - 950$$

where

$P(x) =$ total profit, in dollars

Total profit realized from the sale of 200 units of product is

$$P(200) = -3(200)^2 + 1{,}200(200) - 950 = 119{,}050$$

The sale of 200 units of product should yield a total profit of $119,050. []

Example 17 Studies indicate that the rate of change in the population in a certain trade area is

$$dP/dt = 200 + 30\sqrt{t}$$

where

$P(t) =$ total population
$t =$ time, in years

Population of the area is presently (at $t = 0$) 15,000 persons.

Determine the total-population function. Estimate total population of the area four years from now.

$P(t)$, the total population of the area t years from now, is given by

$$P(t) = \int (200 + 30\sqrt{t})\, dt = 200t + 20t^{3/2} + C$$

To find the value of C we use the information that $P(0) = 15{,}000$; hence, $C = 15{,}000$. Therefore, the population function is

$$P(t) = 20t^{(3/2)} + 200t + 15{,}000$$

Four years from now, the population of the trade area should be approximately

$$P(4) = 20(4)^{(3/2)} + 200(4) + 15{,}000 = 15{,}960$$

In four years, the population of the trade area should be approximately 15,960 persons. []

Example 18 Time-and-motion studies show that, as a worker becomes more experienced at a certain assembly-line task, the length of time t, in minutes, required to perform the task decreases. The rate at which a new employee assembles an electric drill has been determined to be

$$\frac{dt}{dx} = \frac{10}{\sqrt[3]{x}} = 10x^{-1/3}$$

where

$x =$ the number of drills assembled

Find the total time required to assemble 25 drills; 50 drills.

The function $t(x)$ can be determined by finding the antiderivative of dt/dx, as

$$t(x) = \int 10x^{-1/3}\,dx = 10\int x^{-1/3}\,dx = 10(x^{2/3})(3/2) + C = 15x^{2/3} + C$$

Because $t(0) = 0$ (it takes zero time to assemble zero drills), the specific total time function is

$$t(x) = 15x^{2/3}$$

The length of time required to assemble 25 drills is

$$t(25) = 15(25)^{2/3} = 128.25 \text{ minutes}$$

The time required to assemble 50 drills is

$$t(50) = 15(50)^{2/3} = 203.58 \text{ minutes} \qquad []$$

EXERCISES

83. If the marginal revenue for a particular product is $R'(x) = 18$, where x is the number of units sold, what is the total-revenue function?
84. If the marginal revenue for a commodity is $R'(x) = 80 - 0.4x$, where x is the number of units sold, what is the total-revenue function?
85. If the marginal cost for a product is $C'(x) = 0.03x + 18$, where x is the number of units produced and fixed cost of production is \$3,500, what is the total-cost function?
86. If the marginal cost for a product is $C'(x) = 0.001x^2 - 0.1x + 10$, where x is the number of units produced, and fixed cost of production is \$10,000, what is the total-cost function?
87. A manufacturer finds that the marginal cost of production is given by the function $C'(x) = 3x + 9$, where $C'(x)$ is marginal cost, in dollars, and x is the number of units produced. Fixed cost of production is \$6,000. What is the total-cost function?

The marginal cost of producing and selling a product is given by $C'(x) = 240 + 6x$ *where* $C'(x)$ *is marginal cost, in dollars, and* x *is the number of units of product produced and sold. Fixed cost has been determined to be \$5,000.*

The product sells for \$576 per unit; that is, $R'(x) = 576$, *where* $R'(x)$ *is marginal revenue, in dollars, and total revenue is zero when no units are sold.*

88. Find the total-revenue function, $R(x)$, where x is the number of units of product sold.
89. Find the total-cost function, $C(x)$.
90. Find the total-profit function, $P(x)$.

91. Find the sales volume that yields maximum profit. What is the maximum profit, in dollars?

92. The marginal revenue for a company's product is given by

$$R'(x) = 3{,}000 - x$$

where $R'(x)$ is marginal revenue, in dollars, and x is the number of units of product produced and sold. Total revenue is zero when no units are sold. Determine the total-revenue function for the product.

The marginal cost of producing and selling a product is given by

$$C'(x) = 5x + 250$$

where $C'(x)$ *is marginal cost, in dollars, and* x *is the number of units of product produced and sold. Fixed cost has been determined to be $15,000.*

93. Determine the total-cost function for the product.

94. Determine total cost when 10 units are produced and sold.

95. Determine the marginal cost of the 10th unit produced and sold.

96. The marginal profit from producing and selling a product is given by

$$P'(x) = 1{,}000 - 4x$$

where $P'(x)$ is marginal profit, in dollars, and x is the number of units of product produced and sold. Total profit equals $60,000 when 100 units are produced and sold. Determine the total-profit function.

97. The total number of units of a product sold has a slope of $12 + 6t$ at each point t, where t is the number of weeks the product has been on the market. If sales equal zero when $t = 0$, determine the function describing the total number of units sold as a function of time.

98. The rate of change in the population of a certain area is estimated to be

$$dP/dt = 500 + 100\sqrt{t}$$

where $P(t)$ is total population and t is time, in years. Current population (at $t = 0$) is 6,500. Determine the total-population function. Estimate the total population of the area five years hence.

The rate of change of demand for a product is given by

$$D'(x) = -15x + 210$$

where $D'(x)$ *is marginal demand, in units, and* x *is the price per unit, in dollars.*

99. Find the total-demand function, $D(x)$, given that $D(0) = 450$.
100. What quantity will be demanded at a price of \$20? \$25?

The circulation of a certain group of "neighborhood" newspapers is growing in such a way that t *months from now the rate of increase will be* $1 + 6t^{0.8}$ *new subscribers per month.*

101. If the newspapers presently (at month $t = 0$) have a total of 2,500 subscribers, find the function describing the total number of subscribers at time t.
102. How many subscribers will there be in 12 months? in 18 months?

The marginal value of a common stock is $e^{0.06t}$, *where* t *is time in years.*

103. Given that the stock's value at time $t = 0$ is \$150, derive an expression for the value of the stock at any time t.
104. What is the value of the stock at time $t = 5$? at time $t = 10$?

A factory is dumping pollutants into a stream at a rate of $(1/300)t^{0.4}$ *tons per month, where* t *is time, in months, since the dumping first began.*

105. Determine the function $P(t)$ for the total number of tons of pollutants dumped, given that $P(0) = 0$.
106. What quantity of pollutants will have been dumped over the first six months? over the first year?

Geological surveys indicate that t *months from now the number of barrels of oil that a particular oil field yields will be changing at the rate of* $(2e^t - 0.5t^2)$ *thousand barrels per month.*

107. If the field is presently (at $t = 0$) producing 9,000 barrels per month, derive a model giving total production as a function of time.
108. How many barrels will the field be producing at the end of 12 months? at the end of 18 months?

A production engineer has analyzed plant facilities in an effort to determine the effect on the total output of the factory of hiring additional workers. If no additional workers are hired, the total output per day is estimated to be 4,000 units. The marginal productivity of additional workers is estimated to be given by

$$f(x) = 75 - 5\sqrt{x}$$

where $f(x)$ *is the marginal productivity of the* x*th additional worker and* x *is the number of additional workers.*

109. Determine the total-productivity function.

110. What will be total output if 10 additional workers are hired? if 12 additional workers are hired?

13.4 THREE ADDITIONAL RULES OF INTEGRATION

Recall that if $y = [f(x)]^n$, the derivative of y is $y' = n[f(x)]^{n-1} \cdot f'(x)$. Reversing this formula for derivatives, we can see that

$$\int n[f(x)]^{n-1} f'(x)\, dx = [f(x)]^n + C$$

This formula is equivalent to the following Power Rule for Integration.

Rule 7 *Power Rule for Integration:*

$$\int [f(x)]^n f'(x)\, dx = \frac{[f(x)]^{n+1}}{n+1} + C \qquad \text{where } n \neq -1$$

This rule is similar to Integral Rule 2 for a variable carried to a constant power except that Rule 7 allows any *function of the variable* to be carried to a constant power and *requires that $f'(x)$ be present in the integrand.* Of course, if we have $f(x) = x$, meaning that $f'(x) = 1$, we see that the powers-of-x rule is only a special version of the power rule.

Example 19 Evaluate $\int 4(4x - 1)^3\, dx$.

If the power rule is to be used, the integrand must be the product of $[f(x)]^n$ and $f'(x)$. The first step in determining if both these factors are present is the identification of $f(x)$. This is the function which is carried to the power and is here easily identified as $f(x) = 4x - 1$. Once $f(x)$ is defined, $f'(x)$ must be determined. Here $f'(x) = D_x(4x - 1) = 4$.

Noting that $n = 3$, we have verified that the integrand is of the format required for use of the power rule. We proceed by

$$[f(x)]^n f'(x)\, dx$$

$$\int (4x - 1)^3(4)\, dx = \frac{(4x-1)^4}{4} + C$$

[]

Example 20 Evaluate $\int (x^2 + 3)^4(2x)\, dx$.

We begin by identifying $n = 4$ and $f(x) = x^2 + 3$. It follows, then, that $f'(x) = D_x(x^2 + 3) = 2x$. We verify that all factors required for use of Rule 7 are present in the integrand and state

$$\int (x^2 + 3)^4(2x)\, dx = \frac{[x^2 + 3]^5}{5} + C$$

$$[f(x)]^n f'(x)\, dx$$

[]

Example 21 Evaluate $\int \frac{3}{\sqrt{3x+2}}\,dx$.

The integrand can be rewritten as

$$\int (3x+2)^{-1/2}(3)\,dx$$

Then, if $f(x) = 3x + 2$, $f'(x) = D_x(3x + 2) = 3$. The power is $n = -1/2$, so that $n + 1 = +1/2$. Thus, by Rule 7

$$\int (3x+2)^{-1/2}(3)\,dx = \frac{(3x+2)^{1/2}}{1/2} + C = 2\sqrt{3x+2} + C \qquad []$$

Example 22 Evaluate $\int \frac{ln\,x}{x}\,dx$.

If we define $f(x) = ln\,x$, then $f'(x) = 1/x$. With $n = 1$, the integral is

$$\int \frac{ln\,x}{x}\,dx = \int (ln\,x)(1/x)\,dx = \frac{(ln\,x)^2}{2} + C \qquad []$$

Example 23 Evaluate $\int x(2x^2+6)^3\,dx$.

If we define $f(x) = 2x^2 + 6$, then $f'(x) = 4x$. Thus, it would seem that the integrand is not in the proper form for Rule 7 to be applicable. However, since we are missing *only the constant factor* 4, we can make the integrand of the required form by multiplying by $4/4 = 1$, as follows:

$$\int x(2x^2+6)^3\,dx = \int (2x^2+6)^3(4/4)x\,dx = \int (2x^2+6)^3(1/4)(4)x\,dx$$

In multiplying the integrand by $(4/4) = 1$, we altered its appearance but not its value. We now factor the unneeded *constant* $(1/4)$ through the integral, as

$$(1/4)\int (2x^2+6)^3(4x)\,dx$$

We find that the integrand is now of the required format and proceed

$$\int (2x^2+6)^3 x\,dx = (1/4)\int (2x^2+6)^3(4x)\,dx = (1/4)\cdot\frac{(2x^2+6)^4}{4} + C$$

$$= \frac{(2x^2+6)^4}{16} + C \qquad []$$

Example 24 Evaluate $\int (x^2+2x)^3(x+1)\,dx$.

If we let $f(x) = x^2 + 2x$, then $f'(x) = 2x + 2 = 2(x + 1)$. Again it will be necessary to multiply the integrand by a constant and its reciprocal. We obtain

$$\int (x^2 + 2x)^3(x+1)\,dx = (1/2)\int (x^2+2x)^3 2(x+1)\,dx$$
$$= (1/2)\cdot\frac{(x^2+2x)^4}{4} + C$$
$$= \frac{(x^2+2x)^4}{8} + C \qquad []$$

It is important to note that we can introduce only a constant, and not a variable, to get the integrand into the format required by Rule 7. If a variable is needed, some other rule of integration must be used.

Example 25 Evaluate $\int 6(3x^2-5)^3\,dx$.

We identify $f(x) = 3x^2 - 5$ and note that, if this is the case, $f'(x) = 6x$. Since we now have in the integrand only a 6, we might be tempted to multiply by (x/x) and factor $(1/x)$ outside the integral. However we have no integration procedure which allows us to move a variable through the integral sign. Hence, given our current set of rules, we can evaluate this integral only by expanding the parentheses, as

$$\int 6(3x^2-5)^3\,dx = 6\int (27x^6 - 135x^4 + 225x^2 - 125)\,dx$$
$$= (6)[(27x^7/7) - (135x^5/5) + (225x^3/3) - 125x + C]$$
$$= 6x[(27/7)x^6 - 27x^4 + 75x^2 - 125] + C \qquad []$$

EXERCISES

Evaluate each of the following integrals.

111. $\int 3(3x-5)^2\,dx$
112. $\int \sqrt{3+x}\,dx$
113. $\int 2x(x^2-12)^2\,dx$
114. $\int (x^2+4x)^5(x+2)\,dx$
115. $\int (x^2+1)^2\,dx$
116. $\int (3x^3+6)^2 7x^2\,dx$
117. $\int (x+1)\sqrt{x^2+2x}\,dx$
118. $\int x(x^2+7)^4\,dx$
119. $\int x(x+1)^2\,dx$
120. $\int 2(2x-9)^3\,dx$

Rule 8 *Integration of* $f'(x)/f(x)$:

$$\int [f'(x)/f(x)]\,dx = ln\,|f(x)| + C$$

The Power Rule for Integration (Rule 7) applies only if the function is carried to a power n where $n \neq -1$. Rule 8, given above, would be applicable when $n = -1$. That is,

$$\int f(x)^{-1} f'(x)\, dx = \int \frac{f'(x)}{f(x)}\, dx = ln\, |f(x)| + C$$

Example 26 Evaluate $\int \frac{3}{3x+1}\, dx$.

We identify $f(x) = 3x + 1$. Then $f'(x) = D_x(3x + 1) = 3$. Thus, the integrand is in the proper form for use of Rule 8. We have

$$\int \frac{3}{3x+1}\, dx = ln\, |3x+1| + C$$ []

Example 27 Evaluate $\int \frac{x-2}{x^2-4x+5}\, dx$.

We define $f(x) = x^2 - 4x + 5$ and note that, then, $f'(x) = D_x(x^2 - 4x + 5) = 2x - 4 = 2(x - 2)$. The integrand is almost of the proper form. However, if we call again upon the algebraic manipulation of multiplying by a needed constant and its reciprocal, we obtain

$$\int \frac{x-2}{x^2-4x+5}\, dx = (1/2)\int \frac{2(x-2)}{x^2-4x+5}\, dx$$
$$= (1/2)\, ln\, |x^2-4x+5| + C$$ []

Example 28 Evaluate $\int \frac{18x+6}{3x^2+2x}\, dx$.

If we identify $f(x) = 3x^2 + 2x$, then $f'(x) = D_x(3x^2 + 2x) = 6x + 2$. We need to factor the numerator of the integrand, as $3(6x + 2)$, and carry the constant 3 outside the integral sign. Then

$$\int \frac{18x+6}{3x^2+2x}\, dx = 3\int \frac{6x+2}{3x^2+2x}\, dx = 3\, ln\, |3x^2+2x| + C$$ []

Example 29 Evaluate $\int \frac{1}{x\, ln\, x}\, dx$.

If we define $f(x) = ln\, x$, then $f'(x) = D_x(ln\, x) = 1/x$, and the integrand is of the form $f'(x)/f(x)$. Then, by Rule 8

$$\int \frac{1}{x\, ln\, x}\, dx = ln\, |(ln\, |x|)| + C$$ []

Example 30 Evaluate $\int \frac{x^4-3x^3+5x^2-13x-3}{x^2-3x}\, dx$.

Clearly, the integrand is not of the format $f'(x)/f(x)$. However, whenever the degree of the numerator is equal to or greater than the degree of the denominator, we may be able to simplify the problem of finding the antiderivative by dividing the denominator into the numerator. Here the result is

$$\int \frac{x^4 - 3x^3 + 5x^2 - 13x - 3}{x^2 - 3x}\,dx = \int \left[x^2 + 5 + \frac{2x - 3}{x^2 - 3x}\right] dx$$
$$= \int x^2\,dx + 5\int dx + \int \frac{2x - 3}{x^2 - 3x}\,dx$$
$$= (x^3/3) + 5x + ln|\,x^2 - 3x| + C \qquad []$$

EXERCISES

Evaluate the following integrals.

121. $\int \frac{3x^2}{x^3+1}\,dx$

122. $\int \frac{2x}{x^2-5}\,dx$

123. $\int \frac{6x^5}{x^6-14}\,dx$

124. $\int \frac{4x+5}{2x^2+5x-3}\,dx$

125. $\int \frac{x^3}{x^4-1}\,dx$

126. $\int \frac{x^2}{x^3-7}\,dx$

127. $\int \frac{10x^4}{x^5+3}\,dx$

128. $\int \frac{2x^2}{x^3+13}\,dx$

129. $\int \frac{x+1}{3x^2+6x+4}\,dx$

130. $\int \frac{4x+8}{x^2+4x+8}\,dx$

Rule 9 $\int e^{f(x)} f'(x)\,dx = e^{f(x)} + C$

Example 31 Evaluate $\int 2e^{2x+1}\,dx$.

We note that the base e is carried to the power $f(x) = 2x + 1$. Thus, $f'(x) = 2$. The integrand is of the format required by Rule 9. Hence,

$$\int 2e^{2x+1}\,dx = e^{2x+1} + C \qquad []$$

Example 32 Evaluate $\int 3x^2 e^{x^3}\,dx$.

With $f(x) = x^3$, $f'(x) = 3x^2$. The integrand has the appropriate form for Rule 9 and

$$\int 3x^2 e^{x^3}\,dx = e^{x^3} + C \qquad []$$

Example 33 Evaluate $\int xe^{x^2}\,dx$.

With $f(x) = x^2$, $f'(x) = 2x$. The integrand is almost of the required format; only the constant 2 is missing. Thus we multiply by (2)(1/2) and transfer the unneeded constant through the integral sign, as

$$\int xe^{x^2}\,dx = (1/2)\int 2xe^{x^2}\,dx = (1/2)(e^{x^2} + C) = \frac{e^{x^2}}{2} + C \qquad []$$

EXERCISES

Evaluate the following integrals.

131. $\int e^{x+1}\, dx$

132. $\int e^{2x+1}\, dx$

133. $\int 2xe^{x^2+1}\, dx$

134. $\int x^2 e^{x^3+5}\, dx$

135. $\int e^{0.5x+3}\, dx$

136. $\int 5e^{-x/2}\, dx$

137. $\int 6x^2 e^{(2x^3+1)}\, dx$

138. $\int (6x+5)e^{(3x^2+5x)}\, dx$

139. $\int xe^{4x^2}\, dx$

140. $\int 3(x+1)e^{(x^2+2x+3)}\, dx$

13.5 GEOMETRIC INTERPRETATION OF INTEGRATION

We can calculate the areas of many regions that are bounded by straight-line segments using the well-known formulas for area of a rectangle ($A = bH$) and for area of a triangle ($A = bH/2$). (See Figure 13.1.) But the problem becomes more complex when the region is bounded by *curved* lines. We shall see that, if the region is surrounded by curves whose equations are known, the calculus can be used to find the area.

To begin, let us consider the function $f(x) = x^2$, which is shown in Figure 13.2, and see how we can find the shaded area A. This is the area under the curve depicting the function $f(x) = x^2$, above the x-

Figure 13.1

Using Rectangles and Triangles to Determine Area of Regions Surrounded by Straight-Line Segments

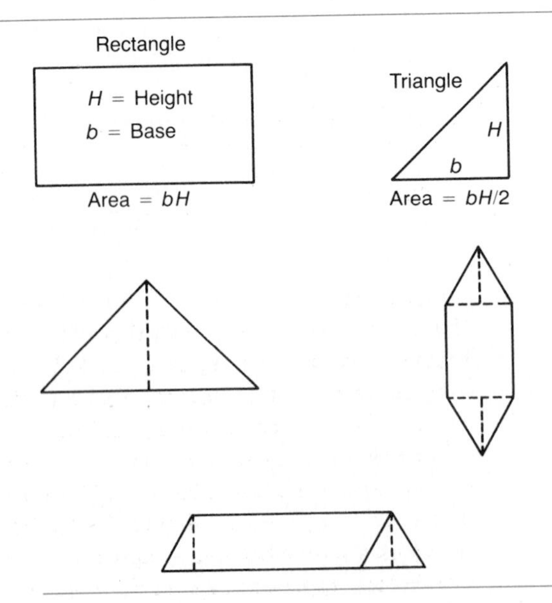

Figure 13.2

Using a Rectangle to Approximate the Area of a Region Bounded by a Curved Line

(A)

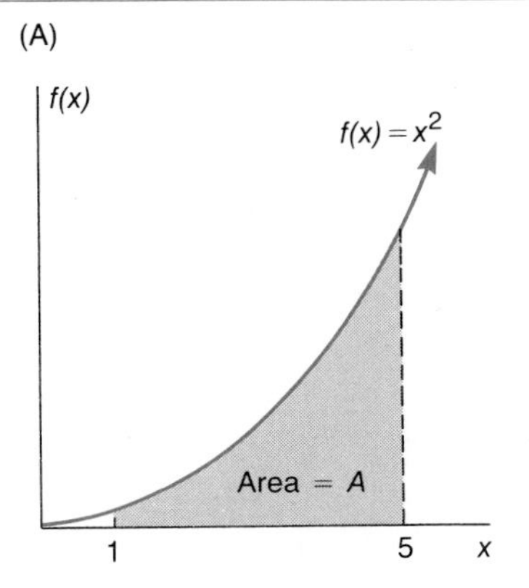

(B)

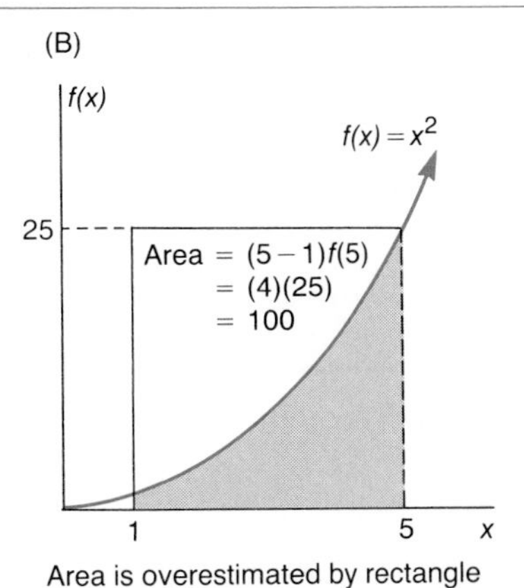

Area is overestimated by rectangle whose height is the maximum value of $f(x)$ on that interval.

(C)

f(x)

f(x) = x²

1

1 5 x

Area = (5 − 1)f(1)
= (4)(1)
= 4

Area is underestimated by rectangle whose height is the minimum value of $f(x)$ on that interval.

(D)

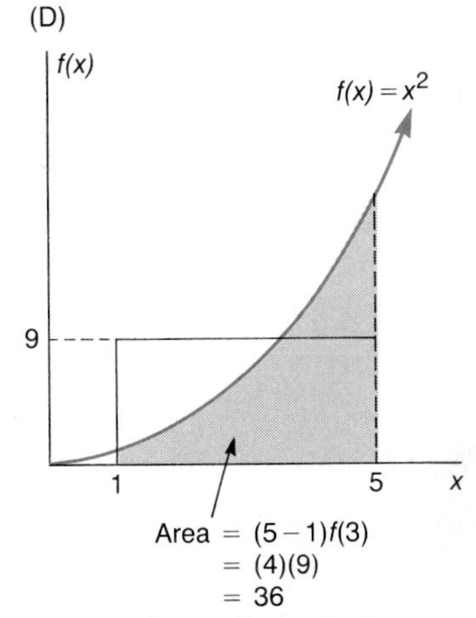

Area = (5 − 1)f(3)
= (4)(9)
= 36

Better estimate obtained using rectangle whose height is the value of $f(x)$ at the midpoint of the interval.

axis, between $x = 1$ and $x = 5$. We shall soon be able to determine that the area is $41\frac{1}{3}$. But because we presently have no geometric formula for the area of this irregularly shaped region, we shall *approximate* the area using rectangles and the fact that the area of a rectangle is given by $A = bH$, where b is the base and H is the height of the rectangle.

Clearly the area we are concerned with is overestimated by the area of the rectangle R_1 in Figure 13.2B and is underestimated by the area of the rectangle R_2 in Figure 13.2C. The rectangle in Figure 13.2D provides a somewhat better approximation. Still, even this is not a very satisfactory approximation of the area A.

Figure 13.3

Using a Series of Rectangles to Approximate the Area of a Region Bounded by a Curved Line

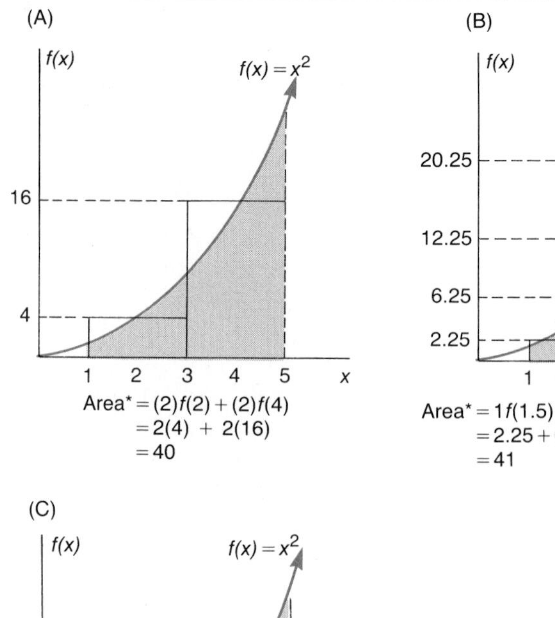

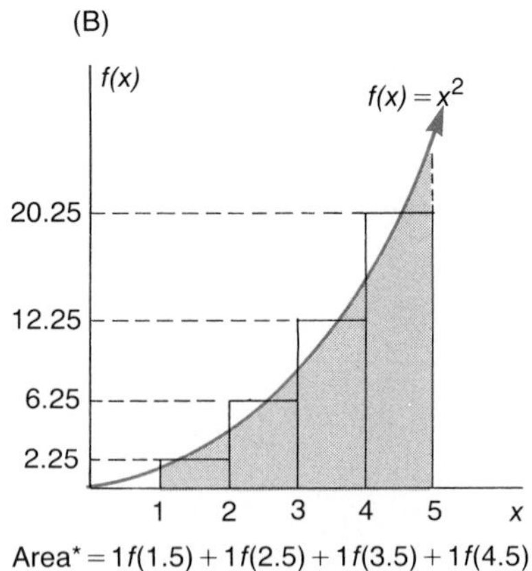

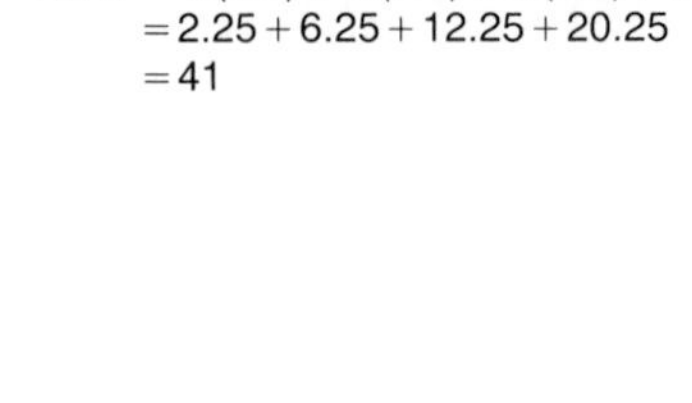

A better approximation is obtained if we divide the interval [1,5] on the x-axis in half and use two rectangles. (See Figure 13.3.) The width of each rectangle is $(5 - 1)/2 = 2$, and the heights are $f(2)$ and $f(4)$, respectively. Using the combined area of these two rectangles to approximate the area of interest, we have

$$A^* = 2f(2) + 2f(4) = 2(4) + 2(16) = 8 + 32 = 40$$

where A^* is the approximate area.

And by forming additional rectangles we obtain an even closer approximation, as shown in Figure 13.3B. Now the width of each rectangle is one, and the combined area of the four rectangles gives an approximation of the area of interest of

$$\begin{aligned} A^* &= 1f(1.5) + 1f(2.5) + 1f(3.5) + 1f(4.5) \\ &= 2.25 + 6.25 + 12.25 + 20.25 \\ &= 41 \end{aligned}$$

In Figure 13.3C, eight rectangles (each of width 0.5) have been drawn to approximate the area of interest. The combined area of these rectangles is computed by

$$A^* = 0.5f(1.25) + 0.5f(1.75) + \cdots + 0.5f(4.75) = 41.25$$

This approximation is even closer to the exact area $A = 41\frac{1}{3}$. In fact, if we continue to subdivide the interval between $x = 1$ and $x = 5$ so as to use more and more rectangles, the approximation will continue to move even closer to the actual area.

Now let us put this discussion on a more general framework. Suppose we are concerned with a function $f(x)$ which is continuous on the closed interval $[a,b]$. We are interested in determining the area between the curve representing the graph of f and the x-axis, and between $x = a$ and $x = b$. We proceed by dividing the interval $[a,b]$ into a number n of subintervals, which will be used to form n rectangles over the area of interest. Let us designate the width of rectangle i by Δx_i and the height of the rectangle by $f(x_i)$. The area A of interest can then be approximated by the sum of the areas of the n rectangles, as

$$A^* = f(x_1)\Delta x_1 + f(x_2)\Delta x_2 + \cdots + f(x_n)\Delta x_n = \sum_{i=1}^{n} f(x_i)\Delta x_i$$

(See Figure 13.4.)

Figure 13.4

Area $= \sum_{i=1}^{n} f(x_i)\Delta x_i$

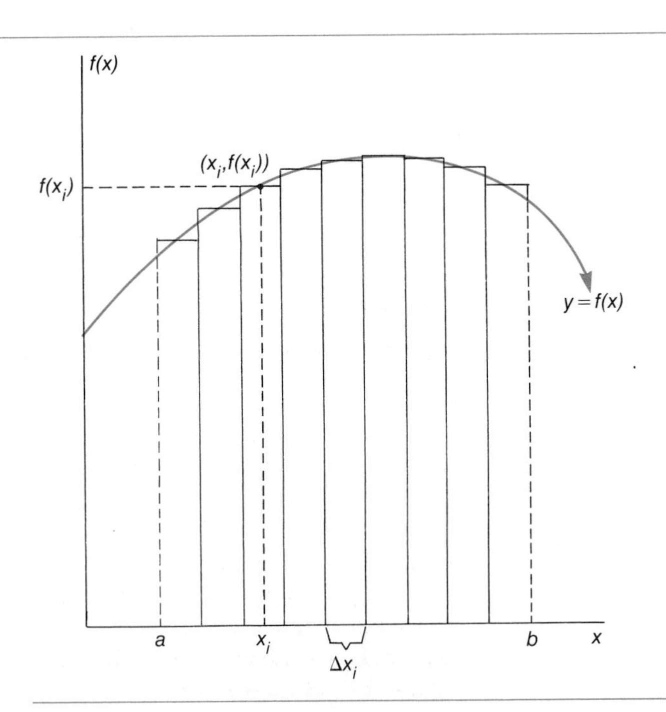

The Definite Integral

The above approximation of area becomes more and more accurate as the number n of rectangles becomes larger and larger (and concurrently the width of each rectangle becomes smaller and smaller). Thus, to obtain the actual area, we let n approach infinity, obtaining

$$A = \lim_{n \to \infty} \sum_{i=1}^{n} f(x_i)\Delta x_i$$

That is, the actual area A is the limiting value of the sum of the areas of the rectangles as the number of rectangles approaches infinity.

This limit, if it exists, is the area A under the curve $y = f(x)$, above the x-axis, between $x = a$ and $x = b$. The sum $f(x_1)\Delta x_1 + f(x_2)\Delta x_2 + \cdots + f(x_n)\Delta x_n$ is referred to as the RIEMANN SUM for the function $f(x)$ on $[a,b]$. The limit is called the RIEMANN INTEGRAL, or the DEFINITE INTEGRAL, from a to b of $f(x)$.

> Given a function f, continuous on the interval $[a,b]$, the DEFINITE INTEGRAL of f from a to b is
>
> $$\int_a^b f(x)\,dx = \lim_{n \to \infty} \sum_{i=1}^{n} f(x_i)\Delta x_i$$
>
> (maximum $\Delta x \to 0$)
>
> if the limit exists

Hence, just as the Greek symbol Σ is used to denote the sum of a finite number of discrete elements, the definite integral denotes summation for continuous functions. The symbol $\int_a^b f(x)\,dx$ is read "the definite integral of $f(x)$ between a and b." The a and b values which appear above and below the integral symbol are called the LIMITS OF INTEGRATION, with a being the LOWER LIMIT OF INTEGRATION and b the UPPER LIMIT OF INTEGRATION.

The Fundamental Theorem of Integral Calculus

The evaluation of area using the limit of a Riemann sum is possible but is tedious and time consuming. The following theorem provides a relatively straightforward technique for evaluating the definite integral.

> Fundamental theorem of integral calculus *If* y = f(x) *is a continuous function on the closed interval* [a,b] *and if it has an antiderivative* F(x) *on this interval, then*
>
> $$\int_a^b f(x)\,dx = F(b) - F(a)$$
>
> *where* F(b) *is the value of the antiderivative at* x = b *and* F(a) *is the value of the antiderivative at* x = a.

According to this theorem, the definite integral is evaluated by first determining an antiderivative, $F(x)$, of $f(x)$. Then the antiderivative is evaluated at $x = b$; that is, $F(b)$ is computed by replacing x by b in the expression. Next, the antiderivative is evaluated at $x = a$ by replacing x by a to obtain $F(a)$. Finally, $F(b) - F(a)$ is computed.

We introduce the notation for the definite integral

$$\int_a^b f(x)\,dx = F(x)\Big|_a^b = F(b) - F(a)$$

In computing the definite integral, the arbitrary constant of integration C will always drop out of the computation, as

$$[F(b) + C] - [F(a) + C] = F(b) - F(a)$$

Hence, there is no need to include the constant C when evaluating definite integrals.

It is also important that we distinguish between an INDEFINITE INTEGRAL of a function and a DEFINITE INTEGRAL. An indefinite integral is a symbol for all the antiderivatives of the function; each antiderivative is itself a function. A definite integral, on the other hand, is a single number.

Let us return to the preceding example and evaluate the shaded area of Figure 13.2, using the definite integral. We compute

$$\int_1^5 x^2\,dx = x^3/3\Big|_1^5 = [5^3/3] - [1^3/3] = 124/3 = 41\ 1/3$$

Certain conditions must be met in order for the above procedure to be applicable. First, f must be continuous on the interval $[a,b]$. Second, f must be nonnegative over the interval $[a,b]$; that is, $f(x) \geq 0$ for $a \leq x \leq b$. Last, f must have an antiderivative on $[a,b]$.

For a function f, continuous on the interval $[a,b]$, and whose value is positive over the interval from $x = a$ to $x = b$ (that is, $f(x)$ lies above the x-axis), the area A which is bounded by the function f, the x-axis, $x = a$, and $x = b$ is determined by

$$A = \int_a^b f(x)\,dx$$

Example 34 Find the area bounded by $f(x) = 10/(x + 1)^2$, the x-axis, $x = 0$, and $x = 3$, as shown in Figure 13.5.

We note that the function is continuous between $x = 0$ and $x = 3$ and that the region of interest lies entirely above the x-axis.

Figure 13.5

Area under the Curve $f(x) = 10/(x + 1)^2$, above the x-Axis, between $x = 0$ and $x = 3$

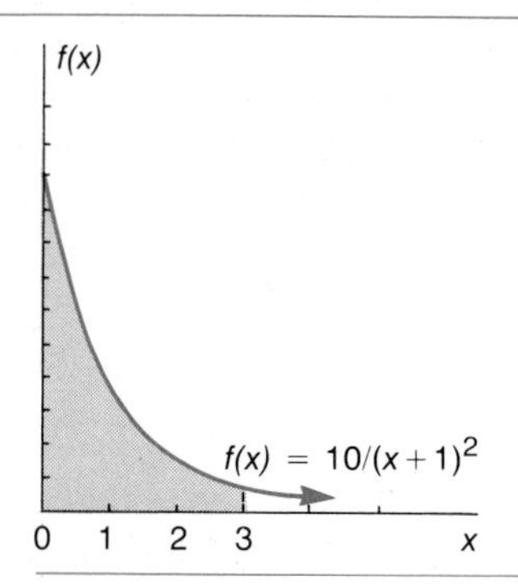

To determine the area, we evaluate the definite integral of $f(x)$ from $x = 0$ to $x = 3$, as

$$\int_0^3 10/(x+1)^2\,dx = -10/(x+1)\Big|_0^3$$
$$= [-10/(3+1)] - [-10/(0+1)]$$
$$= -10/4 + 10/1$$
$$= 30/4 = 7.5$$

[]

Example 35

Find the area of the region bounded by $f(x) = 4 - x^2$, the x-axis, $x = -2$, and $x = 2$, as shown in Figure 13.6.

To determine the area outlined we evaluate

$$\int_{-2}^{2} (4 - x^2)\,dx = [4x - (x^3/3)]\Big|_{-2}^{2} = [4(2) - (2)^3/3] - [4(-2) - (-2)^3/3]$$
$$= 10\ 2/3$$

[]

Figure 13.6

Area under $f(x) = 4 - x^2$, above the x-Axis, and between $x = -2$ and $x = 2$

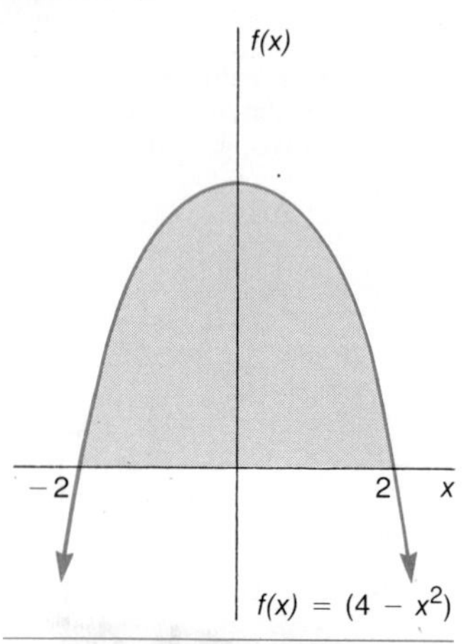

A function must be continuous over the interval of integration in order for these procedures to work. The next example will illustrate this point.

Example 36

Consider the function $f(x) = 1/x^2$ as shown in Figure 13.7. The function is not defined at $x = 0$; in fact, it has an infinite discontinuity at $x = 0$. Hence, the definite integral cannot be evaluated over an interval that includes the point $x = 0$. Let us see what happens when we attempt to evaluate the integral over the interval [0,1]. We obtain

$$\int_0^1 1/x^2\,dx = -1/x \Big|_0^1$$

and $-1/x$ cannot be evaluated at the lower limit of integration, $x = 0$. []

Figure 13.7

The Function Must Be Continuous over the Interval of Integration in Order for the Procedures Outlined in This Chapter to Work

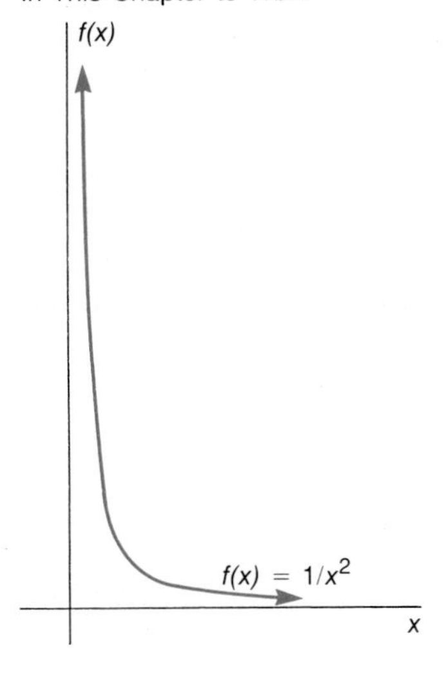

In evaluating area using the definite integral, we specify that the area lie *below* the curve representing $y = f(x)$, *above* the x-axis, between $x = a$ and $x = b$, where $a < b$. One property of area is that it is, by definition, nonnegative. However, the definite integral $\int_a^b f(x)\,dx$ will always be negative when the graph of $f(x)$ is below the x-axis for $[a,b]$. In such a case, area is given by $-\int_a^b f(x)\,dx$.

> For a function f continuous on the interval $[a,b]$ and whose value is *negative over the interval* from $x = a$ to $x = b$ (that is, $f(x)$ lies *below* the x-axis), the area A which is bounded by the function, the x-axis, $x = a$, and $x = b$ is given by
>
> $$A = -\int_a^b f(x)\,dx$$

Example 37

Determine the area shown in Figure 13.8.

Figure 13.8

Area of Region Bounded by $f(x) = -x^2$, the x-Axis, $x = 1$, and $x = 3$

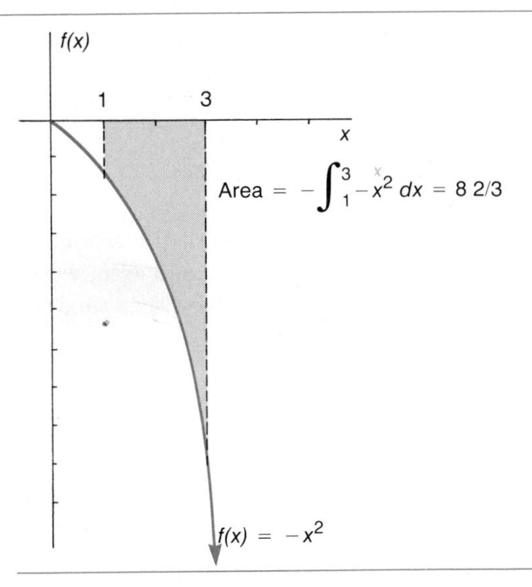

Because the area lies below the x-axis, we compute

$$\text{Area} = -\int_1^3 -x^2\,dx = -\left[\frac{-x^3}{3}\Big|_1^3\right]$$
$$= -\left\{\left[\frac{-(3)^3}{3}\right] - \left[\frac{-(1)^3}{3}\right]\right\}$$
$$= -[(-8\ 2/3)]$$
$$= 8\ 2/3$$

Thus, the shaded region has an area of $8\frac{2}{3}$ square units. []

If $f(x)$ is positive over part of the interval $[a,b]$ and is negative over the remainder of the interval, a portion of the area of interest will lie above the x-axis, while another portion will fall below the x-axis. Then $\int_a^b f(x)\,dx$ will yield the *net* area. That is, area above the x-axis will evaluate positive, while area below the x-axis will evaluate negative, and the two will be combined algebraically.

Example 38

Evaluate $\int_0^8 (x-4)\,dx$.

This definite integral will yield the *net* area shown in Figure 13.9. We compute

$$\int_0^8 (x-4)\,dx = [(x^2/2) - 4x]\Big|_0^8 = [(8^2/2) - 4(8)] - [(0^2/2) - 4(0)]$$
$$= [32 - 32] - [0]$$
$$= 0$$
[]

Figure 13.9

Area of Region Bounded by $f(x) = x - 4$, the x-Axis, $x = 0$, and $x = 8$

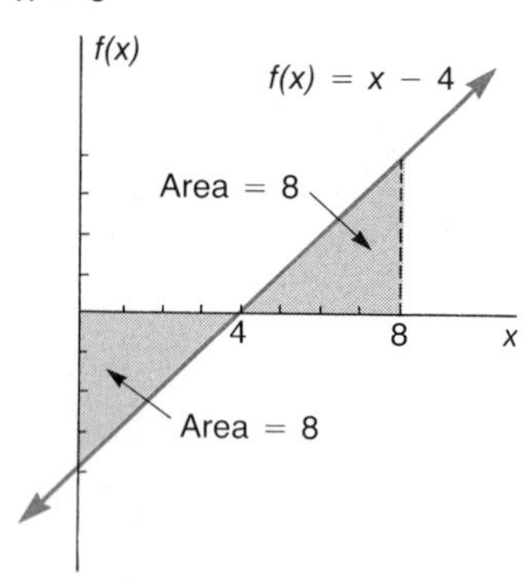

We are able to find the area of a region, a part of which lies above the x-axis and a part of which lies below the x-axis, using the ADDITIVE PROPERTY OF DEFINITE INTEGRALS.

> **Additive property of definite integrals** If f is continuous on some interval containing a, b, and c, where $a \leq b \leq c$, then
>
> $$\int_a^c f(x)\,dx = \int_a^b f(x)\,dx + \int_b^c f(x)\,dx$$

Example 39

Find the total area of the shaded region shown in Figure 13.9.

To find the total shaded area, we must compute the definite integral for the area below the x-axis separately from the definite integral for the area above the x-axis. Hence, we compute

$$\text{Area} = -\int_0^4 (x-4)\,dx + \int_4^8 (x-4)\,dx$$
$$= -[(x^2/2) - 4x]\Big|_0^4 + [(x^2/2) - 4x]\Big|_4^8$$
$$= -[-8] + [8]$$
$$= 16$$
[]

Example 40 Determine the area bounded by $f(x) = x^2 - 6x + 8$ and the x-axis, between $x = 0$ and $x = 6$, as shown in Figure 13.10.

To determine the combined area, we must compute

$$\begin{aligned}\text{Area} &= \int_0^2 (x^2 - 6x + 8)\,dx - \int_2^4 (x^2 - 6x + 8)\,dx + \int_4^6 (x^2 - 6x + 8)\,dx\\ &= 6\ 2/3 - (-1\ 1/3) + 6\ 2/3\\ &= 14\ 2/3\end{aligned}$$

[]

Although the definite integral has been developed with the assumption that the lower limit of integration is less in numerical value than the upper limit, the following property of definite integrals permits us to evaluate the integral even when this is not the case.

> If f is integrable on $[a,b]$ then
>
> $$\int_a^b f(x)\,dx = -\int_b^a f(x)\,dx$$

Another important property of definite integrals is as follows:

> $$\int_a^a f(x)\,dx = 0$$

Figure 13.10

Area of Region Bounded by $f(x) = x^2 - 6x + 8$, the x-Axis, $x = 0$, and $x = 6$

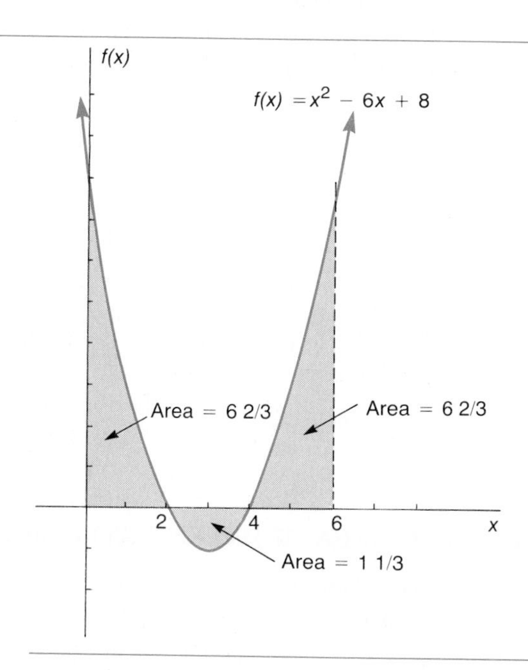

EXERCISES

Evaluate each of the following definite integrals.

141. $\int_1^3 7\,dx$
142. $\int_2^5 (x+4)\,dx$
143. $\int_1^4 (2x+1)\,dx$
144. $\int_1^2 (2x^2-3x+9)\,dx$
145. $\int_1^3 (4x^3+2x)\,dx$
146. $\int_0^5 2x^3\,dx$
147. $\int_1^5 (x/4)\,dx$
148. $\int_1^4 (3x^2+5)\,dx$
149. $\int_1^3 x^{(3/2)}\,dx$
150. $\int_1^2 1/x^3\,dx$
151. $\int_1^5 1/x\,dx$
152. $\int_1^3 (x^3-1)\,dx$
153. $\int_2^3 (x+1)^2\,dx$
154. $\int_0^3 2xe^{x^2}\,dx$
155. $\int_0^1 e^{2x}\,dx$
156. $\int_{-2}^2 (x^2+2x+2)\,dx$
157. $\int_0^3 \frac{1}{x+1}\,dx$
158. $\int_2^4 3\sqrt{x}\,dx$
159. $\int_0^2 \frac{x}{x^2+4}\,dx$
160. $\int_3^3 \sqrt{x^2-1}\,dx$

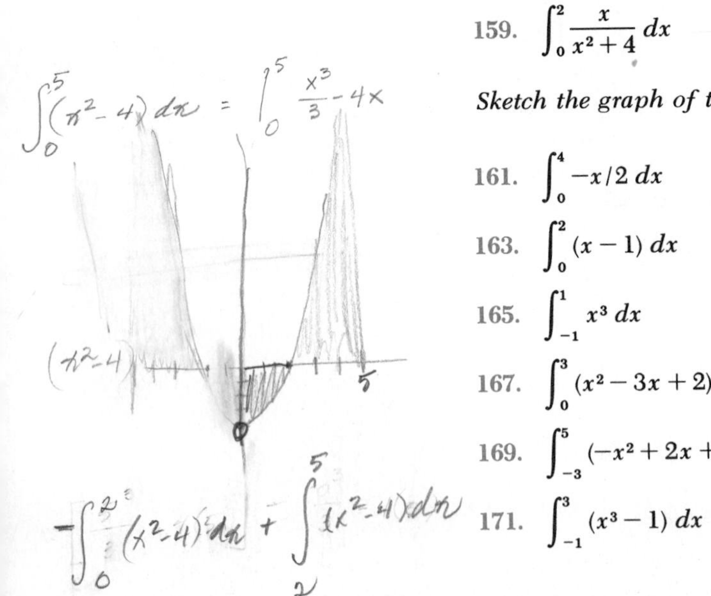

Sketch the graph of the function and determine the area specified.

161. $\int_0^4 -x/2\,dx$
162. $\int_{-2}^2 (2x+5)\,dx$
163. $\int_0^2 (x-1)\,dx$
164. $\int_0^5 (x^2-4)\,dx$
165. $\int_{-1}^1 x^3\,dx$
166. $\int_0^3 (x^2-8x-3)\,dx$
167. $\int_0^3 (x^2-3x+2)\,dx$
168. $\int_{-2}^2 (x^2-x-2)\,dx$
169. $\int_{-3}^5 (-x^2+2x+3)\,dx$
170. $\int_0^5 (x^4-16)\,dx$
171. $\int_{-1}^3 (x^3-1)\,dx$
172. $\int_{-1}^4 (x^2-2x)\,dx$

13.6 THE AREA BETWEEN TWO CURVES

The definite integral can be used to determine the area between two curves.

Figure 13.11

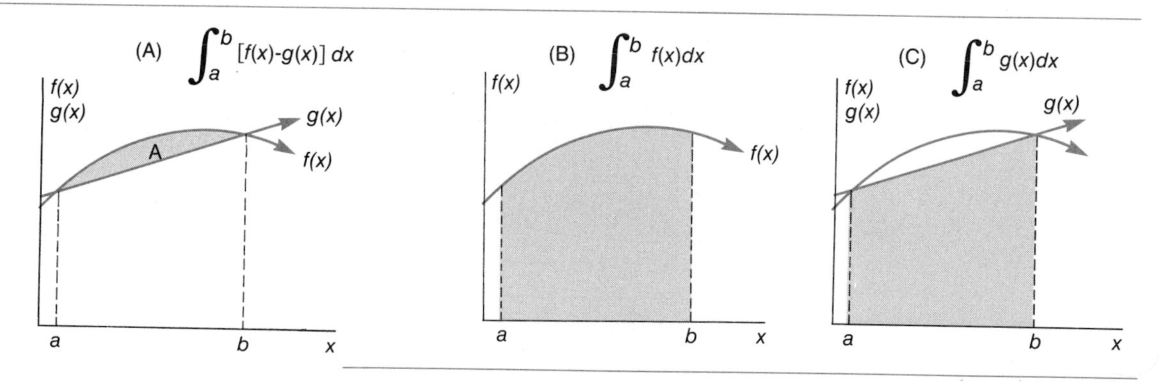

Area between two curves If f and g are continuous and $f(x) \geq g(x)$ over the interval $[a,b]$, the area of the region bounded by $f(x)$, $g(x)$, $x = a$, and $x = b$ is given by

$$A = \int_a^b [f(x) - g(x)]\,dx$$

Consider two functions, f and g, such as those pictured in Figure 13.11, and the region bounded by $f(x)$, $g(x)$, $x = a$ and $x = b$. We have seen that the area of the region bounded by $f(x)$, the x-axis, $x = a$, and $x = b$ is given by $\int_a^b f(x)\,dx$. (See Figure 13.11B.) The resulting area includes that of region A, but it also includes the area of a region which is not a part of A. This means that the area has been overestimated by an amount equivalent to the shaded region in Figure 13.11C. This "surplus" area—that is, the area bounded by $g(x)$, the x-axis, $x = a$, and $x = b$—is given by $\int_a^b g(x)\,dx$. Hence, the area of the region bounded by $f(x)$, $g(x)$, $x = a$, and $x = b$ is given by $\int_a^b f(x)\,dx - \int_a^b g(x)\,dx$.

A property of definite integrals which can greatly facilitate such computations as these is as follows.

$$\int_a^b f(x)\,dx \pm \int_a^b g(x)\,dx = \int_b^b [f(x) \pm g(x)]\,dx$$

That is, if two or more definite integrals are to be evaluated *over the same limits of integration* and then added or subtracted, the integrands may be algebraically combined before they are integrated.

Example 41 Determine the area of the region bounded by $f(x) = 0.05x^2 + 6$, $g(x) = 0.03x^2 + 2$, $x = 1$, and $x = 4$, as shown in Figure 13.12.

Figure 13.12

Area of Region Bounded by $f(x) = 0.05x^2 + 6$, $g(x) = 0.03x^2 + 2$, $x = 1$, and $x = 4$

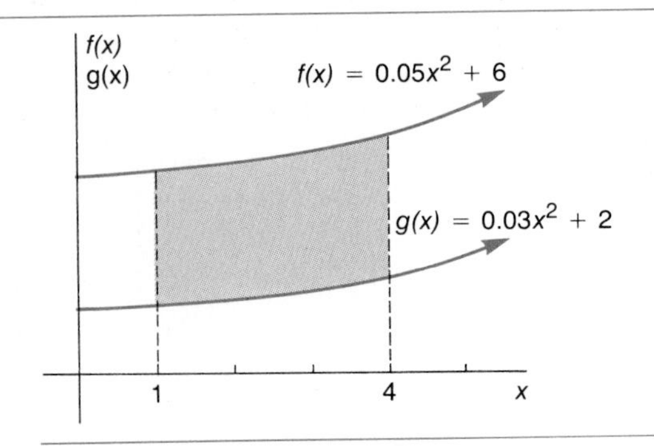

We determine the area by computing

$$
\begin{aligned}
A &= \int_1^4 [(0.05x^2 + 6) - (0.03x^2 + 2)]\, dx \\
&= \int_1^4 (0.02x^2 + 4)\, dx = \frac{0.02x^3}{3} + 4x \Big|_1^4 \\
&= \left[\frac{0.2(4)^3}{3} + 4(4)\right] - \left[\frac{0.2(1)^3}{3} + (4)(1)\right] \\
&= \left[\frac{1.28}{3} + 16\right] - \left[\frac{0.02}{3} + 4\right] \\
&= 12.42
\end{aligned}
$$

The enclosed region has an area of 12.42 square units. []

Example 42

Determine the area of the region bounded by the two functions $f(x) = 2x$ and $g(x) = x^2$, as shown in Figure 13.13.

Figure 13.13

Area of the Region Bounded by $f(x) = 2x$ and $g(x) = x^2$

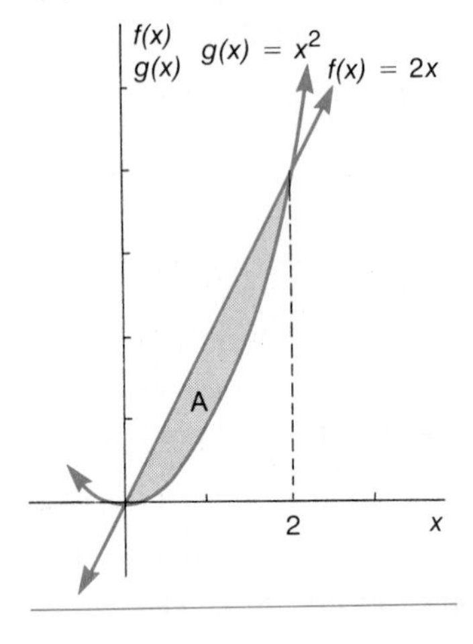

The limits of integration are those values for x for which the two functions are equal. If we equate the functions, we obtain

$$
\begin{aligned}
f(x) &= g(x) \\
2x &= x^2 \\
-x^2 + 2x &= 0 \\
x(-x + 2) &= 0 \\
x &= 0,\ x = 2
\end{aligned}
$$

The area of the region enclosed by the two functions is, thus, given by

$$
\begin{aligned}
\int_0^2 [(2x) - (x^2)]\, dx &= [x^2 - (x^3/3)] \Big|_0^2 \\
&= [2^2 - (2^3/3)] - [0^2 - (0^3/3)] \\
&= 1\,1/3
\end{aligned}
$$

[]

Example 43 Find the area of the region bounded by the curves $f(x) = 8 - x^2$ and $g(x) = x^2$ (as shown in Figure 13.14).

We determine the x-values where the functions intersect by setting $f(x) = g(x)$ and solving for x. We obtain

$$f(x) = g(x)$$
$$8 - x^2 = x^2$$
$$8 - 2x^2 = 0$$
$$2(4 - x^2) = 0$$
$$x = -2,\ x = +2$$

Thus, the area is given by

$$\int_{-2}^{2} [(8 - x^2) - (x^2)]\, dx = \int_{-2}^{2} (8 - 2x^2)\, dx = [8x - (2x^3/3)] \Big|_{-2}^{2}$$
$$= [8(2) - 2(2)^3/3] - [8(-2) - 2(-2)^3/3]$$
$$= [16 - (16/3)] - [-16 + (16/3)]$$
$$= 21\ 1/3$$

[]

Actually, this procedure is valid even if f or g or both take on negative values on $[a,b]$, as the next example will illustrate.

Figure 13.14

Area of the Region Bounded by $f(x) = 8 - x^2$ and $g(x) = x^2$

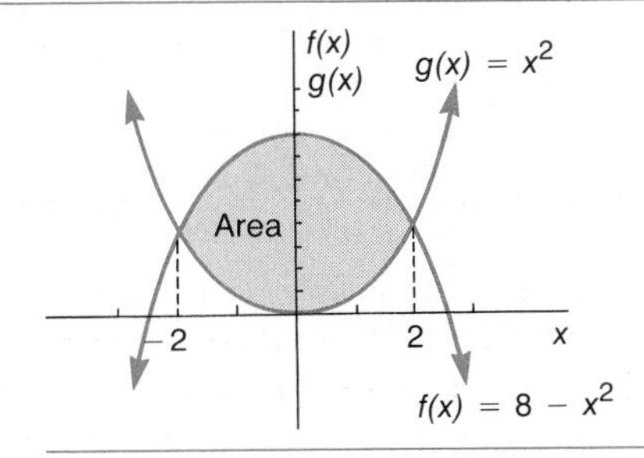

Example 44 Find the area bounded by $f(x) = -x^2 + 4x + 4$ and $g(x) = x^2 - 4x + 4$, as shown in Figure 13.15A.

The shaded area is given by

$$\int_{0}^{4} [(-x^2 + 4x + 4) - (x^2 - 4x + 4)]\, dx = \int_{0}^{4} (-2x^2 + 8x)\, dx$$
$$= [(-2x^3/3) + 4x^2] \Big|_{0}^{4}$$
$$= 21\ 1/3$$

Now let us shift both functions vertically down by four places—as shown in Figure 13.15B. We have

$$s(x) = f(x) - 4 = (-x^2 + 4x + 4) - 4 = -x^2 + 4x$$

Figure 13.15

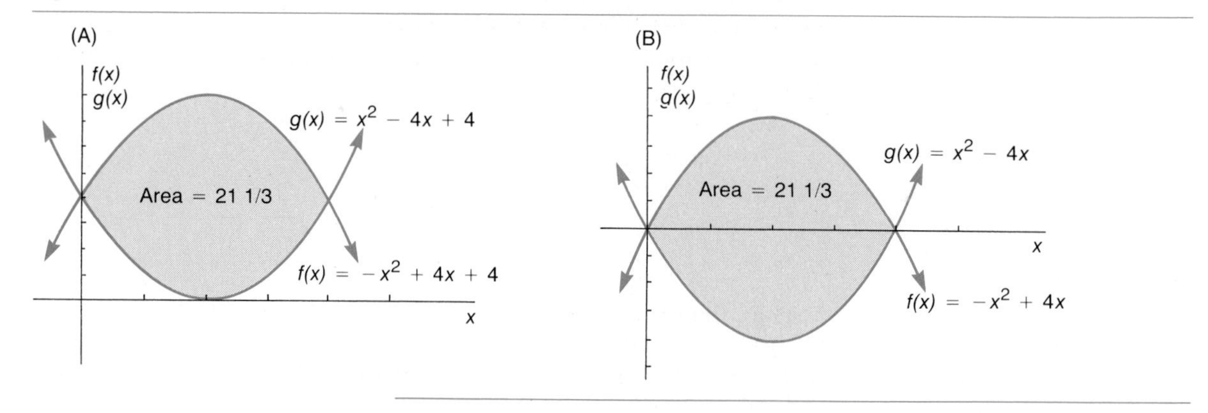

and

$$t(x) = g(x) - 4 = (x^2 - 4x + 4) - 4 = x^2 - 4x$$

Then

$$\int_0^4 [(-x^2 + 4x) - (x^2 - 4x)]\, dx = \int_0^4 (-2x^2 + 8x)\, dx = 21\ 1/3 \quad []$$

It should be noted that in using $\int_a^b [f(x) - g(x)]\, dx$ to determine the area between $f(x)$ and $g(x)$, it is necessary that $f(x) \geq g(x)$ for all x between a and b, inclusive. The next examples indicate how to find area between two curves when this is not the case.

Example 45 Find the area of the region enclosed between the graphs of $f(x) = x^3$ and $g(x) = x$, as shown in Figure 13.16.

Figure 13.16

Region Enclosed by $f(x) = x^3$ and $g(x) = x$

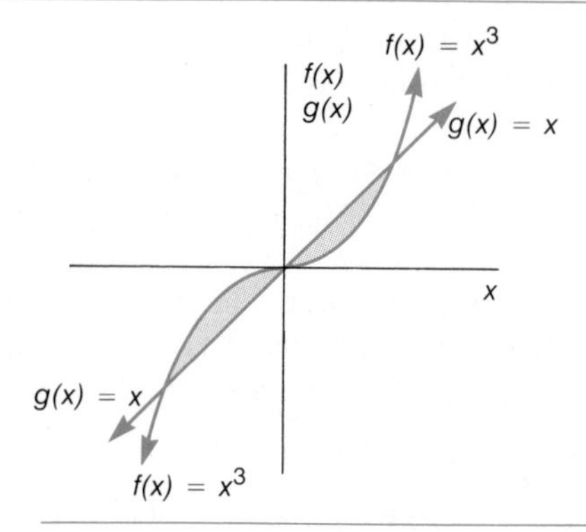

We first determine the limits of integration by finding where the curves intersect. We do this by setting $f(x) = g(x)$ and solving for x, as

$$x^3 = x$$
$$x^3 - x = 0$$
$$x(x^2 - 1) = 0$$
$$x(x + 1)(x - 1) = 0$$
$$x = 0,\ x = -1,\ x = 1$$

Hence, the curves intersect at $x = -1$, at $x = 0$, and again at $x = +1$.

Notice that for x between $x = -1$ and $x = 0$, $f(x) > g(x)$. However, between $x = 0$ and $x = +1$, $g(x) > f(x)$. Thus, to determine the total area, we must evaluate two separate definite integrals and sum the results. We compute

$$A = \int_{-1}^{0} (x^3 - x)\,dx + \int_{0}^{1} (x - x^3)\,dx = [(x^4/4) - (x^2/2)]\Big|_{-1}^{0} + [(x^2/2) - (x^4/4)]\Big|_{0}^{1}$$

$$= \{[0] - [(-1)^4/4 - (-1)^2/2]\} + \{[1^2/2 - 1^4/4] - [0]\}$$
$$= 1/4 + 1/4$$
$$= 1/2$$

[]

This example illustrates the importance of sketching the graph of the two functions when using the definite integral to determine the area of the enclosed region between the curves. Having a picture of the region of interest facilitates identification of pertinent boundaries and underscores the logic required to define the areas.

Example 46 Refer to Figure 13.17 and determine the combination of definite integrals required to compute the area of each of the labeled regions.

a. The upper boundary of region R_1 is formed by $f(x)$, the lower boundary by the x-axis; the region extends from $x = a$ to $x = 0$. Thus, we compute its area A_1 by

$$A_1 = \int_{a}^{0} f(x)\,dx$$

b. The upper boundary of region R_2 is provided by the function $f(x)$, while the lower boundary is $g(x)$. The interval extends from $x = 0$ to $x = b$. Thus, the area A_2 is given by

$$A_2 = \int_{0}^{b} [f(x) - g(x)]\,dx$$

Figure 13.17

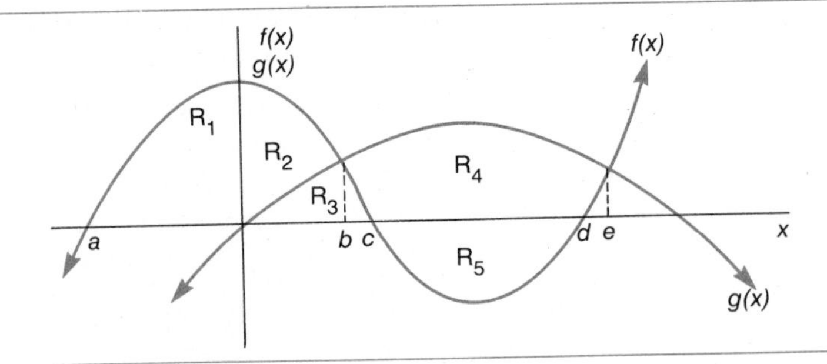

c. Because the upper boundary of region R_3 is formed by $g(x)$ from $x = 0$ to $x = b$ but is formed by $f(x)$ from $x = b$ to $x = c$, two separate definite integrals must be computed. The lower boundary for the region is given by the x-axis. To determine the area, we compute

$$A_3 = \int_0^b g(x)\,dx + \int_b^c f(x)\,dx$$

d. Region A_4, too, must be partitioned. Between $x = b$ and $x = c$, the upper boundary is $g(x)$, the lower boundary is $f(x)$. Between $x = c$ and $x = d$ the region has an upper boundary formed by $g(x)$ and has the x-axis as its lower boundary. Then from $x = d$ to $x = e$, $g(x)$ forms the upper boundary and $f(x)$ the lower boundary. To determine the area, we compute

$$A_4 = \int_b^c [g(x) - f(x)]\,dx + \int_c^d g(x)\,dx + \int_d^e [g(x) - f(x)]\,dx$$

e. Because all of region R_5 lies below the x-axis, with its lower boundary formed by $f(x)$, we compute

$$A_5 = -\int_c^d f(x)\,dx$$

[]

EXERCISES

Sketch the region defined and determine its total area.

173. $f(x) = 6 - x$; $g(x) = 6 + x$; $x = 0$; $x = 6$
174. $f(x) = 5$; $g(x) = x$; $x = 0$; $x = 5$
175. $f(x) = x^2$; $g(x) = x + 2$

176. $f(x) = x^2;\ g(x) = 6 - x$

177. $f(x) = -x^2 + 8x - 7;\ g(x) = 2x - 2$

178. $f(x) = x + 3;\ g(x) = 9 - x^2$

179. $f(x) = x^2;\ g(x) = 32 - x^2;\ x = 0$

180. $f(x) = -x^2;\ g(x) = x^2 - 18$

181. $f(x) = x^3;\ g(x) = 2 - x;\ x = 0;\ x = 1$

182. $f(x) = x;\ g(x) = \sqrt{x}$

183. $f(x) = x^2 - 6x + 9;\ g(x) = 5 - x$

184. $f(x) = -x^2 + 4x + 8;\ g(x) = x^2 - 2x$

185. $f(x) = 3x - x^2;\ g(x) = x - 3$

186. $f(x) = x^3 - x^2 - x;\ g(x) = 1/x;\ x = 1;\ x = 2$

187. $f(x) = x^4;\ g(x) = 20 - x^2$

188. $f(x) = x^3 - 6x^2;\ g(x) = 8x - 32;\ x = 0;\ x = 6$

Determine the combinations of definite integrals which would compute the area of each of the labeled regions.

189.

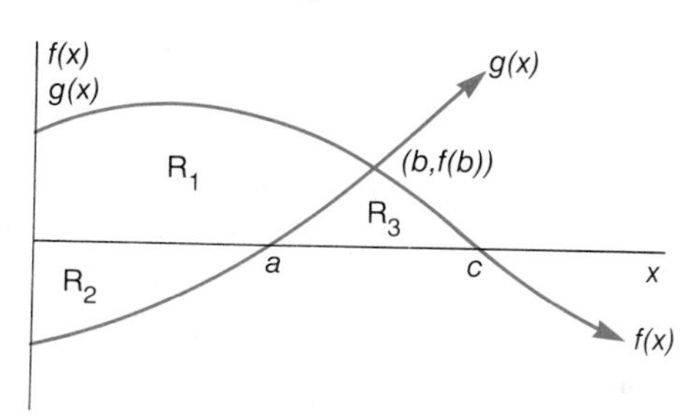

190.

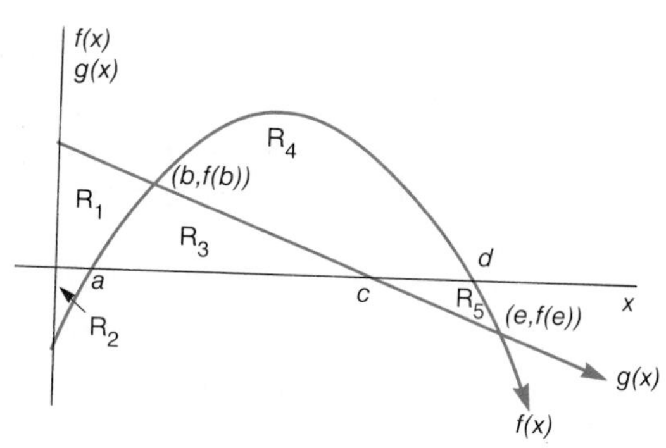

C·H·A·P·T·E·R 14

Additional Methods and Applications of Integral Calculus

The definite integral has many important applications in business and economics.

14.1 CONSUMERS' AND PRODUCERS' SURPLUS

The demand function for a commodity associates a price p of that commodity with a quantity q demanded by consumers. In general, the function $D(q)$ is a decreasing function, indicating that high prices and low demand, low prices and high demand, go hand in hand.

In a somewhat similar manner, the supply function for a commodity associates a price p with a quantity q that producers will be willing to supply. In general, the function $S(q)$ is a increasing function, indicating that larger quantities are supplied at higher prices than at lower prices.

Both the demand function and the supply function for a commodity are shown in Figure 14.1. The economic model of pure competition is founded on the assumption that all consumers pay the same (per-unit) price for the commodity. This price, which comes about by the free interplay of competitive market forces, is the price which equates the quantity consumers are willing to buy with the quantity producers are willing to supply. This EQUILIBRIUM PRICE, which we denote p^*, occurs where $D(q) = S(q)$. The EQUILIBRIUM QUANTITY, q^*, is the quantity that will be supplied, and demanded, at the equilibrium price.

The total amount paid by consumers for the commodity at market equilibrium is p^*q^*—the price per unit times the number of units sold. This total revenue can be interpreted geometrically as the area of the rectangle $0q^*Ep^*$ (the shaded area) in Figure 14.1.

Notice from the demand function (shown again in Figure 14.2) that some consumers would be willing to pay more than the market equilibrium price for the commodity. The quantity q_1 would be demanded at price $p(q_1)$; the quantity q_2 would be demanded at price $p(q_2)$; and so on. These consumers enjoy a special benefit attributable to the fact that they are trading in a competitive marketplace (as opposed to a market wherein the seller could demand from the consumer the *maximum amount* that the consumer would be willing to pay for the commodity). This benefit to the consumers—that is, the difference between what consumers actually pay and the maximum amount they would be willing to pay—is called CONSUMERS' SURPLUS (which we denote CS).

Note the region in Figure 14.3 labeled consumers' surplus. We can use this illustration as a guide in developing a formula for consumers' surplus. The definite integral $\int_0^{q^*} D(q)\,dq$ gives the total area under the demand function between $q = 0$ and $q = q^*$ and represents the total revenue that would have been generated had these consumers paid the maximum they had been willing to pay for the commodity. If we subtract the total amount actually paid, p^*q^*, the result is the CONSUMERS' SURPLUS.

Figure 14.1

Demand and Supply Functions for a Commodity

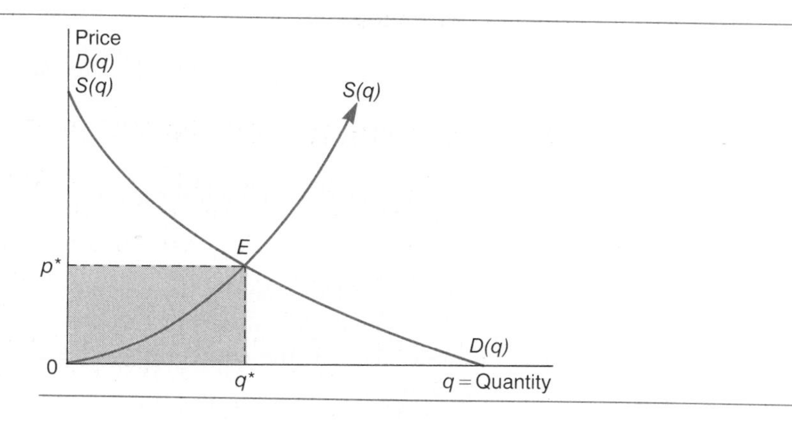

Figure 14.2

Some Consumers Would Be Willing to Pay More than the Equilibrium Price

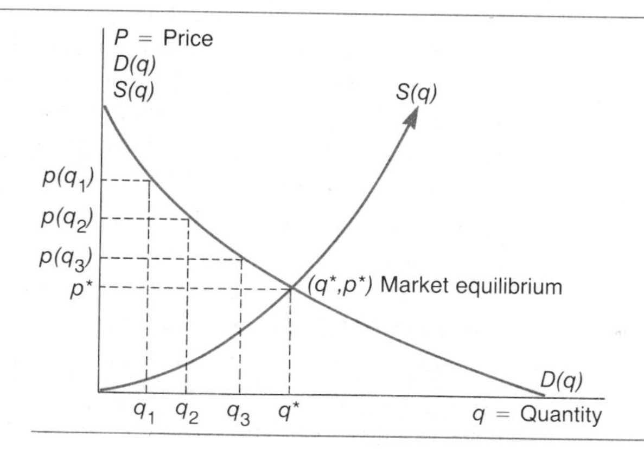

Figure 14.3

Consumers' Surplus

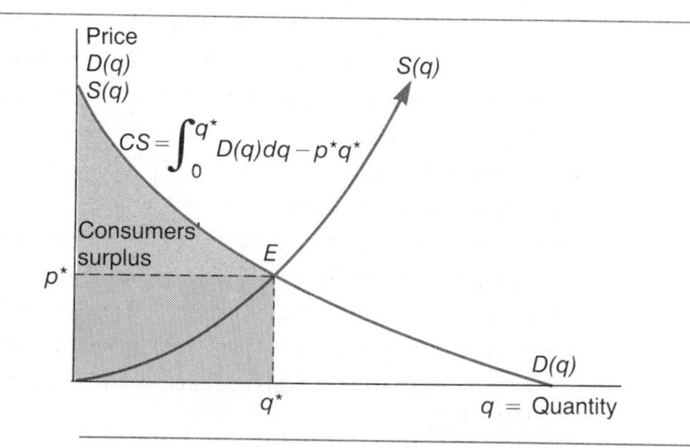

Consumers' surplus

$$CS = \int_0^{q^*} D(q)\,dq - p^*q^*$$

Example 1 The demand function for a commodity is

$$p = D(q) = 20 - 2q$$

where q is the quantity demanded and p is price, per unit, in dollars. (This demand function is pictured in Figure 14.4.)

Assume that the equilibrium price is $p^* = \$10$. Thus the equilibrium quantity is

$$10 = 20 - 2q$$
$$q^* = 5 \text{ units}$$

The revenue actually generated by the sale of the commodity is

$$p^*q^* = (10)(5) = \$50$$

The consumers' surplus is determined by solving

$$CS = \int_0^5 (20 - 2q)\,dq - (10)(5) = (20q - q^2)\Big|_0^5 - 50$$
$$= 75 - 50$$
$$= 25$$

The consumers' surplus is \$25. []

When a commodity is sold at the market equilibrium price, some producers will also enjoy an added benefit, for they would have been

Figure 14.4

Consumers' Surplus

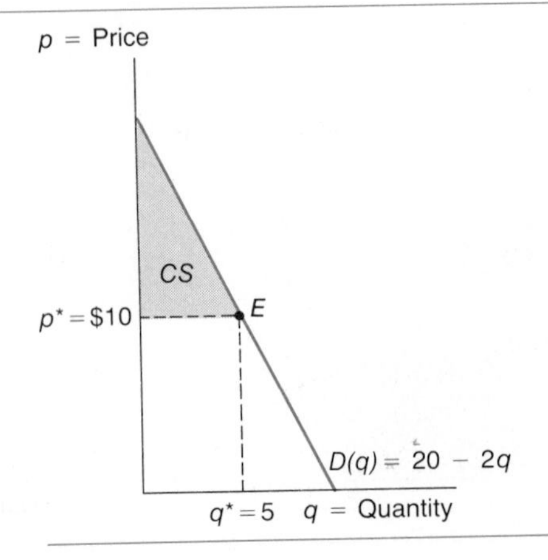

Figure 14.5

Some Producers Would Be Willing to Supply Units of the Commodity at Less than the Equilibrium Price

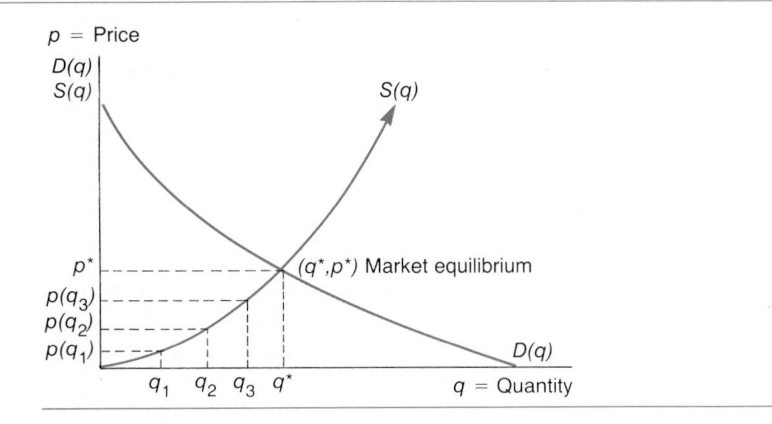

willing to sell the product at a lower price. Notice in Figure 14.5 that q_1 units would have been supplied at price $p(q_1)$, q_2 units would have been supplied at price $p(q_2)$, and so on.

Producers who are willing to supply the commodity at a price less than the market equilibrium price enjoy a PRODUCERS' SURPLUS. This producers' surplus (denoted PS) is the excess of the revenue producers actually receive in a free-market economy and the minimum they would have been willing to receive.

The supply function $S(q)$ for a commodity is shown in Figure 14.6. The total revenue actually received is given by p^*q^*. The quantity $\int_0^{q^*} S(q)\,dq$ is the area under the supply curve $S(q)$ from $q = 0$ to $q = q^*$, the equilibrium quantity, and represents the total revenue that would have been generated had producers sold at prices lower than equilibrium

Figure 14.6

Producers' Surplus

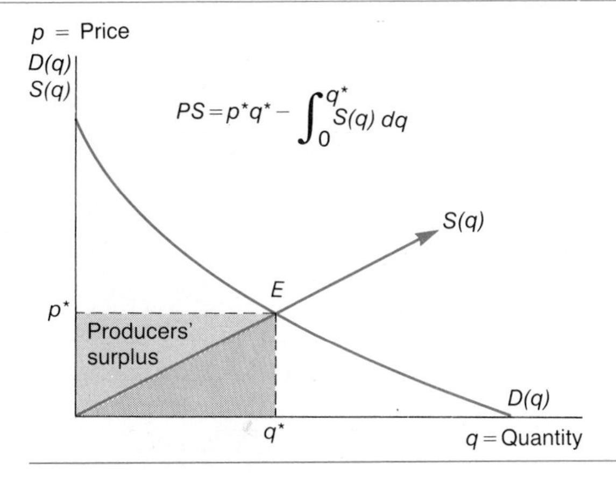

price. If we subtract this latter amount from p^*q^*, the result is the special benefit accruing to producers.

Producers' surplus

$$PS = p^*q^* - \int_0^{q^*} S(q)\, dq$$

Example 2 The supply function for a product is given by

$$S(q) = q^2 + q$$

(See Figure 14.7.) If the market equilibrium occurs where price is \$56, what is the producers' surplus?

Given $p^* = 56$, we can find equilibrium quantity, q^*, as follows:

$$56 = q^2 + q$$
$$q^2 + q - 56 = 0$$
$$(q - 7)(q + 8) = 0$$
$$q = -8,\ q^* = 7$$

Because equilibrium quantity cannot be negative, $q^* = 7$.

At market equilibrium, all producers receive the same per-unit price, $p^* = \$56$, for the product, even though the supply function indicates that some producers would have been willing to supply a quantity of the product at a lesser price.

Figure 14.7

Producers' Surplus

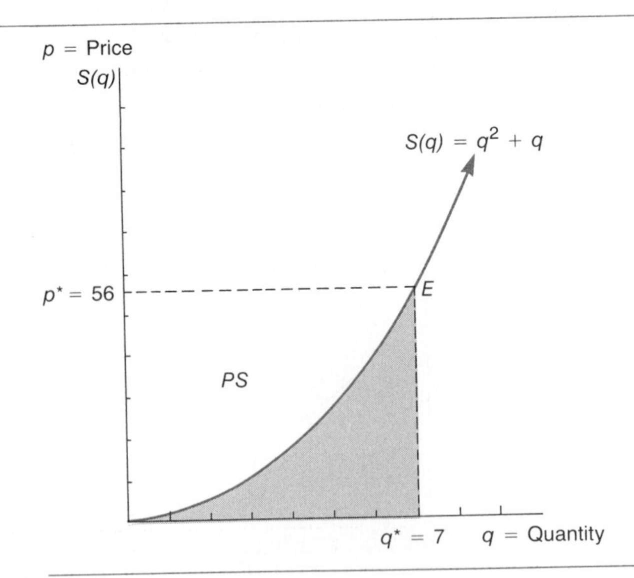

Total revenue actually received by the producers is

$$p^*q^* = (56)(7) = \$392$$

The producers' surplus is given by

$$PS = (56)(7) - \int_0^7 (q^2 + q)\,dq = 392 - [(q^3/3) + (q^2/2)]\Big|_0^7$$
$$= 392 - [343/3 + 49/2]$$
$$= 253.17$$

The producers' surplus is \$253.17. []

Example 3 The demand function for a product is

$$D(q) = \sqrt{124 - 3q}$$

where q is quantity, in units, and $p = D(q)$ is price per unit, in dollars.

The supply function for the same product is

$$S(q) = q + 2$$

Find consumers' surplus and producers' surplus.

The graphs of the demand and supply functions are shown in Figure 14.8.

Figure 14.8

Consumers' and Producers' Surplus

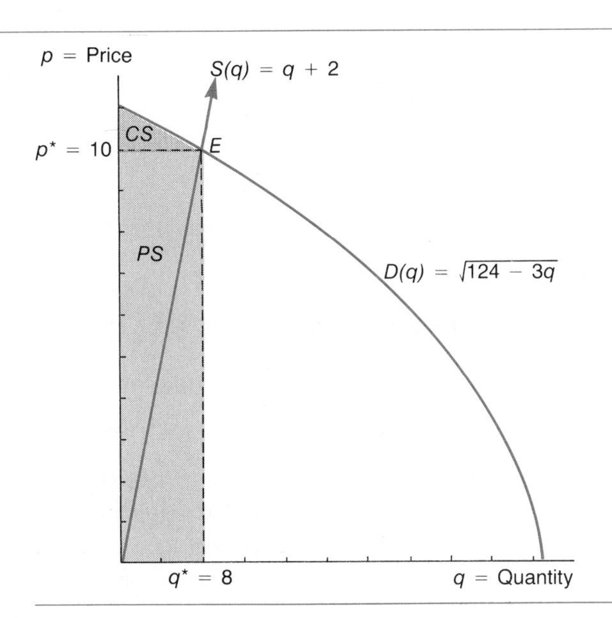

The equilibrium point is determined by solving

$$\begin{aligned}
D(q) &= S(q) \\
\sqrt{124-3q} &= q+2 \\
124-3q &= (q+2)^2 = q^2+4q+4 \\
q^2+7q-120 &= 0 \\
(q+15)(q-8) &= 0 \\
q &= -15,\ q^* = 8
\end{aligned}$$

Because equilibrium quantity cannot be negative, $q^* = 8$.

Given $q^* = 8$, equilibrium price is determined by

$$p = S(8) = 8 + 2 = 10 = p^*$$

(or $p = D(8) = \sqrt{124 - 3(8)} = 10$). Thus, market equilibrium occurs at a price of $p^* = \$10$, with a quantity of $q^* = 8$ units.

The consumers' surplus in this marketplace is given by

$$\begin{aligned}
CS &= \int_0^8 (124-3q)^{1/2}\, dq - (10)(8) \\
&= (-1/3)\int_0^8 (-3)(124-3q)^{1/2}\, dq - 80 \\
&= (-1/3)[(2/3)(124-3q)^{3/2}]\Big|_0^8 - 80 \\
&= (-1/3)(2/3)[(100)^{3/2} - (124)^{3/2}] - 80 \\
&= 4.62
\end{aligned}$$

The consumers' surplus is \$4.62.

The producers' surplus is given by

$$\begin{aligned}
PS = (10)(8) - \int_0^8 (q+2)\, dq &= 80 - [(q^2/2) + (2q)]\Big|_0^8 \\
&= 80 - 48 \\
&= 32
\end{aligned}$$

The producers' surplus is \$32. []

EXERCISES

1. The demand function for a product is

$$p = D(q) = 100 - 5q$$

where q is the quantity demanded, in units, and p is the price per unit, in dollars. If the equilibrium price is \$40, what is the consumers' surplus?

2. The demand function for a product is

$$p = D(q) = 150/(q+1)$$

where q is the quantity demanded, in units, and p is the price per unit, in dollars. If the equilibrium price is \$25, what is the consumers' surplus?

3. The supply function for a product is given by

$$p = S(q) = 0.25q^2$$

where q is the quantity demanded, in units, and p is the price per unit, in dollars. Market equilibrium occurs where price is \$25. Determine producers' surplus in this market.

4. The supply function for a product is

$$p = S(q) = (5 + 0.2q)^{3/2}$$

where q is the quantity demanded, in units, and p is the price per unit, in dollars. What is producers' surplus if the equilibrium price is \$125?

The demand function for a product is

$$p = D(q) = 300 - 6q - q^2$$

and the supply function is

$$p = S(q) = q^2 + 4q$$

where q *is the quantity demanded, in units, and* p *is the price per unit, in dollars.*

5. Determine the price and quantity at market equilibrium.
6. What is the consumers' surplus?
7. What is the producers' surplus?

The supply and demand functions for a product are

$$p = D(q) = 400 - 30q^2$$

and

$$p = S(q) = 10q^2 + 120q$$

where q *is quantity demanded in thousands of tons and* p *is price per ton, in dollars.*

8. Determine the price and quantity at market equilibrium.
9. Determine the amount of consumers' surplus.
10. Determine the amount of producers' surplus.

14.2 THE DEFINITE INTEGRAL AS A MEASURE OF TOTAL CHANGE

Given a "marginal" function $F'(x)$, we determine the associated "total" function $F(x)$ by finding the indefinite integral of $F'(x)$. What is more, we can evaluate the change in that total function over the interval $x = a$ to $x = b$ by finding the definite integral of $F'(x)$ from a to b. The following examples will illustrate this concept.

Example 4 **Marginal and total revenue** Analysts for the Jefferson Company have determined that the marginal revenue function for one of its products is given by

$$R'(x) = 4000 - 8x$$

where x represents the number of units of product sold.

The total revenue function can be found by taking the integral of the marginal revenue function, as

$$R(x) = \int R'(x)\,dx = \int (4{,}000 - 8x)\,dx = 4{,}000x - 4x^2 + C$$

If total revenue is zero when no units are sold, then $C = 0$, and

$$R(x) = 4000x - 4x^2$$

Now we can find total revenue at any level of sales $x = a$ by evaluating $R(a)$. For example, if 40 units were sold, total revenue would be

$$R(40) = 4{,}000(40) - 4(40) = 153{,}600$$

Total revenue from the sale of 40 units would be \$153,600.

Note that we could evaluate the total revenue from the sale of 40 units of product by the computation

$$R(40) = \int_0^{40} R'(x)\,dx = \int_0^{40} (4{,}000 - 8x)dx = (4{,}000x - 4x^2)\Big|_0^{40}$$
$$= [4{,}000(40) - 4(40)^2] - [0]$$
$$= 153{,}600$$

Geometrically this is the area of the region bounded by the marginal revenue function, the x-axis, $x = 0$, and $x = 40$, as shown in Figure 14.9A.

Or, we can determine the change in total revenue as sales change from $x = a$ units to $x = b$ units by finding the definite integral of the marginal revenue function from a to b. For instance, what would be

Figure 14.9

Area beneath Marginal Revenue Function Can Be Interpreted as Total Revenue or Incremental Revenue

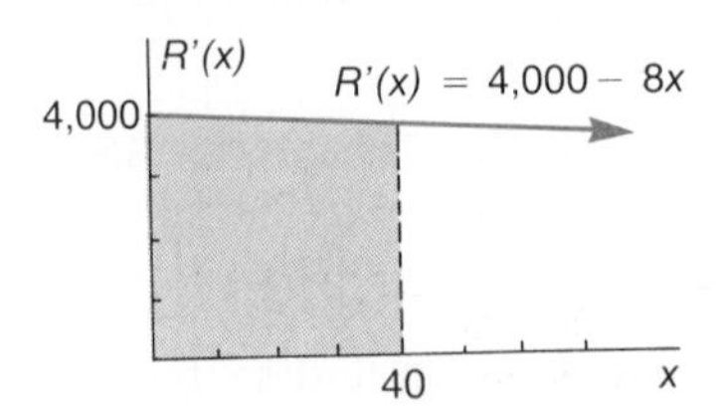

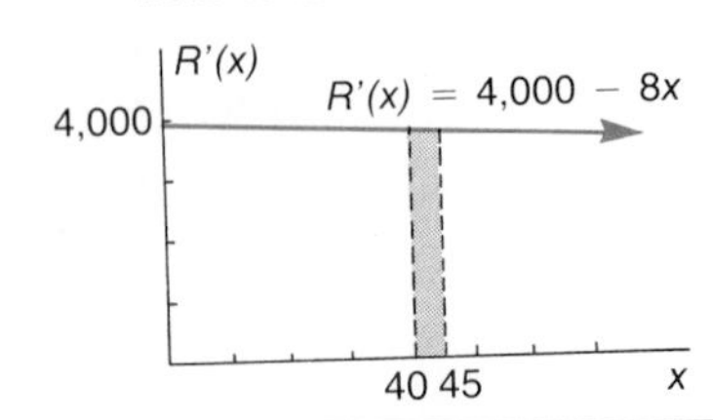

the change in total revenue if sales increased from 40 to 45 units? We compute

$$\int_{40}^{45} (4{,}000 - 8x)\, dx = [4{,}000x - 4x^2] \Big|_{40}^{45}$$
$$= [4{,}000(45) - 4(45)^2] - [4{,}000(40) - 4(40)^2]$$
$$= 171{,}900 - 153{,}600$$
$$= 18{,}300 = \$18{,}300$$

Total revenue would increase by \$18,300 if sales increased from 40 to 45 units. (See Figure 14.9B.) [ ]

Example 5 **Marginal and total cost** The marginal cost of manufacturing a certain product is

$$C'(x) = 3x^2 - 4x + 2$$

where x is the number of units of product manufactured. Determine the increase in total cost that would take place if the production were increased from 20 to 25 units.

The change in total cost can be found by

$$\int_{20}^{25} (3x^2 - 4x + 2)\, dx = [x^3 - 2x^2 + 2x] \Big|_{20}^{25}$$
$$= [(25)^3 - 2(25)^2 + 2(25)] - [(20)^3 - 2(20)^2 + 2(20)]$$
$$= 14{,}425 - 7{,}240$$
$$= 7{,}185$$

Total cost would increase by \$7,185 if production were increased from 20 to 25 units. This may be considered to be the total variable cost associated with the production of the 21st through the 25th unit of product.

Geometrically, this is the area of the region bounded by the marginal-cost function, the x-axis, $x = 20$ and $x = 25$, as shown in Figure 14.10.

Figure 14.10

Incremental Cost when Units Produced Increase from 20 to 25

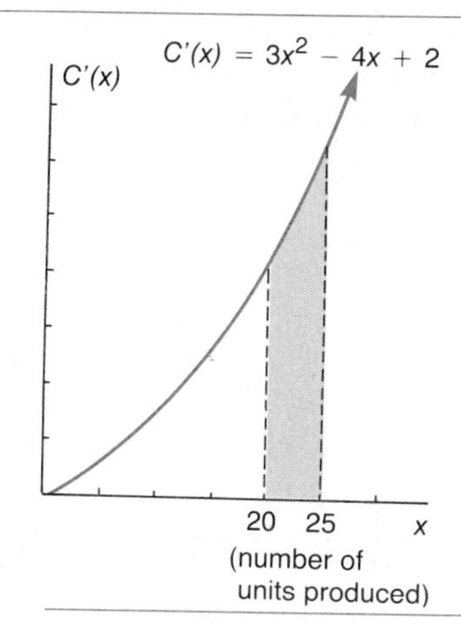

Example 6 **Net earnings (excess of revenue over variable cost)** Foote-Sparks Company has recently purchased new equipment which will generate additional revenue at the rate of

$$f(t) = 70 + 300/(t+1)$$

where $f(t)$ is marginal revenue in thousands of dollars per year, t years hence.

Maintenance and repair costs for the equipment are expected to increase at the rate of

$$g(t) = 50 + 20t$$

where $g(t)$ is marginal cost in thousands of dollars per year, t years hence.

During which year of use of the equipment will the additional annual revenue generated by the equipment equal the annual cost of maintenance and repair? What will be the total net savings over this period of time?

We set

$$f(t) = g(t)$$
$$70 + 300/(t+1) = 50 + 20t$$
$$70(t+1) + 300 = 50(t+1) + 20t(t+1)$$
$$-20t^2 + 320 = 0$$
$$t = 4 \text{ years}$$

Total net earnings are changing at the rate of $f(t) - g(t)$ per year; thus, total net earnings over the first four years are given by

$$\int_0^4 [70 + 300/(t+1)] - [50 + 20t]\, dt = [300 \ln(t+1) - 10t^2 + 20t] \Big|_0^4$$
$$= 402.831$$

Thus, total net earnings over the four years will be $402,831.

Geometrically, this is the area of the region bounded by the marginal revenue function, the marginal cost function, and the y-axis, as shown in Figure 14.11. []

Figure 14.11

Excess of Additional Revenue Generated by Machinery over Maintenance and Repair Costs

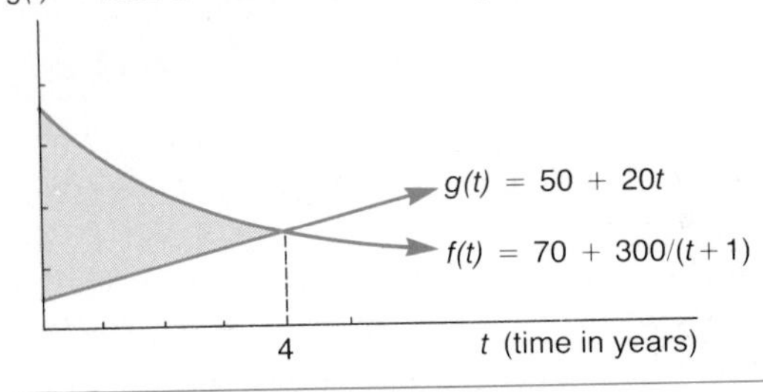

Example 7 **Marginal demand and change in total demand** Marginal demand (that is, the rate of change in total demand) for a product is given by

$$D'(x) = -24x - 36$$

where x represents the per-unit selling price of the product, in dollars. Find the change in total demand when the price changes from $10 to $12.

We evaluate the definite integral of the marginal demand function from $x = 10$ to $x = 12$, as

$$\int_{10}^{12} (-24x - 36)\,dx = [-12x^2 - 36x]\Big|_{10}^{12}$$
$$= [-12(12)^2 - 36(12)] - [-12(10)^2 - 36(10)]$$
$$= [-2{,}160] - [-1{,}560]$$
$$= -600$$

Thus, total demand could decrease by 600 units if price per unit for the product were increased from $10 to $12. (See Figure 14.12.) []

Figure 14.12

Marginal Demand and Change in Total Demand

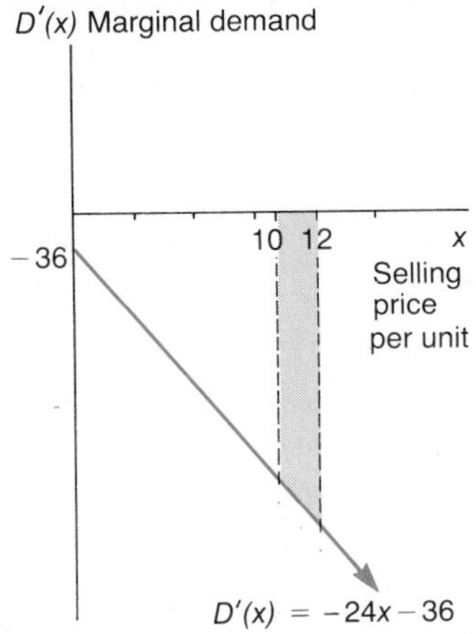

Example 8 **Maximizing net contributions to fund-raising campaign** The Altoona Arts Alliance is conducting its annual fund-raising campaign. Campaign expenditures will be incurred at the rate of $2,800 a day as long as the campaign continues. Experience indicates that daily contributions will be high during the early days of the campaign but will decline as the campaign continues. The rate at which contributions are received is expected to be

$$c(t) = 10{,}000 - 50t^2$$

where t represents the number of days since the campaign began and $c(t)$ is measured in dollars per day.

The alliance wishes to maximize the net proceeds (that is, the excess of contributions over expenditures) from the campaign. How long should the campaign continue?

The rate at which expenses are incurred is

$$e(t) = 2{,}800$$

As long as the rate at which contributions are received exceeds the rate at which expenditures are incurred, net proceeds are positive. Figure 14.13 pictures these two functions. Net proceeds are positive up until the point where the two lines intersect. Beyond this point the daily cost of continuing the campaign exceeds the daily contributions, and net proceeds will be negative.

Figure 14.13

Net Proceeds of Fund-Raising Campaign

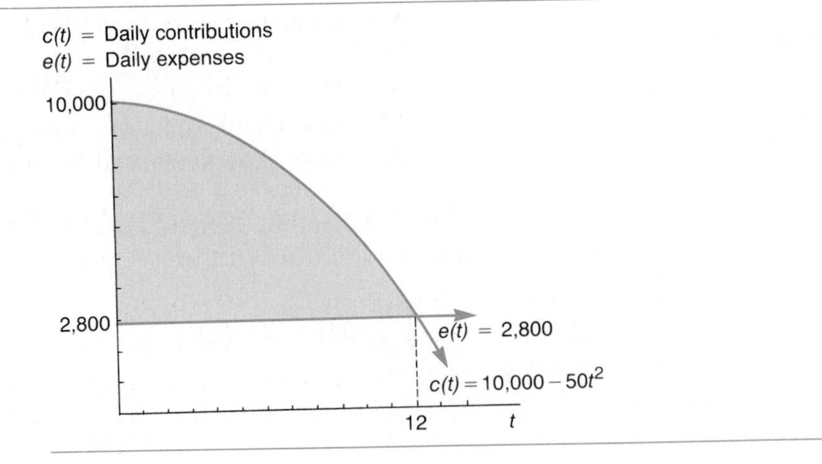

The functions intersect at

$$c(t) = e(t)$$
$$10{,}000 - 50t^2 = 2{,}800$$
$$-50t^2 = -7{,}200$$
$$t^2 = 144$$
$$t = 12$$

Net proceeds will be positive over the first 12 days of the campaign. After that time, daily expenses will exceed daily contributions.

What will the total net proceeds be if the campaign is continued through day 12? Total net proceeds are represented by the region bounded by the $c(t)$ function, the $e(t)$ function, and the y-axis. We evaluate this region by

$$\int_0^{12} [10{,}000 - 50t^2] - [2{,}800]\, dt = \int_0^{12} -50t^2 + 7{,}200\, dt$$
$$= \frac{-50t^3}{3} + 7{,}200t \Big|_0^{12}$$
$$= [(-50/3)(12)^3 + 7{,}200(12)] - [0]$$
$$= 57{,}600$$

The net proceeds from the campaign over the 12 days will be \$57,600. []

EXERCISES

The marginal-revenue function for a product is given by

$$R'(x) = 600 - 0.3x$$

where x *is the quantity of product sold.*

11. What is the total revenue from the sale of 1,000 units?
12. What would be the change in total revenue if sales increased from 1,000 to 1,050 units? Sketch a graph of the marginal-revenue function and shade the region of interest.
13. What would be the change in total revenue if sales increased from 1,050 units to 1,100 units? Sketch a graph of the region of interest.

The marginal-revenue function for a product is given by

$$R'(x) = 5{,}000 - 5x$$

where x *is the number of units of product sold.*

14. Find the total-revenue function.
15. What is the total revenue when 300 units of product are sold? Sketch a graph of the marginal-revenue function and shade the region of interest.
16. What would be the change in total revenue if units sold increased from 300 to 310 units? Sketch a graph of the region of interest.

The marginal profit, in thousands of dollars, for producing and selling x *items is given by*

$$P'(x) = 600 - 5x$$

17. Determine the total-profit function, given that total profit is −\$20,000 when zero units are produced and sold.
18. What is the total profit when 50 items are produced and sold?
19. What is the change in total profit if the level of production changes from 50 to 55 units?

Powers Company has determined that its marginal profit is given by

$$P'(x) = 300 - 2x$$

where x *is the number of items produced and sold.*

20. If the company loses \$2,500 when zero items are produced and sold, what is the total-profit function?
21. What is the company's total profit when 30 items are produced and sold?
22. What is the change in total profit when production and sales increase from 30 to 35 items? Sketch a graph of this region.
23. What is the profit-maximizing level of production? What is maximum profit?

The marginal cost of manufacturing a product is given by

$$C'(x) = 60 - (10{,}000/x^2)$$

where x *represents the number of units of output.*

24. What is the effect on total cost of increasing production from 100 to 125 units?
25. What is the effect on total cost of increasing production from 100 to 150 units?

A firm has determined that its marginal cost at a production level of x units is given by

$$C'(x) = 10 + (160/x^2)$$

26. What is the effect on total cost of increasing production from 30 to 35 units?
27. What is the effect on total cost of increasing production from 30 to 40 units?

The marginal demand for a product is given by

$$D'(x) = 50x - 850$$

where x is the price per unit, in dollars.

28. What is the total-demand function?
29. What is the effect on total demand when price changes from \$4 to \$5? from \$4 to \$5.50? from \$4 to \$6?

A firm has purchased new equipment that will generate revenue at a rate of

$$f(t) = 100 - 15t$$

thousand dollars per year t years hence.

Accumulated maintenance and repair costs for the equipment increase at a rate of

$$g(t) = t^2 + 4$$

thousand dollars per year t years hence.

30. What will be the total revenue generated over the first five years the equipment is used?
31. What will be the total maintenance and repair costs for the equipment over the first five years of use?
32. If the firm decides to sell the equipment when incremental maintenance and repair costs equal incremental revenues, when should the equipment be sold? Compute total net earnings over this period of use.

Maintenance expenses for a certain type of industrial lift truck are estimated to be incurred at a rate given by

$$f(t) = 140 + 15t^2$$

where f(t) *is measured in dollars per year and* t *equals the number of years the lift truck has been in operation.*

33. What are total maintenance expenses expected to be during the first three years of operation?
34. What are total maintenance expenses expected to be during the first five years of operation?

*Maintenance cost on newly purchased heavy-duty construction equipment is expected to be at the rate of 1,500 + 100*t *dollars per year at time* t *years.*

35. Compute total maintenance cost during the first five years.
36. Compute total maintenance cost over the second five years the equipment is in use.
37. How many years will it take for total maintenance cost to amount to $10,000?

The rate of sales of a new product is given by

$$s(t) = 1{,}500 - 900e^{-t}$$

where t *is the number of months the product has been on the market.*

38. Find total sales during the first six months the product is on the market.
39. Find total sales during the second six months the product is on the market.

A production manager finds that the assembly process for a product is appropriately described by the following rate-of-assembly function:

$$r(x) = 800\sqrt{x}$$

where r(x) *represents the rate of labor hours required to assemble the* x*th unit of product.*

40. Determine the total labor requirements for 400 units of output.
41. Determine the additional labor required if a contract is signed to increase production by 25 units (above the 400 units presently produced).
42. The rate at which a certain deposit of zincite is being depleted is given by

$$r(t) = 50{,}000e^{0.04t}$$

where t is time in years. Estimate the amount of zincite that will be used between $t = 3$ and $t = 5$.

A parcel-delivery company estimates that the annual rate of expenditure for maintenance on each of the trucks in its fleet is given by the function

$$r(t) = 350 + 48t$$

where r(t) *is measured in dollars per year and* t *is the age of the truck, in years.*

43. Determine the total maintenance expenditures over the first three years of a truck's service.
44. During what year of service is total maintenance expenditure expected to reach $2,000?

Lassiter Leasing Company leases various types of office machinery and equipment for a fixed monthly rental fee. A certain model word-processing machine is expected to have a useful life of eight years and a rate of maintenance of

$$r(t) = 0.9t^2 + 3t + 12$$

where t *is time in years that the machine is rented and* r(t) *is expressed in dollars.*

45. What will be the total maintenance cost over the eight years?
46. If the company wishes to include a fixed monthly maintenance fee in the leasing contract, what should be the amount of this charge?

Walton Carpet Company is contemplating opening a new regional warehouse. Initial cost of the facility is estimated to be $252,000. The rate of savings generated by the new warehouse is estimated to be

$$r(t) = 6{,}000 + 12{,}000t^{\cdot}$$

where t *is in years.*

47. What will be the total savings over the first five years?
48. How long will it take for the warehouse to pay for itself?

14.3 INTEGRATION BY PARTS

The integration rules we discussed in Chapter 13 apply only to a subset of the functions which we might need to integrate. We now present an additional technique of integration, known as INTEGRATION BY PARTS.

This is a powerful integration technique derived from the product rule of differentiation and is used to determine integrals when the integrand is of the format $u(x) \cdot v'(x)$. Recall that, according to the product rule of differentiation,

$$\frac{d[u(x) \cdot v(x)]}{dx} = v(x) \cdot u'(x) + u(x) \cdot v'(x)$$

If both sides of this expression are integrated, the result is

$$\int \frac{d[u(x) \cdot v(x)]}{dx}\,dx = \int [v(x) \cdot u'(x) + u(x) \cdot v'(x)]\,dx$$

$$u(x) \cdot v(x) = \int [v(x) \cdot u'(x)]\,dx + \int [u(x) \cdot v'(x)]\,dx$$

and

$$\int [u(x) \cdot v'(x)]\,dx = u(x) \cdot v(x) - \int [v(x) \cdot u'(x)]\,dx$$

Integration-by-parts rule

$$\int [u(x) \cdot v'(x)]\,dx = u(x) \cdot v(x) - \int [v(x) \cdot u'(x)]\,dx$$

We must begin by stating that integration by parts can be very successful on certain types of functions but not on all types. Do not attempt to use integration by parts when easier methods are available. Always examine the integrand carefully to determine whether other rules apply; if they do, use them.

The integration-by-parts procedure is a trial-and-error process wherein we partition the integrand into two parts—one part is labeled $u(x)$ and the other part $v'(x)$. Successful use of the formula, if it leads to a solution at all, almost always depends on the choices made for $u(x)$ and $v'(x)$. These choices must be such that it is possible to evaluate $\int [v(x) \cdot u'(x)]\,dx$!

The following examples illustrate successful—and unsuccessful—attempts to use the integration-by-parts approach.

Example 9 Determine $\int xe^x\,dx$.

Observe that the integrand is the product of two factors—and one factor is *not* the derivative of the other—so we will attempt to use integration by parts. (Had one factor been the derivative of the other, we would have used Rule 7 from Chapter 13. Or, had the format been $\int e^{f(x)} f'(x)\,dx$, we would have used Rule 9.)

We must define one factor as $u(x)$ and the other factor as $v'(x)$, the derivative of a function $v(x)$. One key to the successful use of integration by parts is the ability to determine which factor, if either, has the form of the derivative of another function.

With this in mind, let us define

$$u(x) = x \qquad \text{and} \qquad v'(x) = e^x$$

Then,

$$u'(x) = 1 \qquad \text{and} \qquad v(x) = e^x$$

(It is not really necessary to write $v(x) = e^x + C$ at this point. A constant of integration will be added later.)

Thus, we have

$$\int [u(x) \cdot v'(x)]\, dx = u(x) \cdot v(x) - \int [v(x) \cdot u'(x)]\, dx$$

$$\downarrow \quad \downarrow \qquad\qquad \downarrow \quad \downarrow \qquad\quad \downarrow \quad \downarrow$$

$$\int [x \cdot e^x]\, dx = [x \cdot e^x] - \int [e^x \cdot 1]\, dx$$

so that

$$\int xe^x\, dx = xe^x - e^x + C$$

[If we are somewhat unsure of our use of this new rule, we may verify that

$$\frac{d(xe^x - e^x + C)}{dx} = xe^x + e^x \cdot 1 - e^x + 0 = xe^x.]$$

And the integration-by-parts method appears not to be too difficult after all.

But let us start again and see what difficulties we might have encountered had we not been so lucky with our initial definitions of $u(x)$ and $v'(x)$. We thus define

$$u(x) = e^x \qquad \text{and} \qquad v'(x) = x$$

and find that

$$u'(x) = e^x \qquad \text{and} \qquad v(x) = x^2/2$$

Substituting into the integration-by-parts rule, we obtain

$$\int e^x \cdot x\, dx = e^x \cdot (x^2/2) - \int [(x^2/2) \cdot e^x]\, dx$$

But $\int [(x^2/2) \cdot e^x]\, dx$ is more difficult than the integral we originally started with! Whenever this happens, you backtrack and try again! [ ]

How can we select $u(x)$ and $v'(x)$ to make integration by parts work? Unfortunately, there are no general rules for partitioning the integrand. The goal is to select a $v'(x)$ that is integrable and that will result in a $\int v(x) \cdot u'(x)\, dx$ that is also integrable. There are usually just two reasonable choices, and it may be necessary to try both. Practice increases one's insight and will lead to increasingly successful educated guesses.

Example 10 Determine $\int x^2 \ln x\, dx$.

If we define

$$u(x) = \ln x \qquad \text{and} \qquad v'(x) = x^2$$

then

$$u'(x) = 1/x \qquad \text{and} \qquad v(x) = x^3/3$$

Now,

$$\int x^2 \ln x\, dx = (\ln x)(x^3/3) - \int (x^3/3)(1/x)\, dx$$

$$= (x^3/3)(\ln x) - (1/3)\int x^2\, dx$$

$$= (x^3/3)(\ln x) - (1/3)(x^3/3) + C$$

$$= (x^3/3)[\ln x - (1/3)] + C$$

[]

Sometimes it is necessary to repeat the integration-by-parts procedure more than once to complete the evaluation of an integral.

Example 11 Determine $\int x^2 e^x\, dx$.

Let us try

$$u(x) = e^x \qquad \text{and} \qquad v'(x) = x^2$$

which makes

$$u'(x) = e^x \qquad \text{and} \qquad v(x) = (x^3/3)$$

and

$$\int x^2 e^x\, dx = (e^x)(x^3/3) - \int [(x^3/3)(e^x)]\, dx$$

One look at $\int [(x^3/3)(e^x)]\, dx$ causes us to believe we should begin anew. So we redefine

$$u(x) = x^2 \qquad \text{and} \qquad v'(x) = e^x$$

Then

$$u'(x) = 2x \qquad \text{and} \qquad v(x) = e^x$$

and

$$\int x^2 e^x\, dx = (x^2)(e^x) - \int (e^x)(2x)\, dx$$

$$= (x^2)(e^x) - 2\int x e^x\, dx$$

The term $\int x e^x\, dx$, too, looks formidable, and we wonder whether it is possible to determine the original integral by the integration-by-parts method. But look back at Example 9. There we determined—by the integration-by-parts method—that $\int x\, e^x\, dx = x\, e^x - e^x + C$. Therefore,

$$\int x^2\, e^x\, dx = x^2\, e^x - 2(x\, e^x - e^x) + C$$
$$= e^x(x^2 - 2x + 2) + C$$

So, you see, an integral may require the integration-by-parts treatment two—and sometimes even more—times. []

The most "obvious" choices for $u(x)$ and $v'(x)$ are not always the correct ones, as the following example will illustrate.

Example 12 Determine $\int x^3\sqrt{x^2+1}\, dx$.

We probably would begin $u(x) = \sqrt{x^2+1}$ and $v'(x) = x^3$ since x^3 can be easily integrated. However, with these choices, $u'(x) = (1/2)\,(x^2 + 1)^{(-1/2)}\,(2x) = x(x^2 + 1)^{(-1/2)}$ and $v(x) = (x^4/4)$, and we find ourselves faced with $\int (x^5/4)(x^2 + 1)^{(-1/2)}\, dx$. So we will attempt to redefine our factors.

Looking ahead, however, if $\sqrt{x^2+1}$ is to be a part of $v'(x)$, we must use the integral rule $\int [f(x)]^n f'(x)\, dx$ to obtain $v(x)$. This means that $v'(x)$ needs to be defined so as to include $2x\sqrt{x^2+1}$. To obtain the 2, we can multiply by 2 and by its reciprocal 1/2. To obtain the needed x, we factor the x^3 in the original integrand as $x \cdot x^2$.

So, we define

$$u(x) = x^2 \qquad \text{and} \qquad v'(x) = x\sqrt{x^2+1}$$

so that

$$u'(x) = 2x$$

and

$$v(x) = (1/2)\int 2x(x^2+1)^{1/2}\, dx = (1/2)[(2/3)(x^2+1)^{(3/2)}] + C$$
$$= (1/3)(x^2+1)^{(3/2)} + C$$

Now

$$\int x^3\sqrt{x^2+1}\, dx = (x^2)(1/3)(x^2+1)^{(3/2)} - \int (1/3)(x^2+1)^{(3/2)}\,(2x)\, dx$$
$$= (x^2/3)(x^2+1)^{(3/2)} - (1/3)(x^2+1)^{(5/2)}(2/5) + C$$
$$= (1/3)(x^2+1)^{(3/2)}[x^2 - (2/5)(x^2+1)] + C$$
$$= (1/3)(x^2+1)^{(3/2)}[(3/5)x^2 - (2/5)] + C$$
$$= (1/15)(x^2+1)^{(3/2)}[3x^2 - 2] + C$$

[]

EXERCISES

Determine the indefinite integral using integration by parts.

49. $\int x\, e^{2x}\, dx$

50. $\int x \ln x\, dx$

51. $\int \frac{\ln x}{\sqrt{x}}\, dx$

52. $\int \sqrt{x} \ln x\, dx$

53. $\int 4x \ln x\, dx$

54. $\int x^2 \ln 2x\, dx$

55. $\int x\sqrt{x-1}\, dx$

56. $\int 3x\sqrt{2+x}\, dx$

57. $\int \frac{x}{\sqrt{x+1}}\, dx$

58. $\int x(x+2)^5\, dx$

59. $\int x\sqrt[3]{x+2}\, dx$

60. $\int x(x+5)^4\, dx$

61. $\int (3x-2)e^x\, dx$

62. $\int \ln x^2\, dx$

63. $\int \ln 2x\, dx$

64. $\int (\ln x)^2\, dx$

65. $\int (x+2) \ln x\, dx$

66. $\int 2x \ln (x^2)\, dx$

67. $\int 2x \ln 2x\, dx$

68. $\int \frac{\ln x}{x^2}\, dx$

69. $\int x^3\, e^{x^2}\, dx$

70. $\int x^2\, e^x\, dx$

71. $\int x^3 \ln x\, dx$

72. $\int x(\ln x)^2\, dx$

73. $\int x^2(x+4)^5\, dx$

74. $\int x^3\sqrt{3-x^2}\, dx$

75. $\int xe^{-x}\, dx$

76. $\int x^2 e^{2x}\, dx$

77. $\int x^2\, e^{-x}\, dx$

78. $\int x^3\, e^x\, dx$

79. A share of common stock in Billingsley-Clayborne, Inc., increases in value at the rate of $t\sqrt{t+2}$ dollars, where t is time in months since the stock was issued. Find the increase in value between $t = 6$ and $t = 9$.

14.4 INTEGRATION USING TABLES OF INTEGRALS

Determining the integrals of certain functions requires rules and procedures beyond the scope of this text. Many such integrals may, however,

Table 14.1

Table of Integrals*

1. $\int u^n\,du = \frac{u^{n+1}}{n+1} + C$, for $n \neq -1$
2. $\int \frac{du}{u} = \int u^{-1}\,du = ln\,|u| + C$
3. $\int a^u\,du = a^u\,log_a\,e + C$
4. $\int e^u\,du = e^u + C$
5. $\int (a+bu)^n\,du = \frac{(a+bu)^{(n+1)}}{b(n+1)} + C$, for $n \neq -1$
6. $\int \frac{du}{a+bu} = (1/b)\,ln(a+bu) + C$
7. $\int \frac{u\,du}{au+b} = \frac{u}{a} - \frac{b}{a^2}\,ln\,|au+b| + C$
8. $\int \frac{du}{u(au+b)} = (1/b)\,ln\left|\frac{u}{au+b}\right| + C$
9. $\int \frac{du}{u^2-a^2} = (1/2a)\,ln\left|\frac{u-a}{u+a}\right| + C \qquad (u^2 > a^2)$
10. $\int \frac{du}{a^2-u^2} = (1/2a)\,\ln\left|\frac{a+u}{a-u}\right| + C \qquad (a^2 > u^2)$
11. $\int \frac{u\,du}{(au+b)^2} = (1/a^2)\left[ln\,|au+b| + \frac{b}{au+b}\right] + C$
12. $\int u\sqrt{au+b}\,du = \frac{2(3au-2b)(au+b)^{(3/2)}}{15a^2} + C$
13. $\int \frac{u\,du}{\sqrt{au+b}} = \frac{2(au-2b)}{3a^2}\sqrt{au+b} + C$
14. $\int \frac{du}{u\sqrt{au+b}} = \frac{1}{\sqrt{b}}\,ln\frac{\sqrt{au+b}-\sqrt{b}}{\sqrt{au+b}+\sqrt{b}} + C \qquad (b > 0)$
15. $\int \sqrt{u^2+a^2}\,du = (1/2)[u\sqrt{u^2+a^2} + a^2 ln|u+\sqrt{u^2+a^2}|] + C$
16. $\int \sqrt{u^2-a^2}\,du = (1/2)[u\sqrt{u^2-a^2} - a^2 ln|u+\sqrt{u^2-a^2}|] + C$
17. $\int \frac{du}{\sqrt{u^2+a^2}} = ln\,|u+\sqrt{u^2+a^2}| + C$
18. $\int \frac{du}{\sqrt{u^2-a^2}} = ln\,|u+\sqrt{u^2-a^2}| + C$
19. $\int \frac{du}{u\sqrt{a^2+u^2}} = (-1/a)\,ln\left|\frac{a+\sqrt{a^2+u^2}}{u}\right| + C$
20. $\int \frac{du}{u\sqrt{a^2-u^2}} = (-1/a)\,ln\left|\frac{a+\sqrt{a^2-u^2}}{u}\right| + C$
21. $\int \frac{du}{\sqrt{2au+u^2}} = ln|u+a+\sqrt{2au+u^2}| + C$
22. $\int u\,e^u\,du = u\,e^u - e^u + C$
23. $\int u^2 e^u\,du = e^u(u^2 - 2u + 2) + C$

Table 14.1 (*concluded*)

24. $\int u^n \ln u \, du = \frac{u^{(n+1)}}{n+1}\left(\ln u - \frac{1}{n+1}\right) + C$ for $n \neq -1$

25. $\int \frac{\ln u}{u} du = \frac{(\ln u)^2}{2} + C$

26. $\int \frac{du}{u \ln u} = \ln(\ln u) + C$

* In the table, a, b, and n represent constants; C is the constant of integration.

be determined by reference to a table of integrals such as Table 14.1, which contains some of the integral formulas most often needed for business and economic models. (*Standard Mathematical Tables*,* a widely used mathematical reference, contains more than 500 such integral formulas.)

To use a table of integrals, you must be able to match the form of the integrand with the corresponding STANDARD FORM used in the table. To do this, you must understand how to carry out a CHANGE OF VARIABLE in an integral. We went through a similar procedure when using the chain rule to find a derivative. For instance, let us verify that

$$\int \frac{8x+3}{4x^2+3x} dx = \ln(4x^2+3x) + C$$

by differentiating $\ln(4x^2 + 3x) + C$. We will need to use the chain rule, defining

$$y = f(u) = \ln u$$

and

$$u = g(x) = 4x^2 + 3x$$

Observe that we have used the variable u to represent a function of x. This is the key to the change-of-variable procedure.

Now

$$\frac{dy}{du} = \frac{d(\ln u)}{du} = \frac{1}{u}$$

and

$$\frac{du}{dx} = \frac{d(4x^2+3x)}{dx} = 8x + 3$$

* S. M. Selby, ed., *Standard Mathematical Tables*, 22d ed. (Cleveland: Chemical Rubber Company, 1974).

so that

$$\frac{dy}{dx}=\frac{dy}{du}\cdot\frac{du}{dx}=\frac{1}{u}(8x+3)=\frac{1}{4x^2+3x}(8x+3)$$

Hence

$$\frac{d[ln(4x^2+3x)+C]}{dx}=\frac{8x+3}{4x^2+3x}$$

and the integral checks.

If we write the integral using the variable u rather than the variable x, the result is

$$\int\frac{1}{u}\cdot\frac{du}{dx}dx=ln(u)+C$$

$$\downarrow \quad \downarrow \quad \downarrow \quad \downarrow \quad \downarrow$$

$$\int\frac{1}{4x^2+3x}\cdot(8x+3)dx=ln(4x^2+3x)+C$$

Now, while we have used the symbol $\frac{du}{dx}$ to denote the derivative of $u=g(x)$ with respect to x, there are advantages to viewing the symbol as the ratio of two separate quantities, du and dx. When this is done, the quantity du is called the *DIFFERENTIAL OF* u and is defined as

$$du=\left(\frac{du}{dx}\right)\cdot dx \qquad \text{or} \qquad du=f'(x)\,dx$$

This notation allows us to write

$$\int\frac{1}{u}du=\int\frac{du}{u}=ln\,u+C$$

All integral formulas in the table are written in this standard form, using the variable u to represent a function of the variable x. The following examples will illustrate the use of the standard-form integrals.

Example 13 Evaluate $\int(x+1)e^{(x+1)}\,dx$.

We must define u as some function of x and, using this definition, we must determine du. Let us begin with

$$u=x+1$$

so that

$$du=\frac{d(x+1)}{dx}dx=(1)\,dx=dx$$

Thus far we have identified

$$\int\underset{u}{(x+1)}\underset{e^u}{e^{(x+1)}}\underset{du}{dx}=\int ue^u\,du$$

Scanning the Table of Integrals, we see that Integral Rule 22 has this format. That is,

$$\int ue^u\,du = ue^u - e^u + C$$

Thus,

$$\int (x+1)e^{(x+1)}\,dx = (x+1)e^{(x+1)} - e^{(x+1)} + C$$
$$= e^{(x+1)}[x+1-1] + C$$
$$= xe^{(x+1)} + C \qquad []$$

Example 14 Evaluate $\int (x+1)^2 e^{(x+1)}\,dx$.

Again we start by defining u and determining du.

Let us try $u = (x+1)$, so that $du = (1)dx = dx$. Now the integral appears to be of the format

$$\int u^2 e^u\,du$$

and Integral Rule 23 seems to be appropriate. We have

$$\int (x+1)^2 e^{(x+1)}\,dx = e^{(x+1)}[(x+1)^2 - 2(x+1) + 2] + C$$
$$= e^{(x+1)}[x^2 + 2x + 1 - 2x - 2 + 2] + C$$
$$= (x^2+1)e^{(x+1)} + C \qquad []$$

Example 15 Evaluate

$$\int \frac{6x}{(3x^2+5)^3}\,dx$$

We define $u = 3x^2 + 5$ so that

$$du = \frac{d(3x^2+5)}{dx}dx = 6x\,dx$$

Thus, we have identified

$$\int \frac{\overbrace{6x \qquad dx}^{du}}{\underbrace{(3x^2+5)^3}_{u}} = \int \frac{du}{u^3} = \int u^{-3}\,du$$

With $n = -3$, Integration Rule 1 can be used, as

$$\int u^n\,du = \frac{u^{n+1}}{n+1} + C$$

$$\int u^{-3}\,du = \frac{u^{-2}}{-2} + C$$

and, returning to x-notation,

$$\int \frac{6x}{(3x^2+5)^3} dx = \frac{(3x^2+5)^{-2}}{-2} + C = \frac{-1}{2(3x^2+5)^2} + C \qquad []$$

Example 16 Evaluate $\int \sqrt{2x+5}\, dx$

If $u = 2x + 5$,

$$du = \frac{d(2x+5)}{dx} dx = 2\, dx$$

We see that the constant 2 is needed in the integrand if Integration Rule 1 is to be used once again. We can obtain the required constant by following the procedure of multiplying by the constant and its reciprocal as we have done previously. We obtain

$$\int \sqrt{2x+5}\, dx = (1/2) \int \underbrace{\sqrt{2x+5}}_{\sqrt{u}}\, \underbrace{(2)dx}_{du} = (1/2) \int u^{1/2}\, du$$

So, with $n = 1/2$, we have (using Integration Rule 1)

$$(1/2) \int u^{1/2}\, du = (1/2) \cdot \frac{u^{(3/2)}}{(3/2)} + C = \frac{u^{(3/2)}}{3} + C$$

Finally, returning to x-notation, we write

$$\int \sqrt{2x+5}\, dx = \frac{(2x+5)^{(3/2)}}{3} + C \qquad []$$

Example 17 Evaluate $\int \sqrt{x^2-9}\, dx$.

Thinking again of Integration Rule 1, we could define $u = x^2 - 9$ meaning that

$$du = \frac{d(x^2-9)}{dx} dx = 2x\, dx$$

But the required $2x$ in the integrand is missing and, because a variable is involved, unobtainable. So we must abandon these definitions and return to search the table for another formula. Scanning down the list, Integration Rule 16 seems to be a likely candidate. If this formula is to be used, we define $u = x$ so that $u^2 = x^2$, and $a = 3$ so that $a^2 = 9$. Then,

$$du = \frac{d(x)}{du} dx = 1 dx = dx$$

Thus, we have verified correspondence between the two integrands. And, by Integration Rule 16,

$$\int \sqrt{x^2-9}\, dx = (1/2)[x\sqrt{x^2-9} - 9\, ln\, |x + \sqrt{x^2-9}\,|] + C \qquad []$$

EXERCISES

Use a table of integrals to evaluate each of the following integrals.

80. $\int 2e^{(2x+1)}\,dx$

81. $\int (x+1)e^{(x+1)}\,dx$

82. $\int \frac{1}{9-x^2}\,dx$

83. $\int \frac{ln\,(x+5)}{x+5}\,dx$

84. $\int \frac{x}{5x+1}\,dx$

85. $\int \frac{2x}{16-x^4}\,dx$

86. $\int (6x)6^{(x^2+9)}\,dx$

87. $\int \sqrt{x^2-4}\,dx$

88. $\int \frac{x^3}{7-3x^2}\,dx$

89. $\int x\sqrt{3x+2}\,dx$

90. $\int \frac{x\,dx}{x^2\,ln\,x^2}$

91. $\int \frac{dx}{\sqrt{6x+x^2}}$

92. $\int xe^{x^2+1}\,dx$

93. $\int x(3x^2+5)^3\,dx$

94. $\int \frac{dx}{9x^2-4}$

95. $\int \frac{x}{(4x^2+1)^4}\,dx$

96. $\int \frac{dx}{x\sqrt{2x+5}}$

97. $\int \sqrt{x^2+25}\,dx$

98. $\int \frac{2}{\sqrt{4x^2+16}}\,dx$

99. $\int (x+3)^2\,ln\,(x+3)\,dx$

100. $\int \frac{2}{x\sqrt{9+x^2}}\,dx$

101. $\int (x+4)^2\,e^{(x+4)}\,dx$

102. $\int (3x+7)^4\,dx$

103. $\int 4xe^{2x}\,dx$

104. $\int \frac{dx}{\sqrt{9x^2+64}}$

105. $\int \sqrt{x}\,ln\,x\,dx$

106. $\int \frac{x^2}{5-x^3}\,dx$

107. $\int \frac{6}{\sqrt{4x^2-9}}\,dx$

108. $\int \frac{-4}{5x^2+3x}\,dx$

109. $\int \frac{e^x}{e^x+5}\,dx$

14.5 DIFFERENTIAL EQUATIONS

Many important applications of calculus lead to equations involving rates of change, or derivatives. Such equations—equations involving the derivative of an unknown function—are called DIFFERENTIAL EQUATIONS. Examples of typical differential equations are

a. $dy/dx = 4x + 1$ *b.* $y' = 2y$ *c.* $dy/dx = 5x/y$

We *solve* a differential equation by finding all those differentiable functions $y = f(x)$ which, when substituted into the equation, yield an identity. Such functions are called SOLUTIONS to the differential equation. The GENERAL SOLUTION of a differential equation is a solution containing the arbitrary constant of integration. Because

$$\frac{d(x^3 + C)}{dx} = 3x^2$$

for example, $dy/dx = 3x^2$ is a differential equation with the general solution $y = x^3 + C$. A PARTICULAR SOLUTION can be obtained from the general solution when BOUNDARY CONDITIONS are given which allow us to determine specific values for the arbitrary constants. Thus, $y = x^3 - 1$, $y = x^3 + 9$, and so on, are all particular solutions to the differential equation $dy/dx = 3x^2$.

A familiar type of differential equation states that the rate of change of y with respect to x is $f(x)$; that is

$$\frac{dy}{dx} = f(x)$$

The solution to this type of differential equation is given by

$$y = \int f(x)\, dx$$

Example 18 The tangent line to the function $y = f(x)$ at a point (x,y) has slope $2/x^2$. Find the function with this slope that passes through the point (1,5).

This information leads to the differential equation

$$dy/dx = 2/x^2$$

By integrating each side of the equation, we obtain

$$\begin{aligned} y &= \int 2/x^2\, dx \\ &= -2/x + C \end{aligned}$$

From the boundary condition $y = 5$ for $x = 1$, we have

$$\begin{aligned} 5 &= (-2/1) + C \\ 7 &= C \end{aligned}$$

Thus, the function $y = f(x) = (-2/x) + 7$ passes through the point (1,5) and has the slope $2/x^2$ at any point (x,y). []

Example 19 Management of a certain factory has determined that the marginal cost of production (which is a differential equation) is

$$dC/dx = 2e^x + (4/x) - (3/x^2)$$

where C represents total cost and x is the number of units produced.

To find the total cost function $C(x)$, we integrate each side of the differential equation, obtaining

$$C(x) = \int [2e^x + (4/x) - (3/x^2)]\, dx$$
$$= 2e^x + 4 \ln x + (3/x) + C$$

Further, given the boundary condition that total cost is $10,000 when five units are produced, we find the particular solution

$$10{,}000 = 2e^5 + 4 \ln 5 + (3/5) + C$$
$$9{,}696.2 = C$$

Thus, the particular solution to the differential equation is the total cost function

$$C(x) = 2e^x + 4 \ln x + (3/x) + 9{,}696.2$$ []

Another type of commonly encountered differential equation states that the rate of change of y with respect to x is proportional to y; that is,

$$\frac{dy}{dx} = ky$$

where k is a constant

The constant k, called the CONSTANT OF PROPORTIONALITY, may be a *rate of growth* (if $k > 0$) or a *rate of decay* (if $k < 0$).

For this type of differential equation, the solution is that function $y = f(x)$ such that $f'(x) = kf(x)$. The general solution takes the format

$$y = ce^{kx}$$

where c is a constant. (Observe that

$$\frac{dy}{dx} = \frac{d(ce^{kx})}{dx} = kce^{kx} = ky$$

Since $f(0) = ce^0 = c$, each solution of the differential equation can be written in the form $f(x) = f(0)e^{kx}$.

Example 20 Determine the function $y = f(x)$ that satisfies the differential equation $dy/dx = 5y$ if $f(0) = 46$.

The solution to the differential equation is of the form $y = f(x) =$

$ce^{kx} = f(0)e^{kx}$. In this case, k, the rate of proportional change, is 5, and $f(0)$ is 46. Thus

$$f(x) = 46e^{5x}$$ []

The standard model for uninhibited growth is based on the assumption that the rate of change in the size of the population is proportional to the size of the population. That is, when the population is composed of only a few individuals, the rate of growth is small; when the number of individuals is large, the rate of growth of the population is high. If P is the population at time t, such a situation can be described by the differential equation

$$\frac{dP}{dt} = kP$$

for some constant k.

Example 21 The rate of growth in a certain community is proportional to its size. In the five years from 1978 to 1982, it grew from an initial size of 2,500 to 3,200. Derive a model that can be used to estimate population at any arbitrary time t. How fast is the population growing at the end of 10 years? Estimate population 10 years from the initial time. How long will it take for the population to double itself?

The appropriate differential equation is

$$\frac{dP}{dt} = kP$$

where P is the size of the population and t is time, in years. The general solution is

$$P = f(t) = ce^{kt}$$

Since $f(0) = 2{,}500$, we have

$$f(t) = 2{,}500e^{kt}$$

for some constant k. Because $f(4) = 3{,}200$, we can solve for k by

$$\begin{aligned} 3{,}200 &= 2{,}500e^{k(4)} \\ e^{4k} &= 1.28 \\ ln\ e^{4k} &= ln\ 1.28 \\ 4k &= ln\ 1.28 = 0.246860 \\ k &= 0.061715 \end{aligned}$$

The rate of proportional change in population is $0.061715 = 6.1715$ percent. Hence, the model for predicting population size at any time t is

$$P = f(t) = 2500e^{0.061715\,t}$$

Population after 10 years is estimated to be

$$\begin{aligned} f(10) &= 2{,}500^{(0.061715)(10)} \\ &= 4{,}634 \text{ individuals} \end{aligned}$$

At this time, population is growing at a rate that is

$$\frac{dP}{dt} = f'(t) = kf(t) = (0.061715)(4{,}634) = 286$$

individuals per year.

The population will have doubled when it reaches 5,000. This will occur when

$$\begin{aligned} 5{,}000 &= 2{,}500e^{0.061715\,t} \\ 2 &= e^{0.061715\,t} \\ ln\ 2 &= ln\ e^{0.061715\,t} \\ 0.693147 &= 0.061715t \\ t &= 11.23 \end{aligned}$$

The population of the community will double in approximately 11.23 years after 1978. []

A third type of differential equation states that the rate of change of y with respect to x is given by the product of a function of y and a function of x.

$$\frac{dy}{dx} = f(x) \cdot g(y)$$

To find the solution, we must be able to rewrite the equation so that its left side involves only the variable y and its right side involves only the variable x, as

$$g(y)\, dy = f(x)\, dx$$

(A differential equation is called SEPARABLE if it can be written in this form.)

To obtain this format we must consider dy/dx as a quotient of differentials and algebraically separate the variables. Then we integrate both sides of the resulting equation; that is, we determine

$$\int g(y)\, dy = \int f(x)\, dx$$

If we can evaluate the integrals, we have a solution of the differential equation.

Example 22 Find a solution of the differential equation

$$\frac{dy}{dx} = 5xy^2$$

Rewriting the equation so that each side contains only one variable, we get

$$(1/y^2)\,dy = 5x\,dx$$

Then we integrate both sides to obtain

$$\int (1/y^2)\,dy = \int 5x\,dx$$

$$(-1/y) + C_1 = (5/2)x^2 + C_2$$

The constants of integration are combined as $C = C_1 + C_2$, and the expression written

$$-1/y = (5/2)x^2 + C$$

Thus, we find that

$$y = (-2/5x^2) + C$$

is a general solution to the original differential equation. []

Example 23 Determine a general solution of the differential equation

$$\frac{dy}{dx} = \frac{x+1}{y+1}$$

by separating the variables.

We first rewrite the equation as

$$(y+1)\,dy = (x+1)\,dx$$

Then, taking the integral of each side of the equation, we have

$$\int (y+1)\,dy = \int (x+1)\,dx$$

$$(y^2/2) + y + C_1 = (x^2/2) + x + C_2$$

Multiplying by 2 gives

$$y^2 + 2y + 2C_1 = x^2 + 2x + 2C_2$$

Then, combining the constants of integration as $2C_1 - 2C_2 = C$, we obtain

$$y^2 + 2y - x^2 - 2x + C = 0$$

This is the general solution of the differential equation. []

EXERCISES

Find the general solution for each of the following differential equations.

110. $dy/dx = 6x + 5$
111. $dy/dx = 7x + 6$
112. $dy/dx = x^2 - 3x$
113. $dy/dx = 2x^2 + 8x$
114. $dy/dx = e^{3x}$
115. $dy/dx = 4x^2$

116. $dy/dx = 2x - (1/x^2) + 1$

117. $4x^2 - x + dy/dx = 5x^2$

118. $dy/dx = 7y$

119. $dy/dx = -3y$

120. $dy/dx = 3xy^2$

121. $dy/dx = (6 + y)/x$

122. $dy/dx = (4x\sqrt{1 + y^2})/y$

123. $dy/dx = \sqrt{y/x}$

124. $dy/dx = 3y/x$

125. $dy/dx = x^2/\sqrt{y}$

126. $dy/dx = x^3y^3$

127. $dy/dx = x\sqrt{5 + x}/\sqrt{y}$

128. $dy/dx = y/\sqrt{4 + x}$

129. $dy/dx = y(x + 3)^5$

For each of the following differential equations, find the particular solution determined by the given boundary conditions.

130. $dy/dx + 3x = 2x^2$; and $y = 3$ when $x = 0$
131. $dy/dx = 3x + 4$; and $y = -5$ when $x = 0$
132. $dy/dx = 3x^2 - 5x + 1$; and $y = 4$ when $x = 1$
133. $dy/dx = 4x^3 - 6x^2 + 5$; and $y = 0$ when $x = -1$
134. $dy/dx = e^x + 4x - \sqrt{x}$; and $y = 6$ when $x = 0$
135. $dy/dx = (x + 1)(x + 2)$; and $y = -1.5$ when $x = 3$
136. $dy/dx - e^{x-y} = 0$; and $y = 0$ when $x = 0$
137. $e^y(dy/dx) = x^2$; and $y = 0$ when $x = 0$
138. $dy/dx = x/4(1 + x^2)^{3/2}$; and $y = 0$ when $x = 1$
139. $dy/dx = x^2\sqrt{y}/\sqrt{x + 1}$; and $y = 4$ when $x = 0$
140. $dy/dx = 3xye^{x^2}$; and $y = 5$ when $x = 0$
141. A firm's marginal cost function is given by

$$dy/dx = 1{,}200/\sqrt{x - 1} \qquad x > 1$$

where x is the number of items produced. Find the total cost function if $y = 25{,}000$ when $x = 50$.

142. Salvage value of a certain type of industrial machine declines over a 12-year period at a rate which is a function of the age of the machine; that is

$$S'(x) = 5{,}000(x - 12)$$

where $S'(x)$ is stated in dollars and x is the age of the machine, in years. Express the salvage value of the machine as a function of its age, given that the machine was originally purchased for $15,000.

143. The rate at which a radioactive substance disintegrates is proportional to the amount present; that is

$$dQ/dt = -kQ$$

where $Q(t)$ is quantity remaining after t years.

Determine what quantity will remain after 500 years if the original quantity was 20 grams and one half of a supply of the radioactive substance will decompose in 1,000 years.

144. Population of a certain region was 75,000 in 1900 ($t = 0$) and in 1950 was 150,000. Assuming that the growth rate is at any time proportional to the population size—that is, $dP/dt = kP$—find the population projected for 1990.

145. In a certain community the population at any time changes at a rate proportional to the population. If the population in 1970 was 22,000 and in 1980 was 28,000, find an equation giving the population at any time t, where t is the number of years past 1970. What population is expected in 1990?

14.6 IMPROPER INTEGRALS

We have, up to this point, considered the definite integral of a function on the closed interval [a,b] only if the function is continuous over that closed interval. We will now extend our study to include a definite integral on an infinite interval of integration. Such an integral is called an IMPROPER INTEGRAL. Improper integrals have important applications in statistics and in mathematical models which involve projections into the indefinite future.

It would, on the surface, seem that the area of a region lying under a curve whose base is infinitely long would surely be infinite. Sometimes this is the case; but at other times the area of such a region is finite. Consider the area of the region that lies below the graph of $y = e^{-x}$, above the x-axis, to the right of the y-axis, and to the left of a vertical line erected at $x = b$, as shown in Figure 14.14.

The area A of the region, in square units, is given by

Figure 14.14

Finding the Definite Integral over an Infinite Interval of Integration

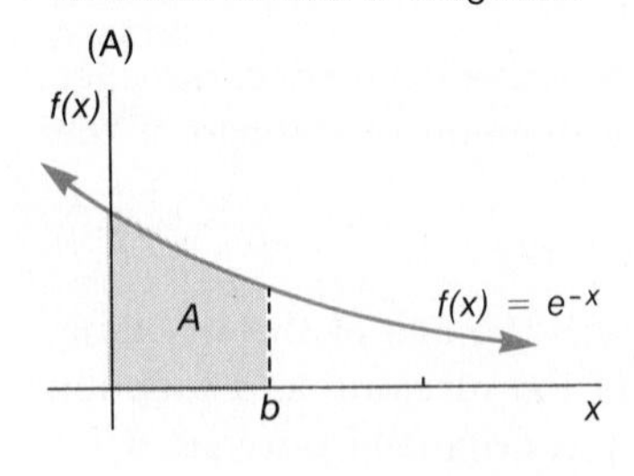

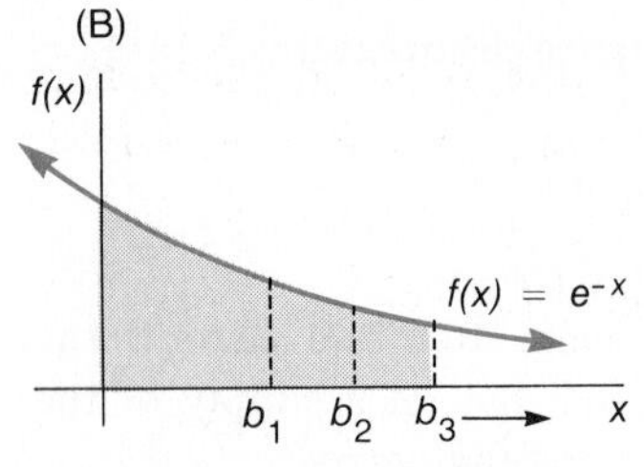

$$A = \int_0^b e^{-x}\,dx = -e^{-x}\Big|_0^b = -e^{-b} - (-e^0) = 1 - e^{-b}$$

If we allow b to increase without bound, the area of the region becomes

$$A = \lim_{b\to+\infty} \int_0^b e^{-x}\,dx = \lim_{b\to+\infty} (1 - e^{-b}) = 1 - 0 = 1$$

Thus, as the upper limit of integration, b, approaches infinity, the area of the region approaches one square unit. We say that

$$\int_0^b e^{-x}\,dx = 1$$

In general, we have the following definition.

> If f is continuous for all $x \geq a$, then
>
> $$\int_a^{+\infty} f(x)\,dx = \lim_{b\to+\infty} \int_a^b f(x)\,dx$$
>
> if this limit exists.

If the lower limit of integration is infinite, the area is given by this definition.

If f is continuous for all $x \leq b$, then

$$\int_{-\infty}^{b} f(x)\,dx = \lim_{a \to -\infty} \int_{a}^{b} f(x)\,dx$$

if this limit exists.

Should both limits of integration be infinite, the following definition applies.

If f is continuous for all x, then

$$\int_{-\infty}^{+\infty} f(x)\,dx = \lim_{a \to -\infty} \int_{a}^{0} f(x)\,dx + \lim_{b \to +\infty} \int_{0}^{b} f(x)\,dx$$

if these limits exist.

If the required limits exist, the improper integral is said to be CONVERGENT. If the limits do not exist, the improper integral is DIVERGENT.

Example 24 Determine whether the improper integral

$$\int_{1}^{\infty} 1/x\,dx$$

is convergent or divergent. If it is convergent, determine its value.

We proceed by

$$\int_{1}^{\infty} 1/x\,dx = \lim_{b \to \infty} \int_{1}^{b} 1/x\,dx = \lim_{b \to \infty} \ln x \Big|_{1}^{b} = \lim_{b \to \infty} [\ln b - \ln 1]$$

Because $\ln 1 = 0$, we have

$$\lim_{b \to \infty} \ln b$$

Now, $\ln b$ increases without bound as b approaches infinity; so the limit does not exist. This improper integral does not exist; it is divergent. [ ]

Example 25 Find, if it exists, the value of the improper integral

$$\int_{-\infty}^{0} e^{x}\,dx$$

We evaluate

$$\int_{-\infty}^{0} e^x\,dx = \lim_{a\to-\infty} \int_a^0 e^x\,dx = \lim_{a\to-\infty} e^x \Big|_a^0 = \lim_{a\to-\infty} [e^0 - e^a]$$

$$\lim_{a\to-\infty} [1 - e^a] = 1 - 0 = 1$$

Thus, this improper integral is convergent; its value is one. []

Example 26 Determine whether the improper integral

$$\int_{-\infty}^{\infty} x^2\,dx$$

is convergent or divergent. If it is convergent, determine its value.

We evaluate

$$\int_{-\infty}^{\infty} x^2\,dx = \lim_{a\to-\infty} \int_a^0 x^2\,dx + \lim_{b\to\infty} \int_0^b x^2\,dx$$

$$= \lim_{a\to-\infty} (x^3/3) \Big|_a^0 + \lim_{b\to\infty} (x^3/3) \Big|_0^b$$

$$= \lim_{a\to-\infty} [0 - (a^3/3)] + \lim_{b\to\infty} [(b^3/3) - 0]$$

Because neither of these limits exists, the improper integral does not exist; it is divergent. []

EXERCISES

Determine whether the following improper integrals are convergent or divergent. If convergent, determine the value of the integral.

146. $\int_1^{\infty} 3/x^2\,dx$

147. $\int_0^{\infty} 5/(x+1)\,dx$

148. $\int_0^{\infty} e^{-x}\,dx$

149. $\int_2^{\infty} (x^2+1)\,dx$

150. $\int_{-\infty}^{2} e^x\,dx$

151. $\int_{-\infty}^{\infty} 3x^2\,dx$

152. $\int_{-\infty}^{\infty} e^x\,dx$

C·H·A·P·T·E·R 15

The Calculus of Multivariate Functions

Until now we have considered the derivatives, and the integrals, of those functions involving one independent and one dependent variable only. It is the purpose of this chapter to present a brief introduction to the calculus of multivariate functions—functions of more than one (independent) variable.

15.1 PICTURING NONLINEAR MULTIVARIATE FUNCTIONS

We have noted in earlier chapters that the number of dimensions required to graph a function is determined by the number of variables in that function. A function of one dependent and one independent variable graphs in two-dimensional space; a function of one dependent and two independent variables graphs in three-dimensional space. No satisfactory method of graphing functions of more than three variables has been devised.

We have also seen that a linear function in one independent variable is pictured as a *straight line* in two-dimensional space, while a linear function in two independent variables is depicted as a *plane* in 3-space. Nonlinear functions of one independent variable graph as *curves* in two-dimensional space while, as we shall soon illustrate, nonlinear functions of two independent variables will be depicted as SURFACES in 3-space. These surfaces are curved—as the undulating surface of rolling hills and valleys—as opposed to the flat plane associated with a linear function.

In Chapter 5 we represented linear functions of two independent variables geometrically in a three-dimensional coordinate system. We came to realize at that juncture that much skill and imagination were required to graph a linear function in 3-space. Still more skill and more imagination will be needed for graphing nonlinear functions in 3-space. Fortunately, in applied problems, it is rarely necessary that we actually construct the graph of such functions; usually visualization of the function as a surface will suffice.

We sketch a surface in 3-space through the aid of TRACES. A TRACE is the intersection of the surface with a coordinate plane. It is the graphical representation of $z = f(x,y)$ when one variable is held constant. The intersection of the surface with the x,y plane is termed the TRACE IN THE x,y PLANE and is obtained by setting $z = c$. Similarly, the intersection of the surface with the x,z plane is termed the TRACE IN THE x,z PLANE, just as the intersection of the surface with the y,z plane is termed the TRACE IN THE y,z PLANE. These are obtained by setting $y = c$ and $x = c$, respectively.

The next few examples will illustrate the procedure for using traces to aid in graphing nonlinear functions in two independent variables.

Example 1 Sketch the graph of $z = x^2$ in 3-space.

The x,z plane is represented by the equation $y = 0$. To find the

Figure 15.1

Sketching the Graph of $z = x^2$ in 3-Space

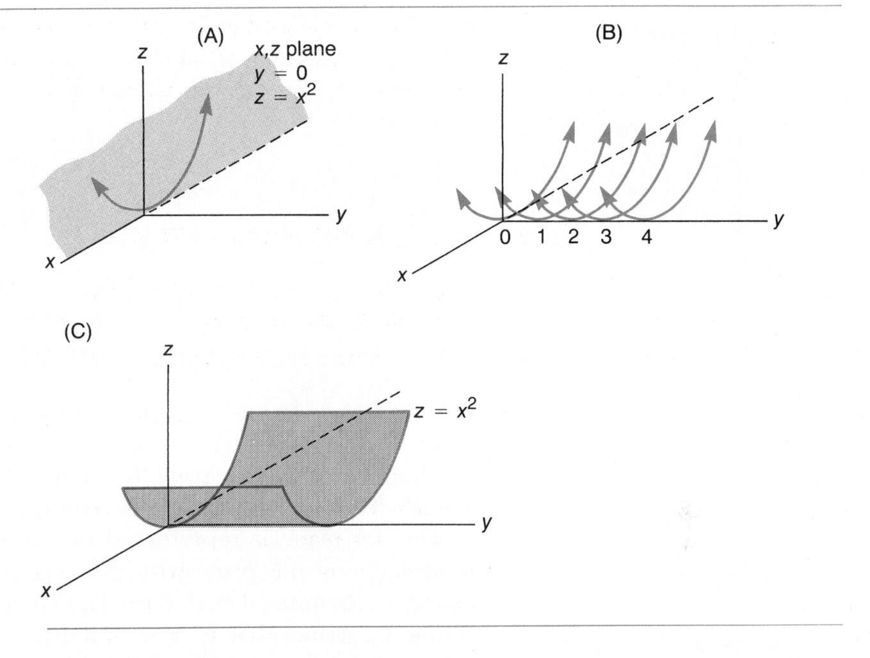

intersection of this plane with the graph (that is, to find the trace in the x,z plane), we substitute $y = 0$ into the equation. The result in the current example is $z = x^2$, which graphs as a concave-upward parabola, as shown in Figure 15.1A.

The x,y plane is represented by the equation $z = 0$. To find the trace in the x,y plane, we substitute $z = 0$ into the equation. This yields $0 = x^2$, which has only one real solution, $x = 0$, and graphs as a point.

The y,z plane is represented by the equation $x = 0$. To find the trace in the y,z plane, we substitute $x = 0$ into the equation. This yields, in the current example, $z = 0$.

These traces, however, represent but a "rib" on the surface which represents the equation. Other "ribs" are obtained by setting the variables, in turn, to other values within their domains. For example, let us return to the x,z plane and set $y = 2$. Again the result is $z = x^2$, but the parabola is located at $y = 2$ on the y-axis. And, setting y at other values in its domain, we obtain other traces, as shown in Figure 15.1B.

Hence, in 3-space the graph of $z = x^2$ would be as illustrated in Figure 15.1C. []

Example 2 Construct the graph of $z = f(x,y) = x^2 + y^2$.

The trace in the x,z plane is obtained by setting $y = 0$. This yields $z = x^2$, which graphs as a concave-upward parabola, as illustrated in Figure 15.2A.

Figure 15.2

Sketching the Graph of $z = f(x,y) = x^2 + y^2$

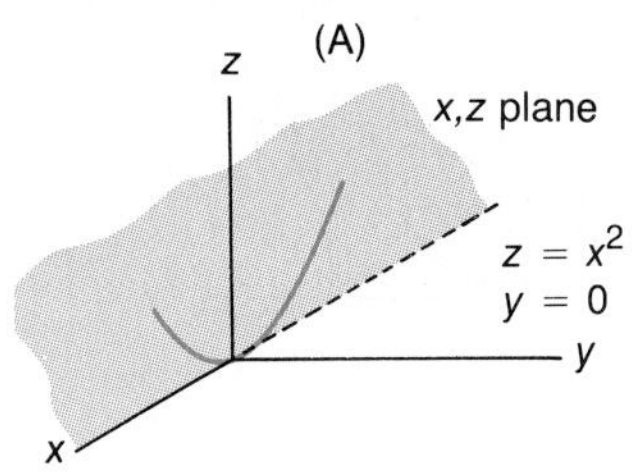

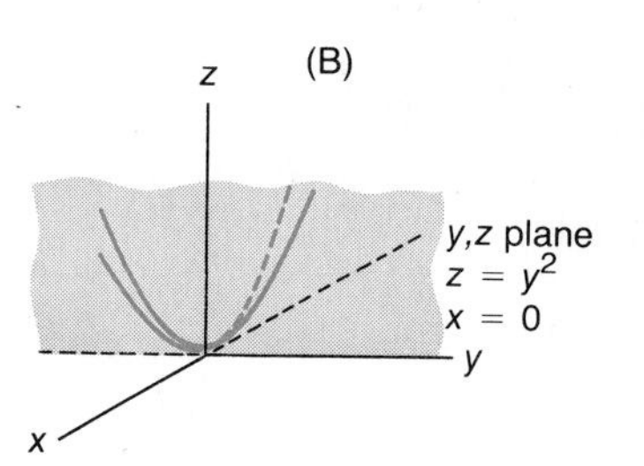

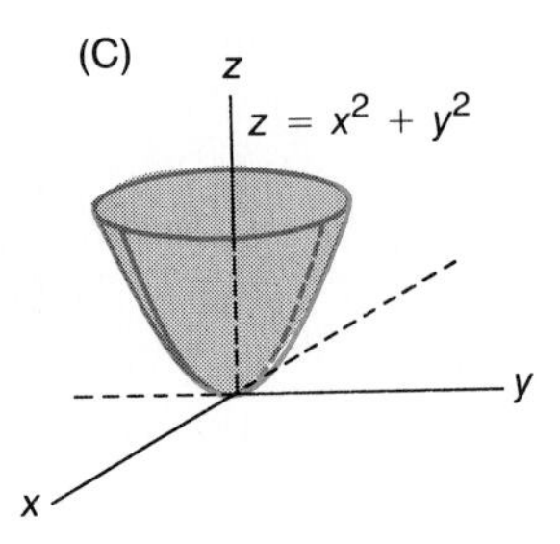

The trace in the y,z plane is obtained by setting $x = 0$, which yields $z = y^2$. This, too, graphs as a parabola, as shown in Figure 15.2B.

The x,y plane is represented by the equation $z = 0$. To find the intersection of this plane with the graph, we substitute $z = 0$ into the equation, obtaining $0 = x^2 + y^2$. This equation has only one real solution in the x,y plane; that is, $x = 0$ and $y = 0$. Hence, the only point of the graph in the x,y plane is (0,0,0).

However, let us find the intersection of the graph with the plane $z = 4$. To do this we substitute $z = 4$ into the equation, to obtain $4 = x^2 + y^2$. This graphs as a circle with radius two. If we substitute $z = 9$, we have $9 = x^2 + y^2$, which graphs as a circle with radius three. Or, for other values of $z = c$, we have $c = x^2 + y^2$, each of which will graph as a circle.

The graph of the equation $z = f(x,y) = x^2 + y^2$, thus, is the surface shown in Figure 15.2C, which is called a CIRCULAR PARABOLOID. []

Example 3 Construct the graph of $z = f(x,y) = y^2 - x^2$.

The trace in the x,z plane is obtained by setting $y = 0$, with the result $z = -x^2$. This graphs as a concave-downward parabola, as shown in Figure 15.3A.

In the y,z plane the trace is obtained by setting $x = 0$, with the result $z = y^2$. This graphs as a concave-upward parabola, as shown in Figure 15.3B.

The trace in the x,y plane is obtained by setting $z = c$, with the result $c = y^2 - x^2$, which graphs as illustrated in Figure 15.3C.

Hence, the graph of $z = f(x,y) = y^2 - x^2$ appears as the surface shown in Figure 15.3D. Such a graph is called a SADDLE SURFACE, and the point (0,0,0) is a SADDLE POINT. []

Figure 15.3

Sketching the Graph of $z = f(x,y) = y^2 - x^2$

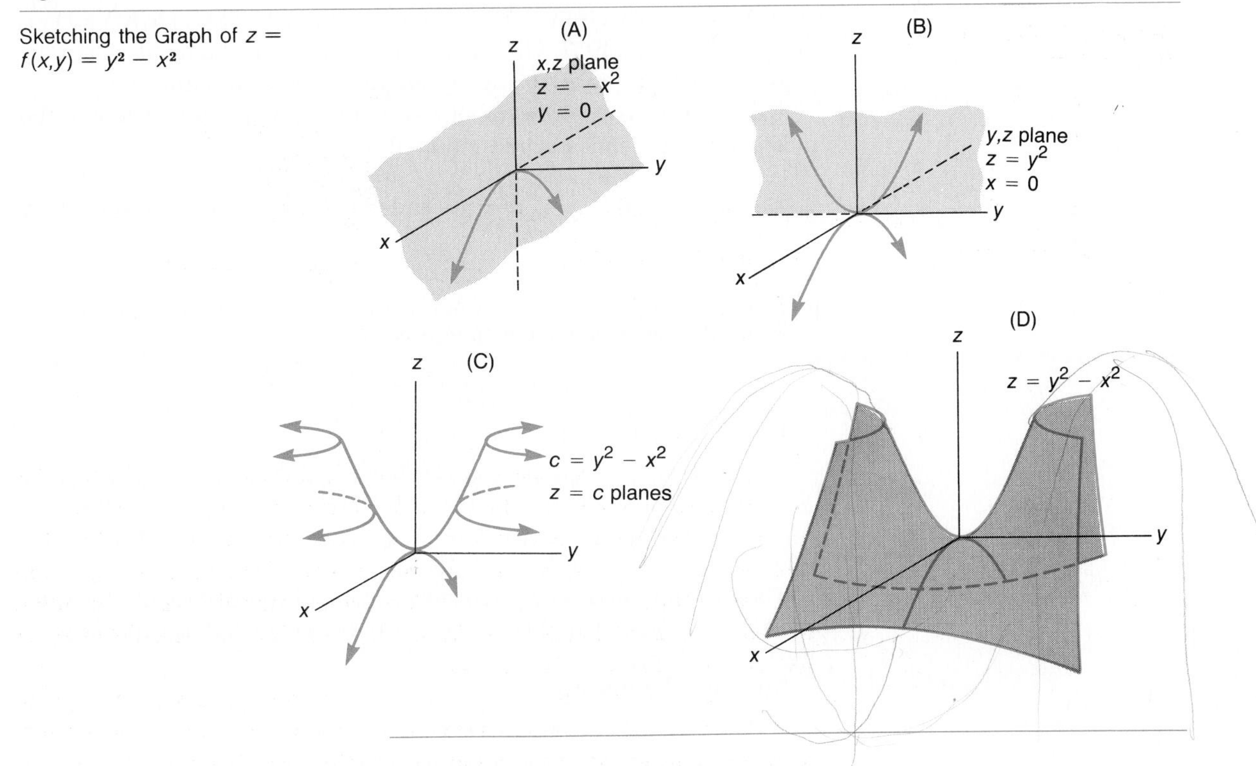

EXERCISES

Sketch a graph of each of the following functions.

1. $z = f(x,y) = 5 - x^2$
2. $z = f(x,y) = 1 + x^2$
3. $z = f(x,y) = x^2 + y^2 + 2$
4. $z = f(x,y) = x^2 + y^2 - 3$
5. $z = f(x,y) = 4 - x^2 - y^2$

15.2 PARTIAL DERIVATIVES

PARTIAL DERIVATIVES are used to analyze the effect on the dependent variable of changes in one of several independent variables *while the other independent variables are held constant.* For a function such as $z = f(x,y)$, for example, each of the variables x and y can be varied independently of the other. One variable, say y, might be set at some fixed value and the other variable, x, then allowed to vary in order to study the net effect of changes in x on the dependent variable, z. The PARTIAL DERIVATIVE OF z WITH RESPECT TO x is found by looking at the function as though it were a function of only one variable, x, with y temporarily being treated as though it were a constant. We interpret this derivative as the rate of change of the function when x alone varies. In precisely the same manner, the PARTIAL DERIVATIVE OF z WITH RESPECT TO y treats the function as though it were only $f(y)$ with x temporarily being considered to be a constant.

Partial derivatives are symbolized in a variety of ways. A few of the most frequently encountered methods are

$f_x(x,y)$ and $f_y(x,y)$ $\qquad \dfrac{\partial f(x,y)}{\partial x}$ and $\dfrac{\partial f(x,y)}{\partial y}$ $\qquad \partial z/\partial x$ and $\partial z/\partial y$

PARTIAL DERIVATIVES of a function $z = f(x,y)$ are defined as follows:
Partial derivative of $f(x,y)$ with respect to x:

$$f_x(x,y) = \lim_{\Delta x \to 0} \frac{f(x + \Delta x, y) - f(x,y)}{\Delta x}$$

Partial derivative of $f(x,y)$ with respect to y:

$$f_y(x,y) = \lim_{\Delta y \to 0} \frac{f(x, y + \Delta y) - f(x,y)}{\Delta y}$$

The rules for differentiation previously stated are used for one variable at a time, all other variables temporarily being treated as constants.

Example 4 Consider the function

$$z = f(x,y) = 3x + xy + 5y$$

This function has two independent variables, x and y, and thus it has two partial derivatives. The partial derivative of $f(x,y)$ with respect to x is found by treating y as though it were a constant (like the 3 or the 5) and applying the derivative rules we have previously learned. Thus,

$$f_x(x,y) = 3 + y$$

In a similar fashion, the partial derivative of $f(x,y)$ with respect to y is found by treating x as though it were a constant and applying the derivative rules we have previously used. Thus,

$$f_y(x,y) = x + 5$$ []

Example 5 Find the partial derivatives of

$$z = f(x,y) = (2x - 5y^2)^3$$

With y treated as a constant, we have (by the extended-power rule) the partial derivative of f with respect to x.

$$f_x(x,y) = (3)(2x - 5y^2)^2(2) = 6(2x - 5y^2)^2$$

Then, with x treated as a constant, we have (again by the chain rule) the partial derivative of f with respect to y,

$$f_y(x,y) = (3)(2x - 5y^2)^2(-10y) = (-30y)(2x - 5y^2)^2$$ []

Example 6 Find the partial derivatives of

$$z = f(x,y) = ln(x^2 + 3y^2)$$

With y treated as a constant, we have the partial derivative of f with respect to x,

$$f_x(x,y) = 2x/(x^2 + 3y^2)$$

Then, with x treated as a constant, we have the partial derivative of f with respect to y,

$$f_y(x,y) = 6y/(x^2 + 3y^2)$$ [ ]

The same approach is used in determining partial derivatives of functions with three or more independent variables. One simply holds constant all variables except one and then defines the derivative of the resulting function of one variable.

Example 7 Find the partial derivatives of

$$w = f(x,y,z) = x^2 + 5y^2 + 10xy + 4z$$

By treating both y and z as constants, we obtain the partial derivative of f with respect to x to be

$$f_x = 2x + 10y$$

Then, if we treat x and z as constants, we obtain the partial derivative of f with respect to y as

$$f_y = 10y + 10x$$

Finally, treating x and y as constants, we obtain the partial derivative of f with respect to z to be

$$f_z = 4$$ [ ]

Example 8 Find the partial derivatives of

$$w = f(x,y,z) = e^{(4x+y^2-2z^3)}$$

With y and z treated as constants, we obtain the partial derivative of f with respect to z to be

$$f_x = 4e^{(4x+y^2-2z^3)}$$

With x and z treated as constants, we obtain the partial derivative of f with respect to y to be

$$f_y = 2ye^{(4x+y^2-2z^3)}$$

Finally, with x and y treated as constants, we obtain the partial derivative of f with respect to z to be

$$f_z = -6z^2e^{(4x+y^2-2z^3)}$$ [ ]

The partial derivative of $z = f(x,y)$ with respect to x evaluated at the point $(x = a, y = b)$ is denoted by $f_x(a,b)$ or $\partial f/\partial x \big|_{(a,b)}$

Example 9 For $f(x,y) = 3x^4y^2 + xy$, find $f_x(2,1)$ and $f_y(-1,3)$.
Holding y constant and differentiating with respect to x yields

$$f_x(x,y) = 12x^3y^2 + y$$

Evaluating this derivative at the point where $x = 2$ and $y = 1$, we obtain

$$f_x(2,1) = 12(2)^3(1)^2 + 1 = 97$$

Holding x constant and differentiating with respect to y yields

$$f_x(x,y) = 6x^4y + x$$

Then,

$$f_y(-1,3) = 6(-1)^4(3) + (-1) = 17$$ [ ]

Figure 15.4

(A) Partial derivative of $z = f(x,y)$ with respect to x

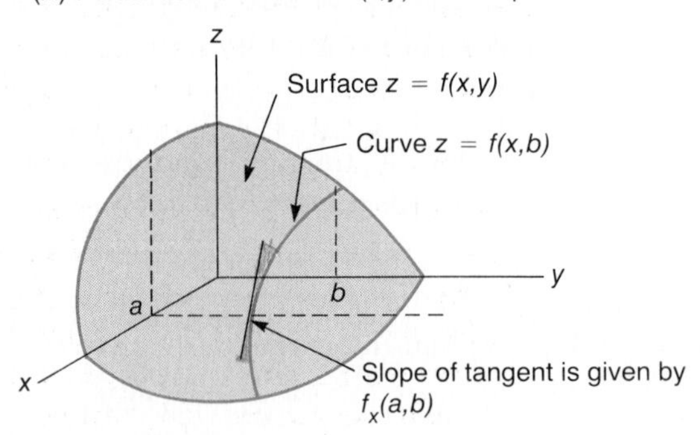

(B) Partial derivative of $z = f(x,y)$ with respect to y

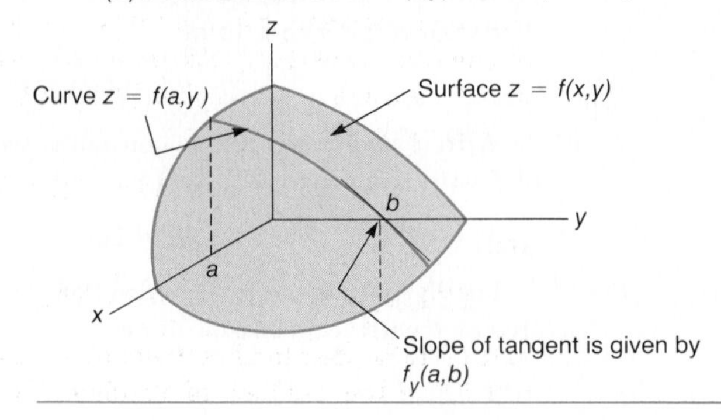

The partial derivatives of $z = f(x,y)$ at a point (a,b) can be interpreted as follows:

$f_x(a,b)$ is the slope of the tangent line drawn to the curve $z = f(x,b)$ at $x = a$ (where the curve $z = f(x,b)$ is the intersection of the surface $z = f(x,y)$ and the plane $y = b$).

$f_y(a,b)$ is the slope of the tangent line drawn to the curve $z = f(a,y)$ at $y = b$ (where the curve $z = f(a,y)$ is the intersection of the surface $z = f(x,y)$ and the plane $x = a$).

These concepts are illustrated in Figure 15.4.

In preceding chapters we have considered cost, revenue, and profit functions of a single-product firm. The next example deals with the JOINT-COST FUNCTION of a firm manufacturing two products.

Example 10 Diverse, Inc., manufactures two types of motorcycle helmets, a standard helmet and a deluxe helmet. The total daily production cost (in dollars) is given by the joint-cost function

$$f(x,y) = 500 + 0.125x^2 + 2xy + 0.18y^2$$

where x represents the number of standard units produced and y represents the number of deluxe units produced.

The MARGINAL COST WITH RESPECT TO STANDARD UNITS is the rate of change in total cost with an increase in the number of standard units produced, the number of deluxe units produced being held constant. The marginal cost with respect to standard units is given by the partial derivative of f with respect to x.

The MARGINAL COST WITH RESPECT TO DELUXE UNITS is the rate of change in total cost with an increase in the number of deluxe units produced, the number of standard units produced being held constant. The marginal cost with respect to deluxe units is given by the partial derivative of f with respect to y.

What is the total cost of producing 50 standard and 30 deluxe helmets? By what amount would total cost increase if the number of standard units produced was increased by 1 while the number of deluxe units was held constant at 30? By what amount would total cost increase if the number of deluxe units produced was increased by 1 while the number of standard units was held constant at 50?

We compute

$$f(50,30) = 500 + 0.125(50)^2 + 2(50)(30) + 0.18(30)^2 = \$3{,}974.50$$

and

$$f(51,30) = 500 + 0.125(51)^2 + 2(51)(30) + 0.18(30)^2 = \$4{,}047.125$$

to determine that total cost would increase by \$4,047.125 − \$3,974.5 = \$72.625 if the number of standard units produced was increased by

1 from 50 to 51 while the number of deluxe units produced was held constant at 30.

This rate of change in the total-cost function when the number of standard units produced increases by one unit but no change takes place in the number of deluxe units produced is approximated by the marginal cost with respect to standard units. To determine this marginal-cost function, we find the partial derivative of f with respect to x, as

$$f_x(x,y) = 0.25x + 2y$$

Evaluating this marginal-cost function at the production level $x = 50$ and $y = 30$, we find

$$f_x(50,30) = 0.25(50) + 2(30) = 72.5$$

Now suppose that the number of deluxe units produced is increased by 1 while the number of standard units produced remains at 50. We determine that total cost will be

$$C = f(50,31) = 500 + 0.125(50)^2 + 2(50)(31) + 0.18(31)^2 = \$4{,}085.48$$

which represents an increase of \$4,085.48 − \$3,974.50 = \$110.98.

This rate of change in the total-cost function when the number of deluxe units produced increases by one unit but no change takes place in the number of standard units produced is approximated by the marginal cost with respect to deluxe units. To determine this marginal-cost function, we find the partial derivative of f with respect to y, as

$$f_y(x,y) = 2x + 0.36y$$

Evaluating this marginal-cost function at the production level $x = 50$ and $y = 30$, we find

$$f_y(50,30) = 2(50) + 0.36(30) = 110.8$$

Thus, the partial derivatives give the effect on the joint-cost function of changes in one variable, when the other variable is held at a constant value. []

The output of a firm may usually be thought of as a function of many economic resources—such as labor, capital, machinery, and so on—which are used in the production process. The partial derivatives of such production functions give the marginal productivity with respect to labor, or the marginal productivity with respect to capital, and so on.

Example 11 The number of units of output for a firm (in units per day) depends upon the number x of units of labor used on that day and the number y of units of machine time that are used. The production function is given by

$$f(x,y) = x^2 + 6xy + 1.5y^2$$

Find the marginal productivity of labor when 40 units of labor and 50 units of machine time are used. Find the marginal productivity of machine time when 40 units of labor and 50 units of machine time are used.

The marginal productivity of labor is given by the partial derivative of f with respect to x and is

$$f_x(x,y) = 2x + 6y$$

The marginal productivity of labor when 40 units of labor and 50 units of machine-time are used is

$$f_x(40,50) = 2(40) + 6(50) = 380$$

That is, if the machine time used were held constant at 50 units while the number of labor units used was increased from 40 to 41 units, total output would be increased by approximately 380 units.

The marginal productivity of machine time is given by the partial derivative of f with respect to y and is

$$f_y(x,y) = 6x + 3y$$

When 40 units of labor and 50 units of machine time are used, the marginal productivity of machine time is

$$f_y(40,50) = 6(40) + 3(50) = 390$$

That is, if the labor units used were held constant at 40 while the number of machine-time units used was increased from 50 to 51, total output would be increased by approximately 390 units. []

Example 12 A company sells soft drinks and snacks through vending machines located in different public buildings. Presently the company has 9 soft drink–vending machines and 6 snack-vending machines in different locations throughout the municipal air terminal. The daily revenue, in dollars, received from these machines is given by

$$f(x,y) = 20\sqrt{x} + 3y^2 + 15xy + e^{x^2/y^2}$$

where x is the number of soft drink–vending machines and y is the number of snack-vending machines in one building.

How much additional revenue will be generated if one additional soft drink–vending machine but no additional snack machines are installed at the terminal?

The extra revenue generated by installing one additional soft-drink machine but no additional snack machines can be represented by the partial derivative of f with respect to x; that is, since x represents the number of soft-drink machines, $f_x(x,y)$ represents the effect on $f(x,y)$ of a change in the number of soft-drink machines and will be termed the marginal revenue with respect to soft-drink machines.

We have

$$f_x(x,y) = 10x^{(-1/2)} + 15y + (2x/y^2)e^{(x^2/y^2)}$$

Evaluated at $x = 9$ and $y = 6$, the partial derivative is

$$\begin{aligned} f_x(9,6) &= (10/\sqrt{9}) + 15(6) + [2(9)/6^2]e^{(9^2/6^2)} \\ &= (10/3) + 90 + 4.74 \\ &= 98.08 \end{aligned}$$

The vending-machine company would increase its total revenue by approximately \$98 by increasing the number of soft drink–vending machines from 9 to 10 while holding the number of snack-vending machines constant at 6.

The marginal revenue with respect to snack-vending machines is

$$f_y(x,y) = 6y + 15x + (-2x^2/y^3)e^{(x^2/y^2)}$$

Evaluated at $x = 9$ and $y = 6$, the partial derivative is

$$\begin{aligned} f_y(9,6) &= 6(6) + 15(9) + [(-2)(9^2)/(6^3)]e^{(9^2/6^2)} \\ &= 36 + 135 - 7.12 \\ &= 163.88 \end{aligned}$$

The vending-machine company would increase its total revenue by approximately \$164 by increasing the number of snack-vending machines in the terminal from 6 to 7 while holding the number of soft drink–vending machines constant at 9. [ ]

Certain products may be related in such a way that changes in the price of one of the products affect the demand *for the other product.* Wherever such a relationship exists between products A and B, the demand for each product is a function of the prices of both products. Let us use the notation

p_A = Price, per unit, for product A
p_B = Price, per unit, for product B

and

$q_A = f(p_A,p_B)$ = Number of units of product A demanded
$q_B = g(p_A,p_B)$ = Number of units of product B demanded

The partial derivatives of q_A can be interpreted as follows:

$\partial q_A/\partial p_A$ is marginal demand for A with respect to the price of A (which reflects the rate of change in total demand for A given an increase in the price of A, with the price of B remaining unchanged).

$\partial q_A/\partial p_B$ is marginal demand for A with respect to the price of B (which reflects the rate of change in total demand for A given an increase in the price of B, with the price of A remaining unchanged).

The partial derivatives of q_B—that is, $\partial q_B/\partial p_A$ and $\partial q_B/\partial p_B$—would have a similar interpretation.

In general, if the price of A increases (while the price of B remains unchanged), the quantity of A demanded will decrease. Thus, $\partial q_A / \partial p_A < 0$ under typical conditions. Similarly, $\partial q_B / \partial p_B < 0$.

However, an increase in the price of one product may result in either an increase or a decrease in the quantity of the other product demanded. If an increase in the price of B (while the price of A remains fixed) causes an increase in the demand for A, or an increase in the price of A (while the price of B remains fixed) causes an increase in the demand for B, A and B are said to be COMPETITIVE (or SUBSTITUTE) PRODUCTS. Butter and margarine, or beef and pork, are examples of competitive products. For competitive products A and B, $\partial q_A / \partial p_B > 0$ and $\partial q_B / \partial p_A > 0$.

If an increase in the price of B (while the price of A remains fixed) causes a decrease in the demand for A, or an increase in the price of A (while the price of B remains fixed) causes a decrease in the demand for B, A and B are said to be COMPLEMENTARY PRODUCTS. Cameras and film, automobiles and gasoline, are complementary products. For complementary products A and B, $\partial q_A / \partial p_B < 0$, and $\partial q_B / \partial p_A < 0$.

Products are said to be INDEPENDENT if changes in the price of one have no direct effect on the quantity of the other product demanded.

Example 13 Two products, A and B, have demand functions

$$q_A = f(p_A, p_B) = 800 - 50p_A + 5p_B$$

and

$$q_B = g(p_A, p_B) = 400 + 3p_A - 15p_B$$

where

p_A = Price per unit of product A
p_B = Price per unit of product B

Are A and B complementary, or competitive, products?

We determine the marginal demand for A with respect to the price of B to be

$$\partial q_A / \partial p_B = +5$$

and the marginal demand for B with respect to the price of A to be

$$\partial q_B / \partial p_A = +3$$

Because both $\partial q_A / \partial p_B > 0$ and $\partial q_B / \partial p_A > 0$, we conclude that A and B are competitive products. An increase in the price of B (with the price of A remaining fixed) results in an increase in the demand for A. Similarly, an increase in the price of A (with the price of B remaining fixed) results in an increase in the demand for B. []

EXERCISES

Find the partial derivatives with respect to x *and* y.

6. $z = f(x,y) = x^2 + xy - y^2$
7. $z = f(x,y) = x^3 + 2x^2y - xy^2 + 4y^3$
8. $z = f(x,y) = x^2y^3 + y^2x^3$
9. $z = f(x,y) = 3x^2y + 4xy^2 - 5y^3$
10. $z = f(x,y) = (x + y)^2(xy)^2$
11. $z = f(x,y) = (x^3 + 2xy + y)^5$
12. $z = f(x,y) = e^{(2x+5y)}$
13. $z = f(x,y) = e^{(x^2+y)}$
14. $z = f(x,y) = x^2e^{(4x+y)}$
15. $z = f(x,y) = xye^{xy}$
16. $z = f(x,y) = ln\, x + ln\, y$
17. $z = f(x,y) = ln(3x^3 + 5y^3)$
18. $z = f(x,y) = 4x \cdot ln(x^2 + y)$
19. $z = f(x,y) = 3e^{x^2} + y^2 + ln(x + 2y)$
20. $z = f(x,y) = 5e^x + 3e^y$

Find the partial derivatives of each of the following functions.

21. $w = f(x,y,z) = x^2 + xy^2z + z^3$
22. $w = f(x,y,z) = x^3y^2z - 3x^2yz^2 + 4xy^2z^3$
23. $w = f(x,y,z) = x^2y + e^{3xy} + ln\, xyz$
24. $w = f(x,y,z) = (x + y)(y + z)$
25. $w = f(x,y,z) = (3x^2 + y^2)/z^3$
26. $w = f(x,y,z) = xe^{yz}$
27. $w = f(x,y,z) = x^2\, ln\, xyz^2$
28. $w = f(x,y,z) = x^3(y + z)^{(1/2)}$
29. $w = f(x,y,z) = (x^2y + z)/(y^2 + z^2)$
30. $w = f(x,y,z) = xyz^2 + e^{xyz}$

Find $f_x(x,y)$, $f_y(x,y)$, $f_x(1,3)$, *and* $f_y(-2,4)$ *for each of the following functions.*

31. $z = f(x,y) = 5x^2y$
32. $z = f(x,y) = 3x^3y^2 - x^2y^3 + 4x + 9$
33. $z = f(x,y) = ln\, xy$
34. $z = f(x,y) = x^3e^y$
35. $z = f(x,y) = e^x\, ln\, y$
36. $z = f(x,y) = x^2e^{(2x+y)}$

The joint-cost function for a firm which manufactures two models of a product is given by

$$f(x,y) = 200 + 1.2x + 2y + 0.001y^2$$

where

x = Number of units of model A produced
y = Number of units of model B produced
$f(x,y)$ = Total production cost in dollars per day

37. Determine total production cost when 20 units of model A and 30 units of model B are produced.
38. Determine the change in total production cost when the number of units of model A produced is increased by 1 while the number of units of model B produced remains at 30.
39. Using partial derivatives, estimate the change expected in total production cost if the number of units of model A produced is increased by 1, from the present level of 20, while the number of units of model B produced is held constant. Compare this with the actual change in total cost as determined in Exercise 38.
40. Determine the change in total production cost when the number of units of model B produced is increased by 1 while the number of units of model A produced remains at 20.
41. Using partial derivatives, estimate the change expected in total production cost if the number of units of model B produced is increased by 1, from the present level of 30, while the number of units of model A produced is held constant at 20. Compare this with the actual change in total cost as determined in Exercise 40.

Municipal Utilities, Inc., employs 18 licensed electricians and 30 electricians' helpers to install utility service for new customers. The number of installations per year is given by

$$f(x,y) = 400x + 5y + x^2y$$

where x *is the number of licensed electricians and* y *is the number of helpers employed.*

42. Determine the marginal productivity with respect to licensed electricians. Interpret.
43. Determine the marginal productivity with respect to electricians' helpers. Interpret.

A company finds that its productivity in units per employee per day is given by

$$f(x,y) = 150xy - 3x^2 - 5y^2$$

where x *is the number of workers in the work force on that day and* y *is the number of machine hours of a certain type that are used.*

44. Determine the marginal product with respect to labor when $x = 50$ and $y = 300$. Interpret.

45. Determine the marginal product with respect to machine hours when $x = 50$ and $y = 300$. Interpret.

A company finds that its annual sales (in thousands of dollars) are a function of the amount of money x *spent on television advertising and the amount of money* y *spent on newspaper and trade journal advertising, where both* x *and* y *are expressed in thousands of dollars, as given by*

$$f(x,y) = 1{,}500x + 4{,}800y - 10x^2 - 6y^2 - 40xy$$

The company presently has an advertising budget which allocates $40,000 to television and $25,000 to newspaper and trade journal advertising.

46. Based on the present advertising budget, what are the expected annual sales?

47. Using partial derivatives, estimate the effect on annual sales if an additional $1,000 is allocated to television advertising while no change is made in the newspaper and trade journal budget.

48. Using partial derivatives, estimate the effect on annual sales if an additional $1,000 is allocated to newspaper and trade journal advertising while no change is made in the television advertising budget.

Given the following sets of demand functions, determine whether the products should be considered to be competitive products, complementary products, or independent products.

49. $q_A = 100p_A/p_B{}^{(2/3)}$ and $q_B = 125p_B/p_A{}^{(1/2)}$

50. $q_A = 40 - p_A - 2p_B$ and $q_B = 80 - 3p_A - 5p_B$

51. $q_A = 200 - 2p_A + p_B$ and $q_B = 100 + 0.5p_A - 2.5p_B$

52. $q_A = 500/(p_A\sqrt{p_B})$ and $q_B = 200/(p_B\sqrt[3]{p_A})$

15.3 SECOND-ORDER PARTIAL DERIVATIVES

Just as there are first-, second-, and higher-order derivatives of a function of only one independent variable, functions of more than one independent variable may have first-order partial derivatives, second-order partial derivatives, and so on.

Second-order partial derivatives are simply first partial derivatives of first partial derivatives. For instance, a function $f(x,y)$ may have four second-order partial derivatives as follows. (Note that the second partial derivatives are symbolized in a variety of ways, but don't let the notation trouble you. It is by far the most confusing aspect of second partial derivatives; the derivatives themselves represent very straightforward concepts.)

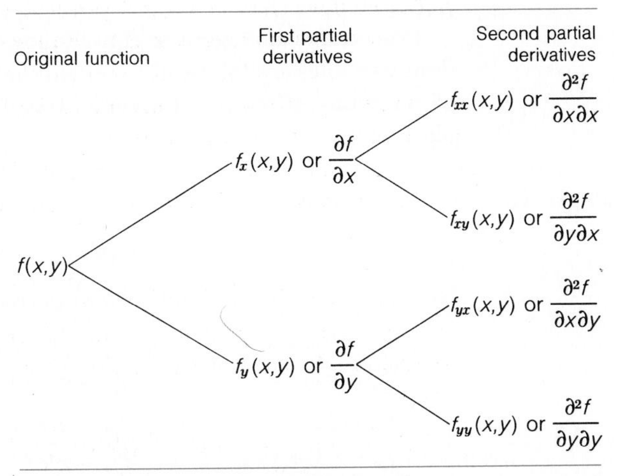

The second partial derivatives f_{xx} and f_{xy} are obtained by differentiating f_x, first with respect to x and then with respect to y. The second partial derivatives f_{yx} and f_{yy} are obtained by differentiating f_y, first with respect to x and then with respect to y.

Example 14 Consider the following function:

$$f(x,y) = x^3 + 27y^2 - xy - 15$$

Its first partial derivatives are

$$f_x(x,y) = 3x^2 - y \qquad \text{and} \qquad f_y(x,y) = 54y - x$$

Two second-order partial derivatives are obtained by differentiating $f_x = 3x^2 - y$, first with respect to x and then with respect to y, as

$$f_{xx}(x,y) = 6x \qquad \text{and} \qquad f_{xy}(x,y) = -1$$

Two additional second-order partial derivatives are obtained by differentiating $f_y = 54y - x$, first with respect to x and then with respect to y, as

$$f_{yx}(x,y) = -1 \qquad \text{and} \qquad f_{yy}(x,y) = 54$$ []

Example 15 The first and second partial derivatives of the function

$$f(x,y) = (x + 4)/y$$

are as follows

$$f_x(x,y) = 1/y \qquad f_y(x,y) = -(x+4)/y^2$$
$$f_{xx}(x,y) = 0 \qquad f_{yx}(x,y) = -1/y^2$$
$$f_{xy}(x,y) = -1/y^2 \qquad f_{yy}(x,y) = 2(x+4)/y^3$$ []

The second-order partial derivatives $f_{xy}(x,y)$ and $f_{yx}(x,y)$ are called CROSS-PARTIAL DERIVATIVES. For a continuous function, $f_{xy}(x,y) = f_{yx}(x,y)$.

Procedures analogous to those followed for functions of two independent variables are followed to obtain the second-order partial derivatives of functions of many independent variables, as the next example will illustrate.

Example 16 The function

$$f(x,y,z) = x^2e^y \ln z$$

has the following first and second partial derivatives:

$f_x = 2xe^y \ln z$	$f_y = x^2e^y \ln z$	$f_z = x^2e^y(1/z)$
$f_{xx} = 2e^y \ln z$	$f_{yx} = 2xe^y \ln z$	$f_{zx} = 2x(e^y/z)$
$f_{xy} = 2xe^y \ln z$	$f_{yy} = x^2e^y \ln z$	$f_{zy} = (x^2/z)e^y$
$f_{xz} = 2xe^y(1/z)$	$f_{yz} = x^2e^y(1/z)$	$f_{zz} = -x^2e^y/z^2$

[]

EXERCISES

Determine the first- and second-order partial derivatives for each of the following functions.

53. $f(x,y) = (2x + 3y)^{(1/2)}$
54. $f(x,y) = e^{(x-3y)}$
55. $f(x,y) = e^{xy}$
56. $f(x,y) = ln(4x + 3y)$
57. $f(x,y) = ln(3x^2 + 2y)$
58. $f(x,y) = 4e^x + 9e^y$
59. $f(x,y) = 2xy^2 - 3e^{x^2y}$
60. $f(x,y) = y\ ln(x + y)$
61. $f(x,y) = (x + 2)/(y + 3)$
62. $f(x,y) = (x + y)^2$

Determine the first- and second-order partial derivatives for each of the following functions.

63. $f(x,y,z) = (2x + 3y + z)^2$
64. $f(x,y,z) = x^2y + xyz + y^2z + z^3$
65. $f(x,y,z) = xe^{yz}$
66. $f(x,y,z) = x^2\sqrt{y} + z^3$
67. $f(x,y,z) = x^2\ ln\ xyz$
68. $f(x,y,z) = (4x^2 + y)/z$
69. $f(x,y,z) = (x + y)^2(y + z)^2$
70. $f(x,y,z) = (x + y)(y + z)(z + x)$

15.4 THEORY OF EXTREMA FOR MULTIVARIATE FUNCTIONS

For functions of more than one independent variable, the tests for extreme points are extensions of the rules given in the one-independent-variable case. However, while the two-independent-variable test can be stated fairly succinctly, if more than two independent variables are involved, the tests involve determinants of matrices and are beyond the scope of the mathematical rigor of this text. We shall thus restrict our discussion to those functions of two independent variables only and to functions which have smooth surfaces, with no edges, breaks, or sharp points.

A relative extremum of a function f of one variable occurs at a if $f(a)$ is either the largest or smallest value of the function in some open interval that contains the point a. We can now give somewhat analogous definitions for relative extrema for bivariate functions.

A function $z = f(x,y)$ is said to have a RELATIVE MAXIMUM at a point (a,b)—that is, where $x = a$ and $y = b$—if there exists in the domain of $z = f(x,y)$ a circular region with (a,b) as its center such that

$$f(a,b) \geq f(x,y)$$

for all (x,y) in the region.

A function $z = f(x,y)$ is said to have a RELATIVE MINIMUM at a point (a,b) if there exists in the domain of $z = f(x,y)$ a circular region with (a,b) as its center such that

$$f(a,b) \leq f(x,y)$$

for all (x,y) in the region.

To say that $z = f(x,y)$ has a relative maximum at (a,b) means geometrically that the point (a,b) on the graph of f is as high as or higher than all points on the surface that are "near" (a,b). In Figure 15.5A, $z = f(x,y)$ has a relative maximum at (a,b).

If $z = f(x,y)$ has a relative maximum at (a,b), as shown in Figure 15.5A, the curve where the plane $y = b$ intersects the surface must have a relative maximum when $x = a$. This means that the slope of the tangent line to the surface, in the x-direction, must be 0 at (a,b). Hence, $f_x(a,b) = 0$. Similarly, if $z = f(x,y)$ has a relative maximum at (a,b), the curve where the plane $x = a$ intersects the surface must have a relative maximum when $y = b$. This means that the slope of the tangent line to the surface, in the y-direction, must be 0 at (a,b). Hence, $f_y(a,b) = 0$.

The same concepts apply for a relative minimum, as indicated in Figure 15.5B.

Hence, a *necessary condition* for $f(a,b)$ to be a relative maximum or a relative minimum is that both $f_x(x,y)$ and $f_y(x,y)$ exist at (a,b) and that $f_x(a,b) = f_y(a,b) = 0$. The point (a,b) is called a CRITICAL (or STATIONARY) POINT on f. This is not to imply, however, that there must be an extremum at a critical point. Just as is the case with functions of only one independent variable, critical points on multivariate functions can represent relative maximums, relative minimums, or neither. In Fig-

Figure 15.5

A Necessary Condition for a Relative Maximum or Relative Minimum is $f_x = f_y = 0$

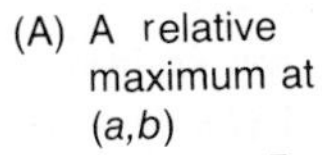
(A) A relative maximum at (a,b)

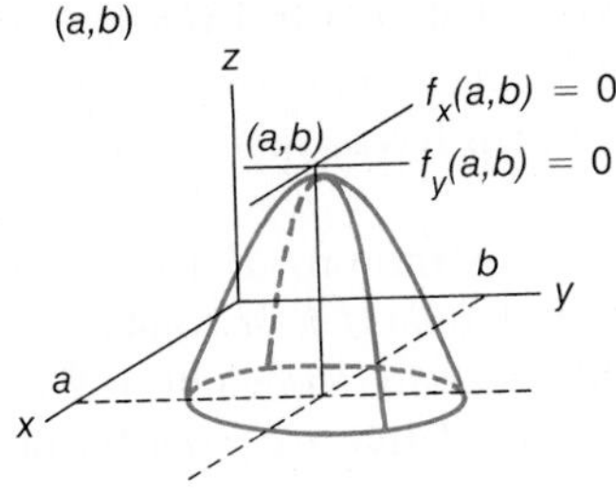

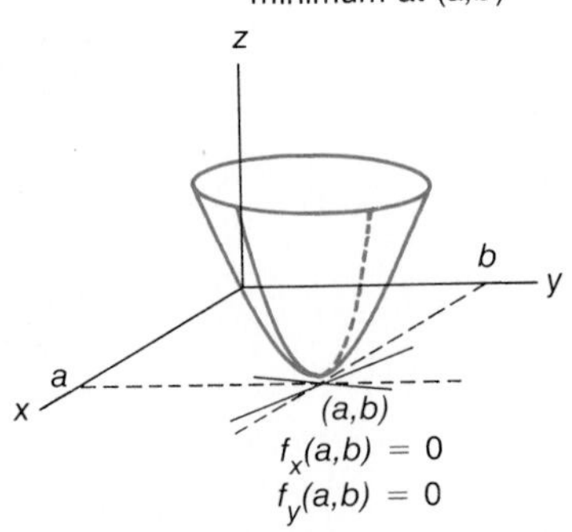

(B) A relative minimum at (a,b)

Figure 15.6

A Critical Point May Represent a SADDLE POINT

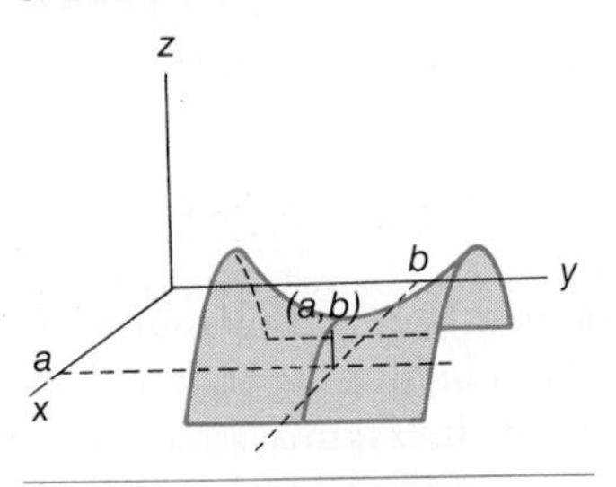

ure 15.6, (a,b) represents a critical point which is neither a maximum nor a minimum but is, instead, termed a SADDLE POINT. (At this point, one of the variables is taking a maximum while the other takes a minimum.)

There exists a second-derivative test, which we shall state without proof, which gives conditions under which a critical point will be a relative maximum or relative minimum.

Second-derivative test for extrema of functions with two independent variables

1. Given $z = f(x,y)$, determine the first partial derivatives, $f_x(x,y)$ and $f_y(x,y)$ and all critical values (a,b); that is, determine all values (a,b) such that $f_x(a,b) = f_y(a,b) = 0$.
2. Determine the set of second partial derivatives, $f_{xx}(x,y)$, $f_{xy}(x,y)$, $f_{yx}(x,y)$, and $f_{yy}(x,y)$. (Note: the cross partial derivatives $f_{xy}(x,y)$ and $f_{yx}(x,y)$ must be equal to one another; otherwise the function is not continuous.)
3. Where (a,b) is a critical point on f, let

$$D = f_{xx}(a,b) \cdot f_{yy}(a,b) - [f_{xy}(a,b)]^2$$

Then

a. *If* $D > 0$ and $f_{xx}(a,b) < 0$, f has a relative maximum at (a,b).
b. *If* $D > 0$ and $f_{xx}(a,b) > 0$, f has a relative minimum at (a,b).
c. *If* $D < 0$, f has neither a relative maximum nor a relative minimum at (a,b).
d. *If* $D = 0$, no conclusion can be drawn; further analysis is required.

Example 17 Examine the function

$$z = f(x,y) = x^2 + y^2 - 4x + 6y$$

for relative maxima or minima by using the second-derivative test.

We begin by finding the first partial derivatives

$$f_x(x,y,) = 2x - 4 \qquad \text{and} \qquad f_y(x,y) = 2y + 6$$

Setting these derivatives equal to 0 and solving for the critical values, we obtain

$$2x - 4 = 0 \qquad \text{and} \qquad 2y + 6 = 0$$
$$x = 2 \qquad\qquad y = -3$$

The critical value of f is $(2,-3)$.

The second partial derivatives are

$$f_{xx}(x,y) = 2 \qquad f_{yx}(x,y) = 0$$
$$f_{xy}(x,y) = 0 \qquad f_{yy}(x,y) = 2$$

Hence, $D = (2)(2) - (0)^2 = 4$.

Because $D > 0$ and $f_{xx} > 0$, a relative minimum occurs on f at $(2,-3)$.

[]

Example 18 Examine the function

$$z = f(x,y) = x^3 + y^2 + 3xy$$

for relative maxima and minima by using the second-derivative test.

We determine that

$$f_x(x,y,) = 3x^2 + 3y \qquad \text{and} \qquad f_y(x,y,) = 2y + 3x$$

Solving $f_x(x,y,) = f_x(x,y,) = 0$, we obtain the critical values $(0,0)$ and $(3/2,-9/4)$ as follows:

$$3x^2 + 3y = 0 \qquad \text{and} \qquad 2y + 3x = 0$$
$$y = -3x/2$$

$$3x^2 + 3(-3x/2) = 0$$
$$3x^2 - (9x/2) = 0$$
$$x(3x - 9/2) = 0$$
$$x^* = 0 \text{ and } x^* = 3/2$$

When $x^* = 0$, $y^* = -3/2(0) = 0$

When $x^* = 3/2$, $y^* = (-3/2)(3/2) = -9/4$

Next, we determine the second derivatives to be

$$f_{xx}(x,y,) = 6x \qquad f_{yx}(x,y,) = 3$$
$$f_{xy}(x,y,) = 3 \qquad f_{yy}(x,y,) = 2$$

At $(0,0)$, $f_{xx}(0,0) = 0$, $f_{yy}(0,0) = 2$, and $f_{xy}(0,0) = 3$. Thus, for this critical point $D = (0)(2) - (3)^2 = -9$. Because $(D = -9) < 0$, f has neither a relative maximum nor a relative minimum at $(0,0)$.

At the critical value $(3/2,-9/4)$, $f_{xx}(3/2,-9/4) = 9$, $f_{yy}(3/2,-9/4) = 2$ and $f_{xy}(3/2,-9/4) = 3$. Hence, for this critical point $D = (9)(2) - (3)^2 = 9$. Because $(D = 9) > 0$ and $(f_{xx} = 9) > 0$, f has a relative minimum at $(3/2,-9/4)$. The value of the function here is

$$z = f(3/2,-9/4) = (3/2)^3 + (-9/4)^2 + 3(3/2)(-9/4) = -27/16$$

[]

Example 19 The total production cost of a product is given by

$$f(x,y) = x^2 + y^2 - 5x - 9y - xy + 90$$

where

$f(x,y)$ = Total cost, in thousands of dollars
x = Number of labor hours used (in hundreds of hours)
y = Number of pounds of raw material used (in hundreds of pounds).

The shop superintendent wishes to determine how many labor hours and how many pounds of raw material should be used in order to minimize total cost.

The first partial derivatives are

$$f_x(x,y,) = 2x - 5 - y$$
$$f_x(x,y,) = 2y - 9 - x$$

The values of x and y where the first derivatives equal zero are determined by solving simultaneously the equations

$$2x - y = 5$$
$$-x + 2y = 9$$

This computation yields $x^* = 19/3$ and $y^* = 23/3$.

Next, the second-order partial derivatives are found to be

$$f_{xx}(x,y,) = 2 \qquad f_{yx}(x,y,) = -1$$
$$f_{xy}(x,y,) = -1 \qquad f_{yy}(x,y,) = 2$$

Thus, at the critical value (19/3,23/3)

$$D = (2)(2) - (-1)^2 = 3$$

Because $(D = 3) > 0$ and $(f_{xx} = 2) > 0$, f has a relative minimum at (19/3,23/3).

Total cost will be at a minimum when 633 labor hours and 767 pounds of raw material are used. Total cost with this production strategy is

$$f(19/3,23/3) = (19/3)^2 + (23/3)^2 - 5(19/3) - 9(23/3) - (19/3)(23/3) + 90$$
$$= 39.677 \text{ or } \$39{,}667$$

[]

Example 20 **An optimization problem involving cost and selling price** Giant Industries operates two manufacturing facilities, both of which are designed for the production of the same product, a small electronic calculator. Cost of production in the two plants is not the same, however. Through a careful analysis of historical cost-volume data, analysts have determined that the cost function for plant A can be appropriately expressed as

$$g(A) = 154.4 - 0.425A + 6.428A^2$$

where

A = Total number of units produced in plant A (in thousands)
$g(A)$ = Total production cost for plant A (in thousands of dollars)

The cost function for plant B can be expressed as

$$h(B) = 202 + 14.5B + 1.25B^2$$

where

B = Total number of units produced in plant B (in thousands)
$h(B)$ = Total production cost for plant B (in thousands of dollars)

Giant Industries wishes to allocate production to the two facilities in such a way as to minimize total cost of production, given some required level of production Q. The total cost function is

$$\begin{aligned} C(A,B) &= [154.4 - 0.425A + 6.428A^2] + [202 + 14.5B + 1.25B^2] \\ &= 356.4 - 0.425A + 6.428A^2 + 14.5B + 1.25B^2 \end{aligned}$$

where

$C(A,B)$ = Total cost of production for both plants (in thousands of dollars).

The total number of units of product manufactured is fixed at Q, where $Q = A + B$. Since Q will be a constant, we can simplify the total cost function by substituting $A = Q - B$ as

$$\begin{aligned} C(Q,B) &= 356.4 - 0.425(Q - B) + 6.428(Q - B)^2 + 14.5B + 1.25B^2 \\ &= 356.4 - 0.425Q + 14.925B + 7.678B^2 - 12.856QB + 6.428Q^2 \end{aligned}$$

The price at which the calculator manufactured by Giant Industries can be sold is dependent upon the number of units offered for sale. Market analysts believe that the functional relationship between price and quantity sold can be expressed as

$$Q = 18 - 0.1P$$

or, alternatively, as

$$P = 180 - 10Q$$

The revenue function under these conditions will be

$$\begin{aligned} \text{Revenue} &= \text{Price} \times \text{Quantity} \\ &= (180 - 10Q)Q \\ &= 180Q - 10Q^2 \end{aligned}$$

Thus, profit expressed as a function of quantity sold and allocation of production to the two plants is

$$\begin{aligned} \text{Profit} &= \text{Revenue} - \text{Cost} \\ f(Q,B) &= [180Q - 10Q^2] - [356.4 - 0.425Q + 14.925B + 7.678B^2 - \\ &\qquad 12.856QB + 6.428Q^2] \\ &= -356.4 - 16.428Q^2 + 180.425Q + 12.856Q - 14.925B - 7.678B^2 \end{aligned}$$

The maximum point on the profit function can be found by taking the first partial derivatives, as

$$f_B(Q,B) = 12.856Q - 14.925 - 15.356B$$
$$f_Q(Q,B) = -32.856Q + 180.425 + 12.856B$$

These first partial derivatives must be set equal to zero and solved simultaneously to obtain the critical values of B and Q, as

$$12.856Q - 14.925 - 15.356B = 0$$
$$B = (12.856Q - 14.925)/15.356$$

Then,

$$-32.856Q + 180.425 + 12.856[(12.856Q - 14.925)/15.356] = 0$$
$$Q^* = 7.601$$

and

$$B^* = \frac{12.856(7.601) - 14.925}{15.356}$$
$$= 5.392$$

That this critical point does represent a maximum profit point should be verified by analyzing the second-order partial derivative conditions, as

$$f_{BB}(Q,B) = -15.356 \qquad f_{QB}(Q,B) = 12.856$$
$$f_{BQ}(Q,B) = 12.856 \qquad f_{QQ}(Q,B) = -32.856$$

Then

$$[D = (-15.356)(-32.856) - (12.856)] > 0$$

while $f_{BB}(Q,B) < 0$, indicating that the point does indeed represent a maximum.

Thus, in order to achieve the maximum possible profit, Giant Industries should manufacture of total of 7,601 calculators. The production at plant B should be 5,392 units, leaving 2,209 units to be produced in plant A.

The selling price when 7,601 units are offered for sale will be

$$P = 180 - 10(7.601) = \$103.99$$

Revenue realized will be (\$103.99)(7,601) = \$790,428. Cost with the optimal allocation of production will total

$$C(7.601, 5.392) = 356.4 - 0.425(7.601) + 14.925(5.392) + 7.678(5.392)^2$$
$$- 12.856(7.601)(5.392) + 6.428(7.601)^2$$
$$= \$501,354$$

The resulting profit will be \$289.074. []

EXERCISES

Find the critical values of the function. Determine, by the second-derivative test, whether these points correspond to a relative maximum, a relative minimum, or neither (or whether the test fails).

71. $z = f(x,y) = x^2 - 4x + y^2 - 6y + 8$
72. $z = f(x,y) = xy + 3x - 2y + 5$
73. $z = f(x,y) = x^2 + y^2 + xy - 8x + 4$
74. $z = f(x,y) = -x^2 + 4x - 6y - y^2 + 15$
75. $z = f(x,y) = 2x^2 - 2xy + 3y^2 - 5x - 6y + 10$
76. $z = f(x,y) = -3x^2 - 2xy + 10x - 15y$
77. $z = f(x,y) = 2x + xy + y^2 + 4y + 6$
78. $z = f(x,y) = x^2y - xy^2$
79. $z = f(x,y) = (x - 1)^2 + y^2$
80. $z = f(x,y) = x^3 + x^2 - xy + y^2$
81. A company produces two products, X and Y. How many of each product should be produced to maximize profit if the profit function is given by

$$P = f(x,y) = -3x^2 + 4xy + 100x - 5y^2 - 30y + 1{,}500$$

where x represents the number of units of product X produced and y represents the number of units of product Y produced. What is the maximum profit?
82. A company finds that its profit (in thousands of dollars) is a function of the amount x (in thousands of dollars) spent on research and development and the amount y (in thousands of dollars) spent on advertising, as

$$P = f(x,y) = 100 - x^2 - 2y^2 + 8x - 3y + 3xy$$

How much should be spent on research and development and how much on advertising in order to maximize profit? What is the maximum profit?
83. A company produces two types of desk lamps. If the revenue and cost functions are

$$R = f(x,y) = 20x + 30y$$

and

$$C = g(x,y) = x^2 - 12xy + 6y^2 + 14x - 18y + 500$$

where x is the number of standard models sold and y is the number of special models sold, how many of each should be produced and sold to maximize profits? What is the maximum profit?
84. A company markets two types of cat food, A and B, and finds that the demand for each depends not only on its own price but also on the price of the other. The joint-demand functions have been determined to be

$$Q_A = 500(-p_A + p_B)$$

and

$$O_B = 400(80 + p_A - 2p_B)$$

where Q_A and Q_B are the quantities of A and B demanded at price p_A for A and p_B for B.

The cost of producing the foods average a constant 50 cents and 75 cents per unit for A and B, respectively.

Determine the selling prices that will maximize the manufacturer's profits. What will be the maximum profit?

15.5 CONSTRAINED OPTIMIZATION AND LAGRANGE MULTIPLIERS

We shall now consider procedures used for locating relative maxima and minima for a multivariate function on which certain constraints are imposed. The need for such procedures arises in many decision-making contexts. Decision makers do not normally have available for their use relatively unlimited resources; or perhaps there may be physical or legal restrictions on the manner in which variables can be used together. For instance, a company may wish to maximize sales resulting from the use of several advertising media while keeping total advertising costs within a budgeted amount. A production engineer may wish to minimize frequency of machine breakdowns while keeping preventive maintenance costs at a certain level. At any rate, in many maximization or minimization decisions, there are constraints which impose restrictions on the values which the controllable variables are allowed to assume. Optimization, in these cases, becomes a matter of finding the largest or smallest value of the function that can be achieved within the permitted range of values of the variables. These are problems of CONSTRAINED OPTIMIZATION. (You will recall that we dealt with constrained optimization when we studied the linear programming model. However, that model requires that all relevant relationships between the variables be linear. With the procedure that we introduce at this point, the requirement of linearity will not be necessary.)

An Illustration: No Constraints

The cost of production for a process can be modeled by

$$f(L,M) = 200 - 0.2\,LM + L^2 - 10M + 2M^2$$

where

$f(L,M)$ = Total cost of production, in thousands of dollars
L = Number of labor units used
M = Number of machine units used

The firm would like to determine that combination of labor units and machine units which would minimize total cost of production.

Cost is given as a function of two variables, labor units and machine units. Hence, partial derivatives are used to find the minimum point on the cost function, as follows:

1. Find the two first-order partial derivatives, set them equal to zero, and solve for the critical values of L and M.

$$f_L(L,M) = -0.2M + 2L = 0 \qquad f_M(L,M) = -0.2L - 10 + 4M = 0$$
$$L = 0.1M \qquad M^* = 2.513$$
$$L^* = 0.1(2.513) = 0.251$$

2. Find the second-order partial derivatives and determine whether or not the critical point is a minimum point on the function.

$$f_{LL}(L,M) = 2 \qquad f_{ML}(L,M) = -0.2$$
$$f_{LM}(L,M) = -0.2 \qquad f_{MM}(L,M) = 4$$

$(D = (2)(4) - (-0.2)^2] > 0$ and $[f_{LL}(L,M) = 2] > 0$; hence, the point is a minimum point.

The cost of the process would be minimized by using 2.513 machine units and 0.251 labor units. The cost with this combination of resources would be $187,437.

A Constraint Exists

However, it is the case that the production process is of such a nature that one labor unit is required for each four machine units. The foregoing solution would not be possible, since the optimum production combination calls for a labor–machine units ratio of 1 to 10.

The constraint on usage of labor and machines specifies that

$$4L = M$$

and the problem becomes one of finding the optimum production combination subject to this restriction.

Substituting the constraint relationship into the original cost function, we obtain

$$C_c = 200 - 0.2L(4L) + L^2 - 10(4L) + 2(4L)^2$$
$$= 200 - 40L + 32.2L^2$$

where C_c indicates that this is a "constrained" cost function. Now, the cost of production is expressed as a function of only one variable, labor units. (Actually, because of their interrelationship, both variables, machine units and labor units, are not truly controllable. When the value of one is determined, the value of the other is automatically fixed.)

Applying the derivative test to the constrained cost function, we compute

$$C_c = f(L) = 200 - 40L + 32.2L^2$$
$$f'(L) = -40 + 64.4L = 0$$

$L^* = 0.621$ is a stationary value on the cost curve. $f''(L) = 64.4$ indicates that the value represents a minimum.

The machine units used will be

$$M = 4(0.621) = 2.484$$

Cost under these conditions will total $187,578 as compared to the unconstrained cost of $187,437. A constraint usually will (but it need not always) change the optimal value of the independent variables and either increase a minimum or decrease a maximum value for the objective function.

Using Lagrange Multipliers

Locating constrained optima by using the constraint to express one of the variables in the original function in terms of the other variable is a widely used technique and often an easy one to implement. However, functions to be optimized and the imposed constraint relationships are not always so uncomplicated as the ones used in the preceding illustration. Therefore, solution by the method of substitution is not always quite so straightforward to effect. And so enters into the picture the technique of LAGRANGE MULTIPLIERS, providing an alternative method for locating extreme values of a function which is subject to one or more constraints.

Given a function to be optimized $f(x,y)$ subject to the constraining condition $g(x,y) = 0$, if we attempted to solve for an optimum directly through the calculus by differentiating partially with respect to each independent variable, setting each of these derivatives equal to zero, and so on, we would have a set of three equations in only two unknowns, x and y, for in addition to the partial-derivatives-equal-to-zero conditions, $f_x(x,y) = 0$ and $f_y(x,y) = 0$, we would also have a constraint condition that must be satisfied, $g(x,y) = 0$. We have seen that whenever a problem is defined by more equations than unknowns, it is said to be an overdefined system and, under normal conditions, without a unique solution.

To overcome this obstacle, we introduce artificial variables (as many as there are constraints). The artificial variables are called Lagrange multipliers and are usually denoted by the Greek symbol λ (read *"lambda"*). The procedure for using the Lagrange multiplier goes like this: The constraint function, $g(x,y)$, is multiplied by the Lagrange multiplier, λ, and the product, $\lambda g(x,y)$, is added to the original function to be optimized, yielding a new function

$$h(x,y,\lambda) = f(x,y) + \lambda g(x,y)$$

It can be shown that if (x^*,y^*) is a critical value of f subject to $g(x,y) = 0$, there exists a value of λ, say λ^*, such that (x^*,y^*,λ^*) is a critical value of h. And, conversely, if (x^*,y^*,λ^*) is a critical value of h, then (x^*,y^*) is a critical value of f subject to the constraint. Thus, to find critical points of f subject to $g(x,y) = 0$, we find critical points on h. These critical points of h are obtained by partially differenting h with respect to each of the three variables, x, y, and λ. Then the system of partial derivatives is set equal to zero and solved simultaneously for the critical values of the variables. Once we obtain a critical point (x^*,y^*,λ^*) of h, we can conclude that (x^*,y^*) is a critical point of f subject to the constraint.

Returning to the previous illustration, we would begin by rewriting the constraint in the format $g(x,y) = 0$, as $4L - M = 0$. Next, we would multiply the expression by the artificial variable λ, as $\lambda(4L - M)$. The result is added to the function which is to be optimized; that is,

$$C_\lambda = h(L,M,\lambda) = 200 - 0.2LM + L^2 - 10M + 2M^2 + \lambda(4L - M)$$

where the subscript indicates that this is a "Lagrangian" expression.

We now have a function of the two decision variables and the artificial variable. Note that because $4L - M = 0$, then $\lambda(4L - M)$ is also a zero-valued expression, and the addition of this expression to the original function does not change the value of the original function.

When we obtain the first partial derivatives, we have a set of three equations in three unknowns:

$$h_L = -0.2M + 2L + 4\lambda$$
$$h_M = -0.2L - 10 + 4M - \lambda$$
$$h_\lambda = 4L - M$$

Solving these equations simultaneously, we obtain $M = 2.484$, $L = 0.621$, and $\lambda = -0.186$. (Note that the values for machine units and labor units are those we found previously by the substitution routine.)

The method of Lagrange multipliers does not directly indicate whether the critical point represents a relative maximum, a relative minimum, or neither of these. We have previously determined that this particular critical point represents a minimum point on the cost curve. Nonetheless, let us develop a second derivative test that can be used directly with the Lagrangian function. The critical value of λ, determined when the first partial derivatives are set equal to zero, should be substituted into the other first partial derivatives, as, in the current illustration,

$$h_L = -0.2M + 2L + 4(-0.186)$$
$$h_M = -0.2L - 10 + 2M - 0.186$$

As a result of this substitution, we have two partial derivatives in two unknowns, and the test outlined previously can be used.

Note that with the optimal values of L and M the constraint $4L = M$ is satisfied. When we set the partial derivative with respect to λ equal to zero, we automatically guarantee that this restriction will be fulfilled. This explains why we multiplied λ times the constraint equation. Since the expression $\lambda(4L - M)$ is equal to zero, the Lagrangian expression for cost is exactly equivalent to the original cost function, and the solution to the Lagrangian problem is necessarily the solution to the original problem as well, subject to meeting the restriction.

But there is additional information available from the Lagrangian multiplier. Its solution value has special significance. We know that the partial derivative of a function is the slope of the function with respect to a particular variable. It is the change in the value of the function associated with an infinitely small change in the value of the variable for which the partial derivative is obtained. If we denote the general model having two decision variables and one equality constraint as

$$\text{Maximize or minimize } f(x,y)$$
$$\text{subject to } g(x,y) = b$$

The value of the Lagrangian multiplier is the negative of the rate of change of the objective function with respect to a change in b; that is,

$$\lambda = -f_b \qquad \text{or} \qquad -\lambda = f_b$$

Thus, λ indicates the approximate amount by which the value of the objective function would be increased if b were decreased or the amount by which the value of the objective function would be decreased if b were increased. Hence, $\lambda = -0.186$ indicates that for a very small change in the constraint from its present relationship, the cost will change by 0.186 units. If the restriction is made more binding, the cost will increase; if the restriction is made less binding, the cost will decrease. It must be emphasized that the Lagrangiam multiplier represents the marginal cost associated with the constraint and that this marginal value of 0.186 applies at this optimum point and is not necessarily applicable over a wide range of values for the variables.

When $\lambda = 0$, the variables are not limited by the constraint.

More than One Constraint

The method of Lagrange multipliers is not restricted to problems involving a single constraint. When more than one constraint is imposed, the function to be optimized is adjusted in this manner:

$$h(x,y,\lambda_1,\lambda_2) = f(x,y) + \lambda_1 g_1(x,y) + \lambda_2 g_2(x,y)$$

and the procedure of partial derivatives is followed as before.

Example 21

Find the critical points for $f(x,y,z) = x^2 + 2y - z^2$ subject to the constraints $2x = y$ and $y + z = 0$.

We use two Lagrange multipliers, λ_1 and λ_2, and set

$$\begin{aligned} h(x,y,z,\lambda_1,\lambda_2) &= (x^2 + 2y - z^2) + \lambda_1(2x - y) + \lambda_2(y + z) \\ &= x^2 + 2y - z^2 + 2\lambda_1 x - \lambda_1 y + \lambda_2 y + \lambda_2 z \end{aligned}$$

Then the first-order partial derivatives are determined to be

$$\begin{aligned} h_x &= 2x + 2\lambda_1 \\ h_y &= 2 - \lambda_1 + \lambda_2 \\ h_z &= -2x + \lambda_2 \\ h_{\lambda_1} &= 2x - y \\ h_{\lambda_2} &= y + z. \end{aligned}$$

Setting these equal to zero and solving for the critical values of the variables, we obtain

$$x = 2/3 \qquad y = 4/3 \qquad z = -4/3$$

and

$$\lambda_1 = -2/3 \qquad \text{and} \qquad \lambda_2 = -8/3$$ []

EXERCISES

Find, by the method of Lagrange multipliers, the critical points of the functions subject to given constraints. If possible, determine whether the critical point represents a relative maximum, a relative minimum, or neither.

85. $f(x,y) = x^2 + 4y^2 + 6$, subject to $x - 4y = 10$
86. $f(x,y) = 2x^2 + 5y^2 + 9$, subject to $x - 2y = 3$
87. $f(x,y) = x^2 + y^2$, subject to $x + y = 5$
88. $f(x,y) = 3x^2 + y^2 + 3xy - 60x - 32y + 600$, subject to $x - y = 8$
89. $f(x,y) = 3x^2 + xy + y^2 - 15x - 7y$, subject to $x = y$
90. $f(x,y) = x^2 + xy + y^2 + 150$, subject to $x + y = 10$
91. $f(x,y) = 6xy - x^2$, subject to $x + y = 2$
92. A company manufactures two products, X and Y, whose joint-cost function is given by

$$f(x,y) = 2x^2 - 6xy + 5y^2 + 10x - 28y$$

where

x = Number of units of product X produced
y = Number of units of product Y produced
$f(x,y)$ = Total cost of production in thousands of dollars

Determine the number of units of each product which should be produced in order to minimize cost of production, given that a total of 12 units must be produced.

93. A steel manufacturer produces two grades of steel. The firm's cost and revenue functions are

$$C = 0.02x^2 + 600x + y^2 - 250y - 0.5xy + 500$$

and

$$R = 3{,}000x - 0.15x^2 + 1{,}000y - y^2 + 0.5xy$$

where

x = number of tons of grade A steel produced
y = number of tons of grade B steel produced

Find the production, in tons of grade A and B, of steel which would maximize the manufacturer's profit, given that four times as many tons of A as of B must be produced.

C·H·A·P·T·E·R 16 The Basic Concepts of Probability Theory

Because we inhabit a world wherein we cannot foresee with certainty what tomorrow—or the next five minutes, for that matter—may bring, we have need for a mechanism for measuring and analyzing the uncertainties associated with possible future events. PROBABILITY THEORY provides such a mechanism.

Each of us has an intuitive feeling for the notion of probability. Words such as *probable, possible, likely, chances,* and *odds* pervade our everyday conversation. We think and speak of the likelihood that we will be given a traffic ticket if we exceed the 55-mile-per-hour speed limit, that Barnard will be a candidate for the city council this year, that the bus drivers will go out on strike, that Clemson will be undefeated at the end of the season, or that Aunt Maudie will come across with a substantial monetary gift at Christmastime. We need now to develop a more formal approach to the matter of quantifying uncertainties.

From a mathematical standpoint, the theory of probability has been rigorously developed and encompasses a very substantial body of knowledge. Actually, the roots of this branch of mathematics are buried deep in games of chance. In the early 1600s Chevalier de Méré, a French nobleman who was also an inveterate gambler, became fascinated with the odds in various gambling situations. Puzzled by the many seeming incongruities which plagued his gambling experiences, de Méré turned to his friend, the renowned mathematician Blaise Pascal (1623–62) for some explanations. Pascal soon became equally intrigued with the whole questions of odds and equivalence between different gambling circumstances. He, in turn, discussed the problems with another well-known mathematician, Pierre de Fermat (1601–65). These discussions between Pascal and Fermat and others spawned an entire branch of mathematics, the theory of probability.

16.1 THE CLASSICAL RANDOM EXPERIMENT

As we have noted, the theory of probability was first developed in a rigorous way by scholars studying games of chance. Accordingly, the traditional mathematician or statistician associates probabilities with what he or she calls "the outcome of a random experiment."

A RANDOM EXPERIMENT is a WELL-DEFINED, THEORETICALLY REPEATABLE, COURSE OF ACTION THAT MAY RESULT IN ANY ONE OF TWO OR MORE POSSIBLE OUTCOMES. Moreover, the specific outcome of the experiment is affected by uncontrollable "chance" factors to such an extent that it cannot be predicted in advance with certainty. Even though the experiment is repeated under essentially constant conditions, different outcomes may occur on different "trials" of the experiment. And, because of the element of chance operating on the process, there is no way to know, before the completion of the particular trial, just which one of the possible outcomes will be experienced.

Example 1 The classical example of a random experiment is the tossing of a fair coin. This experiment can be repeated under essentially unchanging conditions. If the rather unstable "on rim" position is regarded as impossible, and the fact that the coin may roll away and become lost is ignored, there are two possible outcomes—heads or tails—only one of which can occur as the result of a single toss of the coin.

Even though we know what the possible outcomes are, we cannot determine analytically—from the initial position of the coin, from the velocity of the toss, and so on—which of these possible outcomes will result on a specific toss of the coin. []

Example 2 Another of the classic examples of a random experiment is the tossing of a six-sided die. The outcome of the experiment may be considered to be the number of dots on the upturned side of the die when it comes to rest. The die may be tossed over and over again, with different outcomes—one dot on the upturned side, two dots on the upturned side, three dots, and so on—being experienced on different tosses. And we cannot determine, with a "fair" die, which of these possible outcomes will be experienced on a particular toss until after the completion of that toss. []

Still another of the classic examples of a random experiment involves selecting a card *at random* from a well-shuffled deck of playing cards. The term RANDOM SELECTION, or SELECTION AT RANDOM, is used in connection with many random experiments. It describes the selection of elements from a collection in such a way that every element in the collection has the same chance of being selected.

Example 3 A standard deck of playing cards is made up of four suits: hearts, diamonds, spades, and clubs. Within each suit there are 13 cards: 1 ace, 1 king, 1 queen, 1 jack, and 1 card each numbered 2 through 10. All hearts and diamonds are red cards; all clubs and spades are black cards.

Although we know the composition of the deck, we do not know—with a fair deal from a well-shuffled deck—just exactly which of the 52 cards will be obtained on any specific selection. []

Although the terminology *well-defined* and *experiment* may leave the impression that a random experiment is always staged and is controlled step-by-step by a researcher, this is not always the case. Any happening, real or hypothesized, with a known set of possible outcomes, but whose precise outcome is unknown in advance, might be considered a random experiment. Nor are these processes restricted to games of chance. All of the following might be characterized by the requisite uncertainties so that they are, in essence, random experiments.

Sales of a given product during a one-day period.

Passage of automobiles through a given toll station.

Arrival of customers at a service desk.

Defects occurring in a manufactured article.

Durability (lifetime) of flashlight batteries.

Output (in units) from an assembly line.

Sales calls on potential customers.

The terms RANDOM PROCESS and STOCHASTIC PROCESS are often used to refer to these types of processes which have multiple possible outcomes and which are characterized by uncertainties as to which outcome will actually occur. Probabilities are always associated with the outcomes of random processes.

16.2 THE SAMPLE SPACE AND ITS SAMPLE POINTS

A TRIAL of a random experiment is a "once through" of the process, whatever that involves. The OUTCOME of a random experiment is the result that is actually experienced on a trial of that experiment. The SAMPLE SPACE of a random experiment, ordinarily denoted S, is the set whose elements constitute all possible outcomes of the experiment. Each element in the sample space corresponds to exactly one possible outcome, and each possible outcome corresponds to exactly one element in the sample space. The elements in the sample space are termed ELEMENTARY OUTCOMES or SAMPLE POINTS.

For the classical experiment of tossing a fair coin, the sample space is ordinarily designated

$$S = \{\text{heads, tails}\} \qquad \text{or} \qquad S = \{\text{H,T}\}$$

If the situation where uncertainties prevail involves the number of defective washers in a carton containing two dozen washers, the sample space might be designated

$$S = \{0, 1, 2, \ldots, 24\}$$

where 24 is the maximum possible number of defective washers in the carton. If the stochastic procedure involves selecting items from a production process, inspecting them, and noting their quality, the sample space might be written

$$S = \{\text{defective, nondefective}\}$$

Other examples of random processes and possible lists of outcomes are given in Table 16.1.

The elementary outcomes of a sample space have the following properties:

1. They are COMPLETELY EXHAUSTIVE in that each performance of the experiment must result in one of the outcomes listed.
2. They are MUTUALLY EXCLUSIVE in that it is impossible for more than one elementary outcome to occur on the same trial of the experiment.

Table 16.1

Examples of Random Processes and Their Sample Spaces

Random process	List of possible outcomes
Rolling a die and noting the number of dots on the upturned face when it comes to rest	$S = \{1, 2, 3, 4, 5, 6\}$
Contacting a potential customer and noting whether or not a sale is made	$S = \{$sale, no sale$\}$
Asking diners whether they wish to sit in the restaurant's smoking section (S) or in the no-smoking section (N) and recording the responses	$S = \{S,N\}$
Selecting an employee's personnel file and noting the number of years he or she has worked for the firm (rounded to nearest year). (The firm has been in business for only 16 years.)	$S = \{0, 1, 2, \ldots, 16\}$
Tossing two coins, a nickel and a dime, and noting the upturned faces when they come to rest	$S = \{HH, HT, TH, TT\}$
Tossing a coin four times and counting the number of heads	$S = \{0, 1, 2, 3, 4\}$
Observing whether a patron entering a fast-food establishment orders a hamburger (H), french fries (F), both (B), or neither (N)	$S = \{H, F, B, N\}$
Tossing a coin and then rolling a die and noting the upturned side of the coin along with the number of dots on the upturned side of the die	$S = \{(H,1), (H,2), (H,3), (H,4), (H,5), (H,6), (T,1), (T,2), (T,3), (T,4), (T,5), (T,6)\}$
Noting the daily closing quotation for XYZ Amalgamated common stock and recording whether it is higher (H), lower (L), or the same (S) as the previous days closing quotation	$S = \{H, L, S\}$

Note that the terminology is *a* sample space rather than *the* sample space of an experiment, since the outcomes of an experiment might be described in more than one way. The same experiment may be performed with interest centered on different aspects of its results. Because the sample space actually defined depends upon the investigator's interest in the possible outcomes, the list of outcomes in one case could be different from the list of outcomes in another case, even though the same procedure is followed in performing the experiment. For example, an experiment may involve tossing two coins and observing the upturned sides when the coins come to rest. The researcher may be concerned with the faces on the two coins and define a sample space as $S = \{HH, HT, TH, TT\}$. Or, the researcher may be concerned with whether or not the two faces match (M) or do not match (D) and define a sample space accordingly as $S = \{M, D\}$. Or, the researcher may be concerned with the number of heads observed, so that an appropriate sample space is $S = \{0, 1, 2\}$.

Although the appropriate list of outcomes is dictated to a large extent by the decision to be made, good subjective judgment is often required for the specification of a sample space, especially where the amount of detail included in the listing of outcomes is concerned. Note on the sample spaces previously given that the highly improbable "on edge" position for the coin is not ordinarily included as a sample point, and that the quality of items from the production process might have been described in many ways other than "defective-nondefective." Excessive detail serves only to detract from the model's usefulness; yet detail must be sufficient for the decision at hand.

The application of probability concepts begins with the definition of an appropriate sample space.

EXERCISES

Describe an appropriate space for each of the following random experiments.

1. A card is selected from a standard deck of playing cards, and its color—red (R) or black (B)—is noted.
2. A card is selected from a standard deck of playing cards, and its suit—hearts (H), diamonds (D), clubs (C), or spades (S)—is noted.
3. A card is selected from a standard deck of playing cards, and the researcher observes whether the card is a "face card" (F) or not (N).
4. Two cards are selected from a standard deck of playing cards, and the researcher notes whether they are of the same color.
5. Two cards are selected from a standard deck of playing cards, and the researcher notes the color of each card.
6. A Geiger counter is held near an ore sample, and an observer notes whether or not the counter registers any radioactivity.
7. A Geiger counter is held near an ore sample, and the observer notes the number of clicks within a 10-second period.
8. A standard deck of playing cards is shuffled and the cards dealt one at a time until the first red card is dealt. The number of cards dealt prior to the first red card is observed.
9. The numbers 0, 1, 2, 3 are written on separate cards, placed in a box, and mixed. One card is selected, and the number written on the card is noted.
10. The letters A, B, C are written on three separate cards, placed in a box, and mixed. Two cards are drawn, without replacement, and the letters on the first and second cards drawn are noted.

11. A single sock is selected from a box containing red socks, yellow socks, purple socks, and green socks, and the color of the sock is noted.
12. Two socks are selected from a box containing red socks, yellow socks, and purple socks and, if they match, the color of the pair is noted. If the two socks do not match, only this fact is noted.
13. A die is rolled, and then a coin is tossed. An observer notes whether the number of dots on the die is odd or even and the upturned face of the coin.
14. A pair of dice is rolled, and the number of dots on the upturned side of each die is recorded.
15. Three coins are tossed, and the upturned face on each is noted.
16. A coin is tossed until either a head appears or five tosses have been made, and the upturned face on the coin on each toss is recorded.
17. In a public opinion poll of consumer habits, a respondent is asked whether he or she basically agrees (A), basically disagrees (D), or has no opinion (N), with each of two questions, and the responses are recorded.
18. In a taste test, a housewife is asked to rank three brands of peanut butter, X, Y, and Z, by order of preference.
19. A shipment of five generators is examined, and the number of defective generators is noted.
20. The number of customers arriving at a bank teller's window within a five-minute period is observed.
21. A building inspector visits each retail store in Bristol and tabulates the number of building regulations that have been violated.
22. Each person entering a national park is asked the question; "In which of the 50 states do you legally reside?" and the answer is recorded.
23. A field-goal kicker attempts a field goal, and an observer records whether the attempt is successful (S) or not (N).
24. A spool of cable is weighed, and the weight, to the nearest half pound, is recorded. (Note: The maximum possible weight is 6.5 pounds.)

16.3 AXIOMS FOR ASSIGNING PROBABILITIES TO OUTCOMES

Probabilities are, initially, associated with the individual points in the sample space. These probabilities are real numbers that describe the likelihood that the specified outcome will take place as a result of a trial of the experiment. Probabilities are assigned to the sample points following these axioms:

Axiom 1 *The probability assigned to each sample point is a real number between zero and one, inclusive; that is,*

$$0 \leq P(O_i) \leq 1$$

where $P(O_i)$ *is read "the probability of outcome* O_i*."*

Axiom 2 *The sum of the probabilities assigned to the* n *mutually exclusive and completely exhaustive sample points is unity; that is,*

$$\sum_{i=1}^{n} P(O_i) = 1$$

The first requirement of probability assignment is that no probability value can be less than zero or greater than one. If an outcome is certain to occur on each trial of the experiment, its probability is one; if it can never occur, its probability is zero. The second requirement is that the sum of all the probability values associated with the points in the sample space must equal one.

16.4 PHILOSOPHICAL APPROACHES TO PROBABILITY ASSIGNMENT

The two mathematical axioms of probability assignment must be adhered to consistently. Nonetheless, the approach to their application will vary from one type of random process to another. There are, indeed, three prevailing philosophical views of probability assignment. These we may term the CLASSICAL (or AXIOMATIC) VIEW, the RELATIVE-FREQUENCY (or EMPIRICAL) VIEW, and the SUBJECTIVE (or PERSONALISTIC) VIEW.

The Classical (or Axiomatic) Point of View

The classical concept of probability assignment assumes that each of the n possible outcomes of a random experiment is *equally likely* to occur; thus, each of these outcomes has a probability of occurrence of

$$P(O_i) = 1/n \tag{16.1}$$

This rationale leads to the assignment of ½ as the probability of heads on the toss of a fair coin, the assignment of ⅙ as the probability of one dot on the upturned face when a die is tossed, and the assignment of 1/52 that the outcome is the ace of diamonds when a card is selected from a well-shuffled deck of playing cards.

Many times the outcomes of interest in a random experiment are defined in such a way that they are composed of *finer-grain* results. The FINEST-GRAIN RESULT of a random experiment is a result which cannot be further decomposed. If these finest-grain results are equally likely to occur, then the classical definition of probability can be applied to the outcomes of interest by the formula

$$P(\text{outcome of interest}) = \frac{\text{Number of finest-grain results in the outcome of interest}}{\text{Total number of finest-grain results}} \tag{16.2}$$

Example 4

Consider the process of selecting a card from a well-shuffled deck of cards. The finest-grain result would be the individual card. Each of the 52 individual cards in the deck is assumed to have the same likelihood of being selected. If the outcomes of interest were the individual cards, the sample space would consist of 52 outcomes, each with a probability of $\frac{1}{52}$.

However, suppose that the outcomes of interest are "ace" and "not ace," yielding the sample space

$$S = \{\text{Ace, not ace}\}$$

The outcome "ace" consists of four of the finest-grain results; the outcome "not ace" consists of the other forty-eight finest-grain results. The probabilities of these sample points are computed

$$P(\text{Ace}) = \frac{\text{Number of aces in deck}}{\text{Total number of cards in deck}} = \frac{4}{52} = \frac{1}{13}$$

$$P(\text{Not ace}) = \frac{\text{Number of cards that are not aces}}{\text{Total number of cards in deck}} = \frac{48}{52} = \frac{12}{13}$$

[]

Because the classical probability assignments are made on a logical, or self-evident, basis, they are often termed AXIOMATIC PROBABILITIES. Clearly, this classical concept of symmetrical, equally likely, outcomes is most appropriately applied to fair games of chance.

The Relative-Frequency (or Empirical) Point of View

There are many circumstances in which we cannot reasonably apply the assumption of equal likelihood to the outcomes of a random process. If we are thinking about the likelihood of heads on the toss of a *fair* coin, we feel confident about proceeding on the assumption that each of the possible outcomes is just as likely to occur as the other. But if we knew the coin was a *biased* coin, that it had been weighted or tampered with in some way, we would feel much less comfortable with the assumption of equal likelihood. If we were assigning probabilities to the outcomes defective and nondefective when an item was selected from a production process and inspected, we would hope the two outcomes were not equally likely.

In situations such as these, we shall only be able to assign probabilities to the points in a sample space by first carrying out a number of trials of the experiment and observing the results. We would insist upon tossing the weighted coin a number of times and accumulating information about the outcomes actually experienced before we would venture a statement about the probability of heads occurring. Or we would insist upon accumulating information about the proportion of output from the production process that is defective. Probabilities determined in this manner are termed RELATIVE-FREQUENCY, or EMPIRICAL, probabilities.

From the relative-frequency point of view, probability of occurrence of a specified outcome of a random experiment is defined in terms of its LONG-RUN RELATIVE FREQUENCY OF OCCURRENCE. The ratio of a number of times a specified outcome is experienced to the total number of times the experiment is performed is the RELATIVE FREQUENCY OF OCCURRENCE of that outcome. The relative frequency with which a given outcome occurs when the experiment is repeated a very large number of times is called the PROBABILITY of that outcome.

If in n trials of a random experiment, f of which result in outcome O_i, the ratio f/n tends to stabilize at the value $P(O_i)$ as n increased indefinitely, then the ratio

$$P(O_i) = f/n \tag{16.3}$$

is called the PROBABILITY of that outcome.

Unfortunately, the phrase "as n increases indefinitely" is rather vague, and we would prefer to avoid its use. Actually, a determined effort was made early in the development of probability theory to define the probability of an outcome as the MATHEMATICAL LIMIT of its relative frequency of occurrence—that is, as

$$P(O_i) = \lim_{n\to\infty} f/n$$

Such a rigorous definition can only have a theoretical interpretation, however, and not an operational one, since no experiment can be performed an infinite number of times. Thus, we find that we do, after all, tend to think of probability as being numerically equivalent to the "long-run" relative frequency of occurrence of an outcome of a random experiment without giving a precise definition to "long-run."

It seems to be an inherent property of a wide variety of experiments which can be repeated again and again under essentially constant conditions that the proportion of times a particular outcome is observed will become more and more nearly a constant as the number of repetitions of the experiment is increased. This tendency of the relative frequency of occurrence of an outcome to converge to a constant is sometimes termed STATISTICAL CONSTANCY. Given a random experiment that displays this quality, we postulate the existence of a number, $P(O_i)$, which is the true probability for the outcome. For any finite sequence of trials of the experiment, the observed relatve frequency, f/n, is only an estimate of the true probability. However, if the number of trials is very large, we can reasonably expect that the observed relative frequency of an outcome will be close to the true probability of that outcome.

Thus, in spite of the fact that we would obtain almost unanimous agreement that the probability of heads on the toss of a fair coin is one half, we would find very few people who would expect to observe exactly 1 head and 1 tails on 2 tosses of the coin, or exactly 2 heads and 2 tails in 4 tosses, or exactly 5 heads and 5 tails in 10 tosses. But if we tossed the coin 100, 1,000, or 10,000 times, we would expect the observed number of heads to move closer and closer to one half.

In much the same way, when we make a statement such as "the probability that a student will pass the English Proficiency Examination on the first try is 0.85," what we mean is that, of the large number of students who have taken the exam, 85 percent have passed on their first attempt. We are not referring to one specific student taking the examination but to the large collection of students who have taken and who will take the exam. Classical probability concepts are associated with a *large number of trials* of the experiment.

The empirical approach requires that observational data be available regarding the relative frequency of occurrence of the various possible outcomes. Such empirical data serve as the basis for associating probabilities with outcomes.

Example 5 Probabilities may be assigned to the possible number of jets landing at Monroe Airport during a specified time interval based on historical data relative to traffic at this airport. Assume that 200 observations yielded these data: Zero jet landings on 40 of these observations, one jet landing on 70 observations, two landings on 50 observations, three landings on 30 of the observations, and four jet landings on 10 observations. These data could be used to estimate probabilities as:

$$P(\text{zero jets land}) = 40/200 = 0.20$$
$$P(\text{one jet lands}) = 70/200 = 0.35$$
$$P(\text{two jets land}) = 50/200 = 0.25$$
$$P(\text{three jets land}) = 30/200 = 0.15$$
$$P(\text{four jets land}) = 10/200 = 0.05.$$

[]

We see, thus, that axiomatic probabilities are assigned on a logical or self-evident basis, in advance of any actual trials of the random experiment, while empirical probabilities are assigned only after repeated trials of the random experiment. The empirical probability assignments are made on the basis of the information collected from these trials. Both axiomatic and empirical probabilities are associated with the classical random experiment.

The Subjective (or Personalistic) Point of View

It is evident that to define probability only as the long-run relative frequency of occurrence of an outcome of a classical random experiment is to preclude the assignment of probabilities in many situations. Hence, another interpretation of probability is becoming increasingly important. This alternative approach is referred to as a SUBJECTIVE or PERSONAL INTERPRETATION of probability or PROBABILITY BASED ON DEGREE OF RATIONAL BELIEF and is applicable to those real-world situations where the concept of repetitions of a random experiment is not applicable.

Many decision situations do have the properties of a random experiment—except for the fact that they will not be repeated under constant-cause conditions. They are unique processes which have more than one possible outcome, where uncontrollable factors determine just which outcome will be experienced. Because these are situations which have not previously occurred and perhaps will not occur again in precisely the same form, there is no "relative frequency of occurrence" for such events, nor does there exist a clear-cut axiomatic basis for assigning probabilities. Nonetheless, probabilistic statements concerning the likelihood of occurrence of the possible outcomes are both meaningful and useful.

A firm, for instance, must decide whether or not to market a new product. The decision will be based primarily upon the likelihood of good consumer acceptance of the product. Because this is not a repeatable experiment, the relative-frequency concept of probability assignment is not appropriate. Nor is the "self-evident" approach meaningful here. However, the marketing manager should be able to assign subjective probabilities based on knowledge of the product, familiarity with the market structure, and experience with ventures of a similar nature.

It is most important to note that "subjective" probabilities are seldom *purely* subjective. They are ordinarily based on personal assessment of a specific situation, tempered by relevant historical data on roughly similar situations. That is, subjective probabilities are based on the *rational belief of knowledgeable persons.*

No matter which philosophical approach is applicable, the "laws" of probability developed by the classical mathematician are applied in analyzing the probabilistic elements of a decision situation.

EXERCISES

On what philosophical basis—(a) *axiomatic,* (b) *empirical, or* (c) *subjective—would probabilities be assigned in each of these situations?*

25. The likelihood that a new service station built at the intersection of interstate route 109 and state route 64 will be successful.
26. The likelihood that Fas-Pass will win the first race at Belmont tomorrow.

27. The likelihood that a box of washers selected from a production process will meet quality-control specifications.
28. The likelihood of being dealt four aces from a well-shuffled deck of playing cards.
29. The likelihood that there will be more than 700 fatalities on the nation's highways over the next Memorial Day weekend.
30. The likelihood of "double sixes" on the toss of a pair of fair dice.
31. The likelihood that a student will make an A in an introductory literature class.
32. The likelihood that Albert Anderson will make an A in his Literature 101 class.
33. The likelihood that a rat, approaching a corner in a maze, will turn right rather than left.
34. The likelihood that Adams and Brown both are on a committee of two selected at random from a group of five members: Adams, Brown, Cates, Dunnigan, and Ellis.
35. The likelihood of heavy thundershowers in Hazleton tomorrow.
36. The likelihood of selecting a matched pair when two socks are selected at random from a drawer containing two white socks, two black socks, and three red socks.
37. The likelihood that a wildcat oil well drilled on Will Huggins's 40-acre farm will be productive.
38. The likelihood that a student, selected by lottery from the entire student body of Unicorn University, will favor prohibition.
39. The likelihood that a student will default on a Federal Student Assistance loan.
40. The likelihood that, in five tosses of a fair coin, the result will be HTHTH.
41. The likelihood that Lance Remington, going hunting tomorrow (for the first time) in Southeast Paulding County will bag a mountain lion.
42. The likelihood that a taxpayer's income tax return will contain a mathematical error.

43. Gaylord Rationale, notorious riverboat gambler, was betting against the house on the outcome observed—heads or tails—when a coin is tossed. Upon observing five heads in a sequence of five tosses, Gaylord changed his bet from tails to heads, theorizing that the coin was "on a streak." Later in the evening, upon observing five tails in a sequence of five tosses, he changed his bet from tails to heads, theorizing that "heads are due." Is Gaylord rational?

16.5 EVENTS AND THEIR OCCURRENCE

An EVENT is any collection of sample points from one sample space. If an event consists of only one sample point, it is called a SIMPLE EVENT. A COMPOSITE EVENT involves various combinations of simple events and consists of two or more sample points.

Example 6 For the experiment consisting of tossing two coins, the sample space might be specified as

$$S = \{HH, HT, TH, TT\}$$

If we define an event E_1 as "the occurrence of one head and one tail," the event contains two sample points,

$$E_1 = \{HT, TH\}$$

Or, if we define an event E_2 as "the occurrence of heads on the first coin," this event contains the points

$$E_2 = \{HH, HT\}$$

The event E_3, defined as "the occurrence of at least one heads," would consist of three elementary outcomes, as

$$E_3 = \{HH, HT, TH\}$$

The event E_4, defined as "the occurrence of heads on both coins," is a simple event consisting only of the one point HH; that is,

$$E_4 = \{HH\}$$ []

An event is said to "occur" or to "take place" if the outcome which results on the trial of the experiment is any one of the elements included in that event. Because many events may be defined on a sample space, and because several of these may contain the same sample point, it is possible for more than one event to occur on a single trial of the experiment, even though only one outcome is experienced.

Example 7 The number of jet planes arriving at the Municipal Airport during a five-minute time period is observed and recorded. The possible outcomes for the experiment are the nonnegative integers 0 through 10, as

$$S = \{0, 1, 2, \ldots, 10\}$$

where 10 is the maximum possible number of jet arrivals during this length of time.

Any subset of this sample space is an event. For instance, the condition that "at least three jets arrive" would define the event

$$E = \{3, 4, 5, \ldots, 10\}$$

The condition that "fewer than four jets arrive" would define the event

$$F = \{0, 1, 2, 3\}$$

The condition that "exactly one jet arrives" defines the event

$$G = \{1\}$$

Other events of interest could be similarly defined.

Assume that three jets arrive at the airport during one five-minute period. The outcome of the experiment is 3, only one sample point in the sample space. However, because 3 is an element of both subsets E and F, both these events are said to have occurred on this trial of the experiment. []

Special Types of Events

If an event is defined by some condition that is not satisfied by any of the sample points in the sample space, it is called the IMPOSSIBLE, or NULL, EVENT. If in the die-tossing experiment an event were defined as "the number of dots on the upturned side is exactly divisible by 7," this would be an impossible event.

If an event is defined by some condition that is satisfied on each and every trial of the experiment, the event is known as the CERTAIN, or SURE, event. If in the die-tossing experiment an event were defined as "the number of dots on the upturned side of the die is an integer greater than zero but less than seven," this event would be a certain event.

When every outcome contained in one event E is also contained in another event F and, conversely, every outcome contained in F is also contained in E, the events are really the same, and they are said to be EQUAL, or EQUIVALENT, EVENTS. If E and F are equal, either both take place as the result of the outcome of the experiment or neither takes place; the occurrence of one is tantamount to the occurrence of the other.

Two events are said to be MUTUALLY EXCLUSIVE, or DISJOINT, if they cannot both occur on the same trial of the experiment. Not any of the sample points which are elements of one event are elements of the other event. And n events are mutually exclusive if no two of them have any elements in common.

If E is an event associated with the sample space of a random experiment, the event which is the COMPLEMENT of E relative to the sample space is the event, denoted E^c, consisting of all outcomes in the sample space which are not included in E itself. The complement of E is also often referred to as "not E." On each trial of the experiment, either E or its complement, E^c, occurs.

Venn diagrams are used in Figure 16.1 to illustrate disjoint events and the complement of an event.

Figure 16.1

(A) Mutually exclusive (or disjoint) events

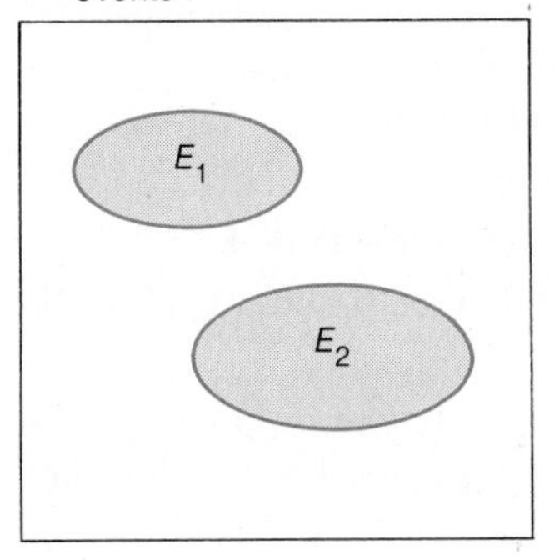

(B) The complement of an event

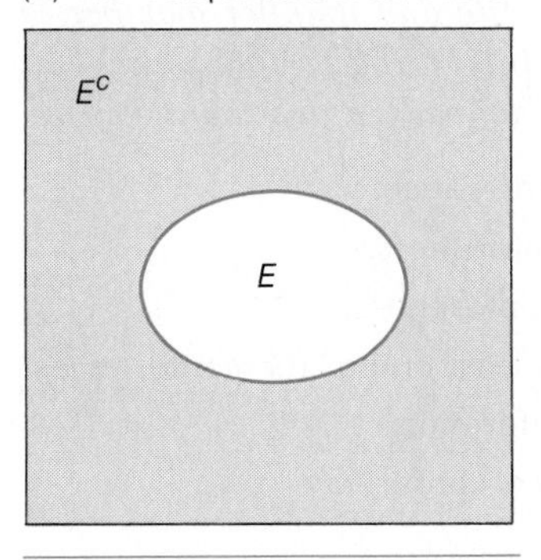

EXERCISES

Three coins are tossed. List the points in the sample space. List the points in each of the following events defined on this sample space.

44. There is exactly one head.
45. There are two heads and one tail.
46. There are at most two heads.
47. There are exactly three heads.
48. Heads and tails alternate.
49. There are more heads than tails.
50. Which of the above events occur if the outcome on a trial of the experiment is HTH?
51. Which of the above events occur if the outcome on a trial of the experiment is TTH?

Two dice—one red and one blue—are tossed. List the points in the sample space. List the points in each of the following events defined on this sample space.

52. The number on the red die is 4.
53. The number on at least one die is 4.
54. The number 4 is on both of the dice.
55. The number on the red die is less than 4.
56. There is a number greater than 4 on either die.
57. The sum of the numbers on the two dice is 4.
58. The sum of the numbers on the two dice is less than 4.
59. Which of the above events occur if the outcome on a trial of the experiment is a two on each of the dice?
60. Which of the above events occur if the outcome on a trial of the experiment is a two on the red die and a three on the blue die?

A committee of five members—A, B, C, D, and E—selects at random one member to attend a conference in Seattle and another member to attend a conference in Boston. List the points in the sample space. List the points in each of the following events defined on this sample space.

61. A is selected to attend the Seattle conference.
62. A is selected to attend one of the conferences.
63. A and B are selected to attend the conferences.
64. Either A or B is selected to attend at least one of the conferences.
65. B is not selected to attend either conference.
66. Neither A nor B is selected to attend a conference.

67. If B is selected to attend the Seattle conference and C is selected to attend the Boston conference, which of the above events occur?
68. If A is selected to attend the Seattle conference and D is selected to attend the Boston conference, which of the above events occur?

Given a sample space S = {1, 2, 3, 4, 5}, explain whether each of the following statements is true or false.

69. $E = \{1, 2, 3\}$ is the sure or certain event.
70. $F = \{5\}$ is an impossible event.
71. If $G = \{1, 3\}$ and $H = \{4, 5\}$, G and H are mutually exclusive events.
72. If $G = \{1, 3\}$ and $J = \{1, 3, 5\}$, G and J are mutually exclusive events.
73. If $G = \{1, 3\}$ and $J = \{1, 3, 5\}$, G and J are equal, or equivalent, events.
74. If $G = \{1, 3\}$ and $K = \{3, 1\}$, G and K are equal, or equivalent events.
75. If $G = \{1, 3\}$, the complement of G is $G^c = \{2, 4\}$.

16.6 CALCULATING THE PROBABILITY OF AN EVENT

In a given experiment, attention may be directed toward events rather than elementary outcomes. Hence, we must be able to associate probabilities with events as well as with individual sample points.

Rule 1 *The probability of an event* E *is the sum of the probabilities assigned to the sample points that comprise the event; that is,*

$$P(E) = \sum_{\text{all } i \text{ in } E} P(O_i) \qquad \textbf{(16.4)}$$

Example 8 In the experiment consisting of tossing a fair die and noting the number of dots on the upturned face, the sample space S has six equally likely sample points, each with a probability of one sixth. If the event of interest is event E that the number of dots on the upturned face of the die is less than three, the event consists of two elementary outcomes:

$$E = \{1, 2\}$$

and the probability associated with the event E is

$$P(E) = P(1) + P(2) = 1/6 + 1/6 = 1/3$$

[]

Example 9 Two fair coins are tossed. The sample space for the experiment is $S = \{HH, HT, TH, TT\}$. Each of these four equally likely outcomes has the probability of occurrence of one fourth.

The event E_1, defined as "exactly one head," contains the sample points HT and TH. Thus,

$$P(E_1) = 1/4 + 1/4 = 1/2$$

The event E_2, defined as "at least one head," contains the sample points HH, HT, and TH and has the probability of occurrence of

$$P(E_2) = 1/4 + 1/4 + 1/4 = 3/4$$

The event E_3, defined as "two tails," contains the sample point TT and has the probability of occurrence of

$$P(E_3) = 1/4$$ []

On each trial of an experiment, either the event E or its complement E^c occurs. Hence the probability of the complement of an event is given by the following rule.

Rule 2 If E and E^c are complementary events, then

$$P(E^c) = 1 - P(E) \qquad \textbf{(16.5)}$$

Example 10 If the sample space consists of the outcomes that may result when one card is selected from a well-shuffled deck of playing cards, there are 52 equally likely sample points, each with a probability of occurrence of 1/52. If the event A is the event that the card selected is an ace, there are four sample points in A and $P(A) = 1/52 + 1/52 + 1/52 + 1/52 = 4/52 = 1/13$.

The event A^c, that the card selected is not an ace, has the probability of occurrence of

$$P(A^c) = 1 - P(A) = 1 - 1/13 = 12/13$$ []

Example 11 Eighty-five percent of the workers at the Maxum Manufacturing Company plant have indicated that they prefer a four-day workweek. If a worker is selected at random from all these plant workers, the probability that the worker selected favors the four-day workweek is

$$P(\text{favor}) = 0.85$$

The probability that a worker selected at random does not favor the four-day workweek is given by

$$P(\text{not favor}) = 1 - P(\text{favor}) = 1 - 0.85 = 0.15$$ []

EXERCISES

A letter is selected at random from the 26 letters of the English alphabet. What is the probability of each of the following events?

76. The letter is Q.
77. The letter comes before E in alphabetical order.
78. The letter is one of the letters in the name *Miss Mamma.*
79. The letter is a vowel (where the vowels are the letters A, E, I, O, U).
80. The letter is not a vowel.
81. The letter is not R or S or T.

An experiment consists of tossing three fair coins. What is the probability of each of the following events?

82. Exactly two heads.
83. At most two heads.
84. At least two heads.
85. Heads on the first coin tossed.
86. Something other than two heads.
87. Either HTH or THT.

Assuming that in a three-child family the eight sample points GGG, GGB, GBG, GBB, BGG, BGB, BBG, and BBB are equally likely, what is the probability of the following events?

88. No boys.
89. All boys.
90. Exactly one boy.
91. At least one boy.
92. At most one boy.
93. Exactly two boys.
94. The oldest child is a girl.

A card is selected at random from a well-shuffled deck of playing cards. What is the probability of each of the following events?

95. Obtaining the two of clubs.
96. Obtaining either the two or the three of clubs.
97. Obtaining a card from the suit of clubs.
98. Obtaining a black card.
99. Obtaining any card other than the two or three of clubs.

Bob, Chuck, David, and Elwood have purchased four tickets to the heavyweight title match but discover, upon reaching the boxing arena, that one ticket has been lost. They decide to write their names on scraps of paper, fold these, and put them in a hat, and then select three of the names at random. The person whose name is not selected will not go to the match. What is the probability that—

100. Elwood does not go?
101. Both Bob and Chuck go?
102. Either Bob or Chuck goes?
103. Chuck, David, and Elwood go?

(Hint: List the points in the sample space and assign a probability to each.)

104. A person wishes to make a telephone call but cannot remember the last digit in the number. What is the probability of getting the correct number if a last digit is simply dialed at random?
105. On some days the receptionist in the doctor's office is coolly efficient, on some days she is rude, and on all other days she is antagonistic. If the probability that she is cooly efficient is 0.2 and the probability that she is rude is 0.55, what is the probability that she is antagonistic?

Of all visitors to a certain national park, only 25 percent are residents of the state in which the park is located. Forty percent of all visitors to the park arrive with a group of at least five people. A visitor is selected at random. What is the probability that—

106. The visitor lives out of state?
107. The visitor arrives with a group of fewer than five people?

A ticket to a football game is to be given as a door prize at a college fraternity dance. The 400 students attending the dance are each given a number, 1 through 400, and one of these numbers is to be selected at random to receive the bowl-game ticket. What is the probability that—

108. The number on the winning ticket is 263?
109. The number on the winning ticket is an even number?
110. The number on the winning ticket is less than 100?

Hardluck Herschel staunchly maintains that, if two dice are tossed and the sum of the dots on the two upturned sides is observed, $P(4) = P(5) = P(9) = P(10)$ *and that* $P(6) = P(7) = P(8)$. *He bases his calculations on the following display:*

	Sum of dots on upturned sides										
	2	3	4	5	6	7	8	9	10	11	12
Number of dots on upturned sides of individual die	1–1	2–1	2–2 or 3–1	3–2 or 4–1	3–3 or 4–2 or 5–1	4–3 or 5–2 or 6–1	2–6 or 3–5 or 4–4	5–4 or 6–3	5–5 or 6–4	6–5	6–6

111. Explain to Herschel the fallacy of his reasoning.

16.7 THE UNION AND INTERSECTION OF EVENTS

The concepts of set theory provide the important tools needed in calculating the probabilities of more complex events. Let us begin by defining the events "E_1 or E_2" and "E_1 and E_2."

Figure 16.2

The UNION and INTERSECTION of Events

(A) The union of events: $E_1 \cup E_2$

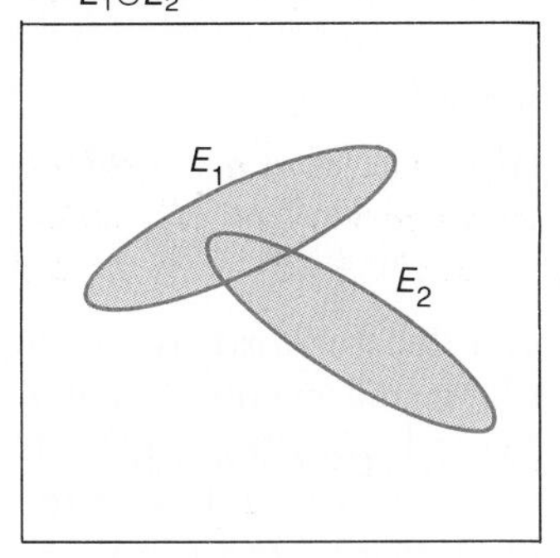

(B) The intersection of events: $E_1 \cap E_2$

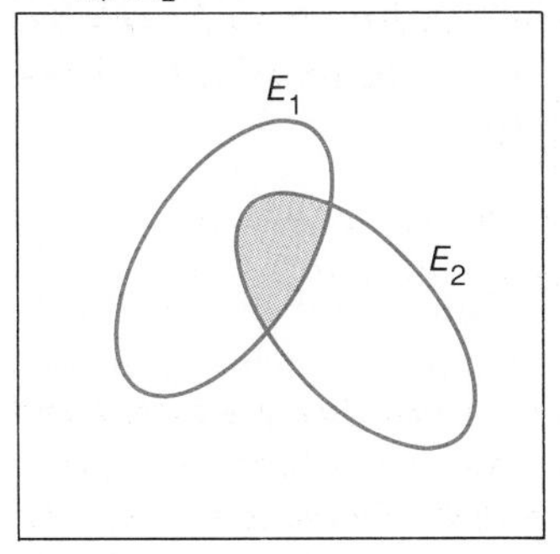

> **The event "E_1 or E_2,"** known as the UNION of E_1 and E_2 and denoted using set theory symbolism as $E_1 \cup E_2$, is the event whose elementary outcomes are either in event E_1 or in event E_2 or in both events. (See Figure 16.2A.)

> **The event "E_1 and E_2,"** known as the INTERSECTION of E_1 and E_2 and denoted using set theory symbolism as $E_1 \cap E_2$, is the event whose elementary outcomes are in event E_1 and in event E_2 as well. (See Figure 16.2B.)

If, as the result of a trial of the experiment, any of the elementary outcomes included either in event E_1 or in event E_2 is experienced, the event $E_1 \cup E_2$ takes place. However, in order for the event $E_1 \cap E_2$ to take place, both events E_1 and E_2 must occur simultaneously as a result of a trial of the experiment.

The following rule may be used for calculating the probability of the union of two events.

> **Rule 3** *For any two events,* E_1 *and* E_2*, the probability of the occurrence of the event* $E_1 \cup E_2$ *is*
>
> $$P(E_1 \cup E_2) = P(E_1) + P(E_2) - P(E_1 \cap E_2) \quad \textbf{(16.6)}$$
>
> *where* $P(E_1 \cap E_2)$ *is the probability of the simultaneous occurrence of the two events (sometimes called their JOINT PROBABILITY).*

Example 12 If A is the event that a card drawn from a deck of cards is an ace, $P(\mathrm{A}) = 4/52 = 1/13$. If H is the event that a card drawn is of the suit hearts, $P(\mathrm{H}) = 13/52 = 1/4$. Now there is one card common to both events—the ace of hearts. Thus, the event "ace and hearts," $\mathrm{A} \cap \mathrm{H}$, has probability $P(\mathrm{A} \cap \mathrm{H}) = 1/52$.

Using the rule for the union of events, we can determine the probability that a card drawn from the deck is either an ace or hearts, as

$$\begin{aligned} P(\mathrm{A} \cup \mathrm{H}) &= P(\mathrm{A}) + P(\mathrm{H}) - P(\mathrm{A} \cap \mathrm{H}) \\ &= 1/13 + 1/4 - 1/52 \\ &= 16/52 \end{aligned}$$

One card, the ace of hearts, was counted as a sample point in the event "the card is an ace" and was counted again in the event "the card is a heart." The last term in the formula could be considered as an adjustment for such double counting. []

Two events are mutually exclusive or disjoint if they both cannot occur on the same trial of the experiment: not any of the sample points which are elements of one of the events are elements of the other. If two events, E_1 and E_2, are mutually exclusive, the probability of their joint occurrence is zero.

Rule 4 *For two mutually exclusive events,* E_1 *and* E_2,

$$P(E_1 \cup E_2) = P(E_1) + P(E_2) \qquad (16.7)$$

Example 13 If A is the event that a card drawn from a deck of cards is an ace and K is the event that a card drawn is a king, the events A and K are mutually exclusive with probabilities $P(\mathrm{A}) = 1/13$, $P(\mathrm{K}) = 1/13$. It is impossible for both events A and K to occur on the same trial of the experiment; thus, their joint probability is zero; that is, $P(A \cap K) = 0$.

Now the event that a card selected is either an ace or a king, $(A \cup K)$, consists of all the points in the two mutually exclusive events and is computed by

$$\begin{aligned} P(A \cup K) &= P(A) + P(K) \\ &= 1/13 + 1/13 \\ &= 2/13 \end{aligned}$$

[]

This addition rule can be extended to any finite number of mutually exclusive events. Thus, if $E_1, E_2, \ldots, E_n$ are all mutually exclusive events, then

$$P(E_1 \cup E_2 \cup \ldots \cup E_n) = P(E_1) + P(E_2) + \ldots + P(E_n) \qquad (16.8)$$

Example 14 The probability that a student in a math course will receive an A, B, C, D, or F for the course are 0.15, 0.25, 0.35, 0.20, and 0.05, respectively.

The probability that a student will receive a C or better is given by

$$\begin{aligned} P(\mathrm{A} \cup \mathrm{B} \cup \mathrm{C}) &= P(\mathrm{A}) + P(\mathrm{B}) + P(\mathrm{C}) \\ &= 0.15 + 0.25 + 0.35 \\ &= 0.75 \end{aligned}$$

The probability that a student will receive at best a C is given by

$$\begin{aligned} P(\mathrm{C} \cup \mathrm{D} \cup \mathrm{F}) &= P(\mathrm{C}) + P(\mathrm{D}) + P(\mathrm{F}) \\ &= 0.35 + 0.20 + 0.05 \\ &= 0.60 \end{aligned}$$

[]

Figure 16.3

Probability of the Union of Mutually Exclusive Events

(A) $P(E) = P(E \cap F^c) + P(E \cap F)$ and $P(E \cap F^c) = P(E) - P(E \cap F)$

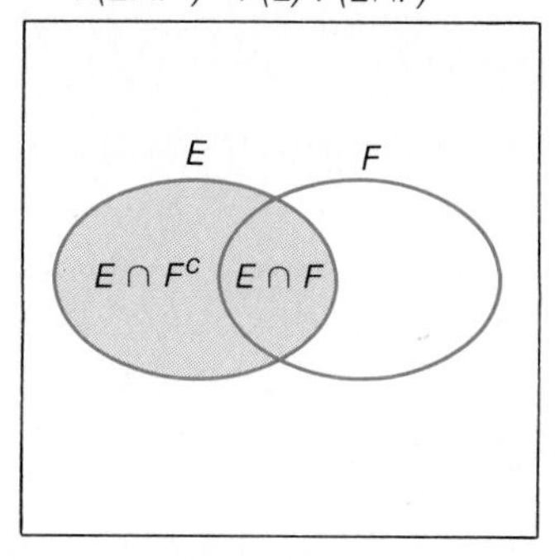

(B) $P(E \cup F) = P(E) + P(F \cap E^c)$

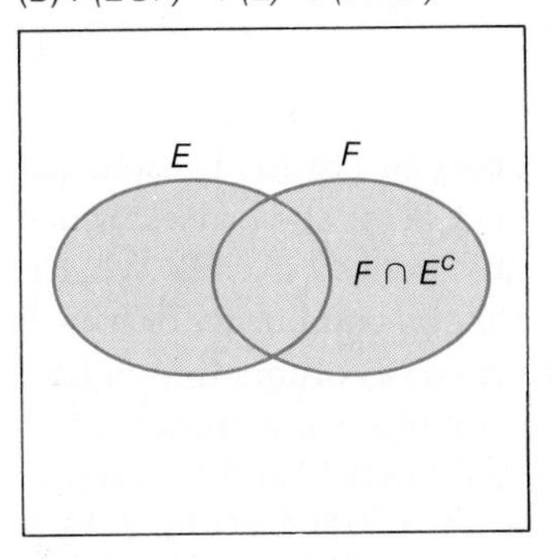

(C) $P(E^c \cap F^c) = 1 - P(E \cup F)$

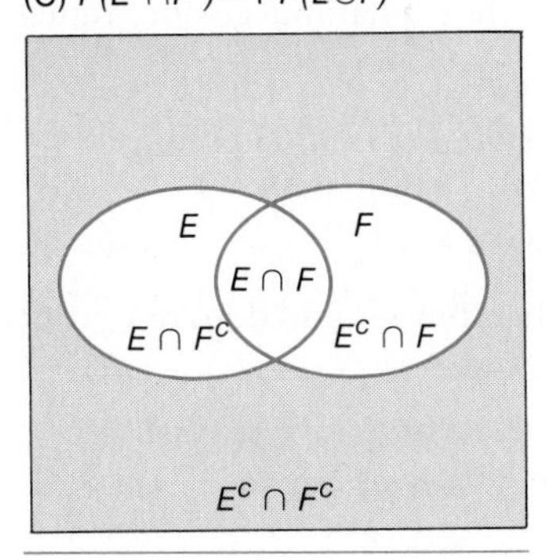

Venn diagrams are very useful tools for visualizing relationships between events and, in particular, for determining the probability of the union of events. The Venn diagram in Figure 16.3 shows that, given two events E and F defined on a sample space, the event E can be viewed as the union of two mutually exclusive events, $(E \cap F)$ and $(E \cap F^c)$; that is,

$$E = (E \cap F^c) \cup (E \cap F)$$

Thus, given the information $P(E \cap F^c)$ and $P(E \cap F)$, the addition rule for determining the probability of mutually exclusive events can be used to find $P(E)$, as

$$P(E) = P(E \cap F^c) + P(E \cap F) \qquad \textbf{(16.9)}$$

Consequently,

$$P(E \cap F^c) = P(E) - P(E \cap F) \qquad \textbf{(16.10)}$$

In the same manner, the Venn diagram in Figure 16.3B shows that the union of events E and F, $E \cup F$, can be represented as the union of two mutually exclusive events, E and $(F \cap E^c)$; that is,

$$E \cup F = E \cup (F \cap E^c)$$

Then

$$P(E \cup F) = P(E) + P(F \cap E^c) \qquad \textbf{(16.11)}$$

Also, as shown in the Venn diagram in Figure 16.3C, the sample space itself can be considered as the union of a set of mutually exclusive events; that is,

$$\begin{aligned} S &= (E \cap F^c) \cap (E \cap F) \cup (E^c \cap F) \cup (E^c \cap F^c) \\ &= (E \cup F) \cup (E^c \cap F^c) \end{aligned}$$

This leads to the proposition that

$$\begin{aligned} P(E \cup F) + P(E^c \cap F^c) &= 1 \\ P(E^c \cap F^c) &= 1 - P(E \cup F) \end{aligned} \qquad \textbf{(16.12)}$$

Example 15 Given two events E and F defined on a sample space and having probability $P(E) = 0.35$, $P(F) + 0.28$, and $P(E \cap F) = 0.06$, determine

a. $P(E \cup F)$ *b.* $P(E \cap F^c)$ *c.* $P(E^c \cap F^c)$

a. From Equation 16.6, the rule for the union of events, we calculate

$$\begin{aligned} P(E \cup F) &= P(E) + P(F) - P(E \cap F) \\ &= 0.35 + 0.28 - 0.06 \\ &= 0.57 \end{aligned}$$

b. From the proposition regarding mutually exclusive events and their union

$$P(E \cap F^c) = P(E) - P(E \cap F)$$

we determine

$$\begin{aligned} P(E \cap F^c) &= 0.35 - 0.06 \\ &= 0.29 \end{aligned}$$

c. Using the relationship

$$P(E^c \cap F^c) = 1 - P(E \cup F)$$

we determine

$$\begin{aligned} P(E^c \cap F^c) &= 1 - 0.57 \\ &= 0.43 \end{aligned}$$

[]

Example 16 The probability that a person entering a certain fast-food restaurant orders a hamburger is 0.76, the probability that he or she orders french fries is 0.69, and the probability that both are ordered is 0.58. What is the probability that such a person (*a*) orders either a hamburger or french fries, (*b*) orders a hamburger but not french fries, (*c*) orders french fries but not a hamburger, (*d*) orders neither a hamburger nor french fries?

Using H to represent the event "orders a hamburger" and F to represent the event "orders french fries," we note that $P(\text{H}) = 0.76$, $P(\text{F}) = 0.69$, and $P(\text{H} \cap \text{F}) = 0.58$.

a. The probability that a person orders either a hamburger or french fries is determined by

$$\begin{aligned} P(\text{H or F}) = P(\text{H} \cup \text{F}) &= P(\text{H}) + P(\text{F}) - P(\text{H} \cap \text{F}) \\ &= 0.76 + 0.69 - 0.58 \\ &= 0.87 \end{aligned}$$

b. The probability that a person orders a hamburger and does not order french fries is determined by

$$\begin{aligned} P(\text{H and "not F"}) = P(\text{H} \cap \text{F}^c) &= P(\text{H}) - P(\text{H} \cap \text{F}) \\ &= 0.76 - 0.58 \\ &= 0.18 \end{aligned}$$

Figure 16.4

Partitioning the Sample Space into Regions

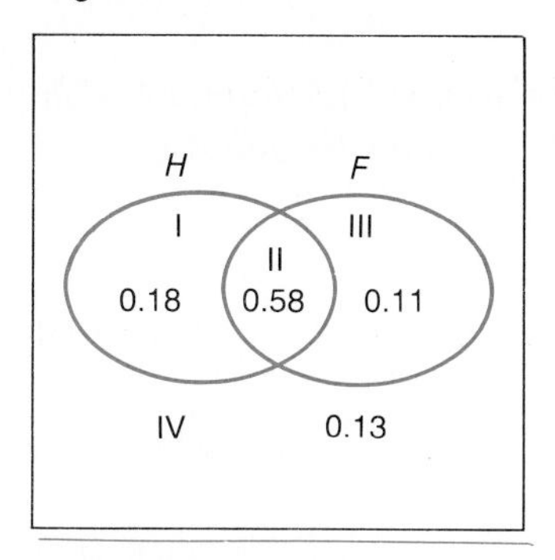

I represents $(H \cap F^c)$
II represents $(H \cap F)$
III represents $(H^c \cap F)$
IV represents $(H^c \cap F^c)$

c. The probability that a person orders french fries and does not order a hamburger is given by

$$P(\text{F and "not H"}) = P(F \cap H^c) = P(F) - P(H \cap F) = 0.69 - 0.58 = 0.11$$

d. Finally, the probability that a person does not order a hamburger and does not order french fries is given by

$$P(\text{"not H" and "not F"}) = P(H^c \cap F^c) = 1 - P(H \cup F) = 1 - 0.87 = 0.13$$

(See Figure 16.4.) []

EXERCISES

A random experiment has the sample space S = {1, 2, 3, 4, 5, 6}, *where each of the outcomes is equally likely to occur. The following events have been defined on this sample space:*

$E_1 = \{1,2,3\}$ $E_2 = \{4,5\}$ $E_3 = \{3,4\}$ $E_4 = \{1,2,3,4\}$

List the sample points in each of the following events and determine the probability of the event.

112. $E_1 \cup E_2$
113. $E_1 \cup E_3$
114. $E_2 \cup E_3$
115. $E_1 \cup E_4$
116. $E_2 \cup E_4$
117. $E_3 \cup E_4$
118. $E_1 \cap E_2$
119. $E_1 \cap E_3$
120. $E_1 \cap E_4$
121. $E_2 \cap E_3$
122. $E_2 \cap E_4$
123. $E_3 \cap E_4$
124. E_1^c
125. E_2^c
126. E_3^c
127. $E_3^c \cup E_4^c$
128. $E_1^c \cap E_3^c$
129. $(E_3 \cup E_4)^c$
130. $(E_1 \cap E_3)^c$
131. $(E_1 \cup E_2) \cap (E_3 \cup E_4)$
132. $(E_1 \cap E_2) \cup (E_3 \cap E_4)$

Given that E *and* F *are two events defined on a sample space* S, *such that* P(E) = *0.51,* P(F) = *0.35, and* P(E ∩ F) = *0.24, find the following:*

133. $P(E^c)$
134. $P(F^c)$
135. $P(E \cup F)$
136. $P(E \cap F^c)$

137. $P(E^c \cap F)$
138. $P(E^c \cup F)$
139. $P(E \cup F^c)$
140. $P(E^c \cup F^c)$
141. $P(E^c \cap F^c)$

Given E *and* F *are two mutually exclusive events defined on a sample space* S *such that* $P(E) = 0.35$ *and* $P(F) = 0.48$*, find the following:*

142. $P(E^c)$
143. $P(F^c)$
144. $P(E \cap F)$
145. $P(E^c \cup F)$
146. $P(E^c \cap F)$
147. $P(E \cup F^c)$
148. $P(E \cap F^c)$
149. $P(E^c \cup F^c)$
150. $P(E^c \cap F^c)$

Given E *and* F *are events defined on a sample space* S *such that* $P(E) = 0.3$, $P(F) = 0.5$, *and* $P(E \cap F) = 0.2$, *find the probability of occurrence of*

151. Either event E or event F or both.
152. Event E but not event F.
153. Event F but not event E.
154. Neither event E nor event F.
155. The probability that Mary will purchase a television set on her current shopping spree is 0.6; the probability that she will purchase a stereo is 0.5; and the probability that she will purchase both a television and a stereo is 0.4. What is the probability that she will purchase either a television or a stereo?
156. The probability that a trainee passes a manual dexterity test is three fifths, and the probability that he or she passes a technical skills test is five eighths. If the probability that the trainee passes at least one of the tests is eight ninths, what is the probability that he or she passes both?
157. The probability that Jim Bob will purchase a sports car this year is 0.8; the probability that he will purchase a dump truck is 0.5; and the probability that he will purchase either a sports car or a dump truck or both is 0.9. What is the probability that Jim Bob will purchase both a sports car and a dump truck?
158. What is the probability of obtaining a number greater than three or an even number on one toss of a fair die?
159. What is the probability of selecting an ace or a red card when one card is selected at random from a well-shuffled deck of playing cards?
160. What is the probability of selecting either a queen or a jack when a single card is selected at random from a well-shuffled deck of cards?

A shipment of 50 one-gallon cans of paint is accidentally submerged in water, and all labels on the cans are lost. The shipment contains 20 cans of red paint, 10 cans of blue paint, 15 cans of green paint, and 5 cans of yellow paint. If a can of paint is selected at random from the shipment,

161. What is the probability that it is red?
162. What is the probability that it is either yellow or green?
163. What is the probability that it is not blue?
164. The probability is 0.5 that a lab-technician job applicant has formal training in the area. The probability is 0.3 that he or she has had both formal training and previous work experience. The probability is 0.8 that the applicant has had formal training or previous work experience or both. What is the probability that the prospective technician has had experience but no formal training?

A district sales manager for a company often travels from the district office in Peoria to the company's home office in Chicago. Sometimes she takes a plane (P), sometimes she takes a train (T), sometimes she takes a bus (B), sometimes she drives the company car (C), and sometimes she drives her husband's car (H). It has been determined that the probabilities for these alternatives are $P(P) = 0.3$, $P(T) = 0.36$, $P(B) = 0.13$, $P(C) = 0.18$, *and* $P(H) = 0.03$. *Find the probability that she—*

165. Will not go by plane.
166. Will go by car.
167. Will catch either a plane or a train.
168. Will catch neither a plane nor a train.

The probability that a shopper making a purchase at Maxi Department Store will use a Maxi's credit card is 0.34; the probability that an AMERICARD credit card will be used is 0.25; the probability that a MASTER CREDIT credit card will be used is 0.22; and the probability that the shopper will pay by personal check is 0.13 and by cash is 0.06. What is the probability that a shopper will—

169. Use a credit card.
170. Not use a credit card.
171. Use a Maxi's credit card or pay by cash.
172. Use neither the AMERICARD nor the MASTER CREDIT card.

A survey of 420 students indicated that 185 are enrolled in an accounting course, 222 are enrolled in a botany course, and 76 are enrolled in both courses. What is the probability that a student selected at random from this group is enrolled in—

173. At least one of the courses.
174. The accounting course but not the botany course.
175. The botany course but not the accounting course.
176. Neither of these courses.

16.8 CONDITIONAL PROBABILITIES

Often partial information about the outcome of an experiment becomes available; and when this happens, the original sample space may well have to be adjusted to take into account this new knowledge. This information may eliminate from consideration certain outcomes which were, without this partial information, believed to be possible, but which are now, with the partial information, known to be impossible. In such a case, the *revised* sample space would omit these impossible outcomes. The probabilities in the revised sample space are said to be CONDITIONAL upon the occurrence of the event defined by the partial information.

Consider an experiment consisting of selecting three cards from a deck containing four cards which are marked 1, 2, 3, and 4 respectively. The sample space for the experiment is

$$S = \{(1,2,3),(1,2,4),(1,3,4),(2,3,4)\}$$

where each of the elementary outcomes is equally likely to occur and thus has the probability of one fourth. (See Figure 16.5.)

Let us define two events, as follows:

$E =$ The card numbered 1 is one of the cards selected
$F =$ The card numbered 4 is one of the cards selected

The points in event E are

$$E = \{(1,2,3),(1,2,4),(1,3,4)\}$$

so that $P(E) = 3/4$.

The points in the event F are

$$F = \{(1,2,4),(1,3,4),(2,3,4)\}$$

so that $P(F) = 3/4$.

Now suppose that we obtain the information that event E has taken place as the result of a trial of the experiment. This information allows us to assign a probability of zero to the outcome (2,3,4) and, in effect, reduces the *relevant* sample space to

$$\{(1,2,3),(1,2,4),(1,3,4)\}$$

The probability associated with each of these elementary outcomes can no longer be considered to be one fourth for, if this were the case, the sum of the probabilities would no longer be unity. Given that event

Figure 16.5

Using Partial Information about the Outcome of an Experiment to Revise the Sample Space

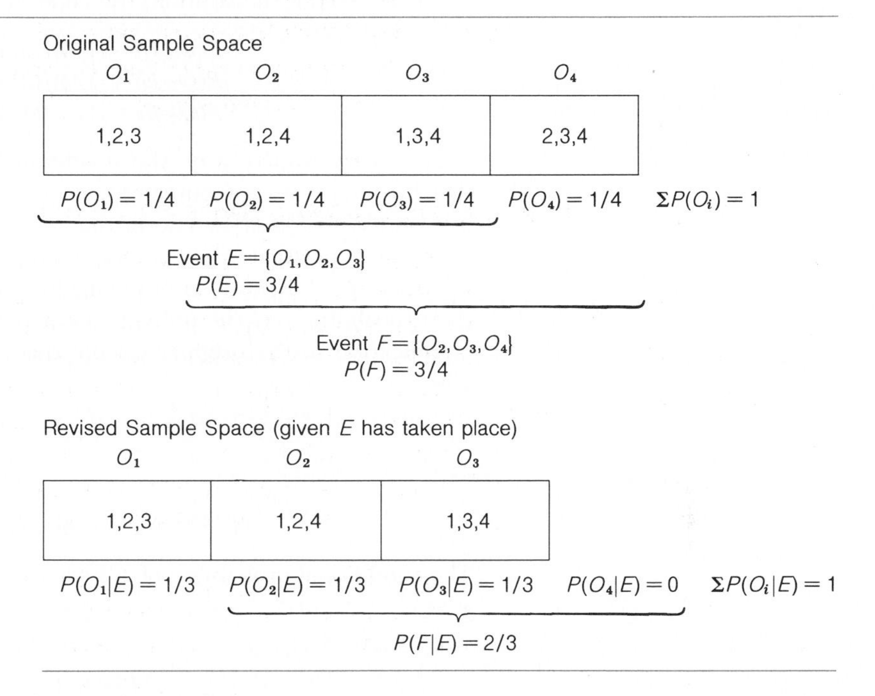

E has occurred, total probability must be redistributed to lie within this event; all elementary outcomes not contained in E must each have zero probability. The new probabilities are CONDITIONAL PROBABILITIES, conditional upon the fact that event E has occurred. The symbolism $P(O_i|E)$ is used and is read "the conditional probability of outcome O_i *given that* event E has occurred."

The conditional probabilities are obtained from the unconditional ones in the original sample space by treating the elements of E as a new sample space. A total probability of one is distributed over this revised sample space in such a way that the new probabilities within E are in the same ratio as in the original sample space; that is,

$$\frac{P(O_j|E)}{P(O_k|E)} = \frac{P(O_j)}{P(O_k)}$$

Thus, the conditional probabilities of the elementary outcomes contained in E are obtained from the original probabilities by multiplying each of them by the factor $1/P(E)$; that is, the conditional probability of an outcome O_i in E is obtained by

$$P(O_i|E) = P(O_i)/P(E)$$

Furthermore, the assignment of these conditional probabilities satisfies the two axioms:

$$0 \leq P(O_i|E) \leq 1 \qquad \text{and} \qquad \Sigma P(O_i|E) = 1$$

In the current illustration, the conditional probabilities are given by

$$P(O_1|E) = (1/4)/(3/4) = 1/3$$
$$P(O_2|E) = (1/4)/(3/4) = 1/3$$
$$P(O_3|E) = (1/4)/(3/4) = 1/3$$

In a very similar way, the conditional probability for an event F, which contains several outcomes, at least one of which is also contained in event E, is found by taking the "unconditional" probability for the set of outcomes that are in common with E (these are in the joint event E and F, $E \cap F$) and multiplying by the same factor used to adjust the probabilities for the individual outcomes, $1/P(E)$.

Referring to the original sample space, we see that the event $E \cap F$ is

$$E \cap F = \{(1,2,3),(1,2,4)\}$$

with

$$P(E \cap F) = (1/4) + (1/4) = 1/2$$

The conditional probability of F's occurring, given that E has occurred, is

$$P(F|E) = \frac{P(E \cap F)}{P(E)} = \frac{1/2}{3/4} = 2/3$$

Rule 5 *If* E *and* F *are events defined on a sample space* S, *the conditional probability of* F, *given the fact that* E *has occurred, written* P(F|E), *is defined as*

$$P(F|E) = \frac{P(E \cap F)}{P(E)} \qquad \text{if } P(E) \neq 0 \qquad \textbf{(16.13)}$$

Similarly,

$$P(E|F) = \frac{P(E \cap F)}{P(F)} \qquad \text{if } P(F) \neq 0 \qquad \textbf{(16.14)}$$

Example 17 Consider the sample space

$$S = \{1,2,3,4,5\}$$

These five outcomes are not equally likely to occur but, instead, have probabilities as follows: $P(1) = 2/12$; $P(2) = 1/12$, $P(3) = 4/12$, $P(4) = 2/12$, and $P(5) = 3/12$.

Two events have been defined on this sample space, as

$$E = \{2,3,5\}$$

and

$$F = \{1,2,4\}$$

The joint event $E \cap F$ consists of the sample points common to both E and F; thus,

$$E \cap F = \{2\}$$

The probabilities for these events can be determined by summing the probabilities associated with the elementary outcomes contained in the events, as

$$\begin{aligned} P(E) &= P(2) + P(3) + P(4) = 1/12 + 4/12 + 3/12 = 8/12 \\ P(F) &= P(1) + P(2) + P(4) = 2/12 + 1/12 + 2/12 = 5/12 \\ P(E \cap F) &= P(2) = 1/12 \end{aligned}$$

Now, suppose that we obtain information which tells us that event E has occurred on a trial of the experiment. The sample space must be revised to reflect this information. If event E has taken place, one of the elementary outcomes 2, 3, or 5 has occurred. These outcomes comprise the new "revised" sample space. Outcomes 1 and 4 are known not to have occurred; they are assigned a probability of zero and are eliminated from further consideration.

The new conditional probabilities of the elementary outcomes in the revised sample space are obtained from the old probabilities by multiplying each of them by the factor

$$\frac{1}{P(E)} = \frac{1}{8/12}$$

Hence,

$$P(2|E) = \frac{1/12}{8/12} = 1/8$$

$$P(3|E) = \frac{4/12}{8/12} = 4/8$$

$$P(5|E) = \frac{3/12}{8/12} = 3/8$$

What is more, the conditional probability of event F, given that event E has occurred, is computed

$$\begin{aligned} P(F|E) &= \frac{P(E \cap F)}{P(E)} \\ &= \frac{1/12}{8/12} \\ &= 1/8 \end{aligned}$$

[]

Example 18 The probability that the purchaser of a new automobile will buy the air-conditioning option is 0.76, the probability that he or she will buy the automatic-transmission option is 0.83, and the probability that both will be selected is 0.69. What is the probability that a purchaser of a new automobile will buy

a. The air-conditioning option, given that he or she also buys the automatic transmission option?
b. The automatic-transmission option, given that he or she also buys the air-conditioning option?
c. Both the air-conditioning and the automatic-transmission option, given that at least one of the two is selected?

Letting A denote the event that the purchaser buys the air-conditioning option and T denote the event that the purchaser buys the automatic transmission option, we see that $P(A) = 0.76$, $P(T) = 0.83$, and $P(A \cap T) = 0.69$. Then, the conditional probability of purchasing air conditioning, given that the automatic transmission is selected, is given by

$$P(A|T) = \frac{P(A \cap T)}{P(T)} = \frac{0.69}{0.83} = 0.8313$$

The conditional probability that the buyer purchase the automatic-transmission option given that he or she also purchases the air-conditioning option is given by

$$P(T|A) = \frac{P(A \cap T)}{P(A)} = \frac{0.69}{0.76} = 0.9079$$

(Note carefully that these two conditional probabilities are not equal in value!)

Finally, the conditional probability that both options are selected given that at least one is purchased is given by

$$P(A \cap T|A \cup T) = \frac{P(A \cap T)}{P(A \cup T)}$$

We can determine the probability of purchasing at least one of the options by calculating

$$\begin{aligned} P(A \cup T) &= P(A) + P(T) - P(A \cap T) \\ &= 0.76 + 0.83 - 0.69 \\ &= 0.9 \end{aligned}$$

Then,

$$P(A \cap T|A \cup T) = \frac{0.69}{0.9} = 0.7667$$

[]

EXERCISES

177. What is the probability of obtaining a 4 on a single toss of a fair die, given that the outcome of the toss is known to be an even number?

178. The probability that Chloe will receive an A in math and an A in English is 0.6. The probability that she will receive an A in English is 0.9. What is the probability that she will receive an A in math given that she receives an A in English?
179. The probability that Mavis will receive an A in English is 0.5. The probability that she will receive an A in math, given that she receives and A in English, is 0.6. What is the probability that Mavis will receive an A in both math and English?
180. The probability that it will sleet and snow tomorrow is 0.25. The probability that it will snow tomorrow given that it also sleets is 0.4. What is the probability that it will not sleet tomorrow?
181. Of all persons who voted in the last local referendum, 30 percent voted negative on the school-bond issue, 40 percent voted negative on the pari-mutuel–betting issue, and 15 percent voted negative on both issues. What is the probability that a person voted negative on the school-bond issue, given that he voted negative on the pari-mutuel–betting issue?
182. If E and F are events defined on a sample space S such that $P(E \cap F) = 0.2$, and $P(E|F) = 0.3$, find $P(F)$.

If E *and* F *are events defined on a sample space* S *such that* $P(E) = 0.4$, $P(F) = 0.3$, *and* $P(E \cap F) = 0.1$,

183. What is the probability of E given F?
184. What is the probability of F given E?

A card is selected at random from a well-shuffled deck of playing cards. What is the probability that—

185. The card is an ace?
186. The card is a black card?
187. The card is a black card or an ace?
188. The card is a black card and an ace?
189. The card is a black card, given that it is an ace?
190. The card is an ace, given that it is a black card?
191. The card is an ace, given that it is not a black card?
192. The card is a black card, given that it is not an ace?

In a survey of 410 salespersons and 350 construction workers, it is found that 164 of the salespersons and 196 of the construction workers were overweight. If a person is selected at random from the group, what is the probability that—

193. This person is overweight?
194. This person is a salesperson, given that the person is overweight?

195. This person is overweight, given that the person is a salesperson?
196. This person is a construction worker, given that the person is not overweight?
197. This person is not overweight, given that the person is a construction worker?

The probability that a grocery shopper makes a purchase in the fresh-produce department is 0.59, while the probability that a shopper makes a purchase in the meat department is 0.71. The probability that a shopper makes a purchase in both the fresh-produce department and the meat department is 0.46.

198. What is the probability that a shopper makes a purchase in the produce department or in the meat department?
199. What is the probability that a shopper makes a purchase in the fresh-produce department, given that he made a purchase in the meat department?
200. What is the probability that a shopper makes a purchase in the meat department, given that he made a purchase in the fresh-produce department?
201. What is the probability that a shopper makes a purchase in both departments, given that he made a purchase in at least one of the departments?

Registration records showed that, of the 800 persons attending a conference, 528 were from the Midwest, 305 were self-employed, and 163 were from the Midwest and self-employed.

202. What is the probability that a person chosen at random from the group is self-employed, given that he is from the Midwest?
203. What is the probability that a person chosen at random from the group is self-employed, given that he is not from the Midwest?
204. What is the probability that a person chosen at random from the group is from the Midwest, given that he is self-employed?
205. What is the probability that a person chosen at random from the group is from the Midwest, given that he is not self-employed?

6.9 JOINT PROBABILITIES

The intersection of two events E and F, denoted $E \cap F$, is the event which contains only those elementary outcomes occurring in *both* E and F. If, as the result of a trial of the experiment, the intersection event $E \cap F$ occurs, *both* events E and F occur. The probability of such simultaneous occurrence of two events is called their JOINT PROB-

ABILITY. The rule for the joint probability of two events is obtained directly from the definition of conditional probability.

Rule 6 *If an experiment can lead to the event* E *and the event* F, *then the probability of the simultaneous occurrence of* E *and* F *is*

$$P(E \cap F) = P(E) \cdot P(F|E) = P(F) \cdot P(E|F) \qquad (16.15)$$

Example 19

A box contains two white marbles and one red marble. Two marbles are selected at random, without replacement, from the box. Let us define the events

W_1 = White marble on the first selection
W_2 = White marble on the second selection

Now, the probability of two white marbles being selected is computed

$$P(W_1 \cap W_2) = P(W_1) \cdot P(W_2|W_1)$$

Before the first selection is made, there are three marbles in the box, two of which are white. Thus, we agree that $P(W_1) = 2/3$. If a white marble is drawn, without replacement, on the first selection, there remain only two marbles—one white and one red—in the box when the second selection is made. Then the conditional probability of a white marble on the second selection, given a white marble on the first selection, is one half; that is, $P(W_2|W_1) = 1/2$.

And the probability of a white marble on both selections is

$$P(W_1 \cap W_2) = (2/3) \cdot (1/2) = 1/3 \qquad []$$

This rule of multiplication to obtain joint probabilities can easily be extended to more than two events.

Example 20

What is the probability of drawing three aces when three cards are drawn, without replacement, from a well-shuffled deck of playing cards?

Let us use A_1 to denote an ace on the first selection, A_2 to denote an ace on the second selection, and A_3 to denote an ace on the third selection.

There are four aces in the deck of 52 cards. Thus, $P(A_1) = 4/52 = 1/13$. Selections are made without replacement, making $P(A_2|A_1) = 3/51 = 1/17$, and $P(A_3|A_1 \cap A_2) = 2/50 = 1/25$.

Thus,

$$P(A_1 \cap A_2 \cap A_3) = (1/13)\ (1/17)\ (1/25) = 1/5{,}525 \qquad []$$

Dependent and Independent Events

Two events are said to be STATISTICALLY INDEPENDENT if the occurrence or nonoccurrence of one in no way affects the probability of occurrence of the other. Their conditional probabilities are equal to their unconditional probabilities. Therefore, the probability of the simultaneous occurrence of two or more independent events reduces to the product of their separate probabilities.

Rule 7 *If* E *and* F *are STATISTICALLY INDEPENDENT EVENTS defined on a sample space* S, *the probability of their simultaneous occurrence is given by*

$$P(E \cap F) = P(E) \cdot P(F) \qquad \textbf{(16.16)}$$

Example 21 If H_1 and H_2 represent getting heads on the first and second tosses of a fair coin, H_1 and H_2 are independent events; the outcome on the second toss is in no way affected by what happened on the first toss. Using T_1 to denote tails on the first toss, we note that

$$P(H_2|H_1) = P(H_2|T_1) = P(H_2) = 1/2$$

Thus,

$$\begin{aligned} P(H_1 \cap H_2) &= P(H_1)\ P(H_2|H_1) = P(H_1) \cdot P(H_2) \\ &= (1/2) \cdot (1/2) \\ &= 1/4 \end{aligned}$$

[]

The definition of independence of two events is symmetric; that is, if an event E is independent of an event F, then the event F is independent of the event E. Also, an event E is independent of an event F if and only if E is independent of the complement of F. Symbolically, for two independent events E and F, if

$$P(E) = P(E|F) = P(E|F^c)$$

then

$$P(F) = P(F|E) = P(F|E^c)$$

If the occurrence or nonoccurrence of one event does in some way affect the probability of occurrence of the other event, the two events are said to be STATISTICALLY DEPENDENT. The conditional probabilities of dependent events are not equal to their unconditional probabilities.

Statistical independence or dependence must be distinguished from causal independence or dependence. Because two events are statistically dependent does not imply that one is caused by the other.

Nor should the concept of independent events be confused with the concept of mutually exclusive, or disjoint, events. In essence, independence is a concept associated with the assignment of probabilities, whereas disjointedness is not defined in terms of probabilities. Venn diagrams can be used to represent the property of disjointedness of events but not to represent independence. Indeed, not only are disjointedness and independence two entirely different concepts, but they are almost incompatible. Disjoint sets are not independent unless one of them has a probability of zero. If E_1 and E_2 are disjoint events, they have no points in common, and their intersection, $E_1 \cap E_2$, is the empty set. If $E_1 \cap E_2 = 0$, then $P(E_1 \cap E_2) = 0$. If two events are independent, their joint probability is given by $P(E_1 \cap E_2) = P(E_1)P(E_2)$. For disjoint events then, $P(E_1 \cap E_2) = 0$ and either $P(E_2) = 0$ or $P(E_2) = 0$. If $E_1 \cap E_2 \neq 0$ so that $P(E_1 \cap E_2) \neq 0$, then E_1 and E_2 are not disjoint.

EXERCISES

206. Two cards are drawn without replacement from a well-shuffled deck of playing cards. What is the probability that they both are spades?
207. Three cards are drawn without replacement from a well-shuffled deck of playing cards. What is the probability that they are all diamonds?
208. A fair coin is tossed three times. What is the probability of heads on all three tosses?
209. What is the probability that heads will first appear on the fourth toss of a fair coin?
210. The probability that an Engineering 409 student will complete the first project on schedule is 0.65. The probability that he or she completes the second project on schedule, given that the first project was completed on time, is 0.80. What is the probability that a student completes both projects on schedule?

The probability that Simone visits Tangier this summer is one third. The probability that Noah visits Tangier this summer is three-fifths. Simone and Noah do not know each other and are totally unrelated. Find the probability that—

211. Simone and Noah will both visit Tangier this summer.
212. Simone will visit Tangier this summer but Noah will not.
213. Noah will visit Tangier this summer but Simone will not.
214. Neither Simone or Noah will visit Tangier this summer.

The probability that Russ Ambrose, novice golfer, gets a good shot if he uses the correct club is three eighths. The probability of a good shot with an incorrect club is only one eighth. There are six clubs in Russ's bag, only one of which is correct for a particular shot. Not knowing which club is the correct one, he selects a club at random.

215. What is the probability that Russ selects the correct club and gets a good shot?

216. What is the probability that he selects the correct club but gets a bad shot?
217. What is the probability that he selects an incorrect club but gets a good shot?
218. What is the probability that Russ selects an incorrect club and gets a bad shot?
219. What is the probability that he gets a good shot?
220. What is the probability that he gets a bad shot?

Given $P(A) = 0.3$, $P(B) = 0.52$, $P(A \cup B) = 0.58$, *find—*

221. $P(A^c)$
222. $P(B^c)$
223. $P(A \cap B)$
224. $P(A^c \cap B)$
225. $P(A \cap B^c)$
226. $P(A^c \cap B^c)$
227. $P(B|A)$
228. $P(B|A^c)$
229. $P(A|B)$
230. $P(A|B^c)$
231. Are the two classifications, A and B, statistically independent?

There are 400 workers employed at the Covington Plant of National Manufacturing Company, 240 on the day shift and 160 on the night shift. Of the 100 plant workers who have joined the company's credit union, 60 are day-shift workers. A worker is selected at random from the work force. Find—

232. *P*(credit union).
233. *P*(not credit union).
234. *P*(credit union and day shift).
235. *P*(credit union and night shift).
236. *P*(credit union, given day shift).
237. *P*(credit union, given night shift).
238. *P*(day shift, given credit union).
239. *P*(night shift, given credit union).
240. *P*(credit union or night shift).
241. *P*(credit union or day shift).
242. Are the two classifications "credit union or not" and "day or night shift" statistically independent?

16.10 SCHEMATIC REPRESENTATION OF RANDOM PROCESSES

We have used Venn diagrams to picture a sample space and events defined on that sample space. Two other devices that are often helpful in picturing a random process and its possible outcomes are the PROBABILITY TREE and the JOINT PROBABILITY TABLE.

The Probability Tree

Many random processes are multiphase processes. A PROBABILITY TREE provides a useful device for structurally defining the relationships within such processes and for computing probabilities associated with possible outcomes. (See Figure 16.6). A CHANCE NODE forms the trunk of the tree and represents the first phase of the process. Extending from this node is a set of BRANCHES, one for each possible result on this phase of the process. At the terminus of each of these branches is another chance node with a new set of branches, these representing the possible results on the second phase of the process. These branches are, in turn, followed by new chance nodes, with another set of branches for the third phase, and so on, until the process is completed.

Figure 16.6

A Probability Tree

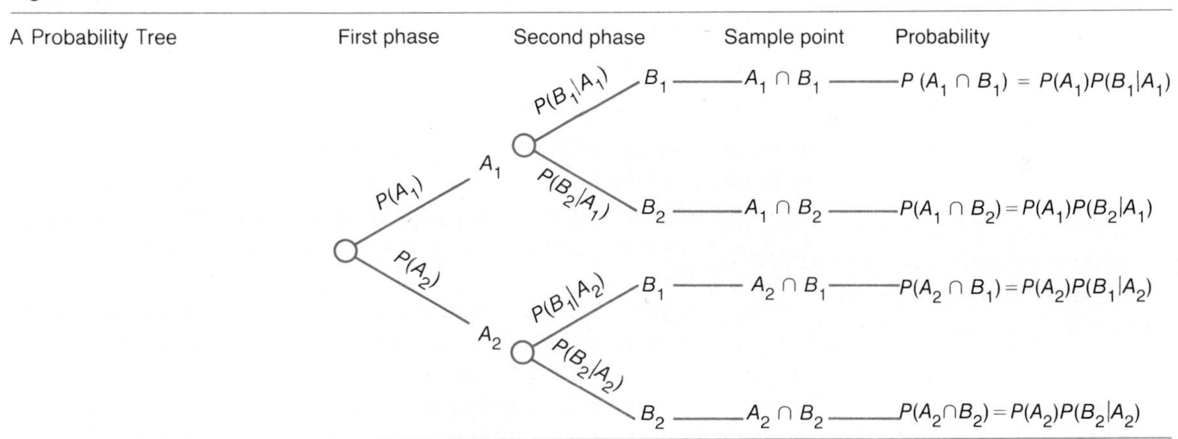

Noted on each of the branches is the probability of that result. After the first set of branches, these probabilities are conditional probabilities, the condition being the occurrence of the result on the preceding phase which leads to this particular chance node. A sequence of branches moving from the trunk of the tree to a tip is called a PATH. Paths through the tree are mutually exclusive and completely exhaustive; they represent all possible outcomes of the multiphase process, and the process can follow one and only one of the paths on a particular trial. The paths thus represent the points in the sample space. Probabilities for these sample points can be computed by multiplying the probabilities listed along the connecting branches of the path.

The probability tree pictured in Figure 16.7 depicts an experiment which involves selecting, without replacement, 2 batteries from a crate containing 10 batteries, all identical except for the fact that 4 of the 10 are defective (D) and 6 are nondefective (N). Note in Figure 16.7 that the possible outcomes on the first step in the process and their probabilities are shown on the first set of branches leading from the

Figure 16.7

A Probability Tree

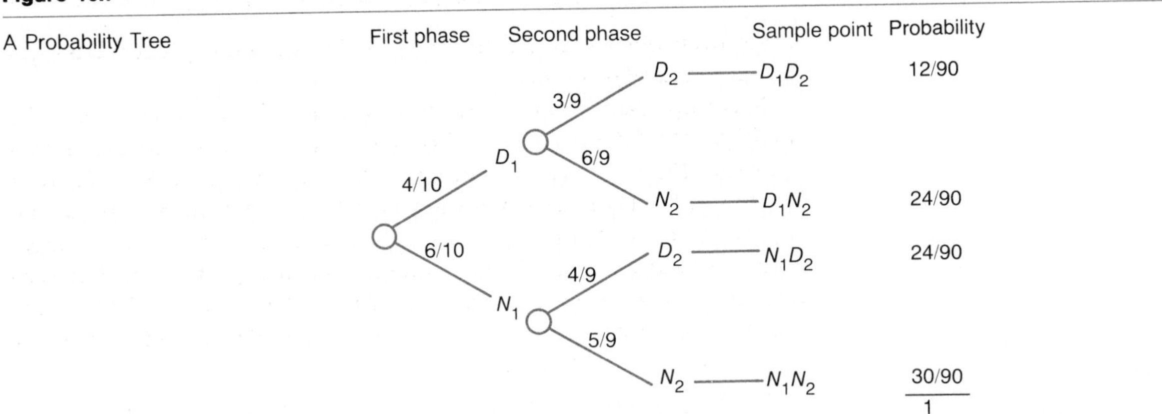

trunk of the probability tree. Because the first selection is made from 10 items, 4 of which are defective (D) and 6 of which are nondefective (N), the probability that the first item selected is defective is $P(D_1) = 4/10$. The probability that the first item selected is nondefective is $P(N_1) = 6/10$.

After the first item is selected, there remain nine items in the crate. The exact number of defectives and nondefectives in this group of items remaining in the crate depends upon the result of the first selection. Hence, the probability of a defective item, or a nondefective item, on the second selection is conditioned by the outcome of the first selection. Leading from each of the branches representing possible results on the first selection is a complete set of branches representing possible results on the second selection. The probabilities on these branches are conditional probabilities. Note the set of branches leading from the chance node following the outcome D_1. If the item selected first was defective, there remain only three defective and six nondefective items in the crate. Accordingly, $P(D_2|D_1) = 3/9$ and $P(N_2|D_1) = 6/9$. The probabilities on the set of branches leading from the chance node following the outcome N_1 are based on the condition that the first selection resulted in a nondefective item. If the item selected first was nondefective, there remain four defective and only five nondefective items in the crate, and the conditional probabilities are $P(D_2|N_1) = 4/9$ and $P(N_2|N_1) = 5/9$.

The joint probability of defective item on the first selection followed by defective item on the second selection is given by

$$P(D_1 \cap D_2) = P(D_1) \cdot P(D_2|D_1) = (4/10) \cdot (3/9) = 12/90 = 2/15$$

The joint probability of two nondefective items on the two selections is given by

$$P(N_1 \cap N_2) = P(N_1) \cdot P(N_2|N_1) = (6/10) \cdot (5/9) = 30/90 = 1/3$$

Other probabilities for the outcomes of the two-phase process are computed in a similar manner.

It is important to note that the total probability, often called the MARGINAL PROBABILITY, of a defective item on the second selection is neither three ninths nor four ninths but is, instead, $12/90 + 24/90 = 36/90 = 2/5$. There are two ways of getting a defective item on the second selection: (1) along with a defective item on the first selection, with probability of $P(D_1 \cap D_2) = 12/90$; or (2) along with a nondefective item on the first selection, with probability of $P(N_1 \cap D_2) = 24/90$. These are mutually exclusive outcomes; hence, the probability of that either one or the other will occur is

$$P(D_2) = P(D_1 \cap D_2) + P(N_1 \cap D_2) = 12/90 + 24/90 = 2/5$$

In a like manner, the total probability of a nondefective item on the second selection is $P(N_2) = 3/5$.

The Joint Probability Table

Another valuable tool for structuring relationships in multifaceted processes and for calculating relevant probabilities is the JOINT PROBABILITY TABLE. This tabular presentation is perhaps most useful when the outcomes are categorized in two ways. It can then be used to display directly all joint probabilities and all marginal probabilities and allows for convenient computation of all conditional probabilities. (See Table 16.2)

Table 16.2

A Joint Probability Table

	B_1	B_2	Total
A_1	$P(A_1 \cap B_1)$	$P(A_1 \cap B_2)$	$P(A_1)$
A_2	$P(A_2 \cap B_1)$	$P(A_2 \cap B_2)$	$P(A_2)$
Total	$P(B_1)$	$P(B_2)$	1

To illustrate, consider an experiment conducted by a large discount department store to study the purchasing patterns of men and women shoppers in that store. The experiment consisted of noting the proportion of weekday and weekend shoppers who made at least one purchase before leaving the store and the proportion who left the store without making a purchase. The information accumulated is displayed in Table 16.3.

Note that shoppers are classified in two ways: (1) weekday (D) or weekend (E) shopper and (2) make a purchase (P) or not make a purchase

Table 16.3

Joint Probability Table

	(D) Weekday shopper	(E) Weekend shopper	Total
Purchaser (P)	0.2	0.1	0.3
Nonpurchaser (N)	0.4	0.3	0.7
Total	0.6	0.4	1.0

(N). The probabilities in the margins are marginal probabilities. (In fact, the term *marginal probability* refers to the fact that these probabilities are found in the margins of a joint probability table.) A marginal probability refers to the probability of the occurrence of an event not conditioned by the occurrence of another event. Thus, if an individual is selected at random from this group, the marginal probabilities are

$$P(\mathrm{P}) = 0.3 \qquad P(\mathrm{D}) = 0.6$$
$$P(\mathrm{N}) = 0.7 \qquad P(\mathrm{E}) = 0.4$$

The probability of the joint occurrence of two events is shown in the cells of the body of the table. For example, the joint probability that a randomly selected shopper is a weekend shopper and makes a purchase is 0.1, read directly from the cell at the intersection of the purchaser row and weekend column. Other joint probabilities that can be read directly from the cells in the body of the table are

$$P(\mathrm{P} \cap \mathrm{D}) = 0.2 \qquad P(\mathrm{N} \cap \mathrm{D}) = 0.4 \qquad P(\mathrm{N} \cap \mathrm{E}) = 0.3$$

Conditional probabilities may be conveniently computed from the joint probability table. For instance, the conditional probability "purchaser" given "weekend" is

$$P(\mathrm{P}|\mathrm{E}) = \frac{P(\mathrm{P} \cap \mathrm{E})}{P(\mathrm{E})} = 0.1/0.4 = 0.25$$

or the conditional probability of "purchaser" given "weekday" is

$$P(\mathrm{P}|\mathrm{D}) = \frac{P(\mathrm{P} \cap \mathrm{D})}{P(\mathrm{D})} = 0.2/0.6 = 0.3\overline{3}$$

We readily see from these computations that the two events are not statistically independent, since $P(\mathrm{P}) \neq P(\mathrm{P}|\mathrm{E}) \neq P(\mathrm{P}|\mathrm{D})$.

EXERCISES

Each job applicant at the Bureau of Buffoons is given a battery of three tests. Sixty percent of all applicants pass the first test. Eighty percent of those who pass the first test also pass the second test, while only 30 percent of those who do not pass the first test pass the second. Ninety percent of those who pass both of the first two tests pass the third; 60 percent of those who pass only one of the first two tests pass the third; and only 10 percent of the applicants who fail the first two tests pass the third. If an applicant is selected at random, what is the probability that he or she—

243. Passes all three tests?
244. Passes none of the tests?
245. Passes the first two and fails the third.
246. Passes the first but fails the second and third?
247. Passes two or more of the tests?
248. Passes exactly one of the tests?

Gravitt Automobile Sales, Inc., has accumulated the following information relative to its sales of automobiles.

	Method of payment		
Type of automobile purchased	Cash (C)	Installment (I)	Total
New (N)	0.15	0.55	0.70
Used (U)	0.25	0.05	0.30
Total	0.40	0.60	1.00

Find:

249. $P(\text{U})$
250. $P(\text{N})$
251. $P(\text{C})$
252. $P(\text{I})$
253. $P(\text{N} \cap \text{C})$
254. $P(\text{N} \cup \text{C})$
255. $P(\text{N}|\text{C})$
256. $P(\text{C}|\text{N})$
257. Are the two methods of classification statistically independent?

Of all persons who voted in the last local referendum, 35 percent voted negative on the school-bond issue, 45 percent voted negative on the rezoning issue, and 15 percent voted negative on both issues. A voter is selected at random.

258. What is the probability that he voted positive on the school-bond issue?
259. What is the probability that he voted positive on the rezoning issue?

260. What is the probability that he voted positive on the school-bond issue, given that he votes negative on the rezoning issue?
261. What is the probability that he voted positive on both issues?
262. What is the probability that he voted positive on at least one issue?
263. What is the probability that he voted negative on at least one issue?
264. Are the two events "voted positive on school-bond issue" and "voted positive on the rezoning issue" statistically independent? Why or why not?

The College of Business Administration of Allover University has analyzed the records of the students enrolled in its master's degree program. Specifically, the students have been classified by undergraduate degree as business (B) or nonbusiness (N) and by scholastic standing as honors student (H) or nonhonors (NH). Twenty-five percent of the students are classified as honors students; 60 percent have undergraduate degrees in business. Fifteen percent both are honors students and have undergraduate degrees in business.

If a student is selected at random from the group, what is the probability that he or she—

265. Has an undergraduate degree in business?
266. Is an honors student and has a business undergraduate degree?
267. Is a nonhonors student, given that he or she has an undergraduate degree in business?
268. Has an undergraduate degree in business, given that he or she is an honors student?
269. Has either an undergraduate degree in business or is an honors student?
270. Has a nonbusiness undergraduate degree?
271. Has a nonbusiness undergraduate degree given that he or she is an honors student?
272. Is an honors student, given that he or she has a nonbusiness undergraduate degree?
273. Is an honors student and has a nonbusiness undergraduate degree?
274. Are the two methods of classification statistically independent? Why or why not?

A large manufacturing plant has determined that 250 of its 400 employees are over 35 years of age. Three hundred of all the employees participate in the company's private retirement plan, but only 125 of those under 35 years of age participate in the retirement plan.

If an employee is selected at random, what is the probability that he or she—

275. Is not over 35 years of age?
276. Does not participate in the company's retirement plan?
277. Is either over 35 years of age or participates in the retirement plan?
278. Participates in the retirement plan and is over 35 years of age?
279. Participates in the retirement plan, given that he or she is over 35 years of age?
280. Participates in the retirement plan, given that he or she is not over 35 years of age?
281. Does not participate in the retirement plan, given that he or she is over 35 years of age?
282. Are the two methods of classification statistically independent? Why or why not?

16.11 BAYES' RULE

BAYES' RULE results from a special application of the definition of conditional probability. Given an event B that can occur only if one of the mutually exclusive and completely exhaustive events $A_1, A_2, \ldots, A_n$ occurs, the probability that B will occur is given by

$$\begin{aligned} P(B) &= P(A_1 \cap B) + P(A_2 \cap B) + \ldots + P(A_n \cap B) \\ &= P(A_1)P(B|A_1) + P(A_2)P(B|A_2) + \ldots + P(A_n)P(B|A_n) \\ &= \sum_{i=1}^{n} P(A_i)P(B|A_i) \end{aligned}$$

(See Figure 16.8.)

The conditional probability of event A_k, given that event B has occurred, is

Bayes' rule

$$P(A_k|B) = \frac{P(A_k)P(B|A_k)}{\sum_{i=1}^{n} P(A_i)P(B|A_i)} = \frac{P(A_k \cap B)}{P(B)} \quad (16.17)$$

Figure 16.8

Event B Can Occur Only if One of the Events A_1, A_2, A_3, or A_4 Occurs

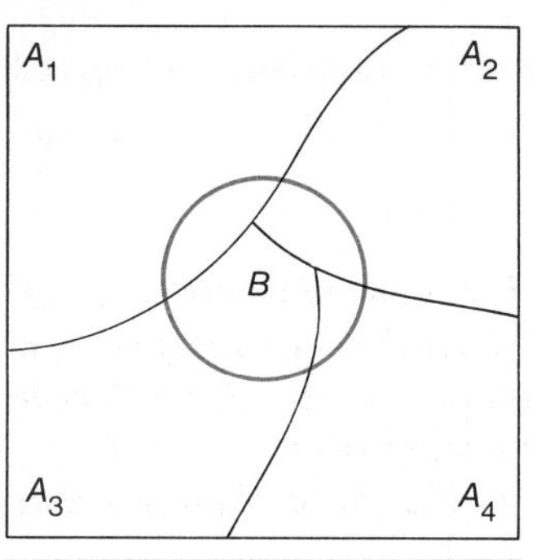

This is Bayes' rule. It occupies a pivotal position in modern decision theory, where it is generally used to adjust a set of probabilities by new evidence which has become available. The events $A_1, A_2, \ldots, A_n$ are often assigned probabilities based on past experience or on the personal belief of the decision maker or a little bit of each. These are commonly called PRIOR PROBABILITIES because they are assigned prior to any direct information that the experiment itself may yield. Then, a trial of the experiment is performed and, as a result, event B occurs. Now we may wish to use the new information made available by the occurrence of B to revise the prior probabilities and obtain POSTERIOR PROBABILITIES. In other words, we adjust the original probabilities, $P(A_1)$, $P(A_2)$, . . . , $P(A_n)$, by the new partial information which has become available as a result of a trial of the experiment and obtain $P(A_1|B)$, $P(A_2|B)$, . . . , $P(A_n|B)$. The significant aspect of Bayes' rule is that it entails a backward sort of reasoning—moving from the tip of the probability tree back to the trunk. It also enables one to start with a prior probability and to update this as new information is provided by the experiment itself.

Example 22 An engineer regularly adjusts a machine. The probability of achieving a correct adjustment is 0.9. If the machine is in proper adjustment, only 1 percent of the items it produces are defective. If, however, the machine is not properly adjusted, 25 percent of the items produced are defective. Given that an item selected at random from the output of the machine is defective, what is the probability that the machine is in proper adjustment?

A probability tree, such as that shown in Figure 16.9, may be useful in picturing the situation. Let us begin by defining the events:

A = The machine was properly adjusted
NA = The machine was not properly adjusted
D = The item is defective
ND = The item is not defective

Event D can occur in either of two ways: (1) A ∩ D, with probability (0.9)(0.01) = 0.009; or (2) NA ∩ D, with probability (0.1)(0.25) = 0.025. The total probability of D is $P(\text{D}) = 0.009 + 0.025 = 0.034$. Given that

Figure 16.9

Probability Tree

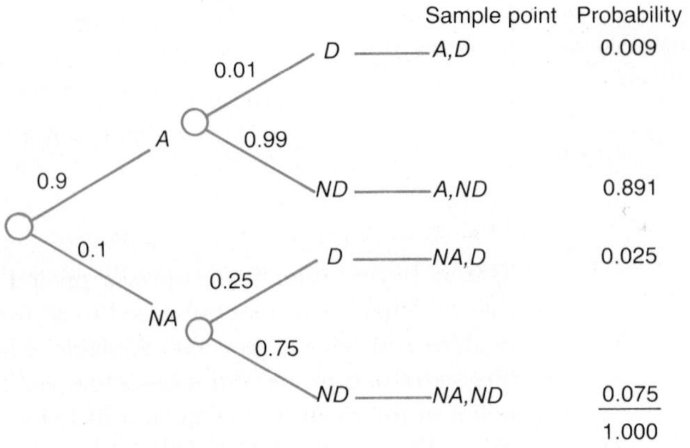

D has occurred (the item selected is defective), the conditional probability of a properly adjusted machine is found by

$$P(\mathrm{A}|\mathrm{D}) = \frac{P(\mathrm{A} \cap \mathrm{D})}{P(\mathrm{D})} = 0.025/0.034 = 0.265$$

The probability of an improperly adjusted machine, given that the item selected is defective, is

$$P(\mathrm{NA}|\mathrm{D}) = \frac{P(\mathrm{NA} \cap \mathrm{D})}{P(\mathrm{D})} = 0.025/0.034 = 0.735$$ []

EXERCISES

An executive of Peppers, Inc., lives in a suburb 25 miles from his office. Each day he may either drive to work or ride the rapid-transit train. He rides the train about 60 percent of the time and drives the other times. If he drives, he gets to work by 9 A.M. 75 percent of the time. If he rides the train he gets to work by 9 A.M. 95 percent of the time.

283. Overall, on what proportion of the time does he get to work by 9 A.M.?
284. If he gets to work late, what is the probability that he drove?
285. If he gets to work by 9 A.M., what is the probability that he drove?

At Edmund Electronics the production machinery is adjusted by maintenance engineers before each production run. The engineers are not infallible and, in fact, fail to achieve a proper adjustment of the machinery 20 percent of the time. If the machinery is properly adjusted, only 5 percent of the items produced are defective. However, if the machinery is not properly adjusted, 30 percent of the items produced are defective.

286. What proportion of all items produced will be defective?
287. Given that an item selected at random is found to be defective, what is the probability that the machinery was not properly adjusted?
288. Given that an item selected at random is found to be nondefective, what is the probability that the machinery was properly adjusted?

Big John Lesser hosts the early morning "Farm and Home Hour" on WRAD radio and must wake up at 4:00 each morning in order to get to the studio on time. He uses an alarm clock that rings properly with probability 0.8. If it rings, the probability is 0.8 that Big John will wake up. If the alarm does not sound, there is a probability of 0.3 that he will wake up on his own.

289. What is the probability that Big John will wake on time?

290. Given that Big John woke up on time, what is the probability that the alarm sounded?

291. Given that Big John did not wake up on time, what is the probability that the alarm sounded?

Three machines, I, II, and III, are used to manufacture the output of Daisy Manufacturing Company. Machine I produces 40 percent of the total output of the company; machine II produces 50 percent of the total output; and machine III produces 10 percent. The percentage of defective items produced by machines I, II, and III, respectively, are 2 percent, 3 percent, and 1 percent. An item is selected at random from the total output of the company.

292. What is the probability that it is defective?

293. What is the probability that it was produced by machine I, given that it is defective?

294. What is the probability that it was produced by machine I, given that it is nondefective?

295. What is the probability that it was produced by machine II, given that it is defective?

C·H·A·P·T·E·R 17

Random Variables and Their Probability Distributions

17.1 THE NATURE OF A RANDOM VARIABLE

In many instances, interest is not focused on the minute details associated with the individual sample points themselves but is directed instead toward some property of these sample points which can be summarized by a set of real numbers. If 10 items were selected from the output of a machine, for example, and each item classified as defective (D) or nondefective (N), there would be $2^{10} = 1{,}024$ elementary sample points in the sample space; that is

$$S = \{NNNNNNNNNN, NNNNNNNNND, NNNNNNNNDN, \ldots, DDDDDDDDDD\}$$

The analysis of the experimental outcomes may not deal with the 1,024 different possible sequences of Ds and Ns but may center around the NUMBER OF DEFECTIVE ITEMS observed in the group of 10 items inspected. Thus, the number of defectives observed—0, 1, 2, . . . , 10—would be associated with each of the sample points. These numbers, assigned to the sample points, are random quantities determined by the outcome of a random experiment. Any such numerical quantity whose value is determined by the outcome of a random experiment is known as a RANDOM VARIABLE. Other frequently used terms for the same concept are CHANCE VARIABLE or STOCHASTIC VARIABLE.

By use of a random variable, we are able to associate a real number with each possible outcome of a random experiment. When the performance of the experiment results in a particular outcome, we say that the random variable *assumes* or *takes on* the numerical value corresponding to that outcome.

Example 1 Consider, again, the experiment of tossing a fair coin three times and noting whether the upturned face is heads (H) or tails (T). The sample space for the experiment is

$$S = \{HHH, HHT, HTH, HTT, THH, THT, TTH, TTT\}$$

Now let us define a random variable as "the number of heads observed" and symbolize this variable by x. This variable is used to associate a real number with each point in the sample space, as

Sample point	Value of the random variable x
HHH	3
HHT	2
HTH	2
HTT	1
THH	2
THT	1
TTH	1
TTT	0

The variable x takes on the value 3 if the outcome of a trial of the experiment is HHH; it takes on the value 2 if the outcome is either HHT, or HTH, or THH; and so on. []

Discrete and Continuous Random Variables

A random variable is termed a DISCRETE RANDOM VARIABLE if it can take on only a finite number of values or if it can take on infinitely many values that are separated one from the other by some nonzero space. A random variable is CONTINUOUS if its possible values form an entire interval of numbers.

In most practical problems, discrete random variables represent count or enumeration data, such as the number of books on a shelf, the number of cars passing a tollgate, or the number of defective items in a shipment of parts. On the other hand, continuous random variables usually represent measurement data, such as weights of items or the distance from one point to another.

17.2 PROBABILITY DISTRIBUTIONS FOR DISCRETE RANDOM VARIABLES

Two of the most important properties of a random variable are (1) the values it can assume and (2) the probabilities that are associated with each of these possible values. Because values assumed by a random variable can be used to symbolize all outcomes associated with a given random experiment, the probability originally assigned to an elementary sample point can now be assigned as the likelihood that the random variable takes on the value corresponding to that outcome.

The set of ordered pairs $(x,P(x))$—where x represents a value of the random variable and $P(x)$ represents the probability associated with that value—is called the PROBABILITY DISTRIBUTION or PROBABILITY FUNCTION of the discrete random variable. Thus a probability function for a discrete random variable is a mutually exclusive and completely exhaustive listing of all values that the random variable can assume along with the probabilities associated with each of these values. In addition to being enumerated as a set of ordered pairs, the probability function may take the form of a table, a graph, or a mathematical equation. Whatever its form, the probability function has these properties:

1. $0 \leq P(x_i) \leq 1$
2. $\sum_i P(x_i) = 1$

Example 2

The eight elementary outcomes of the experiment which involves tossing a fair coin three times are equally likely to occur, each having probability $P(O_i) = 1/8$. The probability distribution for the random variable x, defined on this experiment as the number of heads observed, is obtained as follows:

$P(x=0) = P(\text{TTT}) = 1/8$
$P(x=1) = P(\text{HTT}) + P(\text{THT}) + P(\text{TTH}) = 1/8 + 1/8 + 1/8 = 3/8$
$P(x=2) = P(\text{HHT}) + P(\text{HTH}) + P(\text{THH}) = 1/8 + 1/8 + 1/8 = 3/8$
$P(x=3) = P(\text{HHH}) = 1/8$

The probability function may be expressed as the set of ordered pairs

$$p = \{(0,1/8), (1,3/8), (2,3/8), (3,1/8)\}$$

Or, the probability function may be given in tabular form, as shown in Table 17.1. Or, the function may be given in graphic form as shown in Figure 17.1. []

Table 17.1

Probability Distribution for the Random Variable x, Where x Is the Number of Heads Observed When Three Coins Are Tossed

Value of variable x	Probability $P(x)$
0	1/8
1	3/8
2	3/8
3	1/8
	1

Figure 17.1

Probability Distribution (in Graphic Form) for the Random Variable x, Where x Is the Number of Heads Observed When Three Coins Are Tossed

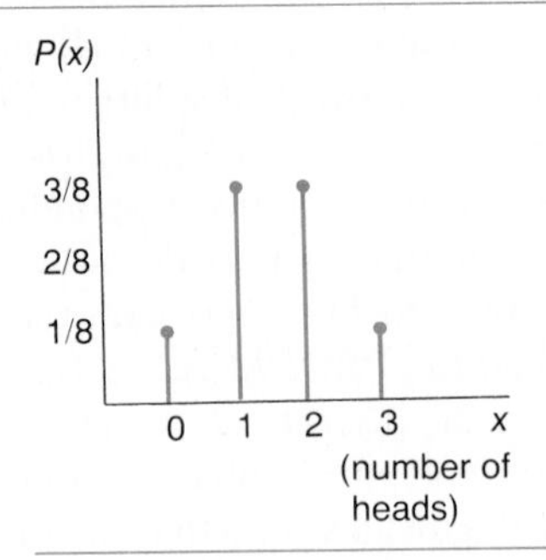

Example 3 An experiment involves selecting at random three items from a production process and observing the number of defective items in the group selected. Ten percent of all items produced by the process are defective, and these defective items occur at random. The other items produced are nondefective.

The sample space consists of $2^3 = 8$ sample points. These are listed in Figure 17.2. The random variable x is defined on this sample space as the number of defective items observed. The value of the random variable and the probability associated with each point are also shown in Figure 17.2. The probability distribution (in tabular form) summarizing this information is shown in Table 17.2.

Figure 17.2

Sample Space and a Random Variable Defined on That Sample Space

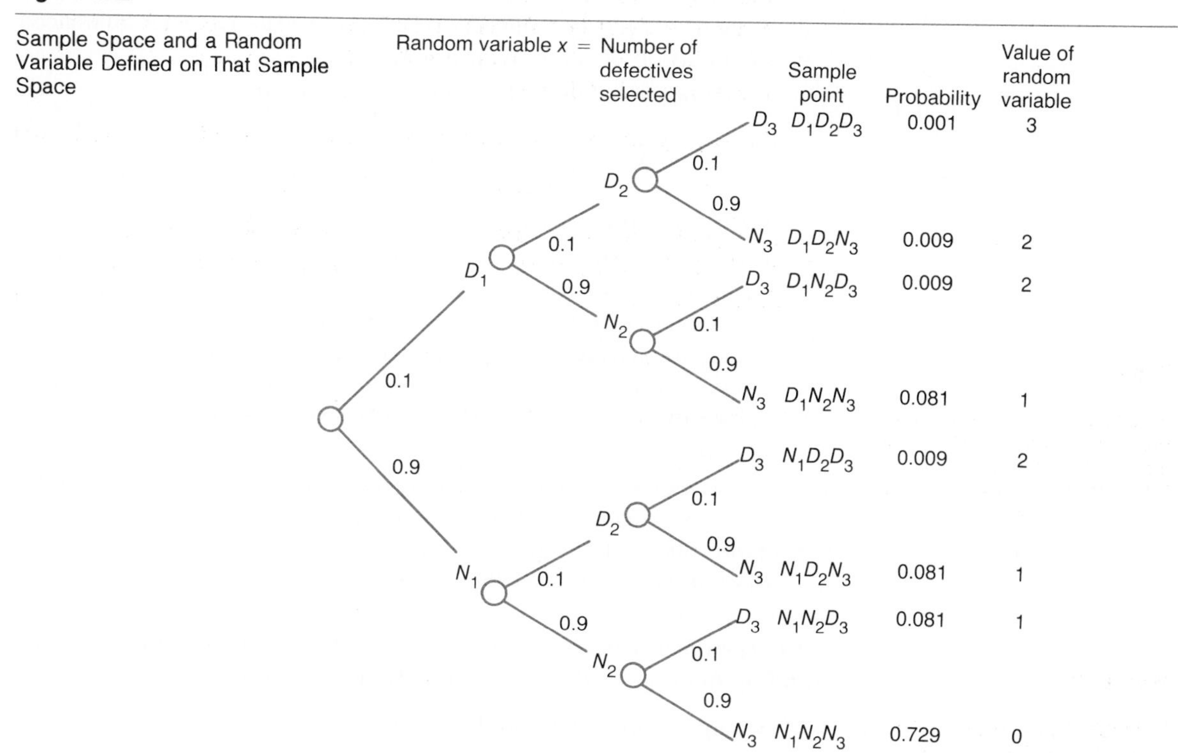

Table 17.2

Probability Distribution for Random Variable x, Where x Is the Number of Defective Items Observed

Value of variable x	Probability $P(x)$
0	0.729
1	0.243
2	0.027
3	0.001
	1.000

EXERCISES

1. Set up, in tabular form, the probability distribution for the random variable X, defined as the number of face cards drawn when two cards are selected at random, without replacement, from a deck of cards containing 16 face cards and 36 cards that are not face cards.
2. Set up, in tabular form, the probability distribution for the random variable X, defined as the number of bull's-eyes scored by a sharpshooter firing three shots at a target, given that the shooter has a

probability of 0.01 of scoring a bull's-eye and has found that when firing more than one shot the result of the nth shot is not influenced by the result of the previous shot.

3. A random variable X is defined by the function

$$P(X) = \frac{4!}{X!(4-X)!}(1/16) \qquad \text{for } X = 0, 1, 2, 3, 4$$

Set up, in tabular form, the probability distribution for X.

4. Records show that 75 percent of the homes in a certain neighborhood are underinsured (that is, do not have insurance coverage sufficient to completely restore the house if it were destroyed by fire). Three homes are selected at random from the neighborhood. Letting the random variable X represent the number of underinsured homes in the group selected, set up, in tabular form, the probability distribution for X.

5. A student is asked to match three historical events with their calendar dates. Not having studied and having no knowledge of the correct dates, the student guesses. What is the probability distribution for the number of answers that are right?

An experiment consists of tossing a pair of dice (one red and one blue) and noting the number of dots on the two upturned sides.

6. Set up the probability distribution, in tabular form, for the random variable X, defined as the sum of the number of dots on the upturned sides.

7. Set up the probability distribution, in tabular form, for the random variable Y, defined as the absolute difference between the number of dots on the upturned side of one die and the number of dots on the upturned side of the other die.

8. Set up the probability distribution, in tabular form, for the random variable Z, defined as follows:

$$Z = \begin{cases} 1 & \text{If the number of dots on the red die is greater than the number of dots on the blue die} \\ 0 & \text{If the number of dots on the red die is equal to the number of dots on the blue die} \\ 2 & \text{If the number of dots on the red die is less than the number of dots on the blue die} \end{cases}$$

17.3 DESCRIPTIVE MEASURES OF A RANDOM VARIABLE

The probability function of a random variable may be summarized using various descriptive measures. The most important of these measures are the EXPECTED VALUE and the VARIANCE of the random variable.

Expected Value of a Discrete Random Variable

The EXPECTED VALUE of a random variable gives a quick picture of the long-run "average" result when the experiment is performed over and over again. It is a weighted average of the possible values of the variable, the weights being the likelihood of occurrence.

> Given a discrete random variable X with possible values $x_1, x_2, \ldots, x_n$, and probability $P(X = x_i) = P(x_i)$, the *expected* (or MEAN) *value* of X, denoted μ or $E(X)$, is defined as
>
> $$\mu = E(X) = \sum_{i=1}^{n} x_i P(x_i)$$

The computations required for determining the expected value of the variable whose probability distribution is given in Table 17.2 are shown in Table 17.3. Note that the expected value of the discrete random variable is not always one of the values the variable can actually take on, nor is it always an integer. It is important to remember that the term *expected value* is an abbreviation for the more-proper term *mathematical expectation.* Mathematical expectation represents that constant around which the values of the random variable oscillate. There should be no interpretation of this measure as a frequent, highly probable, or even possible value of the variable. It is simply the weighted average of the possible values, the weights being the probabilities of occurrence for those values.

Table 17.3

Expected Value and Variance of the Random Variable x

Value of x	Probability	Calculations			
x	$P(x)$	$x\,P(x)$	$x - E(x)$	$[x - E(x)]^2$	$P(x)[x - E(x)]^2$
0	0.729	0	−0.3	0.09	0.06561
1	0.243	0.243	0.7	0.49	0.11907
2	0.027	0.054	1.7	2.89	0.07803
3	0.001	0.003	2.7	7.29	0.00729
		0.300			0.27000

$E(x) = 0.3$
$\mathrm{Var}(x) = 0.27$
$\mathrm{s.d.} = 0.52$

Variance and Standard Deviation for a Discrete Random Variable

The expected value of a variable gives no indication of the dispersion of the values of the variable and does not, in itself, adequately describe the variable. The VARIANCE, usually denoted σ^2 (read "sigma squared") or VAR(X), is one measure of dispersion and represents spread or scatter of the values of the variable around their expected value.

The variance of a discrete random variable is the mathematical expectation of the squares of the deviations of the values of the variable from its expected value. If X is a discrete random variable with possible values $x_1, x_2, \ldots, x_n$ which occur with probabilities $P(x_i)$, and $E(X)$ is the expected value of X, the variance of X is defined as

$$\sigma^2 = \text{VAR}(X) = \sum_{i=1}^{n} [x_i - E(X)]^2 P(x_i)$$

or equivalently, as

$$\text{VAR}(X) = E(X^2) - [E(X)]^2$$

where

$$E(X^2) = \sum_{i=1}^{n} x_i^2 P(x_i)$$

One difficulty with the use of variance as a measure of dispersion is that it does not measure dispersion in the same units as the value of X. The dimensionality of the variance is the square of the dimensionality of the random variable itself. If X represents inches, $E(X)$ is a certain number of inches; but since it is the mean square deviation, VAR(X) is measured in inches squared. In order to have a measure of dispersion in the same units as the original values of X, we define STANDARD DEVIATION of the random variable as the nonnegative square root of the variance.

The standard deviation of a random variable X, usually denoted σ or s.d., is the nonnegative square root of the variance of X; that is,

$$\sigma = \text{s.d.} = \sqrt{\text{VAR}(X)}$$

Variance and standard deviation for the random variable defined on the experiment outlined in Figure 17.2. are computed in Table 17.3.

In mathematical models, variance is often used as an intuitive measure of uncertainty. A random variable that has a variance of zero can take on only one value; that is, there is only one possible outcome for the experiment that generates this random variable. Thus, there is no uncertainty about what this outcome will be. The larger the variance of a random variable, the greater diversity of possible outcomes for the experiment and, correspondingly, the more uncertainty about what that outcome will be.

EXERCISES

9. Records show that the number of accidents occurring on a weekday between 4:30 P.M. and 7:30 P.M. at the intersection of Ivy and Vine is 0, 1, 2, 3, 4, with probabilities 0.01, 0.19, 0.50, 0.20, and 0.10. Find the expected value of the random variable X, defined as the number of accidents occurring during this time period at this location.

10. Find the expected value, variance, and standard deviation of the random variable X whose probability distribution is as follows:

X	$P(X)$
1	0.3
2	0.3
3	0.2
4	0.1
5	0.1
	1.0

11. A random variable X, defined as the number of typing errors per page made by a group of typists, has the following probability distribution:

X	$P(X)$
0	0.80
1	0.12
2	0.04
3	0.03
4	0.01
	1.00

Find the expected value, variance, and standard deviation of the random variable.

12. An urn contains 10 white and 6 red marbles. Three marbles are selected at random, without replacement, from the urn. Set up the probability distribution for the random variable X, defined as the number of red marbles selected. Determine the expected value, variance, and standard deviation of the variable.

13. Three batteries are chosen at random from a lot containing 12 batteries, 2 of which are defective. Let X be defined as the number of defective batteries selected, and set up the probability distribution for X. Determine the expected value and variance of X.

14. A game of chance consists of selecting 4 marbles, without replacement, from an urn containing 100 red, 100 blue, and 100 green marbles. The player is paid $100 if four red marbles are selected, $50 if three red marbles are selected, $10 if two red marbles are selected, and $1 if one red marble is selected. What is the expected value of the random variable X, defined as the payoff for a selection?

15. One thousand tickets are sold in a lottery. Three prizes will be given: first prize is $500, second prize is $200, and third prize is $50. A ticket costs $2. If the random variable X is defined as the net gain on one ticket, find the expected value of X.

Seventy percent of the voters in a certain district are registered as Democrats, 25 percent are registered as Republicans, and 5 percent are registered as Independents. Four voters are selected at random from the district.

16. Set up the probability distribution, in tabular form, for the random variable X, defined as the number of Democrats selected.
17. Display the probability distribution for X in graphic form.
18. Find $E(X)$ and VAR(X).
19. Set up the probability distribution, in tabular form, for the random variable Y, defined as the number of Republicans selected.
20. Set up the probability distribution for Y in graphic form.
21. Find $E(Y)$ and VAR(Y).
22. Set up the probability distribution, in tabular form, for the random variable Z, defined as the number of Independents selected.
23. Find $E(Z)$ and VAR(Z).

17.4 THE BINOMIAL PROBABILITY DISTRIBUTION

The model most frequently encountered among the discrete probability distributions is associated with repeated, but independent, trials of an experiment which has only two possible outcomes. This is the BINOMIAL PROBABILITY DISTRIBUTION.

The properties of the binomial experiment are:

1. There are *only two possible outcomes* on each trial of the experiment: these are commonly labeled "success" and "failure."
2. There are n trials making up the experiment.
3. The trials of the experiment are *statistically independent* of one another; the probabilities associated with the two classes of outcomes are constant from one trial to the next.

The BINOMIAL RANDOM VARIABLE, x, defined on the sample space generated by a binomial experiment, takes on a value that denotes the number of "successes" in n trials of the experiment. It is a discrete variable which can assume the values $0, 1, 2, \ldots, n$.

The probability of "success" on any trial of the experiment is p; the probability of "failure" is $1 - p = q$. The successive trials of the experiment are independent. Hence, the probability p of success is constant for all trials. The quantities n and p are called the PARAMETERS of the binomial experiment because, for any specified experiment, they are constant and they determine the probabilities for all values of the variable x.

Binomial probability function The probability of exactly x "successes" in n trials of a binomial experiment is given by

$$b(x|n,p) = \frac{n!}{x!(n-x)!} p^x q^{(n-x)} \qquad x = 0,1,\ldots,n$$

where p is the probability of success and $q = 1 - p$ is the probability of failure.

The notation $b(x|n,p)$ is read "the binomial probability of x successes, given n trials with probability p of success."

To better understand this function, let us consider a particular sequence of outcomes of n trials for a binomial experiment, each outcome of which is either a success (s) or a failure (f):

$$\underbrace{sss\ldots ss}_{\substack{\text{for } x\\ \text{trials}}}\underbrace{fff\ldots ff}_{\substack{\text{for } (n-x)\\ \text{trials}}}$$

The probability of each success is p and that of each failure is q. Because the trials are independent, we apply the multiplicative law for probability of a sequence of independent events and obtain

$$\underbrace{ppp\ldots pp}_{\substack{\text{for } x\\ \text{trials}}}\underbrace{qqq\ldots qq}_{\substack{\text{for } (n-x)\\ \text{trials}}}$$

For any one sequence of x successes and $(n - x)$ failures, the probability is given by

$$p^x q^{(n-x)}$$

The total number of different sequences of n outcomes, x of which are successes and $(n - x)$ of which are failures, is

$$\frac{n!}{x!(n-x)!}$$

Hence, the probability of x successes in n trials of a binomial experiment is

$$\frac{n!}{x!(n-x)!} p^x q^{(n-x)}$$

where p is the probability of success and q is $1 - p$.

The expected value and variance of the binomial random variable can be found by

$$E(X) = np$$
$$\mathrm{VAR}(X) = npq$$

Example 4

Determine the probability of getting exactly three heads (in any order) on five tosses of a fair coin.

Note that the requisite conditions for the binomial random experiment are met:

1. There are exactly two possible outcomes on each trial: heads ("success") or tails ("failure").
2. Successive trials of the experiment are statistically independent; the outcome of one toss does not affect the outcome of the following toss. Hence, the probability of success is constant from one toss to the next (and, here, is $p = 0.5$ with $q = 1 - 0.5 = 0.5$).
3. There are five trials in the experiment.

The binomial random variable x is defined for this experiment as the number of heads experienced in five tosses. Using the binomial formula we calculate

$$b(3|5,0.5) = \frac{5!}{3!2!}(0.5)^3(0.5)^2 = (10)(0.125)(0.25) = 0.3125$$

Thus, the probability of getting exactly three heads on five tosses of a fair coin is 0.3125. []

Example 5

Construct the probability distribution for the binomial random variable x, defined in the preceding example.

To construct the probability distribution, we must determine the possible values for the random variable and the probabilities associated with each of these values. In five tosses of a coin, we may experience 0, 1, 2, 3, 4, or 5 heads. Probabilities are computed using the binomial formula as:

$$b(0|5,0.5) = \frac{5!}{0!5!}(0.5)^0(0.5)^5 = (1)(1)(0.03125) = 0.03125$$

$$b(1|5,0.5) = \frac{5!}{1!4!}(0.5)^1(0.5)^4 = (5)(0.5)(0.0625) = 0.15625$$

$$b(2|5,0.5) = \frac{5!}{2!3!}(0.5)^2(0.5)^3 = (10)(0.25)(0.125) = 0.3125$$

$$b(3|5,0.5) = \frac{5!}{3!2!}(0.5)^3(0.5)^2 = (10)(0.125)(0.25) = 0.3125$$

$$b(4|5,0.5) = \frac{5!}{4!1!}(0.5)^4(0.5)^1 = (5)(0.0625)(0.5) = 0.15625$$

$$b(5|5,0.5) = \frac{5!}{5!0!}(0.5)^5(0.5)^0 = (1)(0.03125)(1) = 0.03125$$

Table 17.4

Probability Distribution (in Tabular Form) for the Binomial Random Variable x, Defined as the Number of Heads Experienced in Five Tosses of a Fair Coin

Value of x x	Probability $P(x)$
0	0.03125
1	0.15625
2	0.31250
3	0.31250
4	0.15625
5	0.03125
	1.00000

Figure 17.3

Probability Distribution (in Graphic Form) for the Binomial Random Variable x, Defined as the Number of Heads Experienced in Five Tosses of a Fair Coin

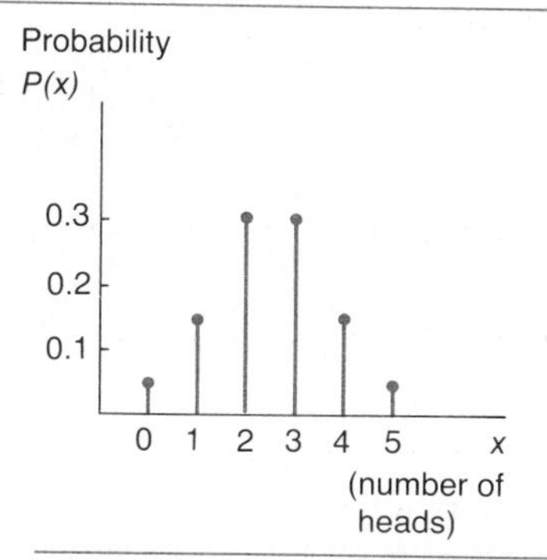

This binomial probability distribution is shown in tabular form in Table 17.4 and is shown in graphic form in Figure 17.3.

The expected value and variance for this binomial random variable are

$$E(x) = np = (5) \cdot (0.5) = 2.5$$

and

$$\text{VAR}(x) = npq = (5) \cdot (0.5) \cdot (0.5) = 1.25$$ []

EXERCISES

24. An experiment consists of rolling a fair die four times. Success is defined as either one dot or two dots on the upturned side of the die when it comes to rest. Construct the probability distribution, in tabular form, for the random variable X, defined as the number of successes experienced.

The parameters of a binomial experiment are: $n = 6$ *and* $p = 0.3$.

25. Set up the probability distribution, in tabular form, for the random variable X.
26. Compute $E(X)$ and VAR(X).
27. Construct a vertical-line graph of the probability distribution.

Five percent of the items coming off a production line have finishing defects. If defects occur at random and six items are packed in a carton,

28. Set up the probability distribution, in tabular form, for the binomial random variable X, defined as the number of items in a carton with finishing defects.
29. Compute $E(X)$ and VAR(X).

30. Construct a vertical-line graph of the probability distribution.

31. Exactly two thirds of the graduate students at a large university are married. What is the probability that in a random sample of 10 of these graduate students, 8 or more are married?

If 80 percent of the single-family residences in a large city contain at least two television sets, what is the probability that in a randomly selected group of nine such homes—

32. All nine have at least two television sets?
33. Exactly five have at least two television sets?
34. No more than two have at least two television sets?

If 30 percent of the workers at a large manufacturing plant drive their personal car to work each day, what is the probability that in a randomly selected sample of eight workers—

35. Exactly two drive their personal car to work each day?
36. More than two drive their personal car to work each day?
37. Fewer than two drive their personal car to work each day?

A manufacturing firm receives a shipment of 12 subassemblies from one of its suppliers. Previous records indicate that 5 percent of the units received from this supplier are defective.

38. What is the probability that all 12 of the subassemblies in this shipment are good?
39. If 10 of these subassemblies are needed urgently for a product being assembled in the plant, what is the probability that there will be enough nondefective subassemblies in this shipment to complete the project?
40. What is the expected number of defective subassemblies in a shipment of 12 items?

Large shipments of an incoming product at a manufacturing plant are inspected for defectives. Ten items are selected at random from the shipment and inspected. The shipment is rejected if two or more defectives are found; otherwise the shipment is accepted.

41. If a shipment contains exactly 5 percent defectives, what is the probability that it is accepted? that it is rejected?
42. If a shipment contains exactly 10 percent defectives, what is the probability that it is accepted? that it is rejected?
43. If a shipment contains exactly 20 percent defectives, what is the probability that it is accepted? that it is rejected?

If imported cars comprise 25 percent of all new car sales in a certain area and if five people who have recently purchased a new car are selected at random from this area, what is the probability that—

44. All five of them purchased an imported car?
45. Three or more of them purchased an imported car?
46. At most one of them purchased an imported car?

Hatfield Huntington III, a headstrong huntsman, plans to go grouse hunting on each of the next six days. He considers a day's outing to be "successful" if he bags at least two birds and has found over the past years that his probability of having a successful hunting day is 0.2. Also, the results of one day's outing seem to be independent of those of the previous day.

47. Set up the probability distribution for the random variable X, defined as the number of successful hunting days out of the six.
48. What is the probability that Hatfield will have at least three successful outings? not more than one successful outing?
49. What is the probability that the first day's hunting trip will be successful? that both the first two days will be successful? that only one of the first two days' trips will be successful?
50. What is the probability that the sixth day's hunting trip will be successful?
51. What is the probability that the sixth day's hunting trip will be successful, given that the first five days were unsuccessful?
52. What is the probability that the first successful day will come on the sixth day's trip?

Haverford Huntington IV, Hatfield's second cousin, is only a half-hearted huntsman and has agreed to go hunting on each of the next six days or *until he has had two successful hunting days. Haverford considers a hunting trip to be "successful" if he bags at least one grouse and has found that he is successful on 60 percent of his hunting trips. Again, the results of one day's hunting seem to be independent of the previous day's results.*

53. What is the probability that Haverford goes hunting on only the first two days?
54. What is the probability that Haverford goes hunting on only the first four days?
55. Set up the probability distribution for the NEGATIVE BINOMIAL RANDOM VARIABLE X, defined as the number of days Haverford goes hunting (or the day, out of the six, on which he experiences his second success). (The negative binomial random variable is gen-

erated by a random experiment where the binomial properties generally hold except that the number of successes is fixed and the number of trials of the experiment is a variable. Although there are mathematical formulas for determining the probability for the negative binomial variable, these probabilities can also be determined easily enough using a probability tree.)

56. What is the probability that Haverford goes hunting on the fifth day?
57. What is the probability that Haverford does not have two successful hunting trips in the six days?

17.5 OTHER DISCRETE PROBABILITY DISTRIBUTIONS

The Multinomial Probability Distribution

One obvious generalization of the binomial function extends the model to cover those experiments whose outcomes fall into three or more distinct categories. This is the MULTINOMIAL PROBABILITY DISTRIBUTION, whose primary characteristics are as follows:

1. The outcome of each trial of the experiment falls into exactly one of k categories, $O_1, O_2, \ldots, O_k$.
2. Each trial of the experiment is independent of all other trials. The probabilities associated with each category of outcome—$p_1, p_2, \ldots, p_k$—remain constant from trial to trial.
3. The experiment consists of exactly n trials.

The MULTINOMIAL RANDOM VARIABLES, $x_1, x_2, \ldots, x_k$, represent the number of times each of the possible outcomes occurs within a fixed number of trials.

Multinomial probability function If any given trial of an experiment can result in one of the k outcomes, $O_1, O_2, \ldots, O_k$, with constant probabilities, $p_1, p_2, \ldots, p_k$, the probability function for the random variables $x_1, x_2, \ldots, x_k$, representing the number of occurrences of the outcomes in n independent trials of the experiment is given by

$$f(x_1, x_2, \ldots, x_k \mid n, p_1, p_2, \ldots, p_k) = \frac{n!}{x_1! x_2! \ldots x_k!} p_1^{x_1} p_2^{x_2} \ldots p_k^{x_k}$$

where

$$\sum_{i=1}^{k} x_i = n$$

and

$$\sum_{i=1}^{k} p_i = 1$$

The expected value and variable of a multinomial random variable will be

$$E(X_i) = np_i \qquad \text{and} \qquad \text{VAR}(X_i) = np_i(1 - p_i)$$

Example 6

A company employs a large number of workers. Fifty percent of the employment force of the company is classified as production workers, one third of the force as salespersons, and the remainder as administrative personnel. A committee of six employees is to be chosen at random from the employment force. What is the probability that three of the six will be production workers, two will be salespersons, and one will be an administrator?

Because the small sample is to be selected from a large population of workers, we can treat the probabilities as though they were constants. In this case, then, we have six independent trials of the experiment, each with three possible outcomes. Probabilities associated with each possible type of outcome are

O_1 = Production worker and has probability $p_1 = 1/2$
O_2 = Salesperson and has probability $p_2 = 1/3$
O_3 = Administrator and has probability $p_3 = 1/6$

The random variable x_1 represents the number of times outcome O_1 occurs in the six trials; the variable x_2 represents the number of times outcome O_2 occurs; and the variable x_3 represents the number of occurrences of outcome O_3. We are interested in the probability that $x_1 = 3$ and $x_2 = 2$ and $x_3 = 1$. This can be determined by

$$f(3,2,1|6,1/2,1/3,1/6) = \frac{6!}{3!2!1!}(1/2)^3(1/3)^2(1/6)^1 = 5/36$$

Hence, the probability that the committee consists of three production workers, two salesmen, and one administrator is 5/36. [ ]

The Hypergeometric Probability Distribution

The binomial probability model is characterized by repeated independent trials of an experiment. Suppose, however, that the trials are not independent, that the probability of success is not constant from one trial of the experiment to the next. In these cases, the HYPERGEOMETRIC PROBABILITY MODEL may be appropriately used.

In general, the hypergeometric probability model is appropriately used when a random sample of n items is drawn without replacement from a finite population of N items. The N items constituting the population are of two classes: k of the N items may be considered "successes," leaving $(N - k)$ items that are "failures." Because the selections are made without replacement, they are not independent, and the probability of success changes with each selection. The hypergeometric random variable x represents the number of successes among the n items selected.

Hypergeometric probability function If the population of N items consists of k items which are labeled "success" and $(N - k)$ items that are labeled "failure," and the random variable X represents the number of successes among n items selected at random without replacement from the population, the hypergeometric probability function is defined as

$$h(x|N,n,k) = \frac{\binom{k}{x}\binom{N-k}{n-x}}{\binom{N}{n}} \qquad \textit{for } x = 0, 1, 2, \ldots, n$$

where

$$\binom{k}{x} = \frac{k!}{x!(k-x)!}$$

$$\binom{N-k}{n-x} = \frac{(N-k)!}{(n-x)!(N-k-n+x)!}$$

and

$$\binom{N}{n} = \frac{N!}{n!(N-n)!}$$

Example 7 Food Products, Inc., employs 15 salespersons, 10 of whom are married and 5 of whom are single. Four salespersons are to be selected at random from the group to develop a new sales campaign. Let the random variable x represent the number of single sales employees selected and calculate the probability that $x = 2$.

This random variable is a hypergeometric random variable. Checking the conditions which characterize this model, we find (1) that a fixed number of items—$n = 4$—is to be selected from a population of fixed size—$N = 15$; (2) that the results of each selection may be classified as either success (single) or failure (married); and (3) that, because the selections are made without replacement, the trials are not independent.

The probability that two of the four salespersons selected are single can be computed as follows:

The total number of ways of selecting two single salespersons from five single sales employees in the group is

$$\binom{5}{2} = \frac{5!}{2!3!} = 10$$

The total number of ways of selecting the remaining 2 people from the 10 married employees in the group is

$$\binom{10}{2} = \frac{10!}{2!8!} = 45$$

From the multiplicative law of probability, we have the total number of ways of selecting two single and two married salespersons as

$$(10)(45) = 450$$

Now this must be compared to the total number of ways of selecting 4 salespersons (without regard to marital status) from the 15. The total number of combinations of 4 items from 15 is

$$\binom{15}{4} = \frac{15!}{4!11!} = 1{,}365$$

Hence, there are 1,365 points in the sample space. Of all these, 450 points represent two married employees and two single employees. The probability of selecting 2 married and 2 single salespersons from the group of 15 is given by

$$h(2|15,4,5) = 450/1{,}365 = 30/91$$ []

The Poisson Probability Distribution

Some events occur not as outcomes of a definite repetition of an experiment but rather at random points in time or space. An important model used to describe phenomena randomly distributed over a given time interval or a specified region is the POISSON PROBABILITY MODEL.

Experience has shown that the Poisson model is an excellent model to use for calculating probabilities associated with the following types of events: the number of telephone calls coming into a switchboard during a fixed time interval, the number of typing errors on a page, the number of defects in a manufactured article, the number of arrivals at a service counter during a specified period of time, the number of breakdowns of a piece of machinery, or the number of claims filed against an insurance company. For events of this kind we are concerned with the number of occurrences of the event, but never with the number of nonoccurrences. We may be interested in the number of vehicles passing a toll station during a certain hour, but we do not attempt to investigate the number of vehicles that do not pass the station.

The POISSON RANDOM VARIABLE x represents the number of occurrences of a given event within an interval of a specified size in a Poisson experiment. It is a discrete variable which may take on any of the values of $x = 0, 1, 2, \ldots$. The probability function for this random variable is the Poisson probability function.

Poisson probability function The probability function of the Poisson random variable x, which represents the number of occurrences of an event during a specified interval, is defined as

$$f(x|\mu) = \frac{e^{-\mu}\mu^x}{x!} \qquad \text{for } x = 0, 1, 2, \ldots$$

where

μ represents the average number of occurrences of the event within the specified interval

e is the mathematical constant 2.71828 . . .

The specified interval may be a time interval of any duration, such as a minute, a day, a week, or a month; it may be a region, such as a line segment, an area, or a volume. For convenience, we generally refer to the interval as if it were a time axis and use the symbol t to denote its duration.

The Poisson process is characterized by these properties: (1) An average for the number of occurrences of the random event during the specified interval exists and remains constant regardless of where the interval begins or how the unit length of the interval is chosen. Therefore, if there is an average number λ of events per unit of interval, the average number of events for t units of interval is $\lambda t = \mu$. (2) The occurrence of the event is completely independent of any other occurrence of the event. (3) The probability of exactly one event's occurring during any small interval is approximately proportional to the length of the interval. Although the assumptions under which random phenomena will obey the Poisson probability laws can be stated theoretically, the value of the constant $\mu = \lambda t$ cannot be deduced theoretically but must be determined empirically. Also, while x, the value of the Poisson random variable, is an integer, λ, because it is an average, need not be an integer; therefore, μ need not be an integer.

The variance of the Poisson distribution is exactly equal to its mean μ, so that the distribution, in effect, has only one parameter. For small values of μ the Poisson distribution is highly skewed, but as the value of μ increases, the curve becomes more symmetrical. (See Figure 17.4.)

Figure 17.4

Poisson Distributions for Various Values of $\mu = \lambda t$

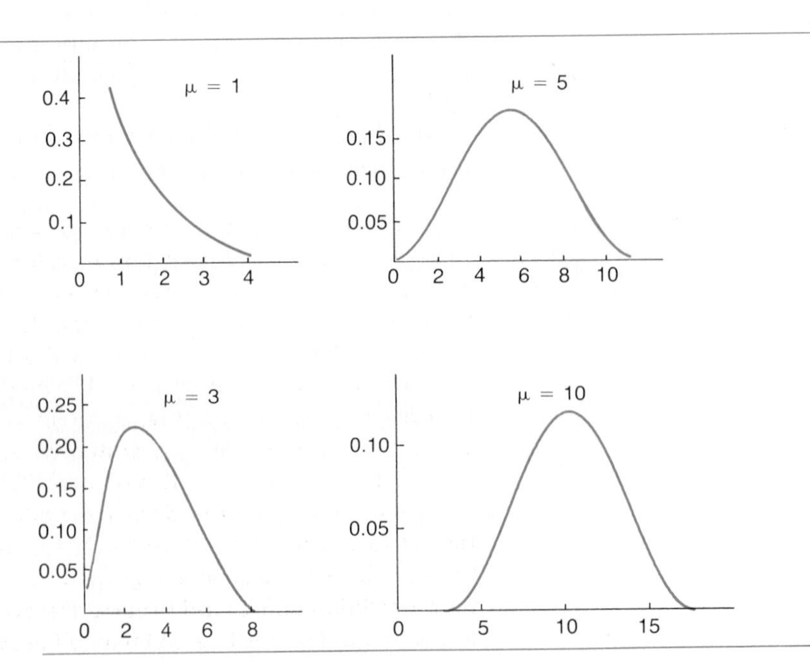

Example 8

The Poisson distribution is widely used in waiting-line, or queueing, models, since it seems to appropriately describe the pattern of arrivals at many types of service facilities. Assume that automobiles are arriving at random times at a teller's window at a drive-in bank but at an overall average rate of 15 an hour and following a Poisson pattern. What is the probability that exactly one car drives up to the window in a five-minute interval? The average number of arrivals during a five-minute interval is

Table 17.5

Probability Distribution for Number of Poisson Arrivals in a Five-Minute Interval When $\mu = 1.25$

Number of arrivals	Probability
0	0.2865
1	0.3581
2	0.2238
3	0.0932
4	0.0291
5	0.0073
6	0.0015
7	0.0003

$$\mu = \lambda t = (15) \times (1/12) = 1.25$$

and the probability of exactly one arrival in a five-minute interval is

$$f(1|1.25) = \frac{e^{-1.25}(1.25)^1}{1!} = 0.3581$$

The complete probability distribution for the number of arrivals in a five-minute interval is given in Table 17.5.

Intervals of different lengths are easy enough to evaluate, since $\mu = \lambda t$. For example, what is the probability of fewer than two arrivals in a 15-minute interval? Now

$$\mu = 15 \times (1/4) = 3.75$$

Since $P(x < 2) = P(x = 0) + P(x = 1)$, we compute

$$f(0|3.75) = \frac{e^{-3.75}3.75^0}{0!} = 0.0235$$

and

$$f(1|3.75) = \frac{e^{-3.75}3.75^1}{1!} = 0.0882$$

Then, $P(x < 2) = 0.0235 + 0.0882 = 0.1117$. [ ]

Example 9

A company tests truck tires by driving them over difficult terrain. They have found that flats due to external causes (causes other than ordinary wear and tear) occur at an average of one flat every 3,000 miles. They have determined that the Poisson function provides an adequate model for the occurrence of flat tires. What is the probability that in a given 1,800-mile test no flats due to external causes will occur?

It is important to note that the Poisson model is completely defined by its mean μ, which is in turn dependent upon the size of the stated interval t. The mean μ changes proportionally whenever the stated interval is changed. Here the distance (1,800 miles) is the interval t, and the random variable of interest is X, the number of flats in 1,800 miles. Because the expected value is proportional to the time interval specified in the definition of X, and because the average originally stated was in terms of one flat tire in a 3,000-mile interval, we make the adjustment

λ = number of flats per mile = 1/3,000
$\mu = \lambda t = (1/3{,}000)(1{,}800) = 0.6$ flats in 1,800 miles, on the average
Using the Poisson formula, we find

$$f(0|0.6) = \frac{e^{-0.6}(0.6)^0}{0} = 0.5488$$

The probability that no flats due to external causes are experienced in an 1,800-mile road test is 0.5488. []

EXERCISES

58. Articles produced on an assembly line are inspected and then characterized as (*a*) having no defects, (*b*) having only minor defects, or (*c*) having major defects. Data indicate that the probabilities of each type of article are as follows:

$$P(\text{no defect}) = 2/3$$
$$P(\text{minor defect}) = 1/4$$
$$P(\text{major defect}) = 1/12$$

What is the probability that for 10 articles selected at random from the assembly line, only 1 will have a major defect, only 1 a minor defect, and the other 8 will have no defects?

59. Forty percent of the doctors attending a medical convention approve of Proposition 815, while 50 percent disapprove and 10 percent are undecided. If a committee of six is selected from all the attendees to further study the proposition, what is the probability that two approve, three disapprove, and one is undecided?

60. Suppose that an experiment consists of tossing a pair of dice six times. Outcomes of interest are: O_1 = A 7 or an 11 occurs; O_2 = A 2 occurs; and O_3 = Any other outcome occurs. What is the probability that in six trials, O_1 occurs twice, O_2 occurs twice, and O_3 occurs twice?

61. Herman, the hot-dog vendor, knows that his customers buy plain dogs, chili dogs, dogs with mustard and relish, and dogs with chili, mustard, and relish with probabilities one sixth, one third, one third, and one sixth respectively. What is the probability that the next five customers, each buying only one hot dog, will purchase two chili dogs, two dogs with mustard and relish, and one dog with chili, mustard, and relish?

62. A discussion panel of six members is to be selected at random from a group of 16 tavern owners and 10 ministers. What is the probability that the panel will contain an equal number of tavern owners and ministers?

63. Two defective motors are accidentally placed on a shelf along with six nondefective motors. If three motors are removed at random

from the shelf, what is the probability that two are nondefective and one is defective?

64. A group of six identical television tubes contain two that are defective and four that are nondefective. Suppose that four of these tubes are selected at random. Letting X represent the number of defective tubes selected, construct the probability distribution for X.
65. Eight vegetable cans, all identical in shape and size, have lost their labels. It is known that four of the cans contain tomatoes and four contain corn. If three are selected at random, what is the probability that all three contain tomatoes? that all three contain corn? that all three contain the same vegetable?

Flaws in the surface of large sheets of metal occur at random on the average of one per section of area 10 square feet. Assume that a Poisson model is appropriate.

66. What is the probability that there will be no flaws in a piece of metal that is 6 by 10 feet?
67. What is the probability that there will be at most one flaw in a 10-by-10-feet sheet of metal?

A used-car lot is lighted at night by 1,000 40-watt light bulbs. The probability that a new bulb will burn for 12 hours is given at 0.995. If all the bulbs are new, what is the probability that—

68. No bulb will burn out during the first 12 hours?
69. At most three bulbs will burn out during the first 12 hours?
70. An average of 12 customers arrives at a service counter each hour. Assuming that the number of arrivals is distributed according to a Poisson probability function, what is the probability that eight customers arrive at the window during any given hour?
71. A large garment-manufacturing plant has many special sewing machines in operation at all times. If the probability that an individual sewing machine breaks down during a given day is 0.08, what is the probability that during a given day two of the machines break down?

If, on the average, calls arrive at the switchboard of Allgood Answering Service at the rate of one every 30 seconds and are distributed by the Poisson probability function,

72. What is the probability that two calls arrive in a 30-second period?
73. What is the probability that no calls arrive in a 30-second period?
74. What is the probability that more than two calls arrive in a given 30-second period?

75. What is the probability that fewer than two calls arrive in a 30-second period?

17.6 CONTINUOUS RANDOM VARIABLES AND THEIR PROBABILITY DISTRIBUTIONS

With a discrete sample space, it is possible to associate some nonzero probability with each of the sample points, although these points may be countably infinite in number, and still have the probabilities assigned sum to unity. Probabilities are thus associated with each possible value of the discrete random variable. With a continuous sample space, because of the uncountably infinite number of sample points, such an assignment of probabilities is an impossible task. When working with a continuous random variable, then, we assign probabilities to the event that the variable assumes a value within an interval. We never speak of the probability that the continuous random variable takes on some specific point value; rather, we speak of the probability that the value of the variable lies between two specified points, and it makes no difference whether the interval is an open or a closed interval.

Suppose that the continuous random variable x has as its range the set of real numbers in the interval [a,b] where a < b. We define a function $f(x)$, called the PROBABILITY DENSITY FUNCTION, with the properties

1. $f(x) \geq 0 \qquad a \leq x \leq b$
2. $\int_{\text{all } x} f(x)\,dx = 1$

Figure 17.5

Probability Density Function for a Continuous Random Variable x and the Area Representing the Probability that x Assumes a Value between c and d

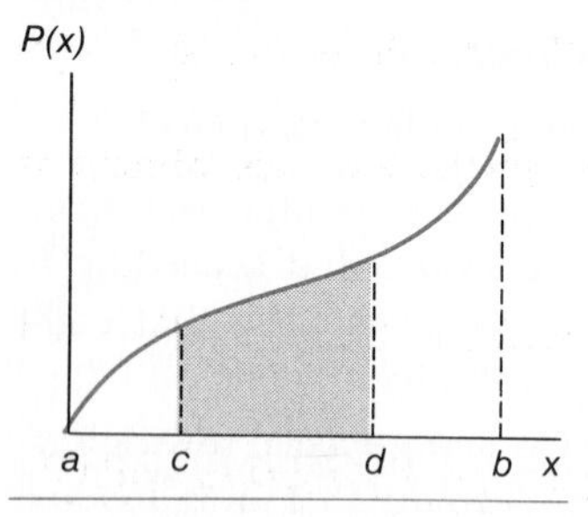

The probability that the continuous random variable takes on a value in the interval $[c,d]$ is given by

$$P(c \leq x \leq d) = \int_c^d f(x)\,dx$$

(See Figure 17.5.)

The expected value of the continuous random variable x with probability density function $f(x)$ is given by

$$E(x) = \int_{\text{all } x} x\,f(x)\,dx$$

Then, variance of the continuous random variable can be obtained by

$$\text{VAR}(x) = E(x^2) - [E(x)]^2$$

where

$$E(x^2) = \int_{\text{all } x} x^2 f(x)\,dx$$

Continuous probability distributions are frequently very easy to work with, and for this reason we often treat a discrete random variable as if it were continuous, particularly when it can take on a very large number of values.

Example 10

City transit buses on a certain route arrive at a certain bus stop every 20 minutes. What is the probability that a person arriving at the bus stop at a random time will have to wait at least five minutes for the bus? What is the probability that a person will have to wait at least 10 minutes for the bus?

The probability density function has a constant value for $0 \le x \le 20$ (where the random variable x represents the number of minutes a person will have to wait), as shown in Figure 17.6. Such a probability density function is called a UNIFORM DISTRIBUTION. Mathematically, the function is given by

Figure 17.6

A Uniform Distribution

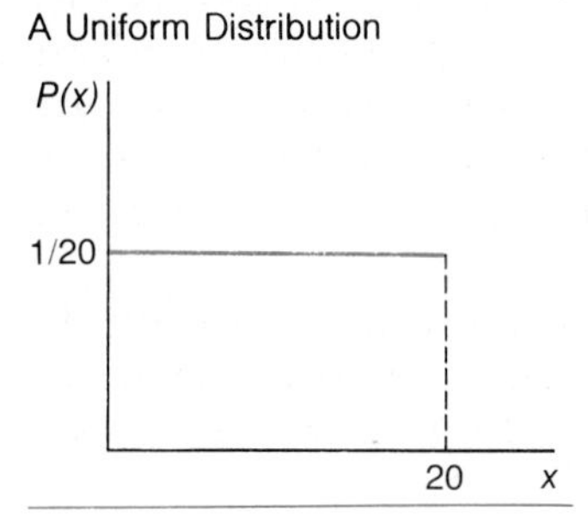

$$P(x) = \begin{cases} \dfrac{1}{b-a} & \text{for } a \le x \le b \\ 0 & \text{elsewhere} \end{cases}$$

and for this bus schedule would become

$$P(x) = 1/20 \qquad 0 \le x \le 20$$

The probability that a person arriving at a random time at the bus stop will have to wait at least five minutes for the bus is given by

$$P(x \ge 5) = \int_5^{20} (1/20)\, dx = \frac{x}{20}\Big|_5^{20} = (20/20) - (5/20) = 3/4$$

The probability that a person will have to wait at least 10 minutes is given by

$$P(x \ge 10) = \int_{10}^{20} (1/20)\, dx = \frac{x}{20}\Big|_{10}^{20} = (20/20) - (10/20) = 1/2 \qquad [\,]$$

Example 11

What are the expected value and the variance of the random variable in the previous example?

The expected length of time a person will have to wait for the bus is given by

$$E(x) = \int_0^{20} x\,(1/20)\, dx = \frac{x^2}{40}\Big|_0^{20} = (400/40) - (0) = 10$$

The variance is determined using

$$E(x^2) = \int_0^{20} x^2(1/20)\, dx = \frac{x^3}{60}\Big|_0^{20} = 133.\overline{3}$$

and, then,

$$\text{VAR}(X) = E(x^2) - [E(x)]^2 = 133.\overline{3} - (10)^2 = 33.\overline{3} \qquad [\,]$$

Example 12 Determine k such that

$$P(x) = \begin{cases} x/k & 0 \le x \le 8 \\ 0 & \text{elsewhere} \end{cases}$$

meets the requirements of a probability density function.

First, $P(x)$ must be nonnegative. This requirement will be met for any positive k.

Second, total probability must sum to one. For the continuous random variable and its probability density function, this means that

$$\int_0^8 (x/k)\, dx = 1$$

$$\frac{x^2}{2k}\bigg|_0^8 = 1$$

$$\frac{(8)^2}{2k} = 1$$

$$64 = 2k$$

$$k = 32$$

Thus, the constant k in the probability density function must take on the value $k = 32$; so

$$P(x) = \int_0^8 x/32\, dx = \begin{cases} x/32 & 0 \le x \le 8 \\ 0 & \text{elsewhere} \end{cases} \qquad []$$

The Normal Distribution

The normal distribution is the most important of the continuous probability distributions. Several mathematicians contributed to the formulation of this probability law. Abraham DeMoivre published, as early as 1733, works containing a derivation of the normal function as the limiting form of the binomial. Pierre S. de Laplace (1749–1827) and Karl Gauss (1777–1855) also published early references to the law. The works of Gauss became especially well known among mathematicians; in fact, the term *Gaussian distribution* is an accepted synonym for *normal distribution.*

The normal distribution rightly occupies a preeminent place in the field of probability theory. In addition to portraying the distribution of many types of measurement phenomena, such as the heights of men, diameters of machined parts, and so on, it also serves as a convenient approximation of many other distributions which are less tractable. Most important, it describes the manner in which certain statistical estimators of population characteristics vary from sample to sample and, thereby, serves as the foundation upon which many inferences from sample statistic to population parameter are made.

A random variable x with mean μ and standard deviation σ is said to be normally distributed if its density function is given by

$$N(\mu,\sigma) = f(x) = \frac{1}{\sigma\sqrt{2\pi}}\, e^{-(x-\mu)^2/2\sigma^2} \qquad -\infty < x < \infty$$

where

$\pi = 3.14159...$
$e = 2.71828...$

The only parameters are the mean and standard deviation, μ and σ. Once these are specified, the normal curve is completely determined. Figure 17.7 illustrates the influence of these parameters on the shape of the curve. Changing the value of μ does not alter the shape of the curve but merely shifts it along the x-axis. A change in σ, however, will alter the shape of the curve. If σ is increased, the maximum ordinate decreases. Because the area bounded by the curve is equal to unity, with increases in σ the center portion of the curve becomes less and less high and the ends of the curve approach the x-axis less and less rapidly. Conversely, with decreases in σ the center portion of the curve is pushed higher and higher up the y-axis and the ends of the curve approach the horizontal axis more and more rapidly.

Figure 17.7

Normal Distributions

(A) The maximum ordinate shifts along the x axis with changes in the value of μ

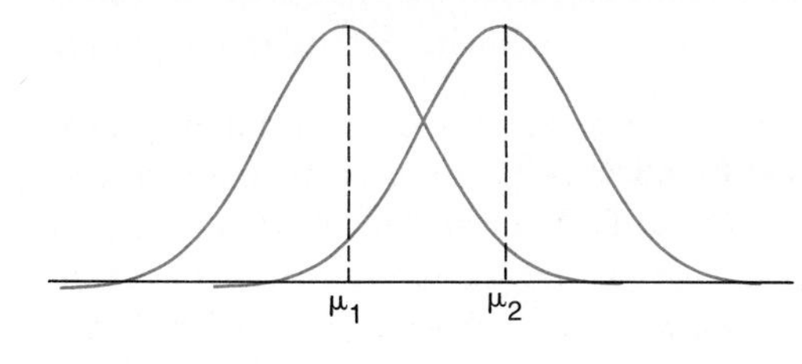

(B) The maximum ordinate moves up and down the x-axis with changes in the value of σ

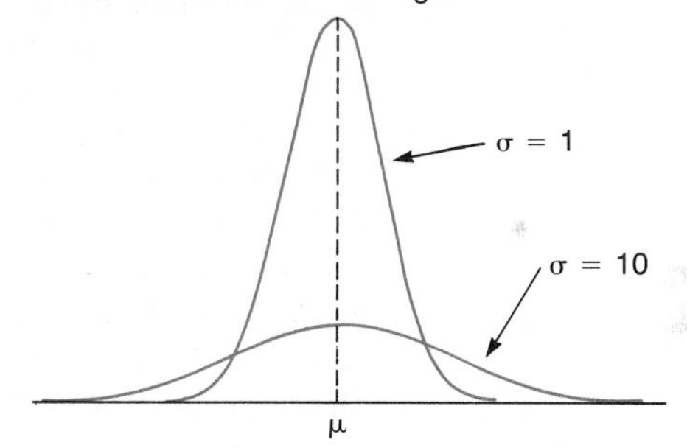

From an inspection of these graphs, we can see that the distribution is bell shaped, unimodal, and symmetrical. The function is defined along the entire x-axis and has a positive value for every value of x; it thus lies entirely above the x-axis.

Use of the Table of Normal Areas

Because the density function cannot be integrated explicitly, the evaluation of probabilities would entail numerical integration were it not for extensive tables of normal areas which are available. These tables are constructed for a *standardized normal distribution.* The ordinates of the normal curve depend upon the mean and standard deviation of the random variable. The area under the curve between any two ordinates, then, depends upon these same values. Hence, all observations of any normal random variable x can be transformed to a set of observations for a *standard normal variable* Z which has mean $\mu = 0$ and standard deviation $\sigma = 1$ by

$$Z = \frac{x - \mu}{\sigma}$$

Example 13 The monthly account balances for the charge-card department of a large financial institution have a normal distribution, with a mean balance of \$100 and a standard deviation of \$25.

a. What proportion of the accounts have a balance in excess of \$100? Since the normal distribution is symmetrical, half the area under the curve is to either side of the mean; thus, $P(x > \$100) = 0.5$ and $P(x < \$100) = 0.5$.

b. What proportion of the accounts have a balance in excess of \$125? It may be convenient to construct a diagram such as shown in Figure 17.8, showing the area under the normal curve. To find the probability that an account balance, represented by x, is greater than \$125, we need to evaluate the area under the normal curve to the right of $x = \$125$. This is accomplished by finding the area for the corresponding value of the standard normal variable Z. Areas in the Table of Normal Areas (Appendix Table III) are shown in standard deviation units from the mean. Hence,

$$Z = (125 - 100)/25 = 1$$

Figure 17.8

Area under the Normal Curve

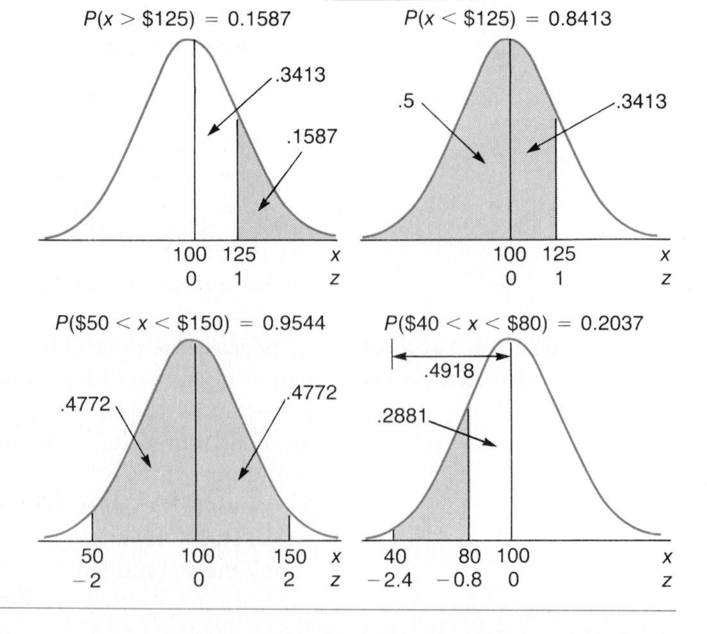

Using the Table of Normal Areas, we find that the area between the mean and $z = 1$ is 0.3413. The entire area to the right of the mean is 0.5. Then the area to the right of $z = 1$ is $0.5 - 0.3413 = 0.1587$, and

$$P(x > \$125) = 0.1587$$

Approximately 16 percent of the accounts have a balance in excess of $125.

c. What proportion of the accounts have a balance which is less than $125? This probability can be determined by

$$P(x < \$125) = 1 - P(x > \$125) = 1 - 0.1587 = 0.8413$$

or

$$P(x < \$125) = P(x < \$100) + P(\$100 < x < \$125)$$
$$= 0.5 + 0.3413 = 0.8413$$

d. What proportion of the accounts have a balance between $50 and $150? The area between $50 and the mean is found by

$$z = (50 - 100)/25 = -2 \qquad \text{Area} = 0.4772 \text{ when } z = -2$$

Thus,

$$P(\$50 < x < \$100) = 0.4772$$

The area between the mean and $150 can be found by

$$z = (150 - 100)/25 = +2 \qquad \text{Area} = 0.4772 \text{ when } z = +2$$

Thus,

$$P(\$100 < x < \$150) = 0.4772$$

Then, combining the two areas we obtain

$$P(\$50 < x < \$150) = 0.4772 + 0.4772 = 0.9544$$

e. What proportion of the accounts have a balance between $40 and $80? All areas must be found using the mean as one reference point. Thus, we find the area under the curve between the mean and $x = \$40$, as follows

$$z = (40 - 100)/25 = -2.4 \qquad \text{Area} = 0.4918 \text{ when } z = -2.4$$

Hence,

$$P(\$40 < x < \$100) = 0.4918$$

Next we find the area under the curve between the mean and $80, as

$$z = (80 - 100)/25 = -0.8 \qquad \text{Area} = 0.2881 \text{ when } z = -0.8$$

Hence,

$$P(\$80 < x < \$100) = 0.2881$$

It follows that

$$P(\$40 < x < \$80) = 0.4918 - 0.2881 = 0.2037$$

f. The 20 percent of the accounts with the largest balances have a balance that exceeds what dollar amount? We are given the area under the curve and want to find the value of the random variable. If the area to the right of $Z = z$ represents 0.20 of the total area under the curve, the area between the mean and $Z = z$ represents 0.30 of the area, since the total area to the right of the mean represents 0.50 of the area under the curve. From the body of the Table of Normal Areas we find that $z \simeq 0.84$ when area = 0.3. Next, if $z = (x - \mu)/\sigma$, then

$$x = \mu + z\sigma$$

Substituting known values, we calculate

$$x = 100 + (0.84)(25) = \$121$$

The largest 20 percent of the account balances exceed \$121.

EXERCISES

Determine k *such that each of the following functions meet the requirements for a probability density function.*

76. $P(x) = k(4 - x)$ for $0 \leq x \leq 4$
77. $P(x) = k(5 + x)$ for $0 \leq x \leq 10$
78. $P(x) = kx^2$ for $0 \leq x \leq 4$
79. $P(x) = (x^2 - 10x)k$ for $0 \leq x \leq 3$

Given the probability density function

$$P(x) = (2 + x)/30 \qquad \text{for } 0 \leq x \leq 6$$

80. Determine $P(x \geq 3)$.
81. Determine $P(2 \leq x \leq 4)$.
82. Determine $P(x \leq 2.5)$.
83. Find $E(X)$.
84. Find VAR(X).

Given the probability density function

$$P(x) = x^2/9 \qquad 0 \leq x \leq 3$$

85. Determine $P(x > 1)$.
86. Determine $P(x < 2)$.
87. Determine $P(0.5 \leq x \leq 2.5)$.

88. Find $E(X)$.

89. Find VAR(X).

Weights of boxes received in a shipping and receiving department are distributed by the density function

$$P(x) = (-1/12)x^3 + (2/3) \qquad \text{for } 0 \leq x \leq 2$$

where x *represent weight in pounds.*

90. What is the probability that a box selected at random will weigh more than one pound?

91. What is the probability that a box selected at random will weigh less than 0.5 pounds?

92. What proportion of the boxes weigh between 0.75 and 1.5 pounds?

93. What is the average weight of the boxes?

All persons applying for admission to a graduate study program at a certain university are given a uniform entrance examination. Scores on this examination have been normally distributed with a mean of 240 and standard deviation of 30.

94. What proportion of the applicants achieves a score of 240 or more?

95. What proportion of the applicants achieves a score of 275 or less?

96. What proportion of the applicants achieves a score of 200 or less?

97. What proportion of the applicants achieves a score of 300 or better?

98. What proportion of the applicants achieves a score of 225 or better?

99. What proportion of the applicants achieves a score within the range from 215 to 285?

100. What proportion of the applicants achieves a score within the range from 250 to 325?

101. The top 10 percent of the scores have been above what level?

102. The lowest 20 percent of the scores have been below what level?

103. The middle 75 percent of the scores have been within what range?

The lifetimes of batteries of a certain type produced by a company are normally distributed with an average of 29 months and standard deviation of 4.

104. What proportion of the batteries will have a lifetime exceeding 30 months?

105. What proportion of the batteries will fail before they have been in service for 24 months?

106. What proportion of the batteries will last between 20 and 36 months?

107. What proportion of the batteries will last more than 25 months?

108. What proportion of the batteries will last less than 32 months?
109. If the batteries are guaranteed to last at least 18 months, what proportion of the batteries will have to be replaced under the guarantee?
110. Five percent of the batteries will have a lifetime exceeding how many months?
111. Twelve percent of the batteries will fail before they have been in use for how many months?
112. If a customer purchases four of these batteries, what is the probability that at least three of the four will last longer than 24 months?

C·H·A·P·T·E·R 18 Mathematics of Finance

THE SAGA OF JARED BEAUCHAMP

Jared was the first male child born into the Beauchamp family in 45 years. To commemorate the momentous occasion, Uncle DuBois Beauchamp felt that he would like to set up an account at the National Bank so that he would be financially able to present Jared with a check for $50,000 on the young man's 21st birthday. DuBois realized that he would not need to deposit the entire $50,000 when he opened the account on the day of Jared's birth, since the money in the account would earn interest at the rate of 12 percent, compounded semiannually, over the 21 years. He did not know, however, just exactly what the dollar amount of the initial deposit should be; so he bought a case of Enfamil, instead.

When Jared was six years old, he started to school. His father contemplated purchasing an automobile so that Mrs. Beauchamp could drive Jared to and from school each day. He located a suitable vehicle at the used-car lot and saw, by the tag on its front windshield, that the purchase price was $3,500. A down payment of $500 was required. The remaining $3,000, plus interest at the rate of 14 percent, compounded monthly, on the unpaid balance, could be paid in 24 equal monthly payments. However, the exact dollar amount of the monthly payment was not stated on the price tag. And lacking this information, Mr. Beauchamp could not determine how the purchase would affect the family budget. So he didn't buy an automobile and, for the next 12 years, Jared walked to school each morning and walked back home again each afternoon.

When Jared graduated from high school and was accepted at the local university, Aunt Melanie considered setting up an annuity which would pay Jared $1,500 each quarter for the four years he would be at the university. She realized she could set up an annuity by making an initial lump-sum deposit into an account and that the funds remaining in the account would draw interest at the rate of 11.5 percent, compounded quarterly. But she did not know what the dollar amount of the lump-sum deposit should be. So she didn't set up the annuity; and Jared worked his way through college by tutoring math students.

Eventually, Jared graduated, with honors, from the university and married his childhood sweetheart, Caroline Cummings. Caroline and Jared wanted very much to own a house with a big backyard and thought that they would be ready for this financial responsibility in five years. They realized that, meanwhile, they needed to save $10,000 for a down payment and that this could be done by making monthly deposits into a savings account which paid 16 percent interest, compounded monthly. But Jared and Caroline were unable to determine how much each monthly deposit into the account should be in order to reach their savings goal in five years. So they didn't open the account. Rather, they lived, all of the remaining years of their lives, in a rented house trailer.

A basic concept of financial transactions is that *money has time value.* The value of money, with respect to a point in time, is described either as PRESENT VALUE or FUTURE VALUE. Present value is the value of money today; future value is the value of money at some point in the future. The charge for exchanging money now for money later is called INTEREST.

18.1 SIMPLE INTEREST

INTEREST is the price paid for the use of a sum of money over a period of time. A savings institution pays interest to a depositor on the money in a savings account since the institution has use of those funds while they are on deposit. A borrower pays interest to a lending agent for use of that agent's funds over the term of the loan.

The original sum of money is the PRINCIPAL. Interest is usually expressed as a percentage of the principal for a specified period of time, which is generally a year. This percentage is termed the INTEREST RATE.

If interest is paid on the initial amount only and not on subsequently accrued interest, it is called SIMPLE INTEREST. Simple interest is given by the formula $I = Prt$.

Simple interest

$$I = Prt \tag{18.1}$$

where

$P =$ Principal amount
$r =$ Simple interest rate per year (stated as a decimal)
$t =$ Time, in years, for which interest is paid

Example 1 The simple interest on a $1,000 loan at 12 percent for six months would be

$$\begin{aligned} I &= Prt \\ &= (1{,}000) \cdot (0.12) \cdot (1/2) \\ &= \$60 \end{aligned}$$

(six months converts to $6/12 = 1/2$ years)

[]

Example 2 A businessman invests $5,000 at 15 percent simple interest for two years. The amount of interest he will earn on the investment is

$$\begin{aligned} I &= Prt \\ &= (5{,}000) \cdot (0.15) \cdot (2) \\ &= \$1{,}500 \end{aligned}$$

[]

In general, at the end of t years, the borrower will owe the lender an amount S that will include the principal plus the interest.

Amount (future value) at simple interest

$$\text{Amount} = \text{Principal} + \text{Interest}$$
$$S = P + Prt = P(1 + rt) \qquad (18.2)$$

where

S = Total amount (or future value)
P = Principal
r = Simple interest rate, per year
t = Time, in years

Example 3 The total amount which will have to be repaid on a $250 loan at 18 percent simple interest for three months is

$$\begin{aligned} S &= P(1 + rt) \\ &= 250[1 + (0.18)(1/4)] \\ &= 250[1.045] \\ &= \$261.25 \end{aligned}$$

The total amount which will have to be repaid is $261.25 []

Example 4 A businessman invests $5,000 at 8 percent simple interest for two years. The amount of interest he will earn on the investment is

$$I = (5{,}000) \cdot (0.08) \cdot (2) = \$800$$

The total amount S that will be available at the end of the two-year period is

$$S = 5{,}000[1 + (0.08) \cdot (2)] = \$5{,}800$$ []

Example 5 How long will it take $10,000 invested at 8 percent simple interest to double in value?

We must find t where S = (2)($10,000) = $20,000, P = $10,000 and $r = 0.08$. We determine

$$\begin{aligned} S &= P(1 + rt) \\ 20{,}000 &= 10{,}000[1 + 0.08t] \\ 2 &= 1 + 0.08t \\ 1 &= 0.08t \\ t &= 12.5 \text{ years} \end{aligned}$$

It will take the principal 12.5 years to double in value. []

Example 6 A borrower receives $4,000 today, agreeing to repay the lender a total of $4,800 at the end of 12 months. What annual simple interest rate is being charged?

Here we wish to determine r given S = $4,800, P = $4,000 and t = 1 year. Thus,

$$S = P(1 + rt)$$
$$4{,}800 = 4{,}000[1 + r(1)]$$
$$1.2 = 1 + r$$
$$r = 0.2 = 20 \text{ percent}$$

The borrower is paying 20 percent interest on the $4,000 loan. [ ]

When time over which interest is paid is given in months, t in either Formula 18.1 or 18.2 is simply the number of months divided by 12. If time is given as a number of days, then one of two methods of computing t may be used. Many lending agencies use a 360-day year, called an ORDINARY-INTEREST YEAR, for computing the time of a loan. Then t is the number of days for which the loan is outstanding divided by 360. Thus, if a loan were for 40 days, t would be $40/360 = 1/9$. When time is determined in this way, the interest is called ORDINARY SIMPLE INTEREST.

EXACT SIMPLE INTEREST is computed based on a 365-day year (or 366 days in a leap year).

In much the same way, if time is given indirectly as time between two specific dates, the number of days may be determined based on a 30-day month (called APPROXIMATE TIME) or on a count of the exact number of days (called EXACT TIME). When counting the exact number of days, it is customary to include either the first day or the last day but not both.

EXACT TIME and the ORDINARY-INTEREST YEAR are widely employed and, unless otherwise specified, will be used in these examples.

Example 7 Find the interest on a $400 loan from August 1 to August 21, using an ordinary-interest year, at 17.5 percent simple interest.

Time from August 1 to August 21 is 20 days. Thus

$$I = Prt$$
$$= (400) \cdot (0.175) \cdot (20/360)$$
$$= \$3.89 \text{ (rounded to the nearest cent)}$$

[ ]

Example 8 Find the exact simple interest on a $900 note for 110 days at 20.5 percent annual interest.

We calculate

$$I = Prt$$
$$= (900) \cdot (0.205) \cdot (110/365)$$
$$= \$55.60$$

[ ]

Simple Discount Notes

A borrower may receive the FACE VALUE (the principal) of a note and repay the face value plus interest at the end of the time period. This is known as a SIMPLE-INTEREST NOTE.

Or, the lender may deduct the interest from the face value of the note at the time the loan is made, so that the borrower initially receives only the remainder, sometimes called the PROCEEDS. At the end of the time period, the borrower repays the face value of the note. This is known as a SIMPLE-DISCOUNT NOTE.

The face value of a simple-discount note (the amount before interest was deducted) is also termed the MATURITY VALUE of the note. The SIMPLE DISCOUNT (or BANK DISCOUNT) is the cost of using the money over the time period. It is based on the amount to be repaid and may be computed by the following formula.

Simple discount

$$D = Mdt \tag{18.3}$$

where

D = Bank discount
M = Maturity value of the note (amount to be repaid)
d = Discount rate, per year
t = Time, in years

The sum initially received, the PROCEEDS, is the amount to be repaid less the discount.

Proceeds on simple discount note

$$\begin{aligned} P &= M - D \\ &= M - Mdt \\ &= M(1 - dt) \end{aligned} \tag{18.4}$$

where

P = Proceeds (amount initially received)
M = Maturity value
d = Discount rate, per year
t = Time, in years

Example 9 A borrower signs a simple discount note for $2,000 at 18 percent discount for eight months. What amount will the loan actually provide today? What discount does the borrower pay?

$$\begin{aligned} P &= M(1 - dt) \\ &= (2{,}000)[1 - (0.18) \cdot (8/12)] \\ &= \$1{,}760 \end{aligned}$$

The borrower will receive proceeds of $1,760 today and will repay (the maturity value of) $2,000 at the end of the eight months. The bank discount is $240. [ ]

Example 10 A borrower needs $1,500 today and will be able to repay a loan at the end of 10 months. A loan company offers a 24 percent simple-discount note. What must the maturity value of the note be in order for the proceeds to be $1,500?

We must determine M given $P = 1{,}500$, $d = 0.24$, and $t = 10/12$. We compute

$$\begin{aligned} P &= M(1 - dt) \\ 1{,}500 &= M[1 - (0.24) \cdot (10/12)] \\ 1{,}500 &= M(0.8) \\ M &= \$1{,}875 \end{aligned}$$

The maturity value of the note should be $1,875 in order for the proceeds to be $1,500. The cost of using $1,500 over 10 months (the simple discount) would be $375. []

Example 11 You may borrow $1,000 from bank A for six months at a simple-interest rate of 18 percent, or you can sign a simple-discount note at bank B at an 18 percent discount for six months. (*a*) What amount would you be required to repay bank A? (*b*) What should be the maturity value of the discount note with bank B in order for the proceeds to be $1,000? (*c*) With which bank would the cost of the money be less?

You would be required to repay bank A $1,000 plus interest, or

$$\begin{aligned} S &= P(1 + rt) \\ &= 1{,}000[1 + (0.18)(6/12)] \\ &= \$1{,}090 \end{aligned}$$

In order to receive $1,000 today, you would need to sign a discount note with bank B with a maturity value of

$$\begin{aligned} P &= M(1 - dt) \\ 1{,}000 &= M[1 - (0.18)(6/12)] \\ M &= \$1{,}098.90 \end{aligned}$$

This includes a $98.90 charge for the use of the $1,000 over six months.

Clearly, the bank discount charge ($98.90) is greater than the simple-interest charge ($90.00). []

Example 12 An investor, A, owns a one-year, 12 percent simple-interest-bearing note with a face value of $6,000. Four months before the due date on the note, A needs money for other investments and decides to discount the note to a lending agent, B, at 15 percent discount. How much will A receive for the note?

First, we must determine the future value of the simple-interest-bearing note. This is

$$\begin{aligned} S &= P(1 + rt) \\ &= (6{,}000)[1 + (0.12)(1)] \\ &= \$6{,}720 \end{aligned}$$

Then, the proceeds from the discounting will be

$$\begin{aligned} P &= M(1 - dt) \\ &= (6{,}720)[1 - (0.15)(4/12)] \\ &= \$6{,}384 \end{aligned}$$

The original holder of the note, investor A, will receive $6,384 for the note today. Then, four months hence, agent B will receive the face value of $6,720. []

EXERCISES

Find the simple interest on each of the following loans.

1. $1,500 at simple interest of 12 percent per year for 4 months.
2. $750 at simple interest of 16.75 percent for 9 months.
3. $900 at simple interest of 14.5 percent for 24 months.
4. $10,000 at simple interest of 15 percent for 6 months.
5. $250 at simple interest of 19.5 percent from April 15 to April 25 (using exact time and ordinary-interest year).

Find the amount (future value) on each of the following:

6. A 90-day, $3,500 loan at 14 percent simple interest.
7. A 2-year, $8,500 loan at 15 percent simple interest.
8. A $900, 18-month loan at 17.5 percent simple interest.
9. A $1,250, 6-month loan at 20 percent simple interest.
10. A $300, 21 percent simple-interest loan from July 1 to July 31 (using exact time and ordinary-interest year).

Find the discount and proceeds on each of the following:

11. $10,000 maturity value, two years, 12.5 percent simple-discount note.
12. $5,000 maturity value, 10 months, 15 percent simple-discount note.
13. $600 maturity value, three months, 18 percent simple-discount note.
14. $8,500 maturity value, six months, 16 percent simple-discount note.
15. $750 maturity value, 15 months, 17.5 percent simple-discount note.
16. If an investor invests $500 at 20 percent simple interest for three months, how much interest will the investment earn?
17. If a person borrows $1,200 at 12 percent simple interest for seven months, how much must be repaid at the end of that time?
18. A person borrowed $800 and repaid $900 at the end of six months. What simple interest rate was paid? 12.5

19. How long will it take for $500 invested at 7 percent simple interest to double in amount? to triple in amount?
20. If you sign a nine-month, $600 note, discounted at 18 percent, how much will you receive? How much must you pay back at the end of nine months?
21. How much should one borrow from a bank that has a 15 percent discount rate in order to receive proceeds of $1,000 if the note is to be repaid in 10 months?
22. If you want to earn 16 percent simple interest on your investments, how much should you be willing to pay for a note that will be worth $2,500 in eight months?
23. If you pay $300 for a note that will pay $350 in four months, what simple interest rate did you earn?
24. An investor decides to buy a $6,000, 12-month, 15 percent simple-interest-bearing note by discounting it at 18 percent 3 months before it is due. What should the investor be willing to pay for the note?
25. You own an 18-month, $7,400, 12 percent simple-interest-bearing note. Six months before the due date of the note, you discount it at 14 percent. How much should you receive for the note?
26. Compare the cost of borrowing $3,000 at 16 percent simple interest for 10 months and the cost on a simple-discount note at 16 percent discount for 10 months which has proceeds of $3,000.

18.2 COMPOUND INTEREST AND EFFECTIVE RATE

If the interest which is due is added to the principal at the end of each interest period, then this interest as well as the principal will earn interest during the next period. In such a case, the interest is said to be COMPOUNDED. The sum of the original principal and all the interest earned is the COMPOUND AMOUNT. The difference between the compound amount and the original principal is the COMPOUND INTEREST.

The time interval between successive conversions of interest into principal is called the INTEREST PERIOD, or CONVERSION PERIOD, or COMPOUNDING PERIOD, and may be any convenient length of time. The interest rate is usually quoted as an annual rate and must be converted to the appropriate rate per conversion period for computational purposes.

A simple example will illustrate the compound-interest procedure. Assume that $1,000 is deposited in an account that pays interest of 12 percent per year, *compounded quarterly*. This means that interest will be computed at the end of each three-month period and added into the principal, thereafter drawing interest itself.

Using the simple-interest formula, we compute the amount in the account at the end of the first quarter to be

$$\begin{aligned} S &= P(1 + rt) \\ &= (1{,}000)[1 + (0.12)(3/12)] = (1{,}000)(1.03) \\ &= \$1{,}030 \end{aligned}$$

With the interest left in the account, the new principal for the second quarter is \$1,030.

At the end of the second quarter, after interest is paid, the amount in the account will be

$$\begin{aligned} S &= (1{,}030)[1 + (0.12)(3/12)] = (1{,}030)(1.03) \\ &= \$1{,}060.90 \end{aligned}$$

Continuing in the same way, at the end of the third and fourth quarters the amount in the account, including both principal and compound interest, will be as follows:

Third quarter:

$$\begin{aligned} S &= (1{,}060.90)[1 + (0.12)(3/12)] = (1{,}060.90)(1.03) \\ &= \$1{,}092.73 \end{aligned}$$

Fourth quarter:

$$\begin{aligned} S &= (1{,}092.73)[1 + (0.12)(3/12)] = (1{,}092.73)(1.03) \\ &= \$1{,}125.51 \end{aligned}$$

At the end of the fourth quarter, the compound amount will be \$1,125.51.

If we summarize the above computations, we will discern a pattern that leads to a general formula for computing compound interest:

$$\begin{aligned} \text{1st quarter: } S &= (1{,}000)(1.03) \\ \text{2nd quarter: } S &= [(1{,}000)(1.03)](1.03) = 1{,}000(1.03)^2 \\ \text{3rd quarter: } S &= [(1{,}000)(1.03)^2](1.03) = 1{,}000(1.03)^3 \\ \text{4th quarter: } S &= [(1{,}000)(1.03)^3](1.03) = 1{,}000(1.03)^4 \end{aligned}$$

In general, the compound amount can be found by multiplying the principal by $(1 + i)^n$ where i is the interest rate *per conversion period* and n is the total number of conversion periods.

Amount—with compound interest

$$S = P(1 + r/m)^{mt} = P(1 + i)^n \qquad \textbf{(18.5)}$$

where

$S =$ Compound amount, after n conversion periods
$P =$ Principal
$r =$ Stated annual rate of interest
$m =$ Number of conversion periods a year
$t =$ Total number of years
$i = r/m =$ Interest rate per conversion period
$n = mt =$ Total number of conversion periods

We may evaluate S in several different ways. Among the possible alternatives are:

1. Use a hand-held calculator with a y^x function key. (This is the procedure most often used in the text.)
2. Restate the equation by finding the logarithm of each side, solving for $log\ S$ (or $ln\ S$) and then finding its antilog, using either a hand-held calculator with logarithmic functions or a table of logarithms.
3. Use specially prepared tables which provide values of $(1 + i)^n$ for selected values of i and n.

The following example illustrates the first two of these computational alternatives.

Example 13 Find the compound amount (to the nearest cent) resulting from the investment of $100 at 8 percent compounded annually for 10 years. What is the compound interest?

Annual compounding will yield an amount

$$\begin{aligned} S &= 100(1 + 0.08)^{10} \\ &= 100(1.08)^{10} \\ &= 100(2.158924997) \quad \text{using } y^x \text{ key on calculator} \\ &= \$215.89 \end{aligned}$$

Or,

$$\begin{aligned} S &= 100(1.08)^{10} \\ \log S &= \log 100 + 10\log 1.08 \\ &= 2 + 10(0.033424) \quad \text{using } \log \text{ function on calculator} \\ \log S &= 2.33424 \\ S &= \text{antilog } 2.33424 = \$215.89 \end{aligned}$$

The compound amount after 10 years will be $215.89. This means that the compound interest on the $100 principal is $115.89. []

Example 14 What would be the compound amount after 10 years if $100 were invested at 8 percent interest, compounded semiannually? compounded quarterly? compounded monthly?

a. **Semiannual compounding** If compounding takes place twice a year, with an 8 percent annual interest rate, the rate per conversion period will be $i = r/m = 0.08/2 = 0.04$, and there will be a total of $n = mt = (2) \cdot (10) = 20$ conversion periods over the 10 years. Thus, the compound amount will be

$$\begin{aligned} S &= 100(1 + 0.04)^{20} \\ &= 100(1.04)^{20} \\ &= 100(2.191123) \\ &= \$219.11 \end{aligned}$$

b. **Quarterly compounding** If compounding takes place quarterly (four times a year), with an 8 percent annual interest rate, the rate per conversion period will be $i = 0.08/4 = 0.02$, and there will be a total of $n = (4) \cdot (10) = 40$ conversion periods over the 10 years. The compound amount will be

$$\begin{aligned} S &= 100(1 + 0.02)^{40} \\ &= 100(1.02)^{40} \\ &= 100(2.208070) \\ &= \$220.80 \end{aligned}$$

c. **Monthly compounding** With monthly compounding, the rate per conversion period will be $i = 0.08/12 = 0.006\overline{6}$, and there will be a total of $n = (12) \cdot (10) = 120$ conversion periods. Under these conditions

$$\begin{aligned} S &= 100(1.006\overline{6})^{120} \qquad \text{using calculator} \\ &= 100(2.219640) \\ &= \$221.96 \end{aligned}$$

Clearly, the amount of interest earned is greater as the number of conversion periods per year becomes greater. []

Example 15 How long will it take to accumulate \$650 if \$500 is invested at 10 percent compounded quarterly?

We must find the number of periods n where $S = 650$, $P = 500$ and $i = 0.10/4 = 0.025$. We compute

$$\begin{aligned} S &= P(1 + i)^n \\ 650 &= 500(1 + 0.025)^n \\ 1.3 &= (1.025)^n \end{aligned}$$

Taking the log of each side of the equation, we obtain

$$\begin{aligned} \log 1.3 &= \log(1.025)^n = n \log 1.025 \\ 0.113943 &= 0.010724n \\ n &= 10.625 \text{ quarters, or approximately } 2\ 2/3 \text{ years} \end{aligned}$$ []

Present Value Frequently it is necessary to determine the principal P which must be invested now at a given rate of interest per conversion period in order that the compound amount S be accumulated at the end of n conversion periods. Under these conditions, P is called the PRESENT VALUE of S. If

$$S = P(1 + i)^n$$

then

$$P = \frac{S}{(1 + i)^n} = S(1 + i)^{-n}$$

Present value of compound amount

$$P = \frac{S}{(1+i)^n} = S(1+i)^{-n} \tag{18.6}$$

where

P = Principal (or present value)
S = Compound amount (or future value)
i = Interest rate per conversion period
n = Total number of conversion periods

Example 16 Find the present value of a loan that will amount to $5,000 in four years if money is worth 10 percent compounded semiannually.

With $S = 5{,}000$, $i = 0.10/2 = 0.05$, and $n = (4)(2) = 8$, we find

$$\begin{aligned} P &= S(1+i)^{-n} \\ &= 5000(1.05)^{-8} \\ &= 5000(0.676839) \\ &= \$3{,}384.20 \end{aligned}$$

The present value of the loan is $3,384.20. []

Example 17 A trust fund for a child's education is being established by a single deposit in a savings account which will pay 16 percent interest compounded quarterly. If the amount required 12 years hence is $30,000, what should the amount of the initial deposit be?

We determine that $i = 0.16/4 = 0.04$ and $n = (4)(12) = 48$ and then compute

$$\begin{aligned} P = S(1+i)^{-n} &= 30{,}000(1.04)^{-48} \\ &= 30{,}000(0.152195) \\ &= \$4{,}565.84 \end{aligned}$$

An initial deposit of $4,565.84 invested in the account will accumulate, at 16 percent compounded quarterly, a total of $30,000 over the 12 years. []

Example 18 If money is worth 14 percent compounded semiannually, would it be better to discharge a debt by paying $500 now or $600 eighteen months from now?

The present value of $600 at 14 percent compounded semiannually over 18 months is

$$\begin{aligned} P = S(1+i)^{-n} &= 600(1.07)^{-3} \\ &= 600(0.816298) \\ &= \$489.78 \end{aligned}$$

Thus, it would be better to wait and pay $600 eighteen months from now than to pay $500 now.

Or, we can see that \$500 invested now at 14 percent compounded semiannually over 18 months would have a future value of

$$\begin{aligned} S = P(1+i)^n &= 500(1.07)^3 \\ &= 500(1.225043) \\ &= \$612.52 \end{aligned}$$

[]

Example 19 Find the amount required to pay off immediately a loan that will amount to \$8,000 in five years if money is worth 12 percent compounded quarterly. Assume a prepayment penalty of 1¼ percent of the difference between the amount of the loan and the present value.

With $i = 0.12/4 = 0.03$ and $n = (5)(4) = 20$, the present value is found to be

$$\begin{aligned} P = S(1+i)^{-n} &= 8{,}000(1.03)^{-20} \\ &= 8{,}000(0.553676) \\ &= \$4{,}429.41 \end{aligned}$$

The difference between the amount of the loan and the present value is \$8,000 − \$4,429.41 = \$3,570.59. The penalty for prepayment then is

$$(0.0125)(3{,}570.59) = \$44.63$$

Thus, the amount required to pay the loan now is

$$\$4{,}429.41 + \$44.63 = \$4{,}474.04$$

[]

Example 20 Mr. Anderson owes Mr. Zetta \$3,000 due in three years and another debt of \$5,000 due in five years. How much should Mr. Zetta be willing to accept in four years to settle both debts if money is worth 12 percent compounded quarterly?

At the end of four years, the \$3,000 debt, which was due after three years, will have earned interest for one year. Its amount will have grown to

$$\begin{aligned} S = P(1+i)^n &= 3{,}000(1+0.03)^4 \\ &= 3{,}000(1.125509) \\ &= \$3{,}376.53 \end{aligned}$$

At the end of four years, the value of the \$5,000 debt, which is due in five years, will be

$$\begin{aligned} P = S(1+i)^{-n} &= 5{,}000(1+0.03)^{-4} \\ &= 5{,}000(0.888487) \\ &= \$4{,}442.44 \end{aligned}$$

Hence, Mr. Zetta should accept \$3,376.53 + \$4,442.44 = \$7,818.97 at the end of four years to settle the debts. []

Effective Rate

Clearly, for a stated annual interest rate, the amount of interest accumulated depends upon the frequency of conversion. This is because interest which has been earned subsequently earns interest itself. When interest is compounded more than once a year, the stated annual rate is called a NOMINAL RATE. The EFFECTIVE RATE corresponding to a given nominal rate r converted m times a year is the simple interest rate that would produce an equivalent amount of interest in one year. Effective rates are particularly useful in evaluating alternative investment opportunities.

The compound-interest amount *for one year* on principal P is

$$I = S - P$$

or with nominal rate r and m conversions periods,

$$I = P(1 + r/m)^m - P = P[(1 + r/m)^m - 1]$$

The *effective rate of interest* is

$$\frac{\text{Compound interest}}{\text{Principal}} = \frac{I}{P}$$

Hence, the effective rate is $(1 + r/m)^m - 1$.

Effective interest rate

$$\text{Effective rate} = (1 + r/m)^m - 1 \qquad (18.7)$$

where

r = Nominal annual rate of interest
m = Number of conversion periods per year

Example 21

What is the effective rate corresponding to a nominal rate of 16 percent compounded quarterly?

$$\begin{aligned} \text{Effective rate} &= (1 + 0.16/4)^4 - 1 \\ &= (1.04)^4 - 1 \qquad \text{(using calculator or} \\ &= 1.169859 - 1 \qquad \text{Appendix Table IV)} \\ &= 16.99 \text{ percent} \end{aligned}$$

[]

Example 22

Bank A offers 12.25 percent compounded semiannually on its savings accounts, while bank B offers 12 percent compounded monthly. Which bank offers the higher effective rate?

For bank A:

$$\begin{aligned} \text{Effective rate} &= (1 + 0.1225/2)^2 - 1 \\ &= (1.06125)^2 - 1 \\ &= 1.126252 - 1 \\ &= 12.6252 \text{ percent} \end{aligned}$$

For bank B:

$$\begin{aligned}\text{Effective rate} &= (1 + 0.12/12)^{12} - 1\\ &= (1.01)^{12} - 1\\ &= 1.126825 - 1\\ &= 12.6825 \text{ percent}\end{aligned}$$

The plan offered by bank B has a slightly higher effective rate than that offered by bank A. []

Example 23 What nominal annual rate of interest converted monthly corresponds to 16 percent converted quarterly?

Quarterly conversion at 16 percent yields an effective rate of $(1 + 0.16/4)^4 - 1$, while monthly conversion at stated interest rate i, *per conversion period,* yields an effective rate of $(1 + i)^{12} - 1$. Equating these amounts gives

$$\begin{aligned}(1 + i)^{12} - 1 &= (1 + 0.04)^4 - 1\\ (1 + i)^{12} &= (1.04)^4\end{aligned}$$

Extracting the 12th root of each side of the equation yields

$$\begin{aligned}[(1 + i)^{12}]^{1/12} &= [(1.04)^4]^{1/12}\\ 1 + i &= (1.04)^{1/3}\\ 1 + i &= 1.013159 \qquad \text{(using calculator)}\\ i &= 0.013159\end{aligned}$$

With $m = 12$ conversion periods a year, the equivalent nominal rate r is $(0.013159)(12) = 0.157908 = 15.7908$ percent.

A stated rate of 15.7908 percent compounded monthly would earn interest equivalent to that earned with a stated rate of 16 percent compounded quarterly. []

EXERCISES

Find the compound amount and compound interest for each of the following investments:

27. $10,000 at 7 percent compounded annually for five years.
28. $10,000 at 7 percent compounded semiannually for five years.
29. $10,000 at 7 percent compounded quarterly for five years.
30. $5,000 at 6 percent compounded annually for 10 years.
31. $5,000 at 6 percent compounded semiannually for 10 years.
32. $5,000 at 6 percent compounded quarterly for 10 years.
33. $7,500 at 8 percent compounded annually for six years.
34. $7,500 at 8 percent compounded semiannually for six years.
35. $7,500 at 8 percent compounded quarterly for six years.
36. $8,000 at 12 percent compounded semiannually for eight years.

37. $8,000 at 12 percent compounded quarterly for eight years.
38. $8,000 at 12 percent compounded monthly for eight years.

Find the present value for each of the following investments:

39. $1,500 due in three years if money is worth 9 percent interest, compounded semiannually.
40. $12,500 due in five years if money is worth 14 percent interest, compounded quarterly.
41. $6,000 due in four years if money is worth 18 percent interest, compounded monthly.
42. $50,000 due in six years if money is worth 15 percent interest, compounded semiannually.
43. $50,000 due in six years if money is worth 14 percent interest, compounded semiannually.
44. $50,000 due in six years if money is worth 13 percent interest, compounded semiannually.

Find the yearly effective rate equivalent to the following nominal rates:

45. 18 percent, compounded semiannually.
46. 16 percent, compounded semiannually.
47. 14 percent, compounded semiannually.
48. 14 percent, compounded quarterly.
49. 12 percent, compounded semiannually.
50. 12 percent, compounded quarterly.
51. 12 percent, compounded monthly.
52. Find the nominal rate of interest, compounded quarterly, that is equivalent to 15 percent compounded semiannually.
53. Find the nominal rate of interest, compounded quarterly, that is equivalent to 15 percent compounded monthly.
54. A borrower obtained a $750, six-month loan by agreeing to pay interest at the rate of 14 percent compounded quarterly. How much had to be repaid at the end of the six-month period?
55. How long will it take $1,000 to double at 12 percent compounded annually?
56. How long will it take $1,000 to amount to $1,500 if invested at 16 percent interest compounded quarterly?
57. How long would $500 have to be invested at 9 percent compounded monthly to earn $150 interest?
58. How much should you invest today at 10 percent compounded quarterly in order to have $9,000 in three years?
59. Find the present value of $10,000 due in five years if money is worth 8 percent compounded quarterly.

60. Assuming money is worth 16 percent compounded quarterly, would it be better to discharge a debt by paying \$5,000 now or \$7,500 in three years?
61. How much money will be required to pay off immediately a loan that will amount to \$3,000 in four years if money is worth 14 percent compounded semiannually and there is a prepayment penalty of 1.5 percent of the difference between the amount of the loan and the present value?
62. Mr. Beecher owes Mr. Woods \$2,500 due in two years and \$6,000 due in five years. How much should Mr. Woods be willing to accept at the end of four years to settle the entire debt if money is worth 15 percent compounded semiannually?
63. Mr. Drake owes Mr. Tucker \$5,000 due in four years and \$4,500 due in seven years. If Mr. Drake agrees to pay \$3,000 now, how much should he have to pay five years from now to settle his entire debt, assuming that money is worth 13 percent compounded semiannually?
64. Mr. Calhoun owes Mr. Young \$3,000 due three years from now and \$5,000 due five years from now. If Mr. Calhoun agrees to pay \$2,500 now, how much should he have to pay four years from now to settle the debt, given that money is worth 14 percent compounded semiannually?
65. What effective rate corresponds to a nominal rate of 10 percent compounded quarterly?
66. To what amount will \$10,000 accumulate in five years if invested at an effective rate of 9 percent?
67. A major credit-card company has a finance charge of 1.5 percent per month on the outstanding indebtedness. What is the nominal annual rate, compounded monthly? What is the effective rate?
68. An investor has a choice between investing a sum of money at 12 percent compounded annually or at 11.75 percent compounded quarterly. Which is the better of the two rates?
69. What nominal rate of interest compounded semiannually corresponds to an effective rate of 9 percent?
70. How long will it take money to double if invested at 12 percent compounded semiannually? if compounded quarterly?
71. What nominal annual rate of interest, converted quarterly, corresponds to 14 percent converted semiannually?

18.3 CONTINUOUS COMPOUNDING

Suppose that P, r, and t are held constant, while m, the number of conversion periods within a year, is increased indefinitely. What happens to the compound amount? Will it, too, increase indefinitely, or will it tend toward some limiting value?

Table 18.1 summarizes the effect of the number of conversion periods within a year on the compound amount when $10,000 is invested at an 8 percent nominal rate. Indeed, it does appear that the compound amount is tending toward an upper limit, something close to $10,833, as the number of conversion periods becomes larger and larger. The limiting case occurs where interest is COMPOUNDED CONTINUOUSLY.

Table 18.1

The Effect of the Number of Conversion Periods on the Compound Amount when $10,000 Is Invested at 8 Percent, with m Conversion Periods a Year

Frequency of compounding	m	Compound amount $S = 10{,}000(1 + 0.08/m)^m$
Annually	1	$10,800.00
Semiannually	2	$10,816.00
Quarterly	4	$10,824.32
Monthly	12	$10,830.00
Weekly	52	$10,832.20
Daily	365	$10,832.78
Hourly	8,760	$10,832.87

To derive a formula for continuous compound interest, we begin by writing

$$(1 + i)^n = (1 + r/m)^{mt}$$

Then, by inserting $1 = r/r$ in the exponent, we obtain

$$(1 + r/m)^{mt(r/r)} = (1 + r/m)^{(m/r)\cdot(rt)}$$

Then, letting $m/r = x$, we have

$$[(1 + 1/x)^x]^{rt}$$

Let us see what happens to the value of the term $(1 + 1/x)^x$ as x increases indefinitely. We compute

x	$(1 + 1/x)^x$
1	2.0
10	2.5937425
100	2.7048138
1,000	2.7169240
10,000	2.7181463
100,000	2.7182723
1,000,000	2.7182818

Clearly, as x increases indefinitely, the term $(1 + 1/x)^x$ approaches the value of the familiar mathematical constant $e = 2.7182818\ldots$. This means that the factor $(1 + i)^n = [(1 + 1/x)^x]^{rt}$ approaches e^{rt} as n increases indefinitely. The resulting formula for the amount under continuous compounding of interest is given below.

Amount with continuous compounding

$$S = Pe^{rt} \tag{18.8}$$

where

S = Amount at end of time t under continuous compounding
P = Principal
r = Annual rate, compounded continuously
t = Time, in years, over which interest is compounded

Values of e^x may be found using a hand-held calculator.

Example 24 If $100 is invested at 8 percent compounded continuously, what amount will be in the account (*a*) after 1 year? (*b*) after 2 years? (*c*) after 10 years?

a. After one year:

$$\begin{aligned} S &= Pe^{rt} \\ &= 100e^{(0.08)\cdot(1)} \\ &= 100e^{0.08} \\ &= 100(1.083287) \\ &= \$108.33 \end{aligned}$$

b. After two years:

$$\begin{aligned} S &= 100e^{(0.08)\cdot(2)} \\ &= 100e^{0.16} \\ &= 100(1.173511) \\ &= \$117.35 \end{aligned}$$

c. After 10 years:

$$\begin{aligned} S &= 100e^{(0.08)\cdot(10)} \\ &= 100e^{0.8} \\ &= 100(2.225541) \\ &= \$222.55 \end{aligned}$$

[]

Example 25 If $500 is invested at 15 percent, compounded continuously, what amount will be in the account after three years?

$$\begin{aligned} S &= Pe^{rt} \\ &= 500e^{(0.15)\cdot(3)} \\ &= 500e^{0.45} \\ &= 500(1.568312) \\ &= \$784.16 \end{aligned}$$

[]

Example 26 (*a*) How long will it take for money invested at 12 percent compounded annually to double in amount? (*b*) How long will it take for money invested at 12 percent compounded continuously to double in amount?

a. For annual compounding, we must find t given $S = P(1 + r)^t$ with $S = 2P$ and $r = 0.12$. We compute

$$2P = P(1 + 0.12)^t$$
$$2 = (1.12)^t$$

To proceed we may take the natural log of both sides of the equation, as

$$ln\ 2 = ln(1.12)^t$$
$$0.693147 = t\ ln\ 1.12$$
$$0.693147 = 0.113329\ t$$
$$t = 0.693147/0.113329 = 6.116 \text{ years}$$

b. For continuous compounding, we must find t given $S = Pe^{rt}$ for $S = 2P$ and $r = 0.12$. We compute

$$2P = Pe^{(0.12)t}$$
$$2 = e^{0.12t}$$
$$ln\ 2 = ln\ e^{0.12t}$$
$$0.693147 = 0.12t$$
$$t = 5.776 \text{ years}$$

It will take 6.116 years (approximately six years, one month, and 12 days) for an amount to double at 12 percent if compounded annually; but it will take only 5.776 years (approximately five years, nine months, and nine days) for an amount to double at 12 percent compounded continuously. []

Example 27

What amount should be invested today at 14 percent compounded continuously in order to have $5,000 in three years?

We must determine P given $S = Pe^{rt}$ and $S = 5{,}000$, $r = 0.14$ and $t = 3$. We calculate

$$5{,}000 = Pe^{(0.14)(3)}$$
$$5{,}000 = Pe^{0.42}$$
$$5{,}000 = P(1.521962)$$
$$P = \$3{,}285.23$$

We should invest $3,285.23 now at 14 percent compounded continuously in order to have $5,000.00 in three years. []

Effective Rate—with Continuous Compounding

Given the nominal (or annual) rate, which is compounded continuously, the annual effective rate can be determined using the following formula.

Effective rate—with continuous compounding

$$\text{Effective rate} = e^r - 1 \qquad \textbf{(18.9)}$$

where r = nominal (or annual) rate, compounded continuously

Example 28 What is the comparable effective rate given that a bank advertises 9.75 percent compounded continuously?

The effective rate is given by

$$\begin{aligned}\text{Effective rate} &= e^{0.0975} - 1\\ &= 0.1024114 \simeq 10.24 \text{ percent}\end{aligned}$$

An interest rate of 10.24 percent compounded annually would be equivalent to a rate of 9.75 percent compounded continuously. []

EXERCISES

Find the compound amount for each of the following:

72. $3,000 at 12 percent compounded continuously for four years.
73. $5,000 at 14 percent compounded continuously for five years.
74. $600 at 16 percent compounded continuously for six years.
75. $1000 at 15 percent compounded continuously for three years.
76. $1000 at 15 percent compounded continuously for six years.
77. $1000 at 15 percent compounded continuously for 10 years.
78. How long will it take for an amount to double if invested at 10 percent compounded continuously?
79. How long would $500 have to be invested at 12 percent compounded continuously to grow in amount to $1,250?
80. How much more will an investor earn on $1,000 invested for five years at 9 percent compounded continuously instead of 9 percent compounded annually?
81. What is the effective rate on $1,000 invested at 11 percent compounded continuously for one year?
82. How much should be invested today at 10 percent interest compounded continuously in order to have $15,000 in five years?
83. How much should be invested today at 13 percent compounded continuously in order to have $20,000 in 10 years?
84. All-Savers Bank advertises 10.25 percent interest, compounded continuously, while Peoples Bank advertises 10.5 percent interest, compounded semiannually. Which bank offers the higher effective rate?

18.4 ANNUITIES

An ANNUITY is a sequence of EQUAL, PERIODIC PAYMENTS. If you make regular deposits of a specified amount to your savings account each month, you are involved with an annuity. Monthly mortgage payments, or car-loan payments, insurance-premium payments, or payments into a retirement account are all examples of annuities.

The payments may be made weekly, monthly, quarterly, annually, or for any fixed period of time. The time between successive payments is called the PAYMENT PERIOD for the annuity. If payments are made at the *end of each payment period,* the annuity is called an ORDINARY ANNUITY. (We shall consider only ordinary annuities in this chapter.)

The time from the beginning of the first payment period to the end of the last payment period is the TERM of the annuity. When the term of an annuity is fixed, that is, when the term starts and ends on definite dates, the annuity is called an ANNUITY CERTAIN. (When the term of an annuity depends upon some uncertain event, such as the death of an individual, the annuity is called a CONTINGENT ANNUITY.)

The AMOUNT (or FUTURE VALUE) of an annuity is the sum of all payments plus all interest earned.

Consider an ordinary annuity of $1,000 a year for four years at 8 percent compounded annually. Since the first payment is not received until the end of the first year, it earns interest only for the next three years. Likewise, the second payment is received at the end of the second year and earns interest only over the third and fourth years. The third payment, received at the end of the third year, earns interest during the fourth year. The last payment, received at the end of the fourth year, earns no interest, since the value of the annuity is computed immediately after this last payment is received.

Table 18.2 outlines the accumulation of interest on these periodic payments. We see that the amount S of the annuity is computed by

$$\begin{aligned} S &= 1{,}000(1.08)^3 + 1{,}000(1.08)^2 + 1{,}000(1.08)^1 + 1{,}000 \\ &= 1259.71 + 1166.40 + 1080.00 + 1000.00 \\ &= \$4506.11 \end{aligned}$$

To derive a general formula for the amount of an ordinary annuity, let us rearrange the terms of S given above from last to first as

$$S = 1{,}000 + 1{,}000(1.08)^1 + 1{,}000(1.08)^2 + 1{,}000(1.08)^3$$

Or, in general, where R represents the amount of the periodic payment, i represents the interest rate per payment period, and n represents the number of payment periods,

$$S = R + R(1+i) + R(1+i)^2 + \cdots + R(1+i)^{n-1} \qquad \textbf{(E1)}$$

Table 18.2

Future Value of Ordinary Annuity with $1,000 Annual Payment over Four Years with 8 Percent Interest Compounded Annually

	Year 1	Year 2	Year 3	Year 4	Amount
Payment 1	$1,000				$1{,}000(1.08)^3$
Payment 2		$1,000			$1{,}000(1.08)^2$
Payment 3			$1,000		$1{,}000(1.08)$
Payment 4				$1,000	1,000

Amount of Annuity at end of four years:

$$S = 1{,}000(1.08)^3 + 1{,}000(1.08)^2 + 1{,}000(1.08) + 1{,}000$$

Multiplying each side of the equation by $(1 + i)$, we obtain

$$S(1+i) = R(1+i) + R(1+i)^2 + R(1+i)^3 + \cdots + R(1+i)^n \quad \text{(E2)}$$

Then, subtracting the first equation (E1) from the second equation (E2), we have

$$S(1+i) - S = R(1+i)^n - R$$
$$S[1+i-1] = R[(1+i)^n - 1]$$
$$S(i) = R[(1+i)^n - 1]$$
$$S = \frac{R[(1+i)^n - 1]}{i}$$

Amount of an ordinary annuity

$$S = R\left[\frac{(1+i)^n - 1}{i}\right] \quad \textbf{(18.10)}$$

where

S = Amount (or future value) of an ordinary annuity at the end of its term
R = Amount of periodic payment
i = Interest rate per payment period
n = Total number of payment periods

Example 29 To find the future value of an ordinary annuity with annual payments of \$1,000 for four years at 8 percent interest per year, we compute, using a hand-held calculator,

$$S = 1{,}000\left[\frac{(1+0.08)^4 - 1}{0.08}\right]$$
$$= 1{,}000\left[\frac{1.360489 - 1}{0.08}\right] \qquad (1.08)^4 = 1.360489$$
$$= 1{,}000(4.506112)$$
$$= \$4{,}506.11$$

[]

Example 30 A firm is paying \$500 each quarter into a fund which pays 12 percent per year interest, compounded quarterly. How much will have accumulated in the fund (*a*) by the end of the first year? (*b*) by the end of the second year? (*c*) by the end of the fifth year?

The 12 percent annual interest rate converts to 3 percent per quarter; there will be four payments period each year.

a. The amount of the annuity at the end of the first year is

$$S = 500\left[\frac{(1.03)^4 - 1}{0.03}\right]$$
$$= 500(4.183627)$$
$$= \$2{,}091.81$$

b. The amount of the annuity at the end of the second year is

$$S = 500\left[\frac{(1.03)^8 - 1}{0.03}\right]$$
$$= 500(8.892336)$$
$$= \$4,446.17$$

c. The amount of the annuity at the end of the fifth year is

$$S = 500\left[\frac{(1.03)^{20} - 1}{0.03}\right]$$
$$= 500(26.870374)$$
$$= \$13,435.19$$

[]

Example 31 Over the past 20 years, Mrs. Zigler has been depositing \$50 each month into a special savings account. For the first 10 years the interest rate was 6 percent, compounded monthly; but for the last 10 years the interest rate has been 9 percent, compounded monthly. Determine the total accumulated value of this annuity at the end of the 20-year period.

At the end of the first 10 years the amount of the annuity was

$$S = 50\left[\frac{(1 + 0.005)^{120} - 1}{0.005}\right] = \$8,193.97$$

Then, over the last 10 years this amount has earned interest at the rate of 9 percent, compounded monthly, to accumulate a sum of

$$S = 8,193.97[1 + 0.0075]^{120} = \$20,086.35$$

In addition, the monthly deposit of \$50 made over the last 10 years will accumulate, with interest at the rate of 9 percent, compounded monthly, to

$$S = 50\left[\frac{(1 + 0.0075)^{120} - 1}{0.0075}\right] = \$9,675.71$$

This gives a total for the annuity of \$20,086.35 + \$9,675.71 = \$29,762.06 at the end of the 20-year period. []

Sinking-Fund Payments

A SINKING FUND is a fund into which equal periodic payments are made. It is designed to accumulate a definite amount of money upon a specified date. Sinking funds are generally established in order to satisfy some financial obligation or to reach some financial goal.

If the payments are to be made in the form of an ordinary annuity, then the required periodic payment into the sinking fund can be determined by reference to the formula for the amount of an ordinary annuity. That is, if

$$S = R\left[\frac{(1 + i)^n - 1}{i}\right]$$

then

$$R = \frac{S}{\left[\frac{(1+i)^n - 1}{i}\right]} = S\left[\frac{i}{(1+i)^n - 1}\right]$$

Required periodic payment (into sinking fund)

$$R = S\left[\frac{i}{(1+i)^n - 1}\right] \tag{18.11}$$

where

R = Amount of periodic payment into sinking fund
S = Amount (or future value) of sinking fund after n payment periods
n = Number of payment periods
i = Interest rate per period

(Payments are made at end of each payment period.)

Example 32 How much will have to be deposited in a fund at the end of each year at 8 percent, compounded annually, to pay off a debt of $50,000 in five years?

We compute the required periodic payment by

$$R = (50{,}000)\left[\frac{0.08}{(1.08)^5 - 1}\right]$$
$$= (50{,}000)(0.174056)$$
$$= \$8{,}522.80$$

A schedule showing how money accumulates in this fund is as follows:

Payment period	Interest earned	Periodic payment	Amount in sinking fund
1	—	$8,522.80	$ 8,522.80
2	$ 681.82	$8,522.80	$17,727.42
3	$1,418.19	$8,522.80	$27,668.41
4	$2,213.47	$8,522.80	$38,404.68
5	$3,072.37	$8,522.80	$49,999.85

[]

Example 33 A firm wishes to establish a sinking fund for the purpose of expanding the production facilities at its Westover Plant. The company needs to accumulate $500,000 over the next five years. A conservative investment policy could earn 8 percent per annum, compounded semiannually. How much should the firm contribute to the fund every six months in order to accumulate $500,000 in five years?

With $i = 0.08/2 = 0.04$ and $n = (2)(5) = 10$, the periodic payment should be

$$P = (500{,}000)\left[\frac{0.04}{(1.04)^{10} - 1}\right]$$
$$= (500{,}000)(0.083291)$$
$$= \$41{,}645.50$$

The periodic payment into the sinking fund should be $41,645.50. []

Example 34 A loan of $10,000 with interest at 10 percent compounded semiannually must be paid off in one payment three years hence. In order to provide for this, quarterly payments are to be placed into a sinking fund which pays 12 percent compounded quarterly. How much should each quarterly payment be?

The amount required to pay off the loan in three years is

$$\begin{aligned} S &= (10{,}000)(1.05)^6 \\ &= 10{,}000(1.340096) \\ &= \$13{,}400.96 \end{aligned}$$

Thus, the sinking fund should accumulate $13,400.96 by the end of three years. The required periodic payment into the fund will be

$$\begin{aligned} R &= (13{,}400.96)\left[\frac{0.03}{(1.03)^{12}-1}\right] \\ &= (13{,}400.96)(0.070462) \\ &= \$944.26 \end{aligned}$$

Quarterly payments of $944.26 should be made into the sinking fund in order to accumulate an amount of $13,400.96 by the end of three years. []

Example 35 Mary Beth Buchanan has a savings goal of $25,000 which she would like to reach 10 years from now. During the first five years she is financially able to deposit only $100 each month into the savings account. What must her monthly deposit over the last five years be if she is to reach her goal? The account pays 10.8 percent interest, compounded monthly.

At the end of the first five years, the amount of the annuity will be

$$S = 100\left[\frac{(1+0.009)^{60}-1}{0.009}\right] = 100(79.0963) = \$7{,}909.63$$

This sum will continue to draw interest at the rate of 10.8 percent, compounded monthly, over the next five years, yielding an accumulated total of

$$S = 7{,}909.63(1.009)^{60} = 7{,}909.63(1.7118668) = \$13{,}540.23$$

This leaves Mary Beth $25,000 − $13,540.23 = $11,459.77 short of her savings goal. Thus, the monthly deposits into the account over the last five years will have to be sufficient to accumulate to this amount. We determine

$$R = (11{,}459.77)\left[\frac{0.009}{(1.009)^{60}-1}\right] = (11{,}459.77)\cdot(0.0126428) = \$144.88$$

Thus, if Mary Beth makes monthly payments of $100 each into the annuity over the first five years and then monthly payments of $144.88

over the next five years, she will reach her savings goal of $25,000 at the end of 10 years. []

EXERCISES

Find the future amount of each of the following annuities.

85. $50 paid every other month for three years earning interest at 12 percent compounded every other month.
86. $100 paid monthly for five years earning 9 percent interest compounded monthly.
87. $500 paid quarterly for 10 years earning 14 percent interest compounded quarterly.
88. $600 paid annually for 25 years earning 8 percent interest compounded annually.
89. $200 paid semiannually for 15 years earning 12 percent interest compounded semiannually.

Find the required periodic payment to an ordinary annuity in each of the following situations:

90. To accumulate $5,000 after three years with 15 percent interest compounded monthly, payments made monthly.
91. To accumulate $25,000 after 12 years with 10 percent interest compounded quarterly, payments made quarterly.
92. To accumulate $10,000 after eight years with 14 percent interest compounded semiannually, payments made semiannually.
93. To accumulate $17,500 after five years with 12 percent interest compounded monthly, payments made monthly.
94. To accumulate $50,000 after 20 years with 7.5 percent interest compounded annually, payments made annually.

A person plans to pay $1,200 per year for six years into an ordinary annuity. If interest is earned at the nominal rate of 12 percent per year, what is the amount to which the annuity will grow by the end of its term under each of the following payment plans:

95. Payments of $600 are made at the end of each six-month period, and interest is compounded semiannually.
96. Payments of $300 are made at the end of each quarter and interest is compounded quarterly.
97. Payments of $100 are made at the end of each month and interest is compounded monthly.
98. Mrs. Grundy deposits $500 in a special savings account at the end of each summer. If the account pays 7.5 percent, compounded

annually, how much money will she have accumulated just after the fifth deposit?

99. What will be the amount of an ordinary annuity of $100 each month for three years if interest at 18 percent is compounded monthly? What will be the amount of the annuity at the end of six years?

100. Parents wish to set up a savings account for their child's education. They plan to deposit $200 every six months. The account earns interest of 16 percent, compounded semiannually. To what amount will the account grow in 18 years? How much interest will be earned during the term of the annuity?

101. A company established a sinking fund to discharge a debt of $120,000 due in five years by making equal semiannual deposits into a fund which pays 14 percent compounded semiannually. What is the required size of each deposit? How much interest will be earned over the five-year period?

102. What amount will have to be invested at the end of each year for 10 years in order to form a sinking fund of $300,000 if interest is 8 percent compounded annually?

103. A bond issue is approved for building a new city library. The city is required to set up a sinking fund by making regular payments every quarter into an account to be used to retire the bonds. The account pays 14 percent interest compounded quarterly. At the end of 10 years the bonds will be retired at a total cost of $1 million. What should the periodic payment into the sinking fund be?

104. A couple would like to have $15,000 for a down payment on a house in five years. What fixed amount should they deposit each month into an account paying 15 percent interest compounded monthly?

105. How much should Jillianne deposit each week (for 50 weeks) in a Christmas club account if she would like to have a total of $500 at Christmastime, given that the account pays 10 percent interest, compounded weekly?

106. Fernando deposits $100 a month for three years into an account that pays interest at an annual rate of 10.8 percent, compounded monthly. Then for the next two years he deposits $150 a month into the same account. If no withdrawals are made from the account, what is the account balance at the end of the five years?

107. Mr. Gabriel wishes to have $150,000 in a special account at the end of 25 years. He deposits in the account $100 each month for the first 15 years. What should be the amount of the monthly deposit over the last 10 years if he is to reach his savings goal? The account pays 10.2 percent interest, compounded monthly.

108. Mrs. Patterson has been making monthly deposits of $150 into a special account for 10 years. The bank paid 9 percent interest, compounded monthly, for the first five years and then raised its interest rate to 12 percent, compounded monthly, for the next five years. How much will Mrs. Patterson have in the account at the end of the 10 years?

109. Determine the accumulated value of a 15-year, $100-a-month annuity if the bank pays 6 percent interest, compounded monthly, over the first five years, 9 percent, compounded monthly, over the next five years, and 12 percent, compounded monthly, over the last five years.

110. The Harold-Jacobs Company purchased a tract of land under a purchase agreement which requires a payment of $250,000 plus 9.5 percent interest, compounded annually, at the end of five years. The company plans to set up a sinking fund to accumulate the amount required to settle the land-purchase debt. What should the quarterly deposit into the fund be if the account pays 14 percent interest, compounded quarterly?

111. Mr. Armistead has borrowed $7,500 for five years at 10 percent interest, compounded continuously. He plans to set up a special account to accumulate the funds needed to settle the debt. How much should he set aside in the account each month if his deposits earn 11.4 percent interest, compounded monthly?

112. Mr. Pugh borrowed $9,000 for four years with interest at 12 percent, compounded quarterly. He set up a special account and deposited $150 each month for the first two years. How much must he deposit each month over the last two years to accumulate the required funds? The account pays 12 percent interest, compounded monthly.

18.5 PRESENT VALUE OF AN ANNUITY

On occasion we are interested in the size of an initial deposit required to give a series of payments in the future. The PRESENT VALUE of an annuity is the sum of the *present values* of all the payments, each discounted to the beginning of the term of the annuity. The present value represents the amount that must be invested now to purchase the payments due in the future.

Assume an annuity of n payments of R (dollars) each, made at the end of each payment period, with interest rate i per payment period. We have seen that the future value of this annuity, after n payment periods, is

$$S = R\left[\frac{(1+i)^n - 1}{i}\right]$$

We have also seen that the value of a lump-sum investment A, after n periods with interest rate i per period, is $A(1 + i)^n$.

The future value of the annuity and the future value of the lump-sum payment should be equal at the end of n periods; thus,

$$A(1+i)^n = R\left[\frac{(1+i)^n - 1}{i}\right]$$

Dividing both sides of the equation by $(1 + i)^n$ gives

$$A = R\left[\frac{1-(1+i)^{-n}}{i}\right]$$

Present value of an annuity

$$A = R\left[\frac{1-(1+i)^{-n}}{i}\right] \qquad \textbf{(18.12)}$$

where

A = Present value of an annuity
R = Periodic payment
i = Interest rate per payment period
n = Number of payment periods

Example 36 What is the present value of an annuity that pays \$400 a month for the next five years if money is worth 12 percent compounded monthly?

To determine the present value, we compute using $i = 0.12/12 = 0.01$ and $n = (12)(5) = 60$,

$$\begin{aligned} A &= (400)\left[\frac{1-(1.01)^{-60}}{0.01}\right] \\ &= (400)(44.955037) \\ &= \$17{,}982.01 \end{aligned}$$

The present value of the series of future payments is \$17,982.01 [ ]

Example 37 What lump sum would have to be invested at 14 percent, compounded quarterly, to provide an annuity of \$1,250 a quarter for four years?

We compute, using $i = 0.14/4 = 0.035$ and $n = (4)(4) = 16$,

$$\begin{aligned} A &= (1{,}250)\left[\frac{1-(1.035)^{-16}}{0.035}\right] \\ &= (1{,}250)(12.094117) \\ &= \$15{,}117.65 \end{aligned}$$

A lump-sum payment of \$15,117.65 would be required to provide the series of future payments described. [ ]

Amortization

AMORTIZING A DEBT means, in general, retiring a debt in a given length of time by equal periodic payments that include compound interest.

Suppose, for example, that you borrow $5000 from a bank and agree to repay the loan in five equal annual installments, including all interest due. The bank's interest charges are 12 percent per year, compounded annually, on the unpaid balance. How much should each annual payment be in order to retire the debt, including interest, in five years?

In actuality the bank has purchased an annuity with $5,000 present value from you. To find the monthly payment R, we refer to the Present Value of an Annuity formula (Formula 18.12); that is,

$$A = R\left[\frac{1-(1+i)^{-n}}{i}\right]$$

Solving for R, we find

$$R = A\left[\frac{i}{1-(1+i)^{-n}}\right]$$

Periodic payment required for amortizing a debt

$$R = A\left[\frac{i}{1-(1+i)^{-n}}\right] \qquad (18.13)$$

where

R = Periodic payment
A = Amount of debt (present value of annuity)
i = Interest rate per period
n = Total number of payment periods

Payments are made at the end of each payment period.

The schedules in Table 18.3 illustrate the difference between the process of establishing a sinking fund and the process of amortizing a debt. Note that with a sinking fund we begin with a fund of zero dollars and make periodic deposits into the fund which, along with the interest earned on these deposits, accumulate to the total amount of a savings goal. But with the amortization of a debt, we begin with a debt balance of x dollars and make periodic payments toward the debt and the interest on the unpaid balance, eventually reducing the debt balance to zero dollars.

Example 38

With an automobile loan of $8,000 which is to be repaid in 48 equal monthly payments at an interest rate of 12 percent, compounded monthly, the periodic payment will be

Table 18.3

Sinking Fund Schedule. Five annual deposits are made into an account paying 12 percent interest, compounded annually, in order to accumulate a total of $5,000

$$R = S\left[\frac{i}{(1+i)^n - 1}\right] = 5{,}000\left[\frac{0.12}{(1.12)^5 - 1}\right] = \$787.05$$

Payment number	Amount in fund at beginning of period	Interest earned during period (12% on beginning balance)	Deposit at end of period	Amount in fund at end of period
1	-0-	-0-	$787.05	$ 787.05
2	$ 787.05	$ 94.45	787.05	1,668.55
3	1,668.55	200.23	787.05	2,655.83
4	2,655.83	318.70	787.05	3,761.58
5	3,761.58	451.39	787.05	5,000.00

Amortization Schedule. Five annual payments are made to settle a $5,000 debt, with interest of 12 percent, compounded annually, on the unpaid balance

$$R = A\left[\frac{i}{1 - (1+i)^{-n}}\right] = 5{,}000\left[\frac{0.12}{1 - (1.12)^{-5}}\right] = \$1{,}387.05$$

Payment number	Remaining principal at beginning of period	Payment amount	Interest paid (12% of remaining balance)	Amount of payment applied to principal	Remaining principal at end of period
1	$5,000.00	$1,387.05	$600.00	$ 787.05	$4,212.95
2	4,212.95	1,387.05	505.55	881.50	3,331.45
3	3,331.45	1,387.05	399.77	987.28	2,344.17
4	2,344.17	1,387.05	281.30	1,105.75	1,238.42
5	1,238.42	1,387.05	148.61	1,238.42	-0-

$$\begin{aligned} R &= (8{,}000)\left[\frac{0.01}{1 - (1.01)^{-48}}\right] \\ &= (8{,}000)(0.02633384) \\ &= \$210.67 \end{aligned}$$

The monthly payments will be $210.67 []

Example 39 You borrow $11,500 from the bank to pay for the construction of a swimming pool and agree to repay the bank in 60 equal monthly installments at 1.5 percent interest per month on the unpaid balance. How much are the monthly payments? How much interest will you pay?

The monthly payment amounts are computed by

$$\begin{aligned} R &= (11{,}500)\left[\frac{0.015}{1 - (1.015)^{-60}}\right] \\ &= (11{,}500)(0.02539343) \\ &= \$292.02 \text{ per month} \end{aligned}$$

Over the next 60 months, the total payments will be

$$60 \times 292.02 = \$17{,}521.20$$

Of this total amount, $6,021.20 represents interest charges. []

Example 40 Darryl Livingstone inherits $50,000. He plans to deposit the money in an account which pays 14 percent interest, compounded quarterly, and make equal quarterly withdrawals from the account over the next 10 years. Determine the amount of the equal quarterly withdrawals.

Withdrawals are made from an annuity in much the same way that payments are made to amortize a debt. We use Formula 18.13 to determine the amount of the withdrawals, as

$$R = 50{,}000\left[\frac{0.035}{1-(1.035)^{-40}}\right] = \$2{,}341.36$$

Darryl will be able to make withdrawals of $2,341.36 each quarter for the next 10 years from the account. []

Example 41 Sara Heller establishes a special retirement fund at age 40 by depositing $100 a month into an account which pays 9 percent interest, compounded monthly. After 20 years, Sara retires. She decides to make equal monthly withdrawals from the retirement fund over the next 15 years. Determine how much money she can withdraw each month from the retirement fund.

The value of the retirement fund when Sara retires will be

$$S = 100\left[\frac{(1.0075)^{240}-1}{0.0075}\right] = \$66{,}788.69$$

From this fund Sara can make equal monthly withdrawals over the next 15 years of

$$R = 66{,}788.69\left[\frac{0.0075}{1-(1.0075)^{-180}}\right] = \$677.42$$

Note that over the 20 years Sara deposited into the fund 240 × $100 = $24,000, but then over the next 15 years she will be able to withdraw from the fund a total of 180 × $677.42 = $121,935.60. The difference represents interest earned on the funds. []

Example 42 Mr. and Mrs. Moore purchase a house for $115,000. They make a 20 percent down payment, with the balance amortized by a 30-year mortgage at an annual interest rate of 12.6 percent, compounded monthly.

a. Determine the amount of their monthly mortgage payment.

The mortgage will be for the purchase price of the house less the down payment, that is, for $115,000 − (0.2)(115,000) = $92.000. The monthly payment required to amortize this debt over 30 years is

$$R = (92{,}000)\frac{0.0105}{1-(1.0105)^{-360}} = \$989.02$$

b. What is the total amount of interest the Moores will pay over the life of the mortgage?

Monthly payments of \$989.02 over a 30-year period will yield a total payment of (\$989.02)(360) = \$356,047.20 against a \$92,000 mortgage. Thus, the amount of interest paid is \$356,047.20 − \$92,000 = \$264,047.20.

c. Determine the amount of the mortgage the Moores will have paid after 10 years.

After 10 years, there will remain (20)(12) = 240 monthly payments of \$989.02 to be made. The amount of the mortgage that is still unpaid at this time is the present value of this series of payments, that is

$$A = (989.02)\frac{1-(1.0105)^{-240}}{0.0105} = \$86{,}513.47$$

Thus, after 10 years, the Moores will have paid \$92,000 − \$86,513.47 = \$5,486.53 against the principal amount of the mortgage. This, plus their down payment, represents their equity in the house.

[]

EXERCISES

Find the present value of an annuity under each of the following conditions:

113. \$600 paid at the end of each year for 15 years with interest rate of 8 percent compounded annually.
114. \$300 paid at the end of each six-month period for 15 years with interest rate of 8 percent compounded semiannually.
115. \$150 paid at the end of each quarter for 15 years with interest rate of 8 percent compounded quarterly.
116. \$150 paid at the end of each quarter for 15 years with interest rate of 10 percent compounded quarterly.
117. \$150 paid at the end of each quarter for 15 years with interest rate of 12 percent compounded quarterly.
118. \$200 paid at the end of each quarter for 15 years with interest rate of 12 percent compounded quarterly.
119. \$200 paid at the end of each quarter for 20 years with interest rate of 12 percent compounded quarterly.
120. You have inherited an annuity paying \$2,000 per quarter for the next 12 years. If money is worth 16 percent interest compounded quarterly, what is the present value of this annuity?
121. Parents wish to set up an annuity which will pay \$1,200 a quarter for four years to a student. What lump sum should they deposit

now at 10 percent interest compounded quarterly to establish this annuity?

122. You borrow $5,000 from the bank and agree to repay the debt in 36 equal installments at 1 percent interest per month on the unpaid balance. How much are your monthly payments?

Determine the amount of the monthly payment on each of the following:

123. $700 to be repaid in 18 equal monthly payments, with 18 percent interest compounded monthly.
124. $2,500 to be repaid in 24 equal monthly payments, with 21 percent interest compounded monthly.
125. $6,200 to be repaid in 48 equal monthly payments, with 15 percent interest compounded monthly.
126. $1,500 to be repaid in 12 equal monthly payments, with 18 percent interest compounded monthly.

On each of the following, determine the amount of the monthly payment, the total amount repaid, and the total interest paid.

127. $2,000 to be repaid in 12 equal monthly payments, with 18 percent interest compounded monthly.
128. $2,000 to be repaid in 15 equal monthly payments, with 18 percent interest compounded monthly.
129. $2,000 to be repaid in 18 equal monthly payments, with 18 percent interest compounded monthly.
130. $2,000 to be repaid in 24 equal monthly payments, with 18 percent interest compounded monthly.

131. A firm purchases a minicomputer for $22,000 and finances it at 15 percent compounded monthly over five years using equal monthly payments. How much will the monthly payments be? What interest will they pay on the loan?
132. A speedboat costs $26,500. You pay one third down and amortize the rest in equal monthly payments over a six-year period with 1.25 percent interest per month on the unpaid balance. What is your monthly payment? How much total interest will you pay?
133. A person pays $310 a month for 40 months for a car, after making an initial down payment of $1,200. If the loan costs 1.25 percent per month on the unpaid balance, what was the original cost of the car? How much total interest will be paid?

Ernie Hoffstetter wants to finance $5,000 on a new car. Determine his monthly payment, and the total amount of interest paid, if the current rate of interest is 14.4 percent, compounded monthly, under each of the following conditions:

134. The loan will be repaid over the next 30 months, with equal monthly payments.
135. The loan will be repaid over the next 36 months, with equal monthly payments.
136. The loan will be repaid over the next 40 months, with equal monthly payments.

137. A vacation cabin is offered for sale for $30,000. The owner will finance the sale at an interest rate of 13.2 percent, compounded monthly. The owner is willing to negotiate the amount of the down payment but insists that the loan be paid off over the next 12 years. If Rachel and Aaron feel that they will be financially able to make $200-a-month payments on the cabin, how much will they have to give as a down payment?
138. Eugene Davidson deposited $450 each quarter into a special retirement fund which paid 8 percent interest, compounded quarterly. These deposits were made over a period of 30 years, after which time Eugene retired. He plans to now withdraw from the fund equal amounts each quarter over the next 20 years. How much can he withdraw each quarter?
139. After he retires, Jackson Baines would like to be able to withdraw $500 a month from a retirement fund over a 15-year period. With this in mind, he sets up a special fund 25 years before he plans to retire. How much should he deposit each month into this account in order to meet his retirement goal? (The account pays interest at the rate of 9 percent, compounded monthly.)

Mr. and Mrs. Casey purchased a $90,000 house, making a $25,000 down payment. They can amortize the balance ($65,000) at 12 percent interest, compounded monthly, over 30 years.

140. What will be the monthly payment on the mortgage?
141. How much total interest will they pay over the 30 years?
142. What will be the remaining mortgage amount at the end of 10 years?
143. What will be the Casey's equity in the house at the end of 10 years?
144. What will be the Casey's equity in the house at the end of 15 years?

Appendix Tables

Appendix Table I

Natural Logarithms

x	ln x	x	ln x	x	ln x	x	ln x
1.0	0.000000	3.5	1.252763	6.0	1.791760	8.5	2.140066
1.1	0.095310	3.6	1.280934	6.1	1.808289	8.6	2.151762
1.2	0.182322	3.7	1.308333	6.2	1.824549	8.7	2.163323
1.3	0.262364	3.8	1.335001	6.3	1.840550	8.8	2.174752
1.4	0.336472	3.9	1.360977	6.4	1.856298	8.9	2.186051
1.5	0.405465	4.0	1.386294	6.5	1.871802	9.0	2.197225
1.6	0.470004	4.1	1.410987	6.6	1.887070	9.1	2.208274
1.7	0.530628	4.2	1.435084	6.7	1.902108	9.2	2.219204
1.8	0.587787	4.3	1.458615	6.8	1.916923	9.3	2.230014
1.9	0.641854	4.4	1.481604	6.9	1.931521	9.4	2.240710
2.0	0.693147	4.5	1.504077	7.0	1.945910	9.5	2.251292
2.1	0.741937	4.6	1.526056	7.1	1.960095	9.6	2.261763
2.2	0.788457	4.7	1.547562	7.2	1.974081	9.7	2.272126
2.3	0.832909	4.8	1.568616	7.3	1.987874	9.8	2.282382
2.4	0.875469	4.9	1.589235	7.4	2.001480	9.9	2.292535
2.5	0.916291	5.0	1.609438	7.5	2.014903	10.0	2.302585
2.6	0.955511	5.1	1.629240	7.6	2.028148		
2.7	0.993252	5.2	1.648659	7.7	2.041220		
2.8	1.029619	5.3	1.667707	7.8	2.054124		
2.9	1.064711	5.4	1.686399	7.9	2.066863		
3.0	1.098612	5.5	1.704748	8.0	2.079442		
3.1	1.131402	5.6	1.722767	8.1	2.091864		
3.2	1.163151	5.7	1.740466	8.2	2.104134		
3.3	1.193923	5.8	1.757858	8.3	2.116256		
3.4	1.223775	5.9	1.774952	8.4	2.128232		

Note: $\ln 120 = \ln [(1.2) \cdot (10^2)] = \ln 1.2 + 2 \ln 10$
$= 0.182322 + 2(2.302585)$
$= 4.787492$

Appendix Table II

Values of e^x and e^{-x}

x	e^x	e^{-x}	x	e^x	e^{-x}
0.1	1.105171	0.904837	5.1	164.021970	0.006097
0.2	1.221403	0.818731	5.2	181.272380	0.005517
0.3	1.349858	0.740818	5.3	200.336848	0.004992
0.4	1.491825	0.670320	5.4	221.406543	0.004517
0.5	1.648721	0.606531	5.5	244.692166	0.004087
0.6	1.822119	0.548812	5.6	270.426253	0.003698
0.7	2.013753	0.496585	5.7	298.867344	0.003346
0.8	2.225541	0.449329	5.8	330.299308	0.003028
0.9	2.459603	0.406570	5.9	365.037329	0.002739
1.0	2.718282	0.367879	6.0	403.428793	0.002479
1.1	3.004164	0.332871	6.1	445.857515	0.002243
1.2	3.320116	0.301194	6.2	492.748947	0.002029
1.3	3.669294	0.272532	6.3	544.572014	0.001836
1.4	4.055198	0.246597	6.4	601.844808	0.001662
1.5	4.481689	0.223130	6.5	665.141633	0.001503
1.6	4.953030	0.201897	6.6	735.094769	0.001360
1.7	5.473946	0.182684	6.7	812.405670	0.001231
1.8	0.049649	0.165299	6.8	897.847463	0.001114
1.9	6.685892	0.149569	6.9	992.274337	0.001008
2.0	7.389056	0.135335	7.0	1096.633158	0.000912
2.1	8.166165	0.122456	7.1	1211.967537	0.000825
2.2	9.025012	0.110803	7.2	1339.430509	0.000747
2.3	9.974184	0.100259	7.3	1480.300210	0.000676
2.4	11.023172	0.090718	7.4	1635.983806	0.000611
2.5	12.182494	0.082085	7.5	1808.042414	0.000553
2.6	13.463743	0.074274	7.6	1998.196657	0.000500
2.7	14.879729	0.067206	7.7	2208.347571	0.000453
2.8	16.444650	0.060810	7.8	2440.602443	0.000410
2.9	18.174138	0.055023	7.9	2697.283872	0.000371
3.0	20.085537	0.049787	8.0	2980.957987	0.000335
3.1	22.197960	0.045049	8.1	3294.469332	0.000304
3.2	24.532526	0.040762	8.2	3640.949613	0.000275
3.3	27.112644	0.036883	8.3	4023.873161	0.000249
3.4	29.964117	0.033373	8.4	4447.069292	0.000225
3.5	33.115452	0.030197	8.5	4914.768840	0.000203
3.6	36.598248	0.027324	8.6	5431.661663	0.000184
3.7	40.447297	0.024724	8.7	6002.916797	0.000167
3.8	44.701193	0.022371	8.8	6634.245272	0.000151
3.9	49.402477	0.020242	8.9	7331.977735	0.000136
4.0	54.598150	0.018316	9.0	8103.083928	0.000123
4.1	60.340311	0.016573	9.1	8955.296120	0.000112
4.2	66.686382	0.014996	9.2	9897.136610	0.000101
4.3	73.699808	0.013569	9.3	10938.021294	0.000091
4.4	81.450915	0.012277	9.4	12088.387647	0.000083
4.5	90.017131	0.011109	9.5	13359.739570	0.000075
4.6	99.484354	0.010052	9.6	14764.787198	0.000068
4.7	109.947256	0.009095	9.7	16317.619647	0.000061
4.8	121.510441	0.008230	9.8	18033.748367	0.000055
4.9	134.289857	0.007447	9.9	19930.381842	0.000050
5.0	148.413301	0.006738	10.0	22026.486801	0.000045

Note: $e^{12} = e^{6+6} = (e^6)(e^6)$

$= (403.428793)(403.428793)$

$= 162{,}754.79$

Appendix Table III

Table of Areas for Normal Curve

An entry in the table is the proportion under the curve between $Z = 0$ and a positive value of Z.

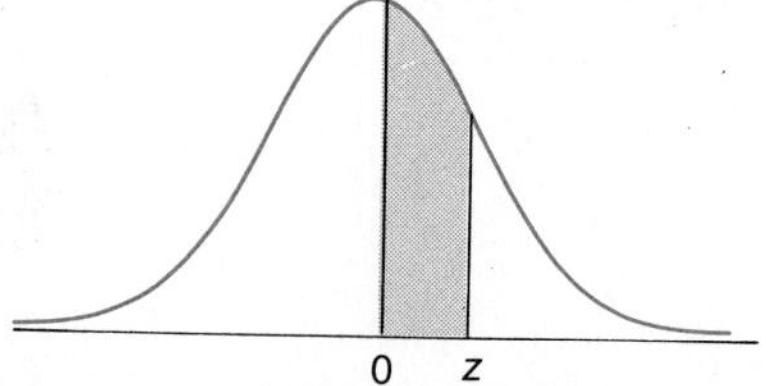

Z	0.00	0.01	0.02	0.03	0.04	0.05	0.06	0.07	0.08	0.09
0.0	0.0000	0.0040	0.0080	0.0120	0.0160	0.0199	0.0239	0.0279	0.0319	0.0359
0.1	0.0398	0.0438	0.0478	0.0517	0.0557	0.0596	0.0636	0.0675	0.0714	0.0753
0.2	0.0793	0.0832	0.0871	0.0910	0.0948	0.0987	0.1026	0.1064	0.1103	0.1141
0.3	0.1179	0.1217	0.1255	0.1293	0.1331	0.1368	0.1406	0.1443	0.1480	0.1517
0.4	0.1554	0.1591	0.1628	0.1664	0.1700	0.1736	0.1772	0.1808	0.1844	0.1879
0.5	0.1915	0.1950	0.1985	0.2019	0.2054	0.2088	0.2123	0.2157	0.2190	0.2224
0.6	0.2257	0.2291	0.2324	0.2357	0.2389	0.2422	0.2454	0.2486	0.2517	0.2549
0.7	0.2580	0.2611	0.2642	0.2673	0.2703	0.2734	0.2764	0.2794	0.2823	0.2852
0.8	0.2881	0.2910	0.2939	0.2967	0.2995	0.3023	0.3051	0.3078	0.3106	0.3133
0.9	0.3159	0.3186	0.3212	0.3238	0.3264	0.3289	0.3315	0.3340	0.3365	0.3389
1.0	0.3413	0.3438	0.3461	0.3485	0.3508	0.3531	0.3554	0.3577	0.3599	0.3621
1.1	0.3642	0.3665	0.3686	0.3708	0.3729	0.3749	0.3770	0.3790	0.3810	0.3830
1.2	0.3849	0.3869	0.3888	0.3907	0.3925	0.3944	0.3962	0.3980	0.3997	0.4015
1.3	0.4032	0.4049	0.4066	0.4082	0.4099	0.4115	0.4131	0.4147	0.4162	0.4177
1.4	0.4192	0.4207	0.4222	0.4236	0.4251	0.4265	0.4279	0.4292	0.4306	0.4319
1.5	0.4332	0.4345	0.4357	0.4370	0.4382	0.4394	0.4406	0.4418	0.4429	0.4441
1.6	0.4452	0.4463	0.4474	0.4484	0.4495	0.4505	0.4515	0.4525	0.4535	0.4545
1.7	0.4554	0.4564	0.4573	0.4582	0.4591	0.4599	0.4608	0.4616	0.4625	0.4633
1.8	0.4641	0.4649	0.4656	0.4664	0.4671	0.4678	0.4686	0.4693	0.4699	0.4706
1.9	0.4713	0.4719	0.4726	0.4732	0.4738	0.4744	0.4750	0.4756	0.4761	0.4767
2.0	0.4772	0.4778	0.4783	0.4788	0.4793	0.4798	0.4803	0.4808	0.4812	0.4817
2.1	0.4821	0.4826	0.4830	0.4834	0.4838	0.4842	0.4846	0.4850	0.4854	0.4857
2.2	0.4861	0.4864	0.4868	0.4871	0.4875	0.4878	0.4881	0.4884	0.4887	0.4890
2.3	0.4893	0.4896	0.4898	0.4901	0.4904	0.4906	0.4909	0.4911	0.4913	0.4916
2.4	0.4918	0.4920	0.4922	0.4925	0.4927	0.4929	0.4931	0.4932	0.4934	0.4936
2.5	0.4938	0.4940	0.4941	0.4943	0.4945	0.4946	0.4948	0.4949	0.4951	0.4952
2.6	0.4953	0.4955	0.4956	0.4957	0.4959	0.4960	0.4961	0.4962	0.4963	0.4964
2.7	0.4965	0.4966	0.4967	0.4968	0.4969	0.4970	0.4971	0.4972	0.4973	0.4974
2.8	0.4974	0.4975	0.4976	0.4977	0.4977	0.4978	0.4979	0.4979	0.4980	0.4981
2.9	0.4981	0.4982	0.4982	0.4983	0.4984	0.4984	0.4985	0.4985	0.4986	0.4986
3.0	0.4987	0.4987	0.4987	0.4988	0.4988	0.4989	0.4989	0.4989	0.4990	0.4990

Answers to Selected Odd-Numbered Exercises

CHAPTER 1 A REVIEW OF THE BASIC CONCEPTS OF MATHEMATICS

1. -2
5. 2
9. -9
13. $13/63$
17. $-23/40$
21. 0
25. 64
29. $1/2$
33. $1/512$
37. 9
41. -1
45. $5\sqrt[3]{3}$
49. 2
53. $2X$
57. $6x^{10} - 2X^6 + 5X^2$
61. $X^3 + 4X^2$
65. $X^3 + 4X + 5$
69. $X^3 - 3X^2 + 3X - 1$
73. $-3X^2 - 17X - 2$
77. $4XYZ(4X^2Y + 2X - Z)$
81. $(3 + X)(2 - X)$
85. $(X + 4)^2$
89. $(X + 2)(X - 2)$
93. $(2X - 1)(4X^2 + 2X + 1)$
97. $(X + 2)(X - 2)^2$
101. $(3X + 2)/[X(X + 1)]$
105. $(12X^2 - 3X + 8)/6X$
109. $(4X - 13)/3X$
113. $6XY/[(X + 1)(4Y - 1)]$
3. -15
7. -2
11. $5/6$
15. $53/30$
19. $27/22$
23. $1/16$
27. 64
31. 25
35. $-1/2$
39. 3
43. $1/125$
47. 5
51. $4X^2 + 9Y^2$
55. $-X^2 + (1/4)X - (1/8)$
59. $3X - 15$
63. $X^2 + 7X + 12$
67. $2X^4 + 3X^3 + 2X^2 - 3X - 4$
71. $3X^3 + 4X^2 - 65X$
75. $-14X^2 - 36X - 18$
79. $(X + 3)(X + 4)$
83. $(6X + 1)(5X - 2)$
87. $X(X - 3)^2$
91. $(X + 5)(X - 5)$
95. $(X + 4)(X^2 - 4X + 16)$
99. $(X + 1)(X - 1)(X^2 + 1)(X^4 + 1)$
103. $(7X - 8)/[(X - 2)(X + 1)]$
107. $(X^4 + X^2 + 10X + 25)/[X^2(5 + X)]$
111. $(XY - X)/(XY - Y)$
115. $-5/7$

117. $[X(X+Y)]/(X+1)$
119. $[X(3X+2)]/(X^3-3)$
121. $X(X+1)(2X-3)$
123. $(X+1)/(X+2)^2$
125. $3Y/X$
127. $X(2X^4+5X^2-1)$
129. $(X-2)(X^2+3)$
131. $Y=(13-2X)/3$; $(0,13/3)$; $(1,11/3)$
133. $X=10$
135. $X=5$
137. $X=1/4$
139. $X=7/3$
141. $X=5$
143. $X=3$ (Note: Do not use $X=4$ since this gives denominator of zero.)
145. $X=7$
147. $X=-4$; $X=5$
149. $X=0$; $X=8$
151. $X \leq -1/3$
153. $X \leq -1/3$
155. $X \geq 2$

157. $X > -1/4$
159. $X > 4$; $X < -12$
161. 3.51×10^8
163. 5.34×10^7
165. 8.21×10^{-5}
167. 1.5×10^3
169. 1.4×10^{-2}
171. $\log_4 64 = 3$
173. $\log_5 1/25 = -2$
175. $\log_{1/4} 4 = -1$
177. $\ln 10 = 2.302585$
179. $X=1/25$
181. $X=3$
183. $X=20.085537$
185. $X=1/3$
187. $X=3$
189. $X=0.1486984$
191. $X=1.7482538$
193. $X=-21.949999$
195. $X=12.283135$
197. $X=3$
199. $X=1.3385662$
201. $X=3$
203. $X=3$
205. $X=5$
207. $X=1/5$
209. $X=1808.0424$
211. $X=316.22618$

CHAPTER 2 A REVIEW OF SET THEORY AND COUNTING

1. $A=C$; $B=D$
3. $G=H$
5. $K=M$
7. Subset of $\mathscr{U}$
9. Subset of $\mathscr{U}$
11. Not a subset of $\mathscr{U}$
13. True
15. True
17. False
19. False
21. True
23. True
25. False
27. True
29. The null set is not an element of the universal set; but the null set is a subset of every universal set and thus is an element of the power set.
31. The subset $\{a,b\}$ is not an element of $\mathscr{U}$, but it is an element of the power set.
33. 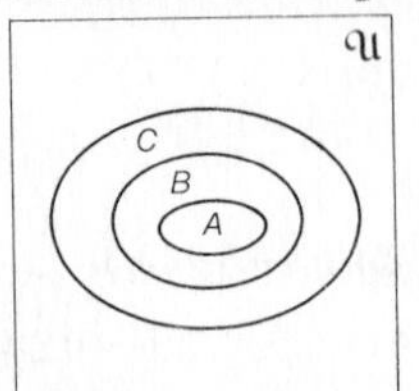

35. 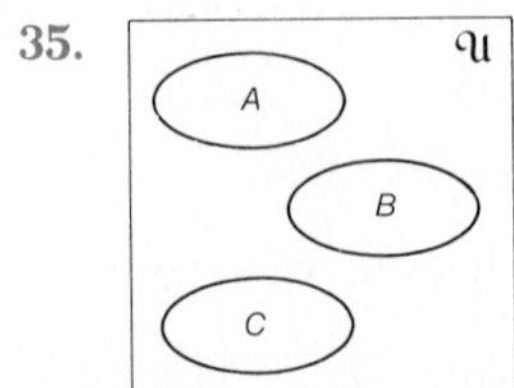

37. {(4,7),(4,8),(5,7),(5,8)}

39. {(7,3,4),(7,6,4),(7,9,4),(8,3,4),(8,6,4),(8,9,4), (7,3,5),(7,6,5),(7,9,5),(8,3,5),(8,6,5),(8,9,5)}

41. {(3,3),(3,6),(3,9),(6,3),(6,6),(6,9),(9,3), (9,6),(9,9)}

43. 12

45. 15

47. 9

49. {(2,1),(2,2),(2,3),(4,1),(4,2),(4,3)}

51. {(1,2),(2,2),(3,2),(1,4),(2,4),(3,4)}

53. {{ },{1},{2},{1,2}}

55. {(1,4),(1,6),(2,4),(2,6)}

57. {(1,4,4,5),(1,4,6,5),(1,6,4,5),(1,6,6,5),(2,4,4,5), (2,4,6,5),(2,6,4,5),(2,6,6,5)}

59. {T,T}

61. {H,H}

63. {(2,2)}

65. {(0,6),(1,4),(1,6),(2,4),(2,6),(3,2),(3,4),(3,6)}

67. R_1 = {(1,2),(2,2),(2,3),(3,2),(3,3),(3,4),(4,2), (4,3),(4,4),(4,5)} is not a function.

69. R_3 = {(1,2),(2,3),(3,4),(4,5)} is a function.

71. There are 27 functions defined on $A \times A$.

73. {5,6}

75. {3,4}

77. {1,2,3,4}

79. {1,2,3,4,5,6}

81. {5,6}

83. { }

85. { }

87. {1,2,5,6}

89. B^c

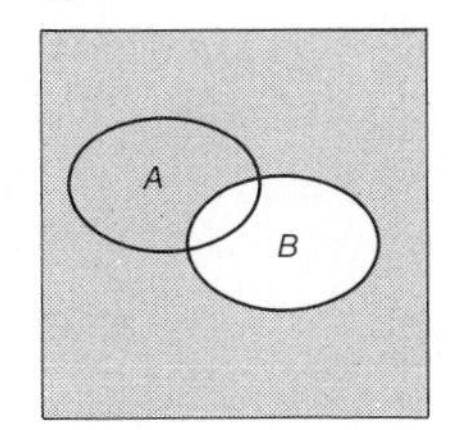

91. $A \cup B$

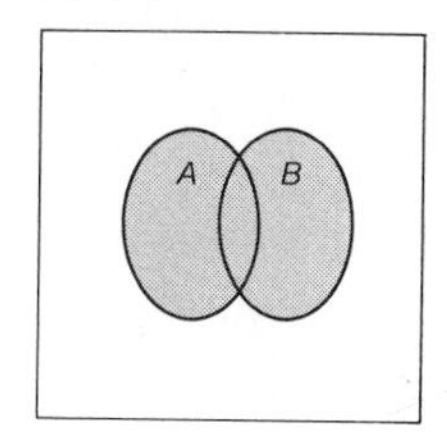

93. $(A \cup B)^c$

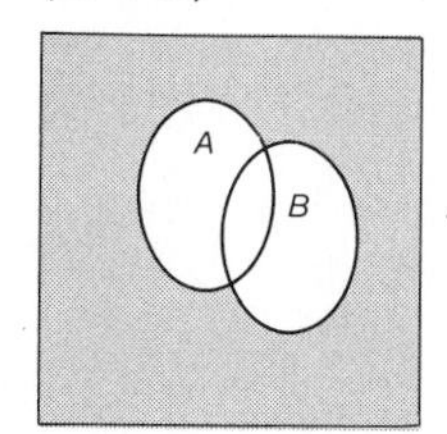

95. $A^c \cap B$

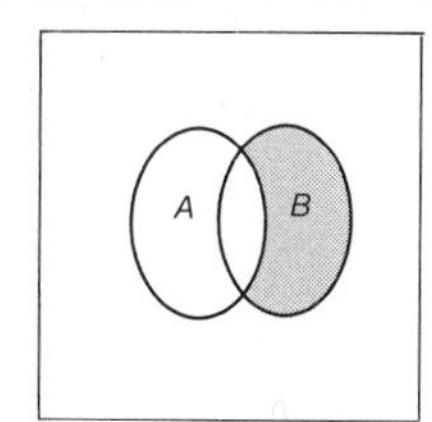

97. $A \cap B^c$

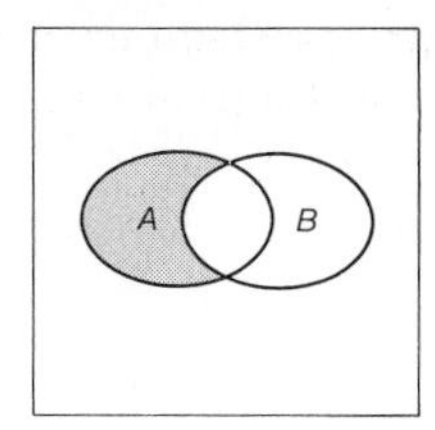

99. $A^c \cap B^c$

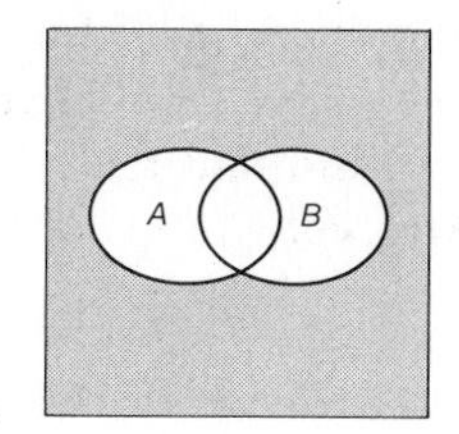

101. D^c is the set of all senators who are not democrats.

103. $E \cap D$ is the set of all Democratic senators whose states are east of the Mississippi River.

105. $E^c \cap D$ is the set of all Democratic senators whose states are not east of the Mississippi River.

107. $E \cap D^c$ is the set of all senators whose states are east of the Mississippi River except those whose are Democrats.

109. $E^c \cup D^c$ is the set of all senators except the Democrats whose states are east of the Mississippi River.

111. $(E \cup D)^c$ is the set of all senators who are neither Democrats nor represent states east of the Mississippi River.

113. {3,4,7,8}

115. {1,2,3,4,5,6}

117. {1,2,3,5,6,7}

119. {1,3}

121. {3,4,5,6,7,8}

123. {1,3,4,5,7,8}

125. {1}

127. {1,2,3,5,7}

129. {1,3,5}

131. {1,2,3,4,5,7,8}

133. {5}

135. {4}

137. {2,3,4,5,6,7,8}

139. {1,2,3,4,5}

141. $A \cap B \cap C$

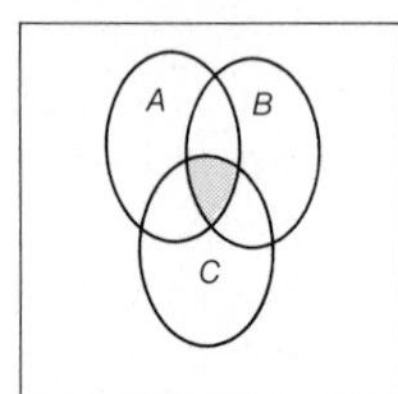

143. $A \cap B \cap C^c$

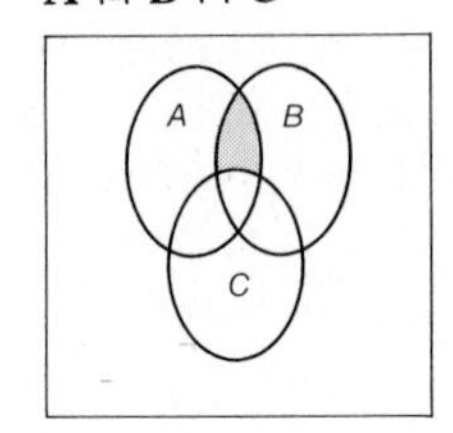

145. $A^c \cap B \cap C$

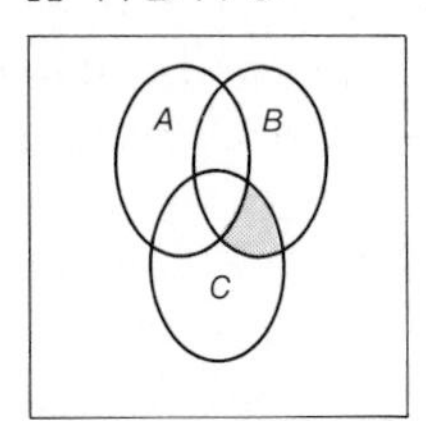

147. $A^c \cap B^c \cap C^c$

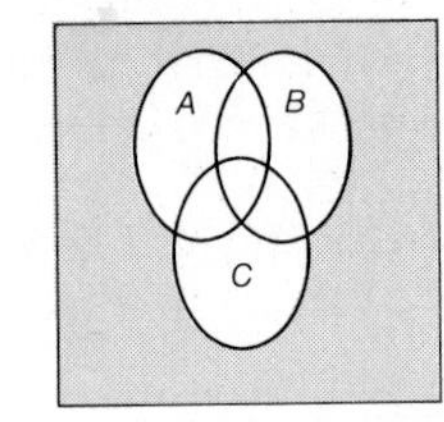

149. $A \cup B \cup C^c$

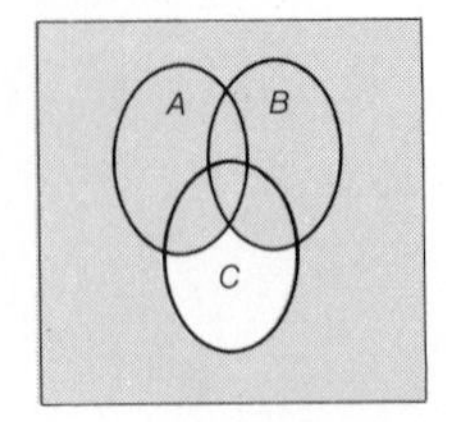

159. S

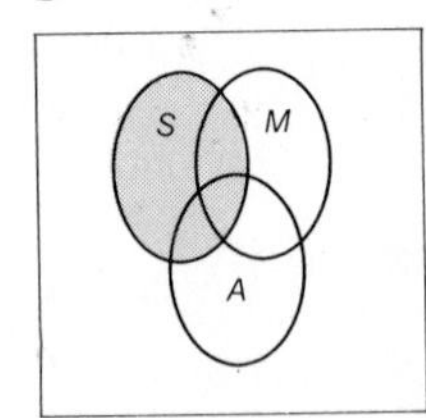

161. M

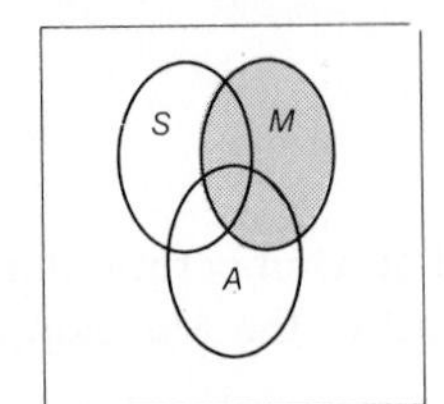

163. M^c

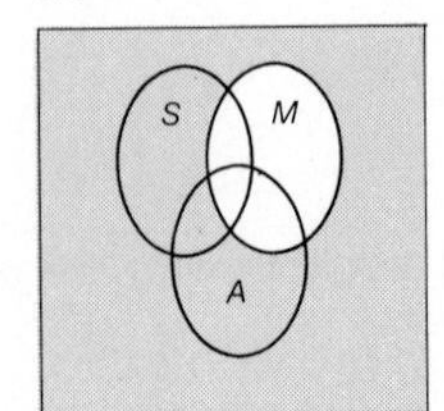

165. $M \cap A^c$

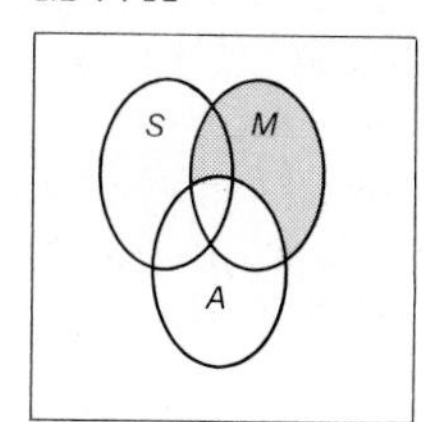

167. $M^c \cap A^c \cap S^c$

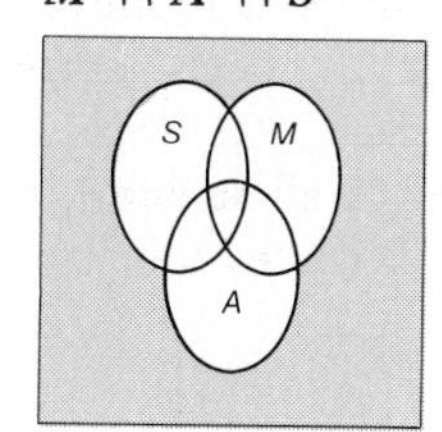

169. A^c is the set of all students who are not taking an accounting course.

171. C^c is the set of all students who are not taking a chemistry course.

173. $B \cap C$ is the set of all students who are taking both a biology and a chemistry course.

175. $A \cup B$ is the set of all students who are taking either an accounting course or a biology course or both.

177. $A \cap B \cap C$ is the set of all students who are taking an accounting course and a biology course and a chemistry course.

179. $A \cup B \cup C^c$ is the set of all students *except* those who are taking just a chemistry course but no accounting and no biology courses.

181. $A^c \cup B \cup C$ is the set of all students except those who are taking just an accounting course but no biology and no chemistry course.

183. $A \cap B^c \cap C$ is the set of all students who are taking both an accounting and a chemistry course but are not taking a biology course.

185. $A \cap B^c \cap C^c$ is the set of all students who are taking an accounting course but not a biology course or a chemistry course.

187. $A \cap (B \cup C)$ is the set of all students who are taking an accounting course along with one (or both) of the other courses, biology or chemistry.

189. $C \cup (A \cap B)$ is the set of all students taking either a chemistry course or both the accounting and biology courses.

191. $A \cap B^c \cap C^c$, $A \cap B \cap C^c$, $A \cap B^c \cap C$, $A \cap B \cap C$

193. $A \cap B^c \cap C^c$, $A \cap B \cap C^c$, $A \cap B \cap C$, $A \cap B^c \cap C$, $C \cap A^c \cap B^c$, $B \cap C \cap A^c$

195. $B \cap A^c \cap C^c$, $A \cap B \cap C^c$, $A \cap B \cap C$, $B \cap C \cap A^c$, $A \cap C \cap B^c$, $C \cap A^c \cap B^c$

197. 6
199. 8
201. 7
203. 10
205. 8
207. 0
209. 6
211. 12
213. 160
215. 80
217. 80
219. 55
221. 40
223. 15
225. 70
227. 38
229. 80
231. 95
233. 220
235. 20
237. 30
239. 35
241. 10
243. 200
245. 43
247. 220
249. 135
251. 81
253. 29
255. 24

257. 30
259. 676,000
261. 520
263. 60
265. 256
267. 120
269. 59,049
271. 840
273. 343
275. 120
277. 15,606,255,000
279. 720
281. 5,040
283. 55,440
285. 24
287. 64
289. 3,840
291. 362,880
293. 30,240
295. 958,003,200
297. 829,440
299. 35
301. 27,720
303. 4
305. 105
307. 3,003
309. 10
311. 4
313. 16
315. 495
317. 7,315
319. 2,940
321. 15
323. 57
325. 4,752
327. 2,496
329. 1,512
331. 4,200
333. 196
335. 2,217,600

CHAPTER 3 MATRIX ALGEBRA

1. 2×3
3. 2×2
5. 2×4
7. 2
9. 10
11. 8
13. (2,2)
15. (4,3)
17. $$\begin{bmatrix} 0 & 1 & 1 & 1 & 1 \\ 2 & 0 & 1 & 1 & 1 \\ 2 & 2 & 0 & 1 & 1 \\ 2 & 2 & 2 & 0 & 1 \\ 2 & 2 & 2 & 2 & 0 \end{bmatrix}$$
19. $$\begin{bmatrix} 2 & 3 & 4 & 5 & 6 \\ 3 & 4 & 5 & 6 & 7 \\ 4 & 5 & 6 & 7 & 8 \\ 5 & 6 & 7 & 8 & 9 \\ 6 & 7 & 8 & 9 & 10 \end{bmatrix}$$
21. $$R = \begin{bmatrix} 225 & 130 \\ 345 & 212 \\ 110 & 48 \end{bmatrix}$$ r_{ij} = Number of respondents of political affiliation i with education level j where $$i = \begin{cases} 1 \text{ is Republican} \\ 2 \text{ is Democrat} \\ 3 \text{ is Independent} \end{cases} \quad \text{and } j = \begin{cases} 1 \text{ is with college degree} \\ 2 \text{ is without college degree} \end{cases}$$
23. $$D = \begin{bmatrix} 25 & 15 \\ 25 & 10 \\ 40 & 5 \end{bmatrix}$$ d_{ij} = Discount percentage on product i granted to account type j where $$i = \begin{cases} 1 \text{ is product A} \\ 2 \text{ is product B} \\ 3 \text{ is product C} \end{cases} \quad \text{and } j = \begin{cases} 1 \text{ is wholesalers} \\ 2 \text{ is retailers} \end{cases}$$
25. $Q = [600 \quad 450 \quad 100 \quad 40]$ q_j = number of units of product j ordered where $$j = \begin{cases} 1 \text{ is product A} \\ 2 \text{ is product B} \\ 3 \text{ is product C} \\ 4 \text{ is product D} \end{cases}$$

27. A, B, C, H

29. E

31. A

33. $A^t = \begin{bmatrix} 1 & 0 \\ 0 & 1 \end{bmatrix}$ $G^t = \begin{bmatrix} 1 & 0 \\ 0 & 1 \\ 1 & 0 \end{bmatrix}$ $E^t = [5 \quad 1/5 \quad 25]$ $D^t = \begin{bmatrix} 8 \\ -9 \\ 7 \end{bmatrix}$

35. $A \neq B$

37. $A \neq B$

39. $\begin{bmatrix} 6 & 8 \\ 10 & 12 \end{bmatrix}$

41. $\begin{bmatrix} 11 & 9 \\ 4 & -1 \\ 6 & -1 \end{bmatrix}$

43. Not conformable for addition

45. $\begin{bmatrix} 7 \\ 4 \end{bmatrix}$

47. $\begin{bmatrix} 2/3 & 3/8 \\ 10/21 & 14.45 \end{bmatrix}$

49. $[4 \quad 4 \quad 3 \quad 1 \quad 1]$

51. $A + B = B + A = \begin{bmatrix} 3 & 4 \\ 5 & 3 \end{bmatrix}$

53. $(A + B)^t = A^t + B^t = \begin{bmatrix} 3 & 5 \\ 4 & 3 \end{bmatrix}$

55. $\begin{bmatrix} -36 & -16 \\ -8 & -12 \\ 0 & -24 \end{bmatrix}$

57. $\begin{bmatrix} 6.3 & 4.3 \\ 0.7 & 2.1 \end{bmatrix}$

59. $[-2 \quad 4 \quad -1]$

61. $1/2$

63. -0.7

65. $(1/12)\begin{bmatrix} 3 & 5 \\ 8 & 10 \end{bmatrix}$

67. $(1/8)[36 \quad 14 \quad 5]$

69. $(x + y)A = xA + yA = \begin{bmatrix} 0 & 52/3 \\ 1/3 & -26/3 \end{bmatrix}$

71. $x(yA) = (xy)A = \begin{bmatrix} 0 & 16/3 \\ 4/3 & -8/3 \end{bmatrix}$

73. $\begin{bmatrix} 180 & 210 & 75 \\ 210 & 290 & 60 \\ 75 & 120 & 40 \end{bmatrix}$

75. $[59 \quad 35 \quad 21.5]$

77. $\begin{bmatrix} -1 & 4 & -3 \\ -1 & 1 & 1 \\ -2 & -9 & -2 \end{bmatrix}$

79. $\begin{bmatrix} -1 & -4 & 1 \\ 1 & -2 & 2 \\ 1 & 9 & -2 \end{bmatrix}$

81. $\begin{bmatrix} 6 & -27 & 12 \\ -9 & 3 & -24 \\ 12 & 27 & 36 \end{bmatrix}$

83. $\begin{bmatrix} -5 & 1/2 \\ 3 & 3 \end{bmatrix}$

85. $\begin{bmatrix} 9/2 & -1/4 \\ -1/2 & -3 \end{bmatrix}$

87. $\begin{bmatrix} -27/5 & -3/10 \\ -7/5 & 22/5 \end{bmatrix}$

89. $\begin{bmatrix} 1 & -7 \\ 96 & 6 \end{bmatrix}$

91. $\begin{bmatrix} 10 & 4 \\ 21 & 9 \end{bmatrix}$

93. $\begin{bmatrix} 7 & 24 & 16 \\ 15 & 63 & 60 \\ 5 & 30 & 40 \end{bmatrix}$

95. $\begin{bmatrix} 10 & 12 \\ 15 & 18 \end{bmatrix}$

97. $\begin{bmatrix} -2 & 0 \\ 7 & -9 \end{bmatrix}$

99. $\begin{bmatrix} 11 & -5.5 & 17.5 \\ 23 & -10 & 37 \\ 35 & -14.5 & 56.5 \end{bmatrix}$

101. $\begin{bmatrix} 18{,}000 \\ 22{,}000 \\ 29{,}200 \end{bmatrix}$

103. $25,550

105. $[4{,}000 \quad 640 \quad 1{,}000 \quad 440]$

107. $\begin{bmatrix} 4{,}145 \\ 4{,}080 \end{bmatrix}$

109. $3,689

111. \$6,250

113. $\begin{bmatrix} 8 & 16 \\ 28 & 16 \end{bmatrix}$

115. $\begin{bmatrix} 15 & 36 \\ 23 & 3 \end{bmatrix}$

117. $\begin{bmatrix} -2 & -2 \\ 3 & 4 \end{bmatrix}$

119. $\begin{bmatrix} 4 & 6 \\ 6 & 10 \end{bmatrix}$

121. $\begin{bmatrix} 7 & 3 \\ 15 & 7 \end{bmatrix}$

123. $\begin{bmatrix} -2 & 0 \\ 3 & 1 \end{bmatrix}$

125. $\begin{bmatrix} 5 & 5 \\ 15 & 5 \end{bmatrix}$

127. $\begin{bmatrix} 5 & 5 \\ 15 & 5 \end{bmatrix}$

129. $AB = [51] \neq BA = \begin{bmatrix} 3 & 6 & 12 \\ 6 & 12 & 24 \\ 9 & 18 & 36 \end{bmatrix}$

131. $AB = AC = \begin{bmatrix} 10 & 14 \\ -20 & -28 \end{bmatrix}$

133. $(AB)^t = \begin{bmatrix} 9 & 18 \\ 8 & -4 \end{bmatrix} \neq A^tB^t = \begin{bmatrix} -10 & 1 \\ 30 & 15 \end{bmatrix}$

135. $(AB)^t = \begin{bmatrix} 9 & 18 \\ 8 & -4 \end{bmatrix} \neq (BA)^t = \begin{bmatrix} -10 & 1 \\ 30 & 15 \end{bmatrix}$

137. $A^2 = \begin{bmatrix} -1 & -2 \\ 4 & -1 \end{bmatrix}$ $A^4 = \begin{bmatrix} -7 & 4 \\ -8 & -7 \end{bmatrix}$ $A^6 = \begin{bmatrix} 23 & 10 \\ -20 & 23 \end{bmatrix}$ $A^8 = \begin{bmatrix} 17 & -56 \\ 112 & 17 \end{bmatrix}$ $A^{16} = \begin{bmatrix} -5{,}983 & -1{,}904 \\ 3{,}808 & -5{,}983 \end{bmatrix}$

139. $(A + B)(A - B) = \begin{bmatrix} -7 & 4 \\ 6 & -8 \end{bmatrix} \neq A^2 - B^2 = \begin{bmatrix} -5 & 3 \\ 4 & -10 \end{bmatrix}$

141. $A^2 = \begin{bmatrix} 0 & 0 \\ 0 & 0 \end{bmatrix}$

143. $\begin{bmatrix} 1 & 0 & 3 \\ 0 & 5 & 6 \\ 7 & 1 & 0 \end{bmatrix}$

145. $\begin{bmatrix} 0 & 0 & 0 \\ 0 & 0 & 0 \\ 0 & 0 & 0 \end{bmatrix}$

147. $\begin{bmatrix} 0 & 0 & 0 \\ 0 & 0 & 0 \\ 0 & 0 & 0 \end{bmatrix}$

149. $[31 \quad -8 \quad 12 \quad 14]$

151. $\begin{bmatrix} 29 \\ 31 \\ 9 \\ 6 \end{bmatrix}$

153. $\begin{bmatrix} 1 \\ 2 \\ 1 \\ 2 \end{bmatrix}$ There is no consensus leader; there is no leader.

155. Barney

157. $\begin{bmatrix} 4 \\ 3 \\ 1 \\ 2 \\ 4 \\ 1 \end{bmatrix}$

159. Team A

161. Brand E

163. $\begin{bmatrix} 0 & 1 & 0 & 0 & 1 \\ 1 & 0 & 1 & 0 & 1 \\ 0 & 1 & 0 & 1 & 0 \\ 0 & 0 & 1 & 0 & 0 \\ 1 & 1 & 0 & 0 & 0 \end{bmatrix}$

165. *A*, 2; *B*, 3; *C*, 2; *D*, 1; *E*, 2

167. $\begin{bmatrix} 0 & 1 \\ 2 & -1 \end{bmatrix}$

169. $(1/11)\begin{bmatrix} 3 & -7 \\ -1 & 6 \end{bmatrix}$

171. $\begin{bmatrix} 2 & -3 \\ -1 & 2 \end{bmatrix}$

173. $(1/60)\begin{bmatrix} 4 & 12 \\ 3 & -6 \end{bmatrix}$

175. $\begin{bmatrix} -2 & -1 & 2 \\ 5 & 2 & -4 \\ 1 & 1 & -1 \end{bmatrix}$

177. $(1/35)\begin{bmatrix} 7 & 2 & -13 \\ -7 & 8 & 18 \\ 7 & -13 & -3 \end{bmatrix}$

179. $\begin{bmatrix} -1 & 3 & 0 \\ 1/2 & -1 & 0 \\ -1 & -1 & 1 \end{bmatrix}$

181. This matrix has no inverse.

183. $(1/4)\begin{bmatrix} -47 & 45 & 51 & -39 \\ -35 & 33 & 35 & -27 \\ 17 & -15 & -17 & 13 \\ 2 & -2 & -2 & 2 \end{bmatrix}$

185. $(A^{-1})^{-1} = \begin{bmatrix} 3 & 7 \\ 2 & 5 \end{bmatrix} = A$

187. $(A^t)^{-1} = (A^{-1})^t = \begin{vmatrix} 5 & -2 \\ -7 & 3 \end{vmatrix}$

189. STUDY EACH DAY TO KEEP THE ZITS AWAY.

CHAPTER 4 MATHEMATICAL FUNCTIONS AND THEIR USE AS MODELS

1. $y = f(x) = 2x + 3$
3. $y = f(x) = 3x + 9$
5. $y = f(x) = 4x$
7. $y = g(x) = x - 1$
9. $f = \{(x,y) | y = x/2\}$
11. $f = \{(x,y) | y = 100 + x^2\}$
13. $f = \{(x,y) | y = |\sqrt{x}|\}$
15. $f(x) = 20x$
17. s = Monthly sales, in dollars; E = Total earnings, in dollars; $E = f(s) = 0.1s + 800$
19. x = Base length, in feet (same as base width); h = Height, in feet = $2.5x$; v = Volume, in cubic feet; $v = f(x) = (x)(x)(h) = x^2(2.5x) = 2.5x^3$
21. x = Number of hours worked; y = Take-home pay, in dollars; $y = f(x) = 7.5x - (0.3)(7.5x) = 5.25x$
23. x = Number of hours worked; y = Number of units of output completed; z = Total weekly earnings, in dollars; $z = f(x) = 8.5x + 1.95y$
25. $12q = 24{,}000$; $q = 2{,}000$ units
27. $f(12{,}000) = \$1{,}100$; $f(18{,}000) = \$1{,}400$
29. $g(12) = \$3.10$
31. v = Sales volume, in units (dependent); p = Selling price, per unit (independent); $v = f(p) = k/(6 + p)$
33. $v(9) = 5{,}33\overline{3}$ units
35. Discrete
37. Continuous
39. All real numbers
41. $x \geq -3$
43. All real numbers except $x = -1$
45. $x \geq 2$
47. $600

49. $$C = f(q) = \begin{cases} q & 0 \le q < 500 \\ 0.9q & 500 \le q < 1{,}000 \\ 0.825q & q > 1{,}000 \end{cases}$$

51.

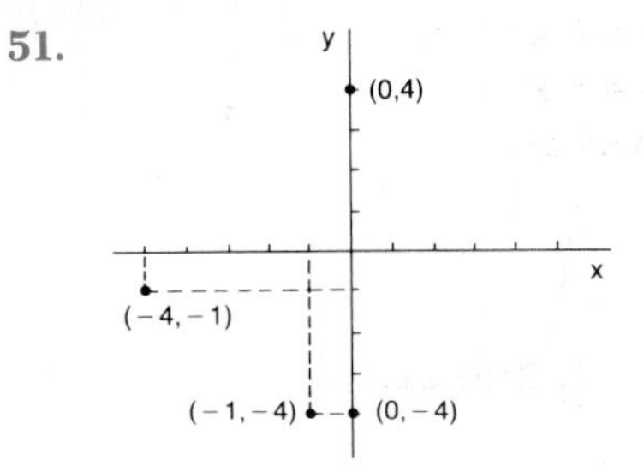

53.

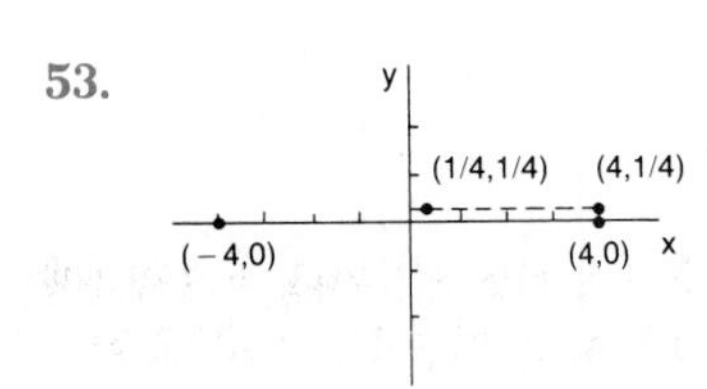

55.

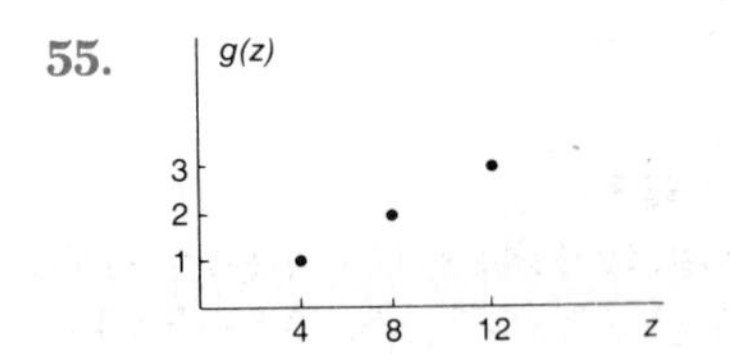

57.

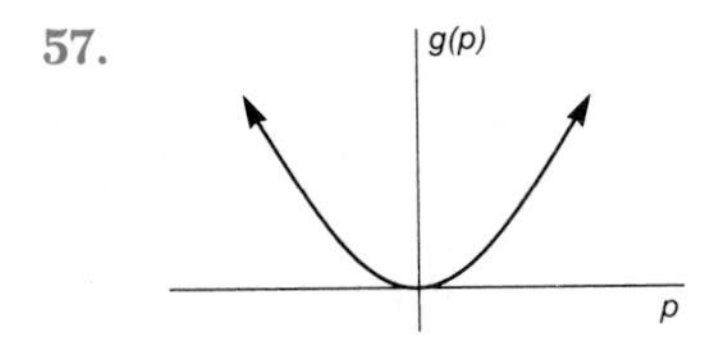

59. Function, one-to-one
61. Relation, not a function
63. Function, one-to-one
65. Function, one-to-one
67. Function, one-to-one

69. Not a function
71. Function, one-to-one
73. Function, one-to-one
75. Not a function
77. Function, one-to-one

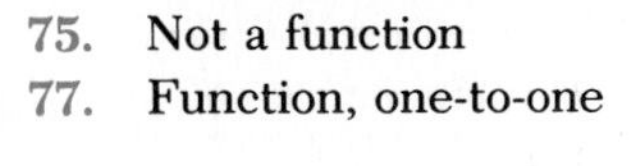

79. $y = f(x) = 2x$
$f^{-1}(x) = (1/2)x$
(a function)

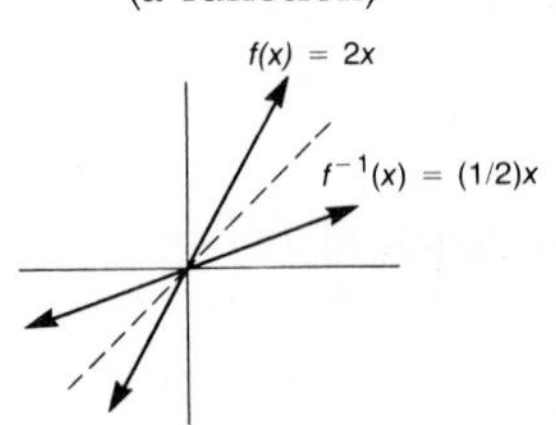

81. $y = f(x) = 5 - 2x$
$f^{-1}(x) = (5 - x)/2$
$= 2.5 - 0.5x$
(a function)

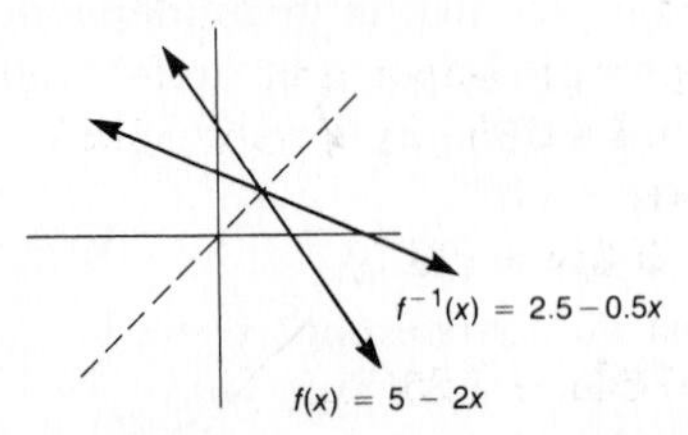

83. $y = f(x) = x^2 - 2$
$f^{-1}(x) = \sqrt{x + 2}$
(not a function)

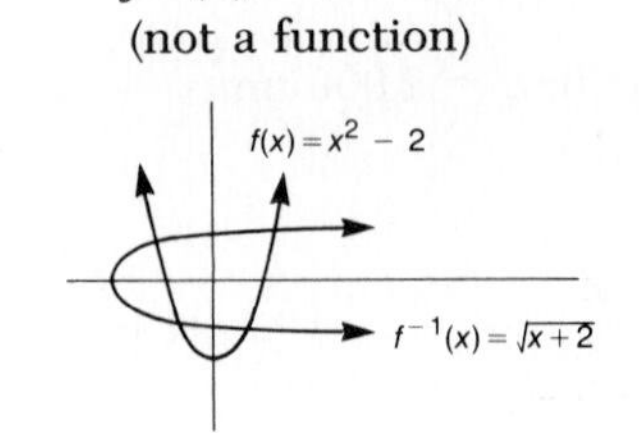

85. $y = f(x) = 4/x$
$f^{-1}(x) = 4/x$
(a function)

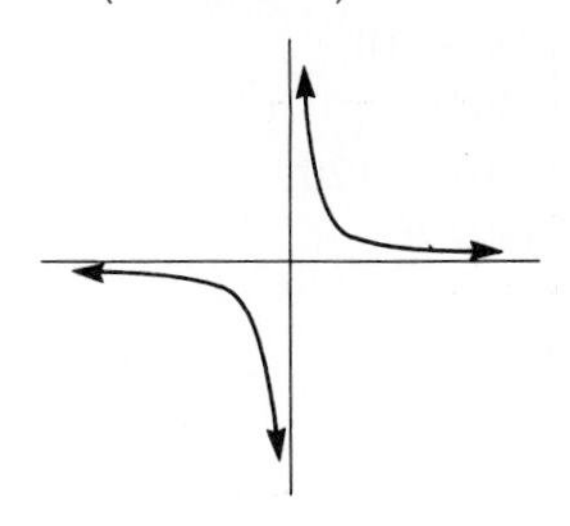

87. $C = (5/9)(F - 32)$
89. $f(x) + g(x) = x^2 + x + 2$
91. $f(x)/g(x) = x^2/(x + 2)$
93. $3f(x) + 2g(x) = 3x^2 + 12x + 1$
95. $f(x)g(x) = 3x^3 - x^2 + x - 35$
97. $g(x)/f(x) = (3x - 7)/(x^2 + 2x + 5)$
99. $f(g(2)) = 27$
101. 195
103. $x + 100 + 1{,}000/x$
105. $-x$
107. $0.1x^2 - 4$
109. $20E - E^2/20$
111. $5{,}000E - 12.5E^2$
113. C = Total cost, in dollars; $C = g(Q) = 4{,}000 + 6.5Q$; $0 \leq Q \leq 8{,}000$; $g(5{,}000) = 36{,}500$; $g(6{,}000) = 43{,}000$; $g(7{,}000) = 49{,}500$
115. $P(Q) = 8.5Q - 4{,}000$; $P(5{,}000) = \$38{,}500$; $P(6{,}000) = \$47{,}000$; $P(7{,}000) = \$55{,}000$
117. $C_n(Q) = 4{,}800 + 6.5Q$; $P_n(Q) = 8.5Q - 4{,}800$; BEP_n: 565 units

CHAPTER 5 LINEAR EQUATIONS, FUNCTIONS, AND INEQUALITIES

1. $2x_1 - x_2 = 5$; linear
3. Not linear
5. Not linear
7. $x - y = 0$; linear
9. Not linear
11. Not linear
13. $r - s = 3$; linear
15. Not linear
17. (0,8),(6,0)
19. (0,3.6),(3,0)
21. (0,2),(2.5,0)
23. (0,12),(4,0)
25. (0,1),(1,0)
27. $x_2 = -4x_1 + 8$; (2,0)
29. $y = (-3/4)x + 3$; (4,0)
31. $y = 1 - x$; (1,0)
33. $y = -3$
35. $x_2 = (13/7) - (3/7)x_1$; (13/3,0)
37.

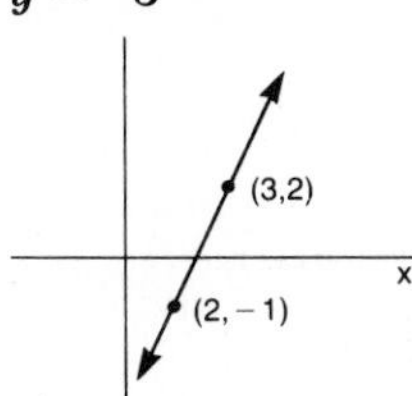

39.

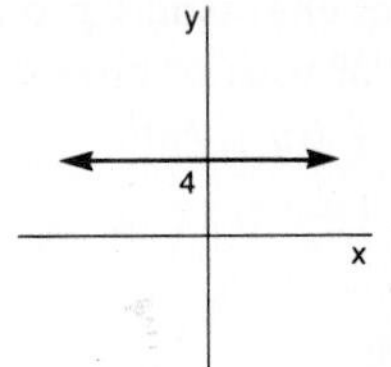

41.

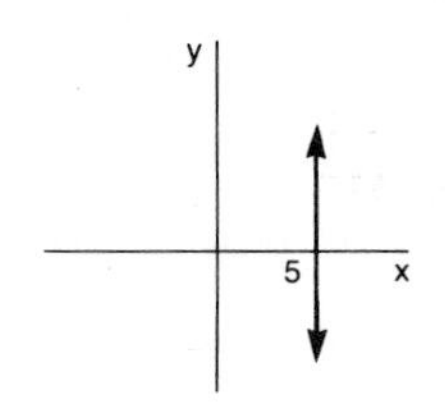

43. A: slope + 4; B: slope + 2
45. A: undefined slope; B: zero slope
47. $y = -3x + 13$
49. $y = (-2/3)x + (26/3)$
51. $y = -6$
53. $y = -1.5\overline{3}x + 1.25$

55. $y = (-3/4)x + 3$

57. $y = 8.\overline{2}x + 180.\overline{6}$

61. $y = 15.868x - 93.176$

63. $y_e = 192.448$

65. Supply function

67. Demand function

69. $P = S(Q) = (-1/20)Q + 30$

71. $P = S(q) = 0.1q + 10$

73. $q^* = 3;\ p^* = 15$

75. $q^* = 10;\ p^* = 80$

77. $q^* = 10;\ p^* = 60$

79. $p = D(q) = -0.2q + 120$

81. $0.6q + 45$

83. Decrease by $507.81

85.

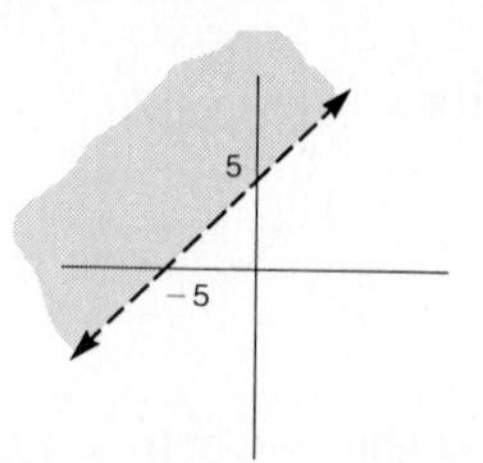

87.

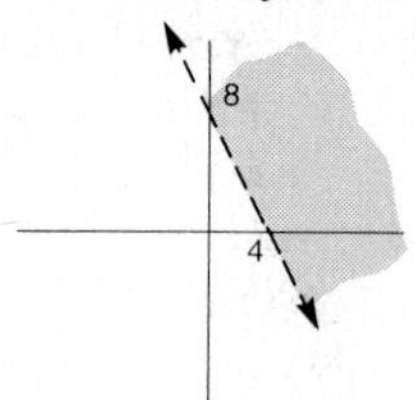

89.

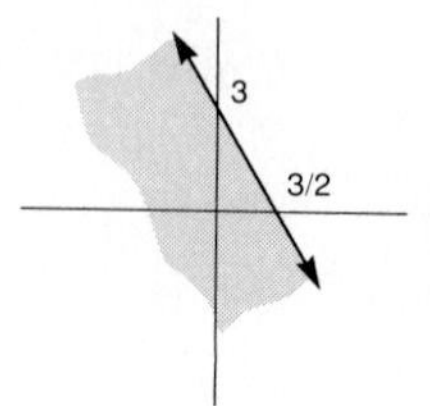

91.

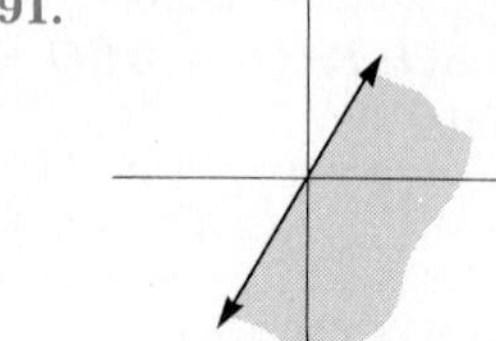

93.

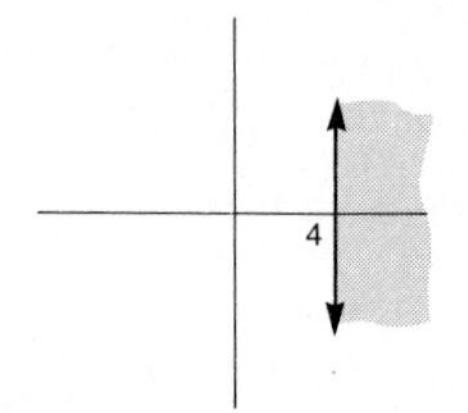

95.

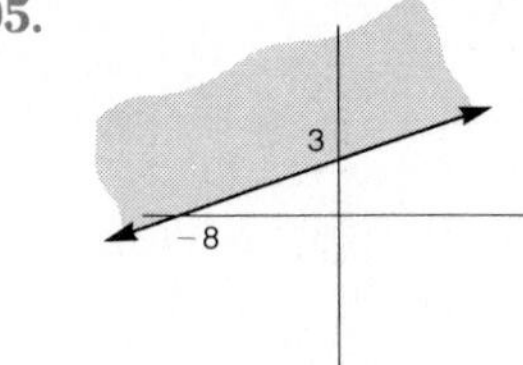

97. x = Number of tons of chemical 1 produced a day; y = Number of tons of chemical 2 produced a day; $4x + 6y \leq 600$

99.

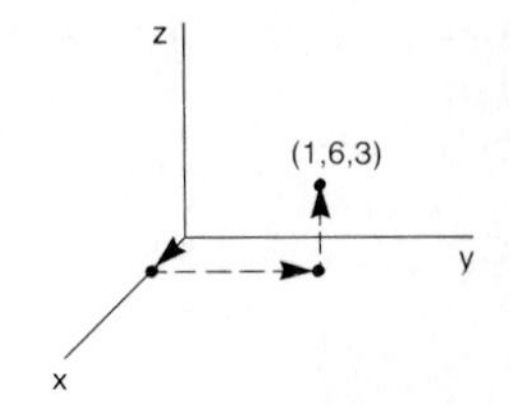

101.

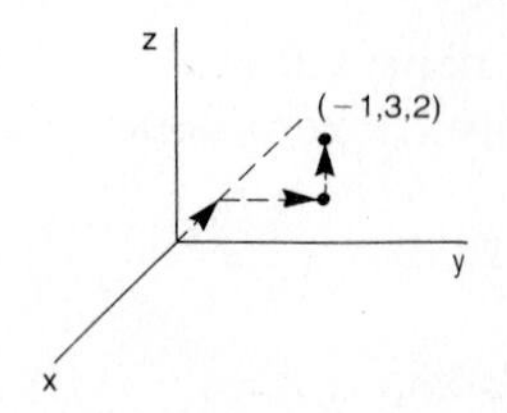

103.

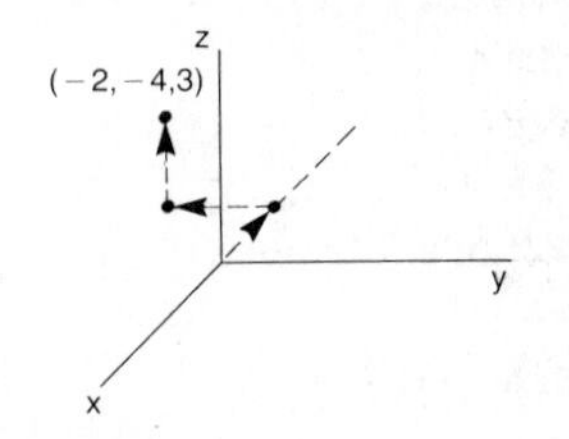

105.

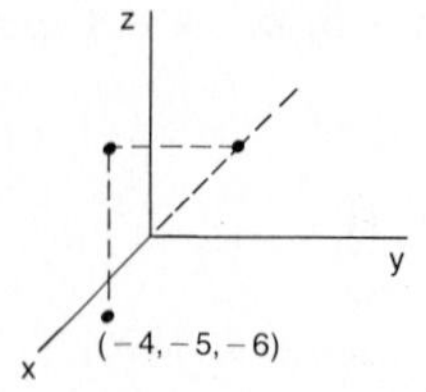

107.

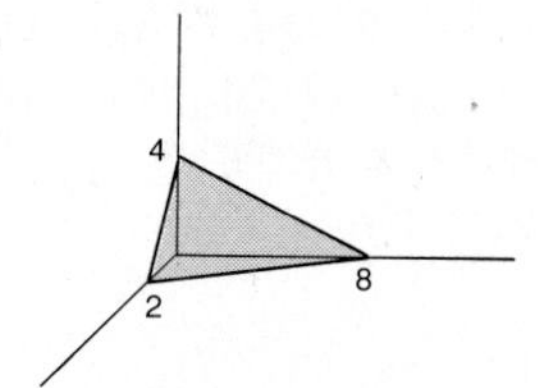

109.

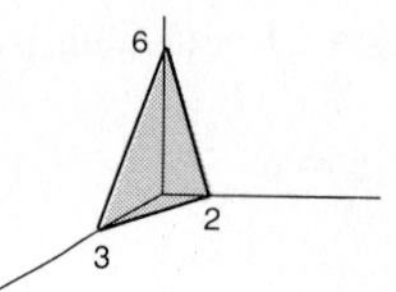

111.

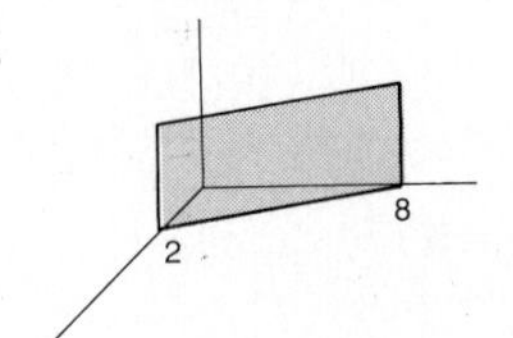

113.

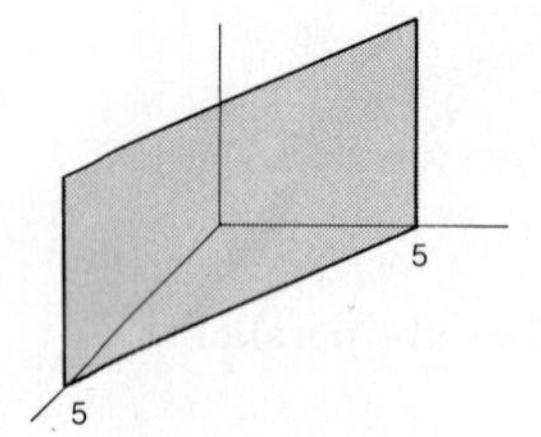

115.

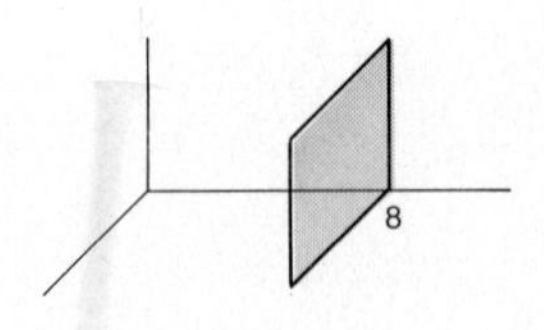

117.

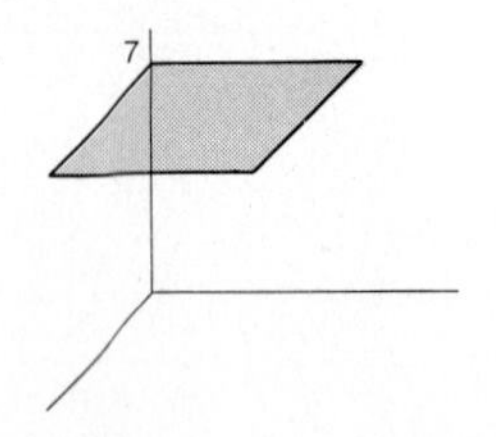

CHAPTER 6 SYSTEMS OF LINEAR EQUATIONS AND INEQUALITIES

1. $x = 3$; $y = 5$
3. $x_1 = 2\ 2/3$; $x_2 = -2/3$
5. $x_1 = -7/11$; $x_2 = 5/11$
7. $x = 0$; $y = -1$
9. No solution
11. $20P + 30Q = 1{,}200$
15. $x + y = 1{,}600$
17. $x = 1{,}200$; $y = 400$
19. $x = 3$; $y = -2$
21. $x = -2$; $y = -3$
23. $x = 4$; $y = 3$
25. $x = 1$; $y = -1$
27. $x_1 = 1$; $x_2 = 2$; $x_3 = 3$
29. $x = 14/9$; $y = 5/9$; $z = 5/9$
31. $x = 1$; $y = 0$, $z = -1$
33. 300 bottles of burgundy, 200 bottles of sauterne; 100 bottles of vin rose
35. $(x_1 = 1,\ x_2 = 1)$, $(x_1 = 2,\ x_2 = 1)$, $(x_1 = 2,\ x_2 = 3)$
37. $(x_1 = 1,\ x_2 = 1,\ x_3 = 1)$, $(x_1 = 3,\ x_2 = -1,\ x_3 = 3)$, $(x_1 = 2,\ x_2 = 2,\ x_3 = 0)$
39. $(A = 3,\ B = 1,\ C = 2)$, $(A = 3,\ B = 0,\ C = 4)$, $(A = 3,\ B = 2,\ C = 3)$
41. $x_1 = 3$, $x_2 = 3$
43. $x_1 = 1$, $x_2 = 3$, $x_3 = -4$
45. $x_1 = 9/10$, $x_2 = 13/10$, $x_3 = 15/10$

47. *a.* $x_1 = 6/14$, $x_2 = 27/14$, $x_3 = 11/14$; *b.* $x_1 = 31/70$, $x_2 = 17/70$, $x_3 = 2/70$; *c.* $x_1 = -1$, $x_2 = -2$, $x_3 = -4$

49. *a.* $x = 3$, $y = 10$, $z = 5$; *b.* $x = 4$, $y = 12$, $z = 6$

51. $\begin{vmatrix} 229.85 \\ 213.79 \end{vmatrix}$

53. $\begin{vmatrix} 66.5 \\ 85.25 \\ 43.24 \end{vmatrix}$

55. 2,825 *E*, 445 *D*, 270 *C*, 80 *B*, 20 *A*

57. Many solutions

59. Many solutions

61. $x_1 = -1 + (3/2)a$; $x_2 = 11 - (5/2)a$; $x_3 = a$

63. No solution

65. No solution

67. $x_1 = 3$, $x_2 = 6$

69. $x = 2 - 3a$, $y = -7$, $z = a$

71. $x_1 = 0$, $x_2 = 5$, $x_3 = 5$

73.

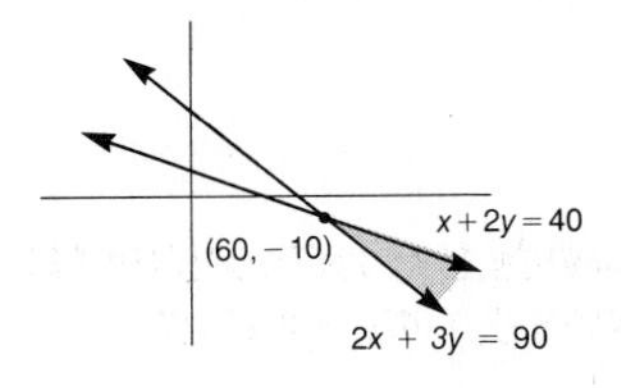

75.

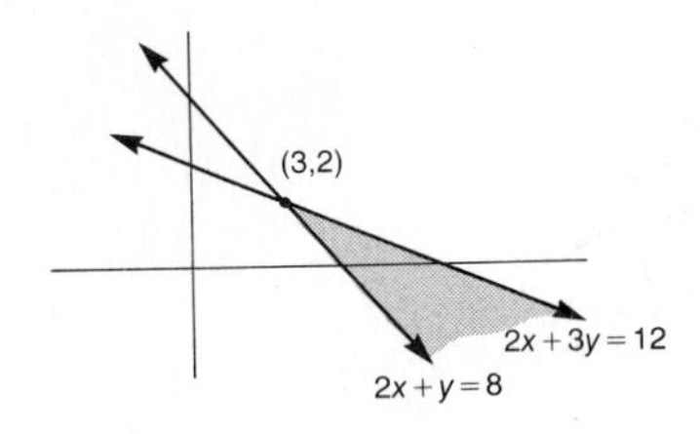

77. No solution; lines are parallel

79. Corner points: (4,8),(6,6)

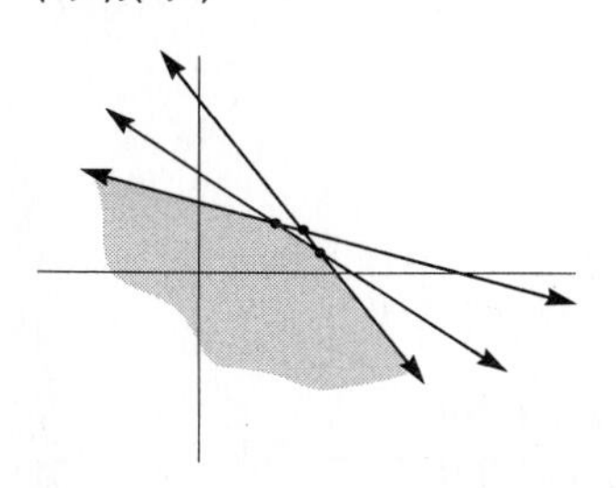

81.

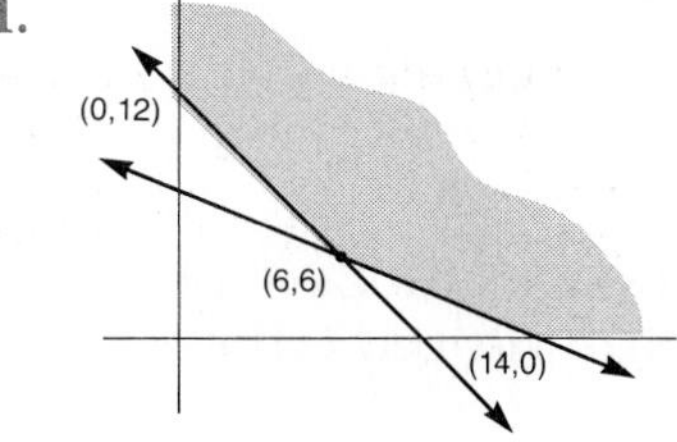

83. Corner points: (5,15),(10,10),(30,0)

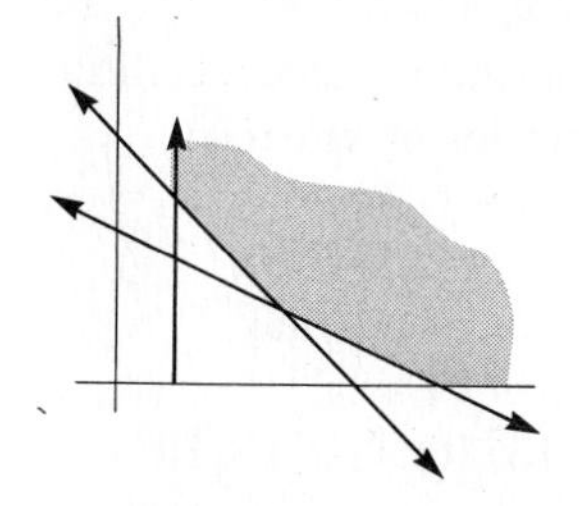

85. Corner points: (0,30),(30/7,120/7),(30,0)

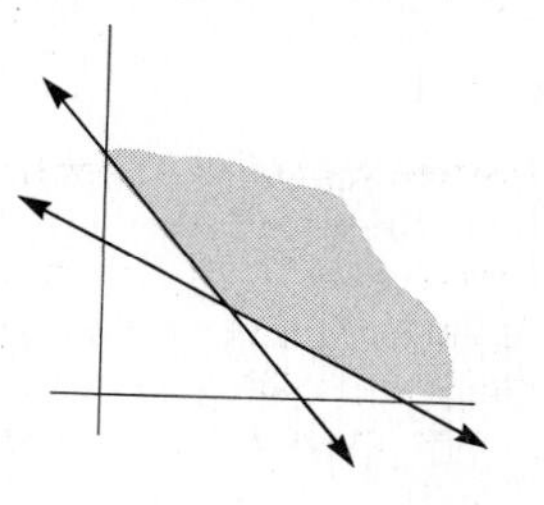

87. No solution

89. Corner points: (8,10),(15,10),(15,3)

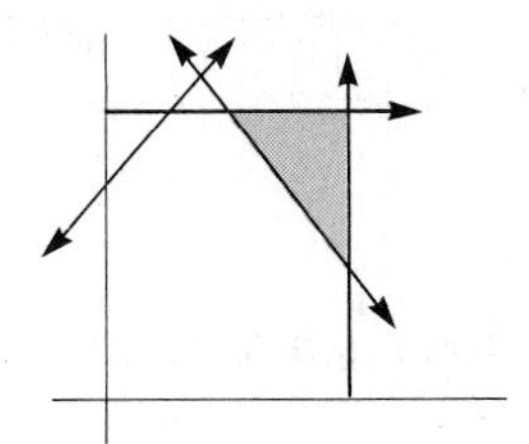

CHAPTER 7 LINEAR PROGRAMMING: AN INTRODUCTION TO THE MODEL

1. Slope of objective function: −0.67

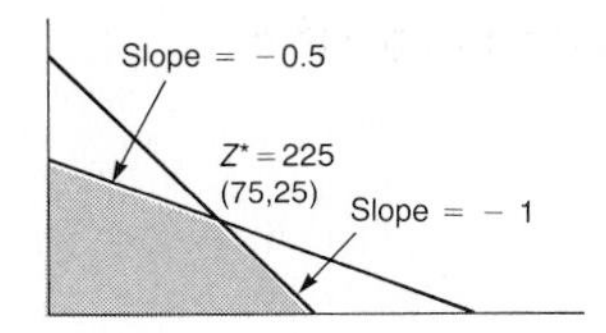

3. Slope of objective function: −0.8

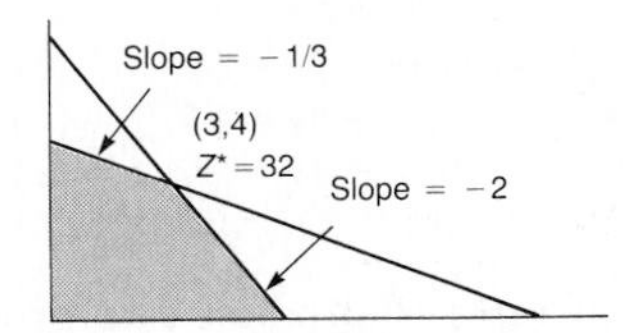

5.

Corner point	Intersecting lines	Coordinates	Value of objective function
A	$y=4;\ x=8$	(8,4)	96
B	$x=8;\ x+y=25$	(8,17)	174
C	$y=4;\ x+y=25$	(21,4)	213**

7.

Corner	Intersecting lines	Coordinates	Value of objective function
A	$x=0;\ y=0$	(0,0)	0
B	$x=0;\ 4x+5y=70$	(0,14)	126
C	$4x+5y=70$	(10,6)	134***
	$x+y=16$		
D	$y=0;\ x+y=16$	(16,0)	128

9.

Corner	Intersecting lines	Coordinates	Value of objective function
A	$x=0;\ y=0$	(0,0)	0
B	$x=0;\ y=12$	(0,12)	108
C	$y=12;\ 2x+y=20$	(4,12)	172**
D	$x=5;\ 2x+y=20$	(5,10)	170
E	$x=5;\ y=0$	(5,0)	80

11. Slope of objective function: −1

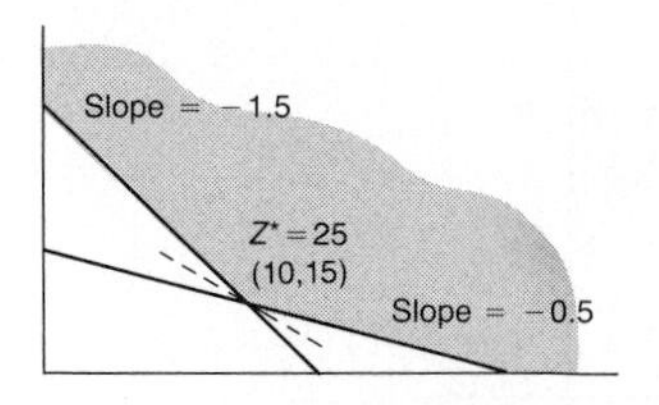

13. Slope of objective function: −0.8.

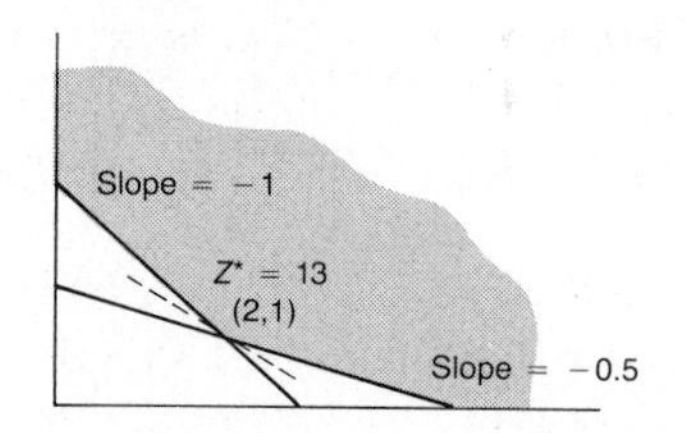

15.

Corner	Intersecting lines	Coordinates	Value of objective function
A	$x=0;\ 1.5x+y=60$	(0,60)	240
B	$1.5xy=60;\ 2x+3y=150$	(12,42)	204***
C	$x+2y=90;\ 2x+3y=150$	(30,30)	210
D	$y=0;\ x+2y=90$	(90,0)	270

17.

Corner	Intersecting lines	Coordinates	Value of objective function
A	$x=0;\ 3x+2y=60$	(0,30)	120
B	$3x+2y=60;\ x+4y=24$	(19.2,1.2)	43.2***
C	$y=0;\ x+4y=24$	(24,0)	48

19. x = Number of units of product 1 to produce each day
y = Number of units of product 2 to produce each day
MAXIMIZE: $Z = 16x + 10y$ = Total profit contribution, in dollars
s.t. $4x + 2y \leq 24$ (raw material supply)
$3x + 3y \leq 21$ (labor supply)
$2x + 5y \leq 30$ (machine-time supply)
and $x, y \geq 0$

OPTIMAL STRATEGY: $x = 5;\ y = 2;\ z = 100$

21. x = Number of units of food F1 to serve
y = Number of units of food F2 to serve
MINIMIZE: $Z = 40x + 5y$ = Total cost, in cents
s.t. $x + 2y \geq 4$ (vitamin A requirement)
$2x + y \geq 5$ (vitamin B requirement)
and $x, y \geq 0$

OPTIMAL STRATEGY: $x = 2;\ y = 1;\ z = 130$

23. x = Number of "standard" suits to make
y = Number of "fashion-wise" suits to make
MAXIMIZE: $z = 160x + 200y$ = Total revenue, in dollars
s.t. $2x + 2.5y \leq 60$ (cotton material available)
$7.5x + 6y \leq 60$ (woolen material available)
$x + 2y \leq 12$ (silk material available)
and $x, y \geq 0$

OPTIMAL STRATEGY: $x = 16/3;\ y = 10/3;$
$z = 1520$
(Note: This decision situation requires integral values for the variables.)

25. x = Tons of coal from source 1
y = Tons of coal from source 2

MINIMIZE: $z = 16x + 12y$ = Cost, in dollars
s.t. $1{,}500x + 750y \leq 6{,}000$ (lbs. of coke required)
$400x + 600y \leq 2{,}400$ (lbs. of tar required)
$140x + 100y \leq 700$ (lbs. of other tar derivatives required)
and $x, y \geq 0$
OPTIMAL STRATEGY: $x = 45/11$; $y = 14/11$; $z = 888/11$

27. x = Number of acres of tomatoes to plant
y = Number of acres of cauliflower to plant
MAXIMIZE: $290x + 200y$ = Profit, in dollars
s.t. $x + y \leq 80$ (total acreage)
$40x + 30y \leq 2{,}500$ (cash expenses)
$10x + 30y \leq 1{,}800$ (labor requirement)
and $x, y \geq 0$
OPTIMAL STRATEGY:
$x = 45$; $y = 35$; $z = 20{,}050$

29. x = Number of packages of Tahitian Tasties to mix
y = Number of packages of Tasty Treat to mix
MAXIMIZE: $z = 2.25x + 2.95y$ = Total revenue, in dollars
s.t. $1.5x + y \leq 320$ (ounces of banana chips)
$1.5x \leq 288$ (ounces of raisins)
$x + 1.5y \leq 288$ (ounces of carob-covered peanuts)
$1.5y \leq 240$ (ounces of almonds)
OPTIMAL STRATEGY:
$x = 153.6$; $y = 89.6$; $z = 609.92$

31. x = Input units of basic ingredient to become grade A plastic
y = Input units of basic ingredient to become grade B plastic
MAXIMIZE: $z = 7.995x + 5.925y$ = Profit contribution, in dollars
s.t. $x + y \leq 200$ (process I capacity)
$0.9x \leq 100$ (process II capacity)
$0.9y \leq 75$ (process III capacity)
$0.72x + 0.855y \geq 75$ (labor union contract)
and $x, y \geq 0$
OPTIMAL STRATEGY:
$x = 111.\overline{1}$; $y = 83.\overline{3}$; $z = 1415$
$z = 1415$

CHAPTER 8 LINEAR PROGRAMMING: THE SIMPLEX ALGORITHM

1. MAXIMIZE: $z = 3x + 4y + 0s_1 + 0s_2$
s.t. $2x + y + s_1 + 0s_2 = 20$
$x + y + 0s_1 + s_2 = 15$
and $x, y, s_1, s_2 \geq 0$

3. MAXIMIZE: $z = x_1 + 4x_2 + 2x_3 + 0s_1 + 0s_2$
s.t. $4x_1 + x_2 + s_1 + 0s_2 = 2$
$-10x_1 + x_2 + 3x_3 + 0s_1 + s_2 = 1$
and $x_1, x_2, x_3, s_1, s_2 \geq 0$

5. MAXIMIZE: $z = 16x + 10y + 0s_1 + 0s_2 + 0s_3$
s.t. $4x + 2y + s_1 + 0s_2 + 0s_3 = 24$
$3x + 3y + 0s_1 + s_2 + 0s_3 = 21$
$2x + 5y + 0s_1 + 0s_2 + s_3 = 30$
and $x, y, s_1, s_2, s_3 \geq 0$
s_1 = Unused units of raw material
s_2 = Unused units of labor
s_3 = Unused machine time

7. MAXIMIZE: $z = 0.8x + 0.15y + 0s_1 + 0s_2 + 0s_3$
s.t. $x + y + s_1 + 0s_2 + 0s_3 = 8{,}000$
$x + 0y + 0s_1 + s_2 + 0s_3 = 7{,}500$
s_1 = Uninvested funds
s_2 = Amount by which investment in type A falls below maximum

$0x + y + 0s_1 + 0s_2 + s_3 = 2{,}500$ s_3 = Amount by which investment in type B falls short of maximum

and $x, y, s_1, s_2, s_3 \geq 0$

9. MAXIMIZE: $z = 20x + 15y + 0s_1 + 0s_2$
s.t. $4x + 2y + s_1 + 0s_2 = 8$ s_1 = Unused machine M1 capacity
$x + 3y + 0s_1 + s_2 = 9$ s_2 = Unused machine M2 capacity
and $x, y, s_1, s_2 \geq 0$

11. MAXIMIZE: $z = 3x + 2y + 0s_1 + 0s_2 + 0s_3$
s.t. $x + 3y + s_1 + 0s_2 + 0s_3 = 150$ s_1 = Unused man-hours
$x + 1.5y + 0s_1 + s_2 + 0s_3 = 90$ s_2 = Unused lathe-hours
$x + 0y + 0s_1 + 0s_2 + s_3 = 50$ s_3 = Unused grinder-hours
and $x, y, s_1, s_2, s_3 \geq 0$

13. $x = 0,\ y = 100,\ s_1 = 0,\ s_2 = 50,\ z = 300$

15. $x = 10,\ y = 0,\ s_1 = 0,\ s_2 = 2,\ s_3 = 5,\ z = 20$

17. $x = 15,\ y = 10,\ s_1 = 0,\ s_2 = 10,\ s_3 = 0,\ z = 65$

19. $x_1 = 0,\ x_2 = 16,\ x_3 = 14,\ s_1 = 0,\ s_2 = 0,\ s_3 = 15,\ z = 228$

21. $x = 5,\ y = 2,\ s_1 = 0,\ s_2 = 0,\ s_3 = 10,\ z = 100$

23. $x = 5{,}500,\ y = 2{,}500,\ s_1 = 0,\ s_2 = 2{,}000,\ s_3 = 0,\ z = 815$

25. $x = 0.6,\ y = 2.8,\ s_1 = 0,\ s_2 = 0,\ z = 54$

27. $x = 50,\ y = 26.\overline{6},\ s_1 = 20,\ s_2 = 0,\ s_3 = 0,\ z = 203.\overline{3}$

29. Assign five salespersons to territory 1, 8/3 salespersons to territory 2, and 4/3 salespersons to territory 3, thus realizing $4,233.33 profit contribution. (Note: Decision situation requires integer values for the variables. Determine the effect on the profit contribution of rounding the values of the variables to the nearest integer.)

31. Plant 200 acres of high-grade land in corn and 100 acres of lower-grade land in corn. Total profit $129,000.

33. $x = 0,\ y = 3,\ s_1 = 1,\ s_2 = 0,\ z' = -3$

35. $x = 0,\ y = 8,\ s_1 = 9,\ s_2 = 2,\ s_3 = 0,\ z' = -8$ (Note; there are many optimal strategies.)

37. $x = 25,\ y = 0,\ s_1 = 0,\ s_2 = 10,\ s_3 = 17,\ z = 25$

39. $x = 3,\ y = 3,\ s_1 = 1,\ s_2 = 0,\ s_3 = 0,\ z' = -9$

41. Serve two units of food F1 and one unit of food F2, thus meeting exactly each vitamin requirement, for a cost of $1.30.

43. Use 45/11 tons of coal from source 1 and 14/11 tons of coal from source 2 for a cost of $888/11. The tar and tar-derivatives constraints will be met exactly, but the coke requirement will be exceeded by 11,300/11 units.

45. Input 111.1 units of basic ingredient to become grade A plastic and 83.3 units to become grade B plastic. Profit contribution will total $1,415. Process II and III capacity will be used, but there will be 5.5 unused units of process I capacity, and the labor-union contract requirement will be exceeded by 76.25 units.

51. The assortment should be made up of 16/15 pounds of sugar cookies and 8/15 pounds of malted milk biscuits, along with 6/15 pounds of gingersnaps. Cost per two-pound tin will be $2.23. All restrictions will be met exactly, except that the quantity of sugar cookies will be less than the maximum allowed.

CHAPTER 9 NONLINEAR FUNCTIONS

1.

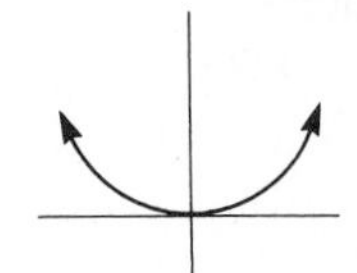

3.

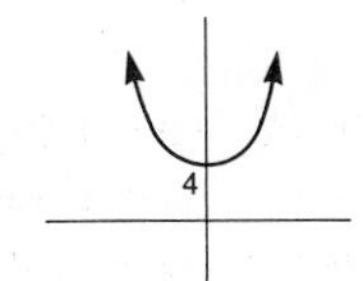

5. x-intercepts: $\pm\sqrt{5/3}$

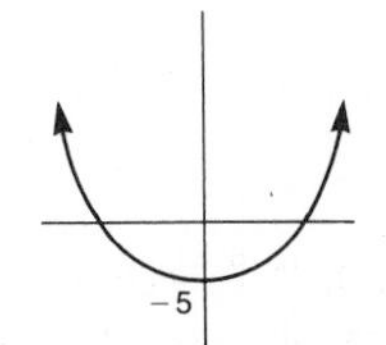

7.

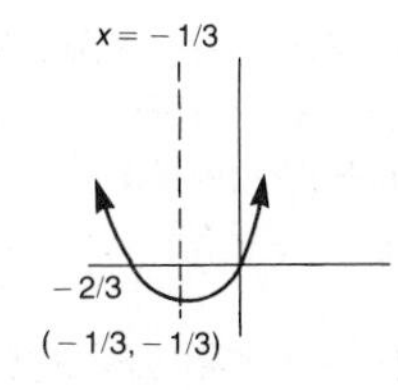

9.

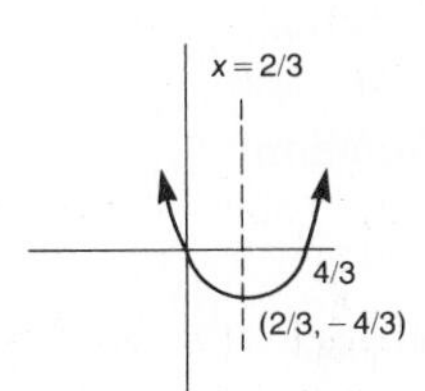

11.

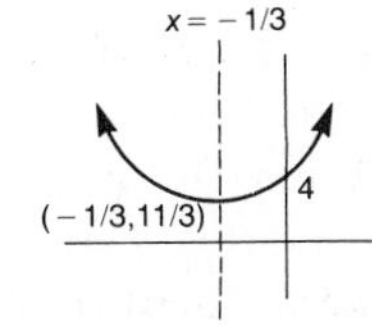

13.

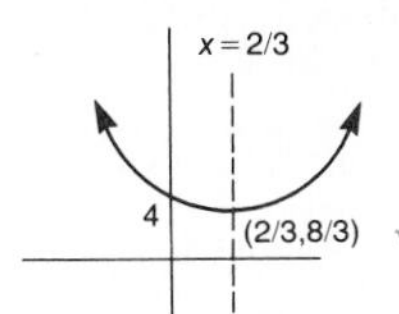

15.

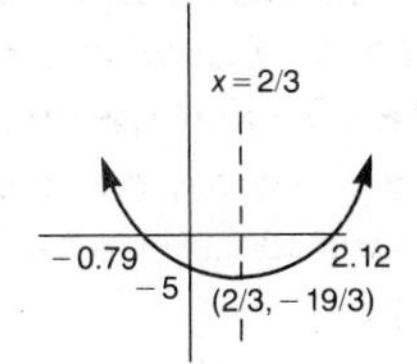

17.

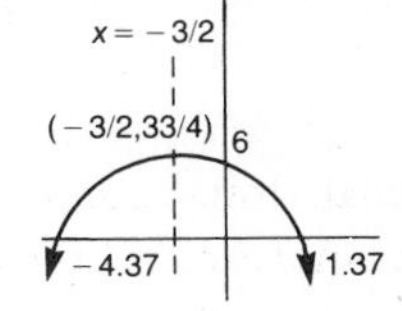

19.

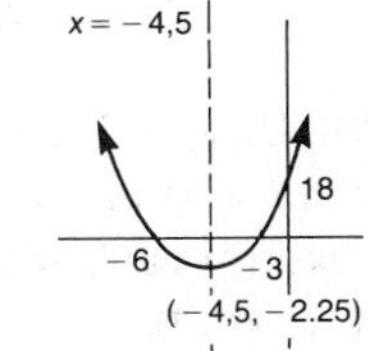

21. x = Number of \$10 increases in rent per apartment $R(x) = (400 + 10x)(100 - x) = 40{,}000 + 600x - 10x^2$

23. 70 apartments

25. x = Number of \$5 increases in selling price; $R(x) = (75 + 5x)(25 - x) = 1{,}875 + 50x - 5x^2$

27. 20 hammocks

29. 89.4 units at \$50; 54.8 units at \$75; 31.6 units at \$85

31. 60 units, \$72

37. $R(x) = 200x - 0.04x^2$

39. \$250,000

41. $P(x) = -0.04x^2 + 140x - 75{,}000$

43. \$47,500

45. $x = 660$ and $x = 2{,}840$ units
47. 2,000 units
49. \$1,000
51. $x = 61.38$; $x = 18.62$ units
53. $x = 7.26$ and $x = 192.74$ units
55. $P(x) = -2.5x^2 + 500x - 3{,}500$
57. \$21,500
59. Second-order, even
61. Third-order, odd
63. Third-order, neither
65. Fourth-order, even
67. Fourth-order, even
69. Fourth-order, even
71. Sixth-order, even
73. Third-order, neither
75. $f(x) = (20/3)x^2 - (25/3)x + 5$
77. $f(x) = x^2$
79. $(x^2/3) + (23/3)x - 6$
81. $D(x) = (2/3)x^2 - 90x + (7{,}300/3)$; 600; $333.\overline{3}$
83. Algebraic; quartic
85. Algebraic; cubic
87. Algebraic; linear
89. Algebraic; linear
91. Algebraic; mixed
93. Algebraic; constant
95. Algebraic; cubic
97. Algebraic; rational
99. Not an algebraic function
101. Algebraic; linear
103. Algebraic; cubic
105. Algebraic; linear
107. Not an algebraic function
109. Algebraic; rational
111. \$3,650
113. \$19
115. \$18 thousand
117. \$33,529; 86.3 percent
119.
121. $C(x) = 18[(9{,}600/x) + x]$
123. Cost is very high for small x because of $9{,}600/x$ term.

125.

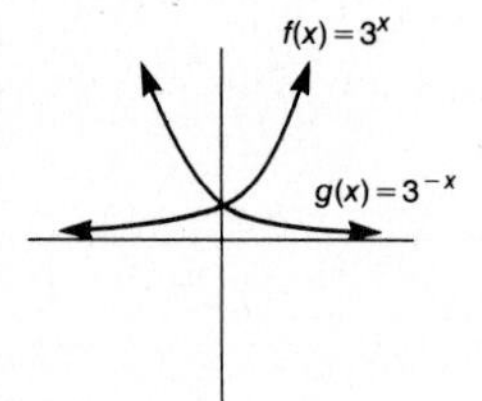

127.

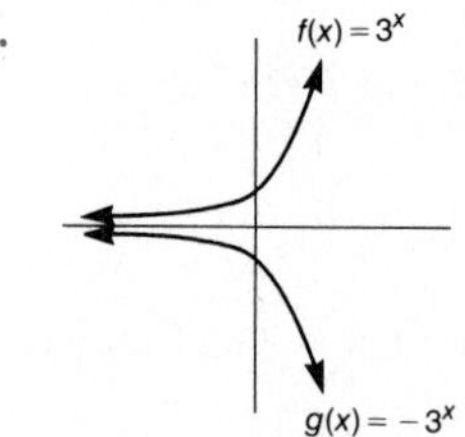

129.

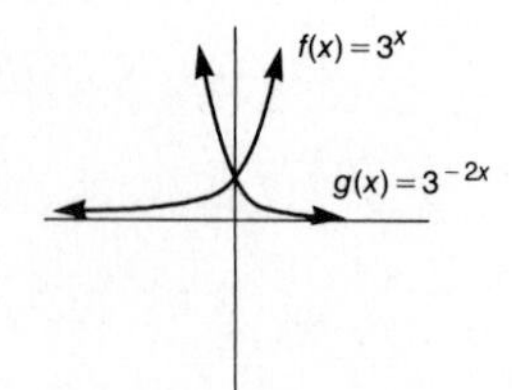

131.

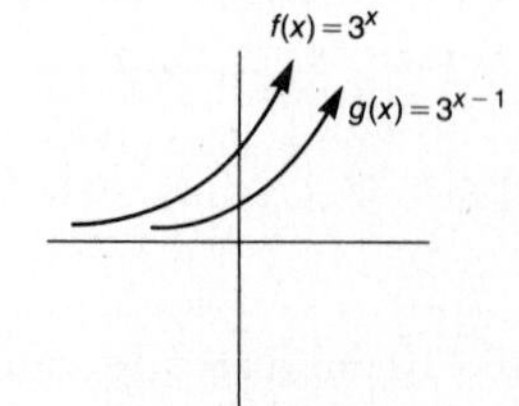

133.

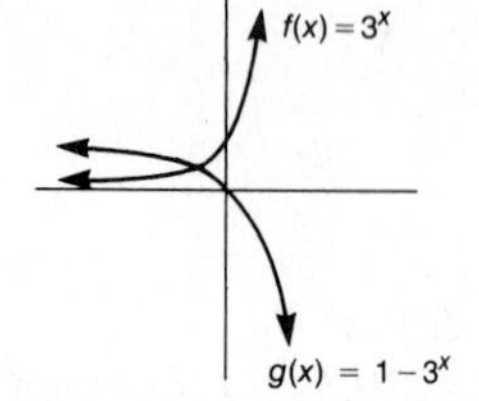

135.

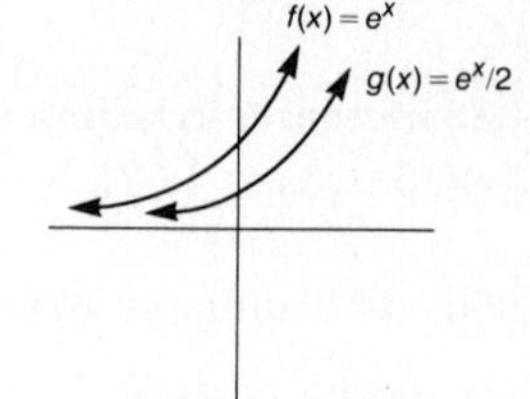

137. $f(t) = 8{,}000(1.12)^t$; \$14,098.73; \$22,184.63
139. $t = 34.5$ in 2015

141. $14,044.21 or 40 percent

143. 13.5 years

145. 20.4 years

147. 36.5 years

149. $k = 0.00784$

151. $f(t) = 5{,}000e^{0.1314t}$

153. 11,552.45 years

155. 9.24 years

157. 40.66 percent

159. 71.35 percent of maximum

161. 416.06 units

167. $P(t) = 380(500{,}000)(1 - e^{-0.15t}) - (40{,}000 + 5{,}000t)$

169. 259.4 units

171. 15 days

173. 6.62 months

175. $x > 0$

177. $x > 5$

179. $x > -3$

181.

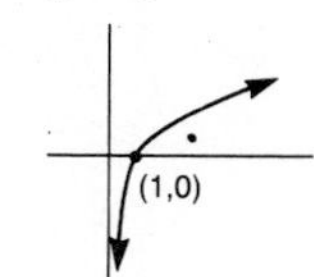

183.

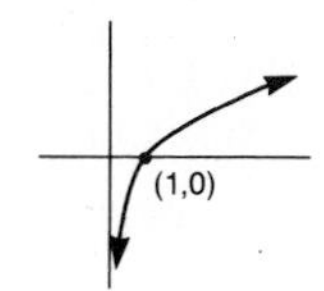

185.

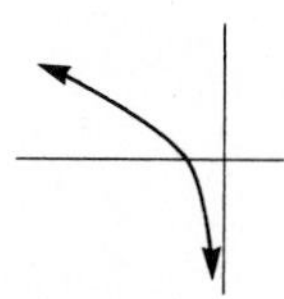

187.

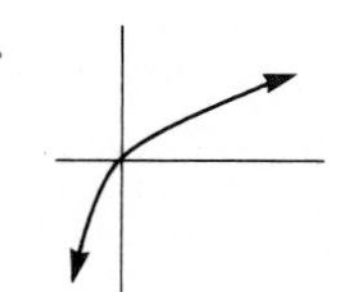

189. $140.63

191. $131.35; $3,940.45

193. $22,922.91

195. $f(x) = 3.67425e^{-0.16347x}$

197. $f(x) = ln(-6.77367 + 4.72091x)$

199. $f(x) = log\,(4.76378 + 2.0743\overline{6}x)$

CHAPTER 10 LIMITS AND CONTINUITY OF A FUNCTION

1. x: $2.9 \to 2.99 \to 2.999 \to 3 \leftarrow 3.001 \leftarrow 3.01 \leftarrow 3.1$
   $f(x) = 3x + 1$: $9.7 \to 9.97 \to 9.997 \to - \leftarrow 10.003 \leftarrow 10.03 \leftarrow 10.3$
   $\lim_{x\to 3}(3x + 1) = 10$

3. x: $-0.01 \to -0.001 \to -0.0001 \to 0 \leftarrow 0.0001 \leftarrow 0.001 \leftarrow 0.01$
   $f(x) = x$: $-0.01 \to -0.001 \to -0.0001 \to - \leftarrow 0.0001 \leftarrow 0.001 \leftarrow 0.01$
   $\lim_{x\to 0} x = 0$

5. x: $1.9 \to 1.99 \to 1.999 \to 2 \leftarrow 2.001 \leftarrow 2.01 \leftarrow 2.1$
   $f(x) = -2x^3$: $-13.718 \to -15.761 \to -15.976 \to - \leftarrow -16.024 \leftarrow -16.241 \leftarrow -18.522$
   $\lim_{x\to 2}(-2x^3) = -16$

7. x: $1.9 \to 1.99 \to 1.999 \to 2 \leftarrow 2.001 \leftarrow 2.01 \leftarrow 2.1$
   $f(x) = 2/x$: $1.0526 \to 1.005 \to 1.0005 \to - \leftarrow 0.9995 \leftarrow 0.995 \leftarrow 0.9523$
   $\lim_{x\to 2}(2/x) = 1$

9. x: $3.9 \to 3.99 \to 3.999 \to 4 \leftarrow 4.001 \leftarrow 4.01 \leftarrow 4.1$
   $f(x) = 2\sqrt{x}$: $3.9497 \to 3.995 \to 3.9995 \to - \leftarrow 4.0005 \leftarrow 4.005 \leftarrow 4.05$
   $\lim_{x\to 4} 2\sqrt{x} = 4$

11. x: $4.9 \to 4.99 \to 4.999 \to 5 \leftarrow 5.001 \leftarrow 5.01 \leftarrow 5.1$
   $f(x) = ln\, x$: $1.589 \to 1.607 \to 1.609 \to - \leftarrow 1.610 \leftarrow 1.611 \leftarrow 1.629$
   $\lim_{x\to 5} ln\, x = 1.609$

13. Does not exist

15. 3

17. 2

19. $L^- = 8$; $L^+ = 8$; $L = 8$

21. $L^- = 0$; $L^+ = 4$; L does not exist.

23. $L^- = 17{,}500$; $L^+ = 36{,}500$; L does not exist.

25. 1

27. 2/3

29. -9

31. 16

33. 16
35. 0.01
37. 3/2
39. 32
41. −108
43. 6
45. 12
47. 6.28
49. 45
51. 3
53. −4
55. 4/7
57. −1/6
59. 6/5
61. 1/2
63. $\sqrt{22}/3$
65. 2
67. −3
69. −1/9
71. ∞
73. ∞
75. 0
77. ∞
79. 1
81. ∞
83. ∞
85. ∞
87. ∞
89. 1
91. 1/2
93. ∞
95. ∞
97. −4/5
99. 0
101. 6,000
103. 1
105. Not continuous
107. Continuous
109. Not continuous
111. Not continuous
113. Continuous
115. Continuous
117. Continuous
119. Continuous
121. Continuous
123. Continuous for all x
125. Discontinuous at $x = 3$
127. Discontinuous at $x = 3$ and $x = -2$
129. Discontinuous for $x = 0$ and $x = 1$

131.

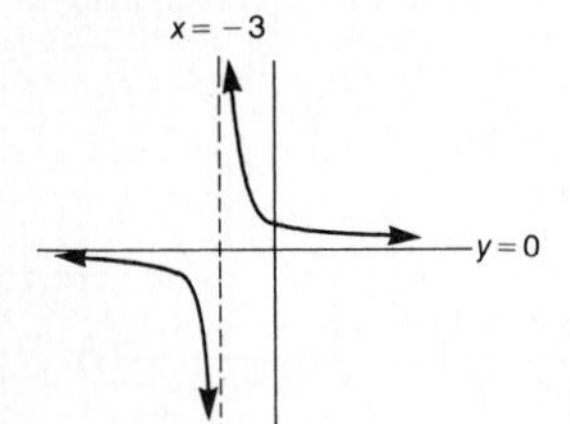

133.

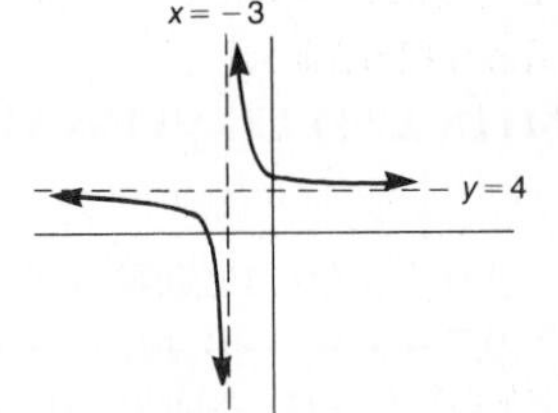

135.

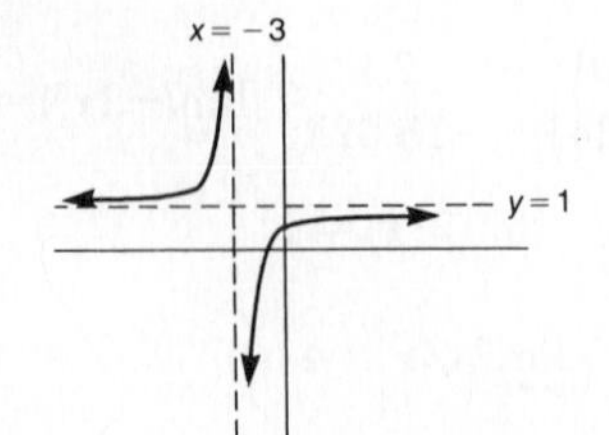

137.

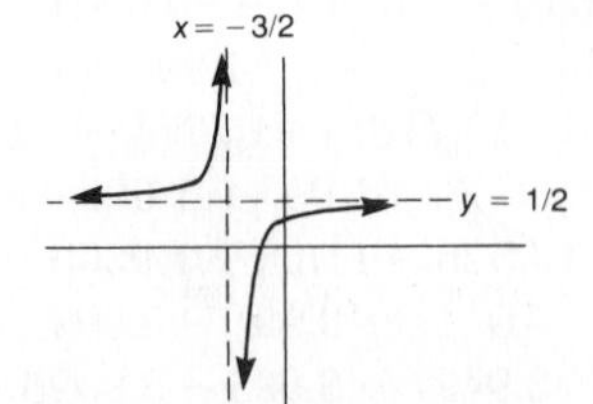

139.

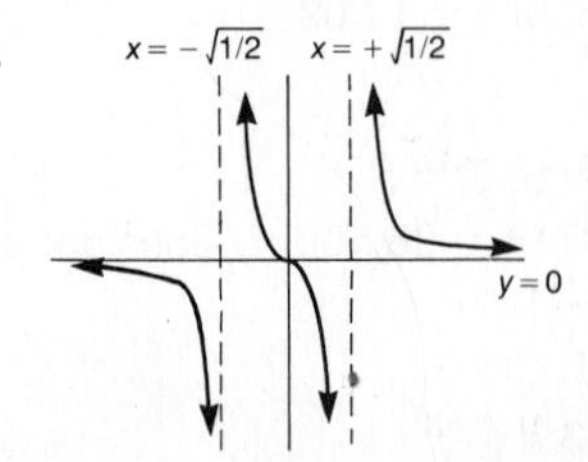

141.

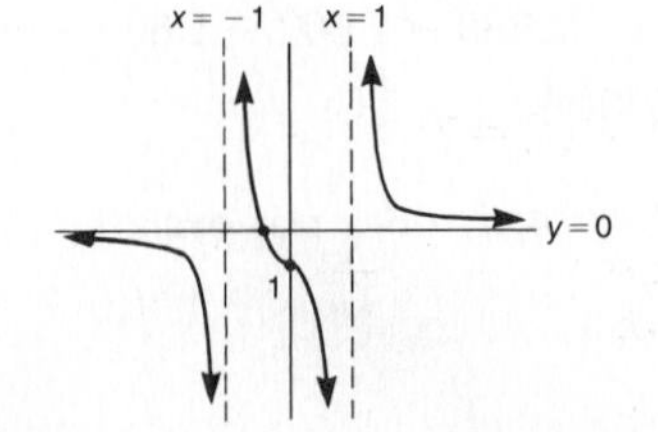

143.

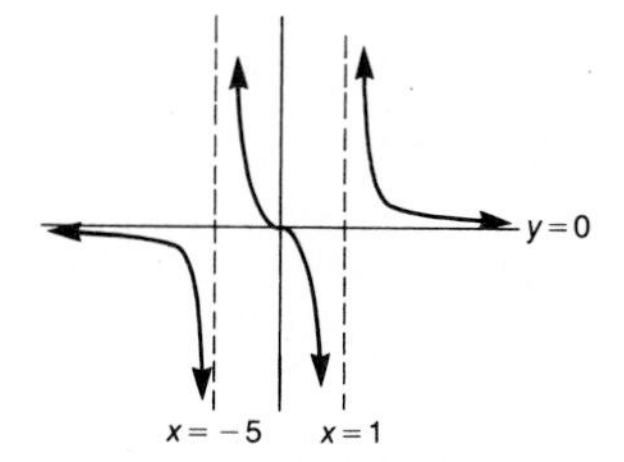

145.

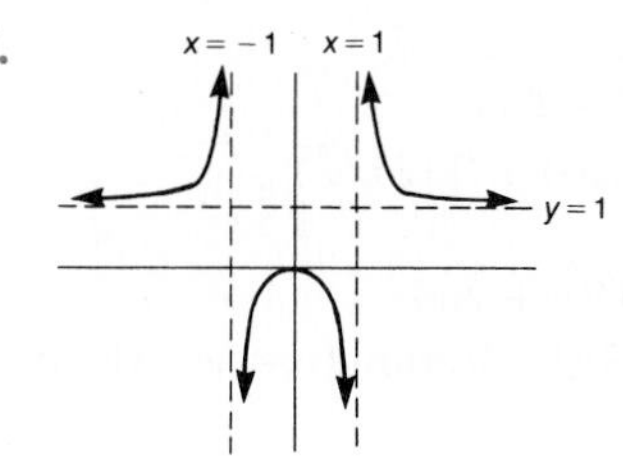

147.

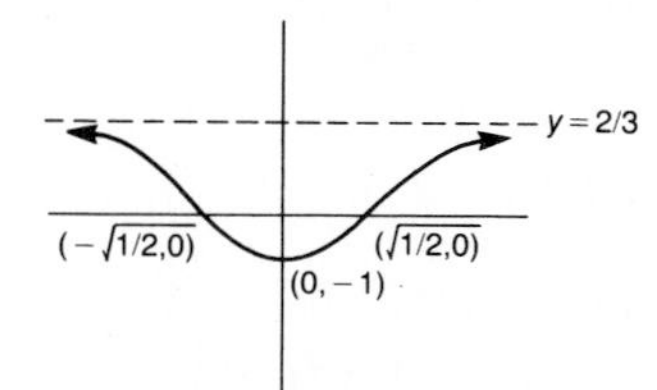

149.

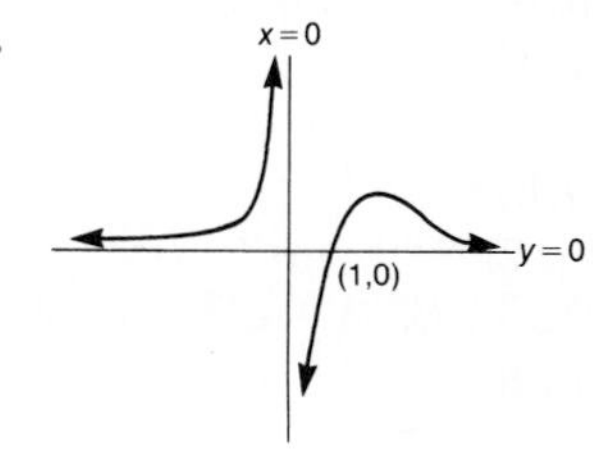

151.

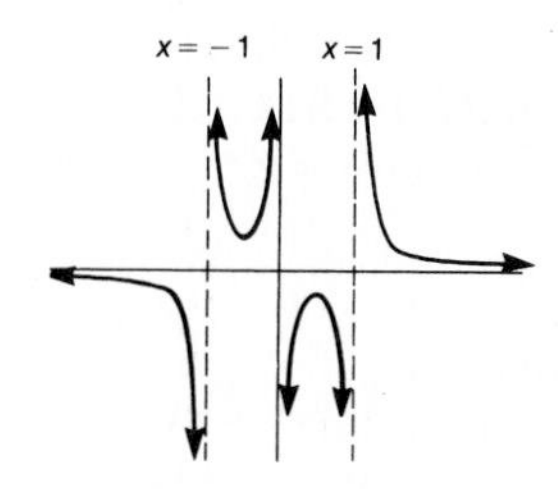

153.

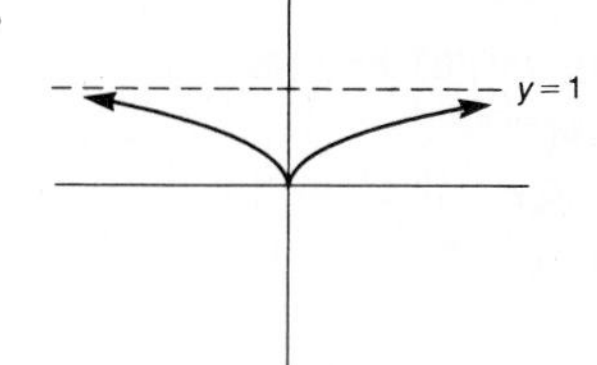

CHAPTER 11 THE DERIVATIVE AND RULES OF DIFFERENTIATION

1. 32; 64
2. 29; 11
3. −5; −11
4. 4; 9
5. 1/12; 1/24
6. −5; −14
7. 8.68; 17.29
9. 46 m.p.h.
11. −12 units; −13 units
13. 130; 155; 182.5
15. $4x$
17. $2x+5$
19. 0
21. −7
23. x^2
25. 6; 12
27. −7; −4
29. 8; −16
31. 0.27; 0.63
33. 80
35. $800t + 400\Delta t + 400$; 1,600; 2,400
39. −64
41. $900 - 32t$
45. No
47. No
49. Yes
53. Yes
55. $y' = 10x^9$
57. $-2/x^3$
59. $(5/3)t^{2/3}$
61. $f'(x) = 7x^6$; $f'(1) = 7$
63. $y' = 3x^2$; $y'\Big|_{x=3} = 27$
65. $f'(x) = 2/3x^{1/3}$
67. $f'(x) = x^{13}$; $f'(1) = 1$
69. $y' = 18x$; 36
71. $0.01/x^{0.9}$
73. −1/5
75. $g'(t) = 32t$; $g'(3) = 96$

77. $f'(x) = x^2/2$; $f'(2) = 2$
79. $f'(x) = x^2 - 1$
81. $y' = (1/3)(2 - 5x)$
83. $(3 - 2x)/x^4$
85. $f'(x) = (-2/x^3) + (1/2\sqrt{x})$
87. $D(3) = 4.2$; $d(5) = 10$
89. 5,750
91. 1,000
93. $N'(x) = -0.2x + 200$
95. $S'(t) = 100 - 32t$
97. At $t = 25/8$ the instantaneous velocity is zero.
99. $f'(x) = 3e^{3x+5}$
101. $f'(x) = 10^x \ln 10$
103. $f'(x) = (2x)(7^{x^2})(\ln 7)$
105. $f'(x) = -9x^2e^{-x^3}$
107. $f'(x) = 2(3x^2 + 2)e^{x^3+2x-1} + (16x^3)(5^{4x^4+3})(\ln 5)$
109. $f'(x) = 5e^{5x} + 5x^4$
111. 4.23; 0.57; 0.03
113. $f'(x) = 2/(x \ln 2)$
115. $f'(x) = 3x^2/(x^3 + 1)$
117. $f'(x) = (1 + 2\sqrt{x})/(2x(1 + \sqrt{x})(\ln 10)$
119. $f'(x) = 9/(3x + 4)$
121. $f'(x) = 4/(x \ln 8)$
123. $f'(x) = 10^x \ln 10 - (1/x \ln 10)$
125. 30; 20
127. 42.5; 46.8; 51.7
129. $R'(x) = 600 - 2x$
131. $R'(x) = -0.45x^2 - 10x + 6{,}545$
133. $C'(50) = 220$; $R'(50) = 234$
135. $C'(70) = 228$; $R'(70) = 214$
137. $P'(t) = -300e^{-0.5t}$
139. $C'(x) = 10/\sqrt{x}$; $C'(16) = 2.5$
141. $R'(x) = 25 - 2x$
143. $y' = 2x + 5$
145. $y' = 9x^2 + 12x - 7$
147. $y' = 6x - 25$
149. $y' = 2(3x^2 - x + 3)$
151. $y' = 15x^2 - 16x - 4$
153. $y' = 10x^4 - 12x^3 - 18x^2 + 1$
155. $y' = [2x^2/(x^2 + 4)] + \ln(x^2 + 4)$
157. $y' = 18x^2 - 86x + 69$
159. $y' = (\ln x)(e^x + 1) + (e^x + x)(1/x)$
161. −10; 10; 40
163. 10,005.64; 12,692.5
165. $y' = 1$
167. $y' = -7/(x - 4)^2$
169. $y' = (x^2 - 2)/2x^2$
171. $y' = -18x^2/(x + x^3)^2$
173. $y' = (-2x^3 - 3)/(x^3 - 3)^3$
175. $y' = -5/(3x + 2)^2$
177. $y' = [5(x^2 + 3) - 2x(5x + 2) \ln(5x + 2)]/[(5x + 2)(x^2 + 3)^2]$
179. $y' = (1 - \ln x)/x^2$
181. $\overline{C}'(x) = (-1{,}200/x^2) + 14 + 2x$
183. $dy/dx = 2x[9(x^2 + 1)^2 - 10(x^2 + 1) + 4]$
185. $y' = -2(2x + 3)/(x^2 + 3x)$
187. $y' = -6x^2/\sqrt{1 - x^3}$
189. $y' = 1/[(x + 1) \ln(x + 1)]$
191. $dy/dx = 2/[x(x + 2)]$
193. $dy/dx = e^x e^{e^x}$
195. $dy/dx = -5/(x + 1)^6$
197. $dy/dx = 2/[x^2(1 + 1/x)^3]$
199. $dy/dx = (8/x)(2 \ln x + 5)^3$
201. $x^{x^2+1}[1 + 2 \ln x]$
203. $y' = x^{3x^2+1}[6x \ln x + (3x^2 + 1)(1/x)]$
205. $y' = (2x + 1)^x[\ln(2x + 1) + (2x/(2x + 1))]$
207. $y' = x^{2x+1}[2 \ln x + (2x + 1)/x]$
209. $y' = (2x + 1)^{x^2+2}[2x \ln(2x + 1) + 2(x^2 + 2)/(2x + 1)]$
211. $y' = x^{\ln x}(2)(1/x)(\ln x)$
213. $y' = 15x^2$; $y'' = 30x$; $y''' = 30$
215. $y' = 0.6x^2 - x$; $y'' = 1.2x - 1$; $y''' = 1.2$
217. $y' = 6x^2 + 2x + 5$; $y'' = 12x + 2$; $y''' = 12$
219. $y' = 1 - (1/x^2)$; $y'' = 2/x^3$; $y''' = -6/x^4$
221. $f' = 4 - (6/x^3)$; $f'' = 18/x^4$; $f''' = -72/x^5$
223. $f' = 10/3x^{1/3}$; $f'' = -10/9x^{4/3}$; $f''' = 40/27x^{7/3}$
225. $y' = (\sqrt{3}/2\sqrt{x}) + 3$; $y'' = -\sqrt{3}/4x\sqrt{x}$; $y''' = 3\sqrt{3}/8x^2\sqrt{x}$
227. $y' = 4(x + 1)^3$; $y'' = 12(x + 1)^2$; $y''' = 24(x + 1)$
229. $y' = 1/x$; $y'' = -1/x^2$; $y''' = 2/x^3$
231. $y' = e^x$; $y'' = e^x$; $y''' = e^x$
233. $y' = 1/(x + 1)^2$; $y'' = -2/(x + 1)^3$; $y''' = 6/(x + 1)^4$
235. $y' = 2x(e^{x^2} + 1)$; $y'' = 2(2x^2e^{x^2} + e^{x^2} + 1)$; $y''' = 4xe^{x^2}(2x^2 + 3)$

237. $f' = (x-1)^3(5x^4)(9x-5)$;
$f'' = (x-1)^2(20x^3)(18x^2-20x+5)$;
$f''' = (x-1)(60x^2)(42x^3-70x^2+35x-5)$

CHAPTER 12 APPLICATIONS OF THE DERIVATIVE

1. $f' < 0$ for all x; f is a strictly decreasing function for all x
3. $(-\infty,0)$ decreasing; $(0,\infty)$ increasing; stationary at $x = 0$
5. $(-\infty,-\sqrt{2})$ increasing; $(-\sqrt{2},\sqrt{2})$ decreasing; $(\sqrt{2},\infty)$ increasing; stationary at $x = -\sqrt{2}$ and $x = \sqrt{2}$
7. $(-\infty,0)$ decreasing; $(0,\infty)$ decreasing; stationary at $x = 0$
9. $(-\infty,0)$ decreasing; $(0,\infty)$ increasing; stationary at $x = 0$
11. $(-\infty,1/3)$ increasing; $(1/3,1)$ decreasing; $(1,\infty)$ increasing; stationary at $x = 1/3$ and $x = 1$
13. $(-\infty,-2)$ decreasing; $(-2/-4/3)$ increasing; $(-4/3,\infty)$ decreasing; stationary at $x = -2$ and $x = -4/3$
15. $(-\infty,0)$ increasing; $(0,\infty)$ decreasing; stationary at $x = 0$
17. $(-\infty,\infty)$ increasing
19. $(-\infty,-2)$ increasing; $(-2,\infty)$ increasing; undefined at $x = -2$
21. $(-\infty,-3/5)$ increasing; $(-3/5,0)$ decreasing; $(0,3/5)$ decreasing; $(3/5,\infty)$ increasing; stationary at $x = -3/5$, $x = 0$, and $x = 3/5$
23. $(-\infty,-1)$ increasing; $(-1,1/2)$ increasing; $(1/2,\infty)$ decreasing; undefined at $x = -1$; stationary at $x = 1/2$
25. $(-\infty,-2/3)$ increasing; $(-2/3,\infty)$ increasing; stationary at $x = -2/3$
27. $(0,24)$ increasing; $(24,50)$ decreasing; stationary at $x = 0$
29. $f' > 0$ for $x > 0$; f always increasing
31. $0 \leq x \leq 25, P' > 0$, P is increasing; $x > 25$, $P' < 0$, P is decreasing
33. Q is increasing for $0 < x < 300$; Q is decreasing for $x > 300$
35. for $0 < p < 50$, R is decreasing; for $p > 50$, R is increasing
37. $t > 50$
39. $(-\infty,\infty)$ concave downward
41. $(-\infty,1/3)$ concave downward; $(1/3,\infty)$ concave upward; $x = 1/3$ is point of inflection
43. $(-\infty,-2/9)$ concave downward; $(-2/9,\infty)$ concave upward; $x = -2/9$ is point of inflection
45. $(-\infty,0)$ concave downward; $(0,\infty)$ concave upward; $x = 0$ is point of inflection
47. $(-\infty,0)$ concave upward; $(0,\infty)$ concave upward; no point of inflection
49. $(-\infty,-\sqrt{3})$ concave downward; $(-\sqrt{3},0)$ concave upward; $(0,\sqrt{3})$ concave downward; $(\sqrt{3},\infty)$ concave upward; point of inflection at $x = -\sqrt{3}$, at $x = 0$, and at $x = \sqrt{3}$
51. $(-\infty,0)$ concave upward; $(0,2/3)$ concave downward; $(2/3,\infty)$ concave upward; point of inflection at $x = 0$ and at $x = 2/3$
53. $(-\infty,\infty)$ concave upward
55. $(-3,\infty)$ concave downward
57. $(-\infty,0)$ concave upward; $(0,\infty)$ concave downward; point of inflection at $x = 0$
59. $(-\infty,1)$ concave upward; $(1,\infty)$ concave upward; no point of inflection

61. a, point of inflection; b, relative maximum; c, point of inflection; d, relative minimum; e, point of inflection; f, relative maximum; g, point of inflection; h, relative minimum; $(-\infty,b)$ increasing; (b,d) decreasing; (d,f) increasing; (f,h) decreasing, (h,∞); (a,c) concave upward; (c,e) concave downward; (e,g) concave upward; (g,∞) concave downward

63. b, relative maximum; $(-\infty,d)$ increasing; (d,b) increasing; (b,e) decreasing; (e,∞) decreasing; $(-\infty,d)$ concave upward; (d,e) concave downward; (e,∞) concave upward

65. $(-\infty,c)$ increasing; (c,d) decreasing; (d,∞) increasing; (c,∞) concave upward; d, relative minimum

67. $x = 2.5$, minimum

69. $x = 0$, minimum

71. $x = 0$, maximum; $x = 2/3$, minimum

73. No relative maximum, no relative minimum

75. $x = 4$, minimum

77. No maximum or minimum point

79. $x = -3$, minimum; $x = 0$ is maximum; $x = 3$, minimum

81. No maximum, no minimum

83. $x = 1$, maximum; $x = 2$, minimum

85. No maximum, no minimum

87. No maximum, no minimum

89. $x = 0$, minimum

91. $x = -4/3$, maximum; $x = 0$, minimum

93. $x = 3$, minimum

95. $x = 0$, maximum

97. No maximum, no minimum

99. $x = -1$, minimum; $x = 1$, maximum

101. $x = -1$, minimum

103. No maximum, no minimum

105. $x = -1.732$, minimum; $x = 0$, maximum; $x = 1.732$, minimum

107. $x = 3$, absolute minimum; $x = 10$, absolute maximum

109. $x = -3$, absolute maximum; $x = 1$, relative minimum

111. $x = -5$, absolute minimum; $x = -2$, relative maximum; $x = 2$, relative minimum; $x = 5$, absolute maximum

113. $x = 0$, absolute minimum

115. $x = 2$, absolute minimum

117. $x = 0$, absolute minimum

119. Dimensions: 300 by 600 feet; area, 180,000 square feet

121. 5.43 by 5.43 inches; volume = 2,612.8 cubic inches; 8 by 8 inches; volume = 8,192 cubic inches

123. Height, 10.4 inches; radius, 3.9 inches; cost, \$23.03

125. Width = 8.94 inches; length = 13.42 inches

127. $x = 62.5$

129. 50,000 miles

131. 3.945

133. \$10,000

135. 4,000

137. 5 additional trees per acre

139. 11 additional \$5 increases, $p =$ \$155

141.

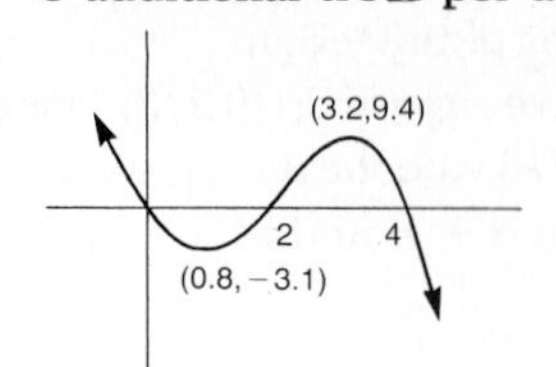

143.

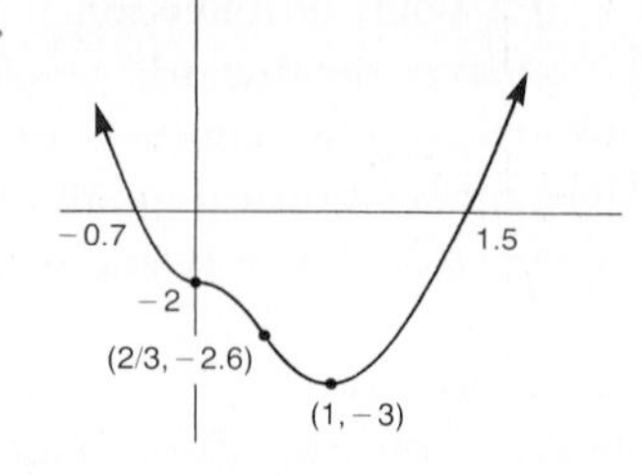

145.

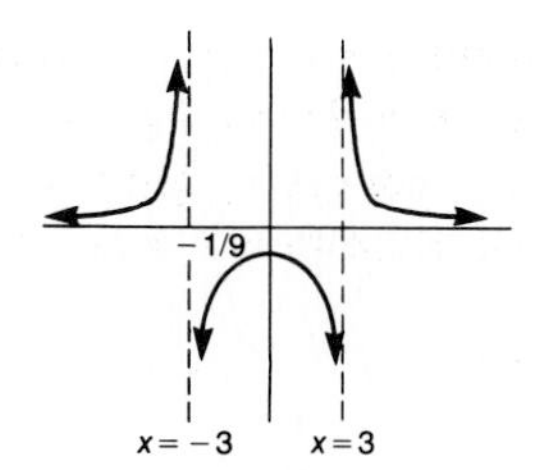

147.

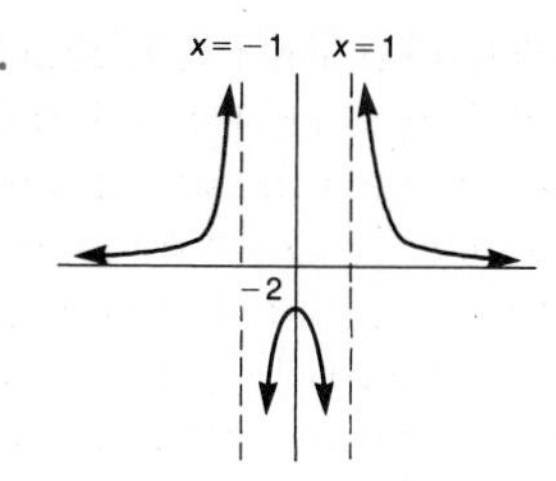

149.

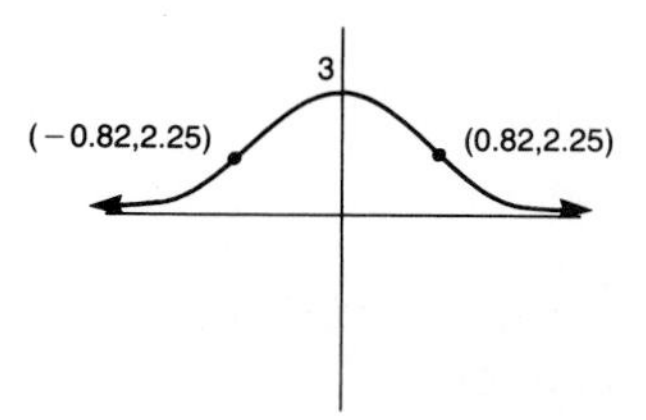

151.

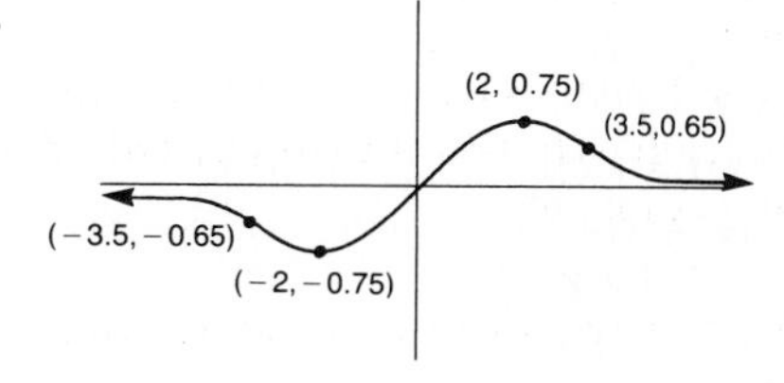

153.

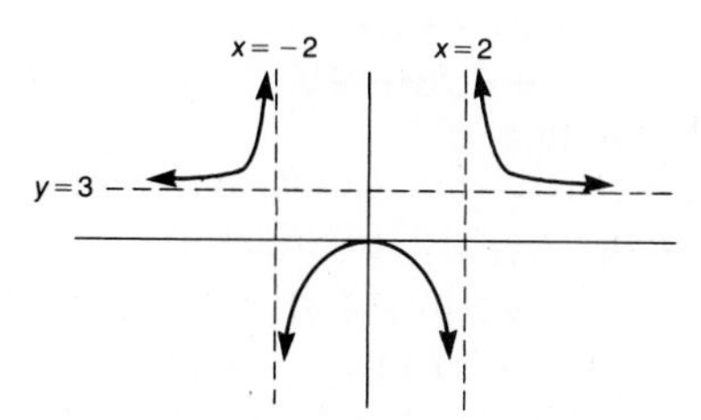

155.

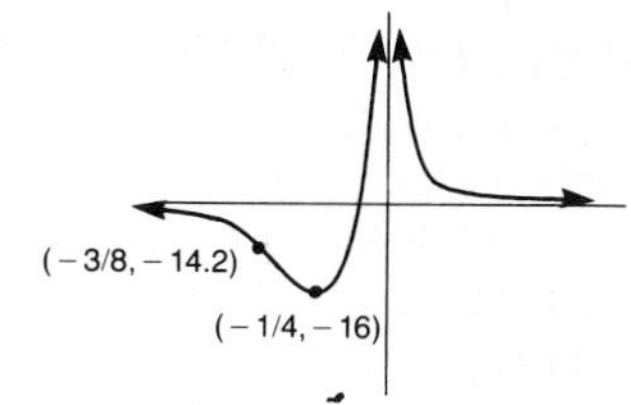

157.

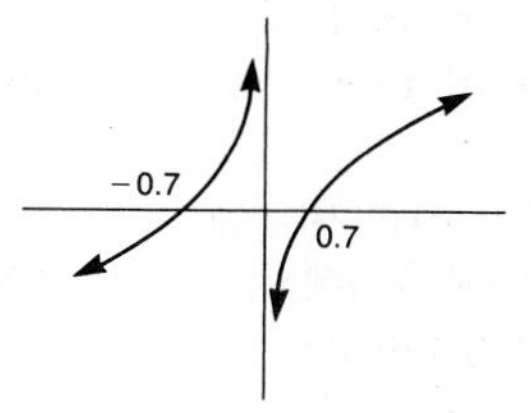

159.

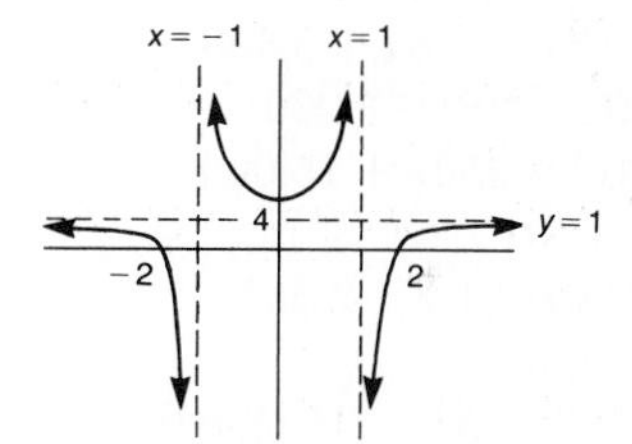

161.

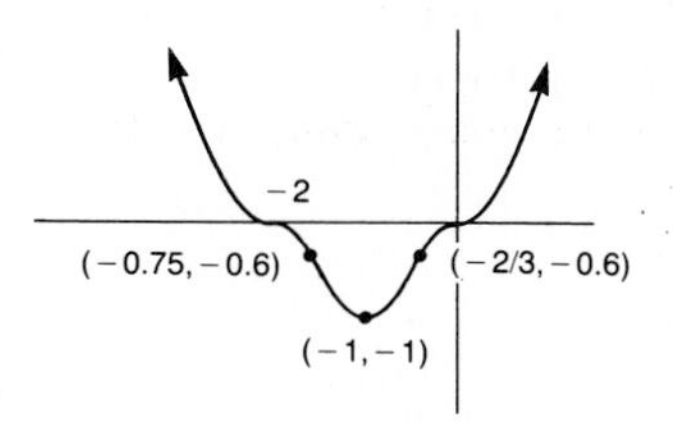

163.

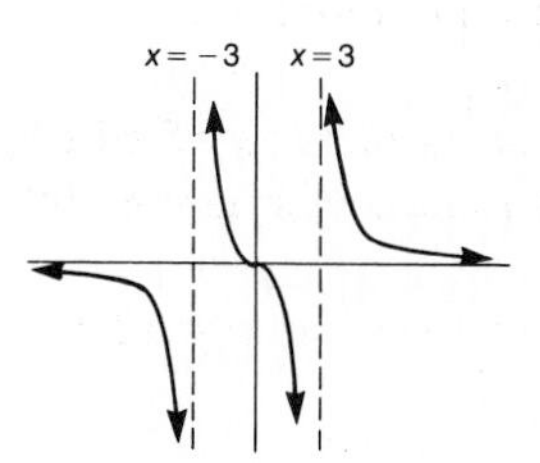

165.

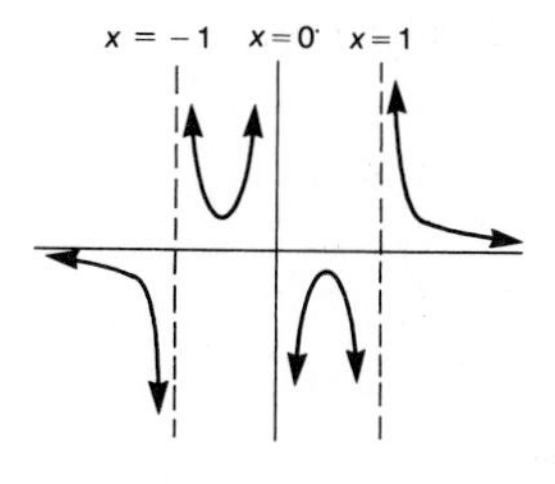

167.

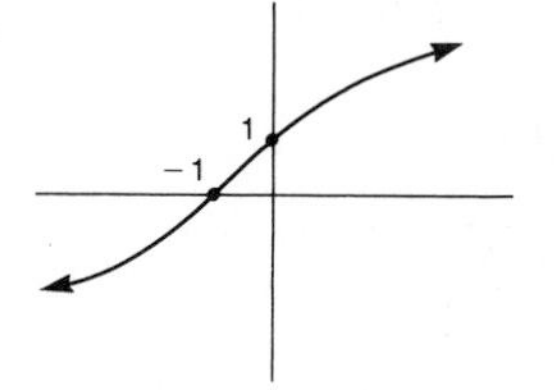

CHAPTER 13 INTEGRAL CALCULUS

1. $5.8x + C$
3. $x^5 + C$
5. $-x^2 + C$
7. $x^2 + 2x + C$
9. $\ln x + C$
11. $9x + C$
13. $y^2/2 + C$
15. $x^4/4 + C$
17. $x^2/2 + 5x + C$
19. $(3/8)t + C$
21. $f(x) = (2/3)x + 8$
23. $f(x) = x^3 + 5x - 20$
25. $f(x) = x^3 + x^2 + 2x - 16$
27. $6x + C$
29. $(4/5)s + C$
31. $\sqrt{3}x + C$
33. $x^5/5 + C$
35. $(4/5)x^{5/4} + C$
37. $x^{1.4}/1.4 + C$
39. $(5/9)x^{9/5} + C$
41. $-4x^2 + C$
43. $(3/2)x^4 + C$
45. $-12\sqrt{x} + C$
47. $(2/5)x^2 + C$
49. $(8/5)x^{5/2} + C$
51. $ex^{e+1}/(e + 1) + C$
53. $(3/2)x^2 + 6x + C$
55. $(x^3/3) + (3/2)x^2 + 5x + C$
57. $3 \ln x + (x^2/6) + C$
59. $(x^5/5) - x^2 + (2/3)x^{3/2} + C$
61. $(x^3/3) + 9x^2 + 81x + C$
63. $3^{2x}/(2 \ln 3) + C$
65. $(1/2)e^{2x} + C$
67. $(-5/3)e^{-3x} + C$
69. $e^{-0.1x} + C$
71. $(8^t/\ln 8) + 5t + C$
73. $x \log 5x - x \log e + C$
75. $x \log_3 4x - x \log_3 e + C$
77. $x[(\ln 6x) - 1] + C$
79. $x \ln 3x - x + x \ln 2x - x + C$
81. $x \ln 7x - x + (7/2)x^2 + C$
83. $R(x) = 18x$
85. $C(x) = 0.015x^2 + 18x + 3{,}500$
87. $C(x) = (3/2)x^2 + 9x + 6{,}000$
89. $C(x) = 240x + 3x^2 + 5{,}000$
91. 56 units; \$4,352
93. $C(x) = 2.5x^2 + 250x + 15{,}000$
95. 300
97. $f(t) = 12t + 3t^2$
99. $D(x) = -7.5x^2 + 210x + 450$
101. $n(t) = (10/3)t^{1.8} + t + 2{,}500$
103. $V(t) = (e^{0.06t}/0.06) + 133.\overline{3}$
105. $P(t) = t^{1.4}/420$
107. $P(t) = 2e^t - t^3/6 + 8{,}998$
109. $F(x) = 4{,}000 + 75x - (10/3)x^{3/2}$
111. $[(3x - 5)^3/3] + C$
113. $[(x^2 - 12)^3/3] + C$
115. $(x^5/5) + (2/3)x^3 + x + C$
117. $[(x^2 + 2x)^{3/2}/3] + C$
119. $(x^4/4) + (2/3)x^3 + (x^2/2) + C$
121. $\ln (x^3 + 1) + C$
123. $\ln (x^6 - 14) + C$
125. $(1/4) \ln (x^4 - 1) + C$
127. $2 \ln (x^5 + 3) + C$
129. $(1/6) \ln (3x^2 + 6x + 4) + C$
131. $e^{x+1} + C$
133. $e^{x^2+1} + C$
135. $2e^{0.5x+3} + C$
137. $e^{2x^3+1} + C$
139. $(1/8)e^{4x^2} + C$
141. 14
143. 18
145. 88
147. 3
149. 5.83538
151. 1.60944
153. $12.\overline{3}$
155. 3.194528
157. 1.38629
159. 0.34657
161. 4
163. 1
165. 1/2
167. 1 5/6
169. 9 1/3
171. 17.5
173. 36
175. 4.5
177. $93.\overline{3}$
179. $85.\overline{3}$

181. 1.25
183. 4.5
185. 10.67
187. $61.8\overline{6}$

CHAPTER 14 ADDITIONAL METHODS AND APPLICATIONS OF INTEGRAL CALCULUS

1. 360
3. $166.6\overline{6}$
5. $q^* = 10$; $p^* = 140$
7. $866.6\overline{6}$
9. 160
11. 450,000
13. 28,500
15. 1,275,000
17. $P(x) = 600x = 2.5x^2 - 20{,}000$
19. 1,687.50
21. $5,600
23. 20,000
25. 1,480
27. 50.76
29. −625; −918.75; −1,200
31. $\$61.\overline{6}$ thousand
33. $555; $1,325
35. $8,750
37. 5.6 years
39. 8,997.78
41. 406,186.4
43. $1,266
45. $345.6
47. $180,000
49. $(e^{2x}/2)(x - 1/2) + C$
51. $2\sqrt{x}\,(ln\, x - 2) + C$
53. $2x^2\,(ln\, x - 1/2) + C$
55. $(2/15)(x-1)^{3/2}(3x+2) + C$
57. $2\sqrt{x+1}\,[x - (2/3)\sqrt{x+1}] + C$
59. $(3/4)(x+2)^{4/3}[x - (3/7)(x+2)] + C$
61. $e^x(3x - 5) + C$
63. $x[ln\, 2x - 1] + C$
65. $[(x^2/2) + 2x](ln\, x) - (x^2/4) - 2x + C$
67. $x^2[ln\, 2x - (1/2)] + C$
69. $(e^{x^2}/2)(x^2 - 1) + C$
71. $(x^4/4)[ln\, x - 1/4] + C$
73. $(x^2/6)(x+4)^6 - (x/21)(x+4)^7 + (1/108)(x+4)^8 + C$
75. $-e^{-x}(x+1) + C$
77. $-x^2e^{-x} - 2e^{-x}(x+1) + C$
79. 51.54
81. $(x+1)e^{(x+1)} - e^{(x+1)} + C$
83. $(1/2)[ln\,(x+5)]^2 + C$
85. $(1/8)\, ln\, [(4+x^2)/(4-x^2)] + C$
87. $(1/2)[x\sqrt{x^2-4} - 4\, ln\, |\, x + \sqrt{x^2+4}\, |] + C$
89. $(2/135)(9x-4)(3x+2)^{3/2} + C$
91. $ln\, |\, x + 3 + \sqrt{6x + x^2}\, | + C$
93. $(3x^2+5)^4/24 + C$
95. $-1/[12(4x^2+1)^3] + C$
97. $(1/2)[x\sqrt{x^2+25} + 25\, ln\, |\, x + \sqrt{x^2+25}\, |] + C$
99. $(x+3)^3/3[ln\,(x+3) - (1/3)] + C$
101. $e^{(x+4)}[(x+4)^2 - 2(x+4) + 2] + C$
103. $(2x)e^{2x} - e^{2x} + C$
105. $(2/3)x^{3/2}[ln\, x - (2/3)] + C$
107. $3[ln\, |\, 2x + \sqrt{4x^2-9}\, |] + C$
109. $ln\,(e^x + 5) + C$
111. $y = (7x^2/2) + 6x + C$
113. $y = (2x^3/3) + 4x^2 + C$
115. $y = (4x^3/3) + C$
117. $y = (x^3/3) + (x^2/2) + C$
119. $y = ce^{-3x}$
121. $y = x - 6$
123. $y = x + C$
125. $y = x^2/2 + C$
127. $y = (3x-2)^{2/3}(x+5)/5 + C$
129. $ln\, y = (-1/4)(x+3)^{-4} + C$
131. $y = (3/2)x^2 + 4x - 5$
133. $y = x^4 - 2x^3 + 5x + 2$
135. $y = (x^3/3) + (3/2)x^2 + 2x - 30$
137. $y = ln\, [(x^3/3) + 1] + C$
139. $y = [(x+1)/32](15x^2 - 2x + 16)^2 - 4$
141. $y = 2{,}400(x-1)^{1/2} + 8{,}200$
143. 70.7 percent
145. 35,636
147. Divergent
149. Divergent
151. Divergent

CHAPTER 15 THE CALCULUS OF MULTIVARIATE FUNCTIONS

1.

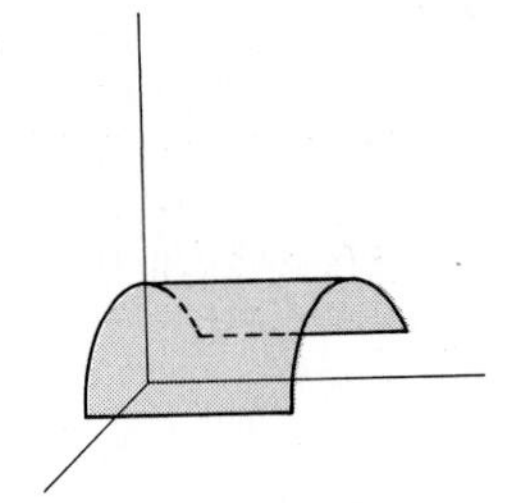

3.

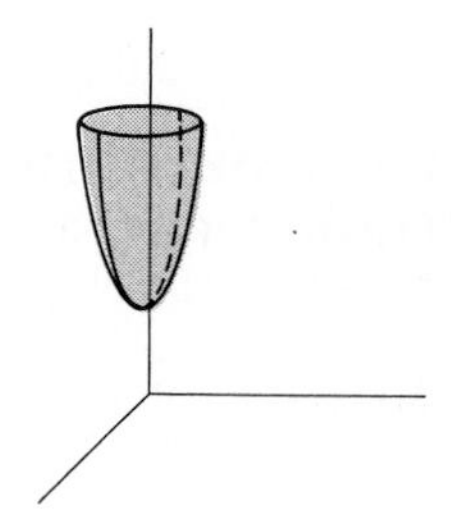

5.

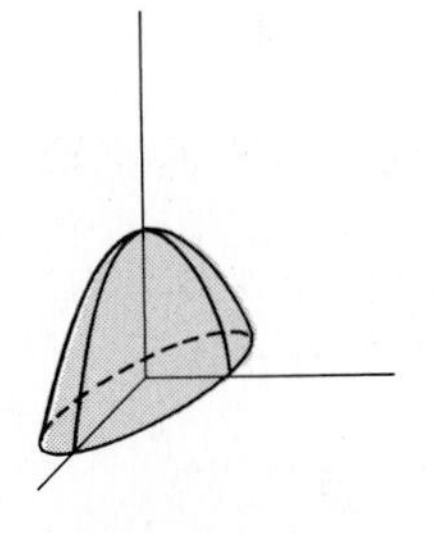

7. $f_x = 3x^2 + 4xy - y^2$; $f_y = 2x^2 - 2xy + 12y^2$
9. $f_x = 6xy + 4y^2$; $f_y = 3x^2 + 8xy - 15y^2$
11. $f_x = 5(3x^2 + 2y)(x^3 + 2xy + y)^4$;
$f_y = 5(2x + 1)(x^3 + 2xy + y)^4$
13. $f_x = (2x)e^{(x^2+y)}$; $f_y = e^{(x^2+y)}$
15. $f_x = ye^{xy}(xy + 1)$; $f_y = xe^{xy}(xy + 1$
17. $f_x = 9x^2/(3x^3 + 5y^3)$; $f_y = 15y^2/(3x^2 + 5y)$
19. $f_x = 6xe^{x^2} + 1/(x + 2y)$;
$f_y = 2y + 2/(x + 2y)$
21. $f_x = 5e^x$; $f_y = 3e^y$; $f_z = xy^2 + 2z^2$
23. $f_x = 2xy + 3ye^{3xy} + (1/x)$;
$f_y = x^2 + 3xe^{3xy} + (1/y)$; $f_z = 1/z$
25. $f_x = 6x/z^3$;
$f_y = 2y/z^3$; $fz = -9x/z^4 - 3y^2/z^4$
27. $f_x = x(1 + 2\ ln\ xyz^2)$; $f_y = x^2/y$; $f_z = x^2/z$
29. $f_x = 2xy/(y^2 + z^2)$;
$f_y = [x^2y^2 + x^2z^2 - 2x^2y^2 - 2yz]/$
$[y^2 + z^2]^2$;
$f_z = [y^2 - 2x^2yz - z^2]/(y^2 + z^2)^2$
31. $f_x = 10xy$; $f_x(1,3) = 30$;
$f_y = 5x^2$; $f_y(-2,4) = 20$
33. $f_x = 1/x$; $f_x(1,3) = 1$;
$f_y = 1/y$; $f_y(-2,4) = 1/4$
35. $f_x = e^x\ ln\ y$; $f_x(1,3) = 2.986$;
$f_y = e^x/y$; $f_y(-2,4) = 0.034$
37. \$284.90
39. 1.2
41. 2.06
43. 329
45. 4,500
47. −300
49. Complementary products
51. Competitive products
53. $f_x = (2x + 3y)^{-1/2}$; $f_{xx} = -1(2x + 3y)^{-3/2}$;
$f_{xy} = (-3/2)(2x + 3y)^{-3/2}$;
$f_y = (3/2)(2x + 3y)^{-1/2}$;
$f_{yx} = (-3/2)(2x + 3y)^{-3/2}$;
$f_{yy} = (-9/4)(2x + 3y)^{-3/2}$
55. $f_x = ye^{xy}$; $f_{xx} = e^{xy}(xy + 1)$; $f_{xy} = y^2e^{xy}$;
$f_y = xe^{xy}$; $f_{yx} = e^{xy}(xy + 1)$; $f_{yy} = x^2e^{xy}$
57. $f_x = 6x/(3x^2 + 2y)$;
$f_{xx} = (-18x^2 + 12y)/(3x^2 + 2y)^2$;

$f_{xy} = -12x/(3x^2+2y)^2$;
$f_y = 2/(3x^2+2y)$;
$f_{yx} = -12x/(3x^2+2y)^2$;
$f_{yy} = -4/(3x^2+2y)^2$

59. $f_x = 2y^2 - 6xye^{x^2}$;
$f_{xx} = 12x^2y^2e^{x^2y} + 6ye^{x^2y}$;
$f_{xy} = 4y - 6x^3ye^{x^2y} - 6xe^{x^2y}$;
$f_y = 4xy - 3xe^{x^2y}$;
$f_{yx} = 4y - 6x^3ye^{x^2y} - 6xe^{x^2y}$;
$f_{yy} = 4x - 3xe^{x^2y}$

61. $f_x = 1/(y+3)$; $f_{xx} = 0$; $f_{xy} = -1/(y+3)^2$;
$f_y = -(x+2)/(y+3)^2$;
$f_{yx} = -1/(y+3)^2$; $f_{yy} = 2(x+2)/(y+3)^3$

63. $f_x = 8x + 12y + 4z$; $f_{xx} = 8$; $f_{xy} = 12$;
$f_{xz} = 4$; $f_y = 12x + 18y + 6z$; $f_{yx} = 12$;
$f_{yy} = 18$; $f_{yz} = 6$; $f_z = 4x + 6y + 2z$; $f_{zx} = 4$;
$f_{zy} = 6$; $f_{zz} = 2$

65. $f_x = e^{yz}$; $f_{xx} = 0$; $f_{xy} = ze^{yz}$; $f_{xz} = ye^{yz}$;
$f_y = xze^{yz}$; $f_{yx} = ze^{yz}$; $f_{yy} = xz^2e^{yz}$;
$f_{yz} = xe^{yz}(yz+1)$;
$f_z = xye^{yz}$; $f_{zx} = ye^{yz}$; $f_{zy} = xe^{yz}(yz+1)$;
$f_{zz} = xy^2e^{yz}$

67. $f_x = x + 2x \ln xyz$; $f_{xx} = 3 + 2x \ln xyz$;
$f_{xy} = 2x/y$; $f_{xz} = 2x/z$; $f_y = x^2/y$;
$f_{yx} = 2x/y$; $f_{yy} = -x^2/y^2$; $f_{yz} = 0$;
$f_z = x^2/z$; $f_{zx} = 2x/z$; $f_{zy} = 0$;
$f_{zz} = (8x^2/z^3) + (2y/z^3)$

69. $f_x = 2(x+y)(y+z)^2$; $f_{xx} = 2(y+z)^2$;
$f_{xy} = 2(y+z)(2x+3y+z)$;
$f_{xz} = 4(x+y)(y+z)$;
$f_y = 2(x+y)(y+z)(x+2y+z)$;
$f_{yz} = 2(y+z)(2x+3y+z)$;
$f_{yy} = 2(x^2+4y^2+z^2+7xy+3xz+6yz)$;
$f_{yz} = 2(x+y)(x+3y+2z)$;
$f_z = (x+y)^2(2)(y+z)$;
$f_{zx} = 4(y+z)(x+y)$;
$f_{zy} = 2(x+y)(x+3y+2z)$;
$f_{zz} = 2(x+y)^2$

71. Relative minimum at $x = 2$, $y = 3$
73. Relative minimum at $x = 16/3$, $y = -8/3$
75. Relative minimum at $x = 2.1$, $y = 1.7$
77. Critical point $x = 0$, $y = -2$ is neither relative maximum nor minimum.
79. Relative minimum at $x = 1$, $y = 0$
81. Relative maximum at $x = 20$, $y = 5$
83. Critical point at $x = -5.4$, $y = -1.4$ is neither maximum nor minimum.
85. Relative minimum where $x = 2$ and $y = -2$
87. Relative minimum where $x = 2.5$ and $y = 2.5$
89. Relative minimum at $x = 2.2$ and $y = 2.2$
91. Critical point at $x = 8/7$ and $y = 6/7$ is neither relative maximum nor minimum.
93. Relative maximum at $x = 7{,}977.94$ and $y = 1{,}994.49$

CHAPTER 16 THE BASIC CONCEPTS OF PROBABILITY THEORY

1. $S = \{R, B\}$
3. $S = \{F, N\}$
5. $S = \{RR, BR, BB\}$
7. $S = \{0, 1, 2, \ldots, n\}$
9. $S = \{0,1,2,3\}$
11. $S = \{\text{red, yellow, purple, green}\}$
13. $S = \{OH, OT, EH, ET\}$
15. $S = \{HHH, HHT, HTH, HTT, THH, THT, TTH, TTT\}$

17. $S = \{$AA, AD, AN, DA, DD, DN, NA, ND, NN$\}$
19. $S = \{0, 1, 2, 3, 4, 5\}$
21. $S = \{0, 0, 2, \ldots, n\}$
23. $S = \{$S, N$\}$
25. Subjective
27. Empirical
29. Subjective
31. Empirical
33. Empirical
35. Subjective
37. Subjective
39. Empirical
41. Subjective
43. No. Outcome of one toss of a fair coin is not influenced by outcome of previous toss.
45. $E_2 = \{$HHT, HTH, THH$\}$
47. $E_4 = \{$HHT$\}$
49. $E_6 = \{$HHH, HHT, HTH, THH$\}$
51. E_1, E_3
53. $E_2 = \{(4,1), (4,2), (4,3), (4,4), (4,5), (4,6), (1,4), (2,4), (3,4), (5,4), (6,4)\}$
55. $E_4 = \{(1,1), (1,2), (1,3), (1,4), (1,5), (1,6), (2,1), (2,2), (2,3), (2,4), (2,5), (2,6), (3,1), (3,2), (3,3), (3,4), (3,5), (3,6)\}$
57. $E_6 = \{(1,3), (3,1), (2,2)\}$
59. E_6, E_4
61. $E_1 = \{$(A,B), (A,C), (A,D), (A,E)$\}$
63. $E_3 = \{$(A,B), (B,A)$\}$
65. $E_5 = \{$(A,C), (A,D), (A,E), (C,A), (C,D), (C,E), (D,A), (D,C), (D,E), (E,A), (E,C), (E,D)$\}$
67. E_4
69. False
71. True
73. False
75. False
77. 4/26
79. 5/26
81. 23/26
83. 7/8
85. 1/2
87. 1/4
89. 1/8
91. 7/8
93. 3/8
95. 1/52
97. 1/4
99. 25/26
101. 1/2
103. 1/4
105. 0.25
107. 0.6
109. 1/2
111. Herschel has not distinguished between such outcomes as 2–1 and 1–2 or 3–1 and 1–3.
113. 2/3
115. 1/2
117. 2/3
119. 1/6
121. 1/6
123. 1/3
125. 2/3
127. 2/3
129. 1/3
131. 2/3
133. 0.49
135. 0.62
137. 0.11
139. 0.89
141. 0.38
143. 0.52
145. 0.65
147. 0.52
149. 1
151. 0.6
153. 0.3
155. 0.7
157. 0.4
159. 7/13
161. 2/5
163. 4/5
165. 0.7
167. 0.66
169. 0.81
171. 0.4

173. 331/420
175. 146/420
177. 1/3
179. 0.3
181. 0.375
183. 1/3
185. 1/13
187. 7/13
189. 1/2
191. 1/13
193. 36/76
195. 164/410
197. 154/350
199. 0.6479
201. 0.5476
203. 142/272
205. 365/495
207. 1,716/140,608
209. 1/16
211. 1/5
213. 2/5
215. 1/16
217. 5/48
219. 1/6
221. 0.7
223. 0.24
225. 0.06
227. 0.8
229. 24/52
231. Not independent
233. 3/4
235. 1/10
237. 1/4
239. 4/10
241. 7/10
243. 0.432
245. 0.048
247. 0.624
249. 0.3
251. 0.4
253. 0.15
255. 3/8
257. Not statistically independent
259. 0.55
261. 0.35
263. 0.65
265. 0.6
267. 2/3
269. 0.7
271. 0.4
273. 0.1
275. 3/8
277. 15/16
279. 7/10
281. 3/10
283. 0.87
285. 0.345
287. 0.6
289. 0.7
291. 16/30
293. 1/3
295. 0.625

CHAPTER 17 RANDOM VARIABLES AND THEIR PROBABILITY DISTRIBUTIONS

1.

Probability x	Distribution $p(x)$
0	1,260/2,652
1	1,152/2,652
2	240/1,652
	1

3.

Probability x	Distribution $p(x)$
0	1/16
1	4/16
2	6/16
3	4/16
4	1/16
	1

5.

Probability x	Distribution $p(x)$
0	2/6
1	3/6
3	1/6
	1

7.

Probability y	Distribution $p(y)$
0	6/36
1	10/36
2	8/36
3	6/36
4	4/36
5	2/36
	1

9. 2.19

11. 0.33; 0.6011; 0.7753

13. 1/2; 15/44

15. −1.25

19.

Probability y	Distribution $p(y)$
0	0.316406
1	0.421875
2	0.210938
3	0.046875
4	0.003906
	1.0

21. 1; 0.75

23. 0.2; 0.18999

25.

Probability x	Distribution $p(x)$
0	0.117649
1	0.302526
2	0.324135
3	0.185220
4	0.059535
5	0.010206
6	0.000729
	1.0

29. 0.3; 0.285

31. 0.299142

33. 0.06606

35. 0.296475

37. 0.255298

39. 0.980432

41. 0.913862; 0.086138

43. 0.375810; 0.624190

45. 0.103516

47.

Probability x	Distribution $p(x)$
0	0.26214
1	0.39322
2	0.24576
3	0.08192
4	0.01536
5	0.00154
6	0.00006
	1.0

49. 0.32

51. 0.2

53. 0.36

55.

Probability x	Distribution $p(x)$
2	0.36000
3	0.28800
4	0.17280
5	0.09216
6	0.08704
	1.0

57. 0.04096
59. 0.03
61. 0.06173
63. 0.53571
65. 1/14; 1/14; 1/7
67. 2.06115×10^{-8}
69. 0.264025
73. 0.367879
75. 0.551819
77. $k = 1/100$
79. $k = -1/36$
81. 1/3
83. 3.6
85. 26/27
87. 0.574
89. 243/45; 27/80
91. 0.33
93. 1
95. 0.8790
97. 0.0228
99. 0.7299
101. 278.4
103. 205.50; 274.50
105. 0.1056
107. 0.8413
109. 0.003
111. 24.3 months

CHAPTER 18 MATHEMATICS OF FINANCE

1. $60
3. $261
5. $1.35
7. $11,050
9. $1,375
11. $2,500; $7,500
13. $27; $573
15. $164.06; $585.94
17. $1,284
19. 14.28 years; 28.57 years
21. $1,142.86
23. 50 percent
25. $8,120.76
27. $14,025.52; $4,025.52
29. $14,147.78; $4,147.78
31. $9,030.56; $4,030.56
33. $11,901.56; $4,401.56
35. $12,063.28; $4,563.28
37. $20,600.66; $12,600.66
39. $1,151.84
41. $2,936.17
43. $22,200.59
45. 18.81 percent
47. 14.49 percent
49. 12.36 percent
51. 12.68 percent
53. 15.1884 percent
55. 6.116 years
57. 2.92 years
59. $6,729.71
61. $1,746.03; $18.81; $1,764.84
63. $3,537.67
65. 10.38 percent
67. 19.56 percent
69. 8.806 percent compounded semiannually
71. 13.75 percent
73. $10,068.76
75. $1,568.31
77. $4,481.69
79. 7.64 years
81. 11.6278 percent
83. $5,450.64
85. $1,070.62
87. $42,275.14
89. $15,811.64
91. $275.15
93. $214.27
95. $10,121.96
97. $10,470.99
99. $4,727.60; $12,807.71
101. $8,685.36; $33,146.40
103. $11,827
105. $9.52
107. $156.50

109. $41,714.40
111. $153.86
113. $5,135.69
115. $5,214.13
117. $4,151.33
119. $6,040.15
121. $15,666
123. $44.66
125. $172.55
127. $183.36; $200.32
129. $127.61; $296.98
131. $523.38; $9,402.80
133. $9,711.35; $2,688.65
135. $171.86; $1,187
137. $15,580.63
139. $43.97
141. $175,696
143. $4,278.14 + $25,000

Index

G–H

I

J–L

M

This book has been set Videocomp, 9 point Caledonia, leaded 3 points. Chapter numbers are 12 and 14 point Souvenir Demi Bold, and chapter titles are 16 point Souvenir Demi Bold. The size of the type area is 36½ by 48 picas.